Dynamics of Complex Quantum Systems

Theoretical and Mathematical Physics

The series founded in 1975 and formerly (until 2005) entitled *Texts and Monographs in Physics* (TMP) publishes high-level monographs in theoretical and mathematical physics. The change of title to *Theoretical and Mathematical Physics* (TMP) signals that the series is a suitable publication platform for both the mathematical and the theoretical physicist. The wider scope of the series is reflected by the composition of the editorial board, comprising both physicists and mathematicians.

The books, written in a didactic style and containing a certain amount of elementary background material, bridge the gap between advanced textbooks and research monographs. They can thus serve as basis for advanced studies, not only for lectures and seminars at graduate level, but also for scientists entering a field of research.

For further volumes:
http://www.springer.com/series/720

Vladimir M. Akulin

Dynamics of Complex Quantum Systems

Springer

Vladimir M. Akulin
CNRS Laboratoire Aimé Cotton
Campus d'Orsay
Orsay Cedex, France

ISSN 1864-5879 ISSN 1864-5887 (electronic)
ISBN 978-94-007-7204-5 ISBN 978-94-007-7205-2 (eBook)
DOI 10.1007/978-94-007-7205-2
Springer Dordrecht Heidelberg New York London

Library of Congress Control Number: 2013956635

Printed on acid-free paper

Springer is part of Springer Science+Business Media (www.springer.com)

Preface

This book is an attempt to put together a large number of similar problems that one encounters in various fields of modern quantum physics and that have common features considering multilevel quantum systems. The main motivation was to present from the same standpoints various models and approaches that have been developed in atomic, molecular, condensed matter, chemical, laser and nuclear physics in various contexts.

The book is based on my lectures in the Moscow Institute of Physics and Technology, in the Aime Cotton Laboratory of CNRS, and some other courses that I have delivered during last two decades. It includes the original results obtained in collaboration with my colleagues V. Aquilanti, I. Averbukh, A. Belousov, M. Blaauboer, E. Borsela, E. Brion, B. Brunetti, C. Brechignac, P. Cahuzac, F. Carlier, I. Dumer, V. Gershkovich, G. Esadze, G. Garsevanishvili, G. Harel, E. Khokhlov, G. Kurizki, R. Larciprete, I. Murachko, A. Nesterenko, A. Orlov, S. Pelegrin, P. Pillet, F. Rebentrost, A. Sarfati, E. Schlag, W. Schleich, F. Vecchiocativi from different scientific centers in the world and with whom I had the pleasure to work on the dynamical aspects of the behaviour of complex quantum systems. I express my deep gratitude to them for their collaboration. The book also contains numerous results of other authors that have, however, been expounded in different notations consistent with the present text, and that sometimes even rely on an alternative derivation as compared to the original version.

In preparing the text I decided to add several results both scientific, yet unpublished, and pedagogical that I feel are necessary for giving the entire picture of the processes in complex quantum systems. Some of these results, presented in Chap. 6, have been obtained in collaboration with V. Kravtsov to whom I express my sincere acknowledgments. I also very much appreciate the discussions with V. Agranovich, E. Bogomolny, B. Chirikov, T. Gallagher, P. Golovinskii, M. Fedorov, Y. Fyodorov, C. Jungen, J. Jortner, L. Maksimov, V. Man'ko, I. Mazets, V. Kac, A. Kazantsev, D. Khmelnitski, A. Kofman, I. Lerner, E. Nikitin, M. Shapiro, D. Shepelyanski, V. Pokrovskii, and A. Prokhorov on the different aspects of the results presented in the book.

I want to worship the memory of G. Askaryan who has influenced my choice of profession, showing me the beauty of physics presented in our life.

Finally, I express my profound gratitude to my teachers Alexandr Dyknne and Nikolay Karlov.

Paris, France Vladimir M. Akulin
December 2003

Preface for the Second Edition

Preparing the second edition, I decided to add the some new results that have been obtained by the scientists working on the broad research field named Complex Quantum Systems after the time when the first edition of the book has been published. The main directions concerned are Quantum Entanglement, Quantum Informatics, Quantum Control and Coherence Protection of open Quantum Systems. Broadening of the scientific field covered in the second edition required to change the title of the book by taking out the word "coherent", since in the present text considerable attention is paid to description of the open systems that may and usually do lose coherence in the course of quantum evolution. It is my pleasure to express in this context my deep gratitude to G. Bensky, A. Mandilara, A. Smilga, L. Viola and add them to the list my colleagues that have participated in obtaining the new results presented in the book, and acknowledge as well the very stimulating discussions that I had with L. Ioffe on Euclidean Random Matrices and Topological Protection of Coherence.

Apart from that, some pedagogically motivated changes of the presentation were done along with corrections of the identified misprints.

My memory keeps alive the brightest moments of my emotional life spent together with my mother Diana Akulina and the intellectual life spent with my teacher Alexandr Dykhne—two persons who have passed away during the last years.

Paris, France
September 2012

Vladimir M. Akulin

Contents

Chapter 1
Complex Systems and Their Statistical Description

We often meet in physics various quantum systems comprising many quantum levels. One may say that the majority of problems of contemporary interest are of this type. Statistics is a natural way to describe large systems. Initially, the statistical approach to individual complex quantum objects had been formulated in the context of the interpretation of nuclear spectra, where the average positions of the energy levels and their average relative distances were the principal concern. Later, the effects of complexity related to disorder were studied in solid state physics and this gave rise to the theory of dynamical localization, focused on the structure of wavefunctions. The rapid development of laser physics, the molecular beam technique, experiments with cold atoms, femtosecond physics, quantum optics, and quantum informatics has formulated a series of theoretical problems centered on the time evolution of multilevel quantum objects. Analytical description of the dynamics of multilevel quantum systems, including the results which can be obtained with the help of the statistical approach, is the main subject of this book.

The dynamics of multilevel systems is a complicated process, which involves population exchange among many quantum states. Though a few types of multilevel problems can be solved analytically, the Schrödinger equation for a multilevel quantum system typically does not possess an exact analytical solution in terms of elementary or special functions nor in terms of integrals over a reasonably small number of variables. Therefore the exact solution implies the numerical integration of the Schrödinger equation. However, the numerical solution of such problems usually encounters difficulties of three different kinds. The first one is rather evident: as the number of states in a system increases, the numerical approach becomes awkward, since it becomes more and more difficult to handle the increasing amount of information. In some cases, this difficulty may arise even earlier, when in order to simply find the matrix elements of the Hamiltonian and to write down the Schrödinger equation in a finite basis representation, one has to perform complicated multidimensional integrations. The second difficulty arises due to the influence of higher-order resonances which may manifest themselves in a hierarchy of interactions that are different by many orders of magnitude. This makes the

V.M. Akulin, *Dynamics of Complex Quantum Systems*, Theoretical and Mathematical Physics, DOI 10.1007/978-94-007-7205-2_1,

dynamics of a multilevel quantum system considerably different at different time-scales. When the influence of the processes developing at exponentially different time-scales becomes important, the numerical approach fails to select an appropriate step-size for the computational algorithms.

There is an inevitable difficulty of a third kind. Provided one has obtained a numerical solution for a large multilevel quantum system and now knows all of the values of the quantum amplitudes for each state and for each moment of time, how is one to gain insight into this mass of information? How does one reach a consistent conclusion, address the main task of theoretical description and find a correspondence law between the input and the output of the calculations? Evidently, the physically meaningful quantities are just a few cumulative characteristics of the quantum system. But how can one identify these characteristics and find comprehensive relations among them? One may say that here we face a conflict between information and insight: the more information we have, the more difficult it is to derive physically interesting conclusions.

To some extent, this conflict is resolved in the context of the field known as "Quantum Informatics", which has rapidly been developing starting from the last decade of the last century. The essence of this resolution is following. One can abandon the idea to trace in all details the state of a large and therefore exponentially complex quantum system but still look for a rather simple, that is, tractable, prescription telling how can one exert control over such a system in such a way that an initial classically tractable quantum state gets transformed to a classically tractable final state. In other words, a tractable finite state is achieved from a tractable initial state along a trajectory in the Hilbert space of a complex system that goes over intractable, exponentially complex, quantum states.

Complexity and the Ensemble Average

A natural question arises when an increasing number of states in a system renders the numerical treatment inefficient; do we really need to know all of the details of the dynamics of a multilevel quantum system as an output of the calculations, or should we rather look for its general properties by considering ensembles of many similar complex quantum objects? The general properties do not depend on all tiny details of the particular realization, but only on a small number of "crude" characteristics of quantum systems. We therefore arrive at the idea of the ensemble average, which after E. Wigner became the key tool of analysis of complex quantum systems. This is in complete analogy with the thermodynamic approach, which summarizes general properties of macroscopically identical systems by ignoring their microscopic differences.

Historically, the random matrix approach to complex quantum systems and the philosophy of ensemble averages associated with this approach were first suggested by E. Wigner for the analysis of the spectra of complex nuclei. In fact, little is known about systems of many interacting nucleons except some general properties such as symmetry with respect to time inversion, permutations of the particles, and other transformations in phase space. Detailed information relating to the structure of the wavefunctions is hidden in the complexity of the system. Nature affords no opportunity to obtain such detailed information and thus provide insight into this structure in a reasonably simple way. The natural approach of mathematical formalization of our unavoidable ignorance is to assume that all matrix elements of the Hamiltonian $H_{i,j}$ are random numbers that satisfy just a small number of general symmetry conditions such as

$$H_{i,j} = H_{j,i} \quad \text{or} \quad H_{i,j} = H^*_{j,i}, \tag{1.1}$$

for real or complex-valued Hamiltonians, respectively. In traditional calculations, the first step is the determination of the eigenvalues of the Hamiltonians. The logical generalization of this concept to the case of a complex system is to employ the distribution function of the eigenvalues. Three principal statistical distributions are usually considered in the analysis of the spectral properties of complex nuclei and their comparison to experimental data: (1) distribution of the eigenvalues, namely the mean density of states $g(E) \equiv \Delta_1(E)$; (2) distribution of the energy spacing between nearest neighbors $\Delta_2(E_n - E_{n-1})$; and (3) the statistics of the deviations $\Delta_3(E_n - n\bar{g})$ of the energy levels E_n from the positions $\bar{g}n$ suggested by the mean local density of states $\bar{g}$. These universal characteristics are usually employed in other fields of physics whenever the quantum system in consideration possesses a complex spectrum.

The idea to employ the ensemble average for the description of complex quantum systems gave rise to a large branch of contemporary theoretical physics and related mathematics that investigates eigenvalue statistics and the properties of the eigenfunctions of complex Hamiltonians. This approach has been successfully adopted in many areas of nuclear physics, in solid state physics, in spectroscopy, and the theory of quantum chaos. In atomic, molecular, and laser physics, a series of theoretical problems has also been formulated where the time evolution of a complex quantum object is the main interest. The frequency domains of the latter range from seconds for cold atoms to femtoseconds for intramolecular physics. The large number of quantum states participating in the dynamical processes allows one to use the ensemble average idea for the analysis of the behavior of these systems as well. However, the main physical characteristics to be found for these systems, such as the net population of a group of levels or the time-dependence of optical polarization, differ essentially from the quantities that one calculates in nuclear or in solid state physics, where the emphasis is usually upon the statistics of eigenstates or on the typical spatial structure of eigenfunctions.

We should clearly understand the basic assumptions underlying the statistical approach, in order to avoid possible misunderstandings concerning the question of what we can and what we cannot pretend to describe with the help of this method. The essence of the ensemble average idea is the following: we are interested in properties that are not sensitive to the details or to the particular realization of a system, and therefore we consider ensembles of many systems that are different in the tiny details but have the same "crude" characteristics. We calculate the physical quantities of interest for each of these systems and take the average over the ensemble. In this way we find the common properties, while individual features of the systems are washed out. We note that technically there is no need to perform the average after the calculations of physical quantities of interest for each of the systems. On the contrary, it is much more convenient to do it at an earlier stage, before the summation of the all-order perturbation series, for example. This usually leads to a considerable simplification of the calculations that can therefore often be performed analytically.

The choice of ensemble is the most delicate matter of this approach. It mainly relies on our physical intuition. We have to decide a priori which details are important and which are not important for the "crude" properties of systems, and then we have to construct the corresponding ensemble. If, after taking the average, the mean value of the quantity of interest is not zero and its dispersion vanishes, one can say that this quantity is indeed "crude" and moreover the ensemble is chosen properly.

Statistical Ensembles

In our consideration we employ a relatively general model of quantum systems, for which the Hamiltonian is not a completely random matrix but consists of two parts: one, $\hat{H}_0$, is regular and the other, the perturbation $\hat{V}$, is random. The matrix of the random part belongs to the so-called Gaussian unitary ensemble where both the real and the imaginary parts of the Hermitian matrix elements obey Gaussian statistics, that is, they are normally distributed around the zero mean value. We shall not give here any proofs that this is the case for each particular problem, but just note that this natural assumption is in the spirit of the central limit theorem of statistics. We also note that this assumption usually allows us to obtain exact analytical expressions for many physical quantities of interest. Therefore this model may serve as a guide for qualitative understanding of the properties of other multilevel systems, for which the applicability of the statistical assumption is questionable, but the exact analytical solution is not available. Though later on we will be mainly concerned with the case where $\hat{H}_0$ is regular and only $\hat{V}$ is random, while all of the physical properties of the complex system result from the interplay of these two parts, here we briefly present the main classical results of random matrix theory that ignore the regular part and treat the Hamiltonian $\hat{H} = \hat{H}_0 + \hat{V}$ of a complex system as a completely random matrix.

Apart from the Gaussian unitary ensemble $H_{i,j} = H^*_{j,i}$, two other standard distributions with Gaussian statistics of matrix elements are often employed in the cases where a certain residual symmetry of the system considered is important. The Gaussian orthogonal ensemble $H_{i,j} = H_{j,i}$ is applicable to time-reversible systems. This implies that all of the matrix element are real, whereas the Gaussian symplectic ensemble $H_{1i,1j} = H^*_{2j,2i}$, $H_{1i,2j} = -H^*_{2j,1i}$ corresponds to the situation where each matrix consists of four blocks, with the diagonal blocks complex conjugated, and the off-diagonal blocks complex conjugated with inverted signs. The last ensemble describes the time-reversible situation in the absence of central symmetry.

All three Gaussian ensembles result in the same distribution of the eigenvalues of the random matrices with mean density $g(E)$ of the eigenstates given by the Wigner semicircle distribution

$$g(E) = \frac{\sqrt{E - N \langle V^2 \rangle}}{\pi \langle V^2 \rangle}, \tag{1.2}$$

where N is the total number of states and $\langle V^2 \rangle$ is the mean squared element of the random matrix. The main difference between these ensembles manifests itself in the statistics of the energy distances $E_{n+1} - E_n$ between the neighboring levels. The probability to find the nearest neighbor on the small energy distance scales as $E_{n+1} - E_n$, $(E_{n+1} - E_n)^2$, and $(E_{n+1} - E_n)^4$ for the orthogonal, unitary, and symplectic ensembles, respectively. We will not consider here the physical properties associated with this difference but just note that all of these ensembles are "rigid" and small energy intervals between the neighboring levels are unlikely. In other words, the levels of complex systems interact and repel each other.

One more important case of the level statistics has to be mentioned. If a system is composed of two or more independent parts that do not interact among themselves, the spectrum of the compound is a direct superposition of the spectra of the parts. In this case, the neighboring levels may belong to different parts. They do not interact and do not repel each other. Therefore the probability to find levels at a small distance is equal to the mean density of states in this energy domain. Such a situation corresponds to the so-called Poisson distribution, typical of Markovian random processes.

We also note another approach to complex quantum systems, which concentrates not on the statistics of the eigenvalues of random Hamiltonians, but on that of the evolution operators $\hat{U} = \exp(-it\hat{H})$ governing the time evolution of the systems. In contrast to $\hat{H}$, the eigenvalues of $\hat{U}$ are not real but occupy the unit circle in the complex plane. Therefore the statistical ensembles introduced for these quantities by Dyson are called circular ensembles, and these ensembles can also be orthogonal, unitary, and symplectic by analogy to the Gaussian ensembles. Note that the circular ensembles had been originally introduced in nuclear physics, not for the time-dependent evolution operator, but for the scattering matrix, which is roughly speaking the operator of evolution during a finite interaction time interval. The probability P to find the eigenvalues of the scattering matrix at the positions $\{e^{-i\varphi_n}\}$ on the unit circle reads

$$P\left(\{e^{-i\varphi_n}\}\right) \sim \prod_{k<l} \left|\exp\left(-i\varphi_k\right) - \exp\left(-i\varphi_l\right)\right|^\beta \prod_{n=1}^{N} d\varphi_n \tag{1.3}$$

with $\beta = 1, 2, 4$ for the orthogonal, unitary, and symplectic Dyson ensembles, respectively.

For circular ensembles, we suggest here a qualitative interpretation of their physical meaning in terms of the time evolution, which is more relevant to the problems that will be considered later on. The eigenvalues of the operator $t\hat{H}$ are simply the eigenenergies multiplied by the elapsed time, and they may be interpreted as phases accumulated by the states during the period of time t. Exponentiation of $-it\hat{H}$ implies that the n-th eigenstate is multiplied by a phase factor $\exp(-itE_n)$, or one can say that the real spectrum of $t\hat{H}$ "winds" on the unit circle. Two energy eigenstates that are separated by an energy interval $2\pi/t$ turn out to be at the same point on the unit circle. If the time interval t is not too long, these states are far from each other and therefore do not experience a significant repulsion, typical of the neighboring energy eigenstates that result in the "rigidity" of the spectrum. However, for a time-dependent Hamiltonian there is a Fourier component of the interaction at the frequency $2\pi/t$ which repels the states separated by the energy of the corresponding quantum as if they were neighboring. This provides us with an idea of how the eigenstates of the scattering matrix can form a "rigid" ensemble on the unit circle.

Simple and Complex Multilevel Systems

A small number of multilevel problems have exact analytical solutions. The corresponding quantum systems usually have a certain, rather high, symmetry, either explicit or hidden in specific relations amongst the parameters of the unperturbed Hamiltonians $\hat{H}_0$ and the perturbations $\hat{V}$. This symmetry is actually the reason why the system dynamics, depending formally on a large number of the matrix elements of these operators, can still be described either in terms of a finite number of algebraic and special functions, or in terms of a few-fold integrals containing a limited number of combinations of these parameters. The symmetry underlying the exact solutions manifests itself in commutation relations between $\hat{H}_0$, the perturbation $\hat{V}$, and their commutators of higher orders, that altogether form a relatively simple operator algebra with the total number of elements much smaller than the total number of levels in the quantum system. In this case, the multilevel quantum system is in fact a multidimensional representation of a relatively small Lie group corresponding to the algebra.

One of the most illustrative examples is the magnetic multiplets of atomic levels with high angular momentum J subject to the action of magnetic fields—the population dynamics of the $2J + 1$ magnetic components is governed by the momentum operators J_+, J_-, and J_z that form a closed three-element algebra $su(2)$. Another example is that of a harmonic oscillator in an external field, where the Hamiltonian $\hat{a}^\dagger\hat{a} + 1/2$, and the creation $\hat{a}^\dagger$ and annihilation operators $\hat{a}$ also form a closed algebra.

However, this is not the case for the generic $N \times N$ matrices $\hat{H}_0$ and $\hat{V}$. Their higher-order commutators are usually linearly independent of the commutators of lower orders and the matrices themselves, and each new commutator yields a new linearly independent matrix until the total number of different matrices reaches the maximum possible number N^2. A statistical description seems the natural approach which allows one to make meaningful statements about the behavior of such systems. Of course, an important role is played by the quantitative relations between the commutators of different orders. When a kind of hierarchic dependence exists, such that the sizes of the commutators rapidly decrease with increasing order of commutation, the same system may behave as a soluble system on a short time-scale but develop complex dynamics for a longer time domain. An example of such a system is a one-dimensional chain of quantum states with disordered energies and the tight binding coupling, which demonstrates diffusive broadening of the population distribution over the states for short times but manifests the effect of quantum localization in the long time limit.

Complexity and Chaos

One encounters complex quantum systems in physical problems of two different kinds. A problem of the first kind concerns a real physical ensemble of multilevel systems, that is, with a set of quantum objects with the Hamiltonian of each object different from the Hamiltonian of another. One can find the ensemble average Hamiltonian and consider the differences between this average and Hamiltonians of particular systems as a random perturbation. We note that even a moderate perturbation of a multilevel system usually results in a large change of eigenstates and in an appreciable transformation of the energy eigenvalues distribution. Therefore many characteristics of a typical representative in the ensemble can be distinct from the mean characteristics over the ensemble, thus yielding a large dispersion.

A problem of the second kind relates to a single system with complex dynamics. In the classical limit such a system manifests chaotic behavior. Dealing with a classical chaotic system one never tries to follow the trajectory in phase space corresponding to a particular initial condition, but rather one takes a set of initial conditions and uses statistical characteristics such as Lyapunov exponents or the Kolmogorov entropy for a description of the system dynamics. In this case the average over the initial conditions plays the role of the ensemble average.

We note, however, that in quantum mechanics one can usually find a natural representation for the Hamiltonian of a system of the second kind, where it has a simple structure and an explicit physical meaning. It might be for instance a product of physical quantities such as the coordinate, the electric field strength, the angular or kinetic momentum, etc. But in spite of this, a relatively small change in the Hamiltonian also results in a drastic change in the eigenstates and eigenvalues of the system. In this respect, a quantum system chaotic in the classical limit behaves like a system perturbed by a random matrix. The statement that all classically chaotic systems behave on the quantum level of consideration like the ensembles of random matrices is known as the Bohigas–Giannoni–Schmit conjecture.

This situation is typical for quantum systems of two or more spatial degrees of freedom. Indeed, if the unperturbed motion is separable, that is, it can be represented as a combination of independent motions along each degree of freedom, the perturbation usually causes mixing of these motions and destroys the separability of the problem. We can say that the perturbation destroys the quantum numbers, or in terms of classical mechanics, it destroys the integrals of motion. Another way to phrase the idea is to say that there are holonom constraints in the phase space of the separable classical system or in the Hilbert space of a quantum system. In classical mechanics these constraints are known as Kolmogorov–Arnold–Moser tori. The perturbation destroys the constraints and makes the system non-holonom.

In the following chapter we discuss a number of examples of complex quantum systems originating from different physical problems. Here we illustrate the analogy between a randomly perturbed quantum system and a system chaotic in the classical limit for one example, known as the Sinai billiard.

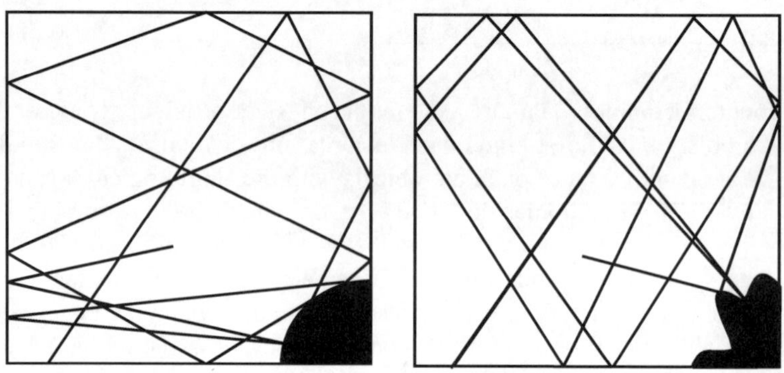

Fig. 1.1 The Sinai billiard and a random system with similar behavior

We consider a particle in a two-dimensional potential well of infinite depth and with a square cross-section of side L. This motion of a particle in such a well is separable and the wavefunctions $\psi_{n,m}$ of the energy eigenstates are the products of the harmonic functions corresponding to each direction $\psi_{n,m} = (2/L)\sin(k_n x)\sin(k_m y)$, where $k_n = 2\pi n/L$. We now place a scattering potential $V(x)$ at the corner of the well as shown in Fig. 1.1. In classical mechanics, for a round scatterer $V(x,y) = 0$ at $x^2 + y^2 > a$ and $V(x,y) = V_0 \to \infty$ at $x^2 + y^2 < a$ this system is known as a Sinai billiard where the motion of the particle becomes chaotic. Apparently it is also the case for a scatterer of an irregular (random) shape and for a large but finite V_0.

Let us now consider the problem quantum mechanically. The particle Hamiltonian reads

$$\hat{H} = \hat{H}_0 + V(x,y) = \frac{1}{2m}\hat{p}^2 + V(x,y), \tag{1.4}$$

and we assume that the potential $V(x,y)$ differs from zero in the region S around the origin $x = y = 0$. We take the eigenstates $\psi_{n,m}$ of the unperturbed Hamiltonian for the basis set of the problem and find the matrix elements of the interaction

$$V_{n,m}^{n',m'} = \frac{4}{L} \int\limits_S \sin(k_n x) \sin(k_m y) V(x,y) \sin(k_{n'} x) \sin(k_{m'} y)\, dx\, dy, \tag{1.5}$$

which turn out to be the Fourier transform of the potential. If $V(x, y)$ is a random function, the matrix elements are apparently random. But for a regular round scatterer they are definitely regular since they have the explicit form

$$\begin{aligned} V_{n,m}^{n',m'} &= \frac{4}{L} \int\limits_0^a r dr \int\limits_0^\pi d\phi \sin(k_n r \cos\phi) \sin(k_m r \sin\phi) \\ &\quad V(x,y) \sin(k_{n'} r \cos\phi) \sin(k_{m'} \sin\phi) \\ &= \frac{2V_0}{L^2} \sum_s c_s \int\limits_0^a r dr J_0(\Delta k_s r) = \frac{2V_0}{L^2} \sum_s c_s \frac{J_1(\Delta k_s a)}{\Delta k_s a}, \end{aligned} \tag{1.6}$$

where $\Delta k_s = \pm\sqrt{(k_n \pm k_{n'})^2 + (k_m \pm k_{m'})^2}$, $J_{0,1}(x)$ are the Bessel functions, and $c_s = \pm 1$.

We know, however, that both problems are equivalent in the classical limit. Therefore it is reasonable to assume that much of this similarity also persists in the quantum case. This implies that many of the predictions made for the systems perturbed by a random matrix will also be valid for a system perturbed by a non-random matrix if such perturbation destroys the separability of the problem and makes the motion chaotic, in agreement with the Bohigas–Giannoni–Schmit conjecture . In algebraic terms, the consecutive commutation of the regular perturbation matrix with the unperturbed Hamiltonian yields in this case the entire set of linearly independent matrices, exactly in the same way as commutation with a random matrix.

Time Dependence and Spectral Properties

Our main concern will be to describe the population dynamics of complex quantum systems under the action of random perturbations. Many of the aspects of the time evolution have specific manifestations in the spectra of the ensemble averaged density of states $g(E)$, or more precisely, in the spectra of the total density of states projected either onto the initial state or onto a given state of the unperturbed system. This is usually the case when the dynamics is governed mainly by a specific unperturbed Hamiltonian, possessing some symmetries or corresponding to a separable motion of different degrees of freedom, while a random perturbation destroys this separability on a longer time-scale. The sharp edges of the state density $g(E)$ that are already present for $E = \pm N \langle V^2 \rangle$ in the Wigner distribution (1.2), a cusp on the dependence $g(E)$ at the point where two Wigner distributions merge, non-analytical dependencies $g(E)$ typical both for fractal structures and for systems corresponding to random walks in the classical limit, and some other spectral signatures, are usually associated with non-exponential decay processes. These processes are also known as coherent damping, which usually corresponds to an intermediate asymptotic regime, that is, to a relatively short time domain where the random perturbation has already started to affect the separability of the unperturbed system, but has not yet destroyed it completely.

Some other processes developing on a longer time-scale do not have signatures in the ensemble averaged density of states but manifest themselves in the statistics of the relative positions of the levels. These processes, known as recurrences and complete or fractional revivals, usually imply complete or partial returns of the population from different groups of quantum levels to the initial state. Interference of the returns from different groups of levels may result in an incomplete decay of the initially populated states, which is one of the manifestations of so-called dynamical localization.

An interesting class of problems is represented by the case where the interaction which destroys the separability of the unperturbed motion changes significantly, by orders of magnitude, for different degrees of freedom. Therefore, in the course of time more and more degrees of freedom are getting involved in the dynamical process. The situation resembles the case known in classical mechanics as Arnold diffusion. For the corresponding quantum case, the recurrences and revivals play a very important role, systematically returning the population back to the initial state while still being unable to prevent this state from complete decay. The spectral density of the states accessible from the initial state is essentially a discontinuous and irregular function of the energy in this case. In the long time limit it corresponds to specific time-dependences of the population of the initial state.

Main Parameters and Quantitative Criterion of Complexity

The commutation relations between $\hat{H}_0$ and $\hat{V}$, generating at each order of commutation new linearly independent operators, cannot yet guarantee that the considered quantum object conforms to the model of a complex system. The quantitative parameters also play a crucial role, by indicating whether the action of the new operators can be considered as a small perturbation or whether the all-order perturbation theory is required for an adequate description. The main parameters that allow one to decide whether the system is complex are the average density of states g of the unperturbed Hamiltonian and the mean squared strength of the perturbation $\langle V^2 \rangle$.

The average density of states in the units where $\hbar = 1$ has the dimensionality of time and represents the so-called Heisenberg time of return, a typical time when the return of the system to an initial state given by a superposition of the eigenstates of $\hat{H}_0$ has to be taken into account. The mean squared interaction enters the transition rate $2\pi g \langle V^2 \rangle$, suggested by the Fermi golden rule for the probability of transition per unit time from a given level to a group of levels with spectral density g that are coupled to this level by an interaction characterized by the mean squared matrix element $\langle V^2 \rangle$. Therefore the natural criterion of complexity reads $2\pi g \langle V^2 \rangle g \gg 1$, which implies that the probability of transition during the return time should be large.

Unless the special algebraic reasons emerging from the hidden symmetry of the problem result in the rapid decrease or vanishing of the commutators with the increase of the commutation order, the criterion $2\pi g^2 \langle V^2 \rangle \gg 1$ also guarantees that the commutators can induce transitions. Since each new commutator is a linearly independent operator, the variety of linear combinations becomes more and more rich with the course of time. The interference of all of these transitions makes the dynamics of the system complex.

Time-Dependent Perturbations and Control

A time-dependent perturbation enriches the system dynamics since each harmonic of the perturbation couples the levels separated by the energy of the corresponding quantum. For the periodic perturbation, the structure of the evolution operator is given by the Floquet theorem, which roughly speaking states that the amplitude $\overline{\psi}_m$ of each harmonic m of the state vector can be considered as an independent variable in a Schrödinger equation with time-independent parameters. The order of this matrix equation is the product of the dimension of the system Hilbert space and the total number of important harmonics. The order therefore approaches infinity when all harmonics are equally important. The only technical question arising for this approach is how to circumvent the ambiguity in relating the initial state of the system to the initial state of the harmonics.

For non-periodic perturbations, the time-dependence can also be taken into account by introducing an additional degree of freedom, which is now associated, not with a discrete index m of $\overline{\psi}_m$, but with a continuous variable τ specifying the corresponding component $\overline{\psi}(\tau)$. In this new representation the Hamiltonians also experience a transformation by acquiring an additional operator term related to the new degree of freedom. Therefore the commutation relations between the unperturbed Hamiltonian and the perturbation may (and usually do) change, most strongly affecting "simple" multilevel systems which possess an underlying algebraic symmetry.

The time-dependence may destroy the holonom constraints in a system including the energy conservation constraint, thus enriching the evolutionary capability of the latter. The question arises: can one employ this destruction and make the system evolve in an arbitrary predescribed way? In other words, can one force the system with a Hamiltonian $\hat{H}_0$ to behave "a la carte", that is, as if it has another predetermined desired Hamiltonian $\hat{H}_d$? The answer to this question is positive for a system that conforms to the Jurdjevic–Sussman theorem and corresponds to the compact algebra generated by the commutators of $\hat{H}_0$ and $\hat{V}(t)$. The main problem is how to construct an effective algorithm that would specify the required time-dependence.

A bunch of new ideas, forming a new field known as Quantum Informatics, has appeared and widely spread during the last decade of the last century. It has brought about a revolutionary change in the vision of the complex quantum systems. When composed and controlled in a special way, a collection of rather simple identical quantum system interacting among themselves in a controlled way comprise a qualitatively new combined quantum system, a quantum computer, which can be employed for quantum calculations. The latter can be viewed as specific control algorithms that rely on the opportunity offered by quantum mechanics to simultaneously process different classical information strings by taking their linear combinations in the Hilbert space. Though the algebraic structure of the Schrödinger equation in the case of quantum computer is almost identical to that typical of the case of Random Systems, the main physical questions addressed in these two cases are rather different and even opposite. One never asks for "typical" or "crude" properties of a quantum computer but looks for a rather involved deterministic control prescriptions realizing the quantum algorithms.

Structure of the Text

The manner of presentation of the material serves our main objective, which is to invite the reader to consider different dynamical problems that belong to various domains of quantum physics from the point of view of complex quantum systems and attribute these problems to one of a number of different cases where exact and physically consistent conclusions can be drawn. In other words, the aim is to develop, as much as possible, the methods of solution rather than to calculate particular physical quantities for the particular physical problems. The latter will be simply impossible considering the vast number of domains where complex quantum systems can be encountered. Moreover, in some cases and for some physical properties, an alternative approach to the problem might be equally efficient, which does not, however, mean that the approach of the problem as a complex system is useless. Sometimes there may be additional insight into the reasons why the system behaves in such a way.

In Chap. 2, a number of examples are given in order to demonstrate the different physical domains where complex quantum systems can be encountered. Both the situation involving a physically existing ensemble of systems, and that involving a single system unknown in detail, are considered from the same standpoint; in other words, we do not specify whether a complex system means a single system with complex dynamics or an ensemble of systems, unless confusion is possible. Some of the examples are well known and widely investigated while others are rather new. Most of them are drawn from the fields of laser physics, physical chemistry, the physics of clusters, and quantum optics, where the dynamical aspects of multilevel quantum systems are the primary concern, whereas examples from solid state physics and nuclear physics—two domains where static aspects like quantum level statistics are explored by the method of random matrices and widely described in the literature—are presented hardly at all. In some examples we also show analogies with classical physics where the different modes of oscillations of a multidimensional system play the role of quantum states. The main purpose of this chapter is to show how one can describe the main features of the dynamical processes in terms of multilevel quantum systems.

In the next chapters we concentrate on the approaches to exactly soluble multilevel problems. Each chapter starts with a simple problem, usually well-known, which serves the principal role of introducing the technical tools and for elucidation of the physical meaning and main parameters governing the dynamics. Technical complexity usually increases in the subsequent parts of the chapters. In these parts we have tried to give as many details as the reader would need for following the narration without doing the calculations, or doing just a few of them.

In Chap. 3 we consider the two-level system and the level–band system, which are problems that can be solved exactly, without an ensemble average. We find the physical parameters responsible for the time evolution of such systems, and demonstrate the important role of coherence, which persists in the large multilevel systems and results in many interesting phenomena, such as quantum revivals, for instance. We discuss the role of the spectrum density and the spectrum edges, allow for the correlations in the matrix elements of interaction in the case of two levels interacting with a band resulting in specific spectral structures like the Fano profile, and analyze the situation when the interaction matrix elements may differ significantly, by orders of magnitude, resulting in non-exponential and incomplete decay of the level.

In Chap. 4 we consider the population dynamics in multilevel quantum systems perturbed by a time-independent random matrix. We illustrate the method of the ensemble average and demonstrate the technique, which can later be used for the description of more complicated problems. It is worth mentioning that in many respects, this technique is an analog of the diagram technique for the many body problem that has been successfully employed in solid state physics for the description of metals at low temperatures. We only employ the representation performed in the spirit of Grotrian diagrams, that are more natural for atomic and molecular physics, where the unperturbed motion is depicted by levels (or dots) and the transitions are shown by lines. The main features of the diagram technique remain the same, including the idea of renormalization of the state energies, better known in atomic physics as "dressed states". We consider the role of the state degeneracy, the coherent damping associated with it, as well as the specific shapes of the spectral lines resulting from the coherence. We pay attention to the effects related to correlations in the perturbation matrices, when the rank of the perturbation matrix is much smaller compared to its order.

In Chap. 5 we consider the effect of a time-dependent perturbation on the population dynamics in "simple", exactly soluble quantum systems such as the two-level system and the level–band system. For a number of cases, the additional dimension associated with the time-dependence can still result in an exactly soluble problem, although it complicates the algebraic relations amongst the Hamiltonian, the perturbation, and their commutators. The most well-known examples are the Landau–Zener problem and the Demkov–Osherov problem. We also discuss the applicability of the semiclassical description also known as non-adiabatic transitions in time-dependent systems conforming to the Dykhne model, and discuss non-adiabatic transitions at the continuum edge.

In Chap. 6 we consider complex quantum systems subject to a time-dependent perturbation that also become exactly soluble following the ensemble average. We focus on the case where the system is perturbed by a random matrix with a perturbation strength with an arbitrary time-dependence. We demonstrate that the shape of the population distribution as a function of energy results from the interplay between the size of the perturbation and the spectral width of the time-dependence. We also consider the harmonic and bi-harmonic perturbation of complex systems, as well as the bi-harmonic perturbation of a two-band system. The main purpose of this consideration is to illustrate how the additional degree of freedom, associated with the time-dependence, manifests itself in the population distribution over the energy scale. We discuss the linear optical response, the delayed coherent polarization or optical echo, and the susceptibility of complex systems, where one can see a difference between the systems corresponding to the unitary and the orthogonal ensembles.

In Chap. 7 we consider the situation where the levels are arranged in a one-dimensional relay-like chain and the interaction occurs only among the closest neighboring levels. We start with the problems that have exact analytical solutions, conforming to one of the low-dimensional algebras of operators. This concerns levels with linearly increasing energy and a constant coupling, the harmonic oscillator in the presence of dipole and Raman harmonic perturbations, angular momentum in an external magnetic field, and a few other examples where the exact solution in terms of hypergeometric functions exists. For the case of smooth variation of the parameters, we describe an approach in the spirit of the WKB approximation. We also consider the one-dimensional chain of irregularly detuned levels, the so-called Lloyd chain, whose spectrum and localized wavefunctions have been widely explored, but whose population dynamics have been much less addressed. In this context we introduce the field-theory methods (also convenient for some other applications) that allow one to calculate ensemble averaged determinants of random matrices, representing them as functional integrals over regular and Grassmann variables.

In Chap. 8 we consider the dynamics of composed complex quantum systems where the perturbation matrix has a block structure. We discuss the diffusion regime in the relay-like structures for various dependencies of the transition rate on the block number that allow an exact description in terms of the hypergeometric functions, and show the important role of the correlations in the perturbation matrix elements. We also concentrate on the dynamics of multilevel systems at long times. We illustrate the dominant role of the quantum interference effects on the population dynamics. In particular, we calculate the probability for a complex system to stay in the initial state at asymptotically long time. This characteristic is complementary to the property of quantum localization in disordered systems.

In Chap. 9 we discuss the manifestation of quantum complexity in the density of states and in the dipole absorption profiles along with a relatively simple way to calculate these profiles for the case when the spectral structures typical of the unperturbed Hamiltonian are still not completely washed out by the random perturbation. We demonstrate the spectral manifestation of the hierarchical structure of the system, originating from the fractal nature of an object and the transformations that this spectrum experiences under a random perturbation. We also consider the density of states of so-called Euclidean Random Matrices that are matrices given in the representation where each basis vector is associated with a point randomly placed in a multidimensional space while the matrix element coupling two states depends on the Euclidian distance between the relative points, exclusively.

In Chap. 10 we concentrate on the essentially quantum characteristic of the composed systems, so-called quantum entanglement. This characteristic cannot be treated as a physical one, since it depends on the partition, that is the way we separate the system in parts. We start the consideration with the simplest examples of entanglement in bipartite systems, then we illustrate the importance of entanglement in the context of quantum informatics algorithms, and finally focus on the classification of entanglement in the multipartite systems in pure quantum states.

In Chap. 11 we consider open quantum systems, paying mainly attention to entanglement of such systems with their environment. We present traditional description based on the Master equation and the Liouville formalism, present an exactly soluble model and trace the dynamics of entanglement between the system and the environment with the help of this model. We conclude this chapter by discussing possible approaches to entanglement classification for mixed quantum states.

In Chap. 12 we return to the complex systems with time-dependent Hamiltonians, but this time in the context of the possibility to exert complete control over such systems and protect them from the loss of coherence with the help of such a control. The intervention algorithms that offer the possibility to achieve complete control over the dynamics are discussed along with the algorithms for the complete protection against unwanted interventions of the environment.

There are no literature references throughout the text, apart from the names of the authors traditionally associated with the well-known results of methodical importance. One of the reasons for this is the fact that many similar or sometimes even identical results have been independently obtained in different domains of physics and are known under different names. For instance, the dipole susceptibility of highly excited molecules has almost the same structure as conductance in metals, while the tunneling transparency of disordered solids has much in common with the cross-section of multiphoton ionization of atoms. The conductivity of narrow-band solids has exactly the same mathematical description as the multiphoton off-resonant excitation of ro-vibronic states of diatomic molecules. The interaction of resonant atoms in a photonic band gap is similar to the Raman transitions between two vibrational states of a molecule. The Anderson localization turns out to be analogous to the Fermi acceleration of the quantized kicked rotor, and the Breit–Wigner distribution in nuclear physics is identical to the Lorentzian profiles of optical line shapes, and also similar to Feshbach resonances for molecular collisions.

The idea of renormalization, widely employed in particle physics, practically coincides with the concept of dressed states for the atoms in strong laser fields, etc. In such a situation, the references to different physical domains throughout the text would be confusing. That is why the review of the literature is performed in a separate, and last, part of the book, Chap. 13, although even there the list of literature cannot be complete and in fact is far from being exhaustive. A relatively short review is presented in the form of problems which the reader can consider with the help of the approaches developed and compare the formulations of the problems and the results with the original papers. More detailed information is available in the monographs listed in accordance with the distribution of topics among the chapters.

However, the reader is assumed to know the main concepts and methods of quantum mechanics and mathematical methods of physics. For the fundamentals of quantum mechanics one can consult the classical series of textbooks by L. D. Landau and E. M. Lifshits [146] as well as *Quantum mechanics* by C. Cohen-Tanoudji, B. Diu, and F. Laloe [58]. The main mathematical tools can be found in the textbook by J. Mathews and R. L. Walker [160] *Mathematical methods of physics.* For more special aspects of the mathematical technique including algebra and group theory, the books *Mathematical methods of classical mechanics* by V. I. Arnold, A. Weinstein, and K. Vogtmann [10], *Theory of group representations and applications* by A. Barut and R. Raczka [14], *Integrable systems of classical mechanics and Lie algebras* by A. M. Perelomov [181], and the article *Universal integrals of motion and universal invariants of quantum systems* by V. V. Dodonov [71] can be recommended.

Chapter 2
Examples of Complex Systems

Molecular and laser physics offer many examples of complex quantum systems that demonstrate the importance of characteristics in disordered spectra other than the density of states or density of states correlations. Here we briefly discuss some of these systems, which have been studied intensively because of their fundamental interest or practical importance.

We start the consideration with problems related to the domain usually known as the interaction of radiation with matter, which is mostly dedicated to atoms and molecules in intense resonant laser fields. In these types of problems, one can usually separate one or a few optically active degrees of freedom that interact with the laser field directly, while the rest of the degrees of freedom are not dipole active, but simply are coupled to the optically active ones by intra-molecular or intra-atomic interactions. The presence of the latter makes the system complex. Highly excited atomic states that belong to the Rydberg series have a rich spectrum and also form a complex system, being perturbed by a strong laser or RF field. However, the multilevel system does not necessarily correspond to highly excited states. Polyatomic molecules, even in the vibrational ground state, possess a rather rich rotational spectrum, which can be strongly affected by an external field. Additional complexity is associated with the presence of multiphoton resonances that correspond to the simultaneous absorption of several laser quanta.

Chemical reactions offer a large number of examples of complex quantum systems. Each molecule is already a complex object in itself, whereas two colliding molecules, even in the absence of chemical reaction, form a complex compound with a rich spectrum of vibrational states, the population of which changes significantly during the collision. An additional source of complexity is the non-adiabatic interaction of the electronic and vibrational degrees of freedom inside a single vibrationally excited molecule which results in intra-molecular energy conversion.

The situation becomes even more involved in the case when the molecules are highly excited electronically, approaching molecular Rydberg states. In such a situation, the adiabatic separation of the fast electronic and slow nuclear motions

V.M. Akulin, *Dynamics of Complex Quantum Systems*, Theoretical and Mathematical Physics, DOI 10.1007/978-94-007-7205-2_2,

fails, since the period of rotation of an electron along a Rydberg orbit becomes comparable to, and sometimes even longer than, the vibrational period. The electronic and vibrational states get strongly entangled and form a common complex spectrum. The collision of a Rydberg molecule with another molecule may result in collisional ionization which adds a continuous component to the system spectrum. An even more complicated situation arises when we consider an ensemble of many interacting Rydberg atoms. Excitation exchange among the atoms results in the motion of populations over the volume occupied by the gas. The spectrum now becomes very rich and involved, corresponding to the mobility of different excitons over the volume.

Another possibility to form complex systems is to put many identical atoms together, such that they form an atomic cluster. Even in the case when each atom has only two quantum states, the spectrum of a compound containing n atoms already has $N = 2^n$ quantum levels and becomes a complex system in the presence of even a very weak interatomic interaction. This spectrum becomes completely disordered if we take into account the interaction of the electrons with atomic vibrations or phonons. This quantum complexity manifests itself in the thermodynamical as well as in the optical properties of the clusters. The complex spectra of clusters are not necessarily formed as the result of atomic forces. The interaction can be mediated by the electromagnetic field. In this case one can see the cooperative interaction of the atoms with the external electromagnetic field.

Quantum complexity is not only typical of atoms and molecules. It has parallels with multidimensional classical systems such as semiclassical translational motion of ions in ionic traps, nonlinear interaction of photons in a resonator, and mobility of electrons on surfaces. In this chapter we show how the equation of motion can be written in a similar form for all of the aforesaid examples. All of the systems, including classical systems, conform to one of many possible models of complex quantum systems.

The statistical description to complex quantum systems is based on the ensemble average. We append the list of examples admitting statistical description by a few examples of the opposite situation, where statistical description is inadequate. One of them is the case of the systems studied in the context of Quantum Information and Quantum Communications—two domains of physics rapidly developing during the last decades. Though the Schrödinger equation describing the dynamics of such systems as well as the underlying algebras, may have very much in common with that for the random systems, the main question one addresses for these systems is completely different—one wants to know whether or not an exhaustive control over these systems can be exerted, such that these system can mimic the dynamics of other quantum systems or to be employed for realization of quantum computations. The other case belongs to the interface between physics and biology. The biological devise responsible for photosynthesis, from the viewpoint of physics, can be seen as a complex quantum network having a huge number of quantum levels. Still, as a result of biological evolution, this system acquires very specific properties maximizing the efficiency of the photon absorption and utilization, which are completely atypical for the statistically most probable realization of similar multilevel systems.

2.1 Molecules and Atoms in Laser Fields

2.1.1 Laser Breaking of a Weakly Bonded Complex

The difference in the vibrational frequencies of a given molecule containing different isotopes is exploited by the technique of laser isotope separation. Let us consider in this context a coordinated compound that is a simple molecule, IsO_2 for example, which is surrounded by a number of relatively weakly bound molecular complexes or ligands as shown in Fig. 2.1. Here Is stands for an element that has

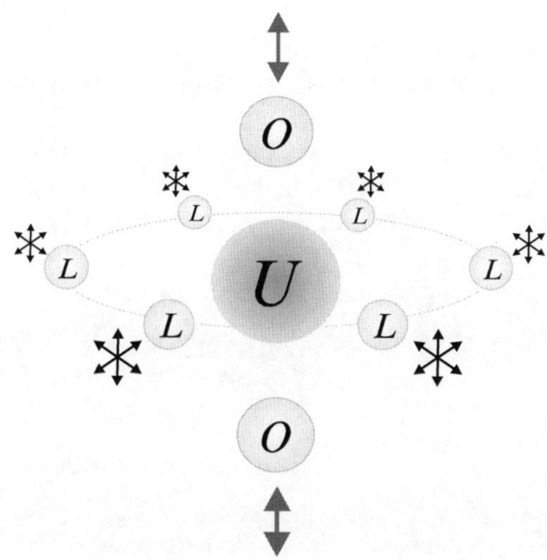

Fig. 2.1 A weakly bound dioxide compound. The dipole-active vibrations of the oxygen atoms are much faster than the chaotic multidimensional motion of the molecular groups surrounding the dioxide

variable isotopic modifications. The central molecule is subjected to the action of the infrared laser which is in resonance with a single dipole-active vibrational mode. The ligands may both alter the frequency of the vibration, and quench the vibrational excitation of the central molecule. The energy of one vibrational quantum of the excited mode is usually sufficient to destroy the weakly bound ligands, and therefore the transfer of the vibrational energy from the molecule to the surrounding complexes immediately results in the dissociation of the compound.

How can one formulate this problem mathematically? We use two indices to describe the quantum state of the system—the index n which enumerates the vibrational state $\psi_n(q)$ of the mode q of the central molecule excited by the radiation, and the index m which enumerates the energy eigenstate of the entire system of the molecular compound provided the vibrational state n of the excited mode is fixed. We denote by $\Phi_{n,m}(Q)$ the corresponding wavefunctions of the compound, where Q stands for all of the set of N coordinates of the compound other than q. We note that the Hamiltonian of motion for the surrounding molecules depends on the vibrational state of the central molecule, since the vibrationally excited central molecule has both a different size and different force constants for the weak bonds compared with the ground vibrational state of the central molecule. We therefore write the Hamiltonian of the entire compound in the form

$$\hat{H} = n\hbar\omega_{IsO_2}\delta_{n,n'}\delta_{m,m'} + H_{m,m'}(n)\delta_{n,n'} \tag{2.1}$$

where we still neglect the influence of the energy transfer from the central molecule to the periphery, and the excitation by the laser field. The eigenenergies

$$E_{n,m} = \hbar\omega n + \Delta_{n,m} \tag{2.2}$$

correspond to the eigenfunctions $|\Psi_{n,m}\rangle = |\psi_n\rangle|\Phi_{n,m}\rangle$.

Now let us allow for the external resonant laser field which induces transitions among the states of the system. The probability amplitudes of the transitions are given by the product of the electric field strength $E = \mathcal{E}\cos\omega t$ and the matrix elements of the dipole moment operator $d_{n,m}^{n',m'} = \langle n, m|e\hat{q}|n', m'\rangle$. The latter is the product of two parts. The first part originates from the excited vibrational mode of the central molecule and is given by the dipole matrix element of the fundamental transition, whereas the second part arises from the overlap integrals of the pairs of wavefunctions of the surroundings, corresponding to the central molecule in the ground vibrational state and in the first excited vibrational state, respectively. It reads

$$\begin{aligned} d_{n,m}^{n',m'} &= \int \psi_n(q)eq\psi_{n\pm1}(q)dq \int \Phi_{n,m}(Q)\Phi_{n\pm1,m'}(Q)d^NQ \\ &= d\sqrt{n}\int \Phi_{n,m}(Q)\Phi_{n\pm1,m'}(Q)d^NQ. \end{aligned} \tag{2.3}$$

Here we have employed the harmonic oscillator model for the vibrational modes of the central molecule which gives the selection rules $n - n' = \pm1$ and have denoted by e the effective charge of the mode q.

Now we assume that the motion of the surroundings is chaotic and therefore a small change in the Hamiltonian of this motion associated with the excitation of the central molecule results in a big change of the eigenfunctions $\Phi_{n,m}(Q)$. In other words the vibrational excitation of the mode q results in a rotation of the ligand energy eigenstates basis in Hilbert space by large angles around many axes. Therefore for an ensemble of molecules, the statistical approach is the natural way to describe the quantum dynamics, and the normal distribution is the logical assumption for the distribution functions of the overlap integrals $\langle\Phi_{n,m}|\Phi_{n',m'}\rangle$ in (2.3).

One can write the Schrödinger equation for the amplitudes $\Psi_{n,m}$ of the states $|\Psi_{n,m}\rangle$ in the form

$$i\hbar\dot{\Psi}_{n,m} = (\hbar\omega_{IsO_2} + \Delta_{n,m})\Psi_{n,m} + \cos\omega t\sum_{m'} V_{n,m}^{n\pm1,m'}\Psi_{n\pm1,m'} \tag{2.4}$$

where $V_{n,m}^{n\pm1,m'} = \mathcal{E}d_{n,m}^{n\pm1,m'}$.

If we now take the resonant approximation, that is, we take the amplitudes in the form

$$\Psi_{n,m} = \psi_{n,m} \exp\{in\omega t\}, \tag{2.5}$$

substitute (2.5) into (2.4), and neglect the rapidly oscillating terms, we arrive at

$$\begin{aligned} i\hbar\dot{\psi}_{n,m} &= (\omega_{IsO_2} - \omega + \Delta_{n,m})\psi_{n,m} \\ &+ \sum_{m'} \left(V_{n,m}^{n+1,m'} \psi_{n+1,m'} + V_{n,m}^{n-1,m'} \psi_{n-1,m'} \right). \end{aligned} \tag{2.6}$$

This is the Schrödinger equation for the slow amplitudes $\psi_{n,m}$. The Hamiltonian of this equation contains two components which we shall consider as random: $\Delta_{n,m}$ gives the random deviations of the level energies from their harmonic positions $\hbar n\omega$, while $V_{n,m}^{n\pm1,m'}$ allows for the random coupling of these states. The problem now is to find some average characteristics of these matrices which govern the population dynamics of the system. That would allow us to identify the common properties of all similar systems, provided such common properties exist.

2.1.2 Laser-Induced Electronic Transitions in Molecules

There is a number of physical systems that are similar to the one just considered. Thus far we were considering the excitation of a vibrational mode of the central molecule and the effect of this vibration on the weak bonds in the compound. The crucial step there was the adiabatic separation of the rapid vibrations and the slow motion of the surrounding molecules. This is completely analogous to the adiabatic separation of the electronic and vibrational motions in molecules, and therefore we face the same situation considering the electronic excitation of polyatomic molecules in laser fields.

This is illustrated in Fig. 2.2. The laser field induces transitions among the manifolds of vibrational states corresponding to two different electronic terms. Transition amplitudes are proportional to the Frank–Condon factors, that is, the overlap integrals $\langle \Phi_m^0 | \Phi_{m'}^1 \rangle$ of the vibrational wavefunctions Φ_m^0 in the ground electronic term, and the vibrational wavefunctions $\Phi_{m'}^1$ in the excited term. The corresponding Schrödinger equation for slow amplitudes $\psi_{n,m}$ of the vibrational states $|\Phi_{n,m}\rangle$ of the n-th electronic term ($n = 0, 1$) reads

$$\begin{aligned} i\hbar\dot{\psi}_{1,m} &= (E_{1,m} - \hbar\omega)\psi_{1,m} + \sum_{m'} V_{1,m}^{0,m'} \psi_{0,m'} \\ i\hbar\dot{\psi}_{0,m} &= E_{0,m}\psi_{0,m} + \sum_{m'} V_{0,m}^{1,m'} \psi_{1,m'}, \end{aligned} \tag{2.7}$$

where $V_{n,m}^{n',m'}$ now stands for the probability amplitudes of the laser-induced transitions.

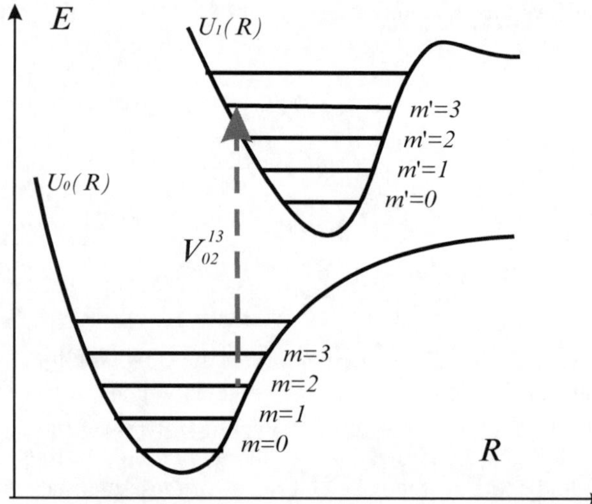

Fig. 2.2 Electronic transition in a molecule

The nuclear motion of vibrationally excited polyatomic molecules is usually chaotic. This implies that the vibrational wavefunction changes drastically and in an irregular way when we move from a given energy eigenstate to a neighboring one. This is even more the case for the vibrational functions corresponding to different electronic terms, which implies that the overlap integrals $\langle \Phi_m^0 | \Phi_{m'}^1 \rangle$ change completely with any change of the indices m and m'. One, therefore, may consider the matrix element of the dipole transition

$$V_{0,m}^{1,m'} = \mathcal{E} d_{0,m}^{1,m'} = \langle \phi_0 | \mathcal{E} \hat{d} | \phi_1 \rangle \langle \Phi_m^0 | \Phi_{m'}^1 \rangle \tag{2.8}$$

as a random value. Here ϕ denotes the electronic wavefunctions, and Φ the vibrational wavefunctions dependent on all the nuclear coordinates.

2.1.3 *Vibrational Excitation of Polyatomic Molecules*

We encounter the same situation for the vibrational motion of a polyatomic molecule excited by an infrared laser field resonant to one of its vibrational modes. In this case the Schrödinger equation formally coincides with (2.6) where n denotes the number of absorbed laser quanta. Let us consider it in more detail. The anharmonicity of vibrations makes the motion of the polyatomic molecule chaotic, and hence the matrix elements of the dipole moment acquire a random character. We illustrate this situation in the most simple example of a threefold degenerate mode of a symmetric polyatomic molecule XY_4 or XY_6. In the harmonic approximation excited vibrational levels of the mode are multiply degenerate. To each eigenstate of the degenerate mode $|I_1, I_2, I_3\rangle$ one can attribute three quantum numbers I_1, I_2, and

I_3, which correspond to numbers of quanta in each of three vibrational degrees of freedom comprising the mode. The energy of the state reads $E_{I_1,I_2,I_3} = \hbar\omega_{vib}(I_1 + I_2 + I_3)$, where the vibrational frequency ω_{vib} coincides with the frequency of the external laser field ω. For light absorption three possible dipole transitions from the state $|I_1, I_2, I_3\rangle$ are allowed by the selection rules. They correspond to an increase by one of the number of vibrational quanta in each degree of freedom:

$$|I_1, I_2, I_3\rangle \to \begin{cases} \to |I_1 + 1, I_2, I_3\rangle \\ \to |I_1, I_2 + 1, I_3\rangle \\ \to |I_1, I_2, I_3 + 1\rangle. \end{cases} \tag{2.9}$$

We now take into account the anharmonicity of the vibrations. In a symmetric, three-dimensional oscillator the anharmonicity consists of three parts. The first part $\hat{H}_1$ is the regular, one-dimensional anharmonicity which shifts the positions of levels of each degree of freedom independently:

$$\hat{H}_1|I_1, I_2, I_3\rangle = \alpha(I_1^2 + I_2^2 + I_3^2)|I_1, I_2, I_3\rangle, \tag{2.10}$$

where by α we denote the corresponding anharmonicity constant. The second part is the cross-anharmonicity. It allows for the change in frequency of one degree of freedom following the excitation of another. We denote the corresponding constant β and write the Hamiltonian in the form

$$\hat{H}_2|I_1, I_2, I_3\rangle = \beta(I_1 I_2 + I_2 I_3 + I_3 I_1)|I_1, I_2, I_3\rangle. \tag{2.11}$$

Neither the first nor the second part allows for energy transfer between the degrees of freedom of the degenerate mode. They do not destroy the separability of the problem and each degree of freedom may thus far be considered as dynamically independent. The third part of the anharmonicity, however, is responsible for the mode mixing. The Hamiltonian $\hat{H}_3$ corresponding to this part is non-diagonal in the representation of the quantum numbers I_1, I_2, I_3, and mixes the states

$$|I_1, I_2, I_3\rangle \leftrightarrow |I_1 - 2, I_2 + 2, I_3\rangle \leftrightarrow |I_1 - 2, I_2, I_3 + 2\rangle \leftrightarrow \text{etc.} \tag{2.12}$$

by transferring pairs of quanta from one degree of freedom to another. It does not commute with $\hat{H}_1$ and $\hat{H}_2$. One can write down this part of the anharmonicity Hamiltonian in terms of the creation $\hat{a}_j^\dagger$ and annihilation $\hat{a}_j$ operators of the vibration of the j-th degree of freedom:

$$\hat{H}_3 = \gamma(\hat{a}_1\hat{a}_1\hat{a}_2^\dagger a_2^\dagger + \hat{a}_1\hat{a}_1\hat{a}_3^\dagger a_3^\dagger + \hat{a}_2\hat{a}_2\hat{a}_3^\dagger a_3^\dagger + \text{c.c.}), \tag{2.13}$$

hereafter c.c. denotes the complex conjugate expression. Note that the third part of the anharmonic Hamiltonian commutes with the total number $I = I_1 + I_2 + I_3$ of the quanta in the mode, and leaves this quantity intact.

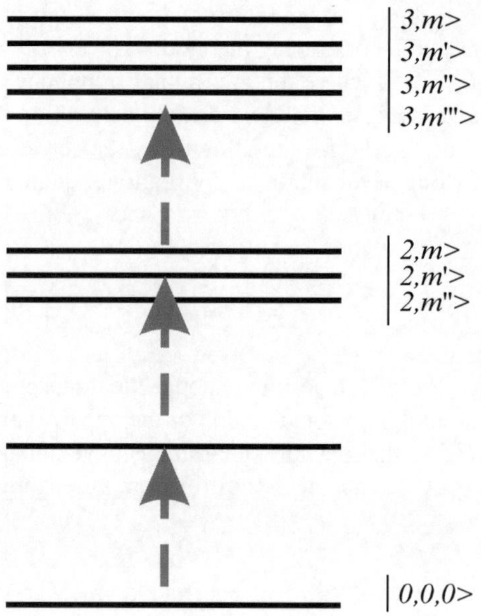

Fig. 2.3 Anharmonic splitting of a threefold degenerate mode

Diagonalization of the sum of the three Hamiltonians $\hat{H}_1 + \hat{H}_2 + \hat{H}_3$ at fixed I gives the eigenstates $|mI\rangle$ of the threefold degenerate mode. Here m enumerates the levels corresponding to a given I. The many-fold degeneracy of the state is now raised, as shown in Fig. 2.3. Usually the results of diagonalization are very sensitive to the values of the anharmonicity constants α, β, and γ, and both the eigenvalues $E_{I,m}$ and the eigenstates $|mI\rangle$ of the Hamiltonian change drastically with a relatively small change in each of the parameters. This is typical of chaotic systems and implies that if we cast the eigenstates $|I, m\rangle$ of the three-dimensional anharmonic oscillator in terms of linear combinations of harmonic states $|I_1\rangle$, $|I_2\rangle$, and $|I_3\rangle$:

$$|I, m\rangle = \sum_{I_2, I_3} c^{I,m}_{I_2,I_3} |I - I_2 - I_3, I_2, I_3\rangle, \tag{2.14}$$

the expansion coefficients $c^{I,m}_{I_2,I_3}$ will be complex functions of I_1, I_2, and m that strongly depend on α, β, and γ. Therefore the matrix of dipole transitions among the states $|Im\rangle$

$$\langle I, m|\hat{d}|I + 1, m'\rangle = \sum_{I_2, I_3} \Big(c^{I,m}_{I_2,I_3} \langle I_1|\hat{d}|I_1 + 1\rangle c^{I+1,m'}_{I_2,I_3} + c^{I,m}_{I_2,I_3} \langle I_2|\hat{d}|I_2 + 1\rangle c^{I+1,m'}_{I_2+1,I_3} + c^{I,m}_{I_2,I_3} \langle I_3|\hat{d}|I_3 + 1\rangle c^{I+1,m'}_{I_2,I_3+1} \Big) \tag{2.15}$$

is an irregular matrix constrained only by the condition $I - I' = \pm 1$. If we replace it by a random matrix we arrive at the Schrödinger equation

$$\begin{aligned} i\hbar\dot{\psi}_{I,m} &= (E_{I,m} - I\hbar\omega)\psi_{I,m} \\ &+ \sum_{m'}\left(V_{I,m}^{I+1,m'}\psi_{I+1,m'} + V_{I,m}^{I-1,m'}\psi_{I-1,m'}\right), \end{aligned} \tag{2.16}$$

which has the same structure as (2.6).

2.1.4 Transitions Among Levels with Fine Structure

We start with the fine structure of the rotational states of spherical-top symmetric polyatomic molecules which also conform with the random matrix model. Spherical-top molecules have three equal components of the inertia tensor, and therefore the rotational states are degenerate with respect to the quantum number K, which gives the projection of the vector of total angular momentum $\boldsymbol{L}$ to the molecular axis. This may be interpreted as the molecule having the same rotational energy regardless of the orientation of the rotation axis in the reference system associated with the atomic positions.

This, however, is not the case for a non-rigid top, if its mechanical elasticity depends on the orientation of the rotation axis. For a given angular velocity $\boldsymbol{\Omega}$ the rotational energy of the top is dependent on the orientation of the rotation axis, and the angular momentum $\boldsymbol{L}$ is a nonlinear vector function of $\boldsymbol{\Omega}$. The direction of $\boldsymbol{L}(\boldsymbol{\Omega})$ may deviate from the direction of $\boldsymbol{\Omega}$ and this disalignment varies with the relative orientation of the molecule. For a classical top the conservation of energy and angular momentum implies that the magnitude and the direction of $\boldsymbol{\Omega}$ changes in time, which also implies that the orientation of the molecule with respect to the angular momentum vector changes. The time evolution of the orientation of a non-rigid molecule has an irregular and chaotic character. At the quantum level of consideration this means that the quantum number K does not correspond to a certain energy eigenstate, that is, the K-degeneracy is raised, and quantum numbers K are destroyed. We mark the quantum states $|n\rangle$ of the resulting multiplets with the index n, which now has nothing to do with the angular momentum projection. The number of these states is given by the number $2L + 1$ of different K-components and can be large for rapidly rotating molecules.

Consider now the dynamics of excitation of vibrational-rotational transitions in spherical-top molecules by an external laser field. The molecule in the ground vibrational state can be considered as rigid, and the rotational levels are K-degenerate. In an excited vibrational state it loses the rigidity and the K-degeneracy is raised. We denote by Δ_n the displacement of the states $|n\rangle$ from their unperturbed positions and write the Schrödinger equations for the slow amplitudes in the form

$$
\begin{aligned}
i\hbar\dot{\psi}_K &= \sum_n V_{K,n}\psi_n \\
i\hbar\dot{\psi}_n &= \Delta_n\psi_n + \sum_K V_{n,K}\psi_K,
\end{aligned}
\tag{2.17}
$$

where ψ_K correspond to the ground vibrational state, and ψ_n denotes the amplitudes of the rotational energy eigenstates of the vibrationally excited molecule. The total momentum L and its projection M are supposed to be conserved in the course of molecular rotations. Note that the displacements Δ_n may also include the detuning of the laser field from the frequency of the unperturbed transition.

For a one-photon process the transition probability amplitudes $V_{n,K}$ are given by the product of the laser field amplitude $\mathcal{E}$, the dipole moment of vibrational transition d_{vib}, and the scalar product of states $\langle n|K\rangle$. Chaotic motion of the non-rigid top suggests the random matrix model for the latter. For a multiphoton excitation we have to replace the transition amplitudes $V_{n,K} = \mathcal{E}d_{vib}\langle n|K\rangle$ by the composite matrix elements

$$
V_{n,K} = \langle n|K\rangle\mathcal{E}d_1 \prod_{l=2}^{\substack{\text{number of}\\ \text{absorbed photons}}} \left\{\frac{\mathcal{E}d_l}{E_l - l\hbar\omega}\right\},
\tag{2.18}
$$

where E_l represents the energies of the intermediate states, and d_l the dipole moments for the transitions $l-1 \to l$.

2.1.5 Excitation of Rydberg States in Atoms

Degeneracy of states is also present in atoms . However, the optical transitions in atoms usually occur among states with relatively small angular momenta, where the number of the magnetic components is relatively small. The population dynamics in atoms can therefore be studied analytically or numerically, where an approach based on statistical ideas is not really needed. Nevertheless they are useful when we consider multiphoton transitions to highly excited Rydberg states. Indeed, multiphoton excitation can be associated with a large acquisition of angular momentum, and thus highly excited states with large angular momenta have rich multiplet structures resulting from the degeneracy of the magnetic quantum numbers M. In the presence of electric and magnetic fields the joint action of Stark and Zeeman effects raises the degeneracy and destroys the quantum numbers M. The situation is identical to that of spherical-top molecules; many components of the multiplets simultaneously fall into a multiphoton resonance with the external laser field, and no selection rules are left. One may use (2.17) with the composed matrix elements similar to those given in (2.18).

Another example of a complex multilevel system is that of Rydberg atoms in an ultra-high-frequency radio field. One can formally describe the system with the help of (2.16) where I denotes the principal quantum number, and m stands for the magnetic quantum number. However, the matrix elements $V_{n,l}^{n',l'}$ cannot a priori be taken as random due to the presence of selection rules. Nevertheless, since the classical motion of an electron in the Coulomb and radio-frequency fields is chaotic, the corresponding quantum problem may manifest some properties of a multilevel system perturbed by a random matrix, although the separation of the Hamiltonian into a regular part and a random part may be different from the separation of the Hamiltonian of the Coulomb problem and the RF-field-dependent perturbation. Such a point of view is complementary to the traditional approach to the problem, which relies on the numerical investigation of quantum maps. In Fig. 2.4 we compare and contrast Rydberg atoms and molecules in an oscillating external field.

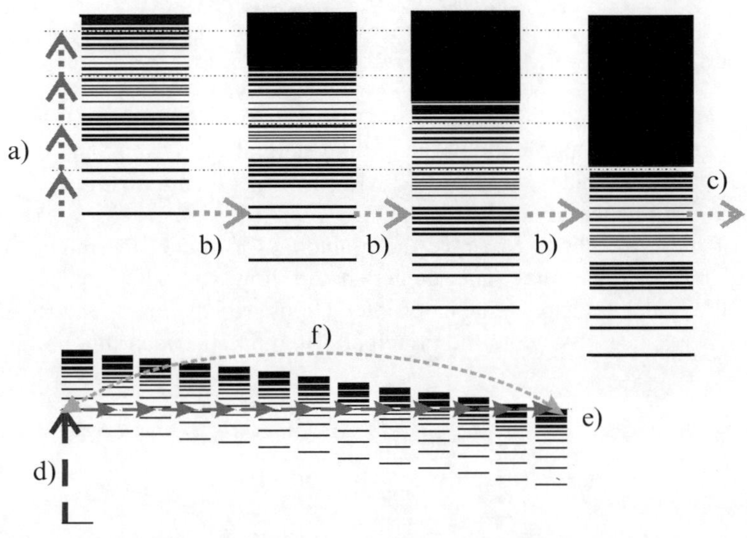

Fig. 2.4 Rydberg atoms versus polyatomic molecules in laser fields. The spectral density of the vibrational states of a polyatomic molecule rapidly increases with energy (**a**). The laser field induces upward transitions. In the compound system of the molecule + quantized field (**b**) the transitions occur among the states of the same energy up to the dissociation limit (**c**). An atom excited by a laser field (**d**) to a Rydberg state in the presence of an RF field performs a series of transitions (**f**) towards the ionization limit (**e**)

2.1.6 *Competition of Multiphoton Processes of Different Orders*

For molecules in strong laser fields one can face competition between multiphoton processes of different orders. Indeed, polyatomic molecules possess rich vibrational spectra. The detuning of the lowest excited level of the vibrational mode from one-photon resonance can be equal or close to the detunings of higher levels from the higher-order multiphoton resonances (Fig. 2.5a). The larger the order of the resonance n, the higher the density of states g_n in the vicinity of the resonant energy $n\hbar\omega$. The excitation dynamics of such a system result from the competition between the increasing density of states and decreasing size of the composed matrix elements (2.18) of the multiphoton transitions. We illustrate such a situation in Fig. 2.5.

In the resonant approximation this system is equivalent to the system of interacting states shown in Fig. 2.5b, where a large matrix element V_1 connects the starting state $|0\rangle$ with a small number of levels, and a weaker interaction V_2 corresponds to a higher density of states g_2, etc. The densest spectrum is hardly coupled to the starting state.

The Schrödinger equation for such a system reads

$$
\begin{aligned}
i\hbar\dot{\psi}_0 &= \sum_m V_{0,m}\psi_m \\
i\hbar\dot{\psi}_m &= \Delta_m\psi_m + V_{m,0}\psi_0,
\end{aligned}
\tag{2.19}
$$

where ψ_0 denotes the amplitude of the ground state of the system, and the excited states with the amplitudes ψ_m are detuned from the resonant positions of one- or multiphoton resonances by the energy Δ_m.

One of the possibilities to tackle the problem is to model the interaction and detunings by random values. This model should, however, allow for large differences in the typical matrix elements of interactions, which correspond to different orders of resonances. We note the principal difference between this problem and

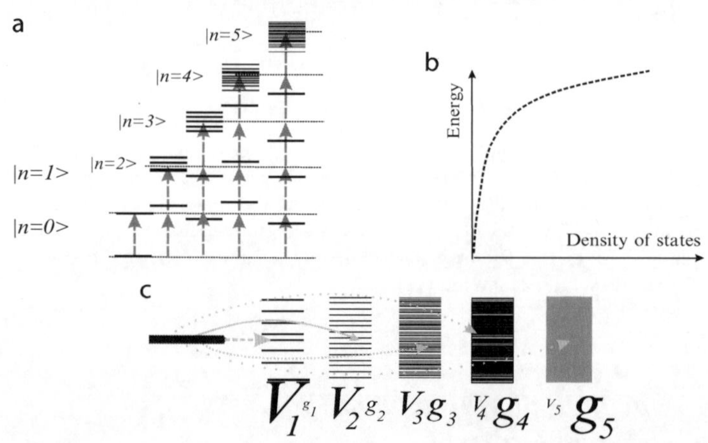

Fig. 2.5 (**a**) Competition between multiphoton processes of different orders. (**b**) Density of resonances as a function of excitation energy. (**c**) Level scheme for resonances of different orders. The increasing density of resonances g_n compensates for the decreasing coupling V_n

the problem of a level decaying to a continuum: two close states $|m\rangle$, and $|m'\rangle$ in the system described by (2.17) which have a small difference in the detunings $\Delta_m - \Delta_{m'}$ may possess transition amplitudes V_m and $V_{m'}$, different by orders of magnitude.

2.2 Collisions and Reactions of Molecules

Interaction of atoms and molecules with laser radiation is not the only domain where consideration of the population dynamics is needed. A great number of similar problems arise in chemistry, in the theory of atomic and molecular collisions, in the theory of Rydberg molecules, associative Penning ionization, etc. In this section we consider some of these examples.

2.2.1 Collisional Redistribution of Energy

Let us first consider the simplest example of a polyatomic molecule colliding with a simple, namely structureless, particle. We assume that the trajectories of the colliding particles are given and concentrate only on the change of the internal state, that is, the vibrational state of the polyatomic molecule. Since vibrationally excited polyatomic molecules move chaotically, we also assume that the interaction

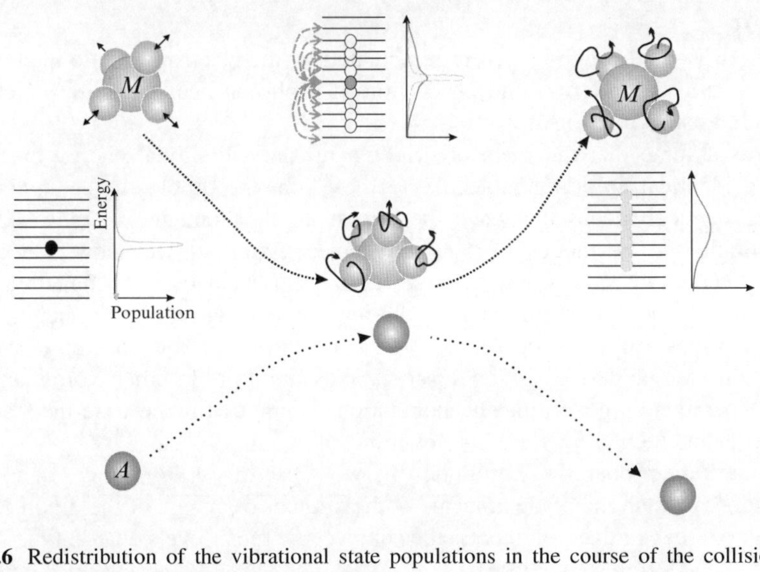

Fig. 2.6 Redistribution of the vibrational state populations in the course of the collision of a polyatomic molecule with a buffer gas molecule

Hamiltonian $\hat{H}_{int}$ in the representation of the vibrational energy eigenstates $|n\rangle$ is complex, and is given by a random matrix $\hat{H}_{int} = G(t)V_{n,n'}$, where the size of the interaction $G(t)$ depends on the distance between the colliding particles, and therefore changes with time. We write the Schrödinger equation for the polyatomic molecule

$$i\hbar\dot{\psi}_n = E_n\psi_n + G(t)\sum_m V_n^m\psi_m, \tag{2.20}$$

which differs from (2.16), since in the interaction term it contains the time-dependence explicitly. The redistribution of the population among the vibrational states of the molecule (see Fig. 2.6) depends on the typical size V and the time-dependence $G(t)$ of the interaction. The solution of (2.16) allows us to find the proper combination of V and $G(t)$ that governs the probability of the energy exchange.

Collisions among vibrationally excited polyatomic molecules also result in vibrational energy exchange. If the distance between two colliding molecules at the position of the closest approach exceeds their sizes, we can assume that the translational and the vibrational degrees of freedom are decoupled from each other and take straight lines for the trajectories of the centers of mass of the molecules. The main contribution to the energy transfer then results from the dipole–dipole interaction among the dipole-active vibrations of different molecules.

If $\boldsymbol{R}(t)$ is the intra-molecular distance, and $\boldsymbol{d}$ and $\boldsymbol{d}'$ are the operators of dipole moments of the molecules, the interaction Hamiltonian reads

$$\hat{V} = -|\boldsymbol{R}(t)|^{-5}\left[3\left(\boldsymbol{d}\cdot\boldsymbol{R}(t)\right)\left(\boldsymbol{d}'\cdot\boldsymbol{R}(t)\right) + \left(\boldsymbol{d}\cdot\boldsymbol{d}'\right)R^2(t)\right]. \tag{2.21}$$

The vibrational exchange is therefore mediated by the electromagnetic interaction, which occurs via emission and absorption of a photon near the point of closest approach, as shown in Fig. 2.7.

Let us assume that the main contribution to the vibrational energy exchange between identical colliding molecules comes from the dipole–dipole interaction of their identical modes that have the largest dipole moment. We may consider for example the ν_3 modes of SF_6 molecules. Later on we show that if the typical interaction time is much longer than the typical time of the vibrational stochastization but much shorter than the inverse spacing among the neighboring quantum states (in units of $\hbar = 1$), the equation for the ensemble average populations can be derived in the quasi-static assumption. In other words, one can find the parameters governing the population dynamics for the time-independent interaction and then allow for their slow time variation.

We mark the probability amplitudes by two subscripts: Latin indices denote the total energy of each molecule in terms of the number of quanta of the dipole-active mode, and Greek indices enumerate the sublevels of these levels emerging from the anharmonic splitting shown in Fig. 2.3.

The pair of colliding molecules comprise a compound system. At large distances when the molecules do not interact with each other, the state of the compound system $|n, m, \alpha, \beta\rangle = |n, \alpha\rangle \times |m, \beta\rangle$ is the direct product of the states of each molecule, and the amplitude of this state $\Psi_{n,m,\alpha,\beta} = \psi_{n,\alpha}\psi_{m,\beta}$ is the product of the amplitudes. At closer distances, for interacting molecules this is no longer the case, since the interaction (2.21) mixes up the states of different molecules and destroys the separability of their individual motions. We therefore have to solve the Schrödinger equation for the entire compound system.

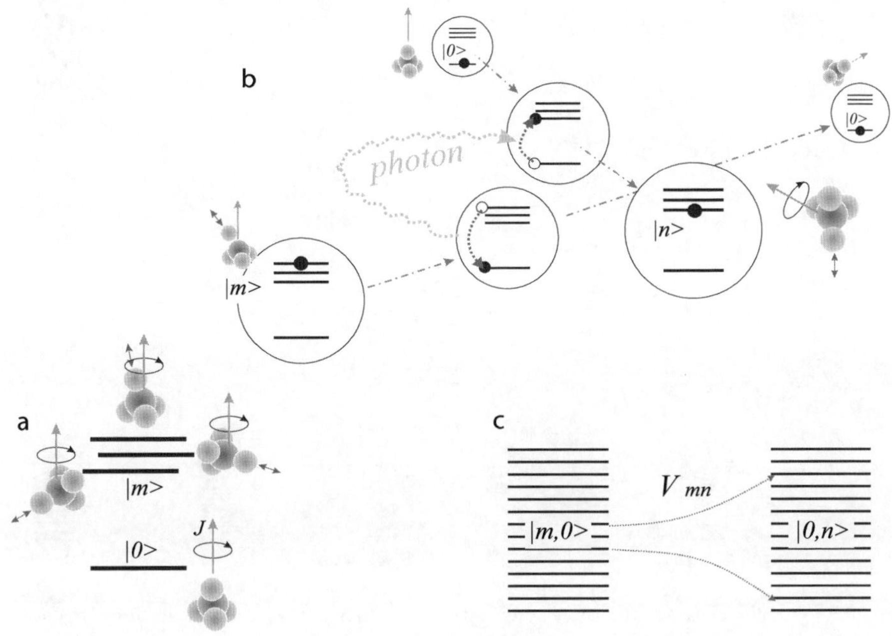

Fig. 2.7 Vibrational energy exchange mediated by a photon. (**a**) Vibrational and rotational spectra of symmetric polyatomic molecules are mixed and form a common ro-vibronic spectrum. (**b**) Sketch of a transition accompanied by change of ro-vibronic states of colliding molecules. The dipole–dipole interaction is equivalent to a photon exchange. (**c**) The compound spectrum of the colliding molecules is a complex multilevel system

For interacting molecules we introduce new variables. By the number $N = (n+m)/2$ we denote the conserving quantity—half of the total energy of the two molecules. Half of the energy difference $M = (n-m)/2$ changes in the course of collision as well as the populations of the sublevels α and β of the anharmonic splitting. In Fig. 2.8 we illustrate the correspondence between the transitions in each of the molecules and the transitions in the compound system. We denote by $\Delta_{N,M,\alpha,\beta} = E_{n,\alpha} - E_{m,\beta}$ the total energy of the non-interacting molecules and introduce the matrix elements of the interaction (2.21)

$$V_{N,M,\alpha,\beta}^{N,M+1,\alpha',\beta'} = V_{n,m,\alpha,\beta}^{n+1,m-1,\alpha',\beta'}; \quad V_{N,M,\alpha,\beta}^{N,M-1,\alpha',\beta'} = V_{n,m,\alpha,\beta}^{n-1,m+1,\alpha',\beta'}. \tag{2.22}$$

In this notation the Schrödinger equation for the compound system reads

$$i\hbar\dot{\Psi}_{N,M,\alpha,\beta} = \Delta_{N,M,\alpha,\beta}\Psi_{N,M,\alpha,\beta} + \sum_{\alpha',\beta'} V_{N,M,\alpha,\beta}^{N,M+1,\alpha',\beta'}\Psi_{N,M+1,\alpha',\beta'}$$
$$+ \sum_{\alpha',\beta'} V_{N,M,\alpha,\beta}^{N,M-1,\alpha',\beta'}\Psi_{N,M-1,\alpha',\beta'}. \tag{2.23}$$

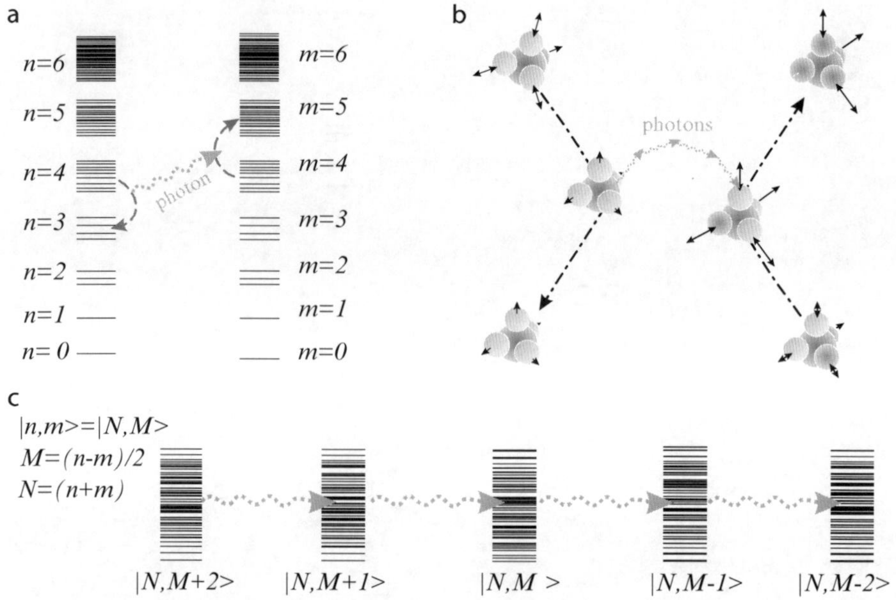

Fig. 2.8 Vibrational exchange in polyatomic molecules. (**a**) A photon exchange results in opposite vibrational transitions in colliding molecules. (**b**) Many photon exchange acts occur during the elementary act of a strong collision. (**c**) Transitions among the states of a compound system of two molecules is a process in a complex quantum system

The random matrix model is a natural selection for the interaction $\hat{V}$ in the chosen representation, since the anharmonic vibrational motion of each molecule is already chaotic, and the interaction would just add more complexity to the motion of the compound system.

2.2.2 *Chemical Reactions*

Chemical reactions of polyatomic molecules in the gas phase are a very important particular case of collisions. From the quantum mechanical point of view they are processes in multilevel quantum systems. We illustrate this by several examples.

Harpoon Reaction

Consider an atom A with a low ionization potential which is at a small distance R from a diatomic molecule BC which has a large electron affinity. The outer atomic electron can transfer to the molecule, which yields an ion–molecular compound $A^+ + BC^-$. In Fig. 2.9 we illustrate this process and show the corresponding change

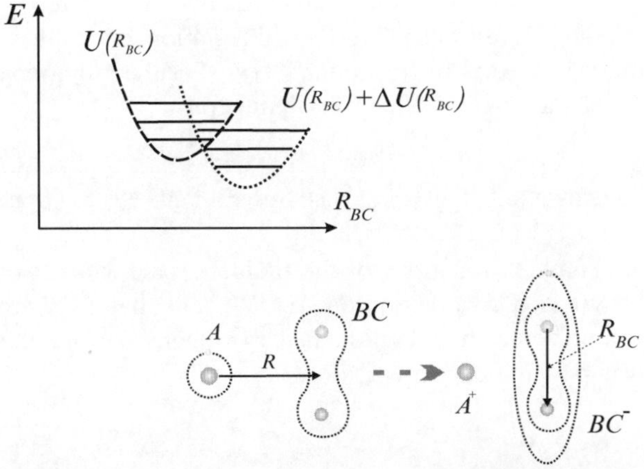

Fig. 2.9 The harpoon reaction. Atom A loses an electron which is captured by the molecule BC at a large distance R. The interatomic potential of the molecular ion differs by a value $\Delta U(R_{BC})$ from the molecular potential $U(R_{BC})$, and this difference serves as a perturbation for the vibrational motion of the ion

$\Delta U(R_{BC})$ of the internuclear potential energy $U(R_{BC})$, which depends on the distance R_{BC} between the atoms B and C. The change of the potential results in a change of the molecular vibrations.

We express the wavefunction of the compound

$$\Psi = \sum_n \psi_n(t)\phi_n \Phi_A + \sum_k \psi_k(t)\phi_k \Phi_{BC} \tag{2.24}$$

in terms of the electronic wavefunctions $\Phi_A(r, R, R_{BC})$ localized at the atom and $\Phi_{BC}(r, R, R_{BC})$ localized at the molecule, and the molecular wavefunctions $\phi_n(R_{AB})$ and $\phi_k(R_{AB})$ of the vibrational energy eigenstates $|n\rangle$ and $|k\rangle$ in the potentials $U(R_{BC})$ and $U(R_{BC})+\Delta U(R_{BC})$, respectively. The coefficients $\psi_n(t)$ stand for the time-dependent probability amplitudes of the states when the electron is associated with the atom A and the molecule BC is at the n vibrational level of the potential $U(R_{BC})$, whereas the coefficients $\psi_k(t)$ correspond to the electron associated with the molecule and the vibrational state k in the potential $U(R_{BC}) + \Delta U(R_{BC})$.

We substitute the ansatz (2.24) into the Schrödinger equation and arrive at the equation for the amplitudes $\psi_n(t)$ and $\psi_k(t)$:

$$i\hbar\dot{\psi}_n = E_n\psi_n + \sum_k V_{n,k}\psi_k$$

$$i\hbar\dot{\psi}_k = E_k\psi_k + \sum_m V_{k,n}\psi_n. \tag{2.25}$$

Here E_n and E_k are the energies of the vibrational states $|n\rangle$ and $|k\rangle$. The probability amplitudes of the transitions are given by the overlap integral

$$V_{n,k} = \int \phi_n(R_{AB})\phi_k(R_{AB})\Phi_{BC}(\mathbf{r}, R, R_{BC})\Phi_A(\mathbf{r}, R, R_{BC})\, d^3r\, dR_{BC} \quad (2.26)$$

where $\boldsymbol{r}$ denotes the coordinates of the transferred electron.

For a polyatomic molecule M one has to replace the integration over the interatomic distance R_{BC} by integration over all N-dimensional space of the internal degrees of freedom $d^N\boldsymbol{R}_M$. For the vibrational energies corresponding to the chaotic motions the overlap integrals

$$V_{n,k} = \int \Phi_M(\mathbf{r}, R, \boldsymbol{R}_M)\Phi_A(\mathbf{r}, R, \boldsymbol{R}_M)\phi_n(\boldsymbol{R}_M)\phi_k(\boldsymbol{R}_M)\, d^N\boldsymbol{R}_M\, d^3r \quad (2.27)$$

are complex functions of the indices n and k and the matrix $V_{n,m}$ can therefore be replaced by a random matrix. We note that if the atom A moves with respect to the molecule, the vibrational energy eigenstates and the transition amplitudes $V_{n,m}$ become time-dependent.

Sub-Barrier Reaction

Another example concerns a chemical reaction which occurs in the framework of a single adiabatic electronic term, that is, it involves only the vibrational motion of molecules. Consider an atom A slowly approaching a rapidly oscillating molecule BC. The presence of the atom effects the frequency of the molecular vibrations, which gradually change as the atom closes with the molecule. After the atom has passed, the vibrational frequency returns to its initial value. This adiabatic change of frequency is a time-dependent perturbation, and in principle it is able to break the molecular bond B–C, although with a low probability. The energy conservation law implies that the breaking of the B–C bond is followed by the formation of another bond, say A–B.

A consistent quantum description of the process should consider the process as a tunneling through the potential barrier along the reaction coordinate, which is shown in Fig. 2.10. Let us consider possible trajectories of the molecular motion in such a term. The trajectory starts at the potential valley that corresponds to the atom A approaching the molecule BC. The oscillations of the trajectory represent the molecular vibrations. A non-reactive process corresponds to the trajectory 1 that leaves the region of interaction via the same potential valley, although probably with a different ratio between the vibrational and the translational energies. The chemical reaction corresponds to the trajectory 2, which escapes the interaction region via the other potential valley. It also includes a tunneling part 3. The oscillations of this trajectory now represent the vibrations of the molecule AB that has been formed in the course of the chemical reaction.

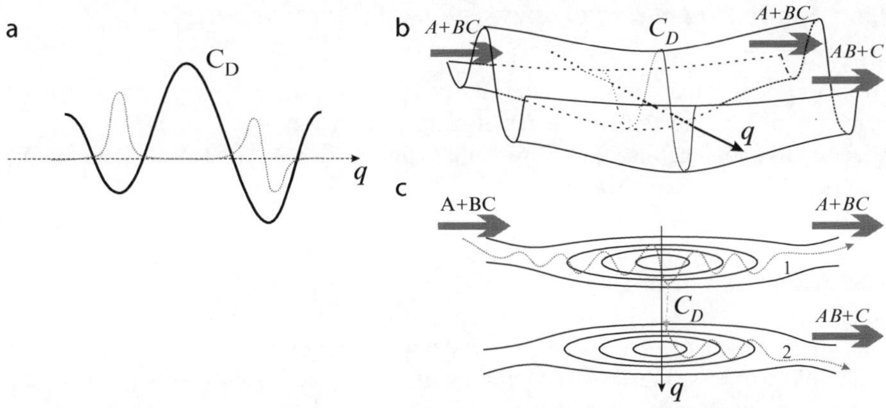

Fig. 2.10 Vibrational wavefunctions (**a**), potential surfaces (**b**), and trajectories (**c**) of a tunneling chemical reaction

We assume that the relative motion of the atom and the molecule is slow and write the Schrödinger equation

$$i\hbar\dot{\psi}_n = E_n(z)\psi_n + \sum_k V_{n,k}(z)\psi_k$$

$$i\hbar\dot{\psi}_k = E_k(z)\psi_k + \sum_m V_{k,n}(z)\psi_n \tag{2.28}$$

for the amplitudes ψ_n of the vibrational states $|n\rangle$ of the molecule BC, and the amplitudes ψ_k of the vibrational states $|k\rangle$ of the reaction product AB. Here $z = z(t)$ is the slowly changing distance between the atom and the molecule, and $V_{n,k}$ represents the probability amplitudes of the tunneling through the potential barrier given by the overlap integrals of the corresponding wavefunctions (Fig. 2.10). The latter evidently depend on the distance z between the atom and the molecule.

When we consider the reactive collisions of polyatomic molecules, the energies $E_n(z)$ and $E_k(z)$ acquire a complex structure as do the overlap integrals $V_{n,k}$. Such reactions occur in a multidimensional space of the configurations which is difficult to depict, although the main features of the process, namely the slow approach of the reagents and the tunneling along the reaction coordinate remain qualitatively the same. This tunneling may, however, occur in many different points of configuration space that correspond to different shapes of the molecules at the moment of the reaction. Moreover, the motion of polyatomic molecules is usually chaotic, and therefore the random matrix model is natural for the transition amplitudes $V_{n,k}$.

In order to calculate the reaction rate or the energy distribution of the products we first have to solve (2.28) and determine which combinations of the spectra $E_n(z)$;$E_k(z)$ and the interaction $V_{n,k}$ govern the reaction properties.

2.2.3 Intermolecular Conversion and Photochemistry

From the conceptual point of view chemical reactions are very similar to the problems of intermolecular transfer of electronic energy to vibrational degrees of freedom, to photochemical processes, and to photon-assisted reactive collisions. We illustrate this with some examples.

Intermolecular Conversion

The adiabatic separation of electronic and nuclear motion is valid only far from the points of term crossing. In the vicinities of these points the nuclear motion may cause electronic transitions, which implies energy exchange between electronic and vibrational degrees of freedom. This perturbation is of a higher order in the Born–Oppenheimer parameter $\Lambda_{BO} = (m_e/m_p)^{1/4}$ and normally is ignored, unless two or more electronic terms happen to be close to each other at a certain position of the nuclei in the molecule. In the latter case the non-adiabatic interactions have to be taken into account.

Let us consider two crossing terms shown in Fig. 2.11. We take the diabatic potentials and construct the basis set from the electronic wavefunctions $\Phi_1(\boldsymbol{r}, \boldsymbol{R})$, $\Phi_2(\boldsymbol{r}, \boldsymbol{R})$ and the vibrational wavefunctions $\phi_n(\boldsymbol{R})$, $\phi_k(\boldsymbol{R})$ in the first and the second potentials, respectively.

Here $\boldsymbol{r}$ denotes the positions of all molecular electrons, and $\boldsymbol{R}$ stands for the position of all nuclei. For the corresponding amplitudes ψ_n and ψ_k we write the Schrödinger equation

$$
\begin{aligned}
i\hbar\dot{\psi}_n &= E_n\psi_n + \sum_k V_{n,k}\psi_k \\
i\hbar\dot{\psi}_k &= E_k\psi_k + \sum_n V_{k,n}\psi_n,
\end{aligned}
\qquad (2.29)
$$

where $V_{n,k}$ now denotes the matrix element of the non-adiabatic interaction:

$$
V_{n,k} = -\int d^p r\, d^N R\, \Phi_1(\boldsymbol{r}, \boldsymbol{R})\phi_n(\boldsymbol{R})\left(\nabla_R\Phi_2(\boldsymbol{r}, \boldsymbol{R}) \cdot \nabla_R\phi_k(\boldsymbol{R})\right). \qquad (2.30)
$$

Here p and N denote the number of electronic and vibrational degrees of freedom, respectively.

For polyatomic molecules the vibrational energies E_n and E_k and the matrix elements $V_{n,k}$ can be modeled, as earlier, by random values.

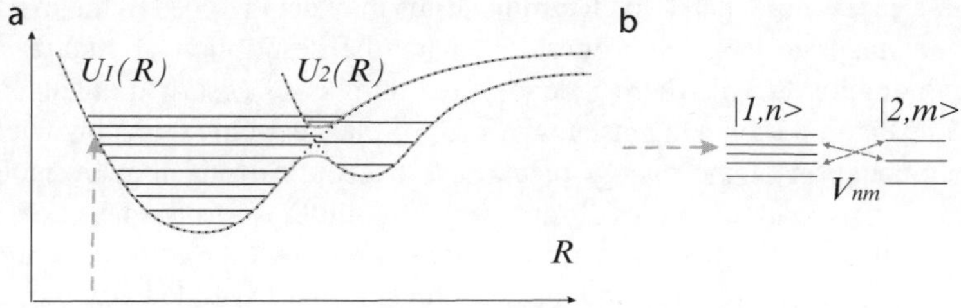

Fig. 2.11 Crossing of molecular terms (**a**). Diabatic terms (*dotted line*) correspond to a certain electronic state of a molecule and serve as potentials for the vibrational motion. Non-adiabatic interactions result in a reconstruction of the electronic terms and yield adiabatic terms (*solid line*) for which the electronic state is entangled with the nuclear position (sometimes also with the momentum). (**b**) Phenomena that occur near the crossing point can be considered in the diabatic vibrational basis as dynamical processes in a complex multilevel system

Photochemistry and Laser-Induced Term Crossing

Optical or ultraviolet laser fields induce transitions among the electronic terms of molecules that usually result in chemical transformations. Strong fields may also affect the vibrational dynamics. In this case one can describe the system by (2.29), where instead of the matrix elements of non-adiabatic interaction (2.30) we have to substitute the probability amplitudes of laser-induced transitions:

$$V_{n,k} = -\mathcal{E}e \int d^p r \, d^N R \, \Phi_1(\boldsymbol{r},\boldsymbol{R})\phi_n(\boldsymbol{R})\boldsymbol{r}\Phi_2(\boldsymbol{r},\boldsymbol{R})\phi_k(\boldsymbol{R}). \tag{2.31}$$

Here $\mathcal{E}$ is the electric field amplitude, and e is the charge of the electron.

This analogy becomes evident when we consider a compound system of the molecule and the quantized laser field. Indeed, at some position of the nuclei, the energy of the electron cloud in the ground electronic term equals the energy of the excited electronic state minus the energy of one photon. One can say that at this point the two potential curves of the compound system cross each other. Coupling occurs due to the dipole interaction with the laser field, which is taken into account by the matrix elements (2.31).

If the typical size of the interaction exceeds the typical energy spacing among the neighboring vibrational states, this process cannot be considered perturbatively. For diatomic molecules it requires a very high laser intensity, at which the typical interaction $\mathcal{E}d$ is of the order of magnitude or larger than the vibrational quantum. However, this requirement is not as restricting for polyatomic molecules, which have a much higher density of vibrational states: even moderately intense laser fields may result in the coupling of electronic and vibrational motions. The complex structure of the vibrational wavefunctions of the excited polyatomic molecules results in an irregular dependence of $V_{n,k}$ on n and on k, and therefore justifies the utilization of a random matrix model.

In some cases more than two electronic terms may get involved in the dynamical process at moderate laser intensities. Then for the description of such a system one has to employ (2.16) where I denotes the number of absorbed quanta, and m enumerates the vibrational eigenstates in each of the electronic terms involved.

Accommodation of the energy of an absorbed optical quantum by molecular vibrations is also an example of processes in multilevel systems. Consider a molecule initially in the ground vibrational state of the lowest electronic term which absorbs a quantum of radiation and makes a transition to an excited vibrational state of the higher electronic term. Thence, due to the non-adiabatic interaction it gets transferred back to the ground electronic term, although in a different vibrational state. This state may also be predissociating. The Schrödinger equation for the amplitudes ψ_k of the vibrational states in the diabatic excited electronic term, the amplitudes ψ_n of the highly excited vibrational states of the ground diabatic term, and for the amplitude ψ_0 of the ground vibrational state read

$$
\begin{aligned}
i\hbar\dot{\psi}_0 &= E_0\psi_0 + \sum_k \mathcal{E}d_{0,k}\psi_k \\
i\hbar\dot{\psi}_n &= E_n\psi_n + \sum_k V_{n,k}\psi_k \\
i\hbar\dot{\psi}_k &= (E_k - \hbar\omega)\psi_k + \sum_n V_{k,n}\psi_n + \mathcal{E}d_{k,0}\psi_0 .
\end{aligned}
\tag{2.32}
$$

Here $V_{m,n}$ represents the matrix elements of the non-adiabatic transitions (2.30), and $\mathcal{E}d_{0,k}$ are the matrix elements (2.31) of the interaction with the electromagnetic field.

2.3 Rydberg Molecules

Multilevel quantum problems arise in a natural way when one considers highly electronically excited or so-called Rydberg molecules, that is, molecules with one electron excited into a high Rydberg state. This electron possesses an energy close to the ionization potential and, in the language of classical mechanics, it moves along a long Kepler orbit in the electrostatic field of the molecular ionic core. Traditional adiabatic separation of the electronic and vibrational motion in molecules implies a large difference in energies, velocities and frequencies of the electrons and nuclei of molecules. This is not completely the case for the Rydberg molecules: the Kepler period might be comparable and even longer than the vibrational or rotational period of the ionic core, although the velocity of the Rydberg electron near the core is still much higher than the velocity of the nuclei. We therefore meet the situation which in a sense is inverse to the regular adiabatic case: the electron is a slowly moving subsystem, while the rotations and vibrations of the ionic core are the fast motion. We can therefore speak of the Rydberg sublevels E_n of each ro-vibrational state $|k\rangle$ of the molecular ion. We give two examples of such molecules.

2.3.1 Subthreshold Photoionization

Consider a molecule in an intense laser field. The outer electron of the molecule makes a multiphoton transition to a highly excited Rydberg state and starts its motion along a Kepler orbit. Returning back to the vicinity of the molecular ion it encounters a potential that differs from the Coulombic case, and therefore experiences scattering at the asymmetric part of the potential. The scattering may be either elastic or non-elastic. The non-elastic scattering is associated with a transfer of energy and angular momentum from the electron to the molecular ion, and hence the electron changes its Rydberg orbit (see Fig. 2.12), whereas the molecular ion finds itself in another rotational or vibrational state. The scattered electron with

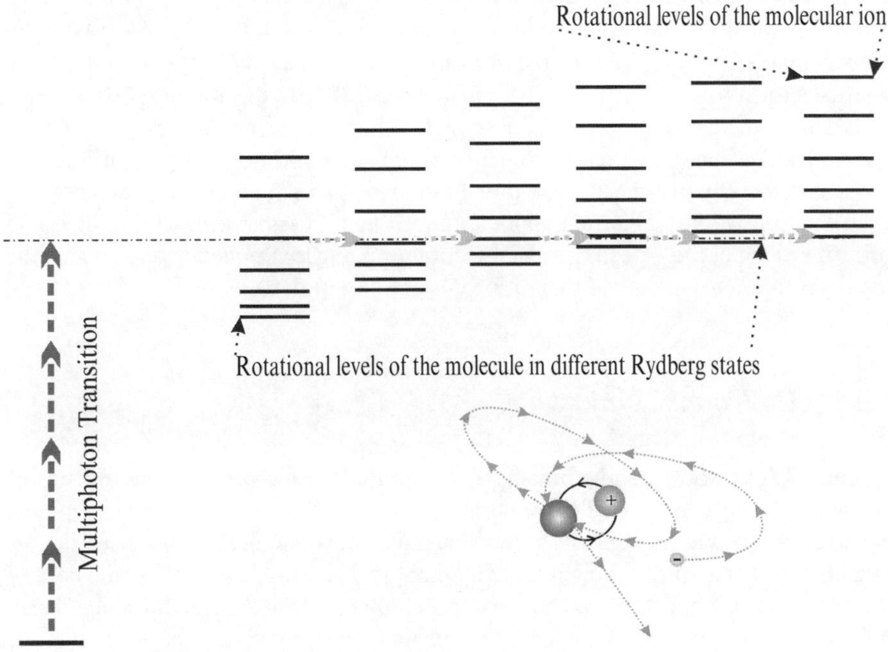

Fig. 2.12 Multiphoton ionization of a molecule via Rydberg states

an energy less than the ionization potential returns back again after a long Kepler period. During this time the molecular ion in the center vibrates and rotates, and there is sufficient time for the position of the nuclei to change. Therefore the returning electron experiences quite a different scatterer as compared to the initial scattering event. Thus the sequence of scattering events is a kind of random process which has a direct analogy to the quantum Sinai billiard.

For the basis set of the problem we employ the products $|n\rangle|k\rangle$ of the Rydberg states $|n\rangle$ of the hydrogen atom and the rotational (and/or vibrational) states $|k\rangle$ of the molecular ion, and write down the Schrödinger equation

$$i\hbar\dot{\psi}_0 = E_0\psi_0 + \sum_{k,n} \tilde{V}_0^{k,n}\psi_{k,n}$$

$$i\hbar\dot{\psi}_{n,k} = (E_n + E_k)\psi_{k,n} + \sum_{k',n'} V_{n,k}^{k',n'}\psi_{k',n'} + \tilde{V}_{k,n}^0\psi_0 \tag{2.33}$$

for the amplitudes $\psi_{k,n}$ of these states. Here $\tilde{V}$ stands for the amplitude (2.18) of multiphoton transitions from the ground state $|0\rangle$. The matrix elements $V_{k,n}^{k',n'}$ allow for the probability amplitudes of scattering from the state $|k'\rangle|n'\rangle$ to the state $|k\rangle|n\rangle$.

We note one interesting peculiarity of this process. The Kepler period may change considerably with the change in the energy of the Rydberg electron: the closer the energy to the ionization limit, the longer the period. In such a situation it is natural to consider the evolution of the system not as a function of time, but as a function of the number of scattering events. One determines the corresponding transformation by multiplying the scattering amplitudes $V_{k,n}^{k',n'}$ in (2.33) by the phase factor $\exp[-i(E_{k'} + E_{n'})T(n')]$, where $T(k')$ is the Kepler period corresponding to the energy $E_{k'}$. This phase factor changes drastically from one state to another, and it makes the matrix elements of the neighboring states very different in phase and thereby justifies the use of the random matrix model for this process.

We also note one important application of this process: by measuring the yield of zero kinetic energy electrons as a function of the laser frequency one obtains the high-resolution rotational and vibrational spectra of molecular ions.

2.3.2 Collisional Ionization

In Fig. 2.13 we show another example of a non-adiabatic process: the transfer of vibrational energy to the Rydberg electron in the course of associative ionization. An excited noble gas atom interacts with a metal atom. At interatomic distances shorter than the orbit size of the excited electron the metal atom sees only the ionic core of the noble atom, since the electron most of the time stays at a longer distance. From the chemical point of view the core resembles a halogen atom, that is, an element with one vacancy in the outer electronic shell. Therefore the reaction between the metal and the noble gas core is of a harpoon type: the outer electron of the metal jumps to the core and fills the vacancy in the outer shell of the noble gas ion. This is essentially the formation of an ionic bond, with the two atoms comprising an ionic molecular core, which generally speaking is in a vibrationally excited state.

At the same time the excited electron of the noble gas atom feels a displacement of the charge center toward the metallic atom. If the distance between the electron and the charge center increases, the electron finds itself in a high Rydberg orbit of the molecular ion. The Kepler trajectory of the electron goes through the space occupied by the molecular core, and therefore the Rydberg electron experiences a sequence of scattering analogous to the case considered just before. In the course of

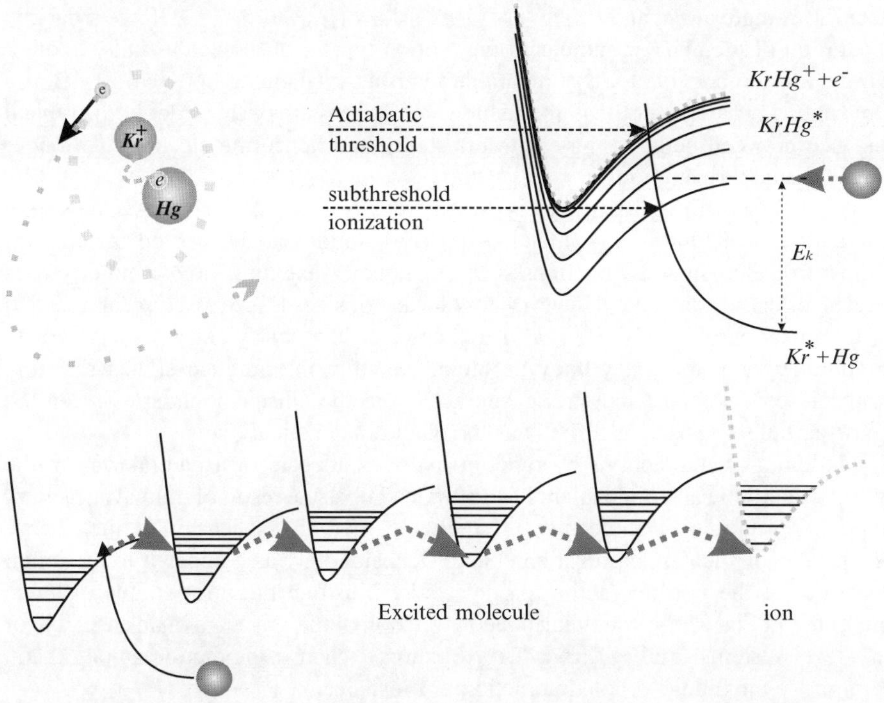

Fig. 2.13 Associative ionization. The outer electron of a metal atom experiences the vacancy in the outer shell of the noble gas core and forms a chemical bond. The excited electron of the noble gas atom finds itself in a high Rydberg orbit of the molecular ion. The vibrational energy of the molecular core is transferred to the Rydberg electron in the course of collisions between the electron and the core. The ionization therefore starts not at the incident energy E_k equal to the adiabatic threshold, but at a lower (subthreshold) energy

non-elastic collisions the vibrational energy of the molecular core can be transferred to the Rydberg electron, which may result in ionization. The Schrödinger equation describing the process

$$i\hbar\dot{\psi}_{n,k} = (E_n + E_k)\psi_{k,n} + \sum_{k',n'} V_{n,k}^{k',n'}\psi_{k',n'} \tag{2.34}$$

coincides with (2.33) when the laser field is at zero strength, $\tilde{V} = 0$.

2.4 Atomic and Molecular Clusters

Clusters are intermediate objects between molecules and solids, and the physics of clusters inherits complexity from both molecular and solid state physics. Indeed, the typical distance among the electronic states of a N-atom cluster is N times smaller than for a single atom, and hence for $N \sim \Lambda_{BO}^{-1} = (m_n/m_e)^{1/4} \sim 10$ the adiabatic separation of the nuclear and electronic motion typical of molecules fails. In other words, for numbers of atoms comparable to or higher than the inverse of the Born–Oppenheimer parameter the non-adiabatic corrections are of the order of the typical distance between neighboring electronic states, and hence the electrons no longer follow the nuclear motion.

The concept of quasi-particles is the key idea of solid state physics. The strong interaction of electrons and nuclei comprising solids can be treated in terms of almost free elementary excitations such as phonons, excitons, holes, and electrons with an effective mass, etc. However, for clusters of sizes less than a typical free path of the quasi-particles ($N \sim 1{,}000$) this concept does not yield any considerable simplification of the many-body problem: quantum interference effects resulting from the confinement, and strong interaction on the cluster boundaries make the quasi-particle approach as difficult as the quantum chemical one.

Modeling of interactions by random matrices appears in a natural way in the physics of atomic and molecular clusters. Indeed if, as the result of mutual attraction, a group of atoms or molecules is stuck together, the spectrum of the energy eigenstates of such an aggregation changes considerably as compared to the initial spectrum of the non-interacting particles. We illustrate this in a simple example presented in Fig. 2.14 by showing a reconstruction of the spectrum of an ensemble of two-level systems resulting from the excitation-exchange interaction among them. It might be the dipole–dipole interaction, for instance.

At finite temperatures various configurations of the cluster are possible, and each of the geometrical arrangements of the atoms corresponds to different interatomic

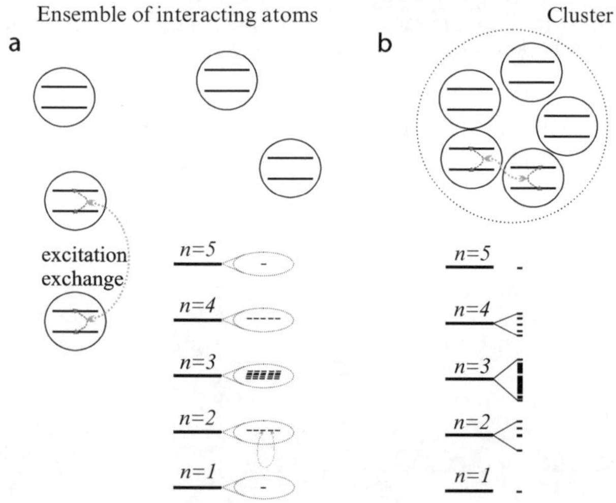

Fig. 2.14 Reconstruction of the energy spectrum of an ensemble of two-level systems as a result of the excitation-exchange interactions. (**a**) Energy spectrum of the ensemble of atoms. The mean interatomic distance is large and the spectrum is almost degenerate. (**b**) Energy spectrum of a cluster formed as a result of the reconstruction of the atomic spectrum due to strong interatomic interaction

distances, and therefore to different sizes of the pair interactions. Let us assume that only one atom was originally excited. Then, as a result of the interaction the excitation can pass to other atoms, and hence the energy eigenstates of the cluster do not correspond to a certain location of the excitation. If nevertheless for the basis set of the problem we choose the states corresponding to all possible locations of the excitation at a single atom, the interaction Hamiltonian becomes a matrix with the elements depending on the particular configuration of the cluster. For a random arrangement of atoms this matrix is random.

In spite of the fact that the interaction modifies considerably the energy eigenstates of the system, the unperturbed Hamiltonian $\hat{H}_0$ still plays an important role. We therefore arrive at the necessity to describe the energy spectrum of a system perturbed by a random matrix $\hat{V}$. Which characteristics have to be employed for the purpose? The density of states $\pi^{-1}\mathrm{Im\ Tr}(E-\hat{H}-i0)^{-1}$ is one of the most representative quantities for a spectrum. It suggests the ensemble averaged density of states

$$g(E) = \left\langle \frac{1}{\pi} \,\mathrm{Im\ Tr} \frac{1}{E - \hat{H}_0 - \hat{V} - i0} \right\rangle \tag{2.35}$$

as one of the key characteristics of a complex system. Indeed, the state density is an ingredient of many experimentally observable properties such as optical absorption, scattering cross-section, etc.

2.4.1 Ground Electronic State of Hot Metallic Clusters

We dwell on the example of metallic clusters which are an important particular case of aggregations. Metallic clusters have common orbitals for the valence electrons. For minute clusters the shell structure of the orbitals effects strongly their basic physical properties: the Fermi energy, work function, electron affinity, and cohesion energy all change abruptly and considerably with an increase of the cluster size. At finite temperatures the thermal motion of atoms perturbs appreciably the shell structure, and thereby it modifies significantly properties of the clusters.

Let us consider this in the framework of the simplest one-particle approximation. The valence electrons of a cluster move in the self-consistent effective potential, which implies that each of the electrons feels an average field of the ionic core and all other electrons. The simplest model of this field is just a rectangular potential well shown in Fig. 2.15. For spherically symmetric clusters the angular momentum $\boldsymbol{l}$ of each of the electrons is conserved, and their angular motion can be taken into account by introducing the centrifugal force potential $U(r) = (\boldsymbol{l})^2/2m_e r^2$, acting on the motion in the radial direction r.

By solving the corresponding Schrödinger equation one finds the energy eigenfunctions

$$\psi_{n,l,m} = \frac{J_{l+1/2}(k_{n,l}r)}{\sqrt{r}}\, Y_{l,m}(\theta, \phi) \tag{2.36}$$

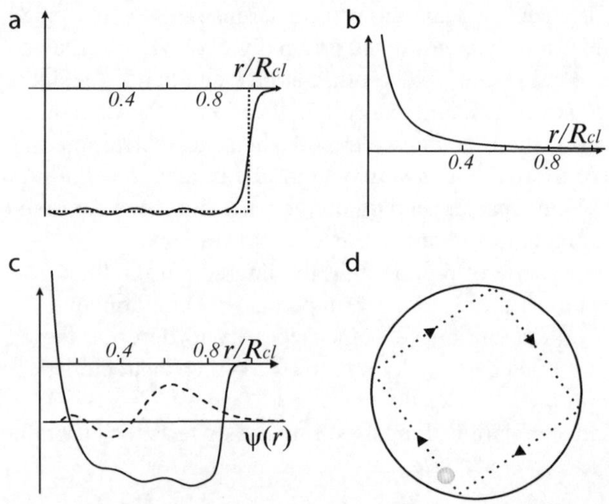

Fig. 2.15 Spherically symmetric potential of cluster electrons. (**a**) Rectangular well model of the work function and self-consistent local density potential. (**b**) Potential of the centrifugal forces. (**c**) Effective potential and a wavefunction. (**d**) Classical trajectory of an electron

in terms of the Bessel functions $J_\nu(x)$, and the spherical harmonics $Y_{l,m}(\theta,\phi)$. For deep potential wells with the help of the boundary condition at the outer border of the cluster

$$\psi_{n,l,m}\Big|_{r=R} = 0 \tag{2.37}$$

one determines the radial wavevectors $k_{n,l}$ and the corresponding energies,

$$E_{n,l} = \frac{\hbar^2 z_{n,l}^2}{2m_e r_S^2 N^{2/3}}, \tag{2.38}$$

provided the number N of atoms in the cluster and the typical size r_S of the atoms (Wigner–Seitz radius) are known. Here $z_{n,l}$ stands for the n-th root of the Bessel function $J_{l+1/2}(z_{n,l}) = 0$. The states are degenerate with respect to the magnetic quantum number m.

More accurate models include a self-consistent consideration of the electronic states based on the local density functional approach. This implies iterative numerical calculation of the electronic wavefunctions and the corresponding effective potential determined as a functional of the electron density distribution. This method results in more accurate values of the cluster electron energies (2.38).

At finite temperatures the thermal motion of atoms comprising the cluster destroys the spherical symmetry and raises the spherical degeneracy of the electron energy eigenstates. Since the atomic motion is much slower than the electron motion we can consider this perturbation as static. We note, however, that this assumption does not mean that the momenta of the electrons play no role in the perturbation. On the contrary, the nuclear momenta bring an important contribution via the non-adiabatic coupling (2.30) of the electronic states. The assumption only means that the time variation of the coupling is slower than its typical size divided by Planck's constant.

The random character of the thermal motion suggests the consideration of the perturbation as random. Indeed, in the classical limit the trajectory of an electron moving in the potential well resembles the trajectory of a particle in Sinai billiard theory (see Fig. 2.16): elastic scattering at thermal phonons at irregularities of the surface and at atomic cores makes the electronic motion chaotic. Therefore the corresponding quantum model relies on the assumption that the thermal motion results in a coupling of the electronic states (2.36) of the cluster via a random matrix with the mean squared matrix element $\langle V^2 \rangle$ depending on temperature T. One can estimate this quantity by noticing that the electron–phonon coupling is the main mechanism of the temperature-dependent conductivity of metals, and the relaxation time of the electronic momentum $\tau(T)$ equals the inverse of the relaxation probability $2\pi \langle V^2 \rangle g_f$, where g_f is a typical one-electron state density near the Fermi energy of the cluster.

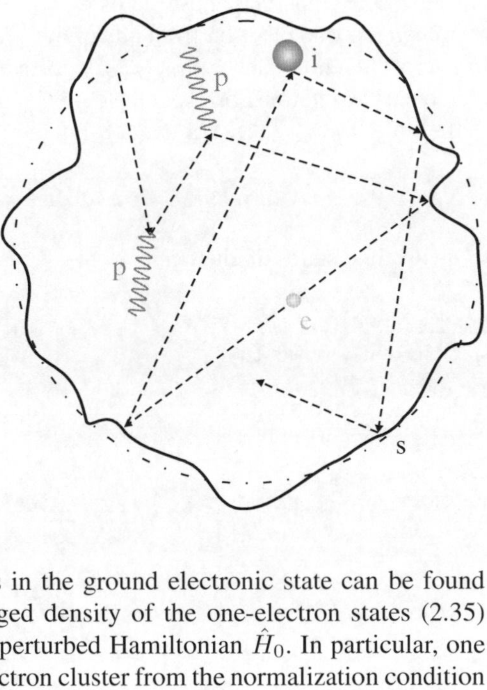

Fig. 2.16 The classical trajectory of an electron (*dashed line*) in a hot cluster. Scattering of the electron occurs at impurities (i), at phonons (p), and at the perturbed surface (s) of the spherical "in average" cluster

Many of the properties of clusters in the ground electronic state can be found with the help of the ensemble averaged density of the one-electron states (2.35) where the energies (2.38) give the unperturbed Hamiltonian $\hat{H}_0$. In particular, one finds the Fermi energy ϵ_f of the N-electron cluster from the normalization condition

$$N = \int_{-\infty}^{\epsilon_f(N, \langle V^2 \rangle)} g(E)\, dE, \tag{2.39}$$

the total energy of the electronic cloud

$$\mathcal{E}(N, \langle V^2 \rangle) = \int_{-\infty}^{\epsilon_f(N, \langle V^2 \rangle)} E g(E)\, dE, \tag{2.40}$$

and the dissociation energies

$$D(N, \langle V^2 \rangle) = E_I + \mathcal{E}(N-1, \langle V^2 \rangle) - \mathcal{E}(N, \langle V^2 \rangle) \tag{2.41}$$

of an ensemble of clusters. Here E_I is the ionization potential of atoms comprising the cluster. Therefore the establishing of a simple recipe for the calculation of the ensemble average density of states $g(E)$ for an arbitrary Hamiltonian H_0 is of practical importance.

2.4.2 Optical Properties of Clusters

The random matrix model can also be applied to the excited states of metallic clusters. One constructs these states by considering elementary excitations, that is, by transferring an electron from one of the occupied one-electron states $|n_e, l_e, m_e\rangle$ to one of the empty states $|n_h, l_h, m_h\rangle$ above the Fermi energy. In other words the elementary excitation corresponds to an electron in the state $|n_e, l_e, m_e\rangle$ and a hole in the state $|n_h, l_h, m_h\rangle$. For the non-interacting electrons it yields the energies

$$E(n_e, l_e, m_e, n_h, l_h, m_h) = E_{n_e, l_e, m_e} - E_{n_h, l_h, m_h} \tag{2.42}$$

of the excited states of clusters.

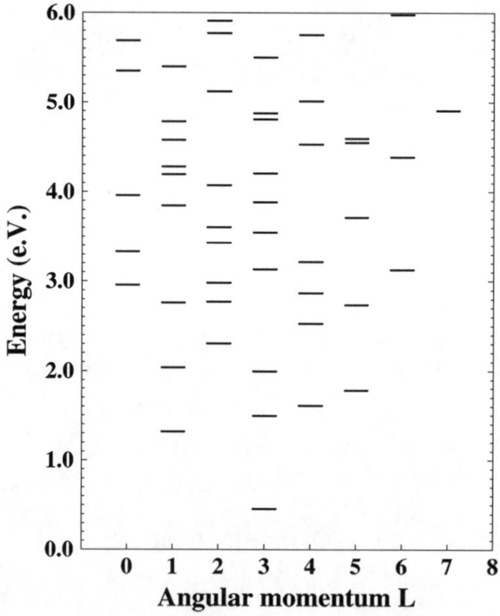

Fig. 2.17 Reconstructed spectrum of the excited states of Na_{21}^+

Electrons, however, interact with each other. Some part of this interaction has already been taken into account when the self-consistent effective potential has been employed as a model for the ground electron term. The remaining part $\hat{V}$ should allow for the deviation of the real interaction from the effective one. The main contribution $\hat{V}^{(1)}$ to this remainder originates from the coupling of the electron–hole pairs with the electromagnetic field. The excited electron is annihilated with the hole and creates an intermediate virtual photon. The cluster therefore returns to the ground electronic term. The photon, in its turn, creates another electron–hole pair by transporting an electron from another occupied one-electron level to a different empty state. Therefore all excited states $|S\rangle = |n_e, l_e, m_e, n_h, l_h, m_h\rangle$ interact among themselves via the ground electronic term and different states of the electromagnetic field. It yields a specific (factorized) structure of the interaction Hamiltonian: each matrix element $V^{(1)}_{S,S'} = T_S \times T_{S'}$ is a product of a quantity related to the starting state $|S'\rangle$ and the same quantity related to the final state $|S\rangle$. One has to include this interaction in the Hamiltonian of the cluster, which yields a considerable modification of the energies (2.42). An example of such a reconstruction of a spectrum is shown in Fig. 2.17.

The rest, $V^{(2)}_{S,S'}$, of the remaining $V_{S,S'}$ has a complicated structure. It originates from the scattering of the excited electron at the hole. The dynamics of the electron resemble the motion in a Sinai billiard with one difference: the scatterer, that is, the hole, does not stay at the same position but also moves along a chaotic trajectory as the result of scattering by the electron. Therefore the matrix elements $V^{(2)}_{S,S'}$ do not have a factorized structure. On the contrary, they depend in a complex and irregular way on the indices S and S' and often are ignored. This is known as the random phase approximation. However, this interaction can also be taken into account within the random matrix model where the energies (2.42) and the interaction $V^{(1)}$ compose the unperturbed Hamiltonian, and the interaction $V^{(2)}$ gives a random perturbation. One can also include in the latter part the random coupling originating from the scattering of the electron and the hole at the atomic core phonons and irregularities of the surface by analogy to the coupling in the ground electronic term discussed in the previous section.

One usually finds cross-sections of the linear absorption from the ground state $|0\rangle$ of a quantum system with Hamiltonian $\hat{H}$ with the help of the expression

$$\sigma(\hbar\omega) = \frac{1}{\pi}\,\mathrm{Im}\langle 0|\,\hat{d}\frac{1}{\hbar\omega - \hat{H} - i0}\hat{d}\,|0\rangle, \tag{2.43}$$

where $\hat{d}$ is the dipole moment operator along the electric field vector. The real part of the quantum average allows for the optical dispersion. For ensembles of clusters (2.43) yields

$$\sigma(\hbar\omega) = \frac{1}{\pi}\,\mathrm{Im}\big\langle\langle 0|\,\hat{d}\frac{1}{\hbar\omega - \hat{H}_0 - \hat{V} - i0}\hat{d}\,|0\rangle\big\rangle, \tag{2.44}$$

where $\hat{H}_0$ is the ensemble average Hamiltonian, and $\hat{V}$ stands for the random perturbation. More sophisticated combinations can be written for multiphoton cross-sections.

2.5 Some Other Examples

We consider here several examples from branches of physics other than atomic and molecular physics, chemistry, or cluster physics. These examples relate to fields with their own conceptual systems and corresponding well-developed mathematical techniques. However, it seems useful to trace the analogies among these problems in terms of interacting quantum states in order to gain a deeper insight into the common properties of complex quantum objects.

2.5.1 *Ion Traps*

Systems such as ions trapped in an oscillating electromagnetic field where classical and quantum effects are of the same order of magnitude represent an interesting branch of contemporary physics. From electrostatics we know that a given number of charged particles can never form an equilibrium configuration in the static electric field. On the other hand, in a time-dependent field a quasi-equilibrium and dynamically stable configurations may exist. Indeed, rapid oscillation of the field results in an effective Kapiza–Dirac potential which may confine the motion of charged particles to a certain region of space. A quasi-equilibrium distribution arises as the result of the interplay between Coulomb repulsion of the ions and the confining action of the effective potential: the particles rapidly oscillate around certain points that either move slowly or are static. This dynamic equilibrium may, however, be destroyed by the influence of the small terms oscillating at double the driving field frequency that result from the nonlinearity of the system. These terms are usually ignored in the first-order approximation. One can take them into account with the help of the random matrix model by the analogy to polyatomic molecules in an external laser field.

Let us illustrate this with the example of N ions in a time-dependent electric field with the potential

$$U(\boldsymbol{r},t) = \alpha(x^2 + z^2 - 2y^2) + A(x^2 \cos \Omega t - Ay^2 \cos \Omega t), \tag{2.45}$$

where $\boldsymbol{r} \equiv (x, y, z)$ is the radius vector and Ω is the oscillation frequency of the electric field vector in the plane (x, y). One sees that both static and time-dependent components of the potential satisfy the Laplace equation $\Delta U = 0$, which means that the chosen potential exists. We start our analysis in terms of classical mechanics and quantize the motion at a later stage, since the result of such a consideration does not differ much from the result of a consistent quantum approach.

The classical Hamiltonian of N ions in the field reads

$$H(\{\boldsymbol{p}_n\},\{\boldsymbol{r}_n\}) = \sum_{n=1}^{N}\left(\frac{\boldsymbol{p}_n^2}{2m} + eU(\boldsymbol{r}_n,t)\right) + \sum_{n\neq m}\frac{e^2}{|\boldsymbol{r}_n - \boldsymbol{r}_m|}, \tag{2.46}$$

where $e, \boldsymbol{p}_n$ are the ion charges and classical momenta, respectively, and m is the ion mass.

In a high-frequency field the motion of each particle consists of two components

$$\boldsymbol{r}_n = \boldsymbol{\rho}_n(t) + \boldsymbol{\xi}_n(t), \tag{2.47}$$

where the $\boldsymbol{\xi}_n(t)$ rapidly oscillates at the frequency of the field, and $\boldsymbol{\rho}_n(t)$ is a slow component. The typical value ξ_t of the rapid component is usually much smaller than the typical value of the slow component ρ_t, and therefore the ratio ξ_t/ρ_t is a small parameter. This is not the case for the velocities

$$\dot{\boldsymbol{r}}_n = \dot{\boldsymbol{\rho}}_n(t) + \dot{\boldsymbol{\xi}}_n(t), \tag{2.48}$$

since ξ rapidly changes with time. In fact, we have $\dot{\boldsymbol{\rho}}_n(t) \sim \dot{\boldsymbol{\xi}}_n(t)$.

We take into account that $\boldsymbol{p}_n = m\dot{\boldsymbol{r}}_n = m\dot{\boldsymbol{\rho}}_n(t) + m\dot{\boldsymbol{\xi}}_n \equiv \boldsymbol{\mathcal{P}}_n + \boldsymbol{\pi}_n$ and substitute (2.47), (2.48) into (2.46), which yields

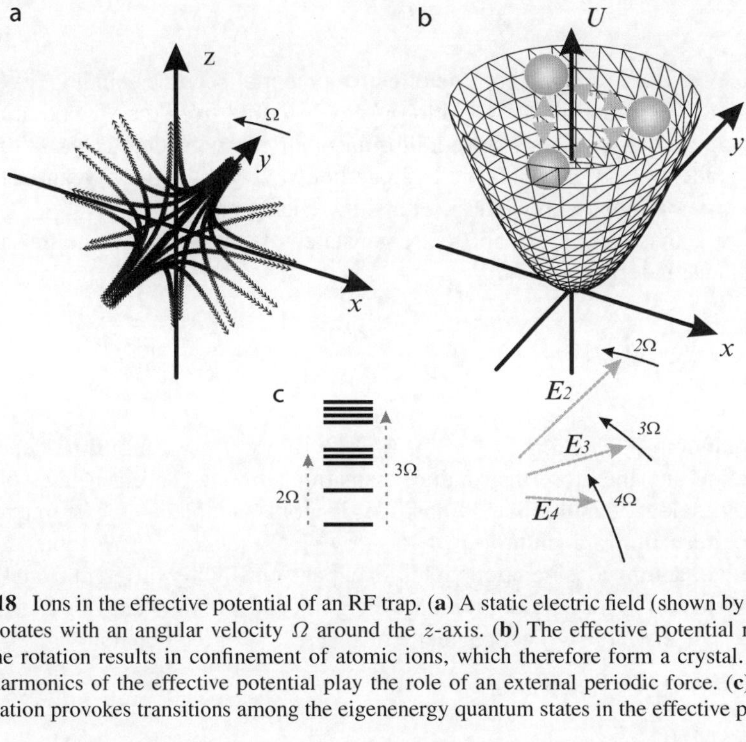

Fig. 2.18 Ions in the effective potential of an RF trap. (**a**) A static electric field (shown by the *field lines*) rotates with an angular velocity Ω around the z-axis. (**b**) The effective potential resulting from the rotation results in confinement of atomic ions, which therefore form a crystal. Higher-order harmonics of the effective potential play the role of an external periodic force. (**c**) Such a perturbation provokes transitions among the eigenenergy quantum states in the effective potential

$$H(\{\boldsymbol{p}_n\},\{\boldsymbol{r}_n\}) = \sum_{n=1}^{N}\left(\frac{\boldsymbol{\mathcal{P}}_n^2+\boldsymbol{\pi}_n^2}{2m}+\frac{(\boldsymbol{\mathcal{P}}_n\cdot\boldsymbol{\pi}_n)}{m}+eU(\boldsymbol{\rho}_n+\boldsymbol{\xi}_n,t)\right)$$
$$+\sum_{n\neq m}\frac{e^2}{|\boldsymbol{\rho}_n+\boldsymbol{\xi}_n-\boldsymbol{\rho}_m+\boldsymbol{\xi}_m|}. \tag{2.49}$$

The rapidly oscillating components of the Hamiltonian (2.49) allow for the rapid motion, and the slowly varying components for the slow motion. We note that the term $\boldsymbol{\pi}_n^2$ contains a slow component, namely the average kinetic energy of the rapid oscillations. This component plays the role of the potential energy for the slow motion, and therefore together with the component of the potential energy depending on the slow coordinates it comprises the Kapiza–Dirac potential. Indeed, the time derivatives of the rapid momenta, say in the x-direction, equal the rapid forces $\boldsymbol{\pi}_n^{(x)} = 2Ax\cos\Omega t$, and hence $\boldsymbol{\pi}_n^{(x)} = 2Ax/\Omega^{-1}\sin\Omega t$, which yields the contribution $\langle(\boldsymbol{\pi}_n^{(x)})^2/2m\rangle_t = 2(Ax/\Omega)^2/m\langle\sin^2\Omega t\rangle_t = (Ax)^2/m$ to the effective potential.

The Hamiltonian for the slow motion reads

$$H(\{\boldsymbol{p}_n\},\{\boldsymbol{r}_n\}) = \sum_{n=1}^{N}\left(\frac{\boldsymbol{\mathcal{P}}_n^2}{2m}+e\alpha(x_n^2+z_n^2-2y_n^2)+\frac{eA^2}{m\omega^2}(x_n^2+y_n^2)\right)$$
$$+\sum_{n\neq m}\frac{e^2}{|\boldsymbol{\rho}_n-\boldsymbol{\rho}_m|}, \tag{2.50}$$

and hence for $2\alpha < A^2/m\omega^2 m$, the effective potential becomes binding, as shown in Fig. 2.18. Therefore the slow motion is confined and from the viewpoint of this motion the system possesses an equilibrium configuration. One can say that the time-dependent electrical field creates a parabolic trap for the ions. Quantization of the slow motion in the trap yields a set of slow-energy eigenstates.

However, these states are not real eigenstates of the system, since the higher-order expansion terms

$$V(\{\rho_n\}) = \sum_{k,l,n\neq m}\frac{1}{l!k!}\frac{\partial^{l+k}}{\partial^l\boldsymbol{\rho}_n\partial^k\boldsymbol{\rho}_m}\frac{e^2}{|\boldsymbol{\rho}_n-\boldsymbol{\rho}_m|}\boldsymbol{\xi}_m^l\boldsymbol{\xi}_n^k \tag{2.51}$$

of the Coulomb interaction in $\boldsymbol{\xi} = \boldsymbol{\rho}(A/m\Omega^2)\sin\Omega t$ result in a time-dependent perturbation and therefore they induce transitions among the eigenstates of slow energy by analogy to multiphoton transitions in atoms or molecules. The frequencies of the perturbation are multiples of the driving frequency Ω. We note that the consistent quantum consideration yields an expansion slightly different from (2.51); however, this difference should not play a significant role for relatively heavy ions and the relatively large typical geometrical size $\alpha^{-1/2}$ of the potential.

Classical motion of the ions in the trap is multidimensional and nonlinear. Therefore at a certain level of the excitation it becomes chaotic. In this case one can consider the matrix elements of perturbation (2.51) as random, and write the Schrödinger equation (2.6) by analogy to polyatomic molecules where one also should allow for the transitions with the energy changing by $2\hbar\Omega$, $3\hbar\Omega$, etc.

2.5.2 *Disordered Solids and Surfaces*

The description of quantum effects in terms of random matrices is a powerful method to tackle a number of problems in solid state physics and surface science. This approach arises in a natural way if one considers the properties of disordered systems like alloys, glasses, non-ideal metals, or systems with chaotic billiard-like dynamics such as quantum dots. It turns out that for such objects interference phenomena play a crucial role by affecting considerably their macroscopic properties. The basic physical processes take place on a limited size-scale which exceeds the atomic sizes, but remains much smaller than a typical macroscopic scale. Therefore these phenomena are called "mesoscopic", and comprise a special field of solid state physics.

Mesoscopic phenomena in solids have as a rule two peculiarities. They usually occur in a coordinate space of three or less dimensions and remain translationally invariant "on average". Therefore the properties of large species result from a kind of superposition of the mesoscopic properties of many statistically equivalent parts of smaller sizes, attached to each other consecutively. It allows one to take advantage of special basis sets, like plane waves for example, to use the scaling hypothesis, or to factorize the propagators of the particles.

Disordered Chains

We illustrate the typical formulation of such a problem by the classical example of the model which considers a single particle in a disordered potential. We take the simplest one-dimensional version of the potential $U(x) = \sum_n k_n \delta(x - na)$, which is a sequence of equidistant Dirac δ-functions shown in Fig. 2.19 of different "depth" k_n, where the k_n values are random. The particle may be located at a given knot n of the potential or tunnel at a distance a to a neighboring knot with a transition amplitude V. The Schrödinger equation for the probability amplitudes ψ_n of the states $|n\rangle$ corresponding to the particle at a given knot n can be written in the form

$$E\psi_n = \delta k_n \psi_n + V\psi_{n+1} + V\psi_{n-1}, \qquad (2.52)$$

where E and δk_n are the deviations of an eigenenergy and the individual depth from the mean depth, respectively, and V is the tunneling amplitude.

The most remarkable property of (2.52) is the so-called Anderson localization phenomenon. Any energy eigenstate $|E\rangle$ of this equation is located mainly at a

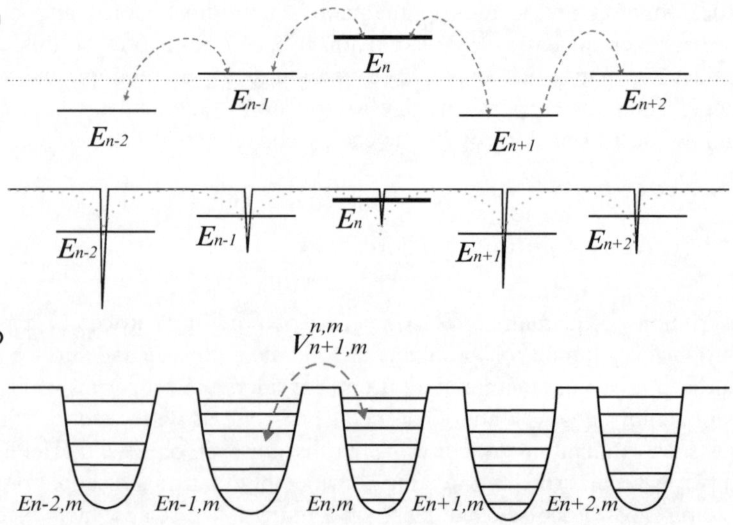

Fig. 2.19 A random sequence (**a**) of δ-potentials (Lloyd chain) and a more complicated potential (**b**) with microstructure

limited number of states $|n\rangle$, which means that the coefficients c_E^n of the expansion $|E\rangle = \sum c_E^n |n\rangle$ drop rapidly when n deviates from the position $n_{max}(E)$ of the knot corresponding to the maximum absolute value of the amplitude ψ_n. The dependence of c_E^n on n is not monotonic and on the contrary is rather random. However, the envelope of the randomly changing absolute values of the amplitudes drops exponentially at $n \to \pm\infty$. As a measure γ of such localization one usually employs the limit

$$\gamma = \lim_{n \to \infty} \frac{1}{n} \langle \log(\psi_n^2 + \psi_{n+1}^2) \rangle \tag{2.53}$$

where the average $\langle \ldots \rangle$ is taken over the ensemble of possible realizations of the random potential, that is, over $\{k_n\}$.

If the deviation δk_n of each of the coefficients k_n from the average is statistically independent of the others and has the probability

$$w(k_n) = \frac{1}{\pi} \frac{\delta}{\delta k_n^2 + \delta_n^2} \tag{2.54}$$

to have a given value δk_n, the problem has an exact solution, which yields

$$\gamma = \text{Arccosh} \left[\frac{1}{4} \sqrt{\left(2 + \frac{E}{V}\right)^2 + \frac{\delta^2}{V^2}} + \frac{1}{4} \sqrt{\left(2 - \frac{E}{V}\right)^2 + \frac{\delta^2}{V^2}} \right]. \tag{2.55}$$

The localization length is $L = 1/\gamma \simeq 2V/\delta$ for small E/V and δ/V.

Quantum States at a Surface

The localization phenomenon is considered as a universal property of one-dimensional disordered quantum systems translationally invariant "on average". The situation may change when we are dealing with a set of level manifolds interacting in a relay-like manner. In this case one more dimension is involved in the problem such as the internal degrees of freedom of particles moving in a spatially limited domain. In Fig. 2.19b we illustrate this by the example of a particle which can move in a domain limited by a binding potential of a complex form. It can also tunnel to a similar potential of the neighboring domain.

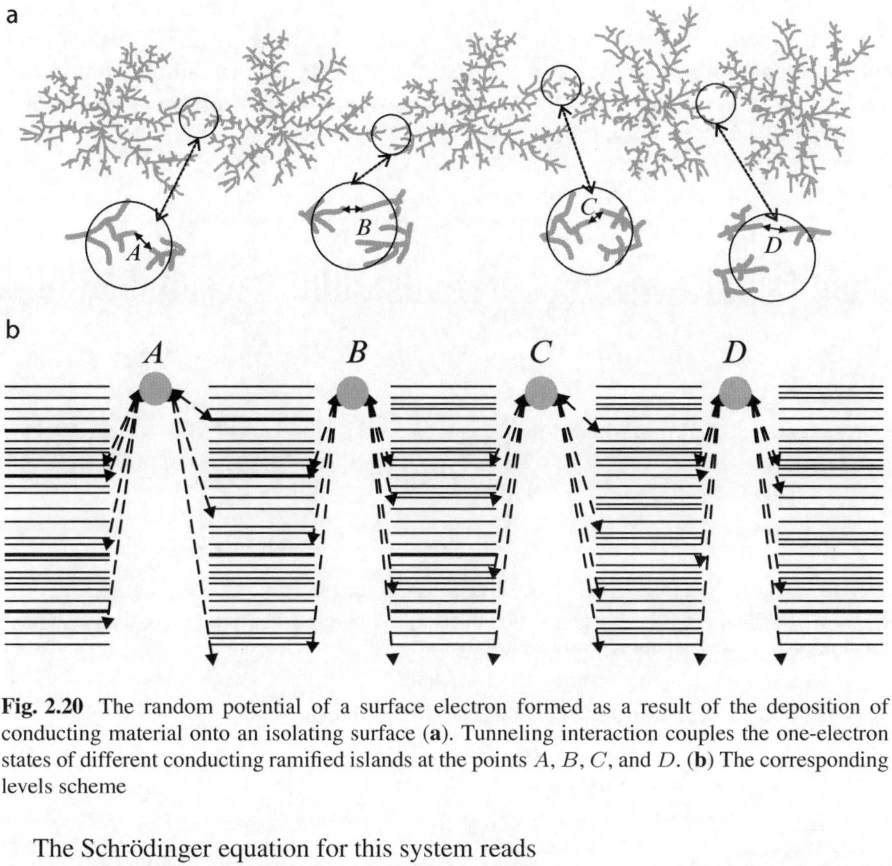

Fig. 2.20 The random potential of a surface electron formed as a result of the deposition of conducting material onto an isolating surface (**a**). Tunneling interaction couples the one-electron states of different conducting ramified islands at the points A, B, C, and D. (**b**) The corresponding levels scheme

The Schrödinger equation for this system reads

$$i\dot{\psi}_{n,k} = E_{n,k}\psi_{n,k} + \sum_m \left(V_{n,k}^{n+1,m}\psi_{n+1,m} + V_{n,k}^{n-1,m}\psi_{n-1,m} \right), \tag{2.56}$$

where $V_{n,k}^{n\pm1,m}$ are the tunneling amplitudes. One sees that this equation has the same form as (2.16) describing the excitation of a polyatomic molecule by an intense laser field. For random potential pits the spectra $E_{n,k}$ are random, and the matrix elements $V_{n,k}^{l,m}$ may also be random for the multidimensional internal motion. The latter is also the case for a particle on a random surface, for which the potential energy resembles a set of conducting fractals as shown in Fig. 2.20. For this case the problem of localization requires special consideration. Examples are known where, in two-dimensional disordered systems uniform "on average", the so-called weak localization phenomenon takes place: the particle does go to infinity, but it takes an exponentially long time.

A Classical Mechanics Analog

It is worth mentioning that one encounters equations of the type (2.52) or (2.56) not only for quantum motion of electrons, but also for classical vibrations of atoms in the lattice, namely phonons. This is the case when atoms of different masses are randomly distributed among the nodes.

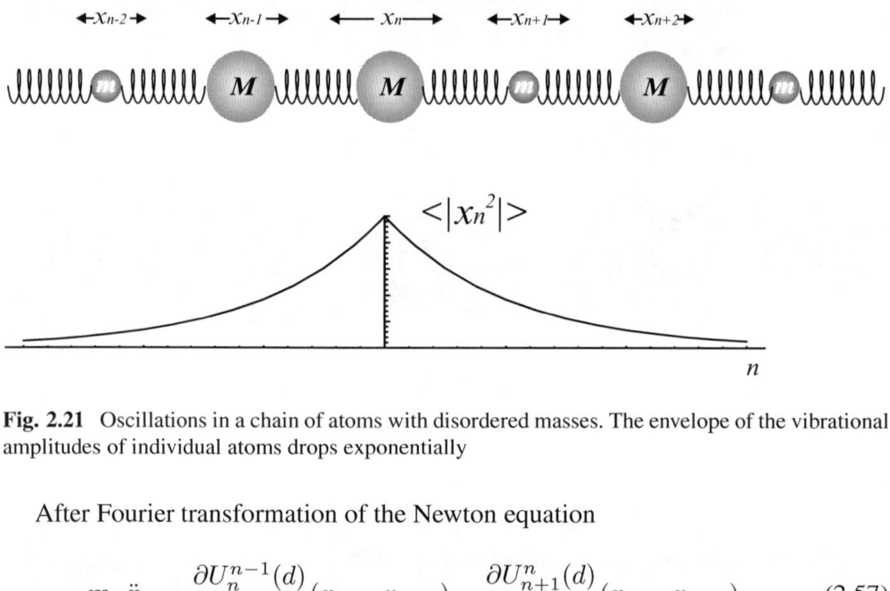

Fig. 2.21 Oscillations in a chain of atoms with disordered masses. The envelope of the vibrational amplitudes of individual atoms drops exponentially

After Fourier transformation of the Newton equation

$$m_n \ddot{x}_n = \frac{\partial U_n^{n-1}(d)}{\partial d}(x_n - x_{n-1}) - \frac{\partial U_{n+1}^{n}(d)}{\partial d}(x_n - x_{n-1}) \tag{2.57}$$

for classical coordinates x_n of the atoms of masses m_n, where $U_n^{n+1}(d)$ is the interaction potential between *the* n-th and $n+1$-th atoms at distance d, we arrive at

$$-\omega^2 m_n x_n = \frac{\partial U_n^{n-1}(d)}{\partial d}(x_n - x_{n-1}) - \frac{\partial U^n n + 1(d)}{\partial d}(x_n - x_{n-1}). \tag{2.58}$$

Equation (2.58) has the same structure as (2.52), since the randomness of masses plays the same role as the distribution of detunings k_n. For a given vibrational mode the individual oscillation amplitudes of atoms change irregularly from one knot to another, but the envelope of this change drops exponentially as sketched in Fig. 2.21. This implies that phonons cannot propagate far in such a medium.

2.5.3 Nonlinear Optics

From the viewpoint of quantum mechanics many of the well-known phenomena described by the theory of nonlinear oscillations and nonlinear optics are processes in multilevel quantum systems. This fact usually does not play an important role in the limit of classical mechanics, that is, when a process involves a large number of quanta. Nevertheless, it may be important for consideration of the threshold and subthreshold regimes of generation, chaotic regimes, or for the processes developing in small confined volumes and on long time-scales.

Let us illustrate this by the example of a nonlinear sample of volume $\mathcal{V}$ placed into a multimode resonator shown in Fig. 2.22. In the empty resonator only one

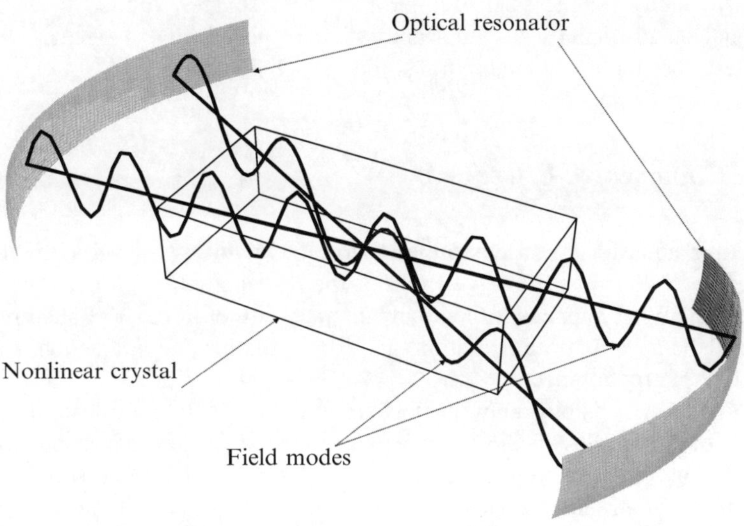

Fig. 2.22 Nonlinear interaction of modes in an optical resonator

mode, say k_0, is excited and contains N photons, whereas all other modes $\{k_n\}$ are in the vacuum state. The electric field amplitude in the cavity $\mathcal{E}(\boldsymbol{r}) = \sqrt{N}\mathcal{E}_0(\boldsymbol{r})$ is therefore given by the number of photons and the mode function $\mathcal{E}_0(\boldsymbol{r})$ of the excited mode. When we introduce the nonlinear medium to the resonator, the radiation of the mode k_0 induces a polarization $\mathcal{P}[\sqrt{N}\mathcal{E}_0(\boldsymbol{r})]$, which couples the excited mode to all other modes. The interaction energy

$$V_{n,0} = -\int_{\mathcal{V}} d^3r\, \mathcal{P}[\sqrt{N}\mathcal{E}_0(\boldsymbol{r})]\mathcal{E}_n(\boldsymbol{r}) \tag{2.59}$$

may change considerably and irregularly with the index n, if the sample is of an irregular shape, or if the nonlinear medium is close to saturation. In the last case the irregularity arises from the spatial structure of the mode function $\mathcal{E}_0(\boldsymbol{r})$. We denote by ψ_0 the amplitude of the state $|N, 0, 0, \ldots, 0\rangle$ with N photons in the mode k_0 and no photons in other modes. By ψ_n we denote the amplitude of the state $|N-1, 0, \ldots, 1, \ldots, 0\rangle$ where one photon has been transported from the mode k_0 to the mode k_n. We neglect the probability of transportation of more than one photon and write the Schrödinger equation

$$\begin{aligned} i\hbar\dot{\psi}_0 &= \sum_n V_{0,n}\psi_n \\ i\hbar\dot{\psi}_n &= \Delta_n\psi_n + V_{m,0}\psi_0, \end{aligned} \tag{2.60}$$

by analogy to (2.19). Here Δ_n is the difference in photon energies of the modes k_0 and k_n.

We mention here one more analogy. A nonlinear medium plays the role of an anharmonicity for the field oscillators in the resonator, similar to the role of vibrational anharmonicity for nuclear motion in polyatomic molecules, whence (2.60) describes the first quantum transition in such a system.

2.5.4 Cooperative Effect

Cooperative emission of an ensemble of non-interacting two-level atoms located in a region smaller than the wavelength of the resonant radiation is known as the Dicke effect. What happens if the atoms interact with each other? Randomness in the mutual distances between the atoms and corresponding pair interactions suggests the random matrix approach.

The following analogy between the ensemble of N two-level atoms and an ensemble of N particles of spin $1/2$ in a magnetic field suggests a convenient means of description. One can consider each of the two-level atoms as a particle of spin $1/2$ in the magnetic field $\boldsymbol{B}$. Transition among the upper and the lower states of an atom corresponds to a spin flip from the state parallel to $\boldsymbol{B}$ to the antiparallel state, whereas the dipole–dipole interaction of atoms correspond to the spin–spin interaction.

The total spin S of the ensemble may take any value from $N/2$ to 0 for N even, or to $1/2$ for N odd. Each value of the total spin apart from $N/2$ can be constructed in many ways by combining the directions of the spins of the individual particles. We mark each such a combination by an index ν. The statistical weight $\mathcal{G}_S$ of a state with spin S is therefore given by the binomial coefficients $\mathcal{G}_S = \begin{pmatrix} S/2 \\ N - S/2 \end{pmatrix}$

for N even and S integer, or $\mathcal{G}_S = \binom{S/2-1/2}{N-S/2+1/2}$ for N odd and S integer plus $1/2$. There are $2S-1$ magnetic sublevels M for any state of a given S. The energy of each sublevel depends linearly on M, and the spectrum of the ensemble is represented in terms of the quantum numbers S and M, as shown in Fig. 2.23. Each energy level corresponds to a certain value of M, and even for a fixed S is $\mathcal{G}_S$-fold degenerate with respect to ν. Optical transitions in the system occur among the states with the same "spin" S and the same number ν.

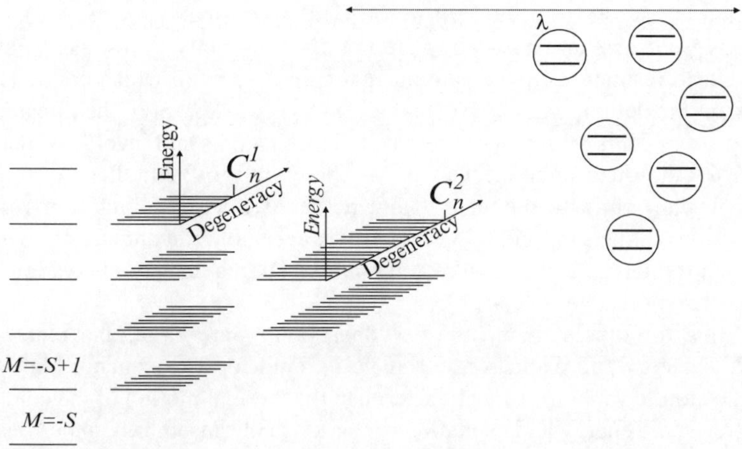

Fig. 2.23 Energy spectrum of an ensemble of two-level systems in the representation of the total "spin" S and its projection M

Interaction of atoms changes considerably the structure of the spectrum, since it implies the possibility of transitions among all $\mathcal{G}_S$-states $|\nu\rangle$ with the same S and M values. The random arrangement of atoms results in a random coupling and yields a splitting of the degenerate levels. The new energy eigenstates $|k\rangle$ are linear combinations $|k\rangle = \sum s_{k,M}^{\nu,M}|\nu, M\rangle$. The radiative transitions induced by the electromagnetic field of amplitude $\mathcal{E}$ now occur in a band-like spectral structure, and therefore conform to the same model [see (2.16)] as anharmonic polyatomic molecules in a resonant laser field. The corresponding Schrödinger equation reads

$$i\hbar\dot{\psi}_{M,k} = \Delta_{M,k}\psi_{M,k} + \mathcal{E}d\sqrt{(M-S)(M+S+1)}\sum_{k'} c_{M,k}^{k',M+1}\psi_{M+1,k'}$$
$$+\mathcal{E}d\sqrt{(M+S)(S+1-M)}\sum_{k'} c_{M,k}^{k',M-1}\psi_{M-1,k'}, \tag{2.61}$$

where $\Delta_{M,k}$ are the interaction-induced random detunings of states from their original positions, $c_{k,M}^{k',M'} = \sum_\nu s_{k,M}^{\nu,M}(s^*)_{k,M'}^{\nu,M'}$, and d is the dipole moment of each atom.

2.5.5 *Many-Body Effects in Cold Rydberg Gas*

Consider an ensemble of cold atoms excited in a high Rydberg state $|np\rangle$. Due to the dipole–dipole interaction, atoms may change their Rydberg state as a result of energy exchange: a couple of neighboring atoms initially in the state $|np, np\rangle$ performs a transition to the state $|ns, n's\rangle$. This energy exchange process may be viewed in the same way as the vibrational exchange in molecules described in Sect. 2.2: one atom emits a photon and performs a transition from the $|np\rangle$ Rydberg state to the $|ns\rangle$ Rydberg state. The photon, being absorbed by the other atom, provokes a transition from the $|np\rangle$ to the $|n's\rangle$ state (with $n' = n + 1$). Atomic transitions $np \to ns$ and $np \to n's$ are not always strictly in resonance, but they may be set to resonance by the external static electric field, which controls by the Stark effect the detuning $\Delta_0 = (E_{np} - E_{ns}) - (E_{np} - E_{n's})$ of the transition.

The process starts with pairs of relatively closed atoms and involves other atoms via the excitation-exchange mechanism, as shown in Fig. 2.24a. Each of the atoms of the pair can return to the state $|np\rangle$ whereas two surrounding atoms perform the transitions to the $|ns\rangle$ and $|n's\rangle$ states. This resembles an autocatalytic process, where the products of a chemical reaction produced in a density fluctuation diffuse out of the reaction zone.

Note that the dipole moments of Rydberg atoms are large, and therefore the process of energy and excitation exchange is rapid for a reasonably high density n_{at} of the atomic vapor. It is much faster than the thermal motion of the cold atoms, and therefore the energy and excitation exchange in the ensemble can be considered as a Hamiltonian process without relaxation, in contrast to collisional relaxation in ensembles of warm atoms and molecules. Viewing this process by analogy to the quasi-particle concept one can say that the couple of states $|ns\rangle$ and $|n's\rangle$ moves coherently to other sites as a result of the excitation-exchange process. We refer to the $|ns\rangle$ and $|n's\rangle$ states as the s and s' states, respectively, and to the $|np\rangle$ state as the p state. The excitation exchange results in the motion of the s-state over the immobile atoms, thus leading to diffusion of these states out of the "active center", that is, the pair of close atoms considered. Therefore, the pair of atoms can produce the next s–s' couple of states.

The level scheme for the elementary act of creation of a single s–s' couple is shown in Fig. 2.24b. The initial state $|0\rangle$ corresponds to all atoms in the $|np\rangle$ state. The first state $|1\rangle$ corresponds to atoms of a close pair (say the atom A and the atom B) in the states $|ns\rangle$ and $|n's\rangle$. The states $|0\rangle$and $|1\rangle$ are coupled by the matrix element of the dipole–dipole interaction $\hat{V}_{1,2} = \mu_{np}^{ns}\mu_{np}^{n's}/R_{AB}^3$ responsible for the energy exchange process. The band corresponds to locations of $|ns\rangle$ and $|n's\rangle$ states on other atoms (say C, D, F, etc.). The state $|1\rangle$ is coupled by the interaction to a combination $|b\rangle = \sum V_{1n}/\langle V^2\rangle^{1/2}\,|n\rangle$, which is a properly normalized superposition of the band states. This combination represents the s–s' couple delocalized over a group of atoms near the initial pair A–B. It is not an eigenstate of the Hamiltonian, and hence is also coupled to another combination of the band states, etc. This hierarchy of couplings of the states corresponds

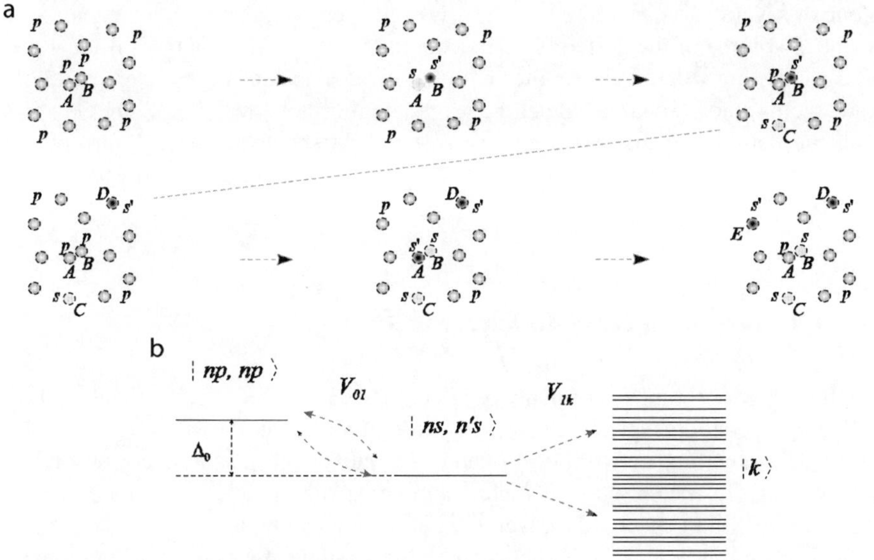

Fig. 2.24 (**a**) A couple of s-states (ss'-couple) is created at a pair of close atoms (A and B). Excitation-exchange processes result in the diffusion of these states out the pair. When the s-states have left the pair can produce the next ss'-couple. (**b**) Level scheme of the elementary act. The initial state $|np, np\rangle$ corresponds to both atoms A and B of the pair in the state $|np\rangle$. It is coupled by the energy exchange matrix element V_{01} to the state where one of the atoms is in the Rydberg state $|ns\rangle$, while the other is in the state $|(n+1)\,s\rangle$. Diffusion of the $|ns\rangle$ and $|(n+1)\,s\rangle$ states over the ensemble of other atoms via the excitation-exchange mechanism is represented by transitions to the band

to a dynamical process, which resembles diffusion with a diffusion coefficient $\simeq 4\pi\mu^2 n_{at} \ln(R_g/a)$ where R_g is the size of the Rydberg ensemble of the density $n_{at} \sim 1/a^3$, and μ is the dipole moment.

The Schrödinger equation reads

$$
\begin{aligned}
i\dot{\psi}_0 &= \Delta_0\psi_0 + V_{01}\psi_1 \\
i\dot{\psi}_1 &= V_{10}\psi_0 + \sum_n V_{1n}\psi_n \\
i\dot{\psi}_n &= \Delta_n\psi_n + V_{n1}\psi_1,
\end{aligned}
\tag{2.62}
$$

where ψ_0, ψ_1 and ψ_n are the amplitudes of the states $|0\rangle$, $|1\rangle$ and $|n\rangle$, respectively, and Δ_n is the detuning of the state $|n\rangle$ in the band from the state $|1\rangle$. Analytical calculation of the energies Δ_n of the eigenstates and the coupling V_{1n} is a challenging task. Indeed, the mean interaction $\int_0^\infty V(R)n_{at}d^3R$ in a gas with $1/R^3$ binary interaction diverges both at small and long distances. This can be interpreted that far atoms and close atoms are equally important for the excitation-exchange process. In such a situation, the spatial structure of stationary states of the quasi-particles is not immediately evident—the stationary states may be localized on one or several atoms, or, on the contrary, each energy eigenstate may extend over the entire volume of the system, like in a solid. It turns out that both well-localized states and strongly delocalized states with multifractal spatial structure are possible, which yields a non-trivial profile of the level detuning statistics $g(\Delta_n)$. On the other hand, the coupling statistics $g(V_{1n})$ similar to that shown in Fig. 2.5 turn out to be statistically independent from detunings and to follow the inverse square law $g(V) \sim V^{-2}$.

2.6 Quantum Circuits and Networks

Quantum system composed of many parts coupled each to another by an interaction depending, generally speaking, on the particular pair of the parts can also be viewed as a quantum network or a quantum circuit. Each part may correspond to a single quantum state or may be a more complex quantum object with its own few-dimensional Hilbert space. Moreover, the interaction can be not only in the form of just an operator acting in the Hilbert space of the coupled parts but also can it have a more sophisticated form of a pseudo-operator implying some internal dynamics of the quantum object intermediating the interaction of the parts. Furthermore, the intermediating objects can be not closed, but open quantum systems, which can be subject to measurements, such that the quantum state of the network or the circuit is conditioned by the measurements outcomes.

Considering dynamics of the circuits and networks, one encounters two types of questions that can be addressed. The first type of questions corresponds to the case where the system depends on a large number of parameters that are not under control. In such a situation one may ask about the ensemble average dynamics of the system, even in the case the system is open and subject by random time-dependent variations of these parameters in complete analogy with the systems considered earlier. The other extreme is when all the parameters are under control, they can be varied in time, and one needs to obtain a certain predetermined outcome of the dynamic process. Thus we are coming to the point where the systems conforming to the random matrix model are contrasted with the systems that cannot be described within this approach, although at the first glance, the corresponding Schrödinger equations have very similar structures. Three of many examples of such systems are the following.

2.6.1 Quantum Communications

Quantum communications suggest an example of the most simple quantum circuit composed just of two parts and a quantum object intermediating their coupling. In Fig. 2.25 we illustrate the idea of quantum communication.

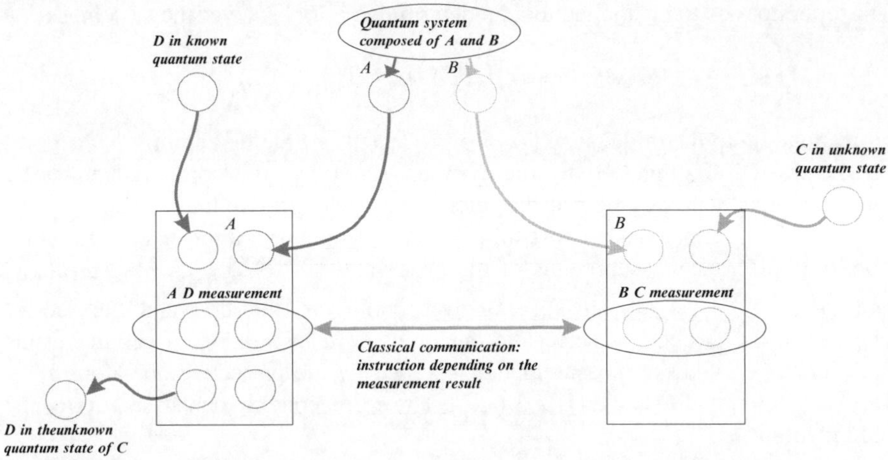

Fig. 2.25 Idea of quantum communication: a quantum system composed of two elements (say a quantum optical field composed of two photons) intermediates interaction between two parts (say two atoms). Being initially in a given quantum state, the intermediating system is spitted into two elements, one of which is send to the first part and the other to the second part. Each of the elements (photons) interacts with the corresponding part (atom) for a while, and then a measurement over the photons is performed. One can specify the initial state of the second part, the protocol of the interactions, and measurements such that as a result of this procedure an unknown state of the first part gets transported to the second part, thus realizing a quantum communication

One can formalize this procedure as a process that occurs in the Hilbert space of the combined system comprising both parts and the intermediating object. Let each of the parts to be a two-level atom, and the intermediating object to be a two-photon quantum field, while each of the photon in its proper mode propagating towards the corresponding atom. The resulting 16 dimensional Hilbert space has the basis vectors $|n_1, n_2, m_1, m_2\rangle$ specified by the numbers of excitations $n_i = 0, 1$ of each of the atoms $i = 1, 2$ and the number of photons $m_j = 0, 1$ in each of the modes $j = 1, 2$. The corresponding Schrödinger equation

$$i\dot{\psi}_{n_1,n_2,m_1,m_2} = \sum_{n'_1,n'_2,m'_1,m'_2} \left({}_pV^{m'_1,m'_2}_{m_1,m_2} + {}_1V^{n'_1,m'_1}_{n_1,m_1} + {}_2V^{n'_2,m'_2}_{n_2,m_2} \right) \psi_{n'_1,n'_2,m'_1,m'_2} \tag{2.63}$$

for the amplitudes of the basis vectors in the case of the time-dependent interactions between the photons, between the first photon and the first atom, and between the second photon and the second atom given by the first, second, and third terms of the Hamiltonian in the parentheses, respectively, yields the evolution operator

$$\hat{U} = \mathrm{T}\exp\left[-i\int\left({}_p\widehat{V}(t) + {}_1\widehat{V}(t) + {}_2\widehat{V}(t)\right)dt\right]. \tag{2.64}$$

This operator has to be applied to a predetermined initial state of the system, say

$$|s,0,0,0\rangle = \alpha\,|1,0,0,0\rangle + \beta\,|0,0,0,0\rangle\,,$$

corresponding to the initial state $|s\rangle = \alpha\,|1\rangle + \beta\,|0\rangle$, of the first atom.

The evolution suggested by the operator (2.64) can be controlled either by the time-dependent variation of the interactions, when possible, or simply by the exposition time if one has no possibility of such variations. At the end of the evolution one performs measurements of the observable $\hat{O}_1 = \sum_{i=1,2} |\omega_{1i}\rangle\,\omega_{1i}\,\langle\omega_{1i}|$ and $\hat{O}_2 = \sum_{i=1,2} |\omega_{2i}\rangle\,\omega_{2i}\,\langle\omega_{2i}|$ corresponding to the first and the second photons, respectively, and select the outcomes corresponding to a certain results (ω_{11} and ω_{12}) of these measurements specified by the projective operators $\hat{o}_1 = |\omega_{11}\rangle\,\omega_{11}\,\langle\omega_{11}|$ and $\hat{o}_2 = |\omega_{12}\rangle\,\omega_{12}\,\langle\omega_{12}|$. The corresponding density matrix of the second atom

$$\mathrm{Tr}_{n_1,m_1,m_2}\left\{\hat{o}_2\hat{o}_1\hat{U}\,|s,0,0,0\rangle\,\langle s,0,0,0|\,\hat{U}^{-1}\right\}$$

depends on the particular realization of the control of the evolution operator. This realization, as well as the projections $\hat{o}_1$ and $\hat{o}_2$, have to be chosen to achieve the control aim, that is, such that the resulting density matrix has rank one and reads $|s\rangle\,\langle s|$. In such a case an unknown initial state of the first atom is transported, or communicated, to the second atom.

2.6.2 *Quantum Computers*

One more example, extremely specific and important realization of complex quantum systems is known as Quantum Computer. By this name one denote a collection of many identical simple quantum systems each of which can be driven by individually chosen unitary evolution and each pair of which can be subject to a standard unitary transformation (gate) in any chosen moment of time. In such a situation, even in the simplest case of two-level systems, so-called qubits, one encounters a quantum system with an exponentially large 2^n-dimensional Hilbert space, where n is the number of qubits. The ideal quantum computer is therefore an extreme case of quantum network, where each part is coupled to each other part by an identical interaction resulting in a standard binary operation applied in a controlled way at a properly chosen moment of time.

Quantum information is thus encoded in a state vector $\sum_{\{n_k\}} \psi_{\{n_k\}}\,|\{n_k\}\rangle$ usually represented in so-called computational basis $|\{n_k\}\rangle$, where $n_k = 0$ corresponds to the ground state of the k-th qubit, and $n_k = 1$ to the excited one. The string $\{n_k\}$ can also be viewed as the binary representation of a natural number written in the computer register. In Fig. 2.26 we illustrate the scheme of the quantum system, which can be employed for quantum computation. The

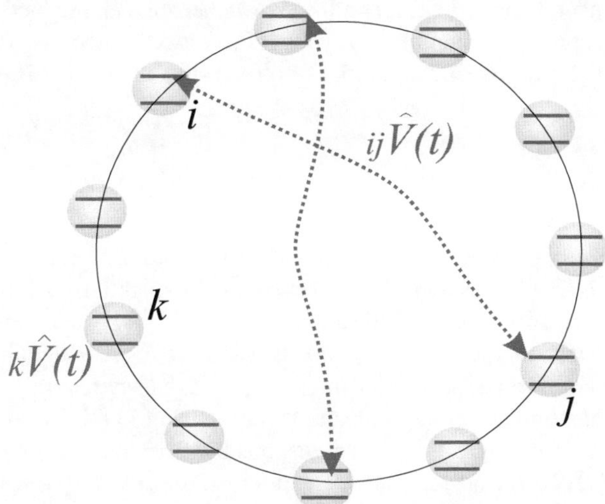

Fig. 2.26 Scheme of a quantum computer: the quantum system is composed of many two-level elements (qubits) each of which can be subject to a local perturbation $\widehat{_kV}(t)$ ensuring universal local transformation $\widehat{_kU} = \mathrm{T}\exp i \int \widehat{_kV}(t)dt$. Pairs of qubits can be subject to a standard binary gate operation $\widehat{_{ij}U} = \mathrm{T}\exp i \int \widehat{_{ij}V}(t)dt$

Schrödinger equation describing dynamics of the quantum computer reads

$$i\dot{\psi}_{\{n_k\}} = \sum_k {_kV}_{n_k}^{n'_k}(t)\psi_{\{n'_k\}} + \sum_{k,k'} {_{k,k'}V}_{n_k,n_{k'}}^{n'_k,n'_{k'}}(t)\psi_{\{n'_{k'}\}}, \tag{2.65}$$

where $\widehat{_kV}(t)$ is the local interaction resulting in the unitary evolution of the qubit k, and $\widehat{_{k,k'}V}(t)$ is the binary interaction coupling qubits k and k' and ensuring the gate transformation.

The quantum dynamics of the computer does not imply any randomness or ensemble average, being an essentially deterministic quantum process of transition from a standard initial quantum state to a final quantum state encoding solution of a computational problem corresponding to the chosen quantum code. Though both the initial and the final states belong to the computational basis, in the middle of the calculation, the state vector of the quantum computer is a generic linear superposition of the basis states given by an exponentially large number of the amplitudes $\psi_{\{n_k\}}$, any of which is as important as the others. Therefore, the quantum computer being the opposite extreme to a random systems cannot be tracked by any reasonable analytical technique based on the state populations and the relevant amplitudes. It worth mentioning that dealing with quantum computers one may need to considerably modify the vocabulary, vision, and the approach to description of physical reality, which can no longer rely on formulas and equations but invokes quantum algorithms as the main conceptual instrument.

There exist a large number of quantum systems suggested as physical realization of quantum computers, starting from the ions in the Paul traps, where the interactions are implemented by laser field and the binary operations are assisted by the vibrational modes of the ionic crystal, and finishing by the spin-spin interactions in molecules in the conditions of the nuclear magnetic resonance setting. Other suggestions include arrays of Josephson junctions, electron spins in diamond dislocations, and a number of other projects.

2.6.3 *Quantum Effects in Photosynthesis*

Two opposite extremes of complex quantum objects considered thus far—random disordered systems and deterministic quantum computers, do not exhaust all interesting cases. A fascinating example of a quantum system, which is neither deterministic nor completely statistical is suggested by biology in the context of the photosynthesis mechanism. One may even say that this is an example of a complex quantum system optimized by Nature itself in the course of Darwin's evolution.

A biological network responsible for the photon absorption is sketched in Fig. 2.27. Speaking in physical terms, the system consists of a number of antennas, each of which is capable to absorb a photon and keep its energy in the form of electronic excitation. Two-level systems offer the simplest model of these antennas. Via the excitation-exchange mechanism, the excitation can be transferred from antenna i to antenna j with a transition amplitude V_{ij}, and a sequence of such transitions occurs until the excitation reaches the root antenna, which transmits the photon energy further to a final recipient—a domain converting it to the chemical form. The intriguing aspect of the process is that in spite of fast dissipation, typical of macromolecular dynamics, the efficiency of the excitation quantum transfer towards the final recipient is very close to unity. Due to this fact, the process is sometimes called "photon harvesting".

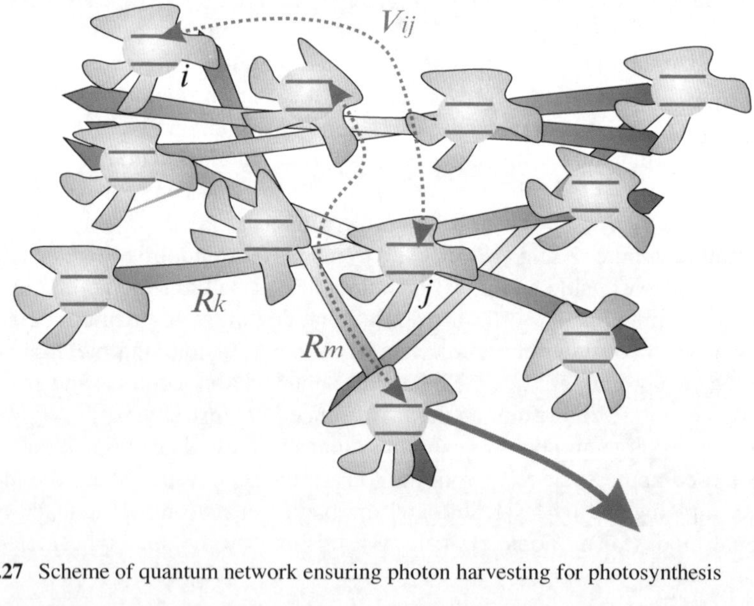

Fig. 2.27 Scheme of quantum network ensuring photon harvesting for photosynthesis

Electronic dissipation in molecules usually occurs via the mechanism of inter-molecular conversion of the electronic energy to vibrations, discussed earlier in Sect. 2.2.3. The network under consideration consists of a very large number of atoms and therefore it has a very reach vibrational spectrum. Some of the vibrational modes R_k correspond to deformation of rather extended molecular species connecting different antennas, and therefore these modes can serve as a common environment of these antennas. Joint dynamics of the electronic excitation and vibration of the common environment can be suggested as a candidate for the mechanism responsible for the photon harvesting.

The Schrödinger equation for the network reads

$$i\dot{\psi}_{\{n_i\},\{m_k\}} = \sum_{n_i',n_j'} V_{n_i n_j}^{n_i' n_j'} \psi_{\{n_j'\},\{m_k\}} + \sum_{m_k'} H_{\{m_k\}}^{\{m_k'\}} \psi_{\{n_i\}\{m_k'\}}$$
$$+ \sum_{n_i',m_k'} \hat{v}_{n_i,m_k}^{n_i',m_k'} \psi_{\{n_i'\}\{m_k'\}} - i\gamma\delta_1^{n_0} \psi_{\{n_i\},\{m_k\}}, \qquad (2.66)$$

where $\hat{V}$ allows for the excitation exchange among the antennas, $\hat{H}$ stands for the Hamiltonian of the vibrational modes R_k, $\hat{v}$ describes the interaction between the electronic and vibrational degrees of freedom, and γ denotes the rate of transition of the excitation from the excited state of the root antenna towards the final recipient. The Hamiltonians entering (2.66) should maximize the probability

$$2\gamma \int \left|\psi_{\{n_0=1;n_i\},\{m_k\}}(t)\right|^2 dt$$

of the transition toward the final recipient for all initial conditions defined at the moment of the photon absorption, that is, this probability should be close to unity whatever is the number of antennas which had absorbed the photon initially.

Chapter 3
Two-Level and Level–Band Systems

We start to consider the dynamics of populations of complex multilevel systems with two well-known examples: a two-level system and a system consisting of a level and a band which is either continuous or composed of discrete levels. The main purpose of such a consideration is to introduce the mathematical methods which are usually employed for the description, and to relate these methods to the main physical processes taking place in the systems, such as Rabi oscillations in two-level systems, repulsion of the levels by interaction, and the exponential decay of the level into the band, as well as to figure out what are the main physical parameters governing the population dynamics. The influence of the continuous band edges and the number of resonant states for the discrete band are discussed along with the process of population injection. We take into account the effects of interference resulting from the structure induced in the continuum by the presence of another interacting level and show how the interference yields special spectral structures known as Fano profiles.

For the discrete bands, we pay a great deal of attention to the asymptotic behavior in the time domain longer than the Heisenberg return time. In this regime, the population returns from the band to the level. In a number of cases, the return can be complete and is then termed a revival. An interesting case of fractional revivals, when the population returns in the form of several short wavepackets, each of which contains a fraction of the complete population, is discussed in detail. The existence of fractional revivals suggests a revision of the Ehrenfest classical limit of quantum mechanics and its extension to the case where, in this limit, one can observe a quantum mixture of two or more classical wavepackets, the situation known as Schrödinger's cats.

We also concentrate on the case where an infinitely dense band has such an irregular structure that it cannot be considered as a continuum. Though this problem has an exact analytical solution in the form of a onefold integral from an algebraic expression containing the sum of contributions of the individual band levels, the dynamics of such a system is not trivial. It consists of multiple returns of the population to the level that occur with different return times, differing by orders

V.M. Akulin, *Dynamics of Complex Quantum Systems*, Theoretical and Mathematical Physics, DOI 10.1007/978-94-007-7205-2_3,

of magnitude. Interference of these returns results in a non-exponential decay of the level population averaged over a time interval. Moreover under certain conditions, the decay of the level can be incomplete, which is the phenomenon related to dynamical localization. We find the parameters of the quantum system governing the population dynamics in this case.

3.1 Two-Level System

Let us concentrate first on a two-level system with the Hamiltonian

$$\hat{H} = \begin{pmatrix} E_1 & 0 \\ 0 & E_0 \end{pmatrix} \tag{3.1}$$

and write down the Schrödinger equation for the amplitudes Ψ_0 of the ground state, corresponding to the energy E_0 and the state vector $|0\rangle$. The amplitude, the energy, and the state vector of the excited state are Ψ_1, E_1, and $|1\rangle$, respectively. The Schrödinger equation for the Hamiltonian (3.1) reads

$$\begin{aligned} i\hbar\dot{\Psi}_1 &= E_1\Psi_1 \\ i\hbar\dot{\Psi}_2 &= E_2\Psi_2. \end{aligned} \tag{3.2}$$

What happens when we switch on an external periodic electromagnetic field of amplitude $\mathcal{E}$ and frequency ω close to the transition frequency $\omega \sim (E_1 - E_0)/\hbar$ if the interaction Hamiltonian is given by the product $\mathcal{E}_\omega \hat{d}$ of the field strength $\mathcal{E}_\omega = \mathcal{E}\cos\omega t$ and the transition dipole moment operator $\hat{d}$? We assume that the dipole moment operator has real non-diagonal matrix elements $d = \langle 0|\,\hat{d}\,|1\rangle$, and the diagonal matrix elements vanish, as is usually the case in optics. The Schrödinger equation now takes the form

$$\begin{aligned} i\hbar\dot{\Psi}_1 &= E_1\Psi_1 + d\mathcal{E}\cos{(\omega t)}\Psi_0 \\ i\hbar\dot{\Psi}_0 &= E_0\Psi_0 + d\mathcal{E}\cos{(\omega t)}\Psi_1. \end{aligned} \tag{3.3}$$

We now apply to (3.3) the so-called rotating wave approximation, which is convenient when the detuning $\Delta = (E_1 - E_0)/\hbar - \omega$ of the field frequency from the transition frequency $(E_1 - E_0)/\hbar$ is relatively small $\Delta \ll \omega$, and which implies that the quantum phases of the states are taken relative to the phase ωt of the field. To this end we substitute the amplitudes of the ground and the excited states in the form

$$\begin{aligned} \Psi_1 &= e^{-i\omega t}\psi_1 \\ \Psi_0 &= \psi_0 \end{aligned} \tag{3.4}$$

to (3.3) and make use of the formula $\cos(\omega t) = \frac{1}{2}(e^{i\omega t} + e^{-i\omega t})$, which yields

$$i\hbar\dot{\psi}_1 e^{-i\omega t} + \hbar\omega\psi_1 e^{-i\omega t} = E_1\psi_1 e^{-i\omega t} + \frac{d\mathcal{E}}{2}(e^{i\omega t} + e^{-i\omega t})\psi_0$$

$$i\hbar\dot{\psi}_0 = E_0\psi_0 + \frac{d\mathcal{E}}{2}(e^{i\omega t} + e^{-i\omega t})\psi_1 e^{-i\omega t}. \tag{3.5}$$

When we set $E_0 = 0$ and replace the combination $d\mathcal{E}/2$ by a shorter notation V, after multiplication of the first equation by $e^{i\omega t}$ the system (3.5) takes the form

$$i\hbar\dot{\psi}_1 = (E_1 - \hbar\omega)\psi_1 + V(1 + e^{2i\omega t})\psi_0$$

$$i\hbar\dot{\psi}_0 = V(1 + e^{-2i\omega t})\psi_1. \tag{3.6}$$

To shorten notation further we also set $\hbar = 1$ hereafter.

We now note that typical time derivatives of the amplitudes ψ_1 and ψ_0 in (3.6) are either of the order of the size of the interaction V or of the order of the detuning $\Delta = (E_1 - \omega)$, when it exceeds V, and therefore these derivatives are much smaller compared to the frequency of the external field ω, provided the condition of resonance $\Delta \ll \omega$ is fulfilled. This enables us to ignore the rapidly oscillating terms $Ve^{2i\omega t}$ in (3.6) that give negligibly small average contributions, and we arrive at

$$i\dot{\psi}_1 = \Delta\psi_1 + V\psi_0$$

$$i\dot{\psi}_0 = V\psi_1. \tag{3.7}$$

Let us now consider the solution of (3.7) corresponding to the initial conditions

$$\psi_0(t = 0) = 1$$

$$\psi_1(t = 0) = 0. \tag{3.8}$$

One of our main tools is a method known either as the Fourier–Laplace transformation or as the Fourier transformation of generalized functions or distributions. The essence of the method is the following. The regular technique of Fourier transformations requires that the subjecting function vanishes at $t \to \pm\infty$, otherwise the Fourier transform of a derivative $\dot{y}(\omega)$, denoted by $F[\dot{y}(t)]$, differs from the Fourier transform of the primitive function multiplied by the frequency $\dot{y}(\omega) \neq i\omega y(t)$. Therefore all the functions ψ are assumed to be zero at $t < 0$ and coincide with the solutions of (3.7) with the initial conditions (3.8) for $t > 0$. This implies that the function $\psi_0(t)$ is discontinuous at $t = 0$ and it experiences a jump from the value $\psi_0 = 0$ to the value $\psi_0 = 1$ at this point. Apparently, the derivative $\dot{\psi}_0$ equals infinity, or in other words it amounts to the Dirac δ-function. Equation (3.7) is no longer valid at $t = 0$, and the correct equation reads

$$i\dot{\psi}_1 = \Delta\psi_1 + V\psi_0$$

$$i\dot{\psi}_0 = V\psi_1 + i\delta(t). \tag{3.9}$$

We note here that the Dirac δ-function in the second equation (3.9) can be interpreted in physical terms. Since we assume that at $t < 0$ there was no particle in the system of two levels, and at $t = 0$ a particle has been supplied to the ground state $|0\rangle$, one can say that the flux of the probability amplitude of the particle supply is infinitely large during an infinitely short time and is given by the δ-function. Such a situation suggests an evident generalization: one can supply a particle with $\Pi = \Pi(t)$, which can be an arbitrary function of time satisfying just a normalization condition. Such a situation is shown schematically in Fig. 3.1.

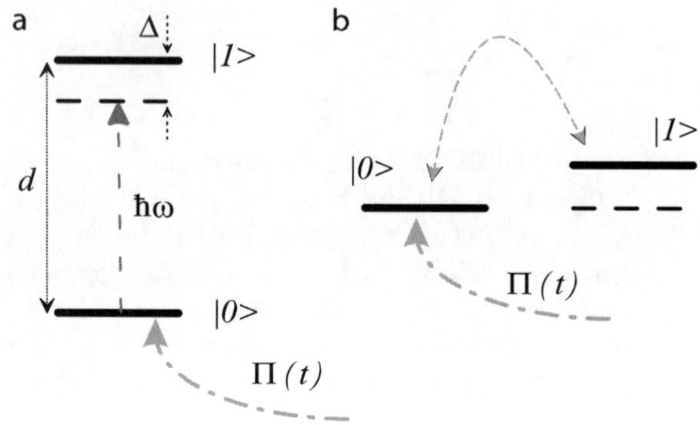

Fig. 3.1 Two-level system in an external resonant field. A quantum particle is supplied at the lower state with the probability amplitude flux $\Pi(t)$. A transition dipole moment d results in the coupling $V = \mathcal{E}d$ between the states. (**a**) Representation in the basis of energy eigenstates. (**b**) Representation in the rotating wave approximation

Let us now perform the Fourier transformation of (3.9) and arrive at

$$\begin{aligned} \varepsilon\psi_1(\varepsilon) - \Delta\psi_1(\varepsilon) - V\psi_0(\varepsilon) &= 0 \\ \varepsilon\psi_0(\varepsilon) - V\psi_1(\varepsilon) &= i, \end{aligned} \tag{3.10}$$

where

$$\psi_{0;1}(\varepsilon) = \int_{-\infty}^{+\infty} e^{i\varepsilon t}\psi_{0;1}(t)dt, \tag{3.11}$$

and we have employed the fact that $\int_{-\infty}^{+\infty} e^{i\varepsilon t}\delta(t)dt = 1$.

The set of two linear algebraic equations (3.10) has the solution

$$\psi_1(\varepsilon) = \frac{\begin{vmatrix} 0, & -V \\ i, & \varepsilon \end{vmatrix}}{\begin{vmatrix} \varepsilon - \Delta, & -V \\ -V, & \varepsilon \end{vmatrix}} = \frac{Vi}{(\varepsilon - \Delta)\varepsilon - V^2}$$

$$\psi_0(\varepsilon) = \frac{\begin{vmatrix} \varepsilon - \Delta, & 0 \\ -V, & i \end{vmatrix}}{\begin{vmatrix} \varepsilon - \Delta, & -V \\ -V, & \varepsilon \end{vmatrix}} = \frac{(\varepsilon - \Delta)i}{(\varepsilon - \Delta)\varepsilon - V^2} = \frac{i}{\varepsilon - \frac{V^2}{(\varepsilon - \Delta)}} \tag{3.12}$$

where vertical brackets around a matrix denote the determinant. We find the time-dependent probability amplitudes $\psi_0(t)$ and $\psi_1(t)$ by performing the inverse Fourier transformation

$$\psi_{0;1}(t) = \frac{1}{2\pi} \int_{-\infty + i\nu}^{+\infty + i\nu} e^{-i\varepsilon t} \psi_{0;1}(\varepsilon) d(\varepsilon), \tag{3.13}$$

that is, by integrating the Fourier transforms $\psi_0(\varepsilon)$ and $\psi_1(\varepsilon)$ in the complex plane along the contour C given by a straight line which goes from $-\infty$ to $+\infty$ at a distance ν above the real axis, as shown in Fig. 3.2.

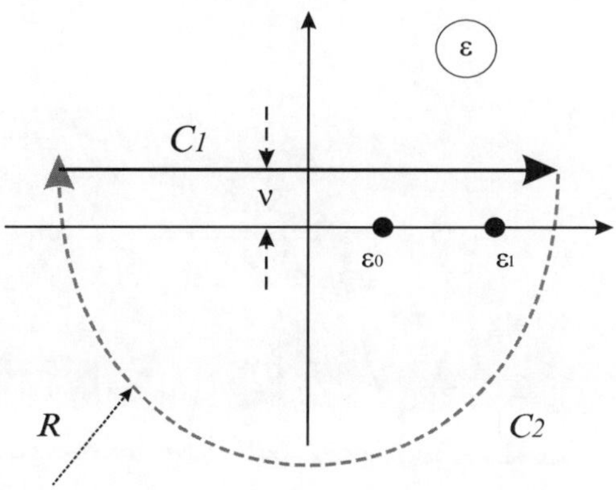

Fig. 3.2 The integration contour of the inverse Fourier transform goes above the real axis. It can be closed in the lower part of the complex plane of ε, where the integrand vanishes

The denominators of the right-hand sides of (3.12) tend to zero at the points

$$\varepsilon_{0;1} = \frac{\Delta}{2} \pm \sqrt{\left(\frac{\Delta}{2}\right)^2 + V^2}; \tag{3.14}$$

we also note that the integrals (3.13) taken along the contour C_2 (a circle of large radius $\varepsilon = R\exp(-i\theta)$; $R \longrightarrow \infty$; $0 \le \theta \le \pi$) vanish in the limit $R \longrightarrow \infty$ since the exponent of the integral acquires a large negative part. We can therefore replace the open integration contour $C_1 = (-\infty - i\nu, \infty - i\nu)$ by the closed integration contour $C = C_1 + C_2$ and make use of the Cauchy theorem by replacing the integral by the sum of the contributions of residuals corresponding to the singularities of the integrand at the points $\varepsilon_{0;1}$ (3.14) in the lower part of the complex plane ε including the real axis. We arrive at

$$\psi_1(t) = \frac{V}{(\varepsilon_0 - \varepsilon_1)} e^{-i\varepsilon_0 t} - \frac{V}{(\varepsilon_0 - \varepsilon_1)} e^{-i\varepsilon_1 t}$$
$$\psi_0(t) = \frac{(\varepsilon_0 - \Delta)}{(\varepsilon_0 - \varepsilon_1)} e^{-i\varepsilon_0 t} - \frac{(\varepsilon_1 - \Delta)}{(\varepsilon_0 - \varepsilon_1)} e^{-i\varepsilon_1 t} \tag{3.15}$$

which yields

$$\psi_1(t) = \frac{\frac{V}{2}}{\sqrt{(\frac{\Delta}{2})^2 + V^2}} e^{-i(\frac{\Delta}{2} + \sqrt{(\frac{\Delta}{2})^2 + V^2})t} - \frac{\frac{V}{2}}{\sqrt{(\frac{\Delta}{2})^2 + V^2}} e^{-i(\frac{\Delta}{2} - \sqrt{(\frac{\Delta}{2})^2 + V^2})t}$$
$$\psi_0(t) = \frac{(\frac{\Delta}{2} + \sqrt{(\frac{\Delta}{2})^2 + V^2})}{2\sqrt{(\frac{\Delta}{2})^2 + V^2}} e^{-i(\frac{\Delta}{2} + \sqrt{(\frac{\Delta}{2})^2 + V^2})t} \tag{3.16}$$
$$- \frac{(\frac{\Delta}{2} - \sqrt{(\frac{\Delta}{2})^2 + V^2})}{2\sqrt{(\frac{\Delta}{2})^2 + V^2}} e^{-i(\frac{\Delta}{2} - \sqrt{(\frac{\Delta}{2})^2 + V^2})t}$$

that is,

$$\psi_1(t) = -\frac{Vi}{\sqrt{(\frac{\Delta}{2})^2 + V^2}} e^{-i\frac{\Delta}{2}t} \sin t\sqrt{\left(\frac{\Delta}{2}\right)^2 + V^2}$$
$$\psi_0(t) = e^{-i\frac{\Delta}{2}t}\left[\cos t\sqrt{\left(\frac{\Delta}{2}\right)^2 + V^2} - \frac{\frac{\Delta}{2} i \sin t\sqrt{(\frac{\Delta}{2})^2 + V^2}}{\sqrt{(\frac{\Delta}{2})^2 + V^2}}\right]. \tag{3.17}$$

The combination $\sqrt{(\Delta/2)^2 + V^2}$ is known as the Rabi frequency, and the corresponding periodic oscillations of the probabilities are referred to as the Rabi oscillations.

Let us now calculate the populations $\rho_1 = |\psi_1|^2$ of the upper and $\rho_0 = |\psi_0|^2$ of the lower levels. Equation (3.17) immediately results in

$$\rho_1(t) = \frac{V^2}{(\frac{\Delta}{2})^2 + V^2} \sin^2 t\sqrt{\left(\frac{\Delta}{2}\right)^2 + V^2}$$
$$\rho_0(t) = \frac{\Delta^2}{\Delta^2 + 4V^2} \sin^2 t\sqrt{\left(\frac{\Delta}{2}\right)^2 + V^2} + \cos^2 t\sqrt{\left(\frac{\Delta}{2}\right)^2 + V^2}. \tag{3.18}$$

Let us now ask for the time average populations $\overline{\rho_{0;1}}$ of the states. Allowing for $\overline{\cos^2 \Omega t} = \overline{\sin^2 \Omega t} = 1/2$ we arrive at

$$\overline{\rho_1} = \frac{2V^2}{\Delta^2 + 4V^2}$$
$$\overline{\rho_0} = \frac{1}{2}\left[\frac{\Delta^2}{\Delta^2 + 4V^2} + 1\right]. \tag{3.19}$$

One can see that for $V \ll \Delta/2$ the mean population ρ_1 of the upper level is small, and the population ρ_0 is of the order of 1. The populations of both levels become of the same order of magnitude only when $V \sim \Delta/2$. At the extreme $V \gg \Delta/2$ we have $\rho_1 \simeq \rho_0 \simeq 1/2$. We can therefore say that the states $|0\rangle$ and $|1\rangle$ are in resonance and the condition of the resonance $V \sim \Delta/2$ is that the typical interaction amplitude becomes larger or of the order of the typical detuning. Later on we will often employ this qualitative criterion analyzing the behavior of multilevel quantum systems.

We also note an important case when the detuning is so small that it can be set zero. In this case

$$\rho_1(t) = \sin^2 Vt$$
$$\rho_0(t) = \cos^2 Vt, \tag{3.20}$$

and the population oscillates between the lower and the upper states with the frequency V. In this case the average populations of the states are apparently equal and amount to $1/2$.

Sometimes it is expedient to represent the time-dependent amplitudes $\psi_{0,1}(t)$ (3.17) in the form of a superposition of exact energy eigenstates $|u\rangle$, $|l\rangle$ of the system corresponding to the upper and lower energy eigenvalues $\varepsilon_{1;0} = \frac{\Delta}{2} \pm \sqrt{(\frac{\Delta}{2})^2 + V^2}$ of (3.14) respectively. The amplitudes $\varphi_{l,u}$ of these states have only the phase time-dependence $e^{-i\varepsilon_{0;1}t}$ whereas the absolute values of the amplitudes and the phase factors at $t = 0$ one finds with the help of the representation

$$|l\rangle = \frac{|0\rangle\left(\frac{\Delta}{2V} + \sqrt{1 + \frac{\Delta^2}{4V^2}}\right)}{\sqrt{1 + \left(\frac{\Delta}{2V} + \sqrt{1 + \frac{\Delta^2}{4V^2}}\right)^2}} + \frac{|1\rangle}{\sqrt{1 + \left(\frac{\Delta}{2V} + \sqrt{1 + \frac{\Delta^2}{4V^2}}\right)^2}}$$

$$|u\rangle = \frac{|0\rangle\sqrt{1 + \left(\frac{\Delta}{2V} - \sqrt{1 + \frac{\Delta^2}{4V^2}}\right)^2}}{2\sqrt{1 + \frac{\Delta^2}{4V^2}}} - \frac{|1\rangle}{\sqrt{1 + \left(\frac{\Delta}{2V} - \sqrt{1 + \frac{\Delta^2}{4V^2}}\right)^2}} \tag{3.21}$$

of the eigenvectors in the basis of states $|0\rangle$ and $|1\rangle$. In particular, for the case $\Delta = 0$ we have two eigenstates

$$|l\rangle = |+\rangle = \frac{1}{\sqrt{2}}|0\rangle + \frac{1}{\sqrt{2}}|1\rangle$$
$$|u\rangle = |-\rangle = \frac{1}{\sqrt{2}}|0\rangle - \frac{1}{\sqrt{2}}|1\rangle \quad (3.22)$$

corresponding to the eigenenergies $\varepsilon_{0;1} = \pm V$, with the relative phase difference $2Vt$ of the amplitudes $\varphi_{u,l} = e^{\pm iVt}/\sqrt{2}$ having identical initial values $\langle l\,|0\rangle = \langle u\,|0\rangle = 1/\sqrt{2}$. Beating of the oscillations of these two amplitudes yields (3.20) for the populations.

3.2 Level–Band System

We continue by tackling a more complex system of a level interacting with a band of levels. We first consider it in the general case and obtain an exact expression for the Fourier transforms of time-dependent amplitudes of the level population amplitudes by the analogy to the results of the previous section. It turns out that this problem possesses an exact analytical solution. However, physically meaningful results emerge from the general expressions only in several limiting cases. The most well-known limit corresponds to an infinite density of states and a constant interaction and yields an expression known as the Fermi golden rule. The other important case is known as the Fano problem. This also implies an infinite density of states but allows for a non-uniform, that is, energy-dependent, amplitude of the transition from the initial level to the band which by itself is a result of the interaction of another discrete level with a continuum. Comparison of the solutions of these two problems shows an important role of the quantum interference that affects strongly the population dynamics. This role becomes even more important when we consider a band of finite state density, that is, a zone composed by a large but finite number of discrete levels. It results in specific effects at long time scales—recurrences, revivals, and fractional revivals. We also study the dynamics of the system in the case, when the coupling among the level and the states of the band can considerably, that is, by orders of magnitudes, vary from state to state. We show that the interference changes completely the dynamics of the system in the long time scale, although this phenomenon does not manifest itself in periodic oscillations typical of interference phenomena.

3.2.1 General Consideration

In the previous section by having employed the rotating wave representation and the resonance approximation we have shown that a system in a periodic external field behaves as a system subject to a time-independent perturbation, provided the detuning of the external frequency from the transition frequency is relatively small.

Considering here a level–band system we make use of the same approximation and by the analogy arrive at the Schrödinger equation

$$i\dot{\psi}_n = \Delta_n\psi_n + V_n^0\psi_0$$
$$i\dot{\psi}_0 = \sum_{n=1}^{N} V_0^n\psi_n. \tag{3.23}$$

Here ψ_0 is a probability amplitude of the system to be in the ground state, and ψ_n denote the amplitudes of the energy eigenstates in the N-level band. The detunings Δ_n and the interaction amplitudes V_0^n now depend on the numbers n of the band levels that are supposed to be non-interacting among themselves. In Fig. 3.3 we illustrate the level scheme of the system.

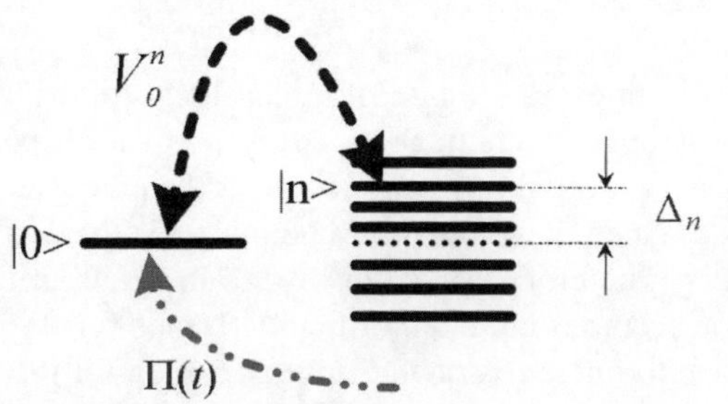

Fig. 3.3 Level–band system. The levels $|n\rangle$ of the band detuned by Δ_n are not coupled among themselves directly, but only via couplings V_0^n with the level $|0\rangle$. The system either was initially in the state $|0\rangle$ with probability amplitude $\psi_0(t=0)=1$, or has been injected into this state with probability amplitude flux $\Pi(t)$

Let us assume that at $t = 0$ only the state ψ_0 is populated. Then by analogy to (3.9) we write

$$i\dot{\psi}_n = \Delta_n\psi_n + V_n^0\psi_0$$
$$i\dot{\psi}_0 = \sum_{n=1}^{N} V_0^n\psi_n + i\delta(t). \tag{3.24}$$

For the Fourier transforms $\psi_n(\varepsilon)$ by analogy to (3.10) we therefore have

$$\varepsilon\psi_n - \Delta_n\psi_n - V_n^0\psi_0 = 0$$
$$\varepsilon\psi_0 - \sum_{n=1}^{N} V_0^n\psi_n. = i. \tag{3.25}$$

We now express $\psi_n(\varepsilon)$ in terms of $\psi_0(\varepsilon)$ with the help of the first equation (3.25), substitute this expression to the second equation, factor out the common multiple $\psi_0(\varepsilon)$ from the sum, and finally arrive at

$$\psi_0(\varepsilon) = \frac{i}{\varepsilon - \sum_{n=1}^{N} \frac{V_0^n V_n^0}{\varepsilon - \Delta_n}}$$
$$\psi_n(\varepsilon) = \frac{V_0^n}{\varepsilon - \Delta_n}\psi_0 = \frac{V_0^n}{\varepsilon - \Delta_n}\frac{i}{\varepsilon - \sum_{n=1}^{N} \frac{V_0^n V_n^0}{\varepsilon - \Delta_n}}, \tag{3.26}$$

which is a formal solution of the problem in the Fourier representation.

3.2.2 Continuous Band Model

Equation (3.26) gives an exact solution of the problem for arbitrary amplitudes of interaction V_0^n and arbitrary detunings Δ_n. It allows for all possible interference phenomena that may occur in the system. However the practical usefulness of these equations is limited, since they contain a complicated function given by the sum in the denominators. Therefore evaluation of the integrals in the inverse Fourier transformation now becomes the main difficulty. Hence for the further analysis of the problem we have to find a reasonable approximation for the sum, which would allow us to find an analytical solution of the problem. One possibility is to replace these sums by an integral which can be evaluated. The following physical example of a level–band system will help us to gain insight into the physical meaning of the approximation.

An Example of the Level–Band Model

Let us consider a quantum particle moving in an external field formed by two potential wells A and B separated by a barrier C shown in Fig. 3.4. The first well is deep and narrow and the second is wide and shallow. The distance between successive levels is of the order of the oscillation frequency, and therefore the spacing among the neighboring levels $|n\rangle$ in the shallow well B is much smaller than that in the deep well A.

Here we restrict our consideration to the energy region around the first excited level in the first well, and allow for the tunneling from this state to the states of the potential well B through the potential barrier C. We therefore assign index 0 to this level, which is the initial state of our level–band problem. The lower state in the pit A serves only as a reservoir for the probability amplitude supply $\Pi(t)$, and will not be taken into account otherwise. Note that energy eigenstates in each of the pits are not the exact eigenstates of the entire problem. However, they can be taken as a good approximative basis set, provided the barrier is

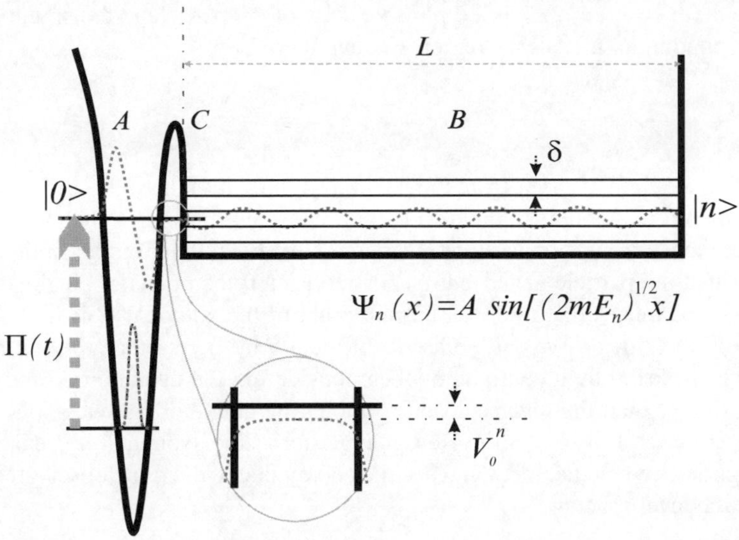

Fig. 3.4 A quantum particle in a two-well potential, which consists of two pits A and B separated by a barrier C, is an example of a level–band system with an amplitude injection to the level. The interaction V_0^n among the energy eigenstates of different pits is of the order of the tunneling exponent. The spacing δ of neighboring levels in the pit of length L is of the order of the inverse period, that is, the inverse time of flight from wall to wall. The population supply $\Pi(t)$ to the excited state in the left pit can be controlled by an external field

sufficiently high. These basis states are not orthogonal, and the Hamiltonian is not diagonal in such a representation. It contains off-diagonal matrix elements given by the overlap integrals, which are of the order of tunneling exponents $V_0^n \sim \exp\left\{-\mathrm{Im}\int_{\text{under barrier } C}\sqrt{E-U(x)dx}\right\}$. Here we take all V_0^n to be identical, since the distances among the levels of interest are much smaller than the height of the barrier, and all tunneling exponents are almost equal.

Let us turn now to the spacing δ among the neighboring levels in the pit B. For a rectangular well the energy eigenfunctions satisfy the Bohr–Sommerfeld quantization condition

$$\int_{\text{left turning point}}^{\text{right turning point}} \sqrt{2mE_n}dx = \sqrt{2mE_n}L = n\pi. \tag{3.27}$$

From (3.27) one immediately finds the energy difference among the neighboring levels:

$$\sqrt{2mE_n} - \sqrt{2mE_{n-1}} = \sqrt{\frac{m}{2E_n}}\left(E_n - E_{n-1}\right) = \frac{\pi}{L}. \tag{3.28}$$

We note that $v = \sqrt{m/2E_n}$ is a typical velocity of the particle in the potential well B corresponding to the energy region E_n, and therefore

$$\delta = (E_n - E_{n-1}) = \frac{\pi}{L} v \simeq \frac{\pi}{T}, \tag{3.29}$$

where T is the time of flight from the left wall of the pit to the right wall. The level spacing therefore amounts to the reciprocal time of return flight in the pit.

We immediately understand the meaning of the continuum model limit $\delta \to 0$: it means that the particle never returns once it has left the potential pit A, since the limit $T \to \infty$ means that the right turning point of the potential B is infinitely far away from the barrier. We can phrase it differently by saying that one can replace a discrete spectrum by a continuum when considering the evolution during a time interval shorter than the inverse spacing among the neighboring levels, that is, the state density $t \ll 1/\delta \sim g$ or the Heisenberg return time. Note that it is exactly the same regime as when the uncertainty of the energy due to the finite observation time exceeds the level spacing δ.

Level and Uniform Continuum

Let us now consider the consequences of the continuous-band approximation. In (3.26) we can now factor out the moduli squared of the interaction amplitudes $V^2 = |V_0^n|^2$, substitute $\Delta_n = n\delta$, and replace summation over n by the integration. We arrive at

$$\psi_0(\varepsilon) = \frac{i}{\varepsilon - V^2 \int \frac{dn}{\varepsilon - n\delta}}$$
$$\psi_n(\varepsilon) = \frac{V}{\varepsilon - n\delta} \frac{i}{\varepsilon - V^2 \int \frac{dn}{\varepsilon - n\delta}}. \tag{3.30}$$

We now remember Fig. 3.2 where the integration contour of the inverse Fourier transformation is shown. It goes above the real axis, which implies that in (3.30) ε is a complex number with a positive imaginary part. We also note that the strongly detuned levels, corresponding to large positive and negative n, are so far that they should not make a physically important contribution to the process. Therefore we can extend our integration over dn to all the real axis from $-\infty$ to $+\infty$. We calculate the integral and find

$$\int_{-\infty}^{+\infty} \frac{dn}{\varepsilon - n\delta} = -\frac{1}{\delta} \int_{-\infty}^{+\infty} \frac{dx}{x - \varepsilon/\delta} = -\frac{1}{\delta} \ln \frac{\infty + i\nu/\delta}{-\infty + i\nu/\delta} = -\frac{i\pi}{\delta} \tag{3.31}$$

where we have employed the explicit expression $\log(z) = \log|z| + i \arg(z)$ for the logarithm of a complex number z. After substitution of (3.31) into (3.30) we perform the inverse Fourier transformation and obtain

$$\psi_0(t) = \frac{1}{2\pi}\int_{-\infty+i\nu}^{\infty+i\nu} \frac{i}{\varepsilon - iV^2\pi/\delta} e^{i\varepsilon t} d\varepsilon$$

$$\psi_n(t) = \frac{1}{2\pi}\int_{-\infty+i\nu}^{\infty+i\nu} \frac{V}{\varepsilon - n\delta}\frac{i}{\varepsilon - iV^2\pi/\delta} e^{i\varepsilon t} d\varepsilon. \tag{3.32}$$

We close the integration contour in the lower plane and apply the Cauchy formula, which yields

$$\psi_0(t) = e^{-tV^2\pi/\delta}$$

$$\psi_n(t) = \frac{e^{itn\delta}}{n\delta - iV^2\pi/\delta} + \frac{e^{-tV^2\pi/\delta}}{iV^2\pi/\delta - n\delta}. \tag{3.33}$$

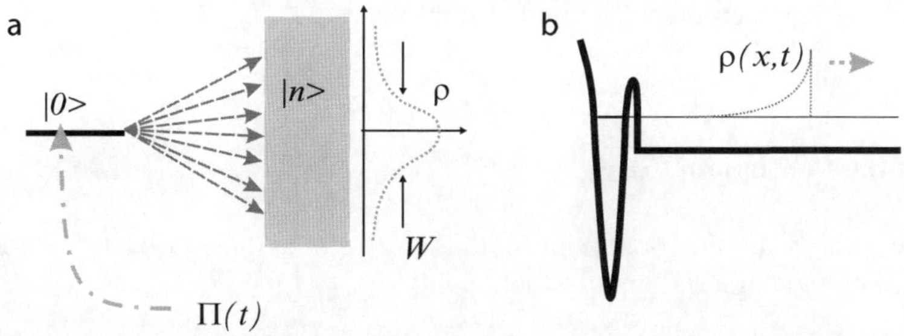

Fig. 3.5 Level–band problem. (**a**) The Lorentzian distribution of the population over the states of the continuum at the long-time limit, when the density of states $g = 1/\delta$ increases and the coupling V decreases, such that the product $W = V^2 g$ remains constant. For the example of Fig. 3.4 such a distribution corresponds to a wavepacket (**b**) which propagates in the continuum to the right and has an exponential envelope

One can see that the population $\rho_0(t) = |\psi_0|^2$ of the state $|0\rangle$ decays exponentially,

$$\rho_0(t) = e^{-t2\pi V^2/\delta} \tag{3.34}$$

with the rate $W = 2\pi V^2/\delta$. It implies that at the time $T \gg t \gg \delta/2\pi V^2$ all the population is on the levels $|n\rangle$ of the band. The second equation (3.33) allows one to find a distribution of the population over these levels. The second term of the equation vanishes at $t \to \infty$ and the first term yields

$$\rho_n(t = \infty) = \frac{V^2}{(n\delta)^2 + (V^2\pi/\delta)^2}, \tag{3.35}$$

which means that the population is distributed over the energy levels of the band according to a Lorentzian profile, as shown in Fig. 3.5. Integration of the right-hand side of (3.35) over n yields unity, as it should be in accordance with the normalization condition.

Let us now rewrite (3.35) in terms of the detuning $\Delta_n = n\delta$ of the state $|n\rangle$ from resonance with the state $|0\rangle$, omit the index n, since it enters the expression only via the detuning, replace $V^2\pi/\delta$ by W, divide both parts by δ, and obtain the spectral density of the population:

$$f(\Delta) = \frac{\rho(\Delta, t = \infty)}{\delta} = \frac{1}{\pi}\frac{W}{\Delta^2 + W^2}. \tag{3.36}$$

This quantity has the same Lorentzian form and the same physical origin as spectral line shapes of width W. Indeed, the Dirac δ-like supply $\Pi(t)$ in (3.24) can be represented as a superposition of harmonics of all frequencies with equal amplitudes. The Schrödinger equation is linear, and hence each harmonic at a frequency Δ brings at $t = \infty$ an independent contribution to the overall population transfer at this frequency. The efficiency of this process depends on the absorption rate at this frequency given by the Lorentzian profile, and thereby the final result coincides with (3.36).

Role of the Continuum Edges

Let us now consider the role of the continuum edges. In (3.30) we have to take into account finite integration limits, which yields

$$\begin{aligned}
\psi_0(\varepsilon) &= \frac{i}{\varepsilon - V^2\int_a^b \frac{dn}{\varepsilon - n\delta}} = \frac{i}{\varepsilon + \frac{V^2}{\delta}\ln(\frac{-\varepsilon+b}{-\varepsilon+a})} \\
\psi_n(\varepsilon) &= \frac{V}{\varepsilon - n\delta}\frac{i}{\varepsilon + \frac{V^2}{\delta}\ln(\frac{-\varepsilon+b}{-\varepsilon+a})}.
\end{aligned} \tag{3.37}$$

The time-dependence of the probability amplitude $\psi_0(t)$ depends on the positions of poles in the complex plane, which are given by roots of the equation

$$\varepsilon + \frac{V^2}{\delta}\ln\left(\frac{-\varepsilon + b}{-\varepsilon + a}\right) = 0. \tag{3.38}$$

For $a <$ Re $[\varepsilon] < b$ the logarithm has an imaginary part and corresponding poles are not disposed on the real axis. Therefore they bring to the probability amplitudes contributions exponentially decreasing in time. Besides on the complex solutions, Eq. (3.38) has two real roots at the points $\varepsilon_a < a$ and $\varepsilon_b > b$ that bring two oscillating contributions. These non-vanishing contributions dominate at the asymptotic regime of long time even in the case their amplitudes are small. In Fig. 3.6 we compare two terms of (3.38), one of which is taken with the opposite sign. Crossing points of the curves correspond to the non-vanishing oscillating contributions

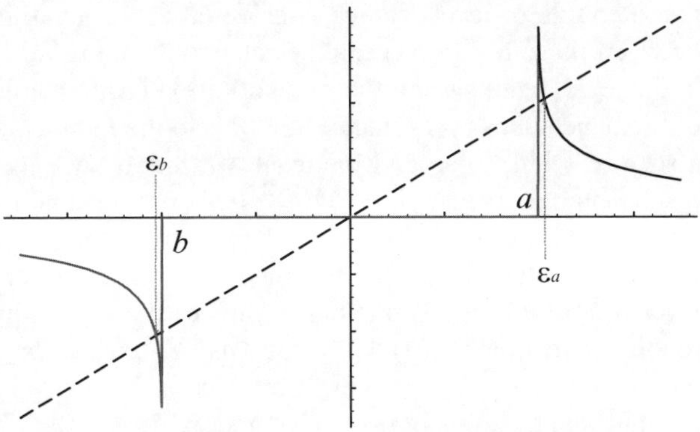

Fig. 3.6 Graphical solution of (3.38). For $\varepsilon < a$ and $\varepsilon > b$ two real roots correspond to oscillating non-vanishing contributions, and their amplitude equal the difference of the slopes at the crossing point

$$\begin{aligned}\psi_0(t) &= -\frac{2\pi\left(\varepsilon_a - b\right)\left(\varepsilon_a - a\right)e^{-i\varepsilon_a t}}{\left(\varepsilon_a - b\right)\left(\varepsilon_a - a\right) + \frac{V^2}{\delta}(b-a)} - \frac{2\pi\left(\varepsilon_b - b\right)\left(\varepsilon_b - a\right)e^{-i\varepsilon_b t}}{\left(\varepsilon_b - b\right)\left(\varepsilon_b - a\right) + \frac{V^2}{\delta}(b-a)} \\ &\simeq -\frac{2\pi\delta}{V^2}\left[\left(\varepsilon_a - a\right)e^{-i\varepsilon_a t} + \left(\varepsilon_b - b\right)e^{-i\varepsilon_b t}\right] \end{aligned} \tag{3.39}$$

of the amplitudes $2\pi\delta\left(\varepsilon_a - a\right)/V^2$ and $2\pi\delta\left(\varepsilon_b - b\right)/V^2$ that are small provided the bandwidth $(b-a)$ considerably exceeds the Lorentzian line width $W = V^2/\delta$. Indeed, for the lower edge of the band $\varepsilon_b \sim b < 0$ Eq. (3.38) yields $\delta\left(b - \varepsilon_b\right)/V^2 \simeq e^{b\delta/V^2}(a-b)\,\delta/V^2$. By analogy $(\varepsilon - a)\,\delta/V^2 \simeq e^{-a\delta/V^2}(a-b)\,\delta/V^2$ implies $a > 0$, and therefore

$$\psi_0(t) \simeq \frac{(a-b)\,\delta}{V^2}\left[e^{b\delta/V^2 - i\varepsilon_b t} - e^{-a\delta/V^2 - i\varepsilon_a t}\right]. \tag{3.40}$$

One sees that the time-independent amplitudes are exponentially small when the edges of the band are disposed at distances larger compared to the width W of the Lorentzian (3.36).

Number of Levels in Resonance

We note that the result (3.35) is in complete agreement with the reasoning based on the time–energy uncertainty principle: the inverse life-time $1/\tau$ given by the decay rate W determines the distribution width. It implies that the population is distributed among the $\mathcal{N} \simeq W/\delta \sim V^2/\delta^2$ states of the band closest to the resonance with the state $|0\rangle$. We also note that the ratio V^2/δ^2 should be a large number, otherwise it is impossible to satisfy the condition that the decay time $\tau = \delta/V^2$ is shorter than the Heisenberg return time $1/\delta$ in the zone, which has been employed for replacing of the summation by integration in (3.30).

The estimate which has just been obtained points to another interesting phenomenon. From the analysis of the two-level system summarized in (3.19) we know that a couple of levels can be considered as resonant, if the interaction strength V between these levels exceeds the relative detuning Δ. Naively we could think that in the level–band system the same criterion is applicable, and therefore the number of the band states in resonance amounts to the number of levels $|n\rangle$ which falls into a strip of width V around the level $|0\rangle$, that is, $\mathcal{N}_{\text{naive}} = V/\delta$. However, the true result obtained reads $\mathcal{N} = (V/\delta)^2$. This implies that the states of the band do not interact with the level $|0\rangle$ independently but rather manifest a cooperative behavior: in order to obtain the correct estimate of the resonant levels with the help of the two-level condition $V > \Delta$ one has to replace the typical interaction V by an effective (cooperative) interaction V_c in the "naive" expression $\mathcal{N} = V_c/\delta$. This effective interaction V_c equals the typical interaction V multiplied by the square root of the true number of levels in resonance, that is, $V \to V_c = V\sqrt{\mathcal{N}}$.

The origin of the $\sqrt{\mathcal{N}}$-factor is in the fact that consideration of the same problem from a different point of view in quantum mechanics usually implies a different choice of the basis sets, and all the transformations of the basis set are given by unitary matrices, which for all operators of physical values preserve sums of the absolute value squares of the matrix elements of any line and any column. Being applied to the operator of the dipole moment it yields the well-known optical theorem, stating that the sum of squares of the transition dipole matrix elements that couple an arbitrary state to all other states is a constant, which remains the same even after reconstruction of the energy spectrum caused by a perturbation. This implies that the sum of squares of the transition matrix elements given by the product of a typical matrix element squared V^2 and the number $\mathcal{N}$ of the resonant states, that is, $V^2 \times (V/\delta)^2$, is a constant, and indeed the decay rate $W = \sqrt{V^2 \times (V/\delta)^2}$ does not depend on the detailed structure of the band spectrum.

One can also look at the situation differently. Imagine that the level–band system originates from a two-level system, where one of the levels, say the level $|1\rangle$, interacts strongly with a $\mathcal{N}$-level band. As a result of this interaction each of the energy eigenstates $|n\rangle$ of the system "level$|1\rangle$+$\mathcal{N}$-level band" contains a contribution from the state $|1\rangle$, and the amplitude of this contribution $\langle 1|n\rangle$ is of the order of $1/\sqrt{\mathcal{N}}$ and hence matrix elements V of the transitions from the state $|0\rangle$ to states $|n\rangle$ are $\sqrt{\mathcal{N}}$ times weaker, compared to the initial matrix element V_c of the transition between the states $|0\rangle$ and $|1\rangle$. But it is clear that this matrix element V_c has to be compared with a typical detuning $\Delta \simeq \delta\mathcal{N}$ in the two-level resonance criterion $V_c \sim \Delta \simeq \delta\mathcal{N}$, that is, $V \sim \delta\sqrt{\mathcal{N}}$, or $\mathcal{N} \sim V^2/\delta^2$. Note that the decay rate $W = 2\pi V^2/\delta$ entering (3.34) can also be given in terms of the cooperative coupling $V_c = V\sqrt{\mathcal{N}}$ divided by a typical width $\Gamma = \delta\mathcal{N}/2\pi$ of the band, that is, $W = V_c^2/\Gamma$.

Wavepacket Representation

Sometimes it is very instructive to consider the motion of the system shown in Fig. 3.4 in the coordinate representation, which allows one to see a deep analogy between the time evolution of the level–band system and the dynamics of wavepackets. For this purpose one has to multiply the state amplitudes $\psi_n(t)$ by the coordinate-dependent eigenfunctions $\Psi_n(x)$ of the particle in the well *B* and sum up over all n in order to obtain the time-dependent wavefunction of the particle $\Psi(t,x) = \sum_n \psi_n(t)\Psi_n(x)$. After substitution of $\psi_n(t) = \frac{1}{2\pi}\int_{-\infty+i\nu}^{+\infty+i\nu} e^{-i\varepsilon t}\psi_n(\varepsilon)d\varepsilon$ and $\psi_n(\varepsilon) = \psi_0(\varepsilon)V_0^n/(\varepsilon - \Delta_n)$ in accordance with (3.26) this yields

$$\Psi(t,x) = \frac{1}{2\pi}\int_{-\infty+i\nu}^{+\infty+i\nu} e^{-i\varepsilon t}K(x,\varepsilon)\psi_0(\varepsilon)d\varepsilon \tag{3.41}$$

where $K(x,\varepsilon) = \sum_n \Psi_n(x)V_0^n/(\varepsilon - \Delta_n)$. We note that the inverse Fourier transform of a product equals the convolution of the inverse Fourier transforms of the cofactors, and write this expression in the more convenient form

$$\Psi(t,x) = \int_{-\infty}^{t} K(x,t-\tau)\psi_0(\tau)d\tau, \tag{3.42}$$

where the time-dependent amplitude $\psi_0(t)$ of the state $|0\rangle$ is convoluted with the kernel $K(x,t) = \sum_n \Psi_n(x)V_0^n \exp\{-i\Delta_n t\}$, resulting from the inverse Fourier transformation of $K(x,\varepsilon)$.

When the spectrum of the band is dense, the total number of levels in the well is large, and the motion is one-dimensional, the semiclassical consideration provides us with a good expression for the wavefunction corresponding to a given energy E_n and any smooth potential $U(x)$:

$$\Psi_n(x) = \frac{1}{\sqrt[4]{E_n - U(x)}}\cos\left[\frac{\pi}{4} - \sqrt{2m}\int_0^x \sqrt{E_n - U(z)}dz\right], \tag{3.43}$$

where $x = 0$ corresponds to the turning point, whereas for the rectangular potential of depth $\hbar\omega - U_0$ shown in Fig. 3.4 one has

$$\Psi_n(x) = \frac{1}{\sqrt[4]{\Delta_n + U_0}}\sin\left[\sqrt{2m}x\sqrt{\Delta_n + U_0}\right]. \tag{3.44}$$

This yields the kernel

$$K(x,t) = \sum_n \frac{V_0^n \exp\{-i\Delta_n t\}}{\sqrt[4]{\Delta_n + U_0}}\sin\left[x\sqrt{2m}\sqrt{\Delta_n + U_0}\right], \tag{3.45}$$

which can be further simplified if we set as earlier all V_0^n identical $V_0^n = V$, expand the root in the argument of sine in a Taylor series around $\Delta_n = 0$, neglect indeed very slow dependence in the denominator and replace the sum by an integral by analogy to (3.31). We arrive at

$$
\begin{aligned}
K(x,t) &= V \int \frac{\exp\{-i\delta t n\}}{\sqrt[4]{U_0}} \sin\left[\frac{\sqrt{m}x\delta}{\sqrt{2U_0}} n\right] dn \\
&= \frac{i\pi V}{\delta \sqrt[4]{U_0}} \left[\delta(t - vx) - \delta(t + vx)\right],
\end{aligned} \tag{3.46}
$$

where $v = \sqrt{2U_0}/\sqrt{m}$ is the particle velocity in the right potential pit, and $\delta(t \pm vx)$ are Dirac δ-functions. Since x and t are both positive only the first of these two δ-functions is important. Substitution of (3.46) into (3.42) yields

$$
\Psi(t,x) = \frac{i\pi V}{\delta \sqrt[4]{U_0}} \psi_0(t - vx), \tag{3.47}
$$

which means that the probability to detect the particle at the distance x at the time moment t equals (apart from a constant factor) the probability to detect the particle at the state $|0\rangle$ at the time $t - vx$, that is, $\rho(t,x) = |\Psi(t,x)|^2 \sim \rho_0(t - vx)$.

The Fano Problem

We now consider a more involved situation, when the continuum is not uniform, but is formed as a result of an interaction between a level and a uniform continuum shown in Fig. 3.7. It turns out that in such a situation the interference phenomena change significantly the Lorentzian line shapes of the resonances. We consider the problem in the representation where the states of the band are not yet reduced to the eigenstate representation, but are still considered as a result of interaction of a state $|1\rangle$ with a uniform band of states $|k\rangle$.

The corresponding Schrödinger equation reads

$$
\begin{aligned}
i\dot{\psi}_n &= \Delta_n \psi_n + V\psi_0 + V'\psi_1 \\
i\dot{\psi}_0 &= V \sum_{n=1}^{N} \psi_n + V''\psi_1 \\
i\dot{\psi}_1 &= V' \sum_{n=1}^{N} \psi_n + \Delta_1 \psi_1 + V''\psi_0,
\end{aligned} \tag{3.48}
$$

where Δ_1 denotes detuning between the states $|0\rangle$ and $|1\rangle$.

We now again proceed in the same way as earlier: after substitution of the Dirac δ-function to the right-hand side of the second equation we perform the Fourier transformation and arrive at

$$
\begin{aligned}
\varepsilon \psi_n &= \Delta_n \psi_n + V\psi_0 + V'\psi_1 \\
\varepsilon \psi_0 &= V \sum_{n=1}^{N} \psi_n + V''\psi_1 + i
\end{aligned}
$$

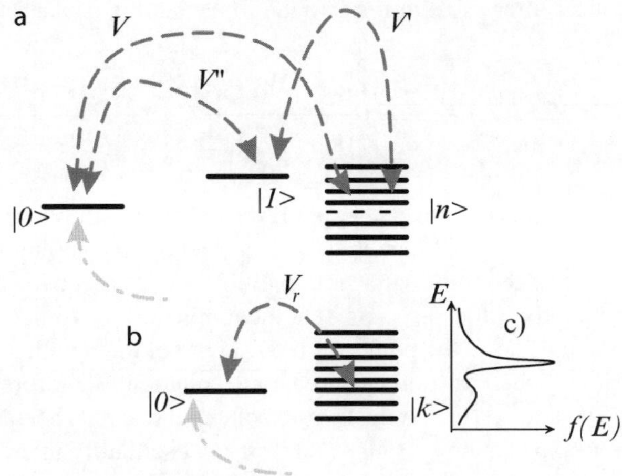

Fig. 3.7 The Fano problem. (**a**) Representation of the band as a level interacting with a zone. The initial state $|0\rangle$ interacts with the state $|1\rangle$ by the coupling V'' and with the zone by the coupling V. The state $|1\rangle$ interacts with the zone by the coupling V'. (**b**) Representation of the band in the energy eigenstate basis. The resulting coupling $V_r(E)$ is not uniform, but has the Fano profile $V_r \sim f(E)$ shown in (**c**)

$$\varepsilon\psi_1 = V'\sum_{n=1}^{N}\psi_n + \Delta_1\psi_1 + V''\psi_0. \tag{3.49}$$

We express ψ_n in terms of ψ_0 and ψ_1 with the help of the first equation and find

$$\varepsilon\psi_0 = V^2\sum_{n=1}^{N}\frac{1}{(\varepsilon-\Delta_n)}\psi_0 + \left(VV'\sum_{n=1}^{N}\frac{1}{(\varepsilon-\Delta_n)} + V''\right)\psi_1 + i \tag{3.50}$$

$$\varepsilon\psi_1 = \left(V'' + VV'\sum_{n=1}^{N}\frac{1}{(\varepsilon-\Delta_n)}\right)\psi_0 + \left((V')^2\sum_{n=1}^{N}\frac{1}{(\varepsilon-\Delta_n)} + \Delta_1\right)\psi_1.$$

We now replace the sums by the integrals (3.50) and obtain

$$(\varepsilon + iW_0)\,\psi_0 - \left(V'' - iW_F\right)\psi_1 = i$$

$$\left(V'' - iW_F\right)\psi_0 + (\Delta_1 - iW_1 - \varepsilon)\,\psi_1 = 0, \tag{3.51}$$

where $W_0 = \pi V^2/\delta;\ W_1 = \pi\,(V')^2/\delta;\ W_F = \pi VV'/\delta = \sqrt{W_0W_1}$. By solving (3.51) one obtains two expressions for ψ_0 and ψ_1:

$$\psi_0(\varepsilon) = \frac{i\,(\Delta_1 - iW_1 - \varepsilon)}{(\varepsilon + iW_0)\,(\Delta_1 - iW_1 - \varepsilon) + (V'' - iW_F)^2}$$

$$\psi_1(\varepsilon) = \frac{-i\left(V'' - iW_F\right)}{(\varepsilon + iW_0)\,(\Delta_1 - iW_1 - \varepsilon) + (V'' - iW_F)^2}, \tag{3.52}$$

which, after substitution into the first equation of the set (3.49), yields

$$\psi_n = i\frac{(\Delta_1 - iW_1 - \varepsilon)\,V - \left(V'' - iW_F\right)V'}{(\varepsilon - \Delta_n)\left[(\varepsilon + iW_0)(\Delta_1 - iW_1 - \varepsilon) + (V'' - iW_F)^2\right]}. \tag{3.53}$$

As usual, the inverse Fourier transformation implies integration along the contour shown in Fig. 3.2, which can be carried out by finding the residuals in the singular points of the right-hand side of (3.52), (3.53) at the lower part of the complex plane ε including the real axis. The denominators of (3.52) are identical and have two singularities at the points ε_1 and ε_2, both of them with negative imaginary parts, that can be found as roots of a quadratic equation. Therefore each of the two amplitudes $\psi_0(t)$ and $\psi_1(t)$ manifests bi-exponential decay, whereas the corresponding populations $\rho_0(t)$ and $\rho_1(t)$ also manifests an oscillatory interference pattern coming from the cross product of different singularities, as shown in Fig. 3.8a. Each amplitude of the band states (3.53) also has an additional pole on the real axis; that is, at $t \to \infty$ it does not decay but brings about the main contribution of the form

$$\psi_n = i\frac{(\Delta_1 - iW_1 - \Delta_n)\,V - \left(V'' - iW_F\right)V'}{(\Delta_n + iW_0)(\Delta_1 - iW_1 - \Delta_n) + (V'' - iW_F)^2}e^{i\Delta_n t}. \tag{3.54}$$

The corresponding population density distribution $f(\Delta) = |\psi_n|^2/\delta$ is no longer of the Lorentzian form (3.36), but has a typical two-hump shape known as the Fano profile, which is shown in Fig. 3.8b. This profile also manifests itself in the dependence of the absorption rate on the frequency ω of a monochromatic excitation.

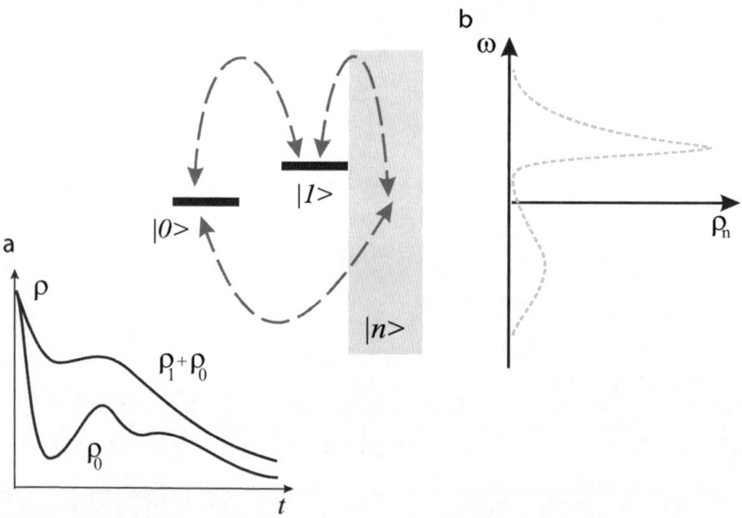

Fig. 3.8 The Fano problem. (**a**) Time-dependent populations of the initial state ρ_1 and the total population of the states $|0\rangle$ and $|1\rangle$. The Fano profile (**b**) of the population distribution over the states of the band

Population Amplitude Injection

Let us now consider the level–band problem assuming that the states of this system were not populated initially, whereas the particle is pumped to the state $|0\rangle$ of the system in the course of an external process of a finite duration τ. In other words the flux $\Pi(t)$ in (3.24) is not the Dirac δ-function but a continuous function localized in a time interval of the order of τ. The results of the consideration will show, on one hand, the relation between population distributions and line shapes, and, on the other hand, will help us later on to understand the phenomena of population recurrences and revivals. Now instead of (3.24) we have

$$i\dot{\psi}_n = \Delta_n \psi_n + V_n^0 \psi_0$$
$$i\dot{\psi}_0 = \sum_{n=1}^{N} V_0^n \psi_n + i\Pi(t), \tag{3.55}$$

which by analogy to (3.26) yields

$$\psi_0(\varepsilon) = \frac{i\Pi(\varepsilon)}{\varepsilon - \sum_{n=1}^{N} \frac{V_0^n}{\varepsilon - \Delta_n}}$$
$$\psi_n(\varepsilon) = \frac{1}{\varepsilon - \Delta_n} V_0^n \psi_0 = \frac{1}{\varepsilon - \Delta_n} \frac{i\Pi(\varepsilon) V_0^n}{\varepsilon - \sum_{n=1}^{N} \frac{V_0^n}{\varepsilon - \Delta_n}}. \tag{3.56}$$

Here $\Pi(\varepsilon)$ is the Fourier transform of $\Pi(t)$, which enters the equations for the amplitudes as an additional factor. We can now also identify the combination $V_0^n \psi_0$ with the probability amplitude supply corresponding to the transition from the state $|0\rangle$ to the state $|n\rangle$.

Let us now consider a particular case of the dependence $\Pi(t)$ by taking the Gaussian function

$$\Pi(t) = \frac{e^{-t^2/\tau^2}}{\sqrt{\pi}\tau}, \tag{3.57}$$

which yields

$$\Pi(\varepsilon) = \frac{1}{\sqrt{\pi}\tau} \int_{-\infty}^{\infty} e^{-t^2/\tau^2 - i\varepsilon t} dt = e^{-\varepsilon^2\tau^2/4}. \tag{3.58}$$

Substitution of (3.58) into (3.56) results, for the continuum,

$$\psi_0(\varepsilon) = \frac{ie^{-\varepsilon^2\tau^2/4}}{\varepsilon - iV^2\pi/\delta}. \tag{3.59}$$

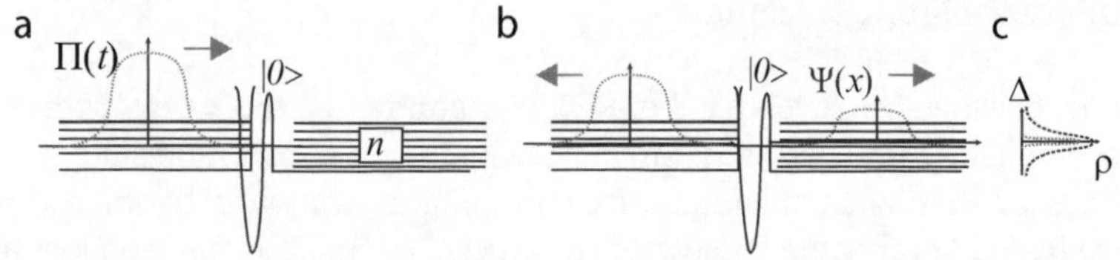

Fig. 3.9 A model of the population amplitude supply. (**a**) A particle with a broad wavepacket arrives from the left and penetrates to the state $|0\rangle$ through the left potential barrier with the penetration rate $\Pi(t)$ proportional to the probability amplitude to be near the barrier. The typical duration of the supply τ considerably exceeds the inverse decay rate $1/W$ of the level. From the state $|0\rangle$ it comes to the states $|n\rangle$ of the right continuum (**b**). Interference of these states, taken with the allowance of the coordinate dependence of their eigenfunctions, results in the propagation of a wavepacket $\Psi(x)$, which is so broad in space (and long in time) that the spectrum width (**c**) of the occupied states in the continuum is much less, compared to the Lorentzian width (3.36) (*dotted line*) given by the transparency of the second barrier W

Let us consider the limiting case, when the decay rate exceeds the rate of particle supply, that is, $V^2/\delta \gg 1/\tau$. The Gaussian dependence makes $\psi_0(\varepsilon)$ vanishing for $\varepsilon > 2/\tau$ and therefore the term ε in the denominator of (3.59) can be omitted, and we arrive at

$$\psi_0(\varepsilon) = -\frac{\delta}{V^2\pi} e^{-\varepsilon^2\tau^2/4}, \tag{3.60}$$

which, after inverse Fourier transformation, reads

$$\psi_0(t) = -\frac{\delta}{V^2 2\pi^2} \int_{\infty - i\nu}^{\infty + i\nu} e^{-\varepsilon^2\tau^2/4 + i\varepsilon t} d\varepsilon = -\frac{\delta}{V^2\pi\sqrt{\pi}\tau} e^{-t^2/\tau^2}. \tag{3.61}$$

One sees that in this case the amplitude of the probability to be at the level $|0\rangle$ follows the supply rate $\Pi(t)$, although it is scaled by the decay rate factor $1/W = \delta/V^2\pi$ and has opposite sign. This means that the probability amplitude of staying in the state $|0\rangle$ is attained as a result of balance of the supply process and the fast decay process corresponding to the rapid transmission of the probability amplitude to the states of the band, natural in the chosen limiting case. We illustrate this in Fig. 3.9 by an instructive analogy. We also note that $\Pi(\varepsilon) = const$ corresponds to a Dirac δ-function-like population amplitude injection, and $\Pi(\varepsilon) \sim 1/\varepsilon$ allows for a steady state flux of the population amplitude.

3.2.3 Measurements and Relaxation as Processes in Level–Continuum Systems

The level–band system can serve as the simplest model of relaxation and of the measurement process in a two-level system. The compound system thus comprises the two-level system and either the meter or the environment. The latter has a very rich spectrum of eigenenergies, originating either from the classical character of the meter, or from the large number of degrees of freedom in the environment. States of the band are therefore related to the environment or the spectrum of the meter. The high band state density implies that the Heisenberg time of return is extremely long, and therefore both the measurement and the relaxation are irreversible processes, and we can consider the band as continuous.

We focus on the simplest case where the relaxation may occur only from the excited state $|1\rangle$ of the two-level system and that this system makes a transition to the ground state $|0\rangle$ as a result. This transition is accompanied by the transition in the environment associated with the energy transfer to one of many degrees of freedom. For a single elementary act of relaxation, we can say that the transition occurs from the vacuum state of the environment $|0,0,0,\ldots,0\rangle$ to an excited state $|k\rangle = |0,\ldots,1,\cdots,0\rangle$ where the k-th degree of freedom gets excited from the ground to the first excited state. By ψ_0 we denote the probability amplitude for the compound system to be in the state $|0\rangle \otimes |0,0,0,\ldots,0\rangle$, while ψ_1 and ψ_k denote the amplitudes of the states $|1\rangle \otimes |0,0,0,\ldots,0\rangle$ and $|0\rangle \otimes |k\rangle$, respectively.

Considering the measurement process, we assume that only the excited state of the two-level system can be detected. In other words, excitation of the two-level system provokes a transition in the meter, such that it leaves its vacuum state for an excited state $|k\rangle$, while the two-level system remains in the upper state $|1\rangle$. The complex variables ψ_0, ψ_1, and ψ_k then denote the amplitudes of the states $|0\rangle \otimes |0\rangle$, $|1\rangle \otimes |0\rangle$, and $|1\rangle \otimes |k\rangle$, respectively.

Population dynamics is a coherent process, characterized by the probability amplitudes as far as the compound system is concerned. These amplitudes satisfy the Schrödinger equation

$$\begin{aligned} i\dot{\psi}_0 &= V\psi_1 \\ i\dot{\psi}_1 &= E_1\psi_1 + V\psi_0 + \sum_k V'\psi_k \\ i\dot{\psi}_k &= E_k\psi_k + V'^*\psi_1, \end{aligned} \tag{3.62}$$

where we have assumed that the upper and lower levels of the two-level system, separated by the energy E_1, are coupled by the interaction V, and that all the band states are coupled to the excited state by an identical interaction V'. With the initial condition $\psi_1(t=0) = 1$ after Fourier transformation this yields

$$\begin{aligned} \psi_0 &= \frac{iV}{(\varepsilon - E_1 + iW(\varepsilon))\,\varepsilon - V^2} \\ \psi_1 &= \frac{i\varepsilon}{(\varepsilon - E_1 + iW(\varepsilon))\,\varepsilon - V^2} \\ \psi_k &= \frac{V'^*}{\varepsilon - E_k}\frac{i\varepsilon}{(\varepsilon - E_1 + iW(\varepsilon))\,\varepsilon - V^2}, \end{aligned} \tag{3.63}$$

where $W(\varepsilon) = i\sum_k |V'|^2 (\varepsilon - E_k)^{-1}$. For the continuum, infinitely large in both sides, the corresponding integration yields $W = \pi g |V'|^2$, which does not depend on ε. After the inverse Fourier transformation we therefore arrive at

$$\psi_0 = e^{-(iE_1+W)t/2}\frac{2V}{\Omega}\sin\frac{t\Omega}{2}$$
$$\psi_1 = e^{-(iE_1+W)t/2}\left[i\cos\frac{t\Omega}{2} + \frac{(E_1 - iW)}{\Omega}\sin\frac{t\Omega}{2}\right]$$
$$\psi_k = \frac{-iE_k V'^* e^{-iE_k t}}{E_1 E_k + V^2 - E_k(E_k + iW)} + \frac{V'^* e^{-(iE_1+W)t/2}}{E_1 E_k + V^2 - E_k(E_k + iW)}$$
$$\times\left[iE_k \cos\Omega/2 + \left(E_1 E_k + 2V^2 - iE_k W\right)\right)\frac{\sin t\Omega/2}{\Omega}\right], \qquad (3.64)$$

where $\Omega = \sqrt{4V^2 + (E_1 - iW)^2}$. Note that the absolute values of the amplitudes ψ_0 and ψ_1 decrease exponentially in time with the rate $(W - \mathrm{Im}\Omega)/2$.

Equation (3.64) provides us with the solution for the compound system. However when concentrating only on the two-level system and ignoring the quantum structure of the environment or of the meter we need to take the trace of the compound system density matrix over all quantum numbers except those of the two levels considered. This yields

$$\begin{aligned}\hat{\rho}_r &= \left(\psi_0 |0\rangle + \psi_1 |1\rangle\right)\left(\psi_0^* \langle 0| + \psi_1^* \langle 1|\right) + \sum \psi_k^*\psi_k |0\rangle\langle 0| \\ \hat{\rho}_m &= \left(\psi_0 |0\rangle + \psi_1 |1\rangle\right)\left(\psi_0^* \langle 0| + \psi_1^* \langle 1|\right) + \sum \psi_k^*\psi_k |1\rangle\langle 1|,\end{aligned} \qquad (3.65)$$

for the environment and the meter, respectively. When we take into account the normalization condition $\psi_0^*\psi_0 + \psi_1^*\psi_1 + \sum \psi_k^*\psi_k = 1$ for the compound system, (3.65) acquires the form

$$\begin{aligned}\hat{\rho}_r &= \psi_0^*\psi_1 |1\rangle\langle 0| + \psi_1^*\psi_0 |0\rangle\langle 1| + \psi_1^*\psi_1 |1\rangle\langle 1| + (1 - \psi_1^*\psi_1)|0\rangle\langle 0| \\ \hat{\rho}_m &= \psi_0^*\psi_1 |1\rangle\langle 0| + \psi_1^*\psi_0 |0\rangle\langle 1| + \psi_0^*\psi_0 |0\rangle\langle 0| + (1 - \psi_0^*\psi_0)|1\rangle\langle 1|.\end{aligned} \qquad (3.66)$$

One sees that, as a result of the elementary act of relaxation, the two-level system comes to the state $|0\rangle$ and the measured system comes to the state $|1\rangle$ after a time $1/(W - \mathrm{Im}\Omega)$, when both amplitudes ψ_0 and ψ_1 vanish.

A Bloch vector $B = (2\mathrm{Re}\rho_{10}, 2\mathrm{Im}\rho_{10}, \rho_{11} - \rho_{00})$ composed of the density matrix elements (3.66) with the allowance of amplitudes (3.64) is a convenient real quantity which helps to trace the time evolution of an open two-level system. Trajectories of B shown in Fig. 3.10a illustrate the dynamics of the two-level system, initially in the upper state $|1\rangle$, which is subjected to either measurement or relaxation. One sees that the vector, initially in the utmost upper position $\rho_{11} = 1$, starts to move toward lower positions as a result of the interaction V but returns along a spiral trajectory as a result of the elementary act of measurement. As a

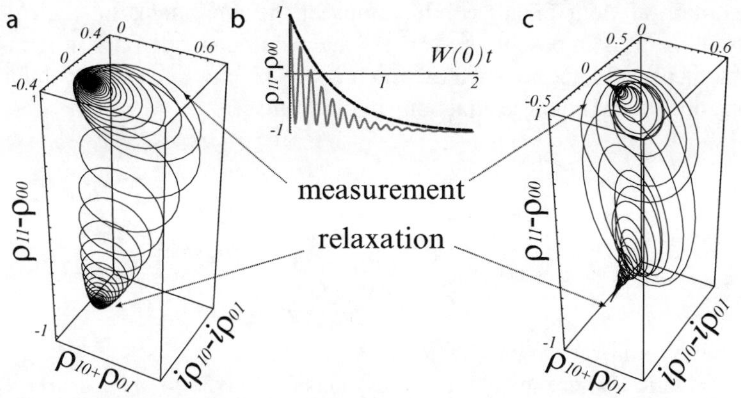

Fig. 3.10 Dynamics of the Bloch vector $B = (\rho_{01} + \rho_{10}, i\rho_{10} - i\rho_{01}, \rho_{11} - \rho_{00})$ representing the density matrix. (**a**) For a relaxing two-level system the vector B starts at the point $(1, 0, 0)$ corresponding to the excited state and goes to the point $(-1, 0, 0)$ following the spiral trajectory associated with Rabi oscillations. For a system measured in the upper states the vector returns to the initial position along a spiral trajectory. (**b**) Exponential relaxation acquires oscillatory behavior in the case of a finite bandwidth. (**c**) The finite bandwidth also affects the trajectory of B

result of the elementary act of relaxation the vector gradually moves to the lowest position $\rho_{00} = 1$, also along a spiral trajectory.

Let us derive differential equations for the density matrices $\hat{\rho}_r$ and $\hat{\rho}_m$. We note that in the case of an infinite band one finds

$$i\dot{\psi}_0 = V\psi_1 \qquad (3.67)$$
$$i\dot{\psi}_1 = E_1\psi_1 + V\psi_0 - iW\psi_1,$$

which results from the Fourier transformation of the last two equations (3.62), substitution of $\psi_k(\varepsilon)$ from the third equation to the second one, and the replacement $\sum_k \frac{|V'|^2}{\varepsilon - E_k} \to -iW$ followed by the inverse Fourier transformation. With the help of (3.67) and their complex conjugate, after direct calculations one obtains the differential equations

$$-i\frac{\partial}{\partial t}\begin{pmatrix} 1-\psi_1^*\psi_1 & \psi_0^*\psi_1 \\ \psi_1^*\psi_0 & \psi_1^*\psi_1 \end{pmatrix} =$$
$$\begin{pmatrix} V\psi_1^*\psi_0 - V\psi_0^*\psi_1 - i2W\psi_1^*\psi_1 & (iW - E_1)\,\psi_0^*\psi_1 - V(\psi_0^*\psi_0 - \psi_1^*\psi_1) \\ (iW + E_1)\,\psi_1^*\psi_0 - V(\psi_1^*\psi_1 - \psi_0^*\psi_0) & V\psi_0^*\psi_1 - V\psi_1^*\psi_0 + i2W\psi_1^*\psi_1 \end{pmatrix}$$
$$-i\frac{\partial}{\partial t}\begin{pmatrix} \psi_0^*\psi_0 & \psi_0^*\psi_1 \\ \psi_1^*\psi_0 & 1-\psi_0^*\psi_0 \end{pmatrix} =$$
$$\begin{pmatrix} V\psi_1^*\psi_0 - V\psi_0^*\psi_1 & (iW - E_1)\,\psi_0^*\psi_1 - V(\psi_0^*\psi_0 - \psi_1^*\psi_1) \\ (E_1 + iW)\,\psi_1^*\psi_0 - V(\psi_1^*\psi_1 - \psi_0^*\psi_0) & V\psi_0^*\psi_1 - V\psi_1^*\psi_0 \end{pmatrix}$$
$$(3.68)$$

which contain all four binary combinations of the amplitudes on the right-hand side. Therefore it is impossible to express the right-hand side only in terms of the matrix elements of $\hat{\rho}_r$ or $\hat{\rho}_m$ each of which contains only three such combinations. However, one can write these equations in the form

$$
\begin{aligned}
i\frac{\partial}{\partial t}\hat{\rho}_r &= [H,\hat{\rho}_r] - iW\begin{pmatrix} -2\rho_{r11} & \rho_{r10} \\ \rho_{r01} & 2\rho_{r11} \end{pmatrix} - V\begin{pmatrix} 0 & 1 \\ -1 & 0 \end{pmatrix}\sum \psi_k^*\psi_k \\
i\frac{\partial}{\partial t}\hat{\rho}_m &= [H,\hat{\rho}_m] - iW\begin{pmatrix} 0 & \rho_{m10} \\ \rho_{m01} & 0 \end{pmatrix} + V\begin{pmatrix} 0 & 1 \\ -1 & 0 \end{pmatrix}\sum \psi_k^*\psi_k
\end{aligned}
\tag{3.69}
$$

involving the sum over the band states.

In the situation where many elementary acts of relaxation or measurements can occur continuously, one can omit the terms $\psi_k^*\psi_k$ on the right-hand side of both equations (3.69), since immediately after the transition to a band state $|k\rangle$ the system starts to be subject to the influence of the next meter or the next set of degrees of freedom of the environment in the vacuum state, as if it were transported back to the states corresponding to the two-level system and the vacuum state of the previous meter or environment. This gives the standard form known as the master equations. However, for the consideration of a single elementary act of measurement or relaxation the sum on the right-hand side of (3.69) is important. In order to express it in terms of the density matrix elements we make use of the normalization condition and take the time derivative of the sum

$$
\begin{aligned}
\sum \psi_k^*\psi_k &= 1 - \psi_0^*\psi_0 - \psi_1^*\psi_1 \\
i\frac{\partial}{\partial t}\sum \psi_k^*\psi_k &= -i\frac{\partial}{\partial t}\left(\psi_0^*\psi_0 + \psi_1^*\psi_1\right) = -i2W\psi_1^*\psi_1
\end{aligned}
\tag{3.70}
$$

which allows one to eliminate the unknown functions and write the second-order differential equations

$$
\begin{aligned}
&-\frac{\partial^2}{\partial t^2}\hat{\rho}_r = \left[H, i\frac{\partial}{\partial t}\hat{\rho}_r\right] - W\frac{\partial}{\partial t}\begin{pmatrix} 2\rho_{r11} & -\rho_{r10} \\ -\rho_{r01} & -2\rho_{r11} \end{pmatrix} + i2WV\begin{pmatrix} 0 & \rho_{r11} \\ -\rho_{r11} & 0 \end{pmatrix}, \\
&\left(\frac{\partial}{\partial t} - 2W\right)\frac{\partial}{\partial t}\hat{\rho}_m = -i\left[H, \left(\frac{\partial}{\partial t} - 2W\right)\hat{\rho}_m\right] \\
&- W\left(\frac{\partial}{\partial t} - 2W\right)\begin{pmatrix} 0 & \rho_{m10} \\ \rho_{m01} & 0 \end{pmatrix} + i2VW\begin{pmatrix} 0 & \rho_{m11} \\ -\rho_{m11} & 0 \end{pmatrix}
\end{aligned}
\tag{3.71}
$$

for the density matrices $\hat{\rho}_r$ and $\hat{\rho}_m$ (3.66). These equations have to be employed in the situation where the environment or the meter saturates after a single elementary act of relaxation or measurement, respectively.

Note the important particular regime of rapid and continuous measurements. After having neglected the term $\psi_k^*\psi_k$, (3.69) for $\hat{\rho}_m$ takes the form

$$i\frac{\partial}{\partial t}\begin{pmatrix}\rho_{00} & \rho_{m10}\\ \rho_{m01} & \rho_{11}\end{pmatrix} =$$
$$\begin{pmatrix} V\rho_{10} - V\rho_{01} & -(iW - E_1)\,\rho_{10} + V(\rho_{00} - \rho_{11}) \\ -(E_1 + iW)\,\rho_{01} - V(\rho_{00} - \rho_{11}) & -V\rho_{m10} + V\rho_{m01}\end{pmatrix}, \tag{3.72}$$

and in the case $W \gg V$ this yields

$$\rho_{01} = V\tfrac{\rho_{11}-\rho_{00}}{E_1+iW}; \qquad \rho_{10} = V\tfrac{\rho_{00}-\rho_{11}}{iW-E_1}$$
$$\frac{\partial}{\partial t}(\rho_{11} - \rho_{00}) = \tfrac{-4WV^2}{E_1^2+W^2}(\rho_{11} - \rho_{00}), \tag{3.73}$$

which means that the off-diagonal elements of the density matrix rapidly assume their stationary values given by the instantaneous value of the difference $\rho_{11} - \rho_{00}$, while the latter change slowly. One sees that for $W \gg E_1$ the relaxation rate $4V^2/W$ of the diagonal matrix elements is inversely proportional to the measurement rate W and tends to zero when the measurements are performed infinitely fast. This implies that the system originally in the state $|1\rangle$ remains in this state indefinitely. This phenomenon is known as the Zeno effect and can be observed in any quantum system subject to frequent measurements, where the relaxation rate of the measured state is inversely proportional to the rate of measurement.

Now let us consider the role of the spectral characteristics of the environment and the meter, which have been taken as an infinite continuum in the previous analysis. If this is not the case, the relaxation rate W is not a constant, but a function which depends on the frequency ε. Therefore in the time-dependent form of the Schrödinger equation it corresponds to an integral operator of convolution, with the kernel given by the Fourier transform of $W(\varepsilon)$. In the simplest case $W(\varepsilon) = w/(\varepsilon - i\Gamma)$, soluble analytically, one finds

$$\psi_0 = i\frac{V}{\varepsilon\,(\varepsilon - E_1 - w/(\varepsilon + i\Gamma)) - V^2},$$
$$\psi_1 = \frac{i\varepsilon}{(\varepsilon - E_1 - w/(\varepsilon + i\Gamma))\,\varepsilon - V^2}. \tag{3.74}$$

These expressions have three simple poles each, in contrast to (3.63) where the same quantities had only two poles. The additional pole and the associated frequency account for the dynamics of the meter or environment resulting from the non-instantaneous character of their motion. The narrower the band width Γ, the larger the influence of this motion on the process of relaxation and measurement. In Fig. 3.10b we illustrate the difference between relaxation to an infinite band and to a band of a finite width for the example of the population difference between the upper and lower levels suggested by the Fourier transforms of (3.74) and the definitions (3.66). The time-dependence of the population difference is exponential for the case of an infinite band, but it experiences dying oscillations when the band is narrow. Other elements of the density matrix represented in Fig. 3.10c in the form of the Bloch vector B are also affected by the band width.

3.3 Long-Time Behavior

We are now in the position to consider dynamics of the level–band system (3.23) at the long-time limit $t \gg 1/\delta$, that is, at time longer compared to the Heisenberg return time from the band, or speaking in terms of the example shown in Fig. 3.4, longer than the period T of oscillations of the particle in the potential pit B. It turns out that in a certain regime one observes at this time scale a number of interesting universal phenomena: quantum recurrences, revivals, and fractional revivals. We start by consideration of the long-time limit with a general formalism, and later we turn to the phenomenon of quantum recurrences, which occurs first, that is, at a time scale shorter than that for two other phenomena—revivals and fractional revivals.

3.3.1 General Consideration of the Long-Time Limit

For the sake of simplicity we restrict ourselves to the case when the population amplitude has been injected to the system during a time τ, which is shorter than the period $T \sim 1/\delta$, but still much longer than the typical decay time $1/W \sim \delta/V^2$ of the state $|0\rangle$. The reason for this is purely technical, since in this case the analytical expressions for the amplitudes are much simpler than for the general case, but nevertheless they manifest all physically important phenomena typical of the level–band system. Full analysis, which can be done by analogy, results in much more cumbersome expressions. For the same reason we restrict ourselves only to the case of identical transition amplitudes $V_n^0 = V_0^n = V$, and an equidistant spectrum of the band $\Delta_n = \delta n$, where n takes all integer values, positive and negative. In such a case one of the states of the band, $n = 0$, has the same energy and the same notation as the level $|0\rangle$, which hopefully will not create any confusion, since this state always enters expressions in combination with other states $|n\rangle$.

Let us concentrate on zeros of the denominator of the first of (3.26) that have to be found from the solution of the algebraic equation which now reads

$$\varepsilon - \sum_{n=-\infty}^{\infty} \frac{V^2}{\varepsilon - \delta n} = 0, \tag{3.75}$$

and can also be written in the form

$$\frac{\varepsilon}{\delta} = \frac{\pi V^2}{\delta^2} \cot\left(\pi \frac{\varepsilon}{\delta}\right) \tag{3.76}$$

since $\sum_{n=-\infty}^{\infty}(x - n\pi)^{-1} = \cot x$. In Fig. 3.11 we depict the right- and the left-hand sides of (3.76), where the roots correspond to the points of crossing. For large V^2/δ^2 the crossing occurs near the points $\varepsilon_n \simeq (n + 1/2)\delta$, where $\cot(\varepsilon\pi/\delta)$ is small. Let us denote by $\alpha_n\delta/\pi$ the deviation of the n-th root ε_n from the point $(n + 1/2)\delta$, and write (3.76) in the form

$$\begin{aligned}\left(n+\frac{1}{2}\right)+\frac{\alpha_n}{\pi} &= \frac{\pi V^2}{\delta^2}\cot\left(\pi\left(n+\frac{1}{2}\right)+\alpha_n\right)\\ &= -\frac{\pi V^2}{\delta^2}\tan(\alpha_n) \simeq -\frac{\pi V^2}{\delta^2}\alpha_n\left(1+\frac{\alpha_n^2}{3}\right).\end{aligned} \tag{3.77}$$

We consider $\alpha_n \sim \arctan\left(\delta^2 n/\pi V^2\right)$ as small, since the typical number $n \sim 1/\tau\delta$ corresponding to a band state populated in the course of the spectrally localized ($\sim 1/\tau$) injection remains small as compared to the total number of states $\pi V^2/\delta^2$ mixed up by the interaction V, as shown in Fig. 3.9.

The third-order equation (3.77) can be resolved exactly with respect to α_n. However, to be consistent with the order of approximation we have to write it also in the approximative form,

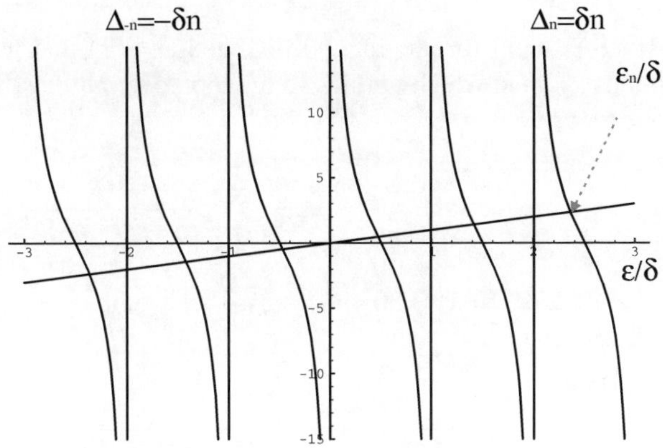

Fig. 3.11 Energy eigenstates of the level–band system corresponding to crossing points of the linear function ε and the band resolvent $\sum_n V^2/(\varepsilon - \Delta_n) = (\pi V^2/\delta)\cot(\pi\varepsilon/\delta)$

$$\alpha_n = \frac{\pi^5 V^2}{\delta^2}\frac{(n+\frac{1}{2})^3}{3\left(1+\frac{\pi^2 V^2}{\delta^2}\right)^4} - \frac{\pi(n+\frac{1}{2})}{1+\frac{\pi^2 V^2}{\delta^2}}, \tag{3.78}$$

which for the same order of approximation yields

$$\varepsilon_n = \left(n+\frac{1}{2}\right)\delta' + \alpha\left(n+\frac{1}{2}\right)^3, \tag{3.79}$$

where $\delta' = \delta\left(1 - \delta^2/\pi^2 V^2\right)$ and $\alpha = \delta^7/3\pi^4 V^6$.

What is the physical meaning of (3.79)? It gives the position of energy eigenstates of the system of the level and the band and implies that the density of states $1/\delta'$ of such a system is slightly larger than the state density $1/\delta$ of the band, since the same spectral domain now contains one more level. Moreover, it shows that the energy eigenstates are not equidistant, as was the case for the band, but experience a third-order correction $\alpha(n+1/2)^3$. One can also notice than n enters the equation in the combination $(n+1/2)$, which implies that for the strong perturbation the energy eigenstates are located in the middle of the intervals among the successive band states.

In order to find the time evolution of the system we have to perform the inverse Fourier transformation, which implies calculation of the residuals at the points of the singularities. One needs to calculate derivatives of the denominator (3.75) for this purpose. We note that $d(\cot yx)/dx = -y/\sin^2 yx$, calculate the derivative and obtain

$$1 - \frac{\pi V^2}{\delta}\frac{\pi}{\delta}\frac{1}{\sin^2\frac{\pi\varepsilon_n}{\delta}} \simeq -\frac{\pi^2 V^2}{\delta^2} \tag{3.80}$$

where we have substituted (3.79) and took into account the relation $\tau V^2/\delta \gg 1$ that is $V^2/\delta^2 \gg n$. We now make use of the formula

$$\frac{1}{f(x)} = \sum_{x_n}\frac{1}{(x-x_n)f'(x_n)}, \tag{3.81}$$

which is valid for any analytical function $f(x)$ that has only simple poles in the points x_n, and which implies summation over all these points. This formula allows one to write (3.56)

$$\psi_0(\varepsilon) = -\frac{i\delta^2 \Pi(\varepsilon)}{\pi^2 V^2}\sum_n \frac{1}{\varepsilon - (n+\frac{1}{2})\delta' - \alpha(n+\frac{1}{2})^3}, \tag{3.82}$$

where (3.79) and (3.80) have also been employed. We also give here without derivation the expression

$$\psi_0(\varepsilon) = \sum_n \frac{-i\delta\Pi(\varepsilon)V^2}{(\pi V^2/\delta)^2 + \left[(n+\frac{1}{2})\delta'\right]^2}\frac{1}{\left[\varepsilon - (n+\frac{1}{2})\delta' - \alpha(n+\frac{1}{2})^3\right]}, \tag{3.83}$$

which is also valid in the case $\tau V^2/\delta < 1$ and relies only on the requirement $V^2/\delta^2 \gg 1$.

Note that expressions (3.81) and (3.82) have a clear quantum mechanical meaning for the case $\Pi(t) = \delta(t)$, that is, $\Pi(\varepsilon) = 1$: we just project the time-dependent quantum state of the system given in terms of the eigenstates $|\varepsilon_n\rangle$ to the initial state $|0\rangle$, which after Fourier transformation yields the amplitude

$$\psi_0(\varepsilon) = \sum_n \frac{\langle 0|\varepsilon_n\rangle}{\varepsilon - \varepsilon_n}, \tag{3.84}$$

whereas the projections $\langle 0|\varepsilon_n\rangle$ are given in terms of the residuals $1/f'_\varepsilon(\varepsilon)$ at the points $\varepsilon = \varepsilon_n$.

Inverse Fourier transformation of (3.82) recovers the time-dependent amplitude $\psi_0(t)$ and yields

$$\psi_0(t) = \frac{2\delta^2}{\pi V^2} \sum_n e^{-\left[(n+1/2)\delta' - \alpha(n+1/2)^3\right]^2 \tau^2/4 - i(n+1/2)\delta' t - i\alpha(n+1/2)^3 t} \tag{3.85}$$

for the case of $\Pi(t)$ given by (3.57). We note that for a typical size of $n \sim 1/\tau\delta$ and for $\alpha \sim \delta^7/V^6$ the combination $\alpha(n+\frac{1}{2})^3 \sim \left(\delta/V^2\tau\right)^3 \tau\delta$ can be omitted as a product of two small parameters, and we are left with

$$\psi_0(t) = \frac{\delta^2}{\pi^2 V^2} \sum_n e^{-(n+1/2)^2 \delta'^2 \tau^2/4 - i(n+1/2)\delta' t - i\alpha(n+1/2)^3 t}. \tag{3.86}$$

If we decide to consider the case $t \ll 1/\delta$, neglect the cubic term in the exponent and replace summation by integration, we immediately recover the result (3.61) which has been obtained directly under the same assumption. Our aim here is to go beyond the limit $t \ll 1/\delta$, which does not allow one to replace sums by integrals, since the phase difference between the neighboring terms of the sum (3.86) is of the order of π.

3.3.2 *Quantum Recurrences*

Generally speaking all three terms in the exponent (3.86) are equally important, since they depend on different combinations of parameters. Nevertheless one can identify a regime $t \sim 1/\delta'$, where the cubic term is still negligible, since $n \sim 1/\tau\delta$, and hence $\alpha(n+\frac{1}{2})^3 t \sim \left(\delta/V^2\tau\right)^3 t\delta \sim \left(\delta/V^2\tau\right)^3 \ll 1$. In terms of the wavepackets shown in Figs. 3.4 and 3.9 this regime corresponds to the situation, where the wavepacket emitted by the level starts to return after either one or several oscillation periods. We therefore have the expression

$$\psi_0(t) = \frac{\delta^2}{\pi^2 V^2} \sum_n e^{-(n+1/2)^2 \delta'^2 \tau^2/4 - i(n+1/2)\delta' t}, \tag{3.87}$$

where summation cannot be replaced by integration.

How do we calculate this sum? Consideration of wavepackets gives an idea of the structure of the sum: it consists of a series of spikes corresponding to the time moments when the wavepacket returns to the left turning point after each oscillation period. A mathematical tool known as the Poisson summation formula turns out to be very convenient for calculation of the sum in such a case. Apparently one can write (3.87) in the form of an integral

$$\psi_0(t) = \frac{\delta^2}{\pi^2 V^2} \int_{-\infty}^{\infty} \sum_n \delta(x-n) e^{-(x+1/2)^2 \delta'^2 \tau^2/4 - i(x+1/2)\delta' t} dx, \tag{3.88}$$

since the sum $\sum_n \delta(x-n)$ of Dirac δ-functions just picks up the values of the integrand in the integer points. Now we note that

$$\sum_n \delta(x-n) = \sum_k e^{i2\pi kx}, \tag{3.89}$$

which can be proved just by casting both sides of this equation, which is periodic in x, into a Fourier series on any interval of length 1 placed around any integer point. Indeed, for such an interval which starts at a point O we find the m-th harmonic of the left-hand side,

$$\int_O^{O+1} e^{i2\pi mx} \sum_n \delta(x-n) dx = \int_O^{O+1} e^{i2\pi mx} \delta(x-n_O) dx = e^{i2\pi mn_O} = 1, \tag{3.90}$$

and of the right-hand side,

$$\int_O^{O+1} e^{i2\pi mx} \sum_k e^{i2\pi kx} dx = \sum_k \int_O^{O+1} e^{i2\pi (m-n)x} dx = \sum_k \delta_n^m = 1, \tag{3.91}$$

where n_O is the integer part of O, and δ_n^m is the Kronecker symbol. Substitution of (3.89) into (3.88) yields

$$\psi_0(t) = \frac{\delta^2}{\pi^2 V^2} \sum_k \int_{-\infty}^{\infty} e^{-(x+1/2)^2 \delta'^2 \tau^2/4 - i(x+1/2)\delta' t + i2\pi kx} dx, \tag{3.92}$$

where we have changed the order of the summation and the integration. The Gaussian integration is straightforward and with the allowance of the integral (3.61) for $\varepsilon = (x+1/2)\,\delta'$ results in the sum

$$\psi_0(t) = \frac{\delta}{2\pi V^2} \sum_k \frac{e^{-i\pi k}}{\sqrt{\pi}\tau} \exp\left\{ -\frac{(t - 2\pi k/\delta')^2}{\tau^2} \right\}, \tag{3.93}$$

where we have neglected the difference between δ' and δ in the pre-exponential factor.

One recognizes in (3.93) a series of Gaussian spikes, which apart from the phase factor $e^{i\pi k} = (-1)^k$ differ only by their arrival times $2\pi k/\delta'$, equally spaced in time by the interval $T_r = 2\pi/\delta'$, which for the model in Fig. 3.4 has the evident meaning of the oscillation period for the particle in the well B. We illustrate the situation in Fig. 3.12. The wavepacket created as the result of Gaussian injection of the population amplitude to the level $|0\rangle$ propagates toward the right turning point. It reaches the point at $t \simeq \pi/\delta$ and after reflection with phase inversion starts to propagate back. At $t \simeq 2\pi/\delta$ it returns and penetrates through the barrier with the extent of penetration given by the ratio $\sim W^{-1}/\tau$ of the decay time $W^{-1} = \delta/\pi V^2$ and the injection time τ.

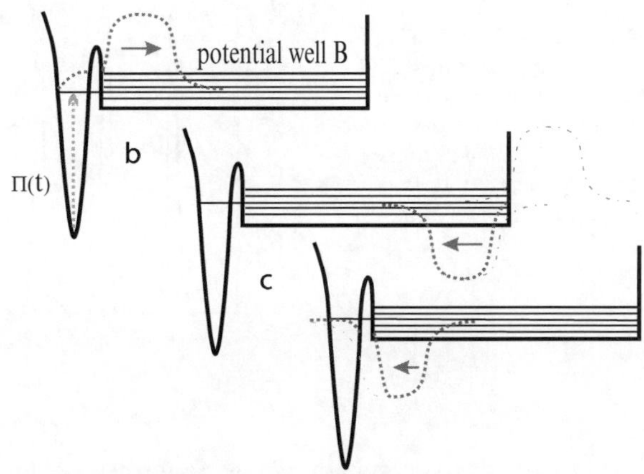

Fig. 3.12 Recurrences in the level–band system corresponding to a two-well potential. (**a**) Injection of the population amplitude to the system and creation of the packet. (**b**) After half of the oscillation period, reflection of the wavepacket with π-shift of the phase. (**c**) The recurrence. The presence of the particle at the initial level is restricted by the necessity of penetration through the barrier

In the extreme case we have just considered the wavepacket of the returned particle has precisely the same shape as it had at the beginning of the evolution. This is not always the case in the presence of anharmonic (that is, nonlinear in n) terms that have been omitted in our consideration. The n^2 term results in dispersion of the wavepacket, which manifests itself in spreading, that is, in the increase of the recurrence duration. It has not been taken into account initially in our band model with equidistant levels. Although the n^3 anharmonicity of the band has also not been taken into account, it arises as a result of the interaction of the band states with the level $|0\rangle$ in the form $\alpha(n + 1/2)^3$, as one can see in (3.86), and results in more significant transformation of the shape of spikes at long time, which is going to be considered now.

The third-order anharmonicity plays the dominating role for the time interval $1/\alpha \gg t \gg 1/\alpha n^3 \sim 1/\alpha'(\tau\delta)^3$, that is, $(V/\delta)^6 \gg \delta t \gg (V^2/\delta)\tau$, where one can neglect the square term compared to the cubic term in the exponent of (3.86). The Poisson summation formula for this case yields

$$\psi_0(t) = \frac{\delta^2}{\pi^2 V^2} \sum_k e^{-i\pi k} \int_{-\infty}^{\infty} e^{-i\alpha(x+1/2)^3 t - i(x+1/2)\delta' t + i2\pi k(x+1/2)} dx, \quad (3.94)$$

where after replacement $x + 1/2 \to x$ and regrouping of terms linear in x in the exponent one recognizes in the integral on the right-hand side an integral representation of the Airy function $2\sqrt{\pi}\mathrm{Ai}(z) = \int_{-\infty}^{\infty} \exp\left(ix^3/3 + ixz\right) dx$ and arrives at

$$\psi_0(t) = \frac{\delta^2}{\pi\sqrt{\pi}V^2\sqrt[3]{3\alpha t}} \sum_k (-1)^k \,\mathrm{Ai}\left(\frac{2\pi k - \delta' t}{\sqrt[3]{3\alpha t}}\right). \tag{3.95}$$

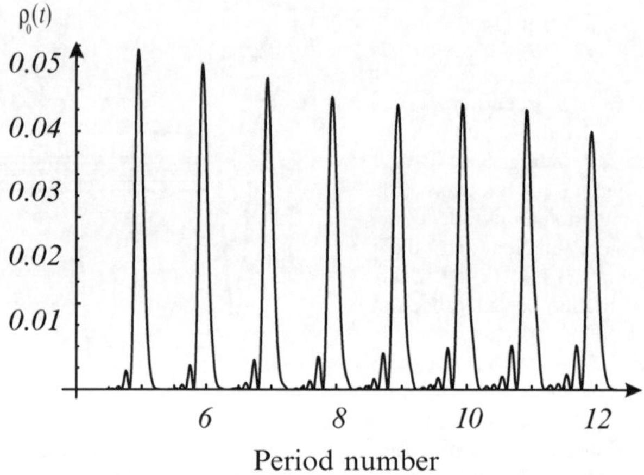

Fig. 3.13 A series of population recurrences for $V^2/\delta \simeq 1.5$ and $\tau\delta \simeq 0.6$ with corresponding anharmonicity $\alpha = 0.001$. The time axis is expressed in the numbers of periods. One clearly sees a cross-over from the bell-shape spikes to the shapes with asymmetric oscillations, typical of the Airy function

Actually, this summation can be performed for the exact expression (3.86), and even in a more general case, when along with the cubic nonlinearity $\alpha(n + 1/2)^3 t$ a quadric term $\beta(n + 1/2)^2 t$ enters the phase of the integrand. One gets rid of the quadric term just by a shift of the integration variable $x \to x - ia$. The final expression reads

$$\psi_0(t) = \frac{\delta^2 e^{2a^3/3}}{\pi\sqrt{\pi}V^2\sqrt[3]{3\alpha t}} \sum_k (-1)^k e^{ab}\,\mathrm{Ai}(b + a^2), \tag{3.96}$$

where $a = (-\delta'^2\tau^2/4 + i\beta t)/(\sqrt[3]{3\alpha t})^2$ and $b = (2\pi k - \delta' t)/\sqrt[3]{3\alpha t}$ enter the argument of the Airy function in (3.95).

In Fig. 3.13 we show a number of sequential recurrences for the population $|\psi_0(t)|^2$, where one clearly sees that the spikes gradually change from the bell-like (Gaussian) shape to a shape with asymmetric oscillations typical of the Airy function, since the relative role of the cubic term increases with time. Note that the spike duration and the period of spike oscillations changes in course of time as is clear from (3.95) where a slow decrease of the spike oscillation frequency with time is associated with the factor $\sqrt[3]{3\alpha t}$ in the denominator of the argument of the Airy function. We also note that this universal Airy-like shape of the spikes is attained in all cases, no matter what the initial profile of the probability amplitude injection, provided the injection was short compared to the wavepacket oscillation period

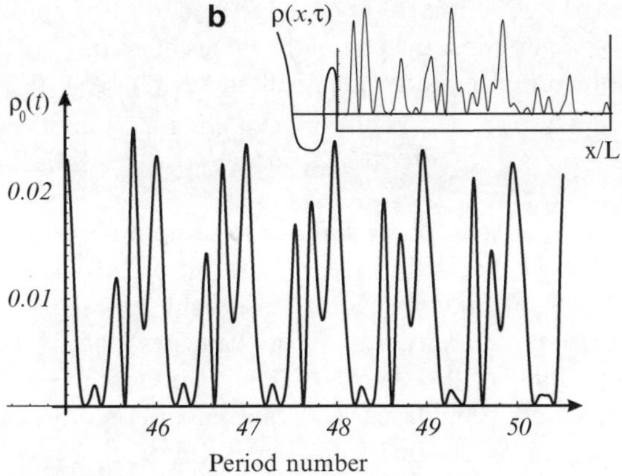

Fig. 3.14 (**a**) A series of population recurrences at longer time. The width of the spikes becomes comparable with the return period and the signal takes on a complicated shape. Corresponding wavepackets (**b**) span all of the potential pit

$T = 2\pi/\delta$ in the potential pit. Destruction of this picture occurs when the spike duration becomes comparable with the period T. In the wavepacket representation this means that at a certain moment the wavepacket becomes so broad that it spans the potential well. Corresponding temporal spikes start to overlap thus entering in the regime of strong interference, where the Poisson summation formula remaining formally valid does not yield a simple expression any longer. The separation of the probability amplitude into a sequence of individual signals no longer makes sense, and the corresponding population acquires a complex irregular shape resembling a random function as shown in Fig. 3.14.

3.3.3 *Quantum Revivals*

Now we show how the interference results in the complete reconstruction of the initial wavepacket, which is followed by the repetition of all the series of recurrences. This phenomenon is called quantum revival. We dwell here mainly on the case of the regular, almost equidistant (that is, slightly anharmonic) spectrum, which we were considering in the previous subsection, and discuss in just a few details another important case of an irregular spectrum, which will be needed later on in this chapter for qualitative estimates.

Regular Spectrum

In order to get an insight into the origin of the quantum revival effect let us consider (3.86) in more detail. The cubic term in the exponent reads

$$i\alpha(n+1/2)^3 t = i\alpha t\left(n^3 + \frac{3}{2}n^2 + \frac{3}{4}n + \frac{1}{8}\right). \tag{3.97}$$

For large t this term is large and differs considerably for the neighboring n. However, for some particular values of αt this big change becomes a multiple of $2\pi i$ and therefore it does not result in any phase shift. Indeed, for $\alpha t = 8\pi$ it amounts to

$$i\alpha(n+1/2)^3 t - i\alpha(n-1/2)^3 t = i\pi(24n^2+2), \tag{3.98}$$

and yields zero phase shift for the amplitudes $\psi_n(t = 8\pi/\alpha)$ whatever the indices n are. Note that in the case $\alpha/(\tau\delta)^3 \ll \delta'$ the cubic term (3.97) changes negligibly if t changes within one period T. This change can be ignored, the overall contribution of the anharmonicity α therefore vanishes, and we return to (3.87) for the probability amplitude of the state $|0\rangle$. This means that the sequence of Gaussian spikes (3.93), which has been destroyed by the anharmonicity α at the time $t > 1/\alpha(\tau\delta)^3$, appears again in its original form at the time $t \simeq 8\pi/\alpha$. The whole scenario of the time evolution repeats again after this revival time $T_r = 8\pi/\alpha$. The wavepacket $\Psi(x, T_r)$ reconstructs itself back from the shape spread over all the potential pit (Fig. 3.14) to the initial Gaussian localized form (Fig. 3.12).

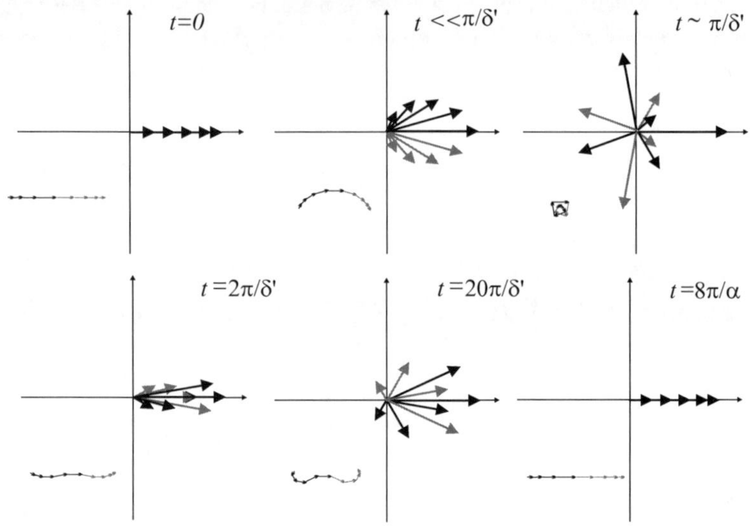

Fig. 3.15 Terms of the sum (3.86) as vectors in the complex plane. The corresponding Cornu constructions for the sum are shown in the inserts in the lower left corners. At $t = 0$ all vectors are aligned and the modulus of the sum is maximum. With the course of time, each of the terms acquires a phase factor, such that the sum decreases. At a time equal to the return period, all of the vectors are almost aligned, since the phase difference comes only from the cubic anharmonicity. Complete alignment occurs at the time of complete revival, when the phase differences of the neighboring terms are multiples of 2π

It is expedient to illustrate the process of recurrence and revivals with the help of vector diagrams in the complex plane, similar to the Cornu geometric construction (Cornu spiral) widely known in diffraction theory. Indeed, each term of the sum (3.86) is a complex number of modulus $e^{-(n+1/2)^2\delta'^2\tau^2/4}$ and phase $-i(n+1/2)\delta' t + i\alpha(n+1/2)^3 t$. We depict these numbers as vectors in the complex plane shown in Fig. 3.15. All the vectors have zero phase at $t = 0$ and therefore their sum has maximum absolute value $\psi_0(0) = 1$. In the course of time the phase difference among the neighboring terms reduces the size of the absolute value of the sum, which almost vanishes at $t \sim \pi/\delta'$, when the neighboring terms are opposite in phase. For the time scale where $\alpha t n^3$ is still a small value for typical $n \sim \delta'\tau$, the sum again has a maximum at the multiples $t = Tk$ of the recurrence time $T = 2\pi/\delta'$, although this maximum is a bit smaller than the value at $t = 0$ due to the influence of the anharmonicity α, which introduce small phase shifts among the neighboring terms resulting in disalignment of the vectors representing these terms. The longer the time, the stronger the disalignment, such that at $t \sim \delta'^3/\alpha$ no revivals can be seen any longer. However, at the time $t \sim T_{rv} = 8\pi/\alpha$ the phase differences resulting from the anharmonicity become multiples of 2π, and therefore complete revivals become again possible. In Fig. 3.16 we show the probability ρ to be in the state $|0\rangle$ at the initial stage of the evolution at $t \sim 0$ and at the time $t \sim T_{rv}$ of the revival. One sees the complete reconstruction of the wavepacket at $t = T_{rv}$ followed by the repetition of the evolution. Note that here we have considered the case of a cubic nonlinearity, although the same phenomenon of revivals takes place also for a square nonlinearity $\alpha t n^2$ for a different T_{rv}.

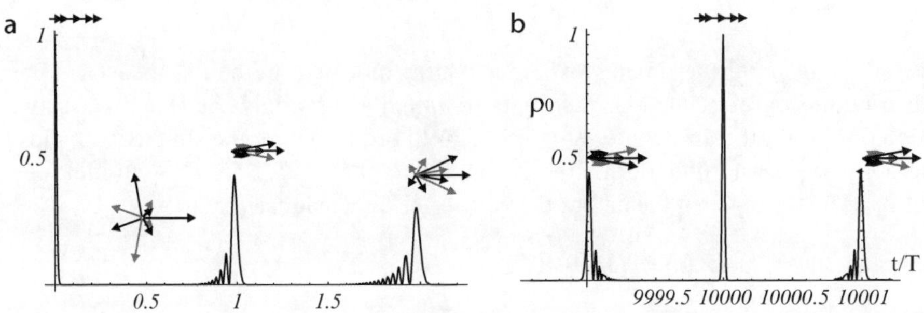

Fig. 3.16 Evolution at the initial stage (**a**) and quantum revival (**b**) at $t = T_{rv} = 10^4 T$. The number of populated levels is $2/\delta'\tau = 40$. The vectors representing the terms of the sum (3.86) become aligned again, as at the beginning, and the evolution repeats

Irregular Spectrum

The important case of the irregular spectrum of eigenstates should also be mentioned here. It is rather typical for a multidimensional motion and does not result in well-pronounced recurrences after a time of the order of the inverse density of states. Indeed, in classical mechanics the motions in different directions are usually nonseparable for two- or higher-dimensional finite systems of general type, as we have already discussed for Sinai billiards (Fig. 1.1). This implies that the classical particle never returns exactly to the initial position, although, according to the Poincaré theorem, it can approach its vicinity with any predetermined accuracy. The smaller the predetermined distance the longer the time needed to get in the vicinity. On the contrary, a finite size of wavepackets in quantum mechanics implies that during such a return a part of the population comes back to the initial state, which means a partial recurrence. Since a long classical time is typically required for return at the vicinity of a size of the initial wavepacket, the packet dispersion during this time is large, and therefore the recurrence amplitude is typically small, resembling rather the regime shown for one-dimensional motion in Fig. 3.14.

In the same time an analog of complete or partial revival can occur in such a system. Revivals of amplitudes of the order of unity are extremely rare events, which resemble giant fluctuations. Let us estimate the time needed to return back to the initial state with an accuracy ϵ. By analogy to (3.86) we take the amplitude in the form

$$\psi_0(t) \sim \sum_n e^{-(n+1/2)^2/N^2 - i\Delta_n t}, \tag{3.99}$$

where N stands for the typical number of the populated energy eigenstates and Δ_n denote detunings of these states from the energy of the level. Actually the simpler expression

$$\psi_0(t) = \frac{1}{2N+1} \sum_{n=-N}^{N} e^{-i\Delta_n t} \tag{3.100}$$

is sufficient for the estimates, where we have replaced the bell-shaped Gaussian distribution by a rectangular distribution, which affects only the shape of revival signals but not the amplitudes at maximum. In order to have the sum (3.100) close to unity we should find a time t such that all the phases $\Delta_n t$ are close to multiples of 2π. This means that the end of the $(2N+1)$-dimensional vector

$$\vec{r}(t) = (\Delta_{-N} t, \Delta_{-N+1} t, \ldots ; \Delta_n t, \ldots \Delta_N t)$$

should be inside of a sphere of radius $\sqrt{2\epsilon N}$ around the point

$$P = 2\pi(m_{-N}, m_{-N+1}, \ldots ; m_n, \ldots m_N)$$

with integer m_n. If all Δ_n are random, one can identify the vector $r(t)$ with coordinates of a particle moving with a random velocity

$$\vec{v} = (\Delta_{-N}, \Delta_{-N+1}, \ldots ; \Delta_n, \ldots \Delta_N)$$

in the $(2N+1)$-dimensional space and the ϵ vicinities of points P with solid spherical scatterers of radii $\sqrt{2\epsilon/N}$ placed in the nodes of a 2π-periodic lattice. By analogy to molecular physics it allows one to say that the revival mean free time T_{rv} equals the inverse relative cross-section $\Gamma(N)\epsilon^{-N}\pi^{-N}$of the $(2N+1)$-dimensional spherical scatterer. With the help of the Stirling formula for the Γ-function we find

$$T_{rv} \sim e^{-N}(N/\epsilon\pi)^N, \tag{3.101}$$

which shows that the revival time rapidly increases with the number N of the energy eigenstates involved. Indeed, the smaller the total mismatch ϵ of the revival is, the smaller should be the mismatches ϵ/N of the individual terms of the sum (3.100). This corresponds to the phase mismatches $\sim \sqrt{\epsilon/N}$ at the position, when the vectors representing the terms are almost aligned and $\cos(x) \simeq 1 - x^2/2$. The probability to have all $2N$ terms aligned with this accuracy is of the order of $\left(\sqrt{\epsilon/N}\right)^{2N}$, which yields (3.101) apart from a numerical factor $1/\sqrt{e\pi}$ for each term of the sum.

3.3.4 *Fractional Revivals*

Let us now return to the almost equidistant spectra. One observes a regular almost periodic time-dependence of the probability amplitude $\psi_0(t)$ not only at the initial stage of the evolution and at the time $T_{rv} = 8\pi/\alpha$ of the complete revival, but also at times $t \sim T_{fr} = (p/q)T_{rv}$, where the fraction is given by the ratio of mutually prime numbers p and q. This phenomenon is known as Averbukh–Perelman fractional revivals, and it results in a more complicated shapes of $\psi_0(t)$ as compared to the simple revivals.

In order to be consistent with the consideration of previous sections we concentrate here on the case of cubic anharmonicity, although the fractional revivals phenomenon can also be observed for quadratic nonlinearity. The wavepacket picture corresponding to the fractional revivals shows that the particle usually is distributed among two or several well-localized wavepackets each of which contains

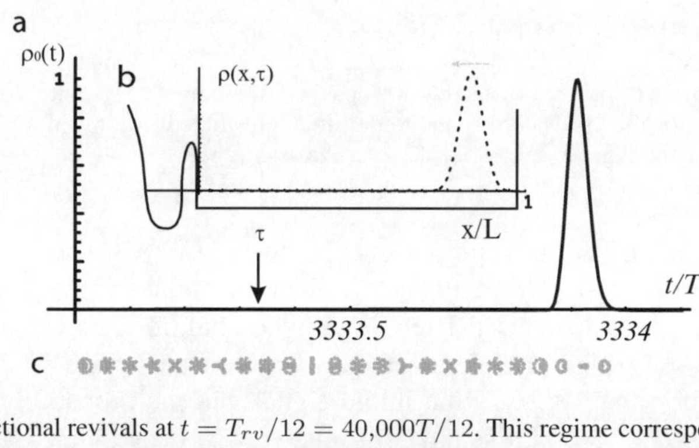

Fig. 3.17 Fractional revivals at $t = T_{rv}/12 = 40{,}000T/12$. This regime corresponds to complete revival that, however, occurs not at the multiple of the return period. The number of the populated levels is $2/\delta'\tau = 18$. (**a**) The time-dependent population of the initial state. (**b**) The wavepacket representation: $\rho = |\Psi(x,t)|^2$ at $\tau = 3{,}333.333$. Only one wavepacket is seen in the potential pit. (v**c**) The ends of vectors representing the terms of the sum (3.86) are shown schematically by points in the complex plane for corresponding times. The revival corresponds to the complete alignment of all the terms

just a fraction of the particle. In other words, the wavefunction $\Psi(x, T_r)$ is a quantum superposition of several well-separated Ehrenfest wavepackets, which allows us to consider this quantum mixture of several classical states as a Schrödinger cat. For some p and q the wavefunction may consist of a single wavepacket which returns to the initial state not at a multiple of the return time nT as it was at the beginning, but at a moment $nT + \Theta T$, where ΘT is a fraction Θ of the return period T. We depict such a situation for $p = 1;\ q = 12$ in Fig. 3.17.

For a shorter time ($p = 1;\ q = 24$) one observes two wavepackets and two corresponding spikes of height $1/2$. We depict such a situation in Fig. 3.18. At the time points corresponding to the position of fractional revivals terms of the sum (3.86) are grouped into two sets with a mutual phase shift of $\pi/2$, as one can see from panel (c).

A more complicated situation takes place at $T_{fr} \sim T_{rv}/32$, when four spikes of different size are seen during one return period. We depict such a situation in Fig. 3.19.

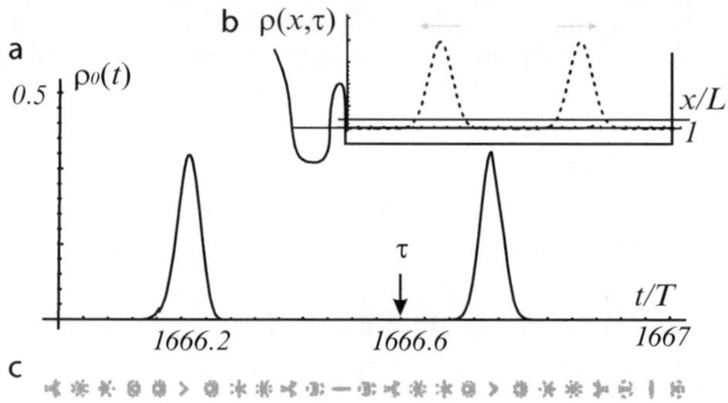

Fig. 3.18 Same as Fig. 3.17 for a more complex situation at $t = T_{rv}/24$. The number of the populated levels is $2/\delta'\tau = 18$ as before. (**a**) The time-dependent population of the initial state has two spikes. (**b**) Two counterpropagating wavepackets are seen at $\tau = 1{,}666.56$ in the potential pit. (**c**) These fractional revivals correspond to the situation where the terms of the sum (3.86) are grouped in two sets of aligned vectors, with a $\pi/2$ phase shift between the sets

The grouping of vectors representing the terms of (3.86) to aligned sets is less evident in this case. However they yield four counterpropagating Gaussian wavepackets for the equivalent particle in the potential pit. One can conclude that in the regime $\alpha T(1/\delta'\tau)^3 \ll 1$ under consideration the spikes of the fractional revivals have the same shape as the initial signal, but the probability of return is shared among two or several time intervals relatively short as compered to the return period T.

Let us dwell now on the distribution law for the return probability amplitudes among the spikes. To this end we consider terms of the sum (3.86) at $t = (p/q)T_{rv} + \vartheta = (8\pi p/\alpha q) + \vartheta$, where $0 < \vartheta < T$ is a time interval of the duration of one return period. The general term of the sum adopts the form

$$\exp\Big[-(n+1/2)^2\frac{\delta'^2\tau^2}{4} + i\Big(\frac{p\,\pi}{q} + \frac{6\,n\,p\,\pi}{q} + \frac{12\,n^2\,p\,\pi}{q} + \frac{8\,n^3\,p\,\pi}{q} - \frac{4\,p\,\pi\,\delta'}{q\,\alpha}$$
$$- \frac{8\,n\,p\,\pi\,\delta'}{q\,\alpha} + \frac{\alpha\,\vartheta}{8} + \frac{3\,n\,\alpha\,\vartheta}{4} + \frac{3\,n^2\,\alpha\,\vartheta}{2} + n^3\,\alpha\,\vartheta - \frac{\delta'\,\vartheta}{2} - n\delta'\,\vartheta\Big)\Big]. \quad (3.102)$$

We neglect the terms proportional to $\alpha\,\vartheta$, since they are small in the regime under consideration, omit the multiples of 2π, and arrive at

$$\exp\Big\{-(n+1/2)^2\frac{\delta'^2\tau^2}{4} + i\Big[\frac{12\,n^2\,p\,\pi}{q} + \frac{8\,n^3\,p\,\pi}{q} + n\,\Big(\frac{12\,p\,\pi}{q} - \frac{8\,p\,\pi\,\delta'}{q\,\alpha} - \delta'\,\vartheta\Big)\Big]\Big\} \quad (3.103)$$

Let us denote by Θ the fractional part of $\left(\frac{12\,p\,\pi}{q} - \frac{8\,p\,\pi\,\delta'}{q\,\alpha} - \delta'\,\vartheta\right)/2\pi$, which allows one to express a moment of time in the interval between two sequential return periods lT and $(l+1)T$ in terms of a fraction of T. Then we set $q = 2q'$ and obtain the sum (3.86) in the form

$$\sum_n \exp\Big\{-(n+1/2)^2\frac{\delta'^2\tau^2}{4} + i\Big(\frac{6\,n^2\,p\,\pi}{q'} + \frac{4\,n^3\,p\,\pi}{q'} + n\,2\,\pi\Theta\Big)\Big\}. \quad (3.104)$$

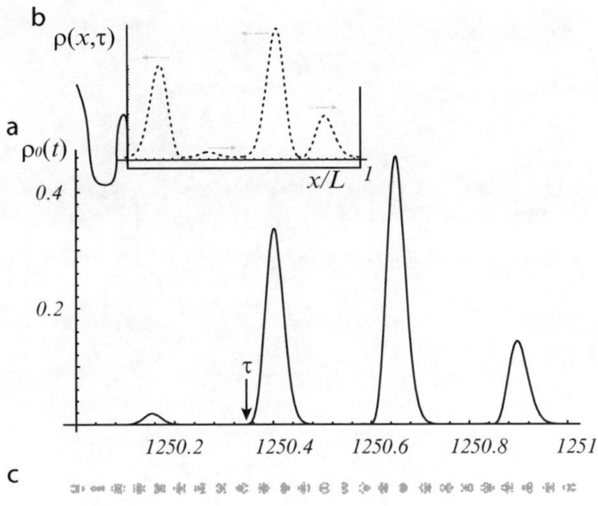

Fig. 3.19 Same as Fig. 3.17 for $t = T_{rv}/32$. (**a**) The time-dependent population of the initial state has four spikes of different size. (**b**) Four counterpropagating wavepackets of different height corresponding to these spikes are seen at $\tau = 1{,}250.333$. (**c**) Grouping of the terms of (3.86) to aligned sets is less apparent

We note that the parts of the phase which depend on n^3 and n^2 yield multiples of 2π for $n = mq$, where m is an integer. We can therefore set $n = mq' + k$ and replace the summation over n by two sums:

$$\sum_{m=-\infty}^{m=\infty}\sum_{k=1}^{k=q'} \exp\Big\{-(mq'+k+1/2)^2\delta'^2\tau^2/4$$
$$+ i\Big[\frac{6\,(mq'+k)^2\,p\,\pi}{q'} + \frac{4\,(mq'+k)^3\,p\,\pi}{q'} + 2\pi(mq'+k)\Theta\Big]\Big\}. \quad (3.105)$$

We replace summation over m by integration over x with the help of the Poisson formula $\sum_m \delta(x-m) = \sum_s e^{i2\pi sx}$ (3.89) and perform this Gaussian integration, which yields

$$\sum_{s=-\infty}^{\infty} \sum_{k=1}^{q'} \frac{2\sqrt{\pi}}{q'\,\delta\,\tau} e^{i\,\pi\,\left(6\,k^2\,p/q'+4\,k^3\,p/q'-2\,k\,s/q'\right)-i\,\pi\left(\,s/q'+\Theta\right)-\frac{4\,\pi^2}{\delta^2\,\tau^2}\left(\,s/q'+\Theta\right)^2}. \tag{3.106}$$

It is convenient to rewrite this equation in the form

$$\sum_{s=-\infty}^{\infty} \frac{2\sqrt{\pi}}{q\,\delta\,\tau} e^{-i\,\pi\left(s/q'+\Theta\right)-\frac{4\,\pi^2}{\delta^2\,\tau^2}\left(s/q'+\Theta\right)^2} \sum_{k=1}^{q'} e^{i\,\pi\,\left(6\,k^2\,p/q'+4\,k^3\,p/q'-2\,k\,s/q'\right)}, \tag{3.107}$$

where one clearly identifies normalized Gaussian spikes with zero phase at the maximum given by terms of the first sum at times $T\Theta$ with $\Theta = -s/q'$ and a universal amplitude function

$$\mathrm{Fr}(p,q',s) = \sum_{k=1}^{q'} e^{i\,\pi\,\left(6\,k^2\,p/q'+4\,k^3\,p/q'-2\,k\,s/q'\right)} \tag{3.108}$$

depending only on the integer numbers, enumerating the revival time (s) and the revival type (p, q).

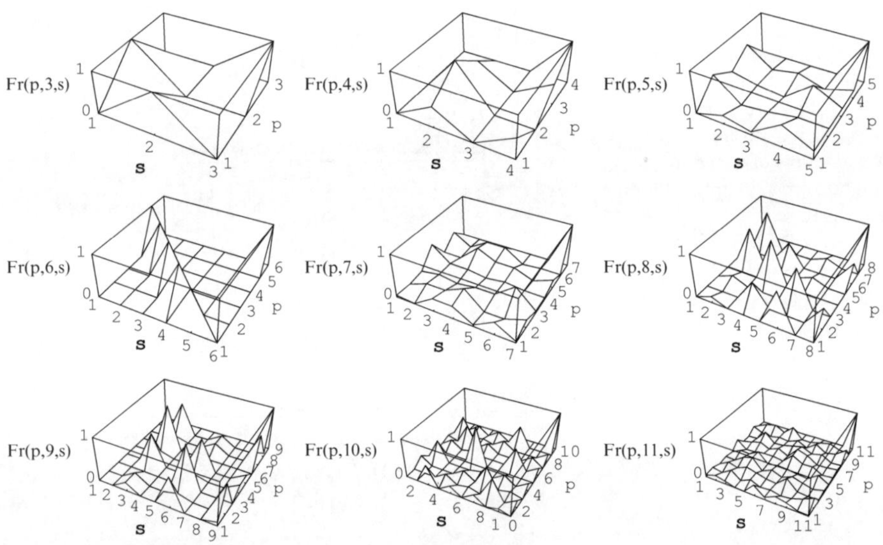

Fig. 3.20 A universal dependence $\mathrm{Fr}(p,q',s)$ which determines the distribution of the amplitudes of the fractional revivals of population as a function of the fractional revival number s for the case of cubic anharmonicity $\alpha(n+1/2)^3$. Normally the fractional revivals in the time region $t \sim \frac{p}{2q'}\frac{8\pi}{\alpha}$ occur at the moments time of $t = lT + \frac{s}{q'}T$, although for some s the revival amplitude may vanish, depending on the factorization properties of the numbers p and $q = 2q'$

For each type of nonlinear dependence of the eigenstate energy on the number n, the function $\mathrm{Fr}(p,q',s)$ of integer variables p, q', and s is determined by (3.108), and yields the probability of fractional revivals at the time $t = lT + st/q'$ for the time region $t \sim lT \sim pT_{rv}/2q'$. Note that an integer l enumerating the number of the return period for which the fractional revivals take place apparently equals the integer part of the ratio $pT_{rv}/2q'T$. For the case of cubic anharmonicity under consideration the function $\mathrm{Fr}(p, q', s)$ is depicted in Fig. 3.20 for several lowest values of q. One sees that for the prime $q' = 5, 7, 11$ all the fractional revivals at different s are present although with different amplitudes, whereas for composed $q' = 4, 6, 8, 9, 10$ some of the spikes disappear. This indicates a hidden relation between the fractional revivals and number theory.

We have considered here the quantum revivals in the simplest version, when all wavepackets were initially of a bell-like Gaussian shape, and we have called revivals and fractional revivals the restoration of such a shape at a given time. This does not always mean that the quantum system returns completely to the initial level, but only implies that the initial state population time-dependence $\rho_0(t)$ is restored at some moment as it was at the beginning, that is, in the course of the population injection. For the original level–band problem, with the initial condition $\psi_0(t = 0) = 1$, when the population does indeed return completely back to the initial state, the wavepackets have different forms—the initially exponential shape gradually transforms to an Airy function-like dependence. In such a situation the variety of possible shapes of fractional revivals is richer, although apart from this circumstance the phenomenon keeps the same general features given by the function $\mathrm{Fr}(p, q', s)$. We also note that for any other type of nonlinearity, like $\alpha n^2, \alpha n^4, \ldots$ etc., the corresponding function $\mathrm{Fr}(p, q', s)$ is different and can be constructed by analogy.

3.3.5 *Revivals and the Classical Limit*

The existence of revivals and fractional revivals suggests to consider more carefully the traditional classical limit of quantum mechanics, at least in the case of the one-dimensional motion where the energy eigenstate spectrum demonstrates a tendency to become equidistant. Indeed, the parameter $n = S/\hbar$ responsible for the transition to the classical limit is the ratio of the mechanical action $S = \int_{\text{period}} p(q)dq$ of the particle and the elementary action given by the Planck constant $\hbar$. It gives the mean $\overline{n}$ of the typical number n of the populated quantum states. The larger the ratio $S/\hbar$, the more classical the system is. The transition frequency $\omega_n = (E_{n+1} - E_n)/\hbar$ among the neighboring states n and $n+1$ is a frequency of classical periodic motion, and it is apparently a classical quantity which may depend on mass and elastic constant but not on $\hbar$. At the same time, the quantum anharmonicity of the motion, that is, the ratio of the anharmonic correction $\omega_{n+1} - \omega_n$ of the transition frequency for the adjacent pairs of the neighboring quantum levels to the frequency ω_n itself, scales as $\hbar/S$. Therefore with increasing n the one-dimensional system becomes more and more harmonic.

The classical limit of quantum mechanics is based on the Ehrenfest theorem, which states that the center of a wavepacket composed as a linear combination of many energy eigenstates follows the classical trajectory. For the harmonic oscillator this is always the case. However the majority of the one-dimensional systems are anharmonic, which results, in particular, in a spread of the wavepacket increasing in the course of time. This quantum dispersion destroys the localization of the Ehrenfest wavepackets. Hence the more classical the system the more equidistant the spectrum is, and the longer is the relative time needed for the spread of a wavepacket composing a localized particle.

In the traditional classical limit of quantum mechanics one first takes $\hbar = 0$ and then considers the limit $t \to \infty$, which implies the time domain of interest remaining shorter than the wavepacket dispersion time $\sim S/\hbar\omega_n$, while implementing the consistent quantum approach one notes that at the long time scale, $t \gg S/\hbar\omega_n$, the anharmonic terms are important, since they wash out the sharp distribution of the quasiclassical particle on the classical trajectory in the manner that the particle becomes distributed along all the length of the trajectory with almost uniform probability, which depends only on the inverse residual time at a chosen point. Figure 3.14, which corresponds to the same situation for a relatively small number of energy eigenstates in the wavepacket, gives an idea about the situation in the classical limit.

From the example of revivals and fractional revivals we have learned that the consistent quantum consideration of the particle propagation problem yields another regime, where the particle is localized in one or a few wavepackets obeying classical mechanics, that is, the particle spread all over the trajectory in the course of evolution reinstates itself again as a single well-localized wavepacket or as a quantum superposition of the well-separated and well-localized wavepackets, each of which follows the same classical trajectory just being displaced in time by a fraction of the oscillation period. In the classical limit such a regime corresponds to an intermediate asymptotic $\hbar \to 0$, $t \to \infty$, but with the constraint $\hbar t = const$ for the case of quadratic anharmonicity αn^2. The size of the constant is of the order of a typical action S multiplied by the oscillation period $T = 2\pi/\omega_n$. For cubic anharmonicity, αn^3, the ratio $(\omega_{n+1} - \omega_n)/\omega_n$ scales as $(\hbar/S)^2$ and this constraint reads $\hbar^2 t = const$, with the constant of the order of $S^2 T$.

The considered asymptotic implies that during the time of evolution the quantum system remains isolated, that is, it obeys Hamiltonian mechanics. For really classical objects the condition $\hbar t = ST$ is difficult to fulfill, since the typical actions are large, and the typical periods are long. Isolating a system for such a long period of time as $t = ST/\hbar$ is unrealistic. But for the case of mesoscopic systems, such as cold atoms or molecules at relatively large distances, this condition can in principle be satisfied.

3.4 Population of Inhomogeneous Bands

The case of an equidistant spectrum coupled to a level by identical matrix elements is one of a few examples of the level–band problem that can be solved exactly and analytically in the traditional manner. It allows to get explicit analytical expressions for a number of different interference effects that may occur in the system and that have been discussed earlier in this chapter. However, no exact analytical solution is available for a more complex quantum system, when the coupling constant V_{0n} differs significantly with the band level number n, and the detunings Δ_n of the band levels deviate from the equidistant positions δn. Of course, the exact expression (3.26) can still be written, but neither the inverse Fourier transformation can be carried out exactly, nor an adequate analysis of this exact expression is possible. We encounter the situation, discussed earlier in Chap. 2, where the idea of the ensemble average becomes indispensable. It yields an analytical expression of the population of the level, although not for a particular quantum level–band system, but for the population averaged over many possible realizations of such a system. We note that in such systems one can encounter non-exponential behavior which is closely related with essentially non-Wigner level statistics.

3.4.1 Statistically Independent Levels

Let us assume that we have an ensemble of level–band systems, each of which has exactly the same numbers $N(V_i)$ of band levels coupled to the state $|0\rangle$ by the matrix elements $V_{0n} = V_i$. Although the numbers of levels with a given coupling V_{0n} are taken to be the same for each system comprising the ensemble, the values of the couplings V_i may change significantly, that is, by orders of magnitude. The positions of the levels, on the contrary, are different for different systems, such that the detuning Δ_n may take any value from a broad strip $\Delta_n \in (-\Gamma, \Gamma)$. An independent distribution of the level energies is called "Poissonian statistics". We assume that the width of the strip $\Gamma \to \infty$, such that the mean density $g_i = N(V_i)/2\Gamma$ of the levels with a fixed coupling remains constant. An example of such a system is shown in Fig. 2.5.

How many band levels are in resonance with the state $|0\rangle$ in such a situation? From the discussion of Sect. 3.2.2 we remember that a two-level system is resonant when the detuning between the levels is less than their coupling. Let us calculate the probability P_r to have a resonance suggested by this criterion. Consider all the levels n of type i that have the matrix elements of coupling $V_{0n} = V_i$. The ratio of the coupling and a typical smallest detuning $\Delta' \sim 1/g_i$ of the level of the group from resonance can be estimated by the product $g_i V_i$. As long as $g_i V_i \ll 1$, this ratio gives the probability to have a resonance in the selected group of levels. In order to get the total probability to have a resonance one has to sum up all the types of level and arrive at $P_r = \sum_i g_i V_i$. In a system with distributed parameters the sum has to be replaced by the integral

$$P_r = \int V g\left(V\right) dV \tag{3.109}$$

where $g(V)\,dV$ is a density of states with $V_{0n} \in (V, V + dV)$. Two distinct cases

$$P_r < \text{a number}$$
$$P_r = \infty \tag{3.110}$$

correspond to different asymptotic behavior of the system.

While P_i is small it does indeed yield the probability to have a resonance. However, the integral (3.109) may be large or may even diverge at the lower limit, which means that the expected number of resonances is much larger than one or is even infinite. Then the probability P_i suggested by the criterion for a two-level system does not represent the expectation value of the number of resonances, since in a multilevel system the resonance criterion is different. From the discussion of Sect. 3.2.2 we know that $N_R \sim 2\pi P_r^2$ gives a better estimate of the number N_R of resonant levels, provided the integral (3.109) converges.

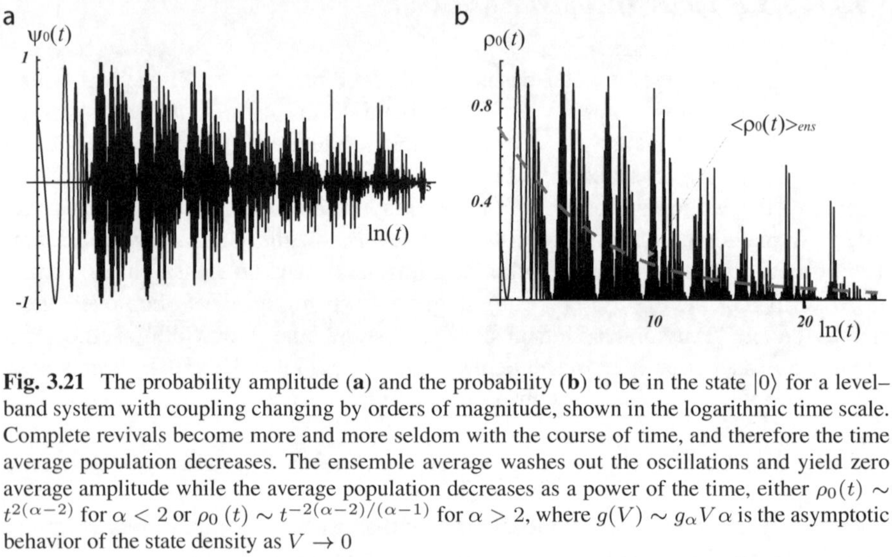

Fig. 3.21 The probability amplitude (**a**) and the probability (**b**) to be in the state $|0\rangle$ for a level–band system with coupling changing by orders of magnitude, shown in the logarithmic time scale. Complete revivals become more and more seldom with the course of time, and therefore the time average population decreases. The ensemble average washes out the oscillations and yield zero average amplitude while the average population decreases as a power of the time, either $\rho_0(t) \sim t^{2(\alpha-2)}$ for $\alpha < 2$ or $\rho_0(t) \sim t^{-2(\alpha-2)/(\alpha-1)}$ for $\alpha > 2$, where $g(V) \sim g_\alpha V\alpha$ is the asymptotic behavior of the state density as $V \to 0$

If the integral (3.109) diverges at the lower limit, the largest contribution is given by the weakly coupled levels. This results in a very special dynamics of the level–band interaction which implies that the number of levels involved in the process increases in the course of time. At the first stage of the process the population rapidly oscillates among a relatively small number of states for which $Vt \sim 1$. In the course of time, more and more weakly coupled levels satisfy this condition and therefore become important, and hence a number of slow Rabi oscillations replace the rapid process. This means that the typical frequency scales down in the course of time, until it becomes zero. One gets an idea about such a regime by considering a function $\phi(a,t) = \prod_{n=0}^{\infty} \cos\left[te^{-an}\right]$, where harmonic oscillations at unit frequency

$\cos[t]$ are modulated by a series of periodically changing amplitudes $\cos[te^{-an}]$ of exponentially decreasing frequencies e^{-an}. In Fig. 3.21 we show the probability amplitude and the probability suggested by the model $\psi_0(t) = \phi(a,t)$. The oscillation frequencies are different for different systems comprising the ensemble, and therefore the ensemble average probability amplitude is not a self-averaging value and already vanishes for relatively short evolution times. But the ensemble average population is well defined, and therefore considering the essentially non-homogeneous bands we concentrate on this very quantity.

Before discussing the technique of the ensemble average it is expedient to consider another aspect of the problem, and turn to the location of the energy eigenvalues by analogy to the approach shown in Fig. 3.11. The divergency of the integral (3.109) at the lower limit means that the poles of the sum $\sum_n V_{0n}V_{n0}/(\varepsilon - \Delta_n + i0)$ in (3.26), which for the equidistant spectrum were placed regularly in the real axis thus giving rise to the cotangent function, are now randomly spread along this axis in such a manner that an infinite number of singularities is present around each real point. Clearly the same is also the case for the zeros of the denominator of (3.26), or, in other words, each point of the real axis is an essential singularity of the Fourier transform $\psi_0(\varepsilon)$ of the type known as a limiting point of poles. The density of poles increases to infinity while their strength $V_{0n}V_{n0}$ decreases. The multitude of the pole positions resembles in some respect the Cantor set.

3.4.2 Factorization of the Level Population and the Ensemble Average

We show now how to perform the ensemble average. The Schrödinger equation for a level–band system in the Fourier representation yields (3.26) in the form

$$\psi_0(\varepsilon) = \frac{i}{\varepsilon - \sum\limits_n V_{0n}V_{n0}/(\varepsilon - \Delta_n + i0)}. \tag{3.111}$$

In the situation where all the levels are similar and each of them is weakly populated and hence never absorbs a significant fraction of the total population, one can replace the sum by the integral $\sum_n \ldots \to \int g(\Delta, V)d\Delta dV \ldots$ at $t \ll g$ as we have done considering the continuous-band model. Here $g(\Delta, V)\,d\Delta dV$ denotes the expectation number of levels in an interval $d\Delta$ around Δ that is coupled to the state $|0\rangle$ by the matrix element in the interval dV around V. This replacement is illegal for matrix elements V_{0n} that are different by orders of magnitude, since different groups of band levels (sorted by the size of $V_{0n}V_{n0}$) result in recurrences, revivals, and fractional revivals in essentially different time scales, which also means that some of the levels can accumulate a significant fraction of the total population more often than the other, and hence during a finite interval of time they contribute to the sum (3.111) with higher weights. In such a situation the statistical spread of various physical values is large, and in order to get a consistent description one has to perform the ensemble average of the quantity of interest, that is, for the population. To this end we take the exact expression for the population

$$\rho_0(t) = \psi_0(t)\,\psi_0^*(t) = \frac{1}{4\pi^2}\int \frac{e^{-i(\varepsilon-\xi)t}}{\varepsilon - \sum\limits_m V_{1m}V_{m1}/\left(\varepsilon - \Delta_m + i0\right)} \frac{1}{\xi - \sum\limits_m V_{1m}V_{m1}/\left(\xi - \Delta_m - i0\right)} d\varepsilon d\xi \qquad (3.112)$$

and perform the ensemble average over $\prod d\Delta_n dV_{0n}$ with a distribution function $g(\{\Delta_n\},\{V_{0n}\})$ depending on the sets of all detunings $\{\Delta_n\}$ and all couplings $\{V_{0n}\}$.

Our assumption implies independent (Poisson) statistics of the position of the band states and their couplings to the level, that is, $g(\{\Delta_n\},\{V_{0n}\}) = \prod g(\Delta_n)\,g(V_{0n})$, with $g(\Delta_n) = \Theta(\Gamma - |\Delta_n|)/2\Gamma$ given by the step function $\Theta(x)$. However, even within so strong a simplifying approximation the ensemble average procedure still requires some effort. For this purpose, one first has to represent the right-hand side of (3.112) in factorized form, that is, as a product of terms, each of which depends only on the detuning Δ_n and coupling V_{0n} of the same level n. Then the ensemble average is reduced to an independent average over the position and the coupling of each level.

Let us first take the average of the population (3.112) over the detunings. It reads

$$\langle\rho_0(t)\rangle = \frac{1}{4\pi^2}\prod_n\left\{\int_{-\Gamma}^{\Gamma}\frac{d\Delta_n}{2\Gamma}\right\}\int_{-\infty}^{\infty} d\varepsilon d\xi \times \frac{e^{-i(\varepsilon-\xi)t}}{\left(\varepsilon - \sum\limits_n \frac{V_{0n}V_{n0}}{(\varepsilon-\Delta_n+i0)}\right)\left(\xi - \sum\limits_n \frac{V_{0n}V_{n0}}{(\xi-\Delta_n-i0)}\right)}. \qquad (3.113)$$

We introduce new variables $\zeta = \varepsilon - \xi$, $\eta = (\varepsilon + \xi)/2$, $x_n = \Delta_n - \eta$ and arrive at

$$\langle\rho_0(t)\rangle = \frac{1}{4\pi^2}\prod_n\left\{\int_{-\Gamma-\eta}^{\Gamma-\eta}\frac{dx_n}{2\Gamma}\right\}\int_{-\infty+i\nu}^{\infty+i\nu}\int_{-\infty}^{\infty} d\zeta d\eta e^{-i\zeta t} \times \frac{1}{\left(\eta + \zeta/2 - \sum\limits_n \frac{|V_{0n}|^2}{(\zeta/2-x_n+i0)}\right)\left(\eta - \zeta/2 - \sum\limits_n \frac{|V_{0n}|^2}{(-\zeta/2-x_n-i0)}\right)}, \qquad (3.114)$$

where the integration contour for ζ is the same as for ε (see C_1 in Fig. 3.2), and the integral over $d\eta$ is calculated along the real axis. We now neglect η with respect to large Γ in the limits of integration over dx_n and perform integration over $d\eta$, which yields

$$\langle \rho_0(t) \rangle = \int_{-\infty+i\nu}^{\infty+i\nu} \frac{d\zeta}{2\pi i} \prod_n \left\{ \int_{-\Gamma}^{\Gamma} \frac{dx_n}{2\Gamma} \right\} \frac{e^{-i\zeta t}}{\zeta - \sum_n |V_{0n}|^2 \left(\frac{1}{\zeta/2 - x_n} + \frac{1}{\zeta/2 + x_n} \right)}. \tag{3.115}$$

Note that on the integration path in the ζ-plane the denominator of the integrand (3.115) has a positive imaginary part, and hence the imaginary part of the fraction is negative. We can therefore make use of the representation $1/x = i \int_0^\infty e^{ix\tau} d\tau$ and write

$$\begin{aligned} \frac{1}{\zeta - \sum_n \frac{|V_{0n}|^2 \zeta}{(\zeta/2)^2 - x_n^2}} &= i \int_0^\infty \exp\left[i\tau\zeta - \sum_n \frac{|V_{0n}|^2 \zeta \tau i}{(\zeta/2)^2 - x_n^2} \right] d\tau \\ &= i \int_0^\infty e^{i\tau\zeta} \prod_n \left\{ \exp\left[\frac{-i|V_{0n}|^2 \zeta\tau}{(\zeta/2)^2 - x_n^2} \right] \right\} d\tau. \end{aligned} \tag{3.116}$$

This representation gives the intended factorized form of the expression and allows one to separate the variables and perform the integration over dx_n independently for each n. Substitution of (3.116) into (3.115) yields

$$\langle \rho_1(t) \rangle = \int_{-\infty+i\nu}^{\infty+i\nu} \frac{d\zeta}{2\pi} e^{-i\zeta t} \int_0^\infty d\tau e^{i\tau\zeta} \prod_n \left\{ \frac{1}{2\Gamma} \int_{-\Gamma}^{\Gamma} dx_n \exp\left[\frac{-i|V_{0n}|^2 \zeta\tau}{(\zeta/2)^2 - x_n^2} \right] \right\}. \tag{3.117}$$

In the limit under consideration of a large number of states and a large width of the band Γ (with a constant state density) the value $\kappa = \frac{1}{2\Gamma} \int_{-\Gamma}^{\Gamma} dx \exp\left[-4i |V_{0n}|^2 \zeta\tau / (\zeta^2 - 4x^2) \right]$ is very close to unity, since the expression in the exponent vanishes for large x, and hence the integrand tends to unity almost everywhere within the integration limits. We make use of this fact and replace κ by $e^{\kappa - 1}$, which for (3.117) yields

$$\begin{aligned} \langle \rho_1(t) \rangle = &\int_{-\infty+i\nu}^{\infty+i\nu} \frac{d\zeta}{2\pi} e^{-i\zeta t} \int_0^\infty d\tau e^{i\tau\zeta} \\ &\exp\left\{ \sum_n \left[\frac{1}{2\Gamma} \int_{-\Gamma}^{\Gamma} \exp\left(\frac{4i|V_{0n}|^2 \zeta\tau}{4x^2 - \zeta^2} \right) dx - 1 \right] \right\}. \end{aligned} \tag{3.118}$$

We note that $\exp\left[4i |V_{0n}|^2 \zeta\tau / (4x^2 - \zeta^2) \right]$ is always 1 along the contour C_2 shown in Fig. 3.2, and hence the expression in the square brackets in (3.118) can be considered as a single integral along the contour $C = C_1 + C_2$ of Fig. 3.2, with the part C_1 coinciding with the real axis:

$$\int_{-\Gamma}^{\Gamma} dx \left[\exp\left\{ \frac{4i |V_{0n}|^2 \zeta\tau}{4x^2 - \zeta^2} \right\} - 1 \right] = \int_C dx \exp\left\{ \frac{4i |V_{0n}|^2 \zeta\tau}{4x^2 - \zeta^2} \right\}. \tag{3.119}$$

In this limit one can also replace $\sum_n 1/2\Gamma$ in (3.118) by the integral over $g\left(V\right)dV$. Note that singularities of the integrand are at different parts of the complex x-plane divided by the real axis, and hence the contour C can be transformed to a small loop around one of these singularities—the point $x = \zeta/2$ at the upper part. After replacement $x \rightarrow \left[1 + 2/(x-1)\right]\zeta/2$ the integral takes the form

$$I = -\int_C \frac{\zeta}{(x-1)^2}\left(\exp\left\{i\frac{\left|V_{0n}\right|^2\tau}{\zeta}\frac{(x-1)^2}{x}\right\}\right)dx, \tag{3.120}$$

and the integration contour C becomes a clockwise oriented circle of large radius centered at the point $x = 1$. We now perform the integration by parts around this point, which results in

$$\begin{aligned} I &= \int_C \frac{\zeta}{1-x}\frac{d}{dx}\left(\exp\left\{i\frac{\left|V_{0n}\right|^2\tau}{\zeta}\left(x-2+\frac{1}{x}\right)\right\}\right)dx \\ &= -i\left|V_{0n}\right|^2\tau\int_C \frac{x+1}{x^2}\left(\exp\left\{i\frac{\left|V_{0n}\right|^2\tau}{\zeta}\left(x-2+\frac{1}{x}\right)\right\}\right)dx. \end{aligned} \tag{3.121}$$

For this integral the radius of the circle C can be taken equal to unity, and therefore by changing variables $x = e^{i\varphi}$ we arrive at

$$I = -\left|V_{0n}\right|^2\tau\exp\left\{\frac{2\left|V_{0n}\right|^2\tau}{i\zeta}\right\}\int_0^{2\pi}\left(1+e^{-i\varphi}\right)\exp\left\{i\frac{2\left|V_{0n}\right|^2\tau}{\zeta}\cos\varphi\right\}d\varphi, \tag{3.122}$$

which can be reduced to a standard integral representation of the Bessel functions $J_k\left(x\right)$ and yields

$$I = -2\pi\left|V_{0n}\right|^2\tau\exp\left\{\tfrac{2\left|V_{0n}\right|^2\tau}{\zeta i}\right\}\left[J_0\left(\tfrac{2\left|V_{0n}\right|^2\tau}{\zeta}\right) + iJ_1\left(\tfrac{2\left|V_{0n}\right|^2\tau}{\zeta}\right)\right], \tag{3.123}$$

where the calculation has been performed assuming a positive τ. The average population (3.118) now takes the form

$$\begin{aligned} \left\langle\rho_0\left(t\right)\right\rangle = \int_{-\infty+i\nu}^{\infty+i\nu}\frac{d\zeta}{2\pi}e^{-i\zeta t}\int_0^\infty d\tau e^{i\tau\zeta}\exp\Bigg\{-\tau\int_0^\infty g\left(V\right)2\pi\left|V\right|^2 \\ \exp\left\{\tfrac{2|V|^2\tau}{\zeta i}\right\}\left\{J_0\left(\tfrac{2|V|^2\tau}{\zeta}\right) + iJ_1\left(\tfrac{2|V|^2\tau}{\zeta}\right)\right\}dV\Bigg\}, \end{aligned} \tag{3.124}$$

and implies evaluation of the integral in the exponent for a given dependence $g(V)$ of the state density on the coupling constant. If the time t is short the typical frequencies $\zeta \sim 1/t$ are big, as suggested by the uncertainty principle, and the combination $2|V|^2 \tau/\zeta$ in the arguments under the integral in the exponent can be set to 0, whence the integral yields the mean decay rate $W = 2\pi \int_0^\infty g(V)|V|^2 dV$. The integration over $d\tau$ yields $1/(\zeta + iW)$, which after the integration over $d\zeta$ results in the exponentially decaying population $\langle \rho_0(t) \rangle = e^{-Wt}$. Now the physical meaning of the exponential and Bessel functions becomes clear: they allow "in average" for the population recurrences from the band to the state $|0\rangle$ that modify strongly the exponential decay process.

3.4.3 The Long-Time Asymptotic

We now consider an asymptotic expression for the population at long times. Note that the main contribution to the integrals comes from the domain where the arguments of the exponents are of the order of unity. For a long time t the slow modes of frequencies $\zeta \sim 1/t$ are important, which means that in (3.124) the main contribution to the integrals come from the domain $\tau \sim 1/\zeta$ and $V \sim \zeta \sim 1/t$. One can therefore replace $g(V)$ by its asymptotic form as $V \to 0$. This is consistent with the fact that at the long-time asymptotic the main part of the population is located at the band levels weakly bound to the state $|0\rangle$ which are present in abundance.

Let us consider the case $g(V) \simeq g_\alpha V^{-\alpha}$ for small V. Note that the power index α should be within certain limits $1 \leq \alpha < 3$, since the state density $g(V)$ is supposed to satisfy two main conditions: (i) the mean transition rate, given by the integral $W = 2\pi \int g(V) V^2 dV$ must be a finite number, which implies that the integral $\int g_\alpha V^{-\alpha} V^2 dV$ converges at the lower limit $V \to 0$, and (ii) the total state density of weakly coupled states given by the integral $\int g(V) dV$ is infinite, which implies that as $V \to 0$ the integral $\int g_\alpha V^{-\alpha} dV$ diverges. On the contrary, the integral $\int g_\alpha V^{-\alpha} V dV$ of (3.109), which gives the expectation value of the number of resonant levels, may diverge or converge, thus yielding two qualitatively different kinds of asymptotic behavior of the population $\langle \rho_0(t) \rangle$.

We also note that for some problems the density of states can indeed be given exactly by a power function of the coupling constant, as is the case for the cold Rydberg atoms with $g(V) \sim V^{-2}$ dependence discussed in Sect. 2.5.5. Such a distribution yields divergence of integrals $\int g(V) V^2 dV$ and $\int g(V) V dV$ at the upper limit $V \to \infty$ and implies a cut-off procedure for strong couplings. In other problems the power dependence is valid just for the asymptotic at $V \to 0$. It turns out that convergence of the integrals is often important for physical conclusions, and therefore the replacement of the density $g(V)$ by its asymptotic dependence has to be done in a way, which excludes these artificial divergencies. One of possible ways is to take $g(V) = g_\alpha V^{-\alpha} e^{-(V/h)^2}$, where $h = \left[W/\pi g_\alpha \Gamma\left(\frac{3-\alpha}{2}\right)\right]^{2/(\alpha-3)}$ is the cut-off parameter consistent with the notation $W = 2\pi \int g(V) V^2 dV$, which can be set to 1 by a proper choice of time units, and $\Gamma(X)$ is the gamma function. This allows us to calculate the integral in the exponent of (3.124):

$$\int_0^\infty g(V)|V|^2 \exp\left\{\frac{2|V|^2\tau}{\zeta i}\right\}\left[J_0\left(\frac{2|V|^2\tau}{\zeta}\right) + iJ_1\left(\frac{2|V|^2\tau}{\zeta}\right)\right] dV, \qquad (3.125)$$

analytically in terms of hypergeometric functions, and obtain the asymptotic expressions for large $\tau/\zeta \sim t^2$ directly from there. However, we present here an approximate albeit physically meaningful approach, which demonstrates directly the nature of the quantities that govern the long-time behavior of the system.

The replacement $\zeta \to z/t$, $\tau \to \theta t$ in (3.124) yields

$$\langle \rho_0(t)\rangle = \int_{-\infty+i\nu}^{\infty+i\nu}\int_0^\infty \frac{dz d\theta}{2\pi} \exp\{-2\pi t\theta \int_0^\infty g(V)|V|^2 e^{-i|V|^2 v}$$
$$\left[J_0\left(|V|^2 v\right) + iJ_1\left(|V|^2 v\right)\right] dV - iz + i\theta z\}, \qquad (3.126)$$

where $v = 2\theta t^2/z$ is a large value for long t. We note that the real part of the integral in the exponent is an even function of v, whereas the imaginary part is an odd one, which makes the entire expression real, as the population should be. One could employ here the asymptotic form of Bessel functions $J_n(x) \simeq \sqrt{2/\pi x}\cos(x - n\pi/2 - \pi/4)$ unless small values of V were important. We therefore divide the integration interval into two parts $\int_0^\infty = \int_0^{1/\sqrt{v}} + \int_{1/\sqrt{v}}^\infty$. The first integral corresponds to small values of $|V|^2 v$ and we set this combination to zero in the integrand, whereas the second integral corresponds to $|V|^2 v \gg 1$ which allows one to make use of the aforesaid asymptotic form of the Bessel functions. This results in

$$\exp\left(-i|V|^2 v\right)\left[J_0\left(|V|^2 v\right) + iJ_1\left(|V|^2 v\right)\right] = e^{i\pi/4}\sqrt{2/\pi|V|^2 v}, \qquad (3.127)$$

and after substitution to (3.126) yields

$$\langle \rho_0(t)\rangle = \int_{-\infty+i\nu}^{\infty+i\nu}\int_0^\infty \frac{dz d\theta}{2\pi} \exp\left\{-t\theta \int_0^{1/\sqrt{v}} 2\pi g(V)|V|^2 dV\right.$$
$$\left. - e^{i\pi/4}\sqrt{\theta z/\pi}\int_{1/\sqrt{v}}^\infty 2\pi g(V)|V| dV - iz + i\theta z\right\}. \qquad (3.128)$$

In the integrals over dV one now recognizes two meaningful quantities. The first integral gives the total decay rate to the states that have coupling matrix elements V less than $1/\sqrt{v}$, that is, less than or of the order of the inverse time $1/t$. For these states $Vt < 1$, and hence they are not populated but work just as an empty reservoir for the exponential decay of the state $|0\rangle$. At large v this integral is small, since the mean decay rate W converges at the lower limit, and for $2\pi g\left(V\right) \simeq g_{\alpha}V^{-\alpha}$ it amounts to $g_{\alpha}v^{(\alpha-3)/2}/\left(3-\alpha\right)$. The second integral gives the expectation value of the number of resonances with the coupling $V > 1/\sqrt{v}$, calculated with the help of the two-level criterion (3.109). We have to consider two physically different cases, $\alpha < 2$ when it converges as $V \to 0$, and $\alpha > 2$ when it diverges. For the first case the integral approaches its asymptotic value $P_r = 2\pi \int_0^{\infty} g\left(V\right) V dV$, with the deviation $\int_{1/\sqrt{v}}^{0} 2\pi g\left(V\right) |V| \, dV \simeq g_{\alpha}v^{(\alpha-2)/2}/\left(\alpha-2\right)$, and for the second case $\int_{1/\sqrt{v}}^{\infty} 2\pi g\left(V\right) |V| \, dV = const + g_{\alpha}v^{(\alpha-2)/2}/\left(\alpha-2\right)$, and it tends to infinity like the power of time $t^{\alpha-2}$. Later on we consider separately the important case $\alpha = 2$. We take into account the specific form of the integrals and write (3.128) in the form

$$\langle \rho_0\left(t\right)\rangle = \int\limits_{-\infty+i\nu}^{\infty+i\nu} \int\limits_{0}^{\infty} \frac{dzd\theta}{2\pi} \exp\Big\{-iz + i\theta z - e^{i\pi/4}\sqrt{\theta z/\pi}P_r$$
$$-z^{\frac{3-\alpha}{2}}\theta^{\frac{\alpha-1}{2}}t^{\alpha-2}A - \theta^{\frac{\alpha-1}{2}}z^{\frac{3-\alpha}{2}}t^{(\alpha-2)}B\Big\}. \tag{3.129}$$

where $A = g_{\alpha}2^{(\alpha-3)/2}/\left(3-\alpha\right)$, $B = e^{-i\pi/4}g_{\alpha}2^{(\alpha-2)/2}/\sqrt{\pi}\left(\alpha-2\right)$, and P_r either is the expectation number of resonances for $\alpha < 2$ or denotes an integration constant for $\alpha > 2$.

Incomplete Decay

We now evaluate the integrals (3.129) taking into account the parity of the exact expression (3.126) with respect to $v \sim \theta/z$ by considering only the real part of the integrals. For $\alpha < 2$ we expand the exponent of the integral (3.129) in powers of $t^{\alpha-2}$ and arrive at

$$\langle \rho_0\left(t\right)\rangle = \mathrm{Re}\left\{\int\limits_{-\infty+i\nu}^{\infty+i\nu} \int\limits_{0}^{\infty} \frac{dzd\theta}{2\pi} e^{-iz+i\theta z-\sqrt{-i\theta z/\pi}P_r}\right.$$
$$\left.\left[1 + z^{\frac{3-\alpha}{2}}\theta^{\frac{\alpha-1}{2}}t^{\alpha-2}C_{\alpha} + \ldots\right]\right\}, \tag{3.130}$$

where $C_{\alpha} = g_{\alpha}2^{(\alpha-3)/2}\left[\sqrt{2/\pi}e^{-i\pi/4}/\left(\alpha-2\right) - 1/\left(3-\alpha\right)\right]$ is a numerical constant. This shows that the population $\langle \rho_0\left(t\right)\rangle$ of the level approaches its asymptotic value

$$\langle \rho_0 (\infty) \rangle = \mathrm{Re} \left[\int\limits_{-\infty+i\nu}^{\infty+i\nu} \int\limits_{0}^{\infty} \frac{dz d\theta}{2\pi} e^{-iz+i\theta z - \sqrt{-i\theta z/\pi} P_r} \right] \tag{3.131}$$

as an inverse power $t^{\alpha-2}$ of time. After the replacement $\theta \to -x/iz$ we perform integration over dz and arrive at

$$\langle \rho_0 (\infty) \rangle = \mathrm{Re} \left[\int\limits_{0}^{\infty} dx e^{-x-\sqrt{x/\pi} P_r} \right] \simeq \frac{2\pi}{P_r^2}, \tag{3.132}$$

where the last equality is the asymptotic form for large P_r of the exact expression

$$\langle \rho_0 (\infty) \rangle = 1 + \left[\mathrm{Erf} \left(\frac{P_r}{2\sqrt{\pi}} \right) - 1 \right] \frac{P_r}{2} \exp \left(\frac{P_r^2}{4\pi} \right) \tag{3.133}$$

depicted in Fig. 3.22a. Here Erf (x) is the error function.

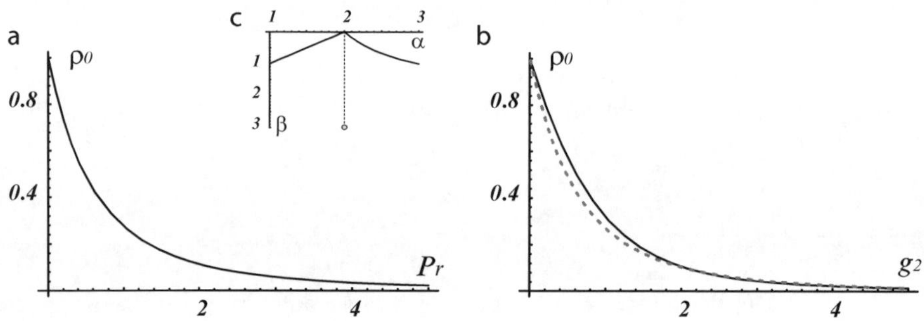

Fig. 3.22 (**a**) Dependence of the asymptotic population on the expectation number P_r of two-level resonances. (**b**) Dependence of the asymptotic value of the population on the density of states for the case $2\pi g(V) = g_2 V^{-2}$ (*solid line*) and for the case $2\pi g(V) = g_2 V^{-2} e^{-V^2}$ (*dotted line*). (**c**) Dependence of the power index β of the asymptotic decay law $\langle \rho_0 (t) - \rho_0 (\infty) \rangle \sim t^{-\beta}$ on the power index α of the asymptotic dependence $2\pi g(V) \simeq g_\alpha V^{-\alpha}$ of the state density on the coupling V in the limit $V \to 0$

Let us consider (3.132) in the context of the discussion of the number of levels in resonance Sect. 3.2.2. For a uniform spectrum and interaction the total number $\mathcal{N}$ of indeed resonant levels is not given by the expectation value $P_r \sim Vg$ calculated with the help of the two-level resonance criterion, but amounts to the square of this number $\mathcal{N} \sim (Vg)^2$. Therefore for large P_r in the case of a uniform population distribution among the resonance levels each of them, including the state $|0\rangle$, has on average $\langle \rho (t) \rangle \sim 1/\mathcal{N} \sim 1/P_r^2$. Equation (3.132) shows that it is also the case for a random spectrum with the total number of resonant levels given by the square of the expectation number of two-level resonances $\mathcal{N} = P_r^2/2\pi$. For a small probability of resonance $P_r \ll 1$ only a small part of the systems comprising the ensemble satisfies the two-level resonance condition $V > \Delta$, which implies that each of the systems has at most one resonance whence the population of the state $|0\rangle$ is of the order of $1/2$. The two-level criterion valid for this case yields $\langle \rho_0 (\infty) \rangle = 1 - P_r/2$ in agreement with (3.133), since $\mathrm{Erf}\,(0) = 0$.

Complete Decay

For $\alpha > 2$ the integral $\int g(V)\, V dV$ diverges, which implies that an infinite number of states are in resonance with the level. In consequence the decay of the level should be complete. In order to show this we keep in the exponent of (3.96) only the leading term $\sim t^{\alpha-2}$ and arrive at

$$\langle \rho_0 (t) \rangle = \mathrm{Re} \left\{ \int_{-\infty+i\nu}^{\infty+i\nu} \int_0^{\infty} \frac{dz d\theta}{2\pi} e^{-iz+i\theta z-\sqrt{-i\theta z/\pi P_r}} \exp\left[z^{\frac{3-\alpha}{2}} \theta^{\frac{\alpha-1}{2}} t^{\alpha-2} C_\alpha \right] \right\}, \tag{3.134}$$

where C_α denotes the same numerical constant as in (3.130). We note that the main contribution to the integral comes from the domain $z \sim 1, \theta \sim t^{-2(\alpha-2)/(\alpha-1)}$, and therefore two terms in the exponent containing the combination θz can be omitted, being small at large t. This yields

$$\langle \rho_0 (t) \rangle = \mathrm{Re} \left\{ \int_{-\infty+i\nu}^{\infty+i\nu} \int_0^{\infty} \frac{dz d\theta}{2\pi} e^{-iz} \exp\left[z^{\frac{3-\alpha}{2}} \theta^{\frac{\alpha-1}{2}} t^{\alpha-2} C_\alpha \right] \right\}, \tag{3.135}$$

which, after the replacement $\theta \to \vartheta t^{-2(\alpha-2)/(\alpha-1)}$, results in

$$\langle \rho_0 (t) \rangle = t^{-2(\alpha-2)/(\alpha-1)} \mathrm{Re} \left\{ \int_{-\infty+i\nu}^{\infty+i\nu} \int_0^{\infty} \frac{dz d\vartheta}{2\pi} e^{-iz} \exp\left[z^{\frac{3-\alpha}{2}} \vartheta^{\frac{\alpha-1}{2}} C_\alpha \right] \right\}, \tag{3.136}$$

where the integral amounts to a time-independent constant. Thus the ensemble average population decays to zero as the power $2(\alpha-2)/(\alpha-1)$ of time, that is, $\langle \rho_0 (t) \rangle \sim t^{-2(\alpha-2)/(\alpha-1)}$ as shown in Fig. 3.21.

We note that in both cases $\alpha > 2$ and $\alpha < 2$ the asymptotic time-dependence of the population becomes less and less rapid when α approaches the value of 2, although the asymptotic population $\langle \rho_0 (\infty) \rangle$ changes abruptly from a finite value to 0. This means that the particular case $\alpha = 2$ when the expectation value of resonances $\int g(V)\, V dV$ diverges, whereas the time-dependence vanishes, requires special consideration.

The Special Case $\alpha = 2$

In the case $\alpha = 2$ the situation is more difficult. Efforts are required for the determination of the long-time asymptotic $\langle \rho_0(t) - \rho_0(\infty)\rangle$ which shows how the population $\langle \rho_0(t)\rangle$ approaches the limiting value, although it is relatively easy to find the asymptotic value $\langle \rho_0(\infty)\rangle$ itself. For $2\pi g(V) = g_2 V^{-2}$ the integral

$$I = 2\pi \int_0^\infty g(V)\,|V|^2 \exp\left(-i\,|V|^2 v\right)\left[J_0\left(|V|^2 v\right) + iJ_1\left(|V|^2 v\right)\right] dV \quad (3.137)$$

entering (3.126) should not be split into two parts as we have done in (3.128), since it has very peculiar behavior. Indeed, after the substitution $2\pi g(V) = g_2 V^{-2}$ and the replacement $V \to x\sqrt{v}$ (3.137) yields

$$\begin{aligned} I &= g_2\sqrt{1/v}\int_0^\infty \exp\left(\mp ix^2\right)\left[J_0\left(x^2\right) \pm iJ_1\left(x^2\right)\right] dx \\ &= g_2\sqrt{z/\theta t^2\pi}\,e^{\mp\pi/4}\left(-2 + C_E \pm \frac{i}{2}\,\pi + 5\,\ln(2)\right), \end{aligned} \quad (3.138)$$

where the upper sign corresponds to positive z in the expression $v = 2\theta t^2/z$, and $C_E \simeq 0.577$ is the Euler constant. After substitution of (3.138) into (3.126) the time-dependence disappears and with the allowance of the symmetry we arrive at

$$\begin{aligned} \langle \rho_0(t)\rangle = \mathrm{Re}\int_0^\infty\int_0^\infty \frac{dz d\theta}{\pi}\exp\Big\{ &- iz + i\theta z - \\ & g_2\sqrt{-iz\theta/\pi}\,\left(-2 + C_E + \tfrac{i}{2}\,\pi + 5\,\ln(2)\right)\Big\}, \end{aligned} \quad (3.139)$$

which after the replacement $\theta \to i\theta/z$ and integration of $\sin z/\pi z$ over dz results in

$$\langle \rho_0(\infty)\rangle = \mathrm{Re}\int_0^\infty d\theta \exp\left\{-\theta - g_2\sqrt{\tfrac{\theta}{\pi}}\,\left(-2 + C_E + \tfrac{i}{2}\,\pi + 5\,\ln(2)\right)\right\}. \quad (3.140)$$

This integral can be expressed in terms of the error function similar to (3.133) which now has a complex-valued argument, and in Fig. 3.22 we show the asymptotic population (3.140) as a function of the constant g_2. One sees that at $g_2 = 0$ the population remains on the level $\langle \rho_0(\infty)\rangle = 1$, as it should if there are no levels present in the band, whereas for larger densities it decreases as a power, $\langle \rho_0(\infty)\rangle \sim g_2^{-2}$.

However, one has to take this result with caution, since it does not yield any time-dependence of the population, or in other words, for such a situation the asymptotic distribution is already attained at $t = 0$. This is an artifact of the model $2\pi g(V) = g_2 V^{-2}$, which yields the expectation number of resonances $\int g(V) V dV$ logarithmically diverging at both upper and lower limits. Thus the strong and the weak couplings V are equally important, and in some way they compensate for each other in the population dynamics. Therefore, in order to determine the asymptotic time-dependence of the level population $\langle \rho_0(t) - \rho_0(\infty) \rangle$ we have to consider the integral

$$I = \int_0^\infty e^{-V^2} \exp\left(-i|V|^2 v\right) \left[J_0\left(|V|^2 v\right) + iJ_1\left(|V|^2 v\right)\right] dV, \tag{3.141}$$

where by introducing the cut-off factor e^{-V^2} we have taken into account that the coupling V cannot have an infinite size. In this way we separate the artificial influence of strongly coupled states ($V \to \infty$) on the long-time behavior of $\langle \rho_0(t) \rangle$, which for the case $\alpha = 2$ by accident has compensated completely the physically important contribution of the weakly coupled states.

Actually now, in the case $\alpha = 2$ under consideration, even the representation (3.141), which was convenient for the splitting of the integral into two parts given in (3.128), requires rather involved analysis. The reason of this complexity becomes clear when we return to an earlier stage of the analysis and consider our initial representation (3.120), with $2\pi g(V) = g_2 V^{-2} e^{-V^2}$, which reads

$$\int_0^\infty \frac{g_2}{V^2} e^{-V^2} \int_C \frac{dx}{2\pi} \exp\left\{\frac{4iV^2 \zeta\tau}{4x^2 - \zeta^2}\right\} dV. \tag{3.142}$$

Earlier, in order to get rid of the diverging singularity $V^{-\alpha}$ of the state density (with $\alpha = 2$ in (3.142)), we have performed, although not explaining the reason, the integration by parts over dx along the contour C:

$$\int_0^\infty dV \frac{g_2}{V^2} e^{-V^2} \left(\frac{x}{2\pi} e^{\frac{4iV^2\zeta\tau}{4x^2-\zeta^2}} \Bigg|_{\text{beg.}C}^{\text{end.}C} - \int_{\text{beg.}C}^{\text{end.}C} dx \frac{x}{2\pi} \frac{d}{dx} e^{\frac{4iV^2\zeta\tau}{4x^2-\zeta^2}} \right). \tag{3.143}$$

As long as the integrand is an analytical function of the variable x the beginning and the end of the contour C correspond to the same point of the complex plane and the first term in parentheses vanishes whereas the second term becomes an integral along a closed loop. But if in the second term we try to change the order of integration and calculate the integral over dV first, we obtain a non-analytic function,

$$I = \frac{g_2}{\pi} \int_{\text{beg}.C}^{\text{end}.C} dx \frac{32i\zeta\tau x^2}{(4x^2 - \zeta^2)^2} \int_0^\infty dV e^{\frac{4iV^2\zeta\tau}{4x^2-\zeta^2} - V^2}$$

$$= \frac{g_2}{\sqrt{\pi}} \int_{\text{beg}.C}^{\text{end}.C} dx \frac{32i\zeta\tau x^2}{(4x^2 - \zeta^2)^{3/2} \sqrt{4i\zeta\tau - 4x^2 + \zeta^2}}, \quad (3.144)$$

which does not return to its initial value after circumvention around the point $x = \zeta/2$ but acquires a phase factor $e^{3i\pi}$. This example shows the typical difficulty which arises when the ensemble average technique is employed—the analytical properties of the functions before and after the average may be different, and hence the change of the order of integration might be illegal. The simplest way to avoid the problem in the case under consideration is to choose a different integration contour. The simplest suitable contour C circumvents both singularities at $x = \pm\zeta/2$ in the opposite directions and thus has an eight-like shape, as shown in Fig. 3.23.

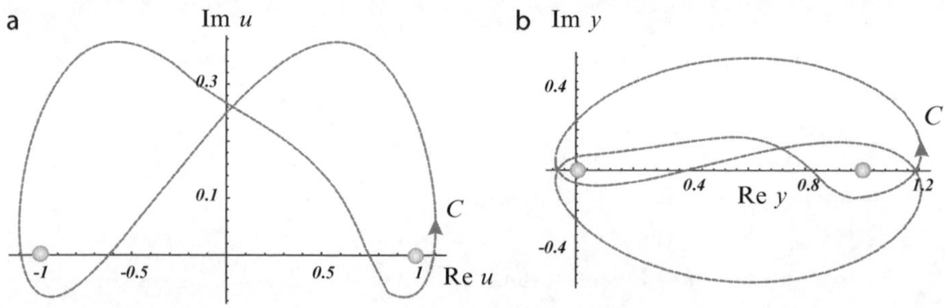

Fig. 3.23 Integration contour C suitable for ensemble average. The averaged function returns to the initial value after circumventing both singularities of (3.126) in the complex plane (**a**) of the variable u. After the replacement $u \to \sqrt{y}$ this contour coincides with the Pochhammer contour (**b**), well-known in the theory of hypergeometric functions

By scaling the variable $x = u\zeta/2$ we change the size of C such that it goes around the points $u = \pm 1$, and after the replacement $u = \sqrt{y}$ this contour takes the Pochhammer shape shown in Fig. 3.23. The integral itself coincides with the integral representation of the hypergeometric function ${}_2F_1(a, b; c; X)$,

$$\int_C \frac{\sqrt{y}\ (y-1)^{-3/2}}{\sqrt{1 - y(1 + 4i\tau/\zeta)^{-1}}} dy = 2\pi i\ {}_2F_1\left(\frac{1}{2}, \frac{3}{2}; 1; \frac{1}{1 + 4i\tau/\zeta}\right) \quad (3.145)$$

and can also be given in different forms either as a hypergeometric function of a different argument or as an elliptic function $E(X)$, which reads

$$\begin{aligned} I &= \frac{-2g_2\sqrt{\pi}\tau}{\sqrt{1+4i\tau/\zeta}}\,{}_2F_1\left(\frac{1}{2},\frac{3}{2};1;\frac{1}{1+4i\tau/\zeta}\right) \\ &= ig_2\sqrt{\pi}\zeta\sqrt{\frac{i\tau}{\zeta}}\,{}_2F_1\left(-\frac{1}{2},\frac{1}{2};1;-\frac{i\zeta}{4\tau}\right) \\ &= \frac{2ig_2}{\sqrt{\pi}}\zeta\sqrt{\frac{i\tau}{\zeta}}E\left(-\frac{i\zeta}{4\tau}\right). \end{aligned} \tag{3.146}$$

Note that when discussing (3.125) earlier we mentioned that for $\alpha \neq 0$ the integral I can also be given in terms of hypergeometric functions, and indeed, by considering (3.144) for this case one obtains a result similar to (3.146), which however cannot be reduced to the elliptic function.

Substitution of (3.146)–(3.118) yields

$$\langle\rho_1(t)\rangle = \int\limits_{-\infty+i\nu}^{\infty+i\nu}\int\limits_0^{\infty}\frac{d\tau d\zeta}{2\pi}\exp\left\{-i\zeta t + i\tau\zeta - i\zeta\frac{g_2}{\sqrt{\pi}}\sqrt{\frac{4\tau}{-i\zeta}}E\left(-\frac{i\zeta}{4\tau}\right)\right\}, \tag{3.147}$$

and after the replacement $\zeta \to -2iz\sqrt{\theta}; \tau \to -\sqrt{\theta}/2z$ results in

$$\langle\rho_1(t)\rangle = \int\limits_{-i\infty-\nu}^{i\infty-\nu}\int\limits_0^{\infty}\frac{d\theta dz}{2\pi z}\exp\left\{-2z\sqrt{\theta}t - \theta - \frac{2g_2}{\sqrt{\pi}}\sqrt{\theta}\,E(z^2)\right\}. \tag{3.148}$$

One can check that such a replacement does not affect the convergence of the integrals.

In Fig. 3.24 we show the absolute value of the integrand (3.148), and the phase of the hypergeometric function ${}_2F_1(-1/2, 1/2; 1; z^2) = 2E(z^2)/\pi$ in the complex plane of the variable z. One clearly sees that the initial contour C can be moved toward large z and can be transformed to a contour C' consisting of two parts. The first part is a loop around the pole at $z = 0$. This yields a time-independent contribution corresponding to the asymptotic population

$$\langle\rho_0(\infty)\rangle = \int\limits_0^{\infty} d\theta\exp\left\{-\theta - \frac{g_2}{\sqrt{\pi}}\sqrt{\theta}\right\}, \tag{3.149}$$

which can be given in terms of the error function by analogy to (3.133), and which is similar in all the main features to the population shown in Fig. 3.22. The second part of the contour C' goes around the branching point of the hypergeometric function at $z = 1$ and allows for the deviation $\langle\rho_0(t) - \rho_0(\infty)\rangle$ decreasing in time. Since the hypergeometric function at different sides of the real axis takes complex conjugated values, this contour integral can be transformed to a definite integral along the real axis from $z = 1$ to $z = \infty$ provided we keep only the imaginary part of the integrand. For long t the integral can be evaluated approximately. Indeed, since the limit $t \to \infty$ corresponds to $\sqrt{\theta} \to 0$ the term θ in the exponent of the integral

$$\langle \rho_0(t) - \rho_0(\infty) \rangle = -2\mathrm{Im} \int_1^\infty \int_0^\infty \frac{d\theta dz}{2\pi z} e^{-2z\sqrt{\theta}t - \theta - \frac{2g_2}{\sqrt{\pi}}\sqrt{\theta}\, E(z^2)} \tag{3.150}$$

can be omitted, being small, and the integration over $d\theta$ yields

$$\langle \rho_0(t) - \rho_0(\infty) \rangle \simeq -2\mathrm{Im} \int_1^\infty \frac{dz}{\pi z \left(2zt + \frac{2g_2}{\sqrt{\pi}}\, E(z^2)\right)^2}. \tag{3.151}$$

The integrand tends to a real number $1/4\pi z^3 t^2$ for long t, and the first non-vanishing correction $g_2\, \mathrm{Im}\left[E(z^2)\right]/2\pi\sqrt{\pi} z^4 t^3$ can be determined by the Taylor expansion over g_2 and, after integration, yields

$$\langle \rho_0(t) - \rho_0(\infty) \rangle \simeq \frac{g_2}{\pi\sqrt{\pi t^3}} \int_1^\infty \mathrm{Im}\left[E(z^2)\right] \frac{dz}{z^4} = \frac{2g_2}{9\pi\sqrt{\pi t^3}}. \tag{3.152}$$

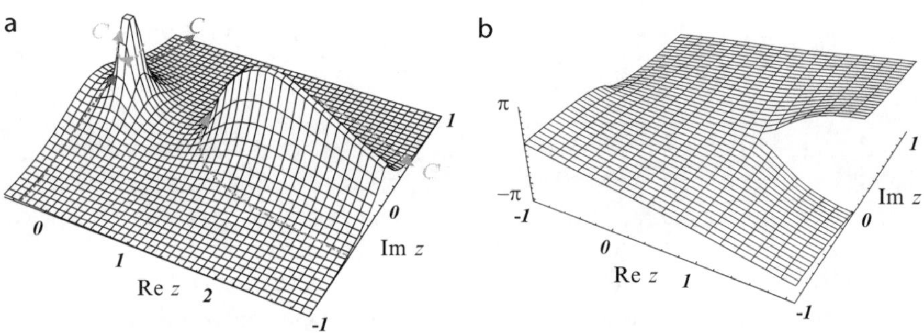

Fig. 3.24 The structure of the integrand (3.148) in the complex plane z. (**a**) The absolute value of the integrand shows that the initial integration contour $C = (-i\infty - \nu, i\infty - \nu)$ can be transformed to the contour C', which consists of two parts and comprises a loop around the pole at $z = 0$ that gives the asymptotic value of $\langle \rho_0(t) \rangle$ and a contour that goes around the branching point $z = 1$ shown in (**b**) which yields the deviation $\langle \rho_0(t) - \rho_0(\infty) \rangle$ vanishing at long t. (**b**) The phase of the hypergeometric function of (3.148) is discontinuous on the real axis, starting from the branching point at $z = 1$

One sees, in spite of the fact that for the case $\alpha = 2$ the expectation number of two-level resonances P_r is infinite, the decay of the level $|0\rangle$ is not complete, although the population of this level approaches its finite asymptotic value $2\pi/g_2^2$ as t^{-3}, which is much faster than for any other α, as shown in Fig. 3.22. It worth mentioning that here the coefficient g_2 plays the role of P_r for other α.

Chapter 4
Two-Band System

Thus far we have been considering the level–band system, for which an explicit analytic expression (3.26) for the Fourier transform of the probability amplitude exists. However, as we have seen, even for such a system, the complexity of the problem does not allow one to directly perform the inverse Fourier transformation in the generic case, and therefore in order to get physically consistent analytic results, one needs to apply a statistical description based on the idea of ensemble averages. The situation becomes more difficult still when we consider the more complex quantum system of two coupled bands. Not only the inverse Fourier transformation, but even the derivation of an explicit, exact expression for the Fourier transforms like (3.26) turns out to be problematic. Therefore, employment of the ensemble average approach becomes a vital part of the analytic approach to this problem.

We start with the consideration of the all-order perturbation series and their graphical representation in the spirit of Feynman diagrams. However, being topologically equivalent to the standard diagrams of many-body theory, where the unperturbed motion of the particle is depicted as straight lines and the perturbations as vortices with the incoming lines of the perturbation, our diagrams more closely resemble Grotrian ones, with interaction shown by lines of transitions among the points denoting the unperturbed energy levels. We believe that such a representation, performed in the spirit suggested by P.W. Anderson in his seminal paper on dynamical localization, is better suited to the case of point (0-dimensional) systems like atoms and molecules, where the free evolution does not imply propagation along a straight trajectory.

We discuss the renormalization associated with the random perturbation, and consider several particular cases of non-degenerate bands, two degenerate levels, and a degenerate level coupled to a non-degenerate band. We consider the dynamics of the total populations of the bands and the degenerate levels as well as the population distribution over the band, and demonstrate the interplay between

V.M. Akulin, *Dynamics of Complex Quantum Systems*, Theoretical and Mathematical Physics, DOI 10.1007/978-94-007-7205-2_4,

the exponential and the coherent damping that yields oscillations in the time-dependencies of the total populations and oscillations in the distribution of the population over the bands.

We dwell on the role of the correlations of the perturbation that manifest themselves in the fact that the rank of the perturbation matrix becomes much smaller relative to its order, and elucidate the role of correlations between the matrix elements of the perturbation and the energy position of the levels to which they belong. We show how these correlations affect the time evolution of the total populations and the distribution of the populations over the bands. We conclude by considering the situation where all of the matrix elements of the perturbation have Gaussian statistics but with an individual width for each transition.

4.1 General Consideration

Under certain conditions, the calculation of the population dynamics in different complex quantum systems can be carried out in the same manner. We first express the Fourier transforms of the probability amplitudes to be at different quantum states in terms of the all-order perturbation series over the coupling. We multiply these series by the similar series for the complex conjugate amplitudes in order to obtain the probabilities. We then perform the ensemble average over the sizes of the couplings and thereby eliminate the majority of terms in the perturbation series. This often allows one to perform an analytic summation of the series for probabilities. We note that after the ensemble average, the particular values of the interaction matrix elements are no longer important for certain regimes, whereas the topological structure of the perturbation series, that is, the number of possible sequences of adjunct transitions induced by the perturbation starts to play a leading role.

4.1.1 Series and Diagrams for the Level–Band Problem

The casting of the probability amplitudes in the all-order perturbation series is the first step of the analysis, and in order to gain a deeper insight into the approach we start with the example of the level–band system, which has already been considered in the previous chapter, but this time with the help of a different technique.

Series and Diagrams for the Probability Amplitudes

One can cast the exact expression (3.26) for the Fourier transform of the probability amplitude of remaining in the initial state $|0\rangle$ in a power series over the coupling,

$$\psi_0(\varepsilon) = \frac{i}{\varepsilon - \sum_n \frac{V_0^n V_n^0}{\varepsilon - \Delta_n}} = i\left[\frac{1}{\varepsilon} + \frac{1}{\varepsilon}\left(\sum_n \frac{V_0^n V_n^0}{\varepsilon - \Delta_n}\right)\frac{1}{\varepsilon}\right.$$
$$+\frac{1}{\varepsilon}\left(\sum_n \frac{V_0^n V_n^0}{\varepsilon - \Delta_n}\right)\frac{1}{\varepsilon}\left(\sum_n \frac{V_0^n V_n^0}{\varepsilon - \Delta_n}\right)\frac{1}{\varepsilon} + \ldots$$
$$\left. +\frac{1}{\varepsilon^{k+1}}\left(\sum_n \frac{V_0^n V_n^0}{\varepsilon - \Delta_n}\right)^k + \ldots\ldots\right] \tag{4.1}$$

$$= i\left[\frac{1}{\varepsilon} + \sum_{n=1}^{N} \frac{1}{\varepsilon}\frac{V_0^n V_n^0}{(\varepsilon - \Delta_n)}\frac{1}{\varepsilon} + \frac{1}{\varepsilon}\sum_{n,m}\frac{V_0^n V_n^0}{(\varepsilon - \Delta_n)}\frac{1}{\varepsilon}\frac{V_0^m V_m^0}{(\varepsilon - \Delta_m)}\frac{1}{\varepsilon}\right.$$
$$\left. +\frac{1}{\varepsilon}\sum_{n,m,k}\frac{V_0^n V_n^0}{(\varepsilon - \Delta_n)}\frac{1}{\varepsilon}\frac{V_0^m V_m^0}{(\varepsilon - \Delta_m)}\frac{1}{\varepsilon}\frac{V_0^k V_k^0}{(\varepsilon - \Delta_k)}\frac{1}{\varepsilon} + \ldots\right]. \tag{4.2}$$

One sees that the generic term of the perturbation series has the structure

$$\frac{1}{\varepsilon}V_0^n\frac{1}{(\varepsilon - \Delta_n)}V_n^0\frac{1}{\varepsilon}V_0^m\frac{1}{(\varepsilon - \Delta_m)}V_m^0\frac{1}{\varepsilon}V_0^k\frac{1}{(\varepsilon - \Delta_k)}V_k^0\frac{1}{\varepsilon} \times \ldots$$
$$\times\frac{1}{\varepsilon}V_0^l\frac{1}{(\varepsilon - \Delta_l)}V_l^0\frac{1}{\varepsilon}V_0^s\frac{1}{(\varepsilon - \Delta_k)}V_s^0\frac{1}{\varepsilon}, \tag{4.3}$$

which can be interpreted in physical terms as a sequence of transitions induced by the interaction V_0^n, and therefore the entire series represents the sum over all transition trajectories. In Fig. 4.1 we show such a trajectory for a sixth order term

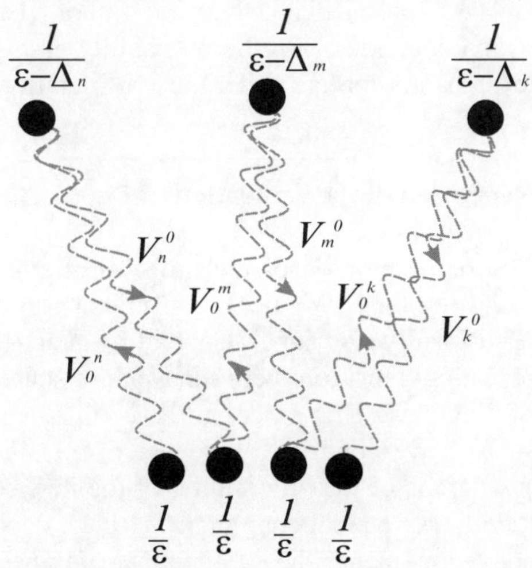

Fig. 4.1

$$\frac{1}{\varepsilon}V_0^n\frac{1}{(\varepsilon-\Delta_n)}V_n^0\frac{1}{\varepsilon}V_0^m\frac{1}{(\varepsilon-\Delta_m)}V_m^0\frac{1}{\varepsilon}V_0^k\frac{1}{(\varepsilon-\Delta_k)}V_k^0\frac{1}{\varepsilon} \tag{4.4}$$

as an example. The system initially in the state $|0\rangle$ makes a transition to the state $|n\rangle$ with the transition amplitude given by the matrix element V_0^n and returns back to the state $|0\rangle$ with the amplitude V_n^0 in order to perform two more "round trip" transitions, to the state $|m\rangle$ and to the state $|k\rangle$ consecutively. The visiting of each level $|l\rangle$ is associated with a fraction $1/(\varepsilon-\Delta_l)$, which corresponds to a pole in the complex plane of ε at the level energy position Δ_l and yields an oscillating exponent $e^{i\Delta_l t}$ after the inverse Fourier transformation, whereas the energy position Δ_0 of the level $|0\rangle$ is taken to be zero and gives the fraction $1/\varepsilon$ corresponding to time integration.

Hereafter, instead of writing down the terms of the perturbation series, which becomes more and more cumbersome as the order of the perturbation increases, we will draw the corresponding diagrams, where each transition between two states $|e\rangle$ and $|s\rangle$ with the amplitude V_e^s is shown by a wave line and points on the ends of these lines denote the fractions $1/(\varepsilon-\Delta_l)$. In this notation, the amplitude (4.2) is given by a sum over all possible "bush-like" trajectories of any possible length that start and end at the point $|0\rangle$, as shown in Fig. 4.2.

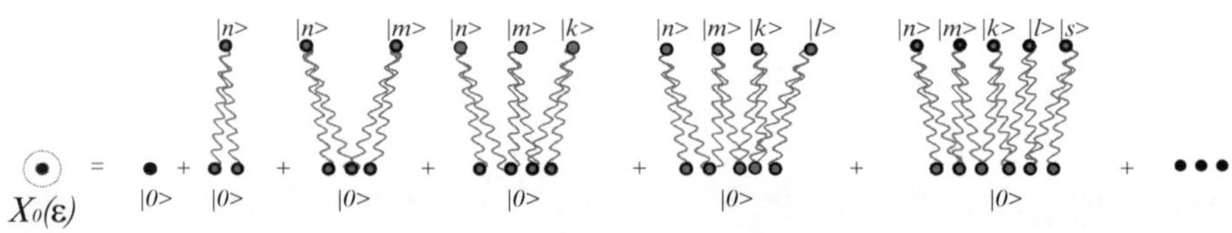

Fig. 4.2

By $X_0(\varepsilon)$ we denote here the sum of all diagrams that start and end with state $|0\rangle$, which coincides with the probability amplitude $\psi_0(\varepsilon)=X_0(\varepsilon)$ for the particular level–band problem under the consideration.

Repetition of the Transitions

The representation of the probability amplitudes in the form of a diagrammatic series not only provides a convenient way to handle cumbersome expressions, but occasionally also suggests a hint for practical calculations. For example, the role of repeating transitions between a pair of states can be readily understood with the help of diagrams. This relies on the comprehension that each term of the perturbation series corresponds to a sequence of transitions, and each transition requires a time τ which is of the order either of the inverse transition probability $\tau\sim 1/W$ or of the inverse typical coupling $\tau\sim 1/V$, depending on which is longer. Therefore at a given moment t the total number of transitions that have occurred in the system is limited, and the typical length of the diagram can be estimated as $N(t)\sim t/\tau$.

If the number $N(t)$ is much less than the total number $\mathcal{N} \simeq W/\delta \sim V^2/\delta^2 \sim P_r^2$ of the levels involved in the process discussed earlier in Sects. 3.2.2 and 3.4.3 (3.133), the probability to encounter a trajectory which arrives twice or more times at the same state $|n\rangle$ of the band is small. Although these trajectories enter into the sum over all diagrams, their relative contribution is negligible for short time-scales. Therefore the ensemble average procedure yields only the mean squared matrix elements $\left\langle V_0^n V_n^0 \right\rangle$, corresponding to the "forward-and-back" transition from the level $|0\rangle$ to a state $|n\rangle$, where the contribution of the higher-order averages remains unimportant. This is in complete agreement with the result of (3.124) and the discussion thereafter, showing that for the short-time regime, while the recurrences or revivals have not yet occurred, only the mean square coupling $\int g\,(V)\,V^2 dV$ is of importance.

This reasoning brings us to the conclusion that by taking into account repeating transitions to the same states, we allow for the influence of recurrences and revivals on the population dynamics of a multilevel system, which definitely play an important role over a long period of time but do not contribute to the short-time dynamics. As we have already seen in Sect. 3.3.2 the revival regime arises at times $t \sim 1/\delta \sim g$ of the order of the inverse spacing $1/\delta$ between the neighboring energy levels, that is, the mean state density g. This means that one can ignore the returns for $t \ll g$ which also corresponds to the regime where one can make use of the continuous-band model.

Diagrams for Populations

As we have already mentioned when considering the level–band system with a complicated and irregular structure of the detunings Δ_n and of the transition matrix elements V_0^n, the value of the probability amplitude $\psi_n(t)$ of a state n may and usually does vanish after average over the ensemble. Therefore, dealing with complex systems we have to consider mainly the probabilities (3.112), which are "good" or "self-averaging" quantities, and draw the corresponding terms of the perturbation series for the populations. We do this by superimposing the diagrams corresponding to the Fourier transform $\psi_0(\varepsilon)$ of the amplitude $\psi_0(t)$ with the diagrams for the Fourier transforms $\psi_0(\xi)$ of the complex conjugate amplitude $\psi_0^*(t)$ as shown in Fig. 4.3. In order to distinguish between these two types of

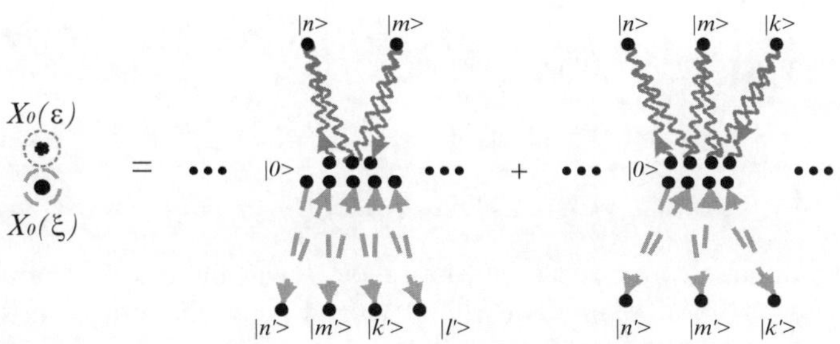

Fig. 4.3

diagrams we use undulating lines for the trajectories of $\psi_0(\varepsilon)$ and dashed lines for those of $\psi_0(\xi)$. It is clear that in a typical term of the perturbation series for populations, the undulating and the dashed trajectories have different lengths. We also note that the dashed line connecting two points, say p and q corresponds to the factor $(V_q^p)^* = V_p^q$ conjugated with the same coupling factor on the wavy line.

The technique under discussion plays no important role in the consideration of the level–band systems; it only brings to light the fact that each term of the perturbation series represents a certain trajectory of sequential jumps, and that the all-order perturbation series itself is just a sum of the amplitudes of all possible trajectories. This property is a particular case of a more general rule stating that in quantum mechanics we have to take into account the interference of all possible channels of the process. It is also a complete analog of the Feynman path integrals, where the integration is carried out in the discrete space of the energy eigenstates of the unperturbed Hamiltonian.

We note, however, that for more complex systems this approach seems to be more convenient as compared with the other methods and to yield the final results in the most straightforward and physically clear way. Indeed for the level–band system, the topological structure of the interaction is such that a transition $0 \to n$ must be followed immediately by the back transition $n \to 0$, and therefore each matrix element V_0^n entering the perturbation series together with its complex conjugate V_n^0 form an irreducible block $V_0^n\,(\varepsilon - \Delta_n)^{-1}\,V_n^0$ in the geometric series (4.2). This is no longer the case for more complex two-band systems, where the transition from a level of the first band to an energy eigenstate of the second band does not imply an immediate back transition. On the contrary, for this problem the generic trajectory travels over different states with little probability of a self-intersection occurring in the regime $t \ll g$.

4.1.2 Series and Diagrams for the Two-Band Problem

Let us now consider the two-band problem. We designate levels of the first band by the index n and levels of the second band by index m, both indices running either from $-\infty$ to ∞ for infinite number of states or from $-N/2$ to $N/2$ and from $-M/2$ to $M/2$ for a finite numbers of states $N+1$ and $M+1$ of the first and the second bands respectively. We assume that at $t=0$ only the state $n=0$ is populated, that is, $\psi_n(t=0) = \delta_n^0$, $\psi_m(t=0) = 0$, and by analogy to (3.10) write down the Fourier transformed Schrödinger equation

$$\begin{aligned} \varepsilon\psi_n(\varepsilon) &= \Delta_n\psi_n(\varepsilon) + V_n^m\psi_m(\varepsilon) + i\delta_n^0 \\ \varepsilon\psi_m(\varepsilon) &= \Delta_m\psi_m(\varepsilon) + V_m^n\psi_n(\varepsilon), \end{aligned} \tag{4.5}$$

where the summation over repeating upper and lower indices is implicit, that is, $V_n^m\psi_m = \sum_m V_n^m\psi_m$. Here, as earlier, Δ_n and Δ_m denote detunings from the resonance of the levels of the first and second bands, respectively, and $\Delta_{n=0}$ is set to zero.

Matrix Form

Two rectangular $(N+1) \times (M+1)$ matrices $\hat{V}$ and $\hat{V}^{+}$ of an interaction with the complex conjugate elements $V_n^m = \left((V^{+})_m^n\right)^*$ together with the unperturbed Hamiltonians of the first $\hat{\Delta}_1$ and second $\hat{\Delta}_2$ bands comprise the Hamiltonian

$$\hat{H} = \begin{pmatrix} \hat{\Delta}_1 & \hat{V} \\ \hat{V}^{+} & \hat{\Delta}_2 \end{pmatrix} \tag{4.6}$$

of the two-band system, which can be separated into two parts: (i) the unperturbed Hamiltonian

$$\hat{H}_0 = \begin{pmatrix} \hat{\Delta}_1 & 0 \\ 0 & \hat{\Delta}_2 \end{pmatrix} \tag{4.7}$$

with $\left(\hat{\Delta}_1\right)_n^{n'} = \delta_n^{n'} \Delta_n$ and $\left(\hat{\Delta}_2\right)_m^{m'} = \delta_m^{m'} \Delta_m$ in the unperturbed energy eigenstate representation, and (ii) the interaction

$$\hat{V}_{int} = \begin{pmatrix} 0 & \hat{V}_{12} \\ \hat{V}_{21}^{+} & 0 \end{pmatrix}, \tag{4.8}$$

where $\hat{H}_0$ and $\hat{V}_{int}$ are both square Hermitian matrices.

Series

In the matrix notation of (4.7)–(4.8) the Schrödinger equation (4.5) reads

$$\left[\varepsilon - \hat{H}_0 - \hat{V}_{int}\right] \bar{\psi}(\varepsilon) = i\overline{\delta_n^0} \tag{4.9}$$

where $\bar{\psi}$ denotes the vector $\left\{\psi_{-N/2}, \ldots, \psi_{N/2}, \psi_{-M/2}, \ldots, \psi_{M/2}\right\}$, and $\overline{\delta_n^0} = \left\{0, \ldots, \delta_n^0, \ldots, 0, \ldots, 0\right\}$ corresponds to the initial condition. The solution of (4.9)

$$\bar{\psi}(\varepsilon) = \frac{i}{\varepsilon - \hat{H}_0 - \hat{V}_{int}} \overline{\delta_n^0} \tag{4.10}$$

given in terms of the resolvent $1/\left(\varepsilon - \hat{H}_0 - \hat{V}_{int}\right)$ means

$$\psi_n(\varepsilon) = \langle n| \frac{i}{\varepsilon - \hat{H}_0 - \hat{V}_{int}} |n = 0\rangle \tag{4.11}$$

$$\psi_m(\varepsilon) = \langle m| \frac{i}{\varepsilon - \hat{H}_0 - \hat{V}_{int}} |n = 0\rangle \,. \tag{4.12}$$

Now we make use of the Taylor expansion of the resolvent operator

$$\begin{aligned}\frac{1}{\varepsilon - \hat{H}_0 - \hat{V}_{int}} &= \frac{1}{\varepsilon - \hat{H}_0} + \frac{1}{\varepsilon - \hat{H}_0}\hat{V}_{int}\frac{1}{\varepsilon - \hat{H}_0} \\ &+ \frac{1}{\varepsilon - \hat{H}_0}\hat{V}_{int}\frac{1}{\varepsilon - \hat{H}_0}\hat{V}_{int}\frac{1}{\varepsilon - \hat{H}_0} + \ldots \end{aligned} \tag{4.13}$$

and take into account the block structure (4.6) of the Hamiltonian. The even terms of (4.13) then contribute to (4.11) and yield the probability amplitudes to find the system at the levels of the first band, whereas the odd terms correspond to (4.12), that is, to the second band. This yields

$$\begin{aligned}\psi_n(\varepsilon) &= \frac{i\delta_n^0}{\varepsilon - \Delta_0} + i\frac{1}{\varepsilon - \Delta_0}V_0^m\frac{1}{\varepsilon - \Delta_m}V_m^n\frac{1}{\varepsilon - \Delta_n} + \ldots \\ \psi_m(\varepsilon) &= i\frac{1}{\varepsilon - \Delta_0}V_0^m\frac{1}{\varepsilon - \Delta_m} \\ &+ i\frac{1}{\varepsilon - \Delta_0}V_0^{m'}\frac{1}{\varepsilon - \Delta_{m'}}V_{m'}^n\frac{1}{\varepsilon - \Delta_n}V_n^m\frac{1}{\varepsilon - \Delta_m} + \ldots, \end{aligned} \tag{4.14}$$

where the summation over repeating indices is implicit.

Diagrams

Comparison of (4.14) and (4.4) suggests that the amplitudes of the states $\psi_n(\varepsilon)$ and $\psi_m(\varepsilon)$ are sums of the contributions of all possible trajectories that start at the state $|0\rangle$, that is, $|n = 0\rangle$ of the first band and end at the states $|n\rangle$ or $|m\rangle$, respectively. In Fig. 4.4 we depict the diagrams representing the sixth order perturbation term for $\psi_n(\varepsilon)$ and the fifth order perturbation term for $\psi_m(\varepsilon)$ with the help of the same notation that has been employed for the level–band problem.

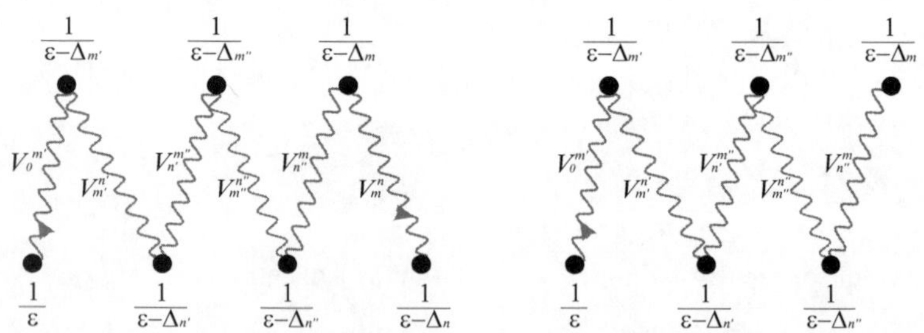

Fig. 4.4

In contrast to the diagrams of Figs. 4.1 and 4.2 the trajectories in the two-band system do not have to return to the same level because of the interaction topology and indeed on the contrary, they travel over all possible intermediate states. Therefore generic terms of the series (4.14) contain the majority of factors V_n^m in the first power. This is also true for terms of the series for populations such as those depicted in Fig. 4.5. This has a dramatic consequence for the ensemble

Fig. 4.5

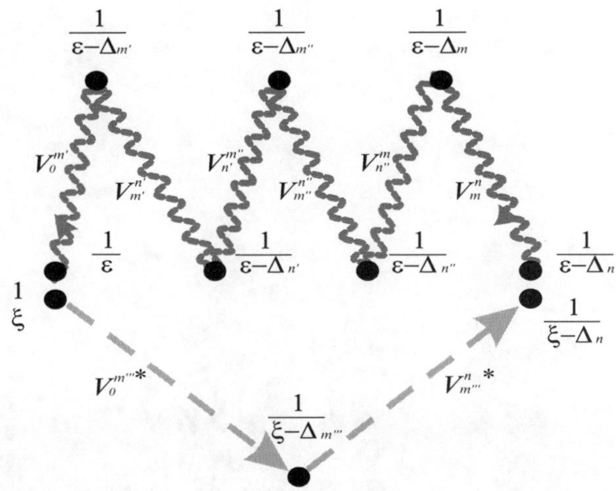

averaged quantities, since after the average over an ensemble of matrices $\hat{V}_{int}$ with a symmetric, say Gaussian, distribution of the matrix elements V_n^m the majority of terms in the perturbation series vanish. Indeed, if a term of the perturbation series contains even a single matrix element V_n^m without its complex conjugate counterpart $(V_n^m)^*$, it is an odd function which yields zero after the ensemble average with an even distribution. It is worth mentioning that if the mean matrix element $\langle V_n^m \rangle$ differs from zero, then the mean interaction $\langle \hat{V}_{int} \rangle$ can be attributed to the unperturbed Hamiltonian $\hat{H}_0 \to \hat{H}_0 + \langle \hat{V}_{int} \rangle$, such that the ensemble average of the rest of the perturbation $\hat{V}_{int} - \langle \hat{V}_{int} \rangle$ would yield zero for the mean values of the matrix elements.

The Diagram Topology

We must now consider the structure of the non-vanishing terms. Generally speaking their structure is complex, once the long-time asymptotic behavior is concerned, as detailed later. Here, however, we concentrate on the regime $t \ll g$ where the structure may be readily determined. Intuitively we understand that during a finite time interval only a finite number of transitions can occur in the two-band system, as was the case in the level–band system previously discussed. If the interval is short enough, the number of transitions is much smaller than the total number of levels involved in the population dynamics, and hence the diagram has enough space for such a location where the number of self-intersections is not larger than that needed to sustain the ensemble average. An example of a diagram of 22-nd order for the amplitude is shown in Fig. 4.6a. In such a diagram each transition

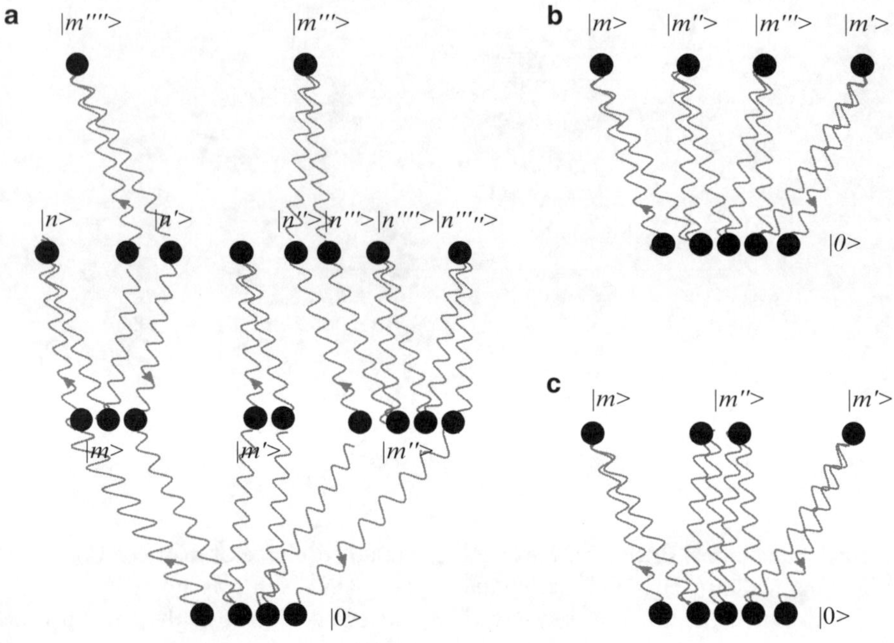

Fig. 4.6

$n \to m$ occurs together with its counterpart—the transition $m \to n$ in the opposite direction, which yields a factor $\langle V_m^n V_n^m \rangle$ after the ensemble average. There is no need to take into account repetition of the transitions, since their contribution is small as long as the total number of transitions remains small compared to the total number $\mathcal{N}$ of the levels involved in the process. We illustrate this in Fig. 4.6b, c where the total number of different diagrams of type (c) is $\mathcal{N}$ times smaller than the number of the diagrams of type (b), whereas a contribution to the perturbation series of each particular eighth order diagram of these types is of the same order of magnitude.

We now depict the diagram for a generic, non-vanishing term of the perturbation series for the amplitude without paying attention to the particular indices of the

Fig. 4.7

visited states. It has the tree-like structure shown in Fig. 4.7. Let us convince ourselves by directly performing the ensemble averaging that the terms of the perturbation series indeed have a tree-like structure. We first consider the matrix elements that are the closest neighbors of their conjugates, that is, the parts of the perturbation series terms where the sequence of factors reads

$$\sum_m \ldots \times V_k^n \frac{1}{(\varepsilon - \Delta_n)} V_n^m \frac{1}{(\varepsilon - \Delta_m)} V_m^n \frac{1}{(\varepsilon - \Delta_n)} V_n^l \times \ldots \tag{4.15}$$

The ensemble average of this part yields

$$\begin{aligned} &\ldots \times \frac{1}{(\varepsilon - \Delta_n)} \sum_m \frac{\langle V_n^m V_m^n \rangle}{(\varepsilon - \Delta_m)} \frac{1}{(\varepsilon - \Delta_n)} \times \ldots \\ &= \ldots \times \frac{1}{(\varepsilon - \Delta_n)} \sum_m \frac{\langle V^2 \rangle}{(\varepsilon - \Delta_m)} \frac{1}{(\varepsilon - \Delta_n)} \times .., \end{aligned} \tag{4.16}$$

where by $\langle V^2 \rangle = \langle V_n^m V_m^n \rangle = \int |V_n^m|^2 g(|V_n^m|) dV_n^m$ we denote the mean square coupling corresponding to the distribution function $g(|V_n^m|)$. One can see that the ensemble average wipes off the matrix structure of the perturbation such that the remaining factor $\sum_m \langle V^2 \rangle / (\varepsilon - \Delta_m)$ can be attributed to its "root" state n, and therefore the next two matrix elements V_k^n and V_n^l separated by this factor can now be considered as neighboring. We repeat this procedure again and again until all of the transition matrix elements become coupled, as shown in Fig. 4.7.

4.1.3 The Renormalization

The tree-like structure of the diagrams after the ensemble average in the short-time regime allows one to perform an analytical summation of the perturbation series and to reduce the problem to the solution of a set of algebraic equations for the ensemble averaged sums and average characteristics of the system that do not contain parameters of particular systems comprising the ensemble. The key idea of the approach employs the fact that each "branch" of a "tree" has exactly the same topology as the entire "tree".

Equations for the Averaged Amplitudes

Indeed, let us denote by $X_m(\varepsilon)$ the sum of all possible tree-like diagrams that start at a level m and return to the level m of the second band, and let us concentrate on the structure of diagrams that start at a specific level n of the first band. The perturbation series for the amplitudes with the allowance of the fact that a branch also has this tree-like topology and therefore can be denoted by $X_m(\varepsilon)$, takes the form shown in Fig. 4.8 where we denote $\langle V^2 \rangle$ by V^2. The first term of the series

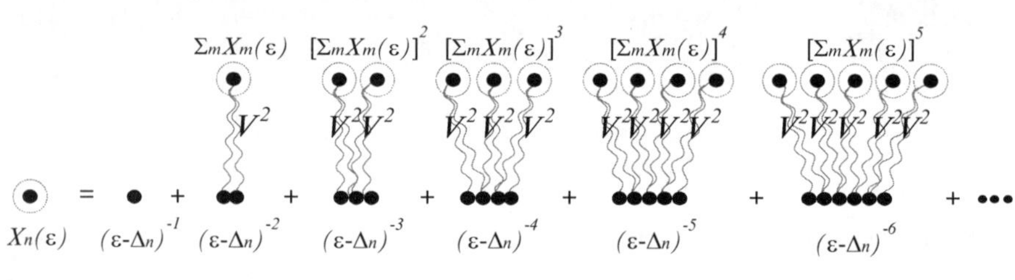

Fig. 4.8

contains no transitions and yields just the factor $1/(\varepsilon - \Delta_n)$ corresponding to the level n of the first band. The first transition starts at the level n, goes to a level m of the second band and returns back. It yields the factor $\langle V^2 \rangle / (\varepsilon - \Delta_m)(\varepsilon - \Delta_n)^2$, which can be combined with all other non-vanishing terms of the perturbation series that pass through the state m and return only once to the state n, thus yielding the combination

$$\frac{\langle V_n^m X_m(\varepsilon) V_m^n \rangle}{(\varepsilon - \Delta_n)^2} = \frac{\langle V^2 \rangle X_m(\varepsilon)}{(\varepsilon - \Delta_n)^2}. \tag{4.17}$$

Summation over m yields the second term of the series shown in Fig. 4.8. The next term combines all the diagrams that return to the state n twice, and thus the contribution reads

$$\frac{1}{\varepsilon-\Delta_n}\left\langle\sum_m V_n^m X_m(\varepsilon) V_m^n\right\rangle \frac{1}{\varepsilon-\Delta_n}\left\langle\sum_{m'} V_n^{m'} X_{m'}(\varepsilon) V_{m'}^n\right\rangle \frac{1}{\varepsilon-\Delta_n}$$

$$=\frac{1}{\varepsilon-\Delta_n}\langle V^2\rangle \sum_m X_m(\varepsilon) \frac{1}{\varepsilon-\Delta_n}\langle V^2\rangle \sum_{m'} X_{m'}(\varepsilon) \frac{1}{\varepsilon-\Delta_n}, \tag{4.18}$$

whereas the generic term of the series corresponding to k returns has the form

$$\frac{1}{(\varepsilon-\Delta_n)^{k+1}}\langle V^2\rangle^k\left[\sum_m X_m(\varepsilon)\right]^k. \tag{4.19}$$

One sees that the generic term (4.19) for the amplitudes in the two-band problem has exactly the same structure as the term (4.1) with the factor $\langle V^2\rangle X_m(\varepsilon)$ playing the role of the factor $V_0^n V_n^0/(\varepsilon-\Delta_n)$ and $\varepsilon-\Delta_n \rightarrow \varepsilon$. The sum $X_n(\varepsilon)$ of the geometric series (4.19) can therefore be given in terms of similar sums $X_m(\varepsilon)$, that is,

$$X_n(\varepsilon)=\sum_k \frac{1}{(\varepsilon-\Delta_n)^{k+1}}\langle V^2\rangle^k\left[\sum_m X_m(\varepsilon)\right]^k$$

$$=\frac{1}{\varepsilon-\Delta_n-\langle V^2\rangle \sum_m X_m(\varepsilon)}. \tag{4.20}$$

By analogy

$$X_m(\varepsilon)=\frac{1}{\varepsilon-\Delta_m-\langle V^2\rangle \sum_n X_n(\varepsilon)}, \tag{4.21}$$

which together with (4.20) yields a set of two nonlinear algebraic equations

$$Q_1(\varepsilon)=\sum_n \frac{1}{\varepsilon-\Delta_n-\langle V^2\rangle Q_2(\varepsilon)}$$

$$Q_2(\varepsilon)=\sum_m \frac{1}{\varepsilon-\Delta_m-\langle V^2\rangle Q_1(\varepsilon)}, \tag{4.22}$$

for the two sums $Q_1(\varepsilon)=\sum_n X_n(\varepsilon)$ and $Q_2(\varepsilon)=\sum_m X_m(\varepsilon)$.

Solutions of (4.22) depend on the structure of bands 1 and 2, and several different examples will be considered later in this chapter. A general conclusion emerging from the consideration of the two-band system should already be mentioned: as the result of the perturbation of a two-band system by a matrix of random couplings, the resonant denominators of the fractions $1/(\varepsilon-\Delta_{n;m})$ that allow for the poles at the positions of energy eigenvalues $\Delta_{n;m}$ adopt the form $1/(\varepsilon - \Delta_{n;m} - \langle V^2 \rangle Q_{2;1}(\varepsilon))$, which means that all of the energy eigenvalues of the same band acquire an imaginary part $\langle V^2 \rangle Q_1(\varepsilon)$ or $\langle V^2 \rangle Q_1(\varepsilon)$, depending on the band number, which are solutions of the set (4.22) of two nonlinear equations. One can refer to this change as renormalization of the energies by a random perturbation that in the regime $t \ll g$ results in the irreversible decay of the states.

Note that by analogy we can obtain the equations

$$X_n(\xi) = \frac{1}{\xi - \Delta_n - \langle V^2 \rangle Q_2(\xi)}; \; X_m(\xi) = \frac{1}{\xi - \Delta_m - \langle V^2 \rangle Q_1(\xi)}$$

$$Q_1(\xi) = \sum_n \frac{1}{\xi - \Delta_n - \langle V^2 \rangle Q_2(\xi)}; \; Q_2(\xi) = \sum_m \frac{1}{\xi - \Delta_m - \langle V^2 \rangle Q_1(\xi)}. \tag{4.23}$$

for the complex conjugate amplitudes that coincide with (4.20)–(4.22). However choosing the solution of these equations we should remember that the integration contour of the inverse Fourier transformation over ξ does not go above the real axis, as in Fig. 3.2, but below it, and therefore the solutions for $X_{n;m}(\xi)$ and for $Q_{1;2}(\xi)$ are the complex conjugates of those for ε.

Expressions for Averaged Populations

Let us now turn to the perturbation series for the populations. A diagram representing a non-vanishing term of the perturbation series for the populations is shown in Fig. 4.9. Comparing this diagram with the diagrams for a level–band system

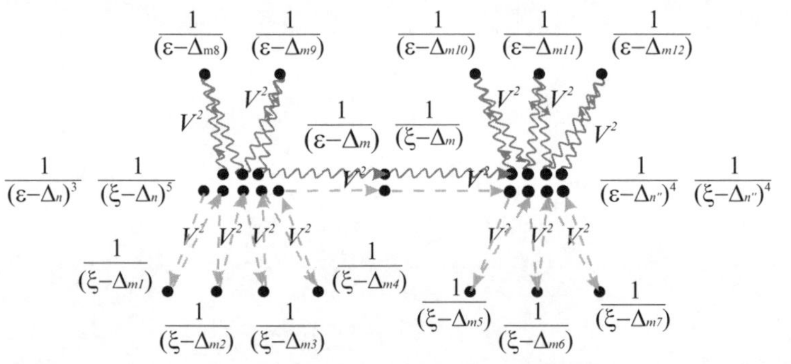

Fig. 4.9

of Fig. 4.3 one sees a new important feature—the non-vanishing contribution may come from an undulating and dashed lines connecting the same levels n and m. Indeed such a diagram contains the product $V_n^m \left(V_n^m\right)^* = V_n^m V_m^n$, since the series for the complex conjugate amplitudes have complex conjugate matrix elements, and hence the parallel transition for two different complex conjugated amplitudes yields exactly the same mean value $\langle V^2 \rangle$ as two opposite transitions for each of them. In Fig. 4.10 we show a higher-order diagram that demonstrates the typical structure of

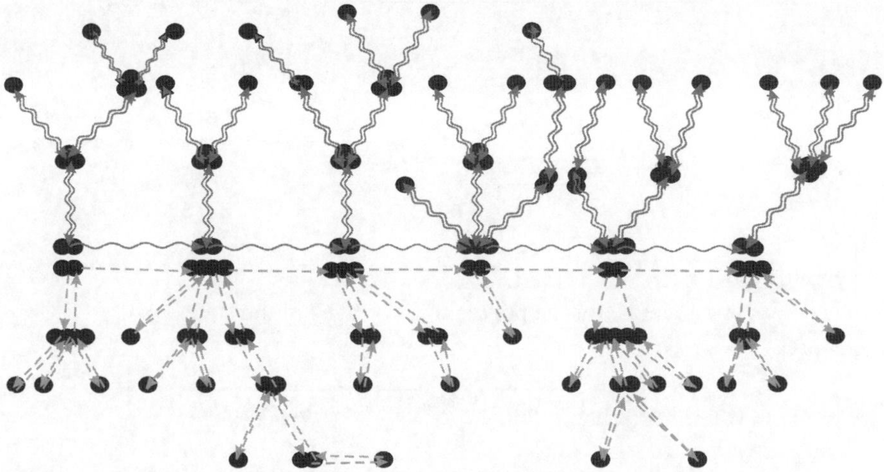

Fig. 4.10

non-vanishing terms—there are two trees "growing" from each level that correspond to the amplitude and conjugate amplitude series, and these trees are connected by coupled lines of different amplitudes.

Employing the notation of (4.20)–(4.23) for the renormalized energies allows one to draw the diagram Fig. 4.10 summed over all possible trees, which takes the form shown in Fig. 4.11.

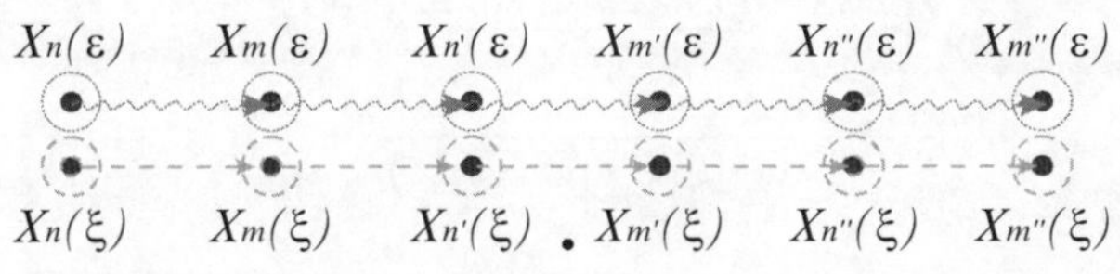

Fig. 4.11

We can now draw the diagram series for the populations shown in Fig. 4.12.

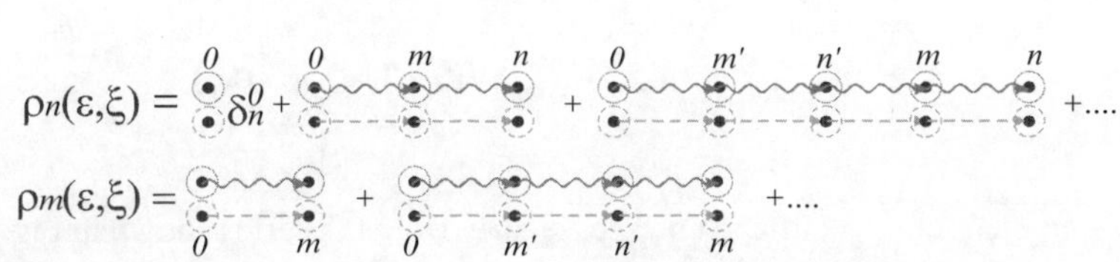

Fig. 4.12

All even terms correspond to band 1, and the odd terms correspond to band 2. One can sum up these perturbation series directly, since they constitute a geometric series with the common factor

$$\Xi(\varepsilon,\xi) = \sum_{n',m'} X_{n'}(\varepsilon)X_{n'}(\xi)\left\langle V^2\right\rangle X'_m(\varepsilon)X_{m'}(\xi)\left\langle V^2\right\rangle \tag{4.24}$$

Fig. 4.13

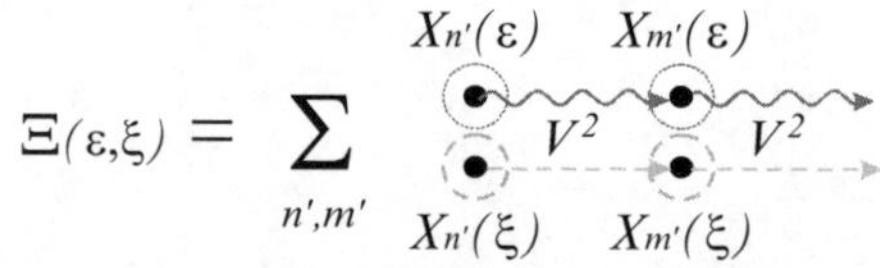

shown in Fig. 4.13.

The summation yields, for the population $\rho_m(\varepsilon,\xi)$ of the states $|m\rangle$,

$$\rho_m(\varepsilon,\xi) = \frac{1}{\varepsilon - \Delta_m - V^2Q_1(\varepsilon)}\frac{1}{\xi - \Delta_m - V^2Q_1(\xi)} \frac{V^2}{1-\Xi(\varepsilon,\xi)}\frac{1}{\xi - V^2Q_2(\xi)}\frac{1}{\varepsilon - V^2Q_2(\varepsilon)}, \tag{4.25}$$

where we write V^2 instead of $\left\langle V^2\right\rangle$ to shorten the notation. We also note an evident albeit remarkable consequence of such a structure of the series

$$\begin{aligned}\rho_m(\varepsilon,\xi) &= \frac{1}{\varepsilon - \Delta_m - V^2Q_1(\varepsilon)}\frac{1}{\xi - \Delta_m - V^2Q_1(\xi)}V^2\sum_n \rho_n(\varepsilon,\xi)\\ \rho_n(\varepsilon,\xi) &= \frac{1}{\varepsilon - \Delta_n - V^2Q_2(\varepsilon)}\frac{1}{\xi - \Delta_n - V^2Q_2(\xi)}V^2\sum_m \rho_m(\varepsilon,\xi)\\ &\quad+\delta_0^n\frac{1}{\xi - V^2Q_2(\xi)}\frac{1}{\varepsilon - V^2Q_2(\varepsilon)},\end{aligned} \tag{4.26}$$

Fig. 4.14

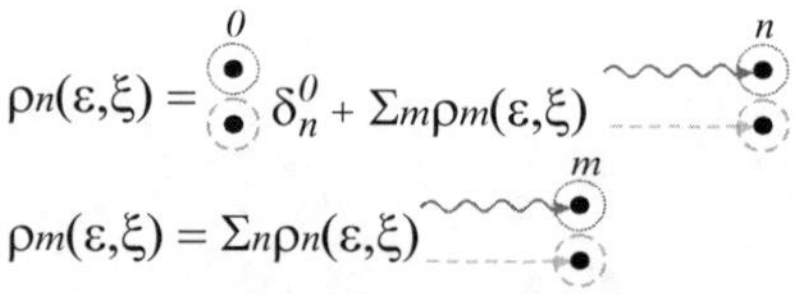

which is depicted in Fig. 4.14 and emerges clearly from the series shown in Fig. 4.12. It has an explicit physical meaning—the population of any state of a band forms as a result of transitions from all populated states of the other band.

Equation (4.26) allows one to find the total populations $\rho_1(\varepsilon,\xi) = \sum_n \rho_n(\varepsilon,\xi)$ of the first and $\rho_2(\varepsilon,\xi) = \sum_m \rho_m(\varepsilon,\xi)$ of the second bands after taking the sums over n for the first equation, and over m for the second one. With the allowance of (4.20)–(4.23) this yields the equations

$$\rho_1(\varepsilon,\xi) = \left(V^2 \sum_n X_n(\varepsilon) X_n(\xi)\right) \rho_2(\varepsilon,\xi) + X_0(\varepsilon) X_0(\xi),$$

$$\rho_2(\varepsilon,\xi) = \left(V^2 \sum_m X_m(\varepsilon) X_m(\xi)\right) \rho_1(\varepsilon,\xi), \qquad (4.27)$$

that have the solutions

$$\rho_1(\varepsilon,\xi) = \frac{X_0(\varepsilon) X_0(\xi)}{1-\left(V^2 \sum_n X_n(\varepsilon) X_n(\xi)\right)\left(V^2 \sum_m X_m(\varepsilon) X_m(\xi)\right)}$$

$$\rho_2(\varepsilon,\xi) = \frac{X_0(\varepsilon) X_0(\xi)\left(V^2 \sum_m X_m(\varepsilon) X_m(\xi)\right)}{1-\left(V^2 \sum_n X_n(\varepsilon) X_n(\xi)\right)\left(V^2 \sum_m X_m(\varepsilon) X_m(\xi)\right)}, \qquad (4.28)$$

and for the population difference $\Delta\rho(\varepsilon,\xi) = \rho_1(\varepsilon,\xi) - \rho_2(\varepsilon,\xi)$ one immediately obtains

$$\Delta\rho(\varepsilon,\xi) = \frac{\left(1 - V^2 \sum_m X_m(\varepsilon) X_m(\xi)\right) X_0(\varepsilon) X_0(\xi)}{1-\left(V^2 \sum_n X_n(\varepsilon) X_n(\xi)\right)\left(V^2 \sum_m X_m(\varepsilon) X_m(\xi)\right)}. \qquad (4.29)$$

4.2 Non-degenerate Bands

We are now in the position to consider a particular realization of the two-band system. We start with the simplest example of two statistically equivalent bands, which are the bands with the same mean density of states, shown in Fig. 4.15, and assume that at $t = 0$ only the state $n = 0$ is populated. We will postpone the important generalization to the case of two different bands until the end of this section, since such a system behaves in qualitatively the same way, although it requires more awkward calculations.

4.2.1 General Remarks and the Main Questions

As earlier, we use the energy of the state $|n = 0\rangle$ as a reference point for the energy scale, that is, we put $\Delta_{n=0} = 0$. Note that in order to find a solution for a different initial condition, the superposition principle for the initial populations can be employed due to the fact that the ensemble averaging procedure wipes out all of the interference effects related to the initial conditions. This circumstance

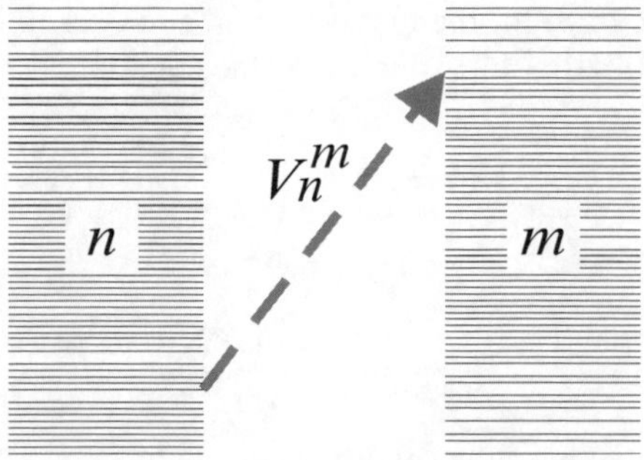

Fig. 4.15 Two-band system with statistically equivalent bands of identical mean density of states $g_1(\Delta) = g_2(\Delta) = const.$

is related directly to the tree-like topology of the terms sustaining the ensemble average, which eliminates completely the diagrams with dashed and undulating lines starting at different levels, namely the diagrams responsible for the interference.

We address here two main questions: (i) what are the total populations of the bands as a function of time, and (ii) in which way are these populations distributed among the levels of the bands? The first question can also be posed differently. One can establish a law, which would govern the total population dynamics, and which would have a simple mathematical form.

The second question may have two qualitatively different answers. Indeed, considering the level–band system in the previous chapter, we have seen that the distribution of the populations among the levels of the band follows the Lorentzian profile (3.35) shown in Fig. 3.5, which has a width given by the level decay rate W. For that case the population of a band state $|n\rangle$ has only one possibility to change, namely to return back to the initial level $|0\rangle$, whereas in the two-band system it has other possibilities—the population arriving at a state $|m\rangle$ of the second band from a state $|n\rangle$ of the first band may, in principle return back to any other state $|n'\rangle$ of the band 1. What happens to the width of the population distribution in this case? Does it become broader by a value W after each subsequent transition from band to band as should be the case if all the phase information is washed away by the ensemble average, or in spite of the ensemble average does the system keep its phase information in such a way that the population distribution over the band remains localized in a finite strip?

4.2.2 Renormalized Energies and the Population Distribution

We start by considering the population distribution over the bands and first solve the Eqs. (4.20)–(4.22) for the case $t \ll g$, which allows one to replace the summation by integration. We also note that for the statistically equivalent bands $Q_1(\varepsilon) = Q_2(\varepsilon) = Q(\varepsilon)$ and $Q_1(\xi) = Q_2(\xi) = Q(\xi)$, and therefore the equations for $Q(\varepsilon)$ and $Q(\xi)$ take the form

$$Q(\varepsilon) = \int_{-\Gamma}^{\Gamma} \frac{g d\Delta}{\varepsilon - \Delta - \langle V^2 \rangle Q(\varepsilon)}$$

$$Q(\xi) = \int_{-\Gamma}^{\Gamma} \frac{g d\Delta}{\xi - \Delta - \langle V^2 \rangle Q(\xi)}, \tag{4.30}$$

where $\Gamma \rightarrow \infty$ is the band width, and g is the state density. We perform the integration and arrive at

$$Q(\varepsilon) = -g \ln \frac{\Gamma - \langle V^2 \rangle Q(\varepsilon) + \varepsilon}{-\Gamma - \langle V^2 \rangle Q(\varepsilon) + \varepsilon}$$

$$Q(\xi) = -g \ln \frac{\Gamma - \langle V^2 \rangle Q(\xi) + \xi}{-\Gamma - \langle V^2 \rangle Q(\xi) + \xi}. \tag{4.31}$$

In the limit $\Gamma \rightarrow \infty$ the numerators of the fractions in the arguments of logarithms are large and positive, and the denominators are large and negative, thus the fraction itself tends to -1. But $\ln(-1)$ may take both values $\pm i\pi$, depending on the direction of the circumvention of the branching point $x = 0$ of $\ln(x)$, which depends in turn on the sign of the imaginary part of the fraction, that is, on the locations of zeros of the denominators of $X(\varepsilon)$ and $X(\xi)$ on the complex plane. We remember that ε has a positive imaginary part, and ξ has a negative one, and therefore the poles of $X(\varepsilon)$ as well as the zeros of $\Gamma - \langle V^2 \rangle Q(\varepsilon) + \varepsilon$ are in the negative imaginary part of the complex plane, whereas for ξ the opposite is the case. This yields in the limit $\Gamma \rightarrow \infty$

$$Q(\varepsilon) = -i\pi g$$
$$Q(\xi) = i\pi g \tag{4.32}$$

and hence, according to (4.20)–(4.23)

$$\begin{aligned} X_n(\varepsilon) = X_m(\varepsilon) = X(\varepsilon) &= \frac{1}{\varepsilon - \Delta_{n;\,m} + i\pi \langle V^2 \rangle g}, \\ X_n(\xi) = X_m(\xi) = X(\xi) &= \frac{1}{\xi - \Delta_{n;m} - i\pi \langle V^2 \rangle g}. \end{aligned} \tag{4.33}$$

One sees that the energy renormalization results in simply adding an imaginary contribution, with the sign changing as a function of the sign of the imaginary part of the variable ε or ξ.

We can now find the sum

$$\sum_n X_n(\varepsilon) X_n(\xi) = \int_{-\Gamma}^{\Gamma} \frac{g d\Delta}{(\varepsilon - \Delta + i\pi \langle V^2 \rangle g)(\xi - \Delta - i\pi \langle V^2 \rangle g)} \tag{4.34}$$

which enters (4.28) and 4.29). For $\Gamma \to \infty$ we calculate the residual at the upper part of the complex plane and arrive at

$$\sum_n X_n(\varepsilon)X_n(\xi) = \frac{-2\pi i g}{\xi - \varepsilon - 2i\pi \left\langle V^2 \right\rangle g}, \tag{4.35}$$

which also gives the identical value of the sum for the other, statistically equivalent band $\sum_n X_n(\varepsilon)X_n(\xi) = \sum_m X_m(\varepsilon)X_m(\xi)$. After substitution of these sums and (4.32) to (4.25) with the allowance of (4.24) one obtains

$$\rho_m(\varepsilon,\xi) = \frac{1}{\varepsilon - \Delta_m + i\pi g V^2} \frac{1}{\xi - \Delta_m - i\pi g V^2}$$
$$\frac{V^2}{1 + \left(\frac{2\pi g V^2}{\xi - \varepsilon - 2i\pi V^2 g}\right)^2} \frac{1}{\xi - i\pi g V^2} \frac{1}{\varepsilon + i\pi g V^2}, \tag{4.36}$$

where as before we write V^2 instead of $\left\langle V^2 \right\rangle$ to shorten the notation. After double inverse Fourier transformation for the time-dependent population of the state $|m\rangle$ of the second band we therefore obtain

$$\rho_m(t) = \frac{1}{4\pi^2} \int_{-\infty - i\nu}^{\infty - i\nu} d\xi \int_{-\infty + i\nu}^{\infty + i\nu} d\varepsilon \; \frac{\exp\left[i(\xi - \varepsilon)t\right]}{(\xi - i\pi g V^2)(\varepsilon + i\pi g V^2)(\xi - \varepsilon)}$$
$$\frac{V^2\left(\xi - \varepsilon - 2i\pi V^2 g\right)^2}{(\xi - \varepsilon - 4i\pi V^2 g)(\varepsilon - \Delta_m + i\pi g V^2)(\xi - \Delta_m - i\pi g V^2)}. \tag{4.37}$$

By analogy to (3.114) we introduce the variables $\zeta = \xi - \varepsilon$, $\eta = (\varepsilon + \xi)/2$, and perform the integration over $d\eta$, which yields the spectral density of the population $\rho(t,\Delta) = \rho_m(t)\,g$ corresponding to the detuning $\Delta_m = \Delta$:

$$\rho(t,\Delta) = \frac{1}{\pi^2} \int_{-\infty + i\nu}^{\infty + i\nu} d\zeta \; \frac{\exp\left[i\zeta t\right]}{\zeta(\zeta - 2iW)} \frac{W(W + i\zeta)}{(\zeta - iW - \Delta)(\zeta - iW + \Delta)}, \tag{4.38}$$

where $W = 2\pi V^2 g$. Performing integration over $d\zeta$ one obtains the answer to question (ii):

$$\rho(t,\Delta) = \frac{W}{\pi\left(W^2 + \Delta^2\right)}\left(1 + e^{-2Wt} - 2e^{-Wt}\cos(\Delta t)\right), \tag{4.39}$$

which indicates clearly that a Lorentzian distribution $W/\pi\left(W^2 + \Delta^2\right)$ of width W attains in the regime $t > 1/W$. Note that for the two-band problem (4.39) gives exactly the same distribution over the levels of the second band as the distribution following from (3.33) for the level–band problem, apart from the fact that the distribution (4.39) is twice as broad and four times smaller than it should be, bearing in mind that for two bands the total sate density is twice as big, and the total population is shared between two bands. We also note that for two statistically equivalent bands the same Lorentzian distribution of populations attains in the first band, although with a slightly different time-dependence:

$$\rho\left(t,\Delta\right)=\frac{W}{\pi\left(W^{2}+\Delta^{2}\right)}\left(1-e^{-2Wt}-2We^{-Wt}\frac{\sin(t\Delta)}{\Delta}\right)+e^{-Wt}\delta(\Delta). \tag{4.40}$$

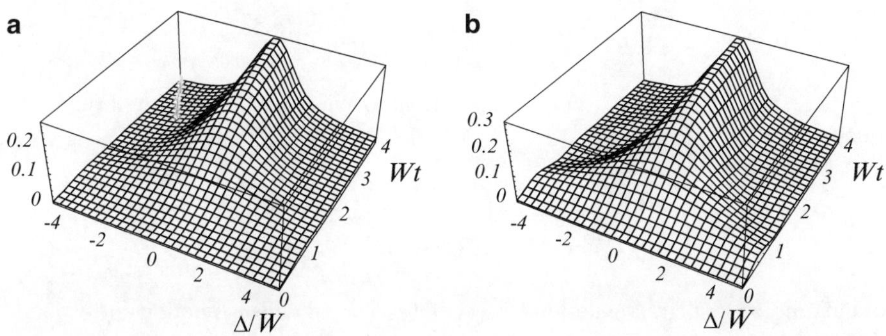

Fig. 4.16 Population distribution over the first (**a**) and the second (**b**) bands. The initial δ-like distribution of the state 0 is not shown

In Fig. 4.16 we depict the distributions (4.39) and (4.40). One sees that both population distributions which can be broad initially become localized in the W-vicinity of the resonance $\Delta = 0$ as time passes.

4.2.3 Dynamics of the Total Populations of Bands

Let us now return to question (i) and consider the dynamics of the population distribution between two bands. Equations (4.27) for the total populations with use of (4.33) and (4.35) read

$$\rho_1\left(\varepsilon,\xi\right)=\frac{-2\pi igV^2}{\xi-\varepsilon-2i\pi V^2g}\rho_2\left(\varepsilon,\xi\right)+\frac{1}{\xi-i\pi gV^2}\frac{1}{\varepsilon+i\pi gV^2},$$

$$\rho_2\left(\varepsilon,\xi\right)=\frac{-2\pi igV^2}{\xi-\varepsilon-2i\pi V^2g}\rho_1\left(\varepsilon,\xi\right), \tag{4.41}$$

As earlier, by analogy with (3.114) we introduce the variables $\zeta = \varepsilon - \xi$, $\eta = (\varepsilon + \xi)/2$, and perform the integration over $d\eta$, which yields

$$\rho_1(\zeta) = \frac{-2\pi i g V^2}{\zeta - 2i\pi V^2 g}\rho_2(\zeta) + \frac{i}{\zeta - i2\pi g V^2},$$
$$\rho_2(\zeta) = \frac{-2\pi i g V^2}{\varsigma - 2i\pi V^2 g}\rho_1(\zeta), \tag{4.42}$$

which for $W = 2\pi V^2 g$ takes the form

$$-i\zeta\rho_1(\zeta) = W[\rho_1(\zeta) - \rho_2(\zeta)] + 1,$$
$$-i\zeta\rho_2(\zeta) = W[\rho_2(\zeta) - \rho_1(\zeta)], \tag{4.43}$$

and after the inverse Fourier transformation over ζ results in the differential equations

$$\dot{\rho}_1(t) = -W[\rho_1(t) - \rho_2(t)] + \delta(t),$$
$$\dot{\rho}_2(t) = -W[\rho_2(t) - \rho_1(t)], \tag{4.44}$$

for the time-dependent populations $\rho_1(t)$ and $\rho_2(t)$ with the δ-function corresponding to the initial condition $\rho_1(t = 0) = 1$.

One immediately recognizes in (4.44) the balance equation, or master equation, and readily finds the population difference

$$\rho_1(t) - \rho_2(t) = e^{-2Wt}, \tag{4.45}$$

whereas for the sum of the populations one obtains

$$\dot{\rho}_1(t) + \dot{\rho}_2(t) = \delta(t), \tag{4.46}$$

which means $\rho_1(t) + \rho_2(t) = 1$ for $t > 0$.

Let us summarize the physical meaning of the results obtained. The total populations of the first and the second band obey the master equation, such that if at $t = 0$ the population was localized at a single level of the first band, the population difference $\rho_1(t) - \rho_2(t)$ decreases exponentially from 1 to 0, with a rate $W = 4\pi V^2 g$ given by the product of the mean squared interaction V^2 and the state density g. After a time $t \sim 1/W$ both bands have a total population close to $1/2$. One may think that this irreversible dynamic is a signature of the loss of coherence in the system. However, the distribution of the populations over the levels of the bands does not become broader with the course of time but on the contrary becomes localized towards a Lorentzian profile of width W which implies that the coherence in the system is present at all times and plays the dominating role in spite of the exponential character of the total population dynamics. The exponential decay is therefore just a transient process that takes place during a limited time interval $t \sim g$, as long as the recurrences do not yet affect the population dynamics. We therefore note an evident condition of the applicability of this approach—there must be many band levels in a strip of the width W of the population distribution. We find this number by multiplying the width $W = 2\pi V^2 g$ by the state density g, so that the condition reads $2\pi V^2 g^2 \gg 1$. This also implies that the recurrence time $T = g$ must be much longer as compared to the typical decay time $1/W$, and that the typical interaction $\sqrt{V^2}$ exceeds the mean distance $\delta \sim 1/g$ among the neighboring band levels. All these conditions allow one to employ the continuous-band model of (4.30).

4.2.4 Different Bands

Let us now consider populations $\rho_n(\varepsilon, \xi)$ and $\rho_m(\varepsilon, \xi)$ of two different bands with the state densities g_1 and g_2, respectively. Equation (4.23) yields

$$
\begin{aligned}
X_n(\varepsilon) &= \frac{1}{\varepsilon - \Delta_n - \langle V^2 \rangle Q_2(\varepsilon)} \\
X_m(\varepsilon) &= \frac{1}{\varepsilon - \Delta_m - \langle V^2 \rangle Q_1(\varepsilon)} \\
X_n(\xi) &= \frac{1}{\xi - \Delta_n - \langle V^2 \rangle Q_2(\xi)} \\
X_m(\xi) &= \frac{1}{\xi - \Delta_m - \langle V^2 \rangle Q_1(\xi)}
\end{aligned}
\tag{4.47}
$$

where by analogy to (4.30) for the quantities Q we have explicit expressions resulting from the set of equations

$$
\begin{aligned}
Q_1(\varepsilon) &= \int_{-\Gamma}^{\Gamma} \frac{g_1 d\Delta}{\varepsilon - \Delta - \langle V^2 \rangle Q_2(\varepsilon)} = -i\pi g_1, \\
Q_2(\varepsilon) &= \int_{-\Gamma}^{\Gamma} \frac{g_2 d\Delta}{\varepsilon - \Delta - \langle V^2 \rangle Q_1(\varepsilon)} = -i\pi g_2, \\
Q_1(\xi) &= \int_{-\Gamma}^{\Gamma} \frac{g_1 d\Delta}{\xi - \Delta - \langle V^2 \rangle Q_2(\xi)} = i\pi g_1, \\
Q_2(\xi) &= \int_{-\Gamma}^{\Gamma} \frac{g_2 d\Delta}{\xi - \Delta - \langle V^2 \rangle Q_1(\xi)} = i\pi g_2,
\end{aligned}
\tag{4.48}
$$

and therefore

$$\sum_n X_n(\varepsilon)X_n(\xi) = \frac{-2\pi i g_1}{\xi - \varepsilon - 2i\pi \langle V^2 \rangle g_2}$$
$$\sum_m X_m(\varepsilon)X_m(\xi) = \frac{-2\pi i g_2}{\xi - \varepsilon - 2i\pi \langle V^2 \rangle g_1}. \tag{4.49}$$

Now (4.28), after integration over $d\eta$, reads

$$\rho_1(\zeta) = \frac{(\zeta + iW_1)}{(\zeta + i(W_1 + W_2))\zeta}$$
$$\rho_2(\zeta) = \frac{iW_2}{(\zeta + i(W_1 + W_2))\zeta}, \tag{4.50}$$

which yields for the time-dependent and density-weighted population difference

$$\rho_1(t)/g_1 - \rho_2(t)/g_2 = e^{-(W_1+W_2)t}/g_1. \tag{4.51}$$

Here $W_{1,2} = 2\pi \langle V^2 \rangle g_{1,2}$. By analogy to (4.39) and (4.40) a Lorentzian distribution of width $(W_1 + W_2)/2$ is attained at $t > 1/(W_1 + W_2)$. We also note that the total populations of the bands satisfy the master equation

$$\dot{\rho}_1(t) = -W_2\rho_1(t) + W_1\rho_2(t) + \delta(t),$$
$$\dot{\rho}_2(t) = -W_1\rho_2(t) + W_2\rho_1(t). \tag{4.52}$$

4.3 Two Degenerate Levels

The coherent character of the population dynamics discussed at the end of Sect. 4.2.3 becomes much more apparent for the case of two degenerate levels, that is, for bands with $\Delta_m = \Delta_n = 0$. Considering such a system we first make use of the results of the general analysis of Sect. 4.1, and then we present an alternative that allows one to obtain deeper insight into the random matrix model.

4.3.1 Degenerate Levels as a Complex System

Equations (4.20)–(4.23), (4.26)–(4.28) remain valid for this system, although the particular algebraic form of (4.20)–(4.23) is now different from (4.32)–(4.33). Let us assume, as earlier in (4.30), that the bands are statistically equivalent both having an identical number N of levels, and write (4.22) for $Q_{1,2}(\varepsilon) = Q(\varepsilon)$ and $Q_{1,2}(\xi) =$

$Q(\xi)$ in the form

$$Q(\varepsilon) = \frac{N}{\varepsilon - \langle V^2 \rangle Q(\varepsilon)}$$
$$Q(\xi) = \frac{N}{\xi - \langle V^2 \rangle Q(\xi)}. \tag{4.53}$$

These second-order algebraic equations yield

$$Q(\varepsilon) = \frac{\varepsilon + \sqrt{\varepsilon^2 - 4N \langle V^2 \rangle}}{2 \langle V^2 \rangle}$$
$$Q(\xi) = \frac{\xi - \sqrt{\xi^2 - 4N \langle V^2 \rangle}}{2 \langle V^2 \rangle}, \tag{4.54}$$

where the signs of the roots are chosen such that the signs of the imaginary parts of $Q(\varepsilon)$ and $Q(\xi)$ coincide with those for ε^{-1} and ξ^{-1} on the integration contours $(-\infty \pm i\nu; \infty \pm i\nu)$ of the inverse Fourier transformation respectively, since otherwise the inverse Fourier transforms of the corresponding probability amplitudes vanish. Equation (4.32) gives a hint for the proper choice.

All states of each of the degenerate bands are apparently statistically equivalent, and therefore we will not consider the population distribution amongst them but dwell on the difference of the total populations of the bands. Substitution of (4.54) into (4.20), (4.23) yields

$$X_n(\varepsilon) = X_m(\varepsilon) = X(\varepsilon) = \frac{2}{\varepsilon - \sqrt{\varepsilon^2 - 4N \langle V^2 \rangle}}$$
$$X_n(\xi) = X_m(\xi) = X(\xi) = \frac{2}{\xi + \sqrt{\xi^2 - 4N \langle V^2 \rangle}}, \tag{4.55}$$

and therefore (4.29) results in

$$\Delta\rho\,(\varepsilon, \xi) = \frac{X(\varepsilon)X(\xi)}{1 + N \langle V^2 \rangle X(\varepsilon)X(\xi)}. \tag{4.56}$$

The inverse Fourier transform of (4.56) reads

$$\Delta\rho\,(t) = \frac{1}{4\pi^2} \int\limits_{-\infty+i\nu}^{\infty+i\nu} \int\limits_{-\infty-i\nu}^{\infty-i\nu} \frac{\exp\left[i\left(\varepsilon - \xi\right)t\right] d\varepsilon d\xi}{\left(\varepsilon - \sqrt{\varepsilon^2 - 4NV^2}\right)\left(\xi + \sqrt{\xi^2 - 4NV^2}\right) + NV^2}, \tag{4.57}$$

whereas earlier we replace $\langle V^2 \rangle$ by V^2.

One can close the integration contours in the complex planes ε and ξ and transform them such that they become the closed loops C_1 and C_2 going around the branching points $\varepsilon_{1,2} = \xi_{1,2} = \pm\sqrt{4NV^2}$ as shown in Fig. 4.17a. After the replacements $\varepsilon = \sqrt{4NV^2}\sin\varphi$ and $\xi = \sqrt{4NV^2}\sin\theta$ these contours become intervals as shown in Fig. 4.17b, whereas the integral (4.57) takes the form

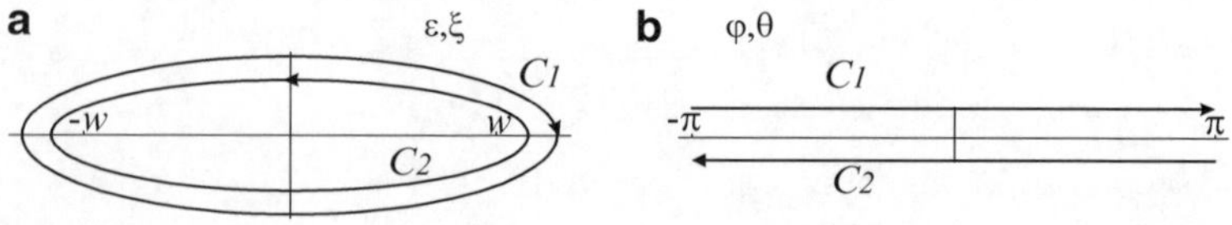

Fig. 4.17 Integration contours C_1 and C_2 in the complex planes ε and ξ respectively (**a**) go around the points $w = \pm\sqrt{4NV^2}$. The replacements $\varepsilon \to \sqrt{4NV^2}\sin\varphi$ and $\xi \to \sqrt{4NV^2}\sin\theta$ transform these contours to the *straight lines* (**b**) in the complex plane of φ and θ

$$\Delta\rho(t) = \int\limits_{C_1}\int\limits_{C_2} \frac{\exp\left[i\sqrt{4NV^2}\left(\sin\varphi - \sin\theta\right)t\right]\cos\varphi\cos\theta d\varphi d\theta}{\pi^2\left(e^{i(\varphi-\theta)}+1\right)}. \tag{4.58}$$

We first perform the integration over one of the variables, say φ, by moving the integration contour C_1 toward $\varphi \to i\infty$ along the strip $(-\pi, \pi)$, by taking the residual at the point $\varphi = \theta + \pi$, and allowing for the fact that as $\varphi \to i\infty$ the integrand vanishes, we arrive at

$$\Delta\rho(t) = \frac{2}{\pi}\int\limits_{-\pi}^{\pi} d\theta \exp\left[-i4t\sqrt{NV^2}\sin\theta\right]\cos^2\theta. \tag{4.59}$$

The integral (4.59) can be given in terms of Bessel functions. Indeed

$$J_n(z) = \frac{1}{2\pi}\int\limits_{-\pi}^{\pi} d\theta \exp\left[in\theta + iz\sin\theta\right], \tag{4.60}$$

and $2\cos^2\theta = 1 + \cos 2\theta$, and therefore

$$\Delta\rho(t) = J_0(4t\sqrt{NV^2}) + J_2(4t\sqrt{NV^2}) = \frac{J_1(4t\sqrt{NV^2})}{2t\sqrt{NV^2}}. \tag{4.61}$$

In Fig. 4.18 we show this dependence. One can see that the population difference of the first and the second bands is an oscillatory and decaying function of time. This means that the populations of these two degenerate levels become equal, that is, they indeed do tend to the limit given by the irreversible master equation, although this process does not have an exponential character, but on the contrary, has strong reminiscences of coherent behavior. Therefore such a dynamics is sometimes called coherent damping.

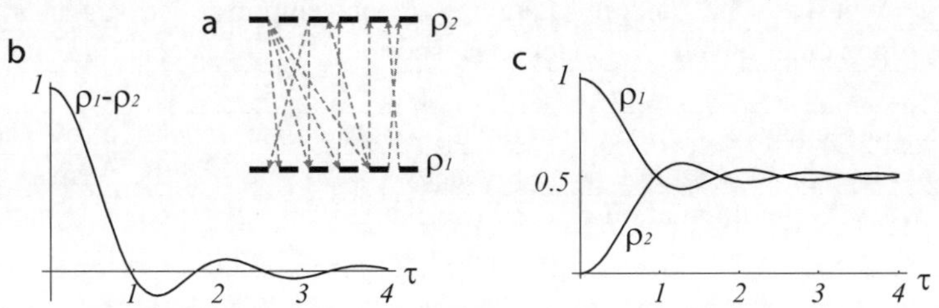

Fig. 4.18 Two degenerate levels (**a**). The *arrows* show possible transitions and correspond to different transition matrix elements V_n^m. (**b**) Population difference of two statistically equivalent degenerate bands as a function of dimensionless time $\tau = t\sqrt{NV^2}$. The population dynamics is not exponential, and is influenced strongly by the coherence, although at the limit of long times it results in the equal distribution of the population among the bands, typical of incoherent decay, as one can see from the total populations ρ_1 and ρ_2 shown in (**c**)

4.3.2 *The Bands as an Ensemble of Two-Level Systems*

Let us consider the results obtained from a different point of view. The Hamiltonian $\hat{H}$ of the system of two degenerate levels coupled to each other by a random matrix has a specific block structure:

$$\hat{H} = \begin{pmatrix} \hat{0} & \hat{V} \\ \hat{V}^+ & \hat{0} \end{pmatrix} = \begin{pmatrix} 0 & V_n^m \\ V_m^n & 0 \end{pmatrix} \tag{4.62}$$

with zero diagonal blocks. The zero matrix has an evident property—it remains zero after multiplication by any other matrix. We can therefore construct a unitary matrix

$$\hat{S} = \begin{pmatrix} \hat{s} & \hat{0} \\ \hat{0} & \hat{s}^+ \end{pmatrix} = \begin{pmatrix} s_n^{n'} & 0 \\ 0 & \left(s^+\right)_m^{m'} \end{pmatrix} \tag{4.63}$$

where the unitary submatrix $s_n^{n'}$ and its conjugate $\left(s^+\right)_m^{m'}$ diagonalize the matrix V_n^m, that is, $\sum_{n'm'} s_n^{n'} V_{n'}^{m'} \left(s^+\right)_{m'}^m = V_n \delta_n^m$, where V_n are the eigenvalues of $\hat{V}$. This implies that the matrix $\hat{V}$ is Hermitian, which is a priori not necessarily the case although we assume it in this section; and in the general case the resulting matrix is not diagonal but a triangular one. After the transformation $\hat{S}\hat{H}\hat{S}^+$ the Hamiltonian takes the form

$$\hat{H} = \begin{pmatrix} 0 & V_n \delta_n^m \\ V_n \delta_n^m & 0 \end{pmatrix} \tag{4.64}$$

with diagonal matrices in the off-diagonal blocks. This form has a physical meaning: we have represented our system of two degenerate levels coupled by a random matrix as a set of N independent two-level systems each of which has the coupling matrix element V_n. The random character of the coupling also suggests that in our new representation all the states corresponding to the level 1 are uniformly populated.

In Chap. 3 we have seen that the population of the upper level of a resonant two-level system with the coupling V is given by (3.20) and has the form $\rho_{up}(t) = \sin^2 Vt$, whereas the population of the lower level is $\rho_{lw}(t) = \cos^2 Vt$, and hence the population difference $\Delta\rho_n(t)$ corresponding to the coupling V_n reads

$$\Delta\rho_n(t) = \cos^2 V_n t - \sin^2 V_n t = \cos 2V_n t. \tag{4.65}$$

The ensemble average population is therefore given by (4.65) averaged with the distribution function of the couplings $g(V_n)$, that is,

$$\Delta\rho(t) = \int dV_n\, g(V_n) \cos 2V_n t. \tag{4.66}$$

From (4.66) we immediately obtain (4.61) by taking the distribution

$$g(V_n) = \frac{\sqrt{(V_n)^2 - 4N\left\langle V^2\right\rangle}}{2\pi N\left\langle V^2\right\rangle} \tag{4.67}$$

of the span $\pm 2\sqrt{N\left\langle V^2\right\rangle}$, which is known as the Wigner semicircular distribution (1.2) of the eigenvalues of a random matrix with Gaussian statistics of the matrix elements $g(V_n^m) = e^{-(V_n^m)^2/\langle V^2\rangle}$. Indeed, when we substitute (4.67) into (4.66) and make the replacement $V_n \to 2\sqrt{N\left\langle V^2\right\rangle}\sin\theta$ we arrive at

$$\Delta\rho(t) = \frac{2}{\pi}\int_{-\pi}^{\pi} d\theta\, \cos^2\theta \cos\left[4t\sqrt{N\left\langle V^2\right\rangle}\sin\theta\right], \tag{4.68}$$

which coincides with (4.59) if we take parity into account. This coincidence is in fact an independent confirmation of the Wigner distribution given by (4.67). We also note that the imaginary part $\mathrm{Im}\left[\frac{X(\varepsilon)}{\pi}\right] = \frac{2}{\pi}\mathrm{Im}\left[\varepsilon - \sqrt{\varepsilon^2 - 4N\langle V^2\rangle}\right]$ $= \sqrt{4N\langle V^2\rangle - \varepsilon^2}/2\pi N\langle V^2\rangle$ of the average resolvent represents the density of eigenstates at the energy $E = \varepsilon$. This part also yields the distribution (4.67), which is quite natural since the two energy eigenstates of a resonant two-level system (3.14) are at the points $E_{1,2} = \pm V_n$ given by the value of the level coupling V_n.

In Sect. 4.1, when we were considering the role of returns, we mentioned that the approach based on the assumption of the tree-like topology of the diagrams may fail when the number of transitions exceeds the number of levels involved in the process and the self-intersections of the diagrams mentioned on p. 136 start to play a role. This is also the case for the degenerate bands. One estimates the number of transitions during the time t by the product $\sqrt{\langle V^2\rangle}t$ and the number of levels is apparently N. This means that the approach is valid as long as $t < N/\sqrt{\langle V^2\rangle}$. This

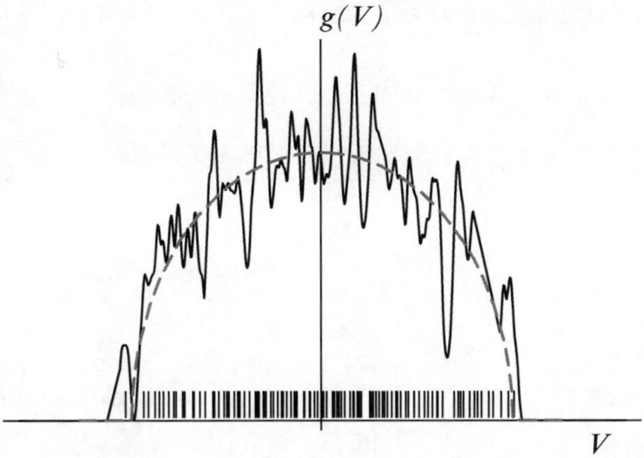

Fig. 4.19 Spectrum of eigenvalues of a 100×100 random Hermitian matrix with Gaussian distribution of the matrix elements shown by ticks on the abscissa, and the convolution of this spectrum with a Gaussian of width 0.5 (*solid line*) that gives the averaged state density. The corresponding Wigner distribution of the eigenvalues of a random matrix is shown by the *dashed line*. One sees that the eigenvalue spectrum is far from being equidistant

time exceeds considerably the mean energy eigenstate density $g \sim \sqrt{N/\langle V^2 \rangle}$ of the coupled degenerate bands, which can be estimated as the total number of states $2N$ divided by the width $2\sqrt{N\langle V^2 \rangle}$ of the distribution of the couplings (4.67).

It is expedient to consider this situation in the context of the quantum recurrences and quantum revivals discussed in Sect. 3.3 that are the physical phenomena destroying the validity of the tree-like topology approximation. If the exact energy eigenstate spectrum is equidistant, or close to equidistant, then at $t \sim g$ one observes a strong recurrence of the population to the initial state that cannot be found in the framework of the continuous-band model. But if the eigenstates are randomly placed on the energy scale, there is no recurrence at $t \sim g$, although a partial revival may take place at a longer time. The latter is rather the case for the system of two degenerate levels, where the spectrum of the coupling matrix elements V_n has no physical reasons to be equidistant for small values of V. We illustrate this in Fig. 4.19, where an example of a random matrix eigenstates is shown along with the Wigner semicircular distribution.

4.4 A Band Coupled to a Degenerate Level

We consider now one more extreme, namely a system of two different bands: an N-fold degenerate level 1 coupled to a band 2 with a constant density of states g. This system has both the features typical of the exponential dynamics of a two-non-degenerate band system and the features typical of the coherent dynamics system of two degenerate levels.

4.4.1 Total Population of the Bands

Equations (4.22) for the case of one degenerate and one uniform band reads

$$
\begin{aligned}
Q_1(\varepsilon) &= \frac{N}{\varepsilon - \langle V^2 \rangle Q_2(\varepsilon)} \\
Q_2(\varepsilon) &= \sum_m \frac{1}{\varepsilon - \Delta_m - \langle V^2 \rangle Q_1(\varepsilon)},
\end{aligned}
\tag{4.69}
$$

and one finds the equations for $Q_{1,2}(\xi)$ analogously. We consider the case $t \ll g$ and by analogy to (4.48) replace the sum by the integral, which immediately yields

$$
\begin{aligned}
Q_1(\varepsilon) &= \frac{N}{\varepsilon + i\pi \langle V^2 \rangle g} \\
Q_2(\varepsilon) &= -i\pi g,
\end{aligned}
\tag{4.70}
$$

and

$$
\begin{aligned}
Q_1(\xi) &= \frac{N}{\xi - i\pi \langle V^2 \rangle g} \\
Q_2(\xi) &= i\pi g.
\end{aligned}
\tag{4.71}
$$

We also note that

$$
\begin{aligned}
X_n(\varepsilon) &= \frac{1}{\varepsilon + i\pi \langle V^2 \rangle g} \\
X_m(\varepsilon) &= \frac{1}{\varepsilon - \Delta_m - \langle V^2 \rangle N \left(\varepsilon + i\pi \langle V^2 \rangle g\right)^{-1}} \\
X_n(\xi) &= \frac{1}{\xi - i\pi \langle V^2 \rangle g} \\
X_m(\xi) &= \frac{1}{\xi - \Delta_m - N \langle V^2 \rangle \left(\xi - i\pi \langle V^2 \rangle g\right)^{-1}},
\end{aligned}
\tag{4.72}
$$

and hence

$$
\begin{aligned}
\sum_n X_n(\varepsilon) X_n(\xi) &= \frac{N}{\left(\varepsilon + i\pi \langle V^2 \rangle g\right)\left(\xi - i\pi \langle V^2 \rangle g\right)} \\
\sum_m X_m(\varepsilon) X_m(\xi) &= \frac{-2\pi i g}{\xi - \varepsilon - \langle V^2 \rangle \left(Q_1(\xi) - Q_1(\varepsilon)\right)},
\end{aligned}
\tag{4.73}
$$

where

$$Q_1(\xi) - Q_1(\varepsilon) = \frac{N\left(\varepsilon - \xi + 2i\pi \left\langle V^2 \right\rangle g\right)}{\left(\varepsilon + i\pi \left\langle V^2 \right\rangle g\right)\left(\xi - i\pi \left\langle V^2 \right\rangle g\right)}. \tag{4.74}$$

We substitute (4.73), (4.74) into (4.29) and after straightforward algebraic transformations arrive at

$$\begin{aligned} \Delta\rho\left(\varepsilon,\xi\right) = &\left(\frac{1}{\varepsilon + i\pi \left\langle V^2 \right\rangle g} - \frac{1}{\xi - i\pi \left\langle V^2 \right\rangle g}\right)\frac{1}{\xi - \varepsilon} \\ &+ \frac{4\pi i g \left\langle V^2 \right\rangle}{\left(\varepsilon + i\pi \left\langle V^2 \right\rangle g\right)\left(\xi - i\pi \left\langle V^2 \right\rangle g\right) + \left\langle V^2 \right\rangle N}\frac{1}{\xi - \varepsilon}. \end{aligned} \tag{4.75}$$

We now note that $i\left(\xi - \varepsilon\right)\Delta\rho\left(\varepsilon,\xi\right)$ is the Fourier transform of the time derivative of the population difference, which we denote $\Delta\rho'\left(\varepsilon,\xi\right)$, and obtain

$$\begin{aligned} \Delta\rho'\left(\varepsilon,\xi\right) = &\frac{i}{\left(\varepsilon + i\pi \left\langle V^2 \right\rangle g\right)} - \frac{i}{\left(\xi - i\pi \left\langle V^2 \right\rangle g\right)} \\ &- \frac{4\pi g \left\langle V^2 \right\rangle}{\left(\varepsilon + i\pi \left\langle V^2 \right\rangle g\right)\left(\xi - i\pi \left\langle V^2 \right\rangle g\right) + \left\langle V^2 \right\rangle N}. \end{aligned} \tag{4.76}$$

The first two terms depend only on one of the two variables ε and ξ, and therefore for $t > 0$ they vanish after the inverse Fourier transformation over both ε and ξ. Hence we have

$$\frac{d}{dt}\Delta\rho\left(t\right) = \frac{-1}{4\pi^2}\int\limits_{-\infty+i\nu}^{\infty+i\nu}\int\limits_{-\infty-i\nu}^{\infty-i\nu}\frac{4\pi i g \left\langle V^2 \right\rangle \exp\left[i\left(\xi - \varepsilon\right)t\right]\, d\varepsilon d\xi}{\left(\varepsilon + i\pi \left\langle V^2 \right\rangle g\right)\left(\xi - i\pi \left\langle V^2 \right\rangle g\right) + \left\langle V^2 \right\rangle N}. \tag{4.77}$$

We now introduce the variables $\zeta = \xi - \varepsilon$, $\eta = \left(\varepsilon + \xi\right)/2$, take into account the relation $\left(\varepsilon + i\pi \left\langle V^2 \right\rangle g\right)\left(\xi - i\pi \left\langle V^2 \right\rangle g\right) = \eta^2 - \left(\zeta/2 - i\pi \left\langle V^2 \right\rangle g\right)^2$, and perform the integration over $d\eta$, which yields

$$\frac{d}{dt}\Delta\rho\left(t\right) = g\left\langle V^2 \right\rangle \int\limits_{-\infty+i\nu}^{\infty+i\nu}\frac{\exp\left[i\zeta t\right]\, d\zeta}{\sqrt{\left(\zeta/2 - i\pi \left\langle V^2 \right\rangle g\right)^2 - \left\langle V^2 \right\rangle N}}. \tag{4.78}$$

One can close the integration path, such that it becomes a closed contour going around the interval $\left(i2\pi \left\langle V^2 \right\rangle g - 2\sqrt{\left\langle V^2 \right\rangle N}, i2\pi \left\langle V^2 \right\rangle g + 2\sqrt{\left\langle V^2 \right\rangle N}\right)$ connecting two branching points in the complex plane ζ, and obtain after the replacement $\zeta \to i2\pi \left\langle V^2 \right\rangle g + 2\sqrt{\left\langle V^2 \right\rangle N}\cos\varphi$

$$\frac{d}{dt}\Delta\rho(t) = -4g\left\langle V^2\right\rangle \exp\left[-2\pi g\left\langle V^2\right\rangle t\right]\int\limits_{-\pi}^{\pi}\exp\left[it2\sqrt{\left\langle V^2\right\rangle N}\cos\varphi\right]d\varphi$$
$$= -4\pi g\left\langle V^2\right\rangle \exp\left[-2\pi g\left\langle V^2\right\rangle t\right] J_0\left(t\sqrt{4\left\langle V^2\right\rangle N}\right), \qquad (4.79)$$

that is,

$$\Delta\rho(t) = 1 - 4\pi g\left\langle V^2\right\rangle \int\limits_0^t \exp\left[-2\pi\left\langle V^2\right\rangle g\tau\right] J_0\left(\tau\sqrt{4\left\langle V^2\right\rangle N}\right) d\tau, \qquad (4.80)$$

where we have taken into account the initial condition $\Delta\rho(t=0) = 1$.

One sees clearly the oscillatory character of the time derivative of the population difference coming from the Bessel function in (4.79), and some reminiscences of these oscillations are seen in the dependence of the population difference $\Delta\rho(t)$ of (4.80) itself, which is shown in Fig. 4.20. This behavior is typical of the coherent

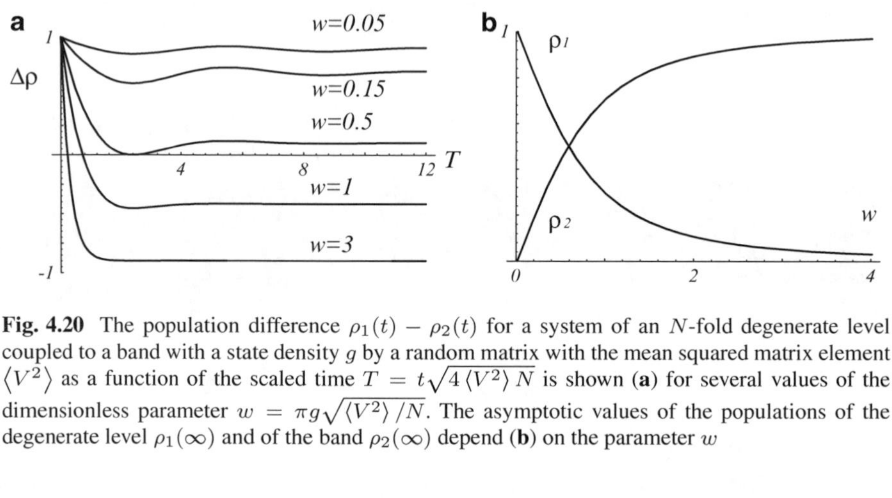

Fig. 4.20 The population difference $\rho_1(t) - \rho_2(t)$ for a system of an N-fold degenerate level coupled to a band with a state density g by a random matrix with the mean squared matrix element $\left\langle V^2\right\rangle$ as a function of the scaled time $T = t\sqrt{4\left\langle V^2\right\rangle N}$ is shown (**a**) for several values of the dimensionless parameter $w = \pi g\sqrt{\left\langle V^2\right\rangle/N}$. The asymptotic values of the populations of the degenerate level $\rho_1(\infty)$ and of the band $\rho_2(\infty)$ depend (**b**) on the parameter w

damping in the system of two non-degenerate bands. At the same time, in (4.79) one also sees the exponential factor $\exp\left[-2\pi g\left\langle V^2\right\rangle t\right]$ typical of the level–band problem and for the two-non-degenerate bands problem. And we see here once more the interplay of the coherent and relaxation processes in the population dynamics of the complex multilevel systems, although we have to remember that the exponential relaxation occurs only until the quantum revivals and recurrences start to play a role at $t > g$.

Let us concentrate on the populations at times $t \gg 1/\sqrt{4N\left\langle V^2\right\rangle}$ and $t \gg 1/2\ \pi g\left\langle V^2\right\rangle$, assuming that $t \ll g$. At this extreme we can extend the integration limit in (4.80) to infinity and obtain

$$\Delta\rho(\infty) = 1 - \frac{4\pi g \left\langle V^2 \right\rangle}{\sqrt{\left(2\pi \left\langle V^2 \right\rangle g\right)^2 + 4 \left\langle V^2 \right\rangle N}}, \tag{4.81}$$

where we have taken into account that $\int_0^\infty \exp\left[-\alpha\tau\right] J_0\left(\tau\beta\right) d\tau = (\alpha^2 + \beta^2)^{-1/2}$. If we make use of the relation $\rho_1(\infty) + \rho_2(\infty) = 1$ we get immediately the explicit asymptotic populations

$$\rho_1(\infty) = 1 - \frac{2\pi g \left\langle V^2 \right\rangle}{\sqrt{\left(2\pi \left\langle V^2 \right\rangle g\right)^2 + 4 \left\langle V^2 \right\rangle N}},$$

$$\rho_2(\infty) = \frac{2\pi g \left\langle V^2 \right\rangle}{\sqrt{\left(2\pi \left\langle V^2 \right\rangle g\right)^2 + 4 \left\langle V^2 \right\rangle N}} \tag{4.82}$$

shown in Fig. 4.20b. The total populations of the degenerate level and the band become equal for a mean squared coupling $\left\langle V^2 \right\rangle = V_{st}^2 = N/\pi^2 g^2$ for which the number of the populated energy eigenstates of the band equals the degeneracy of the level.

4.4.2 Population Distribution over the Band

Let us concentrate on the population distribution among the states of the band. After straightforward, albeit cumbersome transformations, (4.26) with allowance of (4.28) and (4.71)–(4.74) yields

$$\rho_m(\varepsilon, \xi) = \frac{1}{(\varepsilon - \Delta_m)(\varepsilon + iW) - \frac{W^2}{w^2}} \frac{1}{(\xi - \Delta_m)(\xi - iW) - \frac{W^2}{w^2}}$$
$$\left[1 - \frac{2iW^3/w^2}{\left[(\varepsilon + iW)(\xi - iW) + \frac{W^2}{w^2}\right](\xi - \varepsilon)}\right] \frac{W^2}{Nw^2}, \tag{4.83}$$

where we denote $W = \pi \left\langle V^2 \right\rangle g$ and $w = W/\sqrt{\left\langle V^2 \right\rangle N}$. Further analysis of this expression should rely on the search for the positions of poles in the complex planes of the variables ε and ξ followed by the inverse Fourier transformation with respect to these variables, which also require quite cumbersome calculations. However, these calculations become much shorter if we concentrate only on the distribution which is attained in the regime $t \to \infty$, when the only non-vanishing contribution is given by the second term in the parentheses in the second line of (4.83) containing the factor $(\xi - \varepsilon)$ in the denominator. This contribution gives a time-independent term for the population distribution $\rho_m\left(t = \infty\right) = \rho_m$ which reads

$$\rho_m = \frac{1}{\pi}\int_{-\infty}^{\infty} \frac{1}{\left[(\varepsilon-\Delta_m)(\varepsilon+iW)-\frac{W^2}{w^2}\right]}\frac{1}{\left[(\varepsilon-\Delta_m)(\varepsilon-iW)-\frac{W^2}{w^2}\right]}$$

$$\frac{W^5/w^4 N}{\left[(\varepsilon+iW)(\varepsilon-iW)+\frac{W^2}{w^2}\right]}d\varepsilon. \tag{4.84}$$

The integrand (4.84) has six poles at the points of the complex plane ε located symmetrically with respect to the real axis. By taking residuals at three points of the upper part of the complex plane and factoring out W^2 with allowance for the relation $Nw = \pi g W/w$ one obtains

$$\rho_m g = \frac{1}{\pi W}\frac{w}{\sqrt{w^2+1}}\frac{2\sqrt{1+1/w^2}-1}{\left(2\sqrt{1+1/w^2}-1\right)^2+\left(\frac{\Delta_m}{W}\right)^2}. \tag{4.85}$$

In (4.85) one recognizes the total population $\rho_2(\infty) = w/\sqrt{w^2+1}$ of the non-degenerate band (4.82) multiplied by a Lorentzian distribution of width $\Gamma = 2\sqrt{1+1/w^2}-1$ over scaled detunings $\delta = \Delta_m/W$, which reads

$$\rho_m = \frac{\langle V^2\rangle}{\sqrt{\left(\pi\langle V^2\rangle g\right)^2+\langle V^2\rangle N}}$$

$$\frac{2\sqrt{\left(\pi\langle V^2\rangle g\right)^2+\langle V^2\rangle N}-\pi\langle V^2\rangle g}{\left(2\sqrt{\left(\pi\langle V^2\rangle g\right)^2+\langle V^2\rangle N}-\pi\langle V^2\rangle g\right)^2+{\Delta_m}^2}, \tag{4.86}$$

in the initial notation whereas the width is $\Gamma = 2\sqrt{\left(\pi\langle V^2\rangle g\right)^2+\langle V^2\rangle N} - \pi\left\langle V^2\right\rangle g$. This dependence (4.85) is shown in Fig. 4.21a, and we replace hereafter in this subsection Δ_m by Δ. One sees that for large coupling, $\left\langle V^2\right\rangle \to \infty$, when $\sqrt{\langle V^2\rangle N}$ becomes small as compared to $W = \pi g\left\langle V^2\right\rangle$ and as a consequence the number $\pi g^2\left\langle V^2\right\rangle$ of the band levels participating in the population dynamics considerably exceeds the degeneracy N of the level, and the width Γ of the Lorentzian distribution equals $W = \pi\left\langle V^2\right\rangle g$ as is the case for the level–band problem (3.35) discussed in Chap. 3. This is quite natural, since for each state of the degenerate level, there are enough energy eigenstates of the band where its population can go without interfering with the decay of other states of the level. In the opposite limit $\sqrt{\langle V^2\rangle N} \gg \pi g\left\langle V^2\right\rangle$ the distribution width is $\Gamma \simeq 2\sqrt{\langle V^2\rangle N}$, and the number of band levels in resonance amounts to $\pi\Gamma g \simeq 2\pi\sqrt{\langle V^2\rangle N g^2}$. Note that if we assume that the total populations ρ_2 of the band and ρ_1 of the degenerate level are proportional to the corresponding numbers of states in resonance, we arrive at $\rho_2/\rho_1 = 2\pi\sqrt{\langle V^2\rangle N g^2}/N = 2\pi\left\langle V^2\right\rangle g/\sqrt{\langle V^2\rangle N}$, which is twice as large as the value suggested in the same extreme by the exact solution (4.82). This means that the equal distribution principle does not always govern the stationary population distribution among the states of a complex multilevel system.

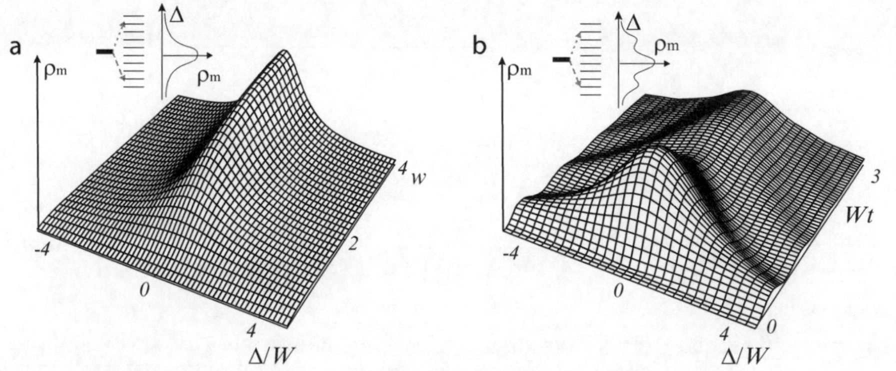

Fig. 4.21 The asymptotic population distribution (**a**) over the levels of the band as a function of the scaled detuning $\Delta/\pi g \left\langle V^2 \right\rangle$ and the dimensionless parameter $w = \pi g \sqrt{\left\langle V^2 \right\rangle / N}$. The asymptotic distribution is attained as a result of a dynamical process that yields a distribution oscillating in time shown in (**b**) as a function of the detuning Δ for $w = 1/\sqrt{3}$

In order to trace the dynamics of the population distribution over the band of levels one needs to perform the inverse Fourier transformation of (4.83), which requires straightforward although cumbersome calculations. We do not present here all the details, but just show the crucial aspects of the main transformation that could serve as guiding lines for such calculations. The unity in the square brackets yields just an absolute value of the inverse Fourier transform of

$$\psi_m(\varepsilon) = \frac{1}{(\varepsilon - \Delta)(\varepsilon + iW) - \frac{W^2}{w^2}} \sqrt{\frac{W^2}{Nw^2}}, \tag{4.87}$$

which is not the total ensemble average population, but "a naive population" $\tilde{\rho}_m(t, \Delta)$ given by the absolute value squared of the ensemble averaged wavefunction. The right-hand side of (4.87) has poles at the points

$$\varepsilon_{1,2} = \frac{1}{2}(\Delta - iW) \pm \sqrt{\frac{1}{2}(\Delta - iW)^2 + \frac{W^2}{w^2}}, \tag{4.88}$$

and the integration performed by calculating the residuals at these points gives the time-dependent average amplitude, which yields the corresponding "probability"

$$\tilde{\rho}_m(t, \Delta) = \frac{W^2}{Nw^2} \frac{2\left|\sin\left(\frac{t}{2}\sqrt{4\frac{W^2}{w^2} + (\Delta - iW)^2}\right)\right|^2 e^{-tW}}{\sqrt{\Delta^4 + \left(W^2 - 4\frac{W^2}{w^2}\right)^2 + 2\,\Delta^2\left(4\frac{W^2}{w^2} + W^2\right)}}. \tag{4.89}$$

The rest of (4.83) is the correlation term $\delta\rho\left(\varepsilon,\xi,\Delta\right)$ to this "naive" population which actually plays the dominant role in the long-time limit. As earlier, to calculate this term, we introduce the variables $\zeta=\xi-\varepsilon$, $\eta=\left(\varepsilon+\xi\right)/2$ and write

$$\delta\rho\left(\zeta,\eta,\Delta\right)=\frac{1}{\left(\eta-\frac{\zeta}{2}-\Delta_m\right)\left(\eta-\frac{\zeta}{2}+iW\right)-\frac{W^2}{w^2}}$$
$$\frac{1}{\left(\eta+\frac{\zeta}{2}-\Delta_m\right)\left(\eta+\frac{\zeta}{2}-iW\right)-\frac{W^2}{w^2}}$$
$$\frac{-2iW^5/Nw^4}{\left[\left(\eta-\frac{\zeta}{2}+iW\right)\left(\eta+\frac{\zeta}{2}-iW\right)+\frac{W^2}{w^2}\right]\zeta}. \tag{4.90}$$

The first and second factors have singularities at the points $\eta=\zeta/2+\varepsilon_{1,2}$ and $\eta=-\zeta/2+\varepsilon_{1,2}^*$, whereas the third factor has poles at $\eta=\pm\sqrt{\left(\zeta/2-iW\right)^2-W^2/w^2}$. It is sufficient to consider only one of the first four singularities, say ε_1, since all the other poles of the first two parts result from ε_1 after complex conjugation associated with the replacement $\zeta\to-\zeta$, and after the replacement $\Delta\to-\Delta$. One has to take the part of the final expression real and even in Δ at the end of the calculations. In the third factor we retain only one pole, which is located in the part of the complex plane which is opposite with respect to the real axis to the part where ε_1 is located, or else the integral over $d\eta$ vanishes. This results in

$$\delta\rho\left(\zeta,\eta,\Delta\right)=\frac{1}{\left(\eta-\frac{\zeta}{2}-\varepsilon_1\right)\left(\zeta+\varepsilon_1-\varepsilon_1^*\right)\left(\zeta+\varepsilon_1-\varepsilon_2^*\right)\left(\varepsilon_1-\varepsilon_2\right)}$$
$$\frac{1}{\sqrt{\left(\frac{\zeta}{2}-iW\right)^2-\frac{W^2}{w^2}}}\frac{iW^5/Nw^4}{\left[\left(\eta+\sqrt{\left(\frac{\zeta}{2}-iW\right)^2-\frac{W^2}{w^2}}\right)\right]\zeta}, \tag{4.91}$$

and yields after integration over $d\eta$

$$\delta\rho\left(\zeta,\Delta\right)=\frac{-1}{\left(\sqrt{\left(\frac{\zeta}{2}-iW\right)^2-\frac{W^2}{w^2}}+\frac{\zeta}{2}+\varepsilon_1\right)\left(\varepsilon_1-\varepsilon_2\right)\zeta}$$
$$\frac{2\pi W^5/Nw^4}{\left(\zeta+\varepsilon_1-\varepsilon_1^*\right)\left(\zeta+\varepsilon_1-\varepsilon_2^*\right)\sqrt{\left(\frac{\zeta}{2}-iW\right)^2-\frac{W^2}{w^2}}}, \tag{4.92}$$

which is equivalent to an expression more convenient for further calculations

$$\delta\rho\left(\zeta,\Delta\right)=\frac{\left(\sqrt{\left(\zeta/2-iW\right)^{2}-W^{2}/w^{2}}-\zeta/2-\varepsilon_{1}\right)}{\left(-\left(\frac{\zeta}{2}-iW\right)^{2}+\frac{W^{2}}{w^{2}}+\left(\frac{\zeta}{2}+\varepsilon_{1}\right)^{2}\right)\left(\varepsilon_{1}-\varepsilon_{2}\right)\zeta}$$

$$\frac{2\pi W^{5}/Nw^{4}}{\left(\zeta+\varepsilon_{1}-\varepsilon_{1}^{*}\right)\left(\zeta+\varepsilon_{1}-\varepsilon_{2}^{*}\right)\sqrt{\left(\frac{\zeta}{2}-iW\right)^{2}-\frac{W^{2}}{w^{2}}}}. \tag{4.93}$$

One sees clearly that (4.92) has poles at the points $\zeta_{1,2,3}=0,\varepsilon_1^*-\varepsilon_1,\varepsilon_2^*-\varepsilon_1$, and at the point $\zeta_4=(W^2+W^2/w^2+\varepsilon_1^2)/(iW-\varepsilon_1)$ corresponding to the first factor in the denominator. Apart from these singularities, the expression has branching points at $\zeta_{5,6}=iW\pm W/w$. We note that the part

$$\delta\rho_1\left(\zeta,\eta,\Delta\right)=\frac{(iW-\varepsilon_1)2\pi W^5/Nw^4\left(\varepsilon_1-\varepsilon_2\right)}{\zeta\left(\zeta-\zeta_2\right)\left(\zeta-\zeta_3\right)\left(\zeta-\zeta_4\right)} \tag{4.94}$$

of the right-hand side of (4.93) does not contain the branching points, and after the inverse Fourier transform over ζ it gives a constant and three exponential time-dependencies, whereas only the remainder

$$\delta\rho_2\left(\zeta,\Delta\right)=\frac{\left(-\zeta/2-\varepsilon_1\right)\left(iW-\varepsilon_1\right)4\pi W^5/Nw^4\left(\varepsilon_1-\varepsilon_2\right)}{\zeta\left(\zeta-\zeta_2\right)\left(\zeta-\zeta_3\right)\left(\zeta-\zeta_4\right)\sqrt{\left(\zeta-\zeta_5\right)\left(\zeta-\zeta_6\right)}} \tag{4.95}$$

which does contain the branching points requires further consideration.

By casting the polynomial fraction factor in (4.95) as a sum of poles

$$\frac{\left(-\zeta/2-\varepsilon_1\right)}{\zeta\left(\zeta-\zeta_2\right)\left(\zeta-\zeta_3\right)\left(\zeta-\zeta_4\right)}=\frac{\varepsilon_1}{\zeta_2\zeta_3\zeta_4}\frac{1}{\zeta}-\frac{\zeta_2/2+\varepsilon_1}{\zeta_2\left(\zeta_2-\zeta_3\right)\left(\zeta_2-\zeta_4\right)}\frac{1}{\left(\zeta-\zeta_2\right)}$$

$$-\frac{\zeta_3/2+\varepsilon_1}{\zeta_3\left(\zeta_3-\zeta_2\right)\left(\zeta_3-\zeta_4\right)}\frac{1}{\left(\zeta-\zeta_3\right)}-\frac{\zeta_4/2+\varepsilon_1}{\zeta_4\left(\zeta_4-\zeta_2\right)\left(\zeta_4-\zeta_3\right)}\frac{1}{\left(\zeta-\zeta_4\right)} \tag{4.96}$$

one reduces (4.95) to the form

$$\delta\rho_2\left(\zeta,\Delta\right)=\sum_k a_k\frac{(iW-\varepsilon_1)4\pi W^5/Nw^4\left(\varepsilon_1-\varepsilon_2\right)}{\left(\zeta-\zeta_k\right)\sqrt{\left(\zeta-\zeta_5\right)\left(\zeta-\zeta_6\right)}} \tag{4.97}$$

where the coefficients a_k emerge from (4.96). The next step is the calculation of the inverse Fourier transform of the expression of type

$$T_F\left(t,\zeta_k,\zeta_5,\zeta_6\right)=\frac{1}{4\pi^2}\int_{-\infty}^{\infty}\frac{\exp\left[it\zeta\right]}{\left(\zeta-\zeta_k\right)\sqrt{\left(\zeta-\zeta_5\right)\left(\zeta-\zeta_6\right)}}d\zeta, \tag{4.98}$$

which is easier to perform by substituting explicitly $\zeta_{5,6} = iW \pm W/w$. After the replacement $\zeta \to xW/w + iW$, equation (4.98) takes the form

$$T_F\left(t, \zeta_k, \zeta_5, \zeta_6\right) = \frac{1}{4\pi^2} \int_{-\infty}^{\infty} \frac{\exp\left[it\left(xW/w + iW\right)\right]}{\left(xW/w + iW - \zeta_k\right)\sqrt{x^2-1}} dx. \tag{4.99}$$

We now multiply both parts of this equation by $\exp\left(-\zeta_k it\right)$, take the time derivative, and arrive at

$$\begin{aligned}\frac{d\left[e^{-it\zeta_k} T_F\left(t, \zeta_k\right)\right]}{dt} &= \int_{-\infty}^{\infty} \frac{\exp\left[it\left(x\frac{W}{w} + iW - \zeta_k\right)\right]}{\left(x\frac{W}{w} + iW - \zeta_k\right)\sqrt{x^2-1}} \frac{dx}{4\pi^2} \\ &= \frac{e^{-t(W+i\zeta_k)}}{4\pi^2} \int_{-\infty}^{\infty} \frac{\exp\left[itx\frac{W}{w}\right]}{\sqrt{x^2-1}} dx = e^{-t(W+i\zeta_k)} \frac{J_0(tW/w)}{2\pi},\end{aligned} \tag{4.100}$$

where we have made use of the same reasoning as for (4.78) and (4.79). This implies that

$$T_F\left(t, \zeta_k\right) = e^{it\zeta_k} \int_0^t e^{-\tau(W+i\zeta_k)} J_0\left(\tau\frac{W}{w}\right) \frac{d\tau}{2\pi}, \tag{4.101}$$

where the lower limit is set to 0, since at $t = 0$ the integration contour (4.99) can be moved to infinity, where the integral vanishes.

Equations (4.94)–(4.98) with allowance for (4.89), (4.101) yield a clumsy, albeit explicit expression for the distribution of the population over the band as a function of time. The result of such a calculation for $w = 1/\sqrt{3}$ is shown in Fig. 4.21b. One sees an important feature of the distribution—it manifests oscillations in time and oscillations as a function of detuning until the asymptotic Lorentzian profile is attained. The larger the ratio $1/w = \sqrt{\langle V^2\rangle N}/\pi g\left\langle V^2\right\rangle$, the more pronounced the oscillations.

4.4.3 Role of the Interaction Rank

Some important aspects of the degenerate level–band problem, including the difference of the factor of 2 in the extreme $\sqrt{\langle V^2\rangle N} \gg \pi g\left\langle V^2\right\rangle$ between the exact asymptotic distribution of the populations and that suggested by the equal distribution principle, are revealed when we consider this system as an ensemble of two-level systems, by analogy with the consideration of two degenerate levels presented in Sect. 4.3.2. As earlier, in (4.62) we concentrate here on the matrix structure of the Hamiltonian

$$\hat{H} = \begin{pmatrix} \hat{H}_2 & \hat{V} \\ \hat{V}^+ & \hat{H}_1 \end{pmatrix} = \begin{pmatrix} \Delta_m \delta_m^{m'} & V_n^m \\ V_m^n & 0 \end{pmatrix} \tag{4.102}$$

and truncate the band such that only the levels that were populated in the process of population transfer are taken into account. The Hamiltonian now also consists of four blocks, a small $M \times M$ block $\hat{H}_2$ of order $M \sim \pi\sqrt{\langle V^2 \rangle N g^2}$, a large $N \times N$ block $\hat{H}_1$ of order N, and two $M \times N$ and $N \times M$ blocks V_n^m and V_m^n of the interaction matrices, which apparently have a rank not larger than M. Since $\hat{H}_1 = 0$, one can perform a linear unitary transformation such that the rectangular conjugated matrices $\hat{V}$ and $\hat{V}^+$ have only non-zero upper and right $M \times M$ blocks, while all of the other matrix elements vanish. The Hamiltonian (4.102) then takes the form

$$\hat{H} = \begin{pmatrix} \hat{H}_2 & \hat{V}' & 0 \\ \hat{V}'^+ & 0 & 0 \\ 0 & 0 & 0 \end{pmatrix} = \begin{pmatrix} \Delta_m \delta_m^{m'} & V_n^m \\ V_m^n & 0 \end{pmatrix} \tag{4.103}$$

where $\hat{V}'$ and $\hat{V}'^+$ are Hermitian conjugate $M \times M$ matrices of rank M' not higher than M. In other words, via the linear unitary transformation applied to the states of the degenerate level we arrive at a representation where only the first M' states of the basis are coupled to the upper band. These states are composed of the properly normalized and orthogonalized linear combinations of M' different states

$$|c_m\rangle = \sum_{n=0}^{N} V_m^n |n\rangle \tag{4.104}$$

interacting with the band, whereas the remaining $N - M'$ states are orthogonal to these M' states. One calls the latter "dark states", as opposed to "bright" states (4.104), since they do not interact with the band. The completely decoupled dark states retain their initial population, and hence the excitation process leaves the $(N - M')/N$ fraction of the initial population of the degenerate level intact. The rest of the population M'/N is equally distributed among M states of band 2 and M' bright states of the degenerate level, provided the matrix $\hat{V}'$ commutes with its Hermitian conjugate $\hat{V}'^+$ and can be diagonalized by a unitary transformation, which we assume to be the case. For $M' = M$ this yields $\rho_2 = M/2N$, $\rho_1 = (N - M)/N + M/2N$, and we arrive at

$$\rho_2/\rho_1 \simeq M/2N = \pi \langle V^2 \rangle g / \sqrt{\langle V^2 \rangle N} \tag{4.105}$$

which coincides with the extreme following from (4.82) and differs by the factor of 2 from the estimate on p. 160.

In the example just considered we have seen that the rank of the interaction matrix, that is, the number of its non-zero eigenvalues, may play an important role in the population dynamics of a complex multilevel quantum system. It affects the number of states that actually participate in the process. The rank of the interaction matrix is closely related to the question of the applicability of the random matrix model to the description of quantum dynamics in a complex system. Indeed, a square $N \times N$ matrix which has rank $R \ll N$ can be cast with the help of the Schmidt expansion into the sum

$$V_j^k = \sum_{s=1}^{R} \langle j \,| c_s \rangle \, V_s \, \langle c_s | \, k \rangle \tag{4.106}$$

over its eigenvalues V_s and eigenvectors $|c_s\rangle$. In the extreme case $R = 1$ this yields

$$V_j^k = \langle j \,| c_1 \rangle \, V_1 \, \langle c_1 | \, k \rangle = C_j V_1 C_k^* \tag{4.107}$$

with $C_j = \langle j \,| c_1 \rangle$, which means that the interaction matrix assumes a factorized form. Note that the matrix elements of $\hat{V}$ can still be random, if the expansion coefficients $\langle c_1 | \, k \rangle$ of its eigenvector $|c_1\rangle$ over the basis set of the energy eigenstates $|k\rangle$ of the unperturbed Hamiltonian are random. In this case, however, as well as in the more general case $R \ll N$, the statistical properties of the products of the matrix elements differ drastically from that for completely random matrices, and in particular this circumstance does not allow one to select from the diagram series of Fig. 4.5 the "tree"-like contributions of Fig. 4.6. This implies a strong statistical correlation among the matrix elements of $\hat{V}$, which one can see explicitly for (4.107) for which we have

$$\left\langle \left| V_j^k V_k^r \right|^2 \right\rangle = (V_1)^4 / N^2 \tag{4.108}$$

and not

$$\left\langle V_j^k V_k^r \right\rangle = \left\langle V^2 \right\rangle \delta_j^r \tag{4.109}$$

as would be the case for completely random matrices.

4.5 The Role of Correlations

The correlations among the interaction matrix elements allow for the interference of different excitation trajectories which not only do not disappear after the ensemble average, but on the contrary strongly affect the population dynamics in the quantum system in general, and the population distribution among the states of different energies in particular. We illustrate this with two examples: that of two levels interacting via a band, and a two-band system with an interaction matrix of rank 1 given in (4.107).

4.5.1 Two Levels and a Band

Let us consider a quantum system of two levels, each of which interacts with a band of levels as shown in Fig. 4.22. This system resembles the Fano system depicted in Fig. 3.7, which was previously considered in Sect. 3.2.2, although here we will not assume equal couplings V and V' of each of the levels to the states of the band, but on the contrary we concentrate on the statistical properties of these matrix elements.

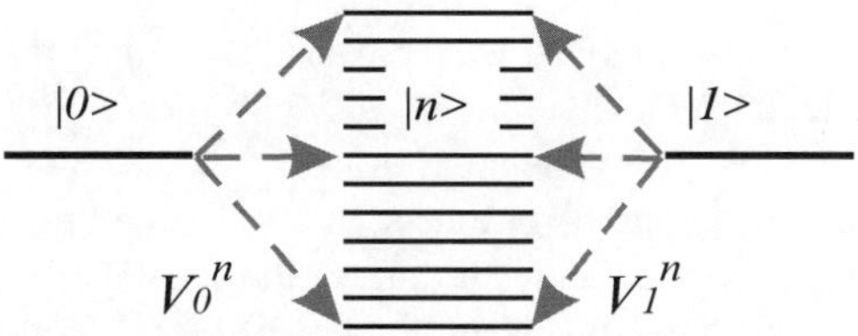

Fig. 4.22 Two levels coupled to a band by different interaction matrix elements V_0^n, and V_1^n

The General Solution

By analogy to (3.49) we can write

$$\begin{aligned}\varepsilon\psi_0 &= \sum_{n=1}^{N} V_0^n \psi_n + i \\ \varepsilon\psi_1 &= \sum_{n=1}^{N} V_1^n \psi_n + \Delta_1\psi_1 + V''\psi_0 \\ \varepsilon\psi_n &= \Delta_n\psi_n + V_n^0\psi_0 + V_n^1\psi_1 \end{aligned} \tag{4.110}$$

where $\psi_0(t = 0) = 1; \psi_1(t = 0) = 0; \psi_n(t = 0) = 0$ are taken as the initial conditions. Substitution of ψ_n from the last equation into the first two equations yields the system of equations

$$\begin{aligned}\left(\varepsilon - \sum_{n=1}^{N} \frac{V_0^n V_n^0}{(\varepsilon - \Delta_n)}\right)\psi_0 - \sum_{n=1}^{N} \frac{V_0^n V_n^1}{(\varepsilon - \Delta_n)}\psi_1 = i \\ -\sum_{n=1}^{N} \frac{V_1^n V_n^0}{(\varepsilon - \Delta_n)}\psi_0 + \left(\varepsilon - \sum_{n=1}^{N} \frac{V_1^n V_n^1}{(\varepsilon - \Delta_n)} - \Delta_1\right)\psi_1 = 0,\end{aligned} \tag{4.111}$$

which have the solutions

$$\psi_0(\varepsilon) = \frac{i\left(\varepsilon - W_1^1 - \Delta_1\right)}{\left(\varepsilon - W_0^0\right)\left(\varepsilon - W_1^1 - \Delta_1\right) - W_0^1 W_1^0}$$
$$\psi_1(\varepsilon) = \frac{iW_1^0}{\left(\varepsilon - W_0^0\right)\left(\varepsilon - W_1^1 - \Delta_1\right) - W_0^1 W_1^0}, \tag{4.112}$$

where we denote $W_p^q = \sum_{n=1}^N V_p^n V_n^q / (\varepsilon - \Delta_n)$ with $p, q = 0, 1$.

Let us now choose the interactions in the form

$$V_n^0 = VC_n$$
$$V_n^1 = VS_n \tag{4.113}$$

similar to (4.107), where V is a coupling constant of energy dimensionality whereas the coefficients $C_n = |C_n|\, e^{i\theta_n}$ and $S_n = |S_n|\, e^{i\chi_n}$ are complex numbers normalized by the conditions $\sum_n |C_n|^2 = \sum_n |S_n|^2 = 1$. Equation (4.113) implies that the state $|0\rangle$ interacts with only one linear combination $|c\rangle = \sum_n C_n |n\rangle$ of the states of the band and that for the state $|1\rangle$ the similar combination $|s\rangle = \sum_n S_n |n\rangle$ of the band levels might be different. Here we assume equal interaction matrix elements, that is, $\sum_n |V_0^n|^2 = \sum_n |V_1^n|^2$, which is actually not a restricting assumption, since the strip of interacting levels can have an arbitrary width.

Substitution of (4.113) allows one to express the coefficients W_p^q in (4.112) in the form

$$W_0^0 = V^2 \mathcal{S}_1 = V^2 \sum_{n=1}^N |C_n|^2 / (\varepsilon - \Delta_n),$$

$$W_1^1 = V^2 \mathcal{S}_2 = V^2 \sum_{n=1}^N |S_n|^2 / (\varepsilon - \Delta_n),$$

$$W_1^0 = (W_0^1)^* = V^2 \mathcal{S}_3 = V^2 \sum_{n=1}^N C_n^* S_n / (\varepsilon - \Delta_n). \tag{4.114}$$

We now assume that both $|C_n|^2$ and $|S_n|^2$ are smooth functions of the index n, and in the vicinity of $\Delta_n = 0$ they amount to $|C|^2$ and $|S|^2$ respectively. For the regime $t < g$ we can therefore factor out these quantities from the sums (4.114) and replace the first two sums by integrations, which yields

$$\mathcal{S}_1 = \sum_{n=1}^N \frac{|C_n|^2}{\varepsilon - \Delta_n} = i\pi g\, |C|^2 ,$$

$$\mathcal{S}_2 = \sum_{n=1}^N \frac{|S_n|^2}{\varepsilon - \Delta_n} = i\pi g\, |S|^2 . \tag{4.115}$$

The sum $\mathcal{S}_3 = V^2 \sum_{n=1}^{N} C_n^* S_n/(\varepsilon - \Delta_n)$ requires a more detailed analysis. Indeed, apart from the moduli $|C_n|$ and $|S_n|$ it contains the phase factors $e^{i\theta_n - i\chi_n}$, which may differ considerably from unity, thus adding another dependence of the summand on the summation index n. However, for $t \ll g$ this problem can be considerably simplified when we make use of the uncertainty principle. The simplest way to employ this principle it is to say that at a time t, the typical frequency ε bringing the main contribution to the integral of the inverse Fourier transformation is of the order of $1/t$, and that generally speaking, this number has a complex value. Denominators of the sums (4.114) do not therefore differ much for two sequential terms, since the difference $\Delta_{n+1} - \Delta_n \sim 1/g$ remains less than the imaginary part of $\varepsilon \sim 1/t$ in the denominator governing the resonance width by the condition $t \ll g$. We can therefore perform the summation for $\mathcal{S}_3$ in two steps: by taking first the average of the phase factors over a number $2K \sim g/t$ of neighboring terms for which the denominators $(\varepsilon - \Delta_n)$ are almost identical, and then by replacing these factors by their average $\left\langle e^{i\theta_n - i\chi_n} \right\rangle = \sum_{m=n-K}^{n+K} e^{i\theta_m - i\chi_m}/2K$, which is likely a smooth function of the index n. Therefore replacing summation by integration one arrives at

$$\mathcal{S}_3 = V^2 \sum_{n=1}^{N} \frac{C_n^* S_n}{\varepsilon - \Delta_n} = i\pi g \left|CS\right| \left\langle e^{i\theta_n - i\chi_n} \right\rangle, \tag{4.116}$$

where, in the last part, the index n corresponds to the states for which $\Delta_n \sim 0$. By taking such an average we assume implicitly that the absolute values of C_n and S_n and their phases are statistically independent.

Substitution of (4.114)–(4.116) into (4.112) yields

$$\psi_0(\varepsilon) = i\frac{\varepsilon - \Delta_1 + i\pi g V^2 \left|C\right|^2}{(\varepsilon - \varepsilon_1)(\varepsilon - \varepsilon_2)}$$

$$\psi_1(\varepsilon) = -\frac{\pi g V^2 \left|CS\right| \left\langle e^{i\theta_n - i\chi_n} \right\rangle}{(\varepsilon - \varepsilon_1)(\varepsilon - \varepsilon_2)}, \tag{4.117}$$

where

$$\varepsilon_{1,2} = \frac{\Delta_1 + gi\pi V^2 \left(\left|C\right|^2 + \left|S\right|^2\right)}{2} \pm \left[\frac{\left(\Delta_1 + gi\pi V^2 \left(\left|C\right|^2 + \left|S\right|^2\right)\right)^2}{4} - g\pi V^2 \left|C\right|^2 \left(i\Delta_1 + \left(\left|\left\langle e^{i\theta_n - i\chi_n} \right\rangle\right|^2 - 1 \right) g\pi V^2 \left|S\right|^2 \right) \right]^{1/2} \tag{4.118}$$

are the roots of the denominators responsible for the time evolution of the system at $t < g$. One sees that the phase correlation of the matrix elements $\left\langle e^{i\theta_n - i\chi_n} \right\rangle$ is an important ingredient in these expressions.

Different Particular Cases

In the particular case of totally random phases, $\left\langle e^{i\theta_n - i\chi_n}\right\rangle = 0$, and hence, according to (4.118) the probability amplitude of being in state $|1\rangle$ vanishes completely. For this case $\varepsilon_1 = \Delta_1 - i\pi g V^2 |C|^2$, and $\varepsilon_2 = -i\pi g V^2 |S|^2$, which yields the result

$$\psi_0(t) = \frac{1}{2\pi}\int\limits_C \frac{ie^{-i\varepsilon t}d\varepsilon}{\left(\varepsilon + i\pi g V^2 |S|^2\right)} = e^{-W_0^0 t} \tag{4.119}$$

coinciding with the corresponding expressions of (3.32)–(3.33). This means that in the absence of correlation among the matrix elements that couple the states $|0\rangle$ and $|1\rangle$ to the band, each of these states interacts with the states of the band independently, thus conforming to the level–band model and yielding all the consequences of this model including (3.35) for the distribution of the population over the band levels. Therefore the state $|1\rangle$, which was not populated initially, remains unpopulated as long as $t < g$, and the continuous-band model remains valid, whereas the population of state $|0\rangle$ prefers to be distributed amongst the more statistically abundant states of the band.

On the other hand, in the case of strongest correlation $\left\langle e^{i\theta_n - i\chi_n}\right\rangle = 1$ the most pronounced effect takes place for the case $\Delta_1 = 0$, when

$$\varepsilon_{1,2} = \begin{cases} 0 \\ -i\pi g V^2(|C|^2 + |S|^2) \end{cases}, \tag{4.120}$$

and therefore (4.117) take the form

$$\psi_0(\varepsilon) = i\frac{\varepsilon + i\pi g V^2 |C|^2}{\varepsilon\left(\varepsilon + i\pi g V^2(|C|^2 + |S|^2)\right)}$$

$$\psi_1(\varepsilon) = i\frac{i\pi g V^2 |CS|}{\varepsilon\left(\varepsilon + i\pi g V^2(|C|^2 + |S|^2)\right)}, \tag{4.121}$$

and yield

$$\psi_0(t) = \frac{|C|^2}{|C|^2 + |S|^2} + \frac{|S|^2}{|C|^2 + |S|^2}e^{-t\pi g V^2(|C|^2+|S|^2)}$$

$$\psi_1(t) = \frac{|CS|}{|C|^2 + |S|^2}\left(1 - e^{-t\pi g V^2(|C|^2+|S|^2)}\right). \tag{4.122}$$

The most important consequence of the complete correlation of the matrix elements is that $\varepsilon_1 = 0$, and hence the probability amplitudes of (4.122) do not vanish as $t \to \infty$. The levels retain a significant part of the total population $\rho_0(t = \infty) = |C|^4/(|C|^2 + |S|^2)^2$, $\rho_1(t = \infty) = |C|^2|S|^2/(|C|^2 + |S|^2)^2$, whereas the total population of the band levels amounts to $\rho_b(t = \infty) = (|S|^4 + |C|^2|S|^2)/(|C|^2 + |S|^2)^2$. In the particular case $|C|^2 = |S|^2 = 1/2$ both of the levels have the asymptotic population $\rho_0(t = \infty) = \rho_1(t = \infty) = 1/4$, while the band takes the other half of the population.

The last case of $|C|^2 = |S|^2 = 1/2$, $\left\langle e^{i\theta_n - i\chi_n}\right\rangle = 1$ and $\Delta_1 = 0$ can be readily understood when we consider the population dynamics of the system taking for the basis set of the problem the combinations $|+\rangle = (|0\rangle + |1\rangle)/\sqrt{2}$, and $|-\rangle = (|0\rangle - |1\rangle)/\sqrt{2}$, that is, the symmetric and antisymmetric combinations of the levels. The state $|+\rangle$ interacts with the band states via a constant matrix element V whereas the state $|-\rangle$ turns out to be completely decoupled. Therefore the state $|-\rangle$ retains the initial population $|\langle 1\,|-\rangle|^2 = 1/2$ and the other half of the population, which was initially in the state $|+\rangle$, moves completely to the levels of the band. We note that in the case where $\left\langle e^{i\theta_n - i\chi_n}\right\rangle = a < 1$, the combination $|-\rangle$ starts to decay as well, since

$$\varepsilon_{1,2} = \frac{1}{2} g\, i\, \pi\, V^2 \pm \frac{1}{2} g\, i\, \pi\, V^2\, |a| \tag{4.123}$$

which yields the amplitudes

$$\begin{aligned} \psi_+(t) &= \frac{1}{\sqrt{2}} e^{-t\pi g V^2 (1+|a|)/2} \\ \psi_-(t) &= \frac{1}{\sqrt{2}} e^{-t\pi g V^2 (1-|a|)/2} \end{aligned} \tag{4.124}$$

and the populations

$$\begin{aligned} \rho_0(t) &= \frac{1}{4}\left(e^{-t\pi g V^2(1+|a|)/2} + e^{-t\pi g V^2(1-|a|)/2}\right)^2 \\ \rho_1(t) &= \frac{1}{4}\left(e^{-t\pi g V^2(1+|a|)/2} - e^{-t\pi g V^2(1-|a|)/2}\right)^2 . \end{aligned} \tag{4.125}$$

The last terms manifest a tri-exponential decay with rates $\pi g V^2(1+|a|)$, $\pi g V^2$, and $\pi g V^2(1 - |a|)$.

One more simple limiting case with $|C|^2 = |S|^2$ (we take both equal to $1/2$), $\left\langle e^{i\theta_n - i\chi_n}\right\rangle = 1$ and $\Delta_1 \neq 0$ corresponds to two resonant levels interacting via a continuum. Equation (4.118) yields for this case

$$\varepsilon_{1,2} = \frac{1}{2}\left[\Delta_1 + g\, i\, \pi\, V^2\right] \pm \frac{1}{2}\left[\Delta_1^2 - \left(g\, \pi\, V^2\right)^2\right]^{1/2}, \tag{4.126}$$

which after substitution into (4.117) and the inverse Fourier transformation results in

$$\psi_0(t) = \frac{-i\Delta_1\, e^{it\left[\Delta_1 + g\, i\, \pi\, V^2\right]/2}}{\sqrt{\Delta_1^2 - (g\,\pi\, V^2)^{\,2}}} \sin\left(\frac{t}{2}\sqrt{\Delta_1^2 - (g\,\pi\, V^2)^{\,2}}\right)$$

$$+ e^{it\left[\Delta_1 + g\, i\, \pi\, V^2\right]/2} \cos\left(\frac{t}{2}\sqrt{\Delta_1^2 - (g\,\pi\, V^2)^{\,2}}\right)$$

$$\psi_1(\varepsilon) = \frac{g\,\pi\, V^2\, e^{it\left[\Delta_1 + g\, i\, \pi\, V^2\right]/2}}{\sqrt{\Delta_1^2 - (g\,\pi\, V^2)^{\,2}}} \sin\left(\frac{t}{2}\sqrt{\Delta_1^2 - (g\,\pi\, V^2)^{\,2}}\right). \quad (4.127)$$

From this equation one sees that the population

$$\rho_1(t) = \frac{\left(g\,\pi\, V^2\right)^{\,2}}{\Delta_1^2 - (g\,\pi\, V^2)^{\,2}} e^{-tg\,\pi\, V^2} \sin^2\left(\frac{t}{2}\sqrt{\Delta_1^2 - (g\,\pi\, V^2)^{\,2}}\right) \quad (4.128)$$

of state $|1\rangle$ manifests two different dynamical regimes of dying oscillations for $|\Delta_1| > \left|g\,\pi\, V^2\right|$, and a bi-exponential decay in the opposite case. For a fixed Δ_1 the decay rate depends on V. For large V, when $\sqrt{(g\,\pi\, V^2)^{\,2} - \Delta_1^2} \simeq g\,\pi\, V^2 - \Delta_1^2/2g\,\pi\, V^2$ and the slowest exponent gives the main contribution to the population, (4.128) reads

$$\rho_1(t) = \frac{1}{4} e^{-t\Delta_1^2/2g\,\pi\, V^2}. \quad (4.129)$$

One sees that the slowest decay rate $\Delta_1^2/2g\,\pi\, V^2$ decreases with the increase of the interaction V. Insight into this paradoxical result is gained by noting that the states $|0\rangle$ and $|1\rangle$ are coupled via the band, and that the stronger the interaction is, the larger the coupling. With an extremely strong coupling, the eigenstates of a two-level system (3.21) are very close to the combinations $|+\rangle$, and $|-\rangle$, and the last state is completely decoupled from the band for the case $\left\langle e^{i\theta_n - i\chi_n}\right\rangle = 1$, as shown above.

Distribution of the Population Over the Band

From the analysis of the two-level–band system we have learned that the correlations of the interaction matrix elements may have an important influence on the population dynamics. Interference phenomena may result in a considerable increase in the decay time. When earlier, in Sect. 3.2.2, considering the Fano problem, we saw that the interference phenomenon in a two-level–band system can manifest itself also in the population distribution over the band. Now we will consider this phenomenon in more detail for the system of two levels and a band. We assume that ε_1 of (4.118) differs from zero and has therefore a non-zero imaginary part, since otherwise the system is equivalent to a level–band system, as has already been considered in Chap. 3.

From the third equation (4.110) with the allowance for (4.113) one obtains

$$\psi_n = \frac{VC_n}{(\varepsilon - \Delta_n)}\psi_0 + \frac{VS_n}{(\varepsilon - \Delta_n)}\psi_1, \tag{4.130}$$

and substitution of ψ_0 and ψ_1 from (4.117) yields

$$\psi_n = i\frac{VC_n}{(\varepsilon - \Delta_n)}\frac{\varepsilon - \Delta_1 + i\pi g V^2 \left|C\right|^2}{(\varepsilon - \varepsilon_1)(\varepsilon - \varepsilon_2)} - \frac{VS_n}{(\varepsilon - \Delta_n)}\frac{\pi g V^2 \left|CS\right| \left\langle e^{i\theta_n - i\chi_n}\right\rangle}{(\varepsilon - \varepsilon_1)(\varepsilon - \varepsilon_2)}. \tag{4.131}$$

In the regime $1/W_m \ll t \ll g$, the main contribution to the inverse Fourier transform comes from the pole $\varepsilon = \Delta_n$, since two other poles $\varepsilon = \varepsilon_{1,2}$ give exponentially small contributions that can be majorized by the exponent e^{-tW_m}, where W_m is the minimum of two imaginary parts of $\varepsilon_{1,2}$. This yields

$$\rho(\Delta) = \frac{W'}{\pi g}\left|\frac{1}{(\Delta - \varepsilon_1)(\Delta - \varepsilon_2)}\right|^2 \left|C_n\left(\Delta - \Delta_1 + iW'\left|C\right|^2\right) + iS_n W' \left|CS\right| \left\langle e^{i\theta_n - i\chi_n}\right\rangle\right|^2 \tag{4.132}$$

for the population $\rho(\Delta_n) = \rho_n = \left|\psi_n\right|^2$ of a state $\left|n\right\rangle$, where by W' we denote the combination $\pi g V^2$.

By taking the average of the last factor in (4.132) over a small energy interval we obtain the mean population spectral density $g\rho(\Delta)$ in the form

$$g\rho(\Delta) = \frac{W}{\pi}\frac{u\left(\Delta - \Delta_1\right)^2 + W^2(u^2 + 2\alpha(1-u))}{\left|(\Delta - \varepsilon_1)(\Delta - \varepsilon_2)\right|^2} \tag{4.133}$$

where by α we denote $\left|a\right|^2 = \left|\left\langle e^{i\chi_n - i\theta_n}\right\rangle\right|^2$, take $u = \left|C\right|^2 / (\left|C\right|^2 + \left|S\right|^2)$, $W = W'(\left|C\right|^2 + \left|S\right|^2)$, and write the roots

$$\frac{\varepsilon_{1,2}}{W} = \frac{\Delta_1}{2W} + \frac{i}{2} \pm \left[\left(\frac{\Delta_1}{2W} + \frac{i}{2}\right)^2 - u(\frac{i\Delta_1}{W} + (\alpha - 1)(1-u))\right]^{1/2} \tag{4.134}$$

in the same notation. The asymptotic population profile of (4.133) is a sum of two Lorentzian profiles

$$g\rho(\Delta) = \frac{1}{\pi}\frac{f_1}{\left|\Delta - \varepsilon_1\right|^2} + \frac{1}{\pi}\frac{f_2}{\left|\Delta - \varepsilon_2\right|^2} + \frac{f_0}{\pi}\left(\frac{\Delta}{\left|\Delta - \varepsilon_1\right|^2} - \frac{\Delta}{\left|\Delta - \varepsilon_2\right|^2}\right) \tag{4.135}$$

that are placed at the positions $\mathrm{Re}\varepsilon_{1,2}$ and have the widths $\mathrm{Im}\varepsilon_{1,2}$ respectively and an asymmetric correction term. The corresponding strengths $f_{0,1,2}$ are algebraic functions of Δ_1, α, and u, that can be found by equating the coefficients of quadratic, linear, and constant terms in the numerators of the right-hand sides of (4.133) and (4.135):

$$f_1 + f_2 - f_0\,(2R_2 - 2R_1) = Wu$$
$$R_2 f_1 + f_2 R_1 - f_0 \frac{|\varepsilon_2|^2 - |\varepsilon_1|^2}{2} = \Delta_1 W u$$
$$f_1\,|\varepsilon_2|^2 + f_2\,|\varepsilon_1|^2 = Wu(\,\Delta_1^2 + W^2(u^2 + 2\alpha(1-u))), \qquad (4.136)$$

where $R_1 = \mathrm{Re}\varepsilon_1$, $R_2 = \mathrm{Re}\varepsilon_2$, which yields

$$f_0 = \frac{2uW(R_2 - R_1)((2\alpha\,(1-u) + u^2)W^2 + \Delta_1^2)}{4R_2^2|\varepsilon_1^2| + 4{R_1}^2|\varepsilon_2^2| + (|\varepsilon_1^2| - |\varepsilon_2^2|)^2 - 4R_1R_2(|\varepsilon_1^2| + |\varepsilon_2^2|)}$$
$$+ \frac{2uW(R_1|\varepsilon_2^2| - R_2|\varepsilon_1^2| - \Delta_1(|\varepsilon_2^2| - |\varepsilon_1^2|))}{4R_2^2|\varepsilon_1^2| + 4{R_1}^2|\varepsilon_2^2| + (|\varepsilon_1^2| - |\varepsilon_2^2|)^2 - 4R_1R_2(|\varepsilon_1^2| + |\varepsilon_2^2|)},$$

$$f_1 = uW\frac{((2\alpha\,(1-u) + u^2)W^2 + \Delta_1^2)(4R_1\,(R_1 - R_2) - |\varepsilon_1^2| + |\varepsilon_2^2|)}{4R_2^2|\varepsilon_1^2| + 4{R_1}^2|\varepsilon_2^2| + (|\varepsilon_1^2| - |\varepsilon_2^2|)^2 - 4R_1R_2(|\varepsilon_1^2| + |\varepsilon_2^2|)}$$
$$+ \frac{uW|\varepsilon_1^2|(4\,(R_2 - R_1)\,\Delta_1 + |\varepsilon_1^2| - |\varepsilon_2^2|)}{4R_2^2|\varepsilon_1^2| + 4{R_1}^2|\varepsilon_2^2| + (|\varepsilon_1^2| - |\varepsilon_2^2|)^2 - 4R_1R_2(|\varepsilon_1^2| + |\varepsilon_2^2|)},$$

$$f_2 = uW\frac{((2\alpha\,(1-u) + u^2)W^2 + \Delta_1^2)(4R_2\,(R_2 - R_1) + |\varepsilon_1^2| - |\varepsilon_2^2|)}{4R_2^2|\varepsilon_1^2| + 4{R_1}^2|\varepsilon_2^2| + (|\varepsilon_1^2| - |\varepsilon_2^2|)^2 - 4R_1R_2(|\varepsilon_1^2| + |\varepsilon_2^2|)}$$
$$+ \frac{uW|\varepsilon_2^2|(4\,(R_1 - R_2)\,\Delta_1 - |\varepsilon_1^2| + |\varepsilon_2^2|)}{4R_2^2|\varepsilon_1^2| + 4{R_1}^2|\varepsilon_2^2| + (|\varepsilon_1^2| - |\varepsilon_2^2|)^2 - 4R_1R_2(|\varepsilon_1^2| + |\varepsilon_2^2|)}. \qquad (4.137)$$

One sees that the asymmetric correction term of the amplitude f_0 vanishes for identical positions of the poles $\varepsilon_1 = \varepsilon_2$, but that in the general case it is of the order of two other terms that give the Lorentzian profiles. However, in the absence of direct interaction between the $|0\rangle$ and $|1\rangle$ levels, this correction term never compensates completely for the Lorentzian profiles, so that the population of the band never becomes zero at a band level with a finite detuning Δ_m, which was the case for the Fano problem where such a direct coupling is present and renders the dependence of the coupling averaged over a small energy interval essentially non-uniform. One can see the resulting profiles in Fig. 4.23, where we show the dependence of the distribution $g\rho(\Delta)$ on the parameters Δ_1, α, and u. In principle, with the help of (4.136) one can find the correlation factor α of the band spectrum by analyzing the asymptotic population distribution $g\rho(\Delta)$ over the band states.

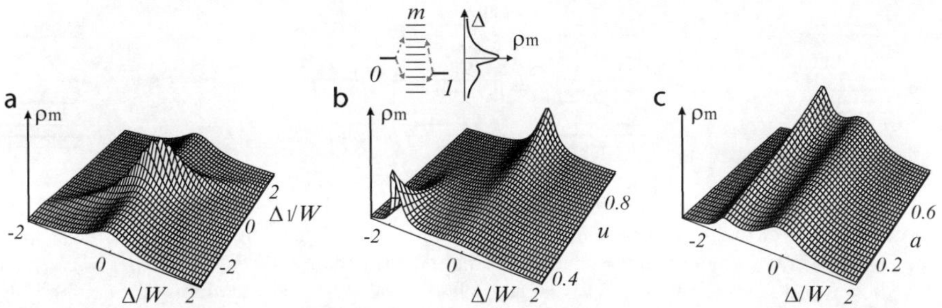

Fig. 4.23 The population distribution over a band as a function of the scaled detuning Δ/W for two levels coupled to a band by different interaction matrix elements $V_0^n = VC_n$ and $V_1^n = VS_n$. The parameters governing the distribution are (**a**) the detuning between the levels Δ_1, divided by $W = i\pi V^2 \langle |C_n|^2\rangle + \langle |S_n|^2\rangle$, (**b**) the relative coupling $u = \langle |C_n|^2\rangle / \langle |C_n|^2\rangle + \langle |S_n|^2\rangle$ of the level 0, and (**c**) the correlation $a = |\langle C_n S_n\rangle|^2 / \langle |C_n|^2\rangle \langle |S_n|^2\rangle$. Two peaks of the population distribution, corresponding to each of the levels, can be seen

From the analysis of the two-level–band system we have learned that the correlation of the matrix elements of interaction may have an important influence on the population dynamics. These correlations may result in a considerable increase in the decay time as was the case for (4.129), which, however, remains finite and roughly amounts to the dispersion of the matrix elements $V^2(1 - |\langle e^{i\chi_n - i\theta_n}\rangle|^2)$ multiplied by the band state density. If the dispersion of the matrix element distribution is of the order of its mean square, the typical decay time does not experience drastic variations, although the distribution of the population over the band can be essentially non-Lorentzian.

4.5.2 Two Bands With a Correlated Coupling

Let us now consider another example of a system with correlated matrix elements, namely a system of two bands (4.5) with a Hamiltonian (4.6) where the coupling matrices have the factorized form $V_n^m = C_n \mathcal{V} C_m^*$ of (4.107). For the sake of simplicity we assume the specific form

$$C_n = \sqrt{\frac{N}{\pi\,(N^2 + n^2)}}$$

$$C_m = \sqrt{\frac{M}{\pi\,(M^2 + m^2)}} \tag{4.138}$$

of the coefficients entering (4.107), which allows one to perform the consideration analytically.

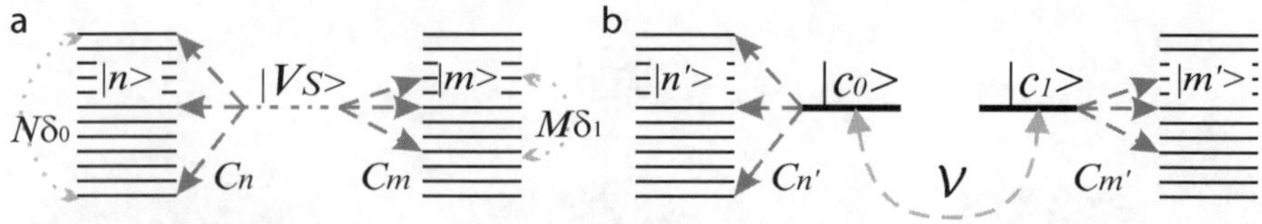

Fig. 4.24 (**a**) Two bands coupled by a factorized interaction. The N levels $|n\rangle$ of the first band located in a strip of width $\sim N\delta_0$ interact with M states $|m\rangle$ of the second band within a strip of width $\sim M\delta_1$ as if the interaction occurs via a single intermediate state $|V_s\rangle = C_n|n\rangle = C_m|m\rangle$, which however does not correspond to an energy level. The mean spacing among the neighboring levels of the first and the second bands are δ_0 and δ_1, respectively. (**b**) The same problem in another representation, where the states $|b1\rangle = \sum C_n|n\rangle$ and $|b2\rangle = \sum C_m|m\rangle$ are included in the basis set together with the orthogonal linear combinations $|n'\rangle = \sum C_n^{n'}|n\rangle$ of the states of the first band and $|m'\rangle = \sum C_m^{m'}|m\rangle$ of the second band. States $|b1\rangle$ and $|b2\rangle$ are coupled by the cooperative interaction $\mathcal{V} = V\sqrt{MN}$

We note that

$$\sum_{n=-\infty}^{\infty} |C_n|^2 = \sum_{m=-\infty}^{\infty} |C_m|^2 \simeq 1 \tag{4.139}$$

for large N and M. This implies that the rank of the perturbation matrix equals one, and that the matrix itself is proportional to a direct product of two normalized vectors $|c_0\rangle\langle c_1|$ that belong to different bands. The only non-zero eigenvalue $\mathcal{V}$ of V_n^m corresponds to the cooperative matrix element for the transition between the bands, which couples $|c_0\rangle$ with $|c_1\rangle$.

What is the physical meaning of (4.138)? It tells us that a number of the order of N levels of the first band interact strongly with a number of the order of M levels of the second band, and that the matrix element of the interaction changes smoothly with the state numbers n and m. Let us denote by δ_0 and δ_1 the mean energy spacing between the neighboring levels of the first and the second bands respectively. Then one can say that the levels located in the energy strip $N\delta_0$ in the vicinity of $\Delta_n = 0$ interact strongly with the levels of the second band located in the energy strip $M\delta_1$ around the position $\Delta_m = 0$ of the second band, as shown in Fig. 4.24. One can say that the interaction of the bands occurs via an intermediate state $|V_s\rangle$, which does not correspond to an energy level, being just the only non-zero eigenvector of the interaction matrix (4.106). Projection of this vector onto the Hilbert space of the first band gives the state vector $|c_0\rangle = \sum_n C_n|n\rangle$, whereas for the Hilbert space of the second band this projection reads $|c_1\rangle = \sum_m C_m|m\rangle$. These states are responsible for the interaction of the bands. When this interaction is a dipole coupling of atomic multiplets mediated by an external electromagnetic field they are called "bright states", in contrast to the "dark states" that are orthogonal to the "bright" ones and do not interact directly with the field.

Formal Solution

The Schrödinger equation (4.5) for the system yields

$$\psi_n(\varepsilon) = \frac{C_n\mathcal{V}}{(\varepsilon - \Delta_n)} \sum_{m=-\infty}^{\infty} C_m\psi_m(\varepsilon) + \frac{i\delta_n^0}{(\varepsilon - \Delta_n)}$$

$$\psi_m(\varepsilon) = \frac{C_m\mathcal{V}}{(\varepsilon - \Delta_m)} \sum_{n=-\infty}^{\infty} C_n\psi_n(\varepsilon). \quad (4.140)$$

We multiply the first equation by C_n, the second equation by C_m, sum up the terms, and obtain the equations

$$\Phi_0(\varepsilon) = \sum_{n=-\infty}^{\infty} \frac{|C_n|^2\mathcal{V}}{(\varepsilon - \Delta_n)}\Phi_1(\varepsilon) + \frac{iC_0}{(\varepsilon - \Delta_0)}$$

$$\Phi_1(\varepsilon) = \sum_{m=-\infty}^{\infty} \frac{|C_m|^2\mathcal{V}}{(\varepsilon - \Delta_m)}\Phi_0(\varepsilon), \quad (4.141)$$

for the combinations

$$\Phi_0(\varepsilon) = \sum_{n=-\infty}^{\infty} C_n\psi_n(\varepsilon)$$

$$\Phi_1(\varepsilon) = \sum_{m=-\infty}^{\infty} C_m\psi_m(\varepsilon). \quad (4.142)$$

We now substitute (4.138) into the sums entering (4.141), substitute $\Delta_n = n\delta_0$, $\Delta_m = m\delta_1$ and replace the summation by integration in the regime $t < (1/\delta_0; 1/\delta_1)$. We arrive at

$$\sum_{n=-\infty}^{\infty} \frac{|C_n|^2}{(\varepsilon - \Delta_n)} = \frac{1}{\pi}\int_{-\infty}^{\infty} dn \frac{N}{N^2 + n^2}\frac{1}{\varepsilon - n\delta_0} = \frac{1}{\varepsilon + iN\delta_0}$$

$$\sum_{m=-\infty}^{\infty} \frac{|C_m|^2}{(\varepsilon - \Delta_m)} = \frac{1}{\pi}\int_{-\infty}^{\infty} dm \frac{M}{M^2 + m^2}\frac{1}{\varepsilon - m\delta_1} = \frac{1}{\varepsilon + iM\delta_1}. \quad (4.143)$$

Note the physical meaning of these sums. They represent the Fourier transformed autocorrelation, that is, the probability amplitude to find the system at the states $|c_0\rangle$ and $|c_1\rangle$, respectively, after a time interval t, provided the system had been in these states at $t = 0$. Substituting (4.143) into (4.141) we obtain

$$\Phi_0(\varepsilon) = \frac{\mathcal{V}}{\varepsilon + iN\delta_0}\Phi_1(\varepsilon) + \frac{i}{\varepsilon\sqrt{\pi N}}$$

$$\Phi_1(\varepsilon) = \frac{\mathcal{V}}{\varepsilon + iM\delta_1}\Phi_0(\varepsilon), \tag{4.144}$$

and solving this system of equations we find

$$\Phi_0(\varepsilon) = \frac{i}{\varepsilon\sqrt{\pi N}}\left(1 - \frac{\mathcal{V}}{\varepsilon + iN\delta_0}\frac{\mathcal{V}}{\varepsilon + iM\delta_1}\right)^{-1}$$

$$\Phi_1(\varepsilon) = \frac{\mathcal{V}}{\varepsilon + iM\delta_1}\frac{i}{\varepsilon\sqrt{\pi N}}\left(1 - \frac{\mathcal{V}}{\varepsilon + iN\delta_0}\frac{\mathcal{V}}{\varepsilon + iM\delta_1}\right)^{-1}, \tag{4.145}$$

which yields, with the allowance for (4.140),

$$\psi_n(\varepsilon) = \frac{i\delta_n^0}{\varepsilon} + \frac{C_n\mathcal{V}}{\varepsilon - \Delta_n}\frac{\mathcal{V}}{\varepsilon + iM\delta_1}\frac{i}{\varepsilon\sqrt{\pi N}} \times\left(1 - \frac{\mathcal{V}}{\varepsilon + iN\delta_0}\frac{\mathcal{V}}{\varepsilon + iM\delta_1}\right)^{-1}$$

$$\psi_m(\varepsilon) = \frac{C_m\mathcal{V}}{\varepsilon - \Delta_m}\frac{i}{\varepsilon\sqrt{\pi N}}\left(1 - \frac{\mathcal{V}}{\varepsilon + iN\delta_0}\frac{\mathcal{V}}{\varepsilon + iM\delta_1}\right)^{-1}. \tag{4.146}$$

Distribution of the Population Over the Band

Let us concentrate on the last equation (4.146) and consider the regime $(1/N\delta_0; 1/M\delta_1) \ll t \ll (1/\delta_0; 1/\delta_1)$ when only the poles at the real points $\varepsilon = \Delta_m; \varepsilon = 0$ are important in the inverse Fourier transformation, whereas two other poles at the imaginary points $\varepsilon = iN\delta_0$ and $\varepsilon = iM\delta_1$ yield in this regime exponentially vanishing contributions.

The inverse Fourier transformation gives the time-dependent amplitude $\psi_m(t)$

$$\psi_m(t) = \frac{-C_m\mathcal{V}}{\Delta_m\sqrt{\pi N}}\left[\frac{N\delta_0 M\delta_1}{N\delta_0 M\delta_1 + \mathcal{V}^2} - \frac{e^{i\Delta_m t}(\Delta_m - iN\delta_0)(\Delta_m - iM\delta_1)}{(\Delta_m - iN\delta_0)(\Delta_m - iM\delta_1) - \mathcal{V}^2}\right] \tag{4.147}$$

and the population $\rho_m(t) = |\psi_m(t)|^2$ reads

$$\rho_m(t) = \frac{1/NM}{1+y^2\Delta^2}\frac{w}{\Delta^2\pi^2}\left|\frac{e^{i\Delta\tau}(\Delta - iy)(\Delta - i/y)}{(\Delta - iy)(\Delta - i/y) - w} - \frac{1}{1+w}\right|^2$$
$$= \frac{1/NM}{1+y^2\Delta^2}\frac{w}{\pi^2}\left|\frac{e^{i\Delta\tau}-1}{\Delta}\frac{(\Delta - iy)(\Delta - i/y)}{(\Delta - iy)(\Delta - i/y) - w}\right.$$
$$\left. + \frac{w}{(1+w)}\frac{\Delta - (iy + i/y)}{((\Delta - iy)(\Delta - i/y) - w)}\right|^2, \tag{4.148}$$

where $w = \mathcal{V}^2/N\delta_0 M\delta_1$; $\Delta = \Delta_m/\sqrt{N\delta_0 M\delta_1}$; $y = \sqrt{N\delta_0/M\delta_1}$; $\tau = t\sqrt{N\delta_0 M\delta_1}$ and we have employed (4.138). One sees the important peculiarity of this system—no stationary distribution over the levels of the second band arises after the relaxation time $t \sim (1/N\delta_0; 1/M\delta_1)$, since the right-hand side of (4.147) remains a function of time.

In the long-time limit, the right-hand side of (4.148) can be considered as consisting of two parts. The well-localized first part arising in the vicinity of the point $\Delta = 0$ from the terms containing the combination $\left(e^{i\Delta\tau} - 1\right)/\Delta$ and its square is given by

$$\rho_m^{(1)}(t) \simeq \frac{1}{NM}\frac{2w^2\left(y + y^{-1}\right)}{\pi^2(1+w)^3}\frac{\sin(t\,\Delta)}{\Delta}$$
$$-\frac{1}{NM}\frac{2w}{\pi^2(1+w)^2}\frac{(\cos(t\,\Delta) - 1)}{\Delta^2} \tag{4.149}$$

whereas $t \to \infty$ one recognizes the Dirac δ-function $\delta(\Delta)$ which originates from the first term on the right-hand side of (4.149) linear in $\left(e^{i\Delta\tau} - 1\right)/\Delta$ and the dominating contribution, $\sim \delta(\Delta)t$, resulting from the second term quadratic in $\left(e^{i\Delta\tau} - 1\right)/\Delta$. The time-independent second part

$$\rho_m^{(2)}(t) \simeq \frac{1/NM\pi^2}{1+y^2\Delta^2}\frac{w^3}{(1+w)}\left|\frac{\Delta - (iy + i/y)}{((\Delta - iy)(\Delta - i/y) - w)}\right|^2 \tag{4.150}$$

emerges from (4.148) when at $\Delta \neq 0$ and $\Delta t \gg 1$ we omit the rapidly oscillating term $\left(e^{i\Delta\tau} - 1\right)/\Delta$.

The time-independent part (4.150) of the population distribution can be represented as a sum of three Lorentzian profiles—one peak of width $1/|y|$ located at $\Delta = 0$, and two other peaks, of width $a = |\mathrm{Re}r_1|$ located at the points $\Delta = \pm b = \pm\mathrm{Im}r_1$, where

$$r_{1,2} = a \pm ib = \frac{y + y^{-1}}{2} \pm \sqrt{\left(\frac{y - y^{-1}}{2}\right)^2 - w} \tag{4.151}$$

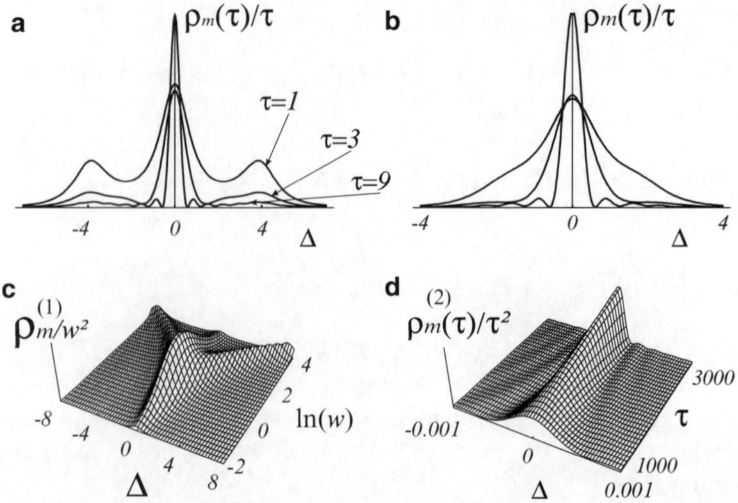

Fig. 4.25 Two bands coupled by a factorized interaction. (**a**) Distribution of population over the second band for $y = 1.2$ for three different times $\tau = 1, 3, 9$ in the regime $\mathrm{Im}\sqrt{r} \neq 0$. One sees that the time-dependent part of the population becomes larger. (**b**) The same as (**a**) for $\mathrm{Im}\sqrt{r} = 0$. (**c**) The time-independent part of the population distribution as a function of w. One sees the cross-over from the regime $\mathrm{Im}\sqrt{r} = 0$ where the distribution is localized around $\Delta = 0$, to the regime $\mathrm{Im}\sqrt{r} \neq 0$ where tree peaks of the distribution are present. (**d**) The time-dependent part of the distribution becomes more and more narrow with the course of time and contains a larger and larger part of the population. The total contribution of this spike increases as $\sim \tau$

is the solution of the bi-quadratic equation $|(ix - iy)(ix - i/y) - w|^2 = 0$ with respect to $\Delta = ix$, corresponding to the poles of (4.150). For real and positive r all three of these peaks are located at the point $\Delta = 0$, but have nevertheless different widths. In Fig. 4.25 we show the population distribution profiles as a function of the detuning for different regimes.

Total Populations of the Bands

Multiplying the probability amplitude $\psi_m(\varepsilon)$ of (4.146) by its complex conjugate one finds the total population of the second band by integrating over $dm = d\Delta/\delta_1$. After substitution of C_m from (4.138) and employing the notation of (4.148)

$$\rho_1(\varepsilon, \xi) = \int \frac{d\Delta}{\pi^2 N} \frac{1/y}{(1/y^2 + \Delta^2)} \frac{w}{(\varepsilon - \Delta)(\xi - \Delta)} \frac{1}{\varepsilon\xi}$$

$$\left(1 + \frac{w}{(\varepsilon - ir_1)(\varepsilon - ir_2)}\right)\left(1 + \frac{w}{(\xi + ir_1)(\xi + ir_2)}\right), \quad (4.152)$$

one obtains

$$\rho_1(\varepsilon,\xi) = \frac{w}{N\pi}\left(1+\frac{2i}{y\,(\varepsilon-\xi)}\right)$$
$$\frac{1}{\varepsilon\left(\varepsilon+\frac{i}{y}\right)}\left(1+\frac{w}{(\varepsilon-b+ai)\,(\varepsilon+b+ai)}\right)$$
$$\frac{1}{\xi\left(\xi-\frac{i}{y}\right)}\left(1+\frac{w}{(\xi-b-ai)\,(\xi+b-ai)}\right). \tag{4.153}$$

By analogy to (4.75) we note that the factor $i/\left(\varepsilon-\xi\right)$ corresponds to the time integration whereas the combinations on the second and the third lines yield, after the inverse Fourier transformation, two complex conjugate time-dependent functions $\phi_\varepsilon(t)$ and $\phi_\xi(t)$, respectively, which read

$$\phi_{\varepsilon,\xi}(t) = \int\limits_C \frac{e^{-i\varepsilon t}}{\varepsilon\left(\varepsilon+\frac{i}{y}\right)}\left(1+\frac{w}{(\varepsilon-b+ai)\,(\varepsilon+b+ai)}\right)$$
$$= y\left(\frac{w}{b^2+a^2}-1\right)-\frac{w\sin(tb\pm\varphi)e^{-at}}{b\,|b-ai|\,\left|b-ai+\frac{i}{y}\right|}$$
$$+y\left(1-\frac{w}{b^2-\left(a-1/y\right)^2}\right)e^{-t/y}, \tag{4.154}$$

where $\varphi=\arg(b-ai)+\arg(b-ai+\frac{i}{y})$. Therefore the time-dependent population $\rho_1(t)$ of the second band can be given in the form

$$\rho_1(\tau) = \frac{w}{N\pi}\left(|\phi_\varepsilon(\tau)|^2+\frac{2}{y}\int\limits_0^\tau |\phi_\varepsilon(t)|^2\,dt\right), \tag{4.155}$$

which after straightforward calculations yields

$$\rho_1(\tau) = \frac{2w}{N\,(1+w)\,\pi^2}\left[\frac{2\tau y}{w}+\frac{w^2+w\left(2+y^2\right)^2-2\left(y^2+y^4\right)}{w\,(1+w)\,(1+y^2)}\right.$$
$$-\frac{2y^2\,(1+w)\left(y^2-1\right)e^{-2a\tau}}{(1+y^2)\,(1-2\,(1+2w)\,y^2+y^4)}$$
$$+\frac{2\left(y^2-w\left(2+y^2\right)\right)}{(1+w)}e^{-a\tau}\cos(b\tau)$$

$$+\frac{2w^2y^2 + y^2\left(y^2-1\right) - w\left(2+3y^2+y^4\right)}{b\left(1+w\right)yw}e^{-a\tau}\sin(b\tau)$$
$$+\frac{2w^2y^2 + y^4(y^2-1) - w(1+y^2+4y^4)}{\left(1+w\right)\left(1-\left(2+4w\right)y^2+y^4\right)}e^{-2a\tau}\cos(2b\tau)$$
$$\left.-\frac{w+2wy^2-y^4}{2b\left(1+w\right)y}e^{-2a\tau}\sin(2b\tau)\right]. \tag{4.156}$$

One sees that the total population demonstrates an exponential decay along with dying oscillations. We should note, however, that for the initial conditions chosen, these terms play a minor role, remaining of the order of inverse numbers $1/N; 1/M$, of levels participating in the population dynamics. The dominating role is played by the linear term

$$\rho_1(\tau) \simeq 4\tau y/N(1+w)\pi^2 = 4t\delta_0/\left(1+\mathcal{V}^2/N\delta_0 M\delta_1\right)\pi^2, \tag{4.157}$$

which however also remains small as compared to unity within the domain $t \ll g \sim 1/\delta_0$ of applicability of the continuous-band model.

Representation by Two Level–Band Systems

The behavior of the two-band system with factorized interaction matrix elements is revealed when one considers the matrix structure of the corresponding Hamiltonian. In Fig. 4.24 we illustrate the principal features of such a structure, by representing the two-band system as a combination of two level–band systems interacting via coupled levels. Indeed, the Hamiltonian of a two band system with the factorized interaction $\hat{V} = \left|C_n\right\rangle \mathcal{V}\left\langle C_m\right|$

$$\hat{H} = \begin{pmatrix} \sum\limits_m \left|m\right\rangle \varDelta_m \left\langle m\right| & \left|c_1\right\rangle \mathcal{V}\left\langle c_0\right| \\ \left|c_0\right\rangle \mathcal{V}\left\langle c_1\right| & \sum\limits_n \left|n\right\rangle \varDelta_n \left\langle n\right| \end{pmatrix} = \begin{pmatrix} \varDelta_m \delta_m^{m'} & C_n^*\mathcal{V}C_m \\ C_m^*\mathcal{V}C_n & \varDelta_n \delta_n^{n'} \end{pmatrix} \tag{4.158}$$

can be rewritten in another representation, which comprises the states $\left|c_0\right\rangle$ and $\left|c_1\right\rangle$ as the basis states together with the other orthogonal basis states $\{\nu\}$ and $\{\mu\}$, with $\mu, \nu = 1, 2, \ldots$. The Hamiltonians of the first $\sum_n \left|n\right\rangle \varDelta_n \left\langle n\right|$ and the second $\sum_m \left|m\right\rangle \varDelta_m \left\langle m\right|$ bands are not diagonal in such a representation, but via an appropriate choice of the sets $\left|\nu\right\rangle$ and $\left|\mu\right\rangle$ one can diagonalize the blocks of the Hamiltonian corresponding to these states. This yields the representation

$$\hat{H} = \begin{pmatrix} \left|\mu\right\rangle \varDelta_\mu' \left\langle \mu\right| & \left|\mu\right\rangle H_\mu \left\langle c_1\right| & 0 & 0 \\ \left|c_1\right\rangle H_\mu^* \left\langle \mu\right| & \left|c_1\right\rangle E_1 \left\langle c_1\right| & \left|c_1\right\rangle \mathcal{V}\left\langle c_0\right| & 0 \\ 0 & \left|c_0\right\rangle \mathcal{V}\left\langle c_1\right| & \left|c_0\right\rangle E_0 \left\langle c_0\right| & \left|\nu\right\rangle H_\nu \left\langle c_0\right| \\ 0 & 0 & \left|c_0\right\rangle H_\nu^* \left\langle n'\right| & \left|\nu\right\rangle \varDelta_\nu' \left\langle \nu\right| \end{pmatrix}, \tag{4.159}$$

in Dirac notation, where summation over ν and μ is implicit, or explicitly in the matrix form

$$\hat{H} = \begin{pmatrix} . & . & . & . & . & . & . & . & . & . \\ . & \Delta'_{\mu-1} & 0 & 0 & H_{\mu-1} & 0 & 0 & 0 & 0 & . \\ . & 0 & \Delta'_{\mu} & 0 & H_{\mu} & 0 & 0 & 0 & 0 & . \\ . & 0 & 0 & \Delta'_{\mu+1} & H_{\mu+1} & 0 & 0 & 0 & 0 & . \\ . & H^*_{\mu-1} & H^*_{\mu} & H^*_{\mu+1} & E_1 & \mathcal{V} & 0 & 0 & 0 & . \\ . & 0 & 0 & 0 & \mathcal{V} & E_0 & H_{\nu+1} & H_{\nu} & H_{\nu-1} & . \\ . & 0 & 0 & 0 & 0 & H^*_{\nu+1} & \Delta'_{\nu+1} & 0 & 0 & . \\ . & 0 & 0 & 0 & 0 & H^*_{\nu} & 0 & \Delta'_{\nu} & 0 & . \\ . & 0 & 0 & 0 & 0 & H^*_{\nu-1} & 0 & 0 & \Delta'_{\nu-1} & . \\ . & . & . & . & . & . & . & . & . & . \end{pmatrix}, \tag{4.160}$$

where $E_0 = \sum_n |C_n|^2 \Delta_n$ and $E_1 = \sum_m |C_m|^2 \Delta_m$.

The precise positions of the energies Δ'_ν and Δ'_μ are not important in the regime $t \ll g$, and can, as earlier, be taken as equidistant with the spacing among the neighboring levels δ_0 and δ_1 respectively, while the new non-diagonal matrix elements H_μ and H_ν can be estimated with the help of the consideration on p. 81 of the number of levels in resonance: We express the number N of levels of the first band accessible from the state $|c_0\rangle$ as $\langle |H_\nu|^2 \rangle / \delta_0^2 = N$ and by analogy $\langle |H_\mu|^2 \rangle / \delta_1^2 = M$ for the second band, which yields $H_\nu \sim \delta_0 \sqrt{N}$ and $H_\mu \sim \delta_1 \sqrt{M}$, respectively.

We note that the combinations $\Phi_0(\varepsilon)$ and $\Phi_1(\varepsilon)$ of (4.142) now have a clear physical meaning, representing the probability amplitudes to be in the states $|c_0\rangle$ and $|c_1\rangle$, whereas (4.144) is the Fourier form of the Schrödinger equation

$$\begin{aligned} \varepsilon\Phi_0(\varepsilon) &= iN\delta_0\Phi_0(\varepsilon) + \mathcal{V}\Phi_1(\varepsilon) + \frac{i(\varepsilon + iN\delta_0)}{\varepsilon\sqrt{\pi N}} \\ \varepsilon\Phi_1(\varepsilon) &= \mathcal{V}\Phi_0(\varepsilon) + iM\delta_1\Phi_1(\varepsilon) \end{aligned} \tag{4.161}$$

for these amplitudes allowing for the irreversible decay with the rates $N\delta_0$ and $M\delta_1$ to the continua of the first and the second bands respectively. In the inhomogeneous term on the right-hand side of (4.161) one also recognizes (see p. 88) a small initial population amplitude $1/\sqrt{\pi N}$ of the state $|c_0\rangle$ at $t = 0$, and a steady state amplitude injection $\Pi(t) = \Theta(t)\sqrt{N}\delta_0/\sqrt{\pi}$ corresponding to the term $\sqrt{N}\delta_0/\varepsilon\sqrt{\pi}$.

The origin of the terms exponentially decaying, oscillating, and linearly increasing with time in (4.156) for the population is now clear: the linearly increasing term originates from the steady state population injection, which is becoming localized with the course of time near the energy $\Delta = 0$, whereas the terms exponentially decreasing in time result from the irreversible (on the scale $t \ll g$) decay of the small initial population of the state $|c_0\rangle$. The decay occurs either with or without oscillations depending on the relation between the coupling strength $\mathcal{V}$ and the typical damping rates $N\delta_0$ and $M\delta_1$ similar to the dynamics of a damped pendulum. Large coupling corresponds to the oscillatory regime, which results in satellite spikes in the population distribution over the band at the positions $\Delta \sim \pm\mathcal{V}$, seen in Fig. 4.25a–c.

But where does the initial population of the state $|c_0\rangle$ and the population amplitude supply come from? To answer this question, one has to note that the initial condition $\psi_n(t = 0) = \delta_n^0, \psi_m(t = 0) = 0$ in the representation of the basis states $|\nu\rangle$ and $|\mu\rangle$ is determined by the amplitudes $\psi_\nu(t = 0)$ and $\psi_\mu(t = 0)$ and reads $\psi_\nu(t = 0) = \langle\nu|n = 0\rangle$, $\psi_\mu(t = 0) = 0$. This yields a small initial amplitude $\langle c_0|n = 0\rangle = C_0 = 1/\sqrt{\pi N}$ for the state $|c_0\rangle$, whereas the main part (~ 1) of the initial population amplitude remains in the band states $|\nu\rangle$ either precisely at the state $|\nu = 0\rangle$ or at a small group of states located in an extremely narrow strip around it. Being coupled to the short-lived state $|c_0\rangle$ by a relatively small matrix element $H_{\nu=0} \sim \delta_0\sqrt{N}$, the state $|\nu = 0\rangle$ supplies a small population flux $\Pi \sim |H_{\nu=0}|^2 / (N\delta_0 + M\delta_1)$ given by the ratio of the squared coupling to the decay rate of $|c_0\rangle$. (Note that this ratio represents the cooperative matrix element of coupling squared divided by the width of the band discussed on p. 82.) For $N\delta_0 \sim M\delta_1$ this yields the estimate $\Pi \sim \rho_1(t)/t \sim \delta_0$ which is consistent with the result of (4.157).

One sees that the correlations of the interaction matrix elements play an important role and may have a significant effect on the population dynamics in a multilevel quantum system. The Fano problem discussed in Sect. 3.2.2 and the problems that have been considered in this section show that the correlations manifest themselves in non-Lorentzian profiles of the population distribution over the band and in the non-exponential time-dependences of the total populations of the bands.

4.5.3 Regime of Stabilization for the Correlated Coupling

An interesting regime exists in a system consisting of an equidistant discrete band coupled to a continuum by a correlated coupling V. It turns out that for time t considerably exceeding the inverse spacing $1/\delta$ between the neighboring discrete levels, the rate of the band population decay to continuum decreases with the increase of V. We consider the continuum with a state density $g = 1/\delta_1$ as a band of N discrete levels in the limit $N \to \infty$, $\delta_1 \to 0$ and employ (4.140) with $C_n = 1/\sqrt{N}$ and $C_m = 1/\sqrt{M}$, where M is the number of levels in the discrete band. This yields

$$\psi_n(\varepsilon) = \frac{\mathcal{V}}{\sqrt{MN}(\varepsilon-\Delta_n)} \sum_{m=-\infty}^{\infty} \psi_m(\varepsilon) + \frac{i\Pi(\varepsilon)}{\sqrt{N}(\varepsilon-\Delta_n)},$$

$$\psi_m(\varepsilon) = \frac{\mathcal{V}}{\sqrt{MN}(\varepsilon-\Delta_m)} \sum_{n=-\infty}^{\infty} \psi_n(\varepsilon), \tag{4.162}$$

where $\Pi(\varepsilon) = \exp(-\varepsilon^2\tau^2/4)$ is the Fourier transformed external injection (3.58) to the continuum and $\mathcal{V} = V\sqrt{MN}$.

By analogy to (4.140)–(4.146) with (4.143) of the form

$$\int_{-\infty}^{\infty} dn \frac{1}{\varepsilon - n\delta_1} = i\pi g,$$

$$\sum_{m=-\infty}^{\infty} \frac{1}{\varepsilon - m\delta} = \frac{\pi}{\delta} \cot \frac{\varepsilon\pi}{\delta} \tag{4.163}$$

suggested by (3.76), we arrive at

$$\psi_m(\varepsilon) = \frac{-\pi g \Pi(\varepsilon) V}{(\varepsilon - \Delta_m)} \left(1 - \frac{i\pi^2 g V^2}{\delta} \cot \frac{\varepsilon\pi}{\delta}\right)^{-1}. \tag{4.164}$$

We determine the level population

$$\rho_m(\varepsilon, \xi) = (\pi g V)^2 \frac{\Pi(\varepsilon)\Pi(\xi)}{(\varepsilon - \Delta_m)(\xi - \Delta_m)} \left(1 - \frac{i\pi^2 g V^2}{\delta} \cot \frac{\varepsilon\pi}{\delta}\right)^{-1} \left(1 + \frac{i\pi^2 g V^2}{\delta} \cot \frac{\xi\pi}{\delta}\right)^{-1}, \tag{4.165}$$

and find the total population $\rho_1(\varepsilon, \xi) = \sum_m \rho_m(\varepsilon, \xi)$ of the discrete band

$$\rho_1(\varepsilon, \xi) = -(\pi g V)^2 \Pi(\varepsilon)\Pi(\xi) \frac{\cot \pi\xi/\delta - \cot \pi\varepsilon/\delta}{(\xi - \varepsilon)} \frac{\pi}{\delta} \left(1 - \frac{i\pi^2 g V^2}{\delta} \cot \frac{\varepsilon\pi}{\delta}\right)^{-1} \left(1 + \frac{i\pi^2 g V^2}{\delta} \cot \frac{\xi\pi}{\delta}\right)^{-1}. \tag{4.166}$$

For large gV^2/δ and for long times $t \gg \pi/\delta$ the main contribution to the inverse Fourier transformation comes from the vicinities of the points where $\xi \simeq \varepsilon \simeq \pi/2 + n\pi$, and after having performed the replacements $\xi \to \xi + \pi/2 + n\pi$ and $\varepsilon \to \varepsilon + \pi/2 + n\pi$ we arrive at

$$\rho_1(\varepsilon, \xi) = -(\pi g V)^2 \sum_n \Pi(n\delta_1)\Pi(n\delta_1) \frac{\pi}{\delta_1} \frac{\pi}{\delta_1} \left(1 - \frac{i\pi^3 g V^2}{\delta^2} \varepsilon\right)^{-1} \left(1 + \frac{i\pi^3 g V^2}{\delta^2} \xi\right)^{-1}, \tag{4.167}$$

and hence

$$\rho_1(t) = \frac{4\delta^2}{V^2} e^{-2t\delta^2/\pi^3 g V^2} \sum_n e^{-(n\delta\tau)^2/2} = \frac{4g\delta\sqrt{2\pi}}{\tau W} e^{-2t\delta^2/\pi^2 W}. \tag{4.168}$$

One sees that the stronger the regular decay rate $2W = 2\pi g V^2$ of the band levels to the continuum at $t < 1/\delta$, the lower is this decay rate $2\delta^2/\pi^2 W$ in the regime $t \gg \pi/\delta$ under consideration. This phenomenon has been observed in Rydberg atoms interacting with a strong laser field and is known as strong field stabilization of atoms.

4.5.4 Correlation Between the Mean Squared Coupling and the Energy Position

Let us now consider a different example of the correlation, when the interaction matrix elements V_n^m do not correlate among themselves, that is, $\langle V_n^m \rangle = 0$, but their mean square values $v_n^m = \left\langle |V_n^m|^2 \right\rangle$ are different for different transitions, and therefore they correlate with the energy positions Δ_m and Δ_n of both the coupled states $|m\rangle$ and $|n\rangle$. All the results of Sect. 4.1 remain valid for this case if one does not put the mean squared interactions in front of the summations symbols, such that (4.20) and (4.21) read

$$X_n(\varepsilon) = \frac{1}{\varepsilon - \Delta_n - \sum\limits_m v_n^m X_m(\varepsilon)}$$

$$X_m(\varepsilon) = \frac{1}{\varepsilon - \Delta_m - \sum\limits_n v_m^n X_n(\varepsilon)}, \tag{4.169}$$

while (4.26) takes the form

$$\rho_m(\varepsilon,\xi) = \frac{1}{\varepsilon - \Delta_m - \sum\limits_n v_m^n X_n(\varepsilon)} \frac{1}{\xi - \Delta_m - \sum\limits_n v_m^n X_n(\xi)} \sum_n v_m^n \rho_n(\varepsilon,\xi)$$

$$\rho_n(\varepsilon,\xi) = \frac{1}{\varepsilon - \Delta_n - \sum\limits_m v_n^m X_m(\varepsilon)} \frac{1}{\xi - \Delta_n - \sum\limits_m v_n^m X_m(\xi)} \sum_m v_n^m \rho_m(\varepsilon,\xi)$$

$$+\delta_0^n \frac{1}{\xi - \sum\limits_m v_0^m X_m(\xi)} \frac{1}{\varepsilon - \sum\limits_m v_0^m X_m(\varepsilon)}. \tag{4.170}$$

We concentrate here on the case of a factorized form of the mean squared interaction $v_n^m = c_n V^2 c_m$ and two identical bands for which one obtains the relations

$$X_n(\varepsilon) = X_m(\varepsilon) = \frac{1}{\varepsilon - \Delta_m - c_m V^2 s(\varepsilon)} \tag{4.171}$$

$$s(\varepsilon) = \sum_n c_n X_n(\varepsilon) \tag{4.172}$$

yielding the renormalization equation

$$s(\varepsilon) = \sum_n \frac{c_n}{\varepsilon - \Delta_n - c_n V^2 s(\varepsilon)}, \tag{4.173}$$

that has much in common with (4.22), although differs from it.

Note that $s(\xi)$ is the complex conjugate function of $s(\varepsilon)$, that is, $s(\xi) = s^*(\varepsilon)|_{\varepsilon=\xi}$. For this case (4.26) reads

$$\rho_m(\varepsilon,\xi) = \frac{1}{\varepsilon - \Delta_m - c_m V^2 s(\varepsilon)} \frac{1}{\xi - \Delta_m - c_m V^2 s(\xi)} c_m V^2 \tilde{\rho}_0(\varepsilon,\xi)$$

$$\rho_n(\varepsilon,\xi) = \frac{1}{\varepsilon - \Delta_n - c_n V^2 s(\varepsilon)} \frac{1}{\xi - \Delta_n - c_n V^2 s(\xi)} c_n V^2 \tilde{\rho}_1(\varepsilon,\xi)$$

$$+\delta_0^n \frac{1}{\xi - c_0 V^2 s(\xi)} \frac{1}{\varepsilon - c_0 V^2 s(\varepsilon)}, \tag{4.174}$$

where $\tilde{\rho}_0(\varepsilon,\xi) = \sum_n c_n \rho_n(\varepsilon,\xi)$ and $\tilde{\rho}_1(\varepsilon,\xi) = \sum_m c_m \rho_m(\varepsilon,\xi)$. Multiplication of the first and the second parts of (4.174) by c_m and c_n respectively followed by summation over all m and n results in the system of equations

$$\tilde{\rho}_1(\varepsilon,\xi) = \tilde{F}(\varepsilon,\xi)\tilde{\rho}_0(\varepsilon,\xi)$$

$$\tilde{\rho}_0(\varepsilon,\xi) = \tilde{F}(\varepsilon,\xi)\tilde{\rho}_1(\varepsilon,\xi) + \tfrac{c_0}{\xi - c_0 V^2 s(\xi)} \tfrac{1}{\varepsilon - c_0 V^2 s(\varepsilon)} \tag{4.175}$$

for the auxiliary quantities $\tilde{\rho}_0(\varepsilon,\xi)$ and $\tilde{\rho}_1(\varepsilon,\xi)$, which has the solution

$$\tilde{\rho}_0(\varepsilon,\xi) = \frac{-c_0}{\left(\tilde{F}(\varepsilon,\xi)^2 - 1\right)(\xi - c_0 V^2 s(\xi))(\varepsilon - c_0 V^2 s(\varepsilon))}$$

$$\tilde{\rho}_1(\varepsilon,\xi) = \frac{-\tilde{F}(\varepsilon,\xi) c_0}{\left(\tilde{F}(\varepsilon,\xi)^2 - 1\right)(\xi - c_0 V^2 s(\xi))(\varepsilon - c_0 V^2 s(\varepsilon))}. \tag{4.176}$$

After substitution into (4.174) this yields, for the total populations,

$$\rho_1(\varepsilon,\xi) = F(\varepsilon,\xi)\tilde{\rho}_0(\varepsilon,\xi)$$

$$\rho_0(\varepsilon,\xi) = F(\varepsilon,\xi)\tilde{\rho}_1(\varepsilon,\xi) + \frac{1}{\xi - c_0 V^2 s(\xi)} \frac{1}{\varepsilon - c_0 V^2 s(\varepsilon)}. \tag{4.177}$$

Here we have employed the notation

$$\tilde{F}(\varepsilon,\xi) = \sum_m \frac{V c_m}{\varepsilon - \Delta_m - c_m V^2 s(\varepsilon)} \frac{V c_m}{\xi - \Delta_m - c_m V^2 s(\xi)}$$

$$F(\varepsilon,\xi) = \sum_m \frac{V^2}{\varepsilon - \Delta_m - c_m V^2 s(\varepsilon)} \frac{c_m}{\xi - \Delta_m - c_m V^2 s(\xi)}. \tag{4.178}$$

Note that the sum $\rho_+(\varepsilon,\xi) = \rho_1(\varepsilon,\xi) + \rho_0(\varepsilon,\xi)$ can be written in the form

$$\rho_+(\varepsilon,\xi) = F(\varepsilon,\xi)\left(\tilde{\rho}_0(\varepsilon,\xi) + \tilde{\rho}_1(\varepsilon,\xi)\right) + \frac{1}{\xi - c_0 V^2 s(\xi)} \frac{1}{\varepsilon - c_0 V^2 s(\varepsilon)}$$
$$= \frac{1 - \tilde{F}(\varepsilon,\xi) - c_0 F(\varepsilon,\xi)}{1 - \tilde{F}(\varepsilon,\xi)} \frac{1}{\xi - c_0 V^2 s(\xi)} \frac{1}{\varepsilon - c_0 V^2 s(\varepsilon)}. \tag{4.179}$$

After the inverse Fourier transformation it should be identical to 1, which implies a hidden functional relation among $\tilde{F}(\varepsilon,\xi)$, $F(\varepsilon,\xi)$, $s(\xi)$ and $s(\varepsilon)$, and indeed, by taking apart the fractional factors in the first part of (4.178) and with allowance for (4.173) one obtains, after straightforward calculations,

$$1 - \tilde{F}(\varepsilon,\xi) = \frac{\varepsilon - \xi}{s(\varepsilon) - s(\xi)} \frac{F(\varepsilon,\xi)}{V^2}, \tag{4.180}$$

which yields

$$\rho_+(\varepsilon,\xi) = \frac{\varepsilon - \xi - c_0 V^2 (s(\varepsilon) - s(\xi))}{\varepsilon - \xi} \frac{1}{\xi - c_0 V^2 s(\xi)} \frac{1}{\varepsilon - c_0 V^2 s(\varepsilon)}$$
$$= \frac{1}{\varepsilon - \xi} \frac{1}{\xi - c_0 V^2 s(\xi)} - \frac{1}{\varepsilon - \xi} \frac{1}{\varepsilon - c_0 V^2 s(\varepsilon)}. \tag{4.181}$$

Indeed, consider the first fraction in the last part of the equation. In the inverse Fourier transformation it corresponds to the term, which as a function of ε, has only one pole at the point $\varepsilon = \xi$. We therefore integrate this term over ε first, thus removing the time-dependence. The analytical properties of the function $s(\xi)$ dictated by the causality principle are such that the remaining factor $(\xi - c_0 V^2 s(\xi))^{-1}$ has no singularities in the lower part of the complex plane, and therefore by transforming the initial integration contour along the real axis into the integration contour C_2, shown in Fig. 3.2 (the circumvention direction is the opposite to that shown in the figure) one gets the numerical contribution $1/2$ implied by the condition $s(\xi) \to 0$ at $|\xi| \to \infty$. The other half comes from the complex conjugate (the second) term in (4.181).

The equation for the population difference is similar to (4.29). It emerges from (4.176) and (4.177) with the allowance for (4.180) and reads

$$\Delta\rho(\varepsilon,\xi) = \left(\frac{\tilde{F}(\varepsilon,\xi) - 1}{\tilde{F}(\varepsilon,\xi) + 1} \frac{V^2 c_0 (s(\varepsilon) - s(\xi))}{(\varepsilon - \xi)} + 1\right)$$
$$\frac{1}{\xi - c_0 V^2 s(\xi)} \frac{1}{\varepsilon - c_0 V^2 s(\varepsilon)}, \tag{4.182}$$

whereas the distribution over the second band results from Eqs. (4.174, 4.176) and (4.180), which yields

$$\rho_m(\varepsilon,\xi) = \frac{1}{\varepsilon - \Delta_m - c_m V^2 s(\varepsilon)} \frac{1}{\xi - \Delta_m - c_m V^2 s(\xi)}$$
$$\frac{(s(\varepsilon) - s(\xi))\, c_0 c_m V^4}{F(\varepsilon,\xi)\,(\varepsilon - \xi)\,(\xi - c_0 V^2 s(\xi))\,(\varepsilon - c_0 V^2 s(\varepsilon))}. \tag{4.183}$$

In the long-time limit, the main contribution comes from the point $\varepsilon = \xi$, where $\tilde{F}(\varepsilon,\xi) = 1$ according to (4.180), and the function $F(\varepsilon,\xi)$ is regular, while the difference $s(\varepsilon) - s(\xi)$ assumes an imaginary value. Therefore (4.183) after the inverse Fourier transformation takes the form

$$\rho_m(\Delta) = \int \frac{1}{\varepsilon - \Delta - c(\Delta) V^2 s(\varepsilon)} \frac{1}{\varepsilon - \Delta - c(\Delta) V^2 s^*(\varepsilon)}$$
$$\frac{2\pi i\,(s(\varepsilon) - s^*(\varepsilon))\, c_0 c(\Delta) V^4}{F(\varepsilon,\varepsilon)\,(\varepsilon - c_0 V^2 s^*(\varepsilon))\,(\varepsilon - c_0 V^2 s(\varepsilon))} d\varepsilon. \tag{4.184}$$

with $s^*(\varepsilon) = s(\xi)|_{\xi=\varepsilon}$. We also note that the asymptotic population distribution over the states of the first band at $t \to \infty$ coincides with that for the second band.

We now have to determine the functions $s(\varepsilon)$, $s(\xi)$, $F(\varepsilon,\xi)$, and $\tilde{F}(\varepsilon,\xi)$, entering (4.183, 4.184). For the model of two continuous bands of an identical state density g they are given by the equations

$$s(\varepsilon) = \int \frac{g d\Delta}{(\varepsilon - \Delta)/c(\Delta) - V^2 s(\varepsilon)} \tag{4.185}$$

$$F(\varepsilon,\xi) = \int \frac{V}{\frac{\varepsilon-\Delta}{c(\Delta)} - V^2 s(\varepsilon)} \frac{V}{\frac{\xi-\Delta}{c(\Delta)} - V^2 s(\xi)} \frac{g d\Delta}{c(\Delta)} \tag{4.186}$$

$$\tilde{F}(\varepsilon,\xi) = \int \frac{V}{\frac{\varepsilon-\Delta}{c(\Delta)} - V^2 s(\varepsilon)} \frac{V}{\frac{\xi-\Delta}{c(\Delta)} - V^2 s(\xi)} g d\Delta.$$

In order to proceed with the calculations we have to choose a model for the dependence $c(\Delta)$ which would allow us to evaluate the integrals (4.185, 4.186) and solve the equation (4.185) for $s(\varepsilon)$ choosing for $s(\xi)$, the complex conjugate solution. However, even for the simplest model of coupling $c(\Delta) = \Gamma/g\pi(\Delta^2+\Gamma^2)$ which has a Lorentzian profile of width Γ, the calculations turn out to be rather cumbersome. We therefore will show here just the main steps of the general analysis and concentrate mainly on the case of strong coupling $V^2 g \gg \Gamma$, which can be performed analytically. We note that the opposite limit $V^2 g \ll \Gamma$ corresponds to the case of two homogeneous bands, as has already been considered in section 4.2.

Replacements $\Delta \to \Gamma x, \varepsilon \to \Gamma\epsilon, s(\epsilon) \to 2\pi g \Gamma^2 r(\epsilon)/V^2, \kappa \to V^2/\pi^2 g \Gamma^3$ in (4.185) yield

$$r(\epsilon) = \frac{\kappa}{2} \int \frac{dx}{(\epsilon - x)(x^2+1) - 2r(\epsilon)}. \tag{4.187}$$

The denominator of the integrand is a third-order polynomial with respect to x, which has roots at the points

$$\begin{aligned}
a &= \frac{\epsilon}{3} + \frac{3 - \epsilon^2 - \Theta^{2/3}}{3\,\Theta^{1/3}}, \\
b &= \frac{\epsilon}{3} + \frac{e^{i\pi/3}\left(\epsilon^2 - 3\right) + e^{-i\pi/3}\,\Theta^{2/3}}{3\,\Theta^{1/3}}, \\
c &= \frac{\epsilon}{3} + \frac{e^{-i\pi/3}\left(\epsilon^2 - 3\right) + e^{i\pi/3}\,\Theta^{2/3}}{3\,\Theta^{1/3}},
\end{aligned} \tag{4.188}$$

where

$$\Theta(\epsilon, r) = 27r - 9\epsilon - \epsilon^3 + 3\sqrt{3}\sqrt{27r^2 + (1+\epsilon^2)^2 - 2r\epsilon(9+\epsilon^2)} \tag{4.189}$$

For $\mathrm{Im}\epsilon > 0$ the points a and b are in the upper part of the complex plane, whereas the point c is in the lower part. Therefore the integral (4.187) can be evaluated by taking the residue at the point $x = c$ which yields the algebraic equation

$$r(\epsilon) = \frac{-\pi i \kappa}{(c-a)\,(c-b)}. \tag{4.190}$$

Let us make the replacement $\Theta \to \left(e^{2i\arccos(j) - 2\pi i/3}\left(\epsilon^2 - 3\right)\right)^{3/2}$, which yields

$$\begin{aligned}
a &= \frac{\epsilon}{3} - \frac{1}{3}(j + \sqrt{3}\sqrt{1-j^2})\sqrt{\epsilon^2 - 3}, \\
b &= \frac{\epsilon}{3} - \frac{1}{3}(j - \sqrt{3}\sqrt{1-j^2})\sqrt{\epsilon^2 - 3}, \\
c &= \frac{\epsilon}{3} + \frac{2}{3}j\sqrt{\epsilon^2 - 3},
\end{aligned} \tag{4.191}$$

for the roots given by (4.188) and thereby allows one to write the expressions (4.189) and (4.190) in a much shorter form:

$$r(\epsilon) = \frac{2\left(\epsilon^2 - 3\right)^{3/2}}{27}\left[4j^3 - 3j + \frac{\epsilon(\epsilon^2 + 9)}{\left(\epsilon^2 - 3\right)^{3/2}}\right] \tag{4.192}$$

$$r(\epsilon) = \frac{3\pi i \kappa}{(\epsilon^2 - 3)\,(4j^2 - 1)}. \tag{4.193}$$

One sees that this system of equations gives an explicit expression for $r(\epsilon)$ provided one has selected the correct branch $j(\epsilon, \kappa)$ from the five possible solutions of the fifth order polynomial equation

$$\frac{81\pi i\kappa}{2\left(\epsilon^2-3\right)^{5/2}}=\left[4j^3-3j+\frac{\epsilon(\epsilon^2+9)}{\left(\epsilon^2-3\right)^{3/2}}\right](4j^2-1) \tag{4.194}$$

and substitutes this solution into (4.192). The correct branch has the asymptotic form $j(\epsilon,\kappa)\to 1$ for $\epsilon\to\infty$, as follows from the direct evaluation of the integral (4.187) in the limit

$$r(\epsilon)\simeq\frac{\pi i\kappa}{\epsilon^2}, \tag{4.195}$$

when the main contribution comes from the pole $x=\epsilon$.

The strong coupling limit $V^2g\gg\Gamma$ corresponds to the extreme $\kappa\to\infty$, where, apart from (4.195), which is valid in the domain $\Delta/\Gamma\sim\epsilon\sim\kappa^{1/3}\gg 1$ containing the main part of the population distribution, the approximate expressions

$$\begin{aligned} j &\simeq 1, \\ c &\simeq \frac{\epsilon}{3}+\frac{2}{3}\sqrt{\epsilon^2-3} \end{aligned} \tag{4.196}$$

can also be found for the auxiliary variable j and for the root c. The main contributions to the integrals of (4.185)–(4.186) and to the inverse Fourier transformation come from the point $\xi\simeq\varepsilon\simeq\Delta,\quad|\varepsilon-\xi|\ll|\varepsilon|$. With allowance for the relations $c_0=1/g\pi\Gamma$, $c(\Delta)\simeq\Gamma/g\pi\Delta^2$ this yields

$$\begin{aligned} s(\varepsilon) &\simeq \frac{2i\Gamma}{\varepsilon^2}, \\ \tilde{F}(\varepsilon,\xi) &\simeq \frac{2i\Gamma^2V^2\left(\varepsilon^{-4}+\xi^{-4}\right)}{g\pi\left(\xi-\varepsilon\right)+2i\Gamma^2V^2\left(\varepsilon^{-4}+\xi^{-4}\right)}. \end{aligned} \tag{4.197}$$

Substitution of (4.197) (4.182) gives the population difference

$$\begin{aligned} \Delta\rho\left(\varepsilon,\xi\right) = &\left(\frac{-V^2 4i\varepsilon^2}{8i\Gamma^2V^2+g\pi\varepsilon^4\left(\xi-\varepsilon\right)}+1\right) \\ &\frac{\varepsilon^2 g\pi}{\varepsilon^3 g\pi-2V^2 i}\frac{\xi^2 g\pi}{\xi^3 g\pi+2V^2 i}. \end{aligned} \tag{4.198}$$

We now perform the inverse Fourier transformation of (4.198) and obtain the time-dependent population $\Delta\rho\left(t\right)$ shown in Fig. 4.26a.

In order to obtain this dependence we consider first the contribution of the pole at $\xi=i\left(2V^2/g\pi\right)^{1/3}$ in the upper part of the complex plane ξ where the integration contour can be displaced for positive t. For this pole the only singularity in the lower part of the complex plane of ε is located at $\varepsilon=-i\left(2V^2/g\pi\right)^{1/3}$, and even at $t=0$ it yields a small $\sim\kappa^{-2/3}$ contribution to the expression (4.198), which becomes yet

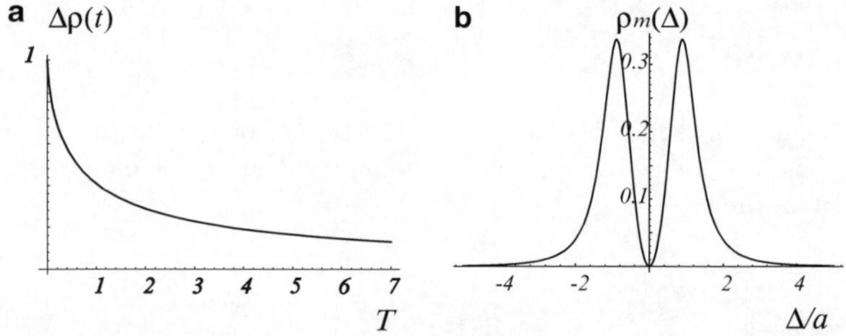

Fig. 4.26 Two identical bands with mean squared interaction depending on the position of the coupled levels: $W_{n,m}(\Delta_n, \Delta_m) = \Gamma^4 \langle V^2 \rangle g / [(\Gamma^2 + \Delta_m^2)(\Gamma^2 + \Delta_n^2)]$ in the regime $\Gamma \ll \langle V^2 \rangle g$. (**a**) The difference of the total populations of the first and the second bands depends on the scaled time $T = 2\pi t \Gamma^2 g / \langle V^2 \rangle$. (**b**) The asymptotic distribution of the populations over the energies of the band states is identical for the first and the second bands and depends only on the scaled detuning Δ/a, where $a = (2\langle V^2 \rangle / g\pi)^{1/3}$

smaller, decreasing exponentially with elapsing time. It can therefore be completely ignored for the given order of approximation. The main contribution to the inverse Fourier transformation for the population difference comes from the point $\varepsilon \simeq \xi$ and reads

$$\Delta\rho\,(t) = \int_C e^{-t8\Gamma^2 V^2/g\pi\varepsilon^4} \frac{2gV^2\varepsilon^2}{\varepsilon^6\,(g\pi)^2 + 4V^4} d\varepsilon. \tag{4.199}$$

The replacements $\varepsilon^3 \to 2\,xV^2/g\pi$ and $t \to T\,\left(2V^2/g\pi\right)^{1/3}/4\Gamma^2$ yield

$$\begin{aligned}\Delta\rho\,(t) &= \frac{1}{\pi}\int_{C_x} e^{-Tx^{-4/3}} \frac{dx}{3\,(x^2+1)} \\ &= \frac{3}{\pi}\int_{-\infty}^{\infty} e^{-Tx^{-4/3}} \frac{dx}{3\,(x^2+1)},\end{aligned} \tag{4.200}$$

where the contour C_x circumvents the point $x = 0$ three times, in order to accumulate the same phase difference of 6π as the function ε^3 after a single circumvention of the point $\varepsilon = 0$ along the contour C. This is the origin of the topological factor 3 in the last equality of (4.200). At $t = 0$ this equation gives $\Delta\rho\,(0) = 1$, and for $t \to \infty$ it yields the asymptotic power-law dependence

$$\Delta\rho\,(t) = \frac{6V^2}{g\pi^2}\left(\frac{g\pi}{8\Gamma^2 V^2 t}\right)^{3/4} \int_{-\infty}^{\infty} e^{-1/u^4} \frac{du}{u^4} \sim t^{-3/4}. \tag{4.201}$$

Let us consider now the population distribution over the energy levels of the bands suggested by (4.184). With the allowance for (4.180)–(4.197), the relations $c_0 = 1/g\pi\Gamma$, $c(\Delta) \simeq \Gamma/g\pi\Delta^2$, and the topological factor 3 mentioned above it reads

$$\rho_m(\Delta) = \frac{6V^2}{(g\pi)^2} \frac{\Delta^2}{\Delta^6 + \left(\frac{2V^2}{g\pi}\right)^2}, \tag{4.202}$$

which corresponds to a two-hump distribution of width $\left(2V^2/g\pi\right)^{1/3}$, shown in Fig. 4.26b. We note that $\int \rho_m(\Delta)\, g d\Delta = 1/2$, since the population is equally distributed amongst the states of the first and the second band of the same energy. We also note that for the regime $V^2 g \gg \Gamma$ under consideration the width Γ of the coupling line $W(\Delta)$ does not affect the population distribution profile (4.202), which is located on the far wings of this line.

Chapter 5
Soluble Time-Dependent Systems

We now consider quantum systems with time-dependent Hamiltonians. The formal expression for the evolution operator for such systems is given by the time-ordered exponent of the integral of the corresponding Hamiltonian,

$$\hat{U}(t) = \hat{T}\exp\left\{-i\int^{t}\hat{H}(\tau)d\tau\right\} = e^{-i\hat{H}(t)d\tau}e^{-i\hat{H}(t-d\tau)d\tau}\ldots e^{-i\hat{H}(\tau+d\tau)d\tau}$$
$$e^{-i\hat{H}(\tau)d\tau}\ldots e^{-i\hat{H}(2d\tau)d\tau}e^{-i\hat{H}(d\tau)d\tau}e^{-i\hat{H}(0)d\tau} \tag{5.1}$$

which simply involves the sequential application of infinitesimal evolution operators $e^{-i\hat{H}(\tau)d\tau}$ with instantaneous Hamiltonians. Here $\hat{T}$ denotes the time-ordering operator. Usually this expression is not very practical when one looks for an exact analytic solution. Moreover, it turns out that for a multilevel quantum system, the possibility to achieve an analytic description is strongly related to the algebraic structure of the time-dependent Hamiltonian suggested by Lie group theory. We therefore start our consideration by introducing some general ideas, remarks, and results of group theory required for our analysis.

We continue with the analysis of a number of exactly soluble problems widely known as the Landau–Zener problem and its generalization to the case of decaying and dephasing levels. We also consider a two-level system for two special cases of the time-dependent perturbation that can be performed are obtained in terms of the hypergeometric functions, which is almost always the case for the exactly soluble problems. Under certain conditions, the hypergeometric functions can be reduced to other special functions, which is the case for exponentially rising coupling. Then we turn to the adiabatic approximation and show that the probability of non-adiabatic transitions is given by the Dykhne formula, also valid for multilevel time-dependent systems.

The next classical example of multilevel time-dependent systems is the Demkov–Osherov system consisting of a level moving across a band. This problem can be solved exactly by the Laplace contour integral method. In this context we address

V.M. Akulin, *Dynamics of Complex Quantum Systems*, Theoretical and Mathematical Physics, DOI 10.1007/978-94-007-7205-2_5,

a question about non-adiabatic processes at the edges of the band. The Laplace method allows one to find a solution in the form of a contour integral for another problem—a level placed at a certain energy difference underneath a continuous band with the coupling exponentially increasing between the level and the band. It turns out that in spite of the fact that, with the course of time, the coupling rapidly exceeds the detuning between the level and the edge of the band, the level keeps the main part of its population, being only repelled toward negative energies.

5.1 Algebraic Structure of Time-Dependent Systems

The dynamics of a quantum system with time-dependent parameters is often more involved compared to the stationary analog. This is always the case when the system Hamiltonians corresponding to different moments in time do not commute. The algebra of their commutators of various orders, namely the rules by which the commutators can be expressed in terms of a linear combinations of operators $\hat{\mathcal{G}}_n$ (where $n = 0 \ldots N_o$, and $\hat{\mathcal{G}}_0 = \hat{I}$) comprising a set of N_o so-called generators of a Lie group, play an important role and shift the consideration of the problem to the domain of Lie group theory. In the following sections, we illustrate the group theory background of several exactly soluble problems, and here we demonstrate it for the simplest example of a system with a Hamiltonian that takes only two values, $\hat{H} = \hat{H}_A$ or $\hat{H} = \hat{H}_B$.

Consider the system after it has been evolving with the Hamiltonian $\hat{H}_A$ during a time interval t_A and with the Hamiltonian $\hat{H}_B$ during a time interval t_B. The evolution operator reads

$$\hat{U}(t_A, t_B) = e^{-it_B\hat{H}_B} e^{-it_A\hat{H}_A}. \tag{5.2}$$

The essence of the problem is in the fact that the regular multiplication rule $e^B e^A = e^{A+B}$ is not valid for non-commuting operators $\hat{A}$ and $\hat{B}$, and must be replaced by the relation known as the Baker–Campbell–Hausdorff formula

$$\begin{aligned} e^{\hat{B}} e^{\hat{A}} &= e^{\hat{C}}, \\ \hat{C} = \; & \hat{A} + \hat{B} + \frac{1}{2}[\hat{A}, \hat{B}] + \frac{1}{12}[\hat{A}, [\hat{A}, \hat{B}]] \\ & - \frac{1}{12}[\hat{B}, [\hat{A}, \hat{B}]] + \ldots + c_{k,s}[A, [\ldots, [A, B] \ldots]] + \ldots, \end{aligned} \tag{5.3}$$

where an explicit albeit non-trivial computation rule exists (see p. 230) for the numerical coefficients $c_{k,s}$ in front of a generic commutator of k-th order. The index $s = 1, \ldots, 2^k$ enumerates different sequences of commutation and has unities and zeros in its binary representation in the order suggested by the sequence of operators $\hat{A}$ and $\hat{B}$, respectively, in the corresponding multiple commutator. For example $[A, [A, [B, [A, B]]]]$ corresponds to $s = 11010$ in the binary representation, that is, $s = 0 + 1 \times 2^1 + 0 \times 2^2 + 1 \times 2^3 + 1 \times 2^4 = 26$. Now one sees that, indeed, the commutation relations generated by two Hamiltonians $\hat{H}_A$ and $\hat{H}_B$ play a crucial role in the dynamics of the quantum system since

$$U(t_A, t_B) = \exp\Big\{ - it_B\hat{H}_B - it_A\hat{H}_A - \tfrac{t_B t_A}{2}[\hat{H}_B, \hat{H}_A] \\ - \tfrac{it_B^2 t_A}{12}[\hat{H}_B, [\hat{H}_B, \hat{H}_A]] + \tfrac{it_B t_A^2}{12}[\hat{H}_A, [\hat{H}_B, \hat{H}_A]] + \dots \Big\}. \quad (5.4)$$

Equation (5.4) gives a clue for the classification of different particular cases—the larger the dimension of the operator space produced by the commutators, that is, the larger the set of operators comprising the algebra, the more involved the quantum dynamics is. For a commuting pair of Hamiltonians the set has only two elements since the evolution operator is given by the exponent of the linear combination $-it_B\hat{H}_B - it_A\hat{H}_A$ of just two operators, and hence the operator space is two-dimensional. One can say that the evolution operator (5.2) belongs to a two-parametric Lie group. Apart from a phase factor $e^{-\frac{t_B t_A}{2}C}$ it is also the case when the commutator is a number, $C = [\hat{H}_B, \hat{H}_A]$. In a more involved case, when all of the commutators of order higher than a certain k vanish, the evolutionary operator turns to be the exponent of a linear combination of a larger, albeit finite number of terms, each of which is a commutator of some order less than k. The dimensionality N_o of the operator space can reach at most 2^k, since this value gives the number of different commutators, unless some of the commutators can be represented as linear combinations of the others, and hence the number of linearly independent elements of the set comprising the algebra is smaller.

Another case, the most interesting from the viewpoint of a comprehensive analytical consideration, corresponds to the situation when no commutators of higher order vanish, but can all be represented as a linear combination of commutators of orders less than a certain order k. In this case, all of the coefficients in (5.3) are different from zero, but nevertheless this infinite sum can be cast in a finite linear combination of commutators with the coefficients $f_n(t_A, t_B)$ given by the infinite series. In this very case, Lie group theory is the most important concept, which allows one to employ generators $\mathcal{G}_n$ of standard groups as a basis set in the space of commutators and thereby express the evolution operator in the form

$$\hat{U}(t) = e^{\sum_{n=0}^{N_o} f_n(t_A, t_B)\hat{\mathcal{G}}_n}, \quad (5.5)$$

which sometimes provides one with an exact analytical solution of the problem. In other words, this is the case where the dimensionality of the operator space generated by the commutators of the Hamiltonians at different moments of time is much smaller than the dimensionality of the Hilbert space of the considered system: $N_o \ll N$. The evolution operator belongs to an N_o-parametric Lie group, although the values of these parameters $f_n(t_A, t_B)$ are possibly rather complex functions of t_A and t_B even for the simplest particular problem (5.2) under consideration. Finally, in the case where all of the commutators of the lower orders turn out to be linearly independent until their total number N_o reaches the maximum possible number N^2 of linearly independent Hermitian matrices in a N-dimensional Hilbert space of the system, we encounter the so-called non-holonom problem. No compact general solution of the problem in terms of analytic or special functions exists for large N, and the analytic description requires statistical hypotheses, as is the case for complex stationary quantum systems.

In order to consider time-dependent Hamiltonians of a general type one could consider time as one of the parameters of the Lie group and treat the operator of the time derivative as one of the group generators. From the formal point of view this consideration relies on the method of characteristics for linear partial deferential equations as well as on the operator of a finite shift. We will not, however, dwell on the mathematical background of the approach but instead illustrate this formal construction by the following physical example. Consider a point quantum system with a time-dependent Hamiltonian $\hat{H}(t)$ for which the Schrödinger equation reads

$$i\left|\dot{\Psi}(t)\right\rangle = \hat{H}(t)\left|\Psi(t)\right\rangle . \tag{5.6}$$

Imagine that the time-dependence of the Hamiltonian originates from the fact that the system moves with a constant velocity v along the axis x, whereas the Hamiltonian $\hat{H}(t) = \hat{H}(x/v)|_{x=vt}$ depends on the position $x(t) = vt$ of the system on this axis. Equation (5.6) describes the evolution of the state vector $|\Psi(t)\rangle$ in the reference frame attached to the moving system, whereas in the reference frame at rest, the Schrödinger equation adopts the form

$$i\left|\dot{\Psi}(t,x)\right\rangle = \left[\hat{H}\left(\frac{x}{v}\right) - iv\frac{\partial}{\partial x}\right]\left|\Psi(t,x)\right\rangle , \tag{5.7}$$

where the last term in the square brackets allows for the Galilean transformation to a reference frame moving with a constant velocity. Once the solution $|\Psi(t,x)\rangle$ of (5.7) is found, one should set $x = vt$. To shorten the notation, we take $x = v\tau$ and rewrite (5.7) in the form

$$i\left|\dot{\Psi}(t,\tau)\right\rangle = \left[\hat{H}(\tau) - i\frac{\partial}{\partial \tau}\right]\left|\Psi(t,\tau)\right\rangle , \tag{5.8}$$

where the new effective Hamiltonian $[\hat{H}(\tau) - i\frac{\partial}{\partial\tau}]$ depends on the new "time" variable τ, but does not depend on "real" time t. Therefore the operator of evolution corresponding to (5.8) reads

$$\hat{U}(t) = e^{-it[\hat{H}(\tau) - i\frac{\partial}{\partial\tau}]}, \tag{5.9}$$

which implies

$$|\Psi(t,\tau)\rangle = e^{-it[\hat{H}(\tau) - i\frac{\partial}{\partial\tau}]}\left|\Psi(0,\tau)\right\rangle , \tag{5.10}$$

given the initial condition $|\Psi(0,\tau)\rangle$ at $t=0$. One can choose an arbitrary initial condition for the state vector $|\Psi(t,\tau)\rangle$ in two-dimensional time space, which is restricted only by the correspondence requirement $|\Psi(0,0)\rangle = |\Psi(0)\rangle$ suggested by the initial condition for the original state vector $|\Psi(t)\rangle$ entering (5.6).

Let us consider first the evolution operator (5.9) in the particular case $\hat{H}(\tau)=0$, that is,

$$\hat{U}(t) = e^{-t\frac{\partial}{\partial\tau}}. \tag{5.11}$$

The application of this operator to an arbitrary function or a state vector, say $|\Psi(0,\tau)\rangle$, with allowance for Taylor expansion formula yields

$$e^{-t\frac{\partial}{\partial\tau}}\,|\Psi(0,\tau)\rangle = \sum_{n=0}^{\infty}\frac{(-t)^n}{n!}\left(\frac{\partial}{\partial\tau}\right)^n|\Psi(0,\tau)\rangle = |\Psi(0,\tau-t)\rangle\,. \tag{5.12}$$

One sees that application of the operator (5.11) results in a shift of the argument τ of an arbitrary function or state vector by a finite value $-t$. Therefore one can refer to $e^{t\frac{\partial}{\partial\tau}}$ as the operator of finite shift by the value t. Note the evident consequence of (5.10) for $\hat{H}(\tau)=0$ and (5.12):

$$|\Psi(t,\tau)\rangle = e^{-t\frac{\partial}{\partial\tau}}\,|\Psi(0,\tau)\rangle = |\Psi(0,\tau-t)\rangle\,, \tag{5.13}$$

therefore

$$|\Psi(t)\rangle = |\Psi(t,\tau)\rangle|_{\tau=t} = |\Psi(0,\tau-t)\rangle|_{\tau=t} = |\Psi(0,0)\rangle = |\Psi(0)\rangle\,, \tag{5.14}$$

as it should for $\hat{H}(\tau)=0$, which simply indicates displacement of a non-evolving system in space with constant velocity.

Let us now consider the evolution operator (5.9) in the general case, rewrite it in the equivalent form

$$\hat{U}(t) = e^{-t\frac{\partial}{\partial\tau}}e^{t\frac{\partial}{\partial\tau}}e^{-it[\hat{H}(\tau)-i\frac{\partial}{\partial\tau}]}, \tag{5.15}$$

and apply the Baker–Campbell–Hausdorff formula (5.3) to the last two factors. This yields

$$\hat{U}(t,\tau) = e^{-t\frac{\partial}{\partial\tau}}e^{-it\hat{H}(\tau)-t\frac{\partial}{\partial\tau}+t\frac{\partial}{\partial\tau}+\frac{1}{2}[-it\hat{H}(\tau)-t\frac{\partial}{\partial\tau},t\frac{\partial}{\partial\tau}]+\dots}. \tag{5.16}$$

One sees that the operator of finite shift $e^{-t\frac{\partial}{\partial\tau}}$ acts on the exponent of the argument

$$\begin{aligned}&-it\hat{H}(\tau)+\tfrac{1}{2}[-it\hat{H}(\tau),t\tfrac{\partial}{\partial\tau}]+\tfrac{1}{12}[t\tfrac{\partial}{\partial\tau},[-it\hat{H}(\tau),t\tfrac{\partial}{\partial\tau}]]\\&-\tfrac{1}{12}[-it\hat{H}(\tau)-t\tfrac{\partial}{\partial\tau},[-it\hat{H}(\tau),t\tfrac{\partial}{\partial\tau}]]+\dots,\end{aligned} \tag{5.17}$$

which, apart from the time-dependent Hamiltonian $\hat{H}(\tau)$ comprises commutators of different orders of the Hamiltonian and the time derivative $\frac{\partial}{\partial\tau}$. From the relation

$$
\begin{aligned}
&[\tfrac{\partial}{\partial\tau}, \hat{H}(\tau)]\,|\Psi(t,\tau)\rangle = \tfrac{\partial}{\partial\tau}\hat{H}(\tau)\,|\Psi(t,\tau)\rangle - \hat{H}(\tau)\tfrac{\partial}{\partial\tau}\,|\Psi(t,\tau)\rangle \\
&= \left[\tfrac{\partial}{\partial\tau}\hat{H}(\tau)\right]|\Psi(t,\tau)\rangle \quad + \hat{H}(\tau)\tfrac{\partial}{\partial\tau}\,|\Psi(t,\tau)\rangle - \hat{H}(\tau)\tfrac{\partial}{\partial\tau}\,|\Psi(t,\tau)\rangle \\
&= \dot{\hat{H}}(\tau)\,|\Psi(t,\tau)\rangle
\end{aligned}
\tag{5.18}
$$

it becomes evident that commutation of a time-dependent operator with the time derivative yields the time derivative of this operator. Equation (5.16) now reads

$$
\begin{aligned}
\hat{U}(t,\tau) = e^{-t\frac{\partial}{\partial\tau}} exp\{-it\hat{H}(\tau) - \frac{it^2}{2}\dot{\hat{H}}(\tau) + \dots \\
+ \frac{it^k a_{k,n,\dots,m,l}}{2}[\hat{H}^{(n)}(\tau), [\dots, [\hat{H}^{(m)}(\tau), \hat{H}^{(l)}(\tau)]\dots] + \dots\},
\end{aligned}
\tag{5.19}
$$

where dots denote commutators of different orders of the time derivatives $\dot{\hat{H}}(\tau)$, $\ddot{\hat{H}}(\tau), \dots, \hat{H}^{(n)}(\tau)\dots$ of the Hamiltonian that enter the series with the coefficients $a_{k,n,\dots,m,l}$ related in a certain way to the coefficients $c_{k,s}$ in the Baker–Campbell–Hausdorff series (5.3).

Note that the approach based on the expansion (5.19) gives a tool which helps us to determine whether or not a quantum system possesses a hidden symmetry, which would allow one to reduce the original problem to a problem of smaller dimension. That is, it allows one to identify the situation when the space spanned by the commutators of $\hat{H}^{(n)}(\tau)$ has a smaller dimension compared to the dimension of the entire Hilbert space. Unfortunately this method does not suggest any practical way to calculate the operator of evolution, since the coefficients $c_{k,s}$ and hence $a_{k,n,\dots,m,l}$ are given by a recurrent computation rule and there is no explicit expression available for these coefficients at the moment.

We also note that the application of the operator of finite shift in (5.16) results in

$$
\hat{U}(t,\tau) = e^{-it\hat{H}(\tau-t) - \frac{it^2}{2}\dot{\hat{H}}(\tau-t) + \dots}, \tag{5.20}
$$

and once we set $\tau = x/v = t$ [that is, we remain on the characteristics, or in other words on the trajectory, see (5.6)–(5.7)] it yields the equation

$$
\begin{aligned}
\hat{U}(t) = \exp\Big\{-it\hat{H}(0) - \frac{it^2}{2}\dot{\hat{H}}(0) - \frac{it^3}{3!}\ddot{\hat{H}}(0) + \dots \\
+ a_{k,n,\dots,m,l}\frac{it^k}{2}[\hat{H}^{(n)}(0), [\dots, [\hat{H}^{(m)}(0), \hat{H}^{(l)}(0)]\dots] + \dots\Big\} \\
= \hat{T}\exp\left\{-i\int^t \hat{H}(\tau)d\tau\right\},
\end{aligned}
\tag{5.21}
$$

which plays the role of the Taylor expansion for the integral in the time-ordered exponent $\hat{T}\exp\left\{\int^t \hat{H}(\tau)d\tau\right\}$ encountered earlier in (5.1). Indeed, in the first line of (5.21) one recognizes the Taylor expansion of the integral $\int^t \hat{H}(\tau)d\tau$ such that (5.21) has the structure

$$\begin{aligned}\hat{U}(t) &= \hat{T}\exp\left\{-i\int^t \hat{H}(\tau)d\tau\right\}\\ &= \exp\left[-i\int^t \hat{H}(\tau)d\tau - i\hat{R}\left(t;\hat{H}^{(n)}(0)\right)\right],\end{aligned} \tag{5.22}$$

where the remainder $\hat{R}(t,\hat{H}^{(n)}(0))$ is a Hermitian operator, which has a complex explicit form, contains the commutators of the Hamiltonian and its derivatives of different orders, and thereby allows for the non-commutativity of the Hamiltonian corresponding to different moments in time.

Consider an important particular case of time-dependence. A Hamiltonian periodic in time of period $2\pi/\omega$ can be represented as a Fourier series $\hat{H}(\tau) = \sum_n \hat{H}_n e^{-i\omega n\tau}$. One can also cast the state vector into a Fourier series

$$|\Psi(t,\tau)\rangle = \sum_n e^{-i\omega n\tau}\,|\Psi_n(t)\rangle \tag{5.23}$$

and rewrite (5.8) in the form

$$i\left|\dot{\Psi}_m(t)\right\rangle = \sum_k \hat{H}_{m-k}\,|\Psi_k(t)\rangle - \omega m\,|\Psi_m(t)\rangle\,, \tag{5.24}$$

which is a system of linear first-order differential equations with time independent coefficients. One can represent the state vector (5.23) as a double sum

$$|\Psi(t,\tau)\rangle = \sum_{n,j} e^{-i\omega n\tau}\psi_{n,j}(t)\,|j\rangle\,, \tag{5.25}$$

where the states $|j\rangle$ form a complete basis set for the system, which yields

$$i\dot{\psi}_{m,j}(t) = \sum_{k,l} H_{m-k,j,l}\psi_{k,l}(t) - \omega m\psi_{m,l}(t). \tag{5.26}$$

The Fourier transformation of this equation over the time t yields

$$E\psi_{m,l} = \sum_{k,l} H_{m-k,j,l}\psi_{k,l} - \omega m\psi_{m,l}. \tag{5.27}$$

The eigenvalues E_ν of this equation are called quasienergies by analogy with the quasimomenta for the systems with Hamiltonians periodic in space, while such a form of the time-dependent Schrödinger equation (5.6) is called the quasienergy representation. This method is also known as the Floquet approach.

The initial condition for (5.26) reads

$$|\Psi(t=0)\rangle = \sum_{n,j} \psi_{n,j}(t=0)\,|j\rangle\,. \tag{5.28}$$

Later on we will see that, for the case where the time-dependence results from the interaction of the system with a quantized external electromagnetic field, the coefficients $\psi_{n,j}(0)$ have the clear physical interpretation of the probability amplitudes to find the system in the quantum state $|j\rangle$ and the field in the state with n photons at $t=0$. However, for a classical electromagnetic field, or in the general case of a periodic Hamiltonian, the initial condition (5.28) alone does not permit one to determine unambiguously all of the coefficients $\psi_{n,j}(t=0)$, as was the case for (5.10). One of the most common ways to choose the initial condition for the amplitudes $\psi_{n,j}(t=0)$ is to take them as identical for all the quasienergy indices $\psi_{n,j} = \psi_{n',j}$, which actually corresponds to the classical limit of the coherent quantum field state.

5.2 Time-Dependent Two-Level Systems

The simplest example of systems with time-dependent Hamiltonians are two-level systems. The corresponding Schrödinger equation has the form of a set of two linear, first-order differential equations

$$\begin{aligned} i\dot{\psi}_0 &= \Delta_0(t)\psi_0 + V(t)\psi_1, \\ i\dot{\psi}_1 &= \Delta_1(t)\psi_1 + V(t)\psi_0, \end{aligned} \tag{5.29}$$

for the probability amplitudes ψ_0 and ψ_1 to be in the states $|0\rangle$ and $|1\rangle$ respectively, and is similar to (3.7) for the time-independent Hamiltonian. It can also be rewritten in the equivalent form of one linear second-order differential equation with time-dependent coefficients

$$\ddot{\psi}_1 + \left(i\Delta_1 + i\Delta_0 - \frac{\dot{V}}{V}\right)\dot{\psi}_1 + \Delta_1\left(i\frac{\dot{\Delta}_1}{\Delta_1} - i\frac{\dot{V}}{V} - \Delta_0 + \frac{V^2}{\Delta_1}\right)\psi_1 = 0, \tag{5.30}$$

which just by interchange of indices yields the analogous equation for ψ_0. Solutions of a large number of second-order linear differential equations are known. They can be found and expressed in terms of special functions, provided a proper replacement and change of the variable equation (5.30) is reduced to one of the standard forms of the second-order equations for the special functions. In this section we consider several examples that give insight into the typical behavior of two-level systems in various conditions.

5.2.1 Landau–Zener Problem

In Fig. 5.1 we show an example of an exactly soluble problem, known as the Landau–Zener problem, considering a two-level system with a linear time dependence αt of the energy distance between the levels and a time-independent coupling V of the levels. The corresponding Schrödinger equation reads

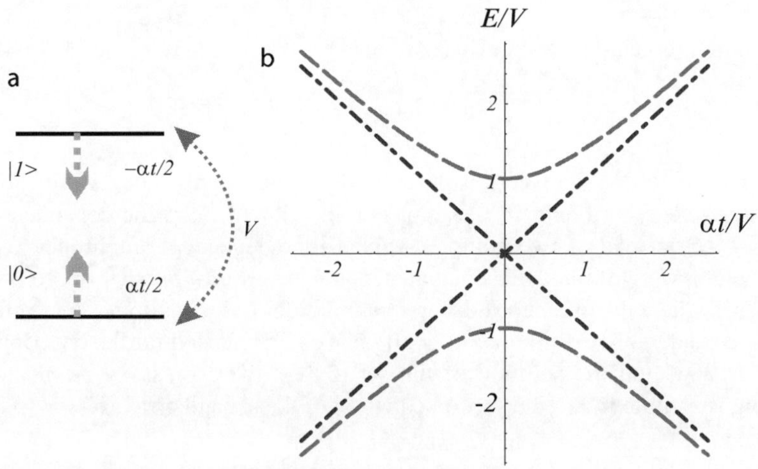

Fig. 5.1 Landau–Zener problem. (**a**) Two states move toward each other with linear dependence of the energy difference with time. At $t = 0$ the level crossing occurs, which means that the positions of the levels on the energy scale coincide, and for positive times the levels continue their motion with the same velocity α. The states are coupled by the interaction V. (**b**) Position of the level energies and the Hamiltonian eigenstates of the system as a function of time. The dot-dash lines show the so-called diabatic levels, that is, the energy positions of the states in the absence of the interaction. The dashed lines show the so-called adiabatic levels, that is, the energy eigenstates of the system corresponding to the instantaneous Hamiltonian

$$
\begin{aligned}
i\dot{\psi}_0 &= \frac{\alpha t}{2}\psi_0 + V\psi_1, \\
i\dot{\psi}_1 &= -\frac{\alpha t}{2}\psi_1 + V\psi_0.
\end{aligned}
\tag{5.31}
$$

In order to illustrate the speculations of the previous section, it is convenient to represent the Hamiltonian

$$\hat{H} = \frac{\alpha t}{2}\hat{\sigma}_z + V\hat{\sigma}_x \tag{5.32}$$

with the help of the three Pauli matrices,

$$\hat{\sigma}_x = \begin{pmatrix} 0 & 1 \\ 1 & 0 \end{pmatrix}; \ \hat{\sigma}_y = \begin{pmatrix} 0 & -i \\ i & 0 \end{pmatrix}; \ \hat{\sigma}_z = \begin{pmatrix} 1 & 0 \\ 0 & -1 \end{pmatrix}, \tag{5.33}$$

which together with the unit matrix $\hat{I}$ generate the $SU(2)$ group and comprise a closed Lie algebra of three elements $\hat{\mathcal{G}}_1 = \hat{\sigma}_x$; $\hat{\mathcal{G}}_2 = \hat{\sigma}_y$; $\hat{\mathcal{G}}_3 = \hat{\sigma}_z$ satisfying the commutation relations

$$[\hat{\sigma}_x, \hat{\sigma}_y] = 2i\hat{\sigma}_z; \ [\hat{\sigma}_z, \hat{\sigma}_x] = 2i\hat{\sigma}_y; \ [\hat{\sigma}_y, \hat{\sigma}_z] = 2i\hat{\sigma}_x. \tag{5.34}$$

The evolution operator (5.5) for this case reads

$$\hat{U}(t) = e^{f_0(t)\hat{I} + f_x(t)\hat{\sigma}_x + f_y(t)\hat{\sigma}_y + f_z(t)\hat{\sigma}_z}. \tag{5.35}$$

Such a representation does not result, however, in any significant simplification of the Landau–Zener problem since it implies utilization of four time-dependent coefficients $f_n(t)$ instead of two complex-valued time-dependent amplitudes $\psi_{0,1}(t)$. Nevertheless, we will trace this equivalence which is useful in other, more complicated cases where the number of different amplitudes (which is sometimes infinite), exceeds considerably the number of coefficients $f_n(t)$ needed for the description.

Let us now find the solution of the set (5.31). We express ψ_1 from the first equation, substitute it into the second equation of the set and arrive at

$$\left(i\frac{\partial}{\partial t} + \frac{\alpha t}{2}\right)\left(i\frac{\partial}{\partial t} - \frac{\alpha t}{2}\right)\psi_0 = V^2\psi_0, \tag{5.36}$$

which has the form of the Schrödinger equation for an eigenfunction $\psi(x)$ of a particle of mass $1/2$ moving in the parabolic potential $-\alpha^2 x^2/4$ with energy $V^2 + i\alpha/2$, provided we have replaced time t by the coordinate x. In the combinations $i\frac{\partial}{\partial t} + \frac{\alpha t}{2}$ and $\frac{\alpha t}{2} - i\frac{\partial}{\partial t}$ one recognizes the harmonic oscillator annihilation and creation operators, respectively, which together with their product comprise generators of W_1 Lie algebra. Due to this fact, the solution of (5.36) can be explicitly expressed in terms of special functions. It can either be found in handbooks on special functions in the chapters dedicated to parabolic cylinder functions, or just calculated directly by substituting a Taylor series into (5.36), which after the replacement $\psi_0(t) = \varphi_0(t)e^{-i\alpha t^2/4}$ takes the form

$$\left(\frac{\partial^2}{\partial t^2} - i\alpha t\frac{\partial}{\partial t} + V^2\right)\varphi_0(t) = 0. \tag{5.37}$$

Two linearly independent solutions correspond to even $\varphi_0(t) = \sum_{n=0}^{\infty} \frac{a_n}{n!} t^{2n}$ and odd $\varphi_0(t) = \sum_{n=0}^{\infty} \frac{b_n}{n!} t^{2n+1}$ functions of time. Direct substitution of these series into (5.37) yields the recurrence relations

$$\left(n+\tfrac{1}{2}\right) a_{n+1} = \tfrac{i\alpha}{2}\left(n - \tfrac{V^2}{2i\alpha}\right) a_n,$$
$$\left(n+\tfrac{3}{2}\right) b_{n+1} = \tfrac{i\alpha}{2}\left(n - \tfrac{V^2}{2i\alpha} + \tfrac{1}{2}\right) b_n, \tag{5.38}$$

for the coefficients a_n and b_n. The same relations among coefficients hold for the confluent (or degenerate) hypergeometric functions

$$_1F_1(\beta;\gamma;z) = 1 + \frac{\beta}{\gamma}\frac{z}{1!} + \frac{\beta(\beta+1)}{\gamma(\gamma+1)}\frac{z^2}{2!} + \ldots, \tag{5.39}$$

when $z = i\alpha t^2/2$, $\beta = -V^2/2i\alpha$, $\gamma = 1/2$ for even and $z = i\alpha t^2/2$, $\beta = 1/2 - V^2/2i\alpha$, $\gamma = 3/2$ for odd functions. Therefore, the general solution of (5.36) reads

$$\psi_0(t) = \left[c_1\, {_1F_1}\Big(-\frac{V^2}{2i\alpha}; \frac{1}{2}; \frac{i\alpha t^2}{2}\Big) + c_2 t\, {_1F_1}\Big(\frac{1}{2} - \frac{V^2}{2i\alpha}; \frac{3}{2}; \frac{i\alpha t^2}{2}\Big)\right] e^{-i\alpha t^2/4}, \tag{5.40}$$

where c_1 and c_2 are two constants. From (5.31) one sees that the replacement $\alpha \to -\alpha$ corresponds to the permutation $\psi_0(t) \rightleftharpoons \psi_1(t)$ and therefore the general solution for $\psi_1(t)$ reads

$$\psi_1(t) = \left[c_3\, {_1F_1}\Big(\frac{V^2}{2i\alpha}; \frac{1}{2}; -\frac{i\alpha t^2}{2}\Big) + c_4 t\, {_1F_1}\Big(\frac{1}{2} + \frac{V^2}{2i\alpha}; \frac{3}{2}; -\frac{i\alpha t^2}{2}\Big)\right] e^{i\alpha t^2/4}, \tag{5.41}$$

where the constants c_n can be given in terms of the amplitudes at $t = 0$, that is, $c_1 = \psi_0(0)$; $c_2 = -iV\psi_1(0)$; $c_3 = -iV\psi_0(0)$; $c_4 = \psi_1(0)$. In other words the evolution operator reads

$$\hat{U}(t) = \begin{pmatrix} {_1F_1}(\frac{iV^2}{2\alpha}; \frac{1}{2}; \frac{i\alpha t^2}{2}) e^{\frac{-i\alpha t^2}{4}} & \frac{Vt}{i}\, {_1F_1}(\frac{1}{2} + \frac{iV^2}{2\alpha}; \frac{3}{2}; \frac{i\alpha t^2}{2}) e^{\frac{-i\alpha t^2}{4}} \\ \frac{Vt}{i}\, {_1F_1}(\frac{1}{2} + \frac{V^2}{2i\alpha}; \frac{3}{2}; \frac{\alpha t^2}{2i}) e^{\frac{i\alpha t^2}{4}} & {_1F_1}(\frac{V^2}{2i\alpha}; \frac{1}{2}; \frac{\alpha t^2}{2i}) e^{\frac{i\alpha t^2}{4}} \end{pmatrix}, \tag{5.42}$$

which, after comparison with (5.35), determines implicitly the four functions $f_n(t)$.

One sees that the evolution operator (5.42) as well as the amplitudes (5.40)–(5.41) depend only on the dimensionless time $\sqrt{\alpha}t$ and the dimensionless adiabaticity parameter V^2/α, and in Fig. 5.2 we show the population

$$\rho_0(t) = |\psi_0(t)|^2 = \left|\left(\hat{U}(t)\hat{U}^{-1}(-\infty)\right)_{0,0}\right|^2 \tag{5.43}$$

of the state $|0\rangle$ as a function of these parameters, calculated for the initial condition $\psi_0(t = -\infty) = 1$ with the help of (5.42). One can identify the two limiting cases of (i) a small and of (ii) a large parameter of adiabaticity V^2/α. In case (i), the coupling V between the states $|0\rangle$ and $|1\rangle$ is small and the system mainly remains in its initial state following the diabatic trajectory shown in Fig. 5.1 while the population of the state $|0\rangle$ experiences just slight, dying oscillations with the phase $\sim \alpha t^2$ after crossing the resonance at $t = 0$. In the opposite extreme (ii), the velocity α of the state displacement along the energy scale is small compared to V^2, such that the system follows the adiabatic trajectory and performs around $t = 0$, an almost complete transition from the initial state $|0\rangle$ to the final state $|1\rangle$, although the population still manifests slight and dying oscillations corresponding to the transitions between the states.

One finds the asymptotic expression for the population

$$\rho_0(\infty) = e^{-2\pi V^2/\alpha}, \tag{5.44}$$

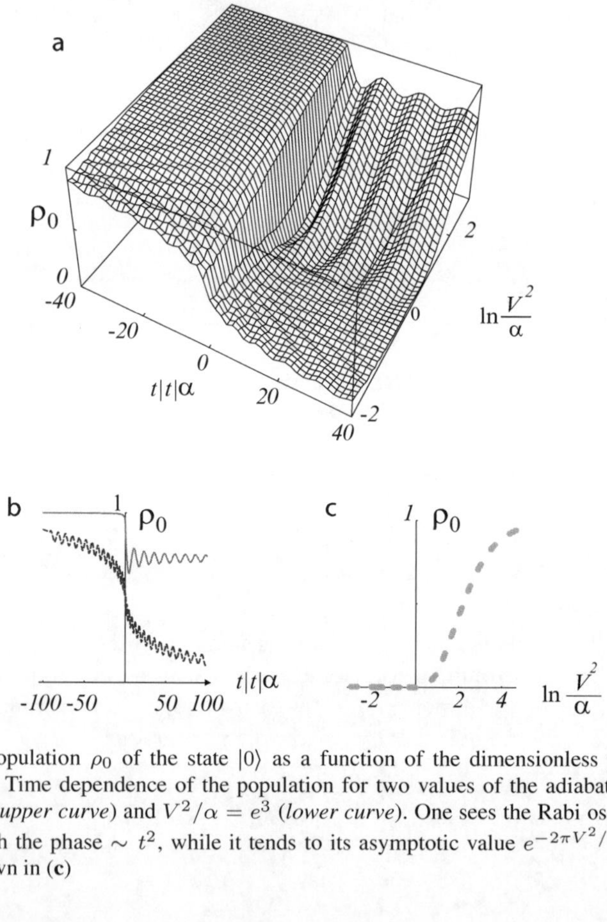

Fig. 5.2 (**a**) Population ρ_0 of the state $|0\rangle$ as a function of the dimensionless parameters $t^2\alpha$ and V^2/α. (**b**) Time dependence of the population for two values of the adiabaticity parameter $V^2/\alpha = e^{-3}$ (*upper curve*) and $V^2/\alpha = e^3$ (*lower curve*). One sees the Rabi oscillations of the probability with the phase $\sim t^2$, while it tends to its asymptotic value $e^{-2\pi V^2/\alpha}$ suggested by (5.44) and shown in (**c**)

provided the asymptotic expression for the degenerate hypergeometric function

$$ {}_1F_1(\beta;\gamma;z) = \tfrac{\Gamma(\gamma)}{\Gamma(\gamma-\beta)}(-z)^{-\beta} \tag{5.45}$$

relevant to the case under consideration ($\mathrm{Re}z > 0$) and the relations

$$\begin{aligned} |\Gamma(iy)|^2 &= \frac{\pi}{y\sinh(\pi y)} \\ \left|\Gamma(\frac{1}{2}+iy)\right|^2 &= \frac{\pi}{\cosh(\pi y)} \end{aligned} \tag{5.46}$$

for Γ-functions has been employed for the expression

$$\begin{aligned} \rho_0(\infty) &= \lim_{t\to\infty}\left|\left(\hat{U}(t)\hat{U}^{-1}(-t)\right)_{0,0}\right|^2 \\ &= \lim_{t\to\infty}\left|{}_1F_1(-\tfrac{V^2}{2i\alpha};\tfrac{1}{2};\tfrac{i\alpha t^2}{2})\,{}_1F_1(\tfrac{V^2}{2i\alpha};\frac{1}{2};-\tfrac{i\alpha t^2}{2})+\right. \\ &\left.(Vt)^2_1\,F_1(\tfrac{1}{2}-\tfrac{V^2}{2i\alpha};\tfrac{3}{2};\tfrac{i\alpha t^2}{2})\,{}_1F_1(\tfrac{1}{2}+\tfrac{V^2}{2i\alpha};\tfrac{3}{2};-\tfrac{i\alpha t^2}{2})\right|^2 \end{aligned} \tag{5.47}$$

following from (5.43) with allowance for (5.42). The probability $\rho_0(\infty) = e^{-2\pi V^2/\alpha}$ to remain in the initial state after the level crossing is shown in Fig. 5.2c.

From (5.44) one concludes that the probability to make a transition to the state $|1\rangle$ is $\rho_1(\infty) = 1 - e^{-2\pi V^2/\alpha}$. We note that for slowly changing level positions, that is, at small α, the probability to remain in the initial state is small, whereas the probability to make a transition is large. At the same time one can consider the problem in another frame, the so-called adiabatic basis (see Fig. 5.1), by making use of the instantaneous energy eigenstates

$$\begin{aligned} |+\rangle &= \frac{V\,|1\rangle - \left(\frac{\alpha t}{2} - \sqrt{(\frac{\alpha t}{2})^2+V^2}\right)|0\rangle}{\sqrt{\left(\frac{\alpha t}{2} - \sqrt{(\frac{\alpha t}{2})^2+V^2}\right)^2+V^2}}, \\ |-\rangle &= \frac{V\,|1\rangle - \left(\frac{\alpha t}{2} + \sqrt{(\frac{\alpha t}{2})^2+V^2}\right)|0\rangle}{\sqrt{\left(\frac{\alpha t}{2} + \sqrt{(\frac{\alpha t}{2})^2+V^2}\right)^2+V^2}}, \end{aligned} \tag{5.48}$$

corresponding respectively to the eigenvalues $E_\pm = \pm\sqrt{(\frac{\alpha t}{2})^2+V^2}$ of the time-dependent Hamiltonian (5.32). Note that $|+\rangle = |0\rangle$ and $|-\rangle = |1\rangle$ at $t = -\infty$, whereas at $t = \infty$ one finds $|+\rangle = |1\rangle$ and $|-\rangle = -|0\rangle$. This means that $\rho_+(t = -\infty) = 1$ and $\rho_-(t = -\infty) = 0$, and the probability of making a transition from the initially populated state $|+\rangle$ to the state $|-\rangle$ is exponentially small in the limit $\alpha \to 0$, or—as one says in this context—adiabatically small, that is,

$$\rho_+(t = \infty) = 1 - e^{-2\pi V^2/\alpha} \tag{5.49}$$

and

$$\rho_-(t = \infty) = e^{-2\pi V^2/\alpha}. \tag{5.50}$$

It is worth mentioning that the so-called Landau–Zener formula given by the expression $\rho_{dp}(\infty) = 2e^{-2\pi V^2/\alpha}(1 - e^{-2\pi V^2/\alpha})$ is valid for the probability of a transition following a double passage (forward and back) of the level crossing, in the absence of the quantum interference between the passages. This means that the resulting probability is just a product of the probabilities of two sequential events—to remain in the initial state after the first passage and to make a transition during

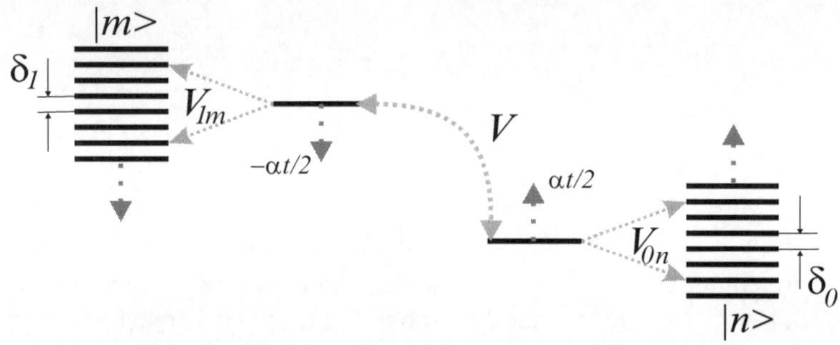

Fig. 5.3 Landau–Zener transition between two decaying levels. The presence of the continua results in the complex shift of the time variable $\alpha t/2 \to \alpha t/2 - i\gamma_{1;2}$ where the decay rate $\gamma_{1,2} = \pi V^2_{0,n;1,m}/\delta_{0,1}$ is in accordance with (3.34)

the second one, while the factor 2 takes into account the possibility for the opposite order of these events. The deviation of the probability from this expression resulting from the quantum interference is known as the Ramsay effect.

5.2.2 Landau–Zener Transition to a Decaying State

The results of the previous section can be generalized in the case of decaying states, where the Schrödinger equation similar to (5.31) reads

$$\begin{aligned} i\dot{\psi}_0 &= \left(\frac{\alpha t}{2} - i\gamma_0\right)\psi_0 + V\psi_1, \\ i\dot{\psi}_1 &= -\left(\frac{\alpha t}{2} + i\gamma_1\right)\psi_1 + V\psi_0. \end{aligned} \tag{5.51}$$

By analogy to the level–band problem discussed in Sect. 3.2 one can consider the decay rates γ_0 and γ_1 as resulting from the interaction of levels $|1\rangle$ and $|0\rangle$ with two continua of states $|n\rangle$ and $|m\rangle$, respectively, that accompany the levels in their translational motion along the energy axis, as shown in Fig. 5.3. The Schrödinger equation for this entire system does not apparently contain non-Hermitian terms proportional to γ_0 and γ_1, and indeed has a more complex form that reads

$$\begin{aligned}
i\dot{\psi}_n &= \left(\frac{\alpha t}{2} + n\delta_0\right)\psi_n + V_{n0}\psi_0, \\
i\dot{\psi}_0 &= \frac{\alpha t}{2}\psi_0 + V\psi_1 + \sum_{n=-\infty}^{\infty} V_{0n}\psi_n, \\
i\dot{\psi}_1 &= -\frac{\alpha t}{2}\psi_1 + V\psi_0 + \sum_{m=-\infty}^{\infty} V_{1m}\psi_m, \\
i\dot{\psi}_m &= \left(-\frac{\alpha t}{2} + m\delta_1\right)\psi_m + V_{m1}\psi_1.
\end{aligned} \tag{5.52}$$

However, since the parts of the total Hamiltonians corresponding to the interaction of the levels with their continua commute with each other as well as with the part responsible for the level coupling, the problem can actually be separated into three independent parts: the results of the levels–continua interaction yield according to (3.34) the decay rates $\gamma_0 = \pi |V_{0n}|^2/\delta_0$ and $\gamma_1 = \pi |V_{1m}|^2/\delta_1$ for the probability amplitudes, while for the interaction of the states $|1\rangle$ and $|0\rangle$ one finds (5.51). This means that the algebraic structure of the problem in the case of decaying levels remains the same as in the former case, and therefore the analytical solution can be found in terms of the same special functions. Indeed, after the replacement $\psi_{0;1}(t) \to \tilde{\psi}_{0;1}(t)e^{-(\gamma_0+\gamma_1)t/2}$ (5.51) takes the form

$$\begin{aligned}
i\dot{\tilde{\psi}}_0 &= \left(\frac{\alpha t - i\gamma_0 + i\gamma_1}{2}\right)\tilde{\psi}_0 + V\tilde{\psi}_1, \\
i\dot{\tilde{\psi}}_1 &= -\left(\frac{\alpha t - i\gamma_0 + i\gamma_1}{2}\right)\tilde{\psi}_1 + V\tilde{\psi}_0,
\end{aligned} \tag{5.53}$$

which coincides with (5.31) for non-decaying states after the substitution $t \to t' + i(\gamma_0 - \gamma_1)/\alpha$. The evolution operator therefore reads

$$\hat{\tilde{U}}(t') = e^{-(\gamma_0+\gamma_1)t/2}\hat{U}\left(t - i\frac{\gamma_0 - \gamma_1}{\alpha}\right), \tag{5.54}$$

where the evolution operator $\hat{U}(t)$ is given by (5.42).

The exponential pre-factor in (5.54) takes into account the exponential decrease of the total population, whereas the complex argument of the evolution operator (5.42) allows for the relative populations of the states $|1\rangle$ and $|0\rangle$:

$$\rho_0(t) \sim \left| \left(\hat{U}(t - ig)\hat{U}^{-1}(-\infty - ig) \right)_{0,0} \right|^2,$$
$$\rho_1(t) \sim \left| \left(\hat{U}(t - ig)\hat{U}^{-1}(-\infty) - ig \right)_{1,0} \right|^2, \tag{5.55}$$

where $g = (\gamma_0 - \gamma_1)/\alpha$. Therefore, there is a third dimensionless parameter $(\gamma_0 - \gamma_1)/\sqrt{\alpha}$ which along with dimensionless time $\sqrt{\alpha}t$ and the dimensionless adiabaticity parameter V^2/α govern the time evolution of the decaying Landau–Zener system. In Fig. 5.4 we show the dependence of the population of the state $|0\rangle$ normalized to its value $e^{-2\gamma_0 t}$ at $V = 0$ as a function of time and the adiabaticity parameter for a positive and a negative value of g.

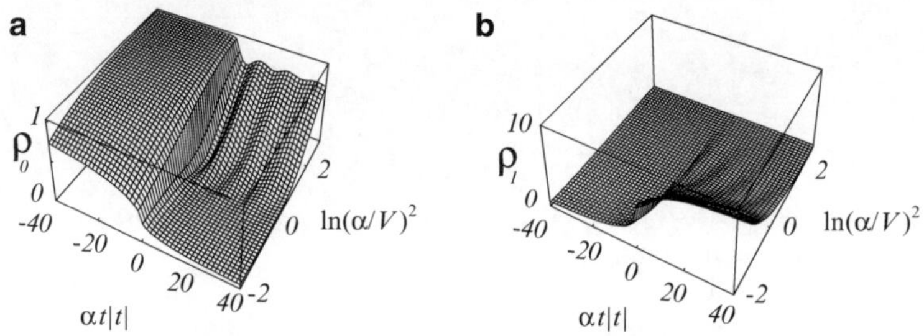

Fig. 5.4 Landau–Zener transition between two decaying levels. (**a**) The population of the state $|0\rangle$ for the case $\gamma_0 < \gamma_1$ (normalized to the population of the decaying state at zero coupling $V = 0$). The asymptotic value of the population is the same, as for the case of non-decaying states, although the time evolution is somewhat different. (**b**) The population of the state $|1\rangle$ for the case of $\gamma_0 > \gamma_1$ manifests completely different behavior and exceeds sometimes considerably ρ_1 for the case of non-decaying levels, although it vanishes in the long-time asymptotic (5.58)

Asymptotic expressions for the degenerate hypergeometric functions

$$_1F_1(\beta;\gamma;z) = \frac{\Gamma(\gamma)}{\Gamma(\beta)} z^{\beta-\gamma} e^z \tag{5.56}$$

and (5.45) that are valid in the case ($\mathrm{Re}z < 0$) allow one to find the population ρ_0 (5.55) in the limit $t \to \infty$, which reads

$$\rho_0(t \to \infty) = \begin{cases} e^{-2\gamma_0 t} e^{-2\pi V^2/\alpha}; & g < 0 \\ e^{-2\gamma_1 t} \left(\frac{V^2}{\alpha^2 t^2} \right)^2 e^{2\pi V^2/\alpha}; & g > 0. \end{cases} \tag{5.57}$$

One sees that apart from the exponential factor allowing for the decay of the state $|0\rangle$, the asymptotic population for $g < 0$ coincides with that of (5.44) which is typical of non-decaying levels. On the contrary, if the decay rate of the state $|0\rangle$ exceeds that of the state $|1\rangle$, that is, $g < 0$, the asymptotic behavior is completely different, since the population ρ_0 exists mainly because of ρ_1, which decays more slowly and feeds ρ_0 via the interaction V with a rate proportional to the fourth power of the ratio of the coupling to the instantaneous detuning.

5.2.3 Landau–Zener Transition in the Presence of Transversal Relaxation

The decay of quantum state populations resulting from the interaction with a continuum or continua is just one of many examples of irreversible processes that may occur in quantum systems. Another type of relaxation takes place when a system is subject to a perturbation which we do not know in all details, such as white noise for instance. This makes the system evolve in Hilbert space, along an unknown and unpredictable trajectory. This process can no longer be considered in terms of the Hamiltonian dynamics of a quantum state, but requires the density matrix formalism incorporating a statistical model for this unknown action. The description is usually given in terms of the density matrix $\hat{\rho}(t) = \overline{|\Psi\rangle\langle\Psi|}$ of the system, which is the direct product of the state vector by its conjugate averaged over all possible realizations of the unknown external action. The simplest model takes into account the relaxation by introducing the so-called relaxation operator $\hat{\hat{\mathcal{R}}}$ acting on $\hat{\rho}(t)$, such that the master equation for the density matrix reads

$$i\dot{\hat{\rho}}(t) = [\hat{H}, \hat{\rho}(t)] + \hat{\hat{\mathcal{R}}}\hat{\rho}(t), \tag{5.58}$$

as we have already seen on p. 92. The action of the relaxation operator to the $n \times n$ density matrix does not have the form of multiplication by another matrix of the same size, but it is a convolution with the fourth rank $n \times n \times n \times n$ tensor $\hat{\hat{\mathcal{R}}}$, that is, $\left(\hat{\hat{\mathcal{R}}}\hat{\rho}(t)\right)_{ij} = \sum_{kl} R_{ijkl}\rho_{kl}$. For the two-level system with the Landau–Zener Hamiltonian (5.22)

$$\hat{H} = \begin{pmatrix} \alpha t/2 & V \\ V & -\alpha t/2 \end{pmatrix} \tag{5.59}$$

(5.58) has the form

$$\begin{aligned}
i\dot{\rho}_{00} &= V(\rho_{10} - \rho_{01}) + i\rho_{11}/T_1 \\
i\dot{\rho}_{01} &= \alpha t\rho_{10} + V(\rho_{11} - \rho_{00}) - i\rho_{10}/T_2 \\
i\dot{\rho}_{10} &= -\alpha t\rho_{10} + V(\rho_{00} - \rho_{11}) - i\rho_{10}/T_2 \\
i\dot{\rho}_{11} &= V(\rho_{01} - \rho_{10}) - i\rho_{11}/T_1,
\end{aligned} \tag{5.60}$$

where T_1 and T_2 are two parameters of the relaxation matrix called the longitudinal and transversal relaxation times.

By adding the first and the fourth equations in the system (5.60) one immediately finds the integral of motion $\rho_{11} + \rho_{00} = const$, and the constant has to be unity as follows from the normalization condition. The remaining three linearly independent equations rewritten in terms of three new variables: the polarization $\mathcal{P} = \rho_{01} + \rho_{10}$, dispersion $\mathcal{Q} = \rho_{01} - \rho_{10}$, and population of the upper level $\rho_{11} \equiv \frac{1}{2}\mathcal{Z}$, read

$$\begin{aligned}\dot{\mathcal{Z}} &= 2V\mathcal{P} - \gamma\mathcal{Z},\\ \dot{\mathcal{P}} &= -2V\mathcal{Z} + \alpha t\mathcal{Q} - \gamma\mathcal{P} + 2V,\\ \dot{\mathcal{Q}} &= -\alpha t\mathcal{P} - \gamma\mathcal{Q}.\end{aligned} \tag{5.61}$$

Here we have assumed that the constant $\gamma = 1/T_{1,2}$ is identical for the transverse and longitudinal relaxation.

The initial conditions corresponding to the case $\rho_{00}(-\infty) = 1$ are

$$\mathcal{Z}(-\infty) = \mathcal{P}(-\infty) = \mathcal{Q}(-\infty) = 0. \tag{5.62}$$

For large positive time, the two-level system is out of resonance, and hence in the asymptotic $t \to \infty$ due to the longitudinal relaxation, the system returns to its initial state $|0\rangle$. This gives the boundary condition

$$\mathcal{Z}(\infty) = \mathcal{P}(\infty) = \mathcal{Q}(\infty) = 0. \tag{5.63}$$

Since the asymptotic population distribution between the states is evident, we concentrate here on a non-trivial characteristic of the process—the longitudinal relaxation yield I, namely on the mean number of the elementary acts of the longitudinal relaxation given by the expression

$$I = \gamma \int_{-\infty}^{\infty} \rho_{11}(t)dt = \frac{\gamma}{2} \int_{-\infty}^{\infty} \mathcal{Z}(t)dt. \tag{5.64}$$

Equations (5.62)–(5.63) allow us to perform a Fourier transformation of (5.61) which yields

$$\begin{aligned}i\omega\mathcal{Z} &= 2V\mathcal{P} - \gamma\mathcal{Z},\\ i\omega\mathcal{P} &= -2V\mathcal{Z} + i\alpha\frac{\partial}{\partial\omega}\mathcal{Q} - \gamma\mathcal{P} + 4\pi V\delta(\omega),\\ i\omega\mathcal{Q} &= -i\alpha\frac{\partial}{\partial\omega}\mathcal{P} - \gamma\mathcal{Q}.\end{aligned} \tag{5.65}$$

Now we express $\mathcal{Z}$ and $\mathcal{Q}$ in terms of $\mathcal{P}$ with the help of the first and the third equations of (5.65) and substitute them into the second equation of this set. This yields

$$\frac{\alpha^2}{\omega - i\gamma}\frac{\partial}{\partial\omega}\frac{1}{\omega - i\gamma}\frac{\partial}{\partial\omega}\mathcal{P} + \mathcal{P} - \frac{4V^2}{(\omega - i\gamma)^2}\mathcal{P} = 4\pi\frac{V}{\gamma}\delta(\omega). \tag{5.66}$$

The Dirac δ-function on the right-hand side of the equation implies that at the point $\omega = 0$ the Fourier transform of the polarization $\mathcal{P}(\omega)$ has a discontinuity in the first derivative. When we integrate (5.66) over the small vicinity ϵ of $\omega = 0$ we find

$$\frac{\partial\mathcal{P}}{\partial\omega}|_{\omega=0+\epsilon} - \frac{\partial\mathcal{P}}{\partial\omega}|_{\omega=0-\epsilon} = -4\pi\frac{\gamma V}{\alpha^2}. \tag{5.67}$$

Thus, we have to find the solution of the homogeneous equation (5.66) that satisfies the boundary condition, (5.67), and corresponds to a real $\mathcal{P}(t)$, that is,

$$\mathcal{P}(\omega) = \mathcal{P}^*(-\omega). \tag{5.68}$$

We may now obtain the exact analytical solution of (5.66) in terms of Whittaker functions

$$\begin{aligned} W_{\lambda,\mu}(z) = {} & \frac{\Gamma(-2\mu)z^{\frac{1}{2}+\mu}e^{-\frac{z}{2}}}{\Gamma(\frac{1}{2}-\mu-\lambda)}\,{}_1F_1(\tfrac{1}{2}+\mu-\lambda; 1+2\mu; z) \\ & + \frac{\Gamma(2\mu)z^{\frac{1}{2}-\mu}e^{-\frac{z}{2}}}{\Gamma(\frac{1}{2}+\mu-\lambda)}\,{}_1F_1(\tfrac{1}{2}-\mu-\lambda; 1-2\mu; z), \end{aligned} \tag{5.69}$$

which are another form of degenerate hypergeometric function of (5.39), and which has an integral representation

$$W_{\lambda,\mu}(z) = \frac{z^{\frac{1}{2}+\mu}e^{-\frac{z}{2}}(1+e^{2i\pi(\mu-\lambda)})^{-1}}{\Gamma(\frac{1}{2}+\mu-\lambda)}\int_C \tau^{\mu-\frac{1}{2}-\lambda}(\tau+1)^{\mu-\frac{1}{2}+\lambda}e^{-\tau x}d\tau, \tag{5.70}$$

with the integration contour C going from ∞ to ∞ around the point 0.

In order to simplify the notation, we put $2V = 1$, $\gamma = \tilde{\gamma}/(2V)$, $\alpha = \tilde{\alpha}/(2V)^2$, $\omega = \tilde{\omega}/(2V)$, and introduce the new variable $x = (i\omega + \gamma)^2/2\alpha$. Then we have

$$x\frac{\partial^2}{\partial x^2}\mathcal{P} + \frac{1}{2\tilde{\alpha}}\mathcal{P} + x\mathcal{P} = 0. \tag{5.71}$$

Solutions of the differential equations with the coefficients that are first-order polynomials in the independent variable, can be found with the help of the so-called Laplace contour integral method, which suggests that the differential equation

$$\sum_n (a_n + b_n x) \frac{\partial^n}{\partial x^n} \mathcal{Y} = 0 \tag{5.72}$$

has the solution

$$\mathcal{Y}(x) = \int_C \frac{d\tau}{\sum_n b_n \tau^n} \exp\left[x\tau + \int^{\tau} \frac{\sum_n a_n y^n}{\sum_n b_n y^n} dy\right], \tag{5.73}$$

where the integration contour C has to be chosen such that the exponent in the integrand has the same values at each end. Different contours satisfying this condition usually yield different, linearly independent solutions, provided they circumvent differently the singularities of the function $\left(\sum_n b_n \tau^n\right)^{-1}$. In particular for (5.71) one has $\left(\sum_n b_n \tau^n\right)^{-1} = \left(\tau^2 + 1\right)^{-1}$; $\int^{\tau} \sum_n a_n y^n / \sum_n b_n y^n dy = \frac{1}{4\tilde{\alpha}} \ln\left(\tau^2 + 1\right)$, and this method yields

$$\mathcal{P} = \frac{A'}{2i\alpha} \int_{C_W} \frac{\exp(\tau x)}{(\tau + i)^{1 - i/4\tilde{\alpha}} (\tau - i)^{1 + i/4\tilde{\alpha}}} d\tau. \tag{5.74}$$

The contour C_W arrives from $\tau = i\infty$, goes around the point $\tau = i$, returning along the direction $\tau = i\infty$. This integration path ensures convergence of the integral for $\omega > 0$. The integral representation for $\omega < 0$ follows from the condition (5.68). We note that another possible contour C_M is the loop which circumvents both points i and $-i$. The corresponding polarization $\mathcal{P}$ accounts for the homogeneous solution which is regular at the point $\omega = 0$ and therefore does not contribute to the final result.

The substitution $\tau = i + i2z$ transforms the integral (5.74) to the integral representation of the Whittaker function (5.70),

$$u_1(\omega) = W_{i/4\alpha, -1/2}\left(-i\frac{(i\omega + \gamma)^2}{\alpha}\right). \tag{5.75}$$

Here, we have made use of the explicit expression for x. Hence, the Whittaker function $u_1(\omega)$ is a solution of (5.71) for $\omega > 0$, that is, $\mathcal{P}(\omega)|_{\omega > 0} = Au_1(\omega)$. The other linearly independent solution of (5.66),

$$\mathcal{P}(\omega) = A^* u_2(\omega) = A^* W_{-i/4\alpha, -1/2}\left(i\frac{(i\omega + \gamma)^2}{\alpha}\right), \tag{5.76}$$

accounts for the case $\omega < 0$.

Now we can find the proportionality constant A with the help of the condition given in (5.68) at $\omega = 0$ and the condition (5.67). We come then to the set of equations

$$A u_1(0) - A^* u_2(0) = 0$$
$$A \frac{\partial u_1}{\partial \omega}|_{\omega=0} - A^* \frac{\partial u_2}{\partial \omega}|_{\omega=0} = -2\pi \frac{\gamma}{\alpha^2}. \tag{5.77}$$

The determinant D of this set is proportional to the Wronskian $W(u_1; u_2) \equiv (u_1 \frac{\partial}{\partial x} u_2 - u_2 \frac{\partial}{\partial x} u_1)$ of the Whittaker functions and reads

$$D = \frac{2\gamma}{\alpha} W(u_1; u_2) = \frac{2\gamma}{\alpha} \exp\left(\frac{\pi}{4\alpha}\right). \tag{5.78}$$

The factor $\frac{2\gamma}{\alpha}$ results from the derivative of the arguments $\frac{\partial x}{\partial \omega}$. Solving (5.77) we obtain the constant A and find for the polarization

$$\mathcal{P}(\omega)|_{\omega>0} = \pi \frac{u_2(0) u_1(\omega)}{\alpha} \exp\left(-\frac{\pi}{4\alpha}\right). \tag{5.79}$$

Now we are able to calculate the longitudinal relaxation yield (5.64) that is $I = \frac{\gamma}{2} \int_{-\infty}^{\infty} \mathcal{Z}(t) dt = \frac{\gamma}{2} \mathcal{Z}(\omega = 0)$. Since, according to the first line of (5.65), we have $\frac{1}{2} \mathcal{Z}(\omega = 0) = (2\gamma)^{-1} \mathcal{P}(\omega = 0)$ we find from Eqs. (5.75)–(5.76) and (5.79) for $\omega = 0$

$$I = 2\pi \frac{V^2}{\alpha} e^{\frac{-\pi V^2}{\alpha}} \left| W_{\frac{iV^2}{\alpha}, -\frac{1}{2}} \left(-\frac{i\gamma^2}{\alpha} \right) \right|^2. \tag{5.80}$$

In Fig. 5.5 we show the dependence of the longitudinal relaxation yield I on the adiabaticity parameter V^2/α and on the quenching parameter γ^2/α suggested by (5.80). At $\gamma = 0$, the yield is given by the regular Landau–Zener dependence (5.44) for $\rho_1 = 1 - \rho_0$, which simply means that the relaxation yield amounts to the probability to make the transition. It changes with the increase of the quenching parameter γ^2/α, as the relaxation yield increases with increasing relaxation rate.

We now consider the asymptote of (5.80). We start with the case of slow passage through the resonance, that is, $\alpha \to 0$. We use the integral representation of the Whittaker function and find the following expression with the help of the stationary phase method:

$$W_{\frac{iV^2}{4\alpha}, -\frac{1}{2}} \left(-\frac{i\gamma^2}{\alpha}\right) = \frac{e^{i\gamma^2/2\alpha}}{\Gamma(-\frac{iV^2}{\alpha})} \int_0^\infty \frac{\exp\{i\frac{\gamma^2}{\alpha} t - i\frac{V^2}{\alpha} \ln \frac{t}{1+t}\}}{t(t+1)} dt$$
$$\simeq \frac{e^{i\Phi}}{\Gamma(-iV^2/\alpha)} \frac{\sqrt{\pi\alpha}}{[1 + (2V/\gamma)^2)]^{1/4} V}, \tag{5.81}$$

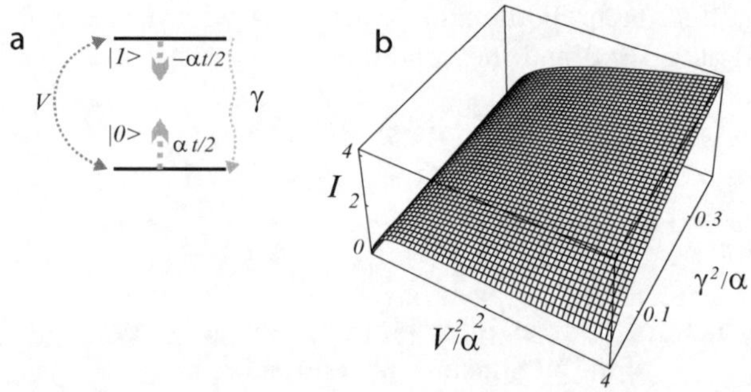

Fig. 5.5 Landau–Zener transition to a decaying level. (**a**) Level scheme. (**b**) Expectation value of the relaxation yield I given by (5.80) as a function of the Landau–Zener parameter V^2/α and quenching parameter γ^2/α. At $\gamma = 0$ the dependency recovers the probability of the Landau–Zener transition $\rho_1 = 1 - \rho_0$

where Φ is a phase factor that does not affect the final result, (5.80). Here we have taken into account that in the limit $\alpha \to 0$ the only stationary point on the integration path is $t = -\frac{1}{2} + \frac{1}{2}\sqrt{1 + \frac{4V^2}{\gamma^2}}$. We also have assumed $V^2\gamma^2/\alpha^2 \gg 1$, allowing us to neglect derivatives higher than second order at the stationary phase point and thus satisfy the requirement of the stationary phase method. Substitution of (5.81) into (5.80) results in

$$I = 2\pi \frac{\gamma V^2}{\alpha} \frac{1 - \exp\{\frac{-2\pi V^2}{\alpha}\}}{(\gamma^2 + 4V^2)^{1/2}}, \tag{5.82}$$

where we have made use of (5.46). The exponential term, reminiscent of the Landau–Zener expression, arises from the Γ-function and not from the stationary phase analysis of (5.81). However, in order to be consistent within the order of approximation it should be neglected. Note that this result also follows from the rate equation. Indeed, one can obtain the result (5.82) directly from (5.61) in the quasi-stationary limit, namely when all time derivatives are equal to zero. This results in $\mathcal{Z} = 2V\mathcal{P}/\gamma$, and $\mathcal{Q} = -\alpha t\mathcal{P}/\gamma$, which gives $\mathcal{P} = 2\gamma V[4V^2 + (\alpha t)^2 + \gamma^2]^{-1}$ and hence

$$I = V \int \mathcal{P} dt = 2\pi \frac{\gamma V^2}{\alpha\sqrt{\gamma^2 + 4V^2}}.$$

The limit of weak interaction, $V^2/\alpha \to 0$, follows from the condition $|W_{0,-1/2}(x)| = |e^{ix}| = 1$, and (5.80) simplifies to $I \simeq 2\pi V^2/\alpha$. This implies that the asymptotic formula (5.82) is also valid for weak interactions.

We now consider the limit of slow decay, $\gamma^2/\alpha \ll 1$, with the help of the asymptotic expression for the Whittaker function

$$\lim_{z\to 0} W_{ia,-\frac{1}{2}}(z) = \frac{1}{\Gamma(1 + ia)} = \frac{1}{ia\Gamma(ia)}, \tag{5.83}$$

which, after substitution into (5.80), results in

$$I = 1 - \exp\left(-\frac{2\pi V^2}{\alpha}\right). \tag{5.84}$$

We hence recover the probability of the Landau–Zener transition, (5.44), which indeed is identical to the probability of emission of a spontaneous photon in the case of a small decay rate. We note that a simple expression for the limit $V^2\gamma^2/\alpha^2 \ll 1$ is difficult to obtain due to the non-analytical behavior of the Whittaker function at small values of the argument, as shown in Fig. 5.2.

We conclude by noting that Rabi oscillations of population between the states (Fig. 5.2) that take place while the levels are in resonance, do not manifest themselves in any oscillations of the asymptotic populations nor in the relaxation yield. At first sight, one might think that the relaxation would bring them to light, since it "brakes" the population oscillations in the middle of the Rabi cycle, that is, before the passage through the resonance is complete. However, they remain hidden even in the case when relaxation processes are taken into account: the irreversible decay does not affect the probability of the transition at all as long as the population of the initially populated state decays faster than the population of the other state. The longitudinal relaxation of the upper level accounts for the multiple returns of the particle to the initial state in the middle of the Rabi cycle. However, it is also the case that these returns do not bring about any oscillations to the expectation value of the relaxation yield. The gradual change of the quenching parameter results instead in the gradual transformation of the Landau–Zener dependence to the smooth dependence given by the rate equation.

5.2.4 Excitation by a Pulse

Another example of an exactly soluble two-level problem corresponds to the case of a constant detuning $\Delta_0 = 0$, $\Delta_1 = const$ and the time-dependent interaction of a very special profile $V(t) = V/\cosh \alpha t$. The Schrödinger equation (5.29) after the replacements

$$z = \frac{1}{1+e^{-2\alpha t}}; \quad \Delta = \frac{\Delta_1}{2\alpha}; \quad \theta = \frac{V\pi}{\alpha} \tag{5.85}$$

yields the second-order differential equation (5.30) in the form

$$-z(z-1)\frac{\partial^2\psi_0}{\partial z^2} + \left(\frac{1}{2} - z + i\Delta\right)\frac{\partial\psi_0}{\partial z} + \frac{2\theta}{\pi}\psi_0 = 0, \tag{5.86}$$

which for $\psi_0(t \to -\infty) = \psi_0(z \to 0) \to 1$ has an explicit solution in terms of the hypergeometric function

$$\psi_0(z) = {}_2F_1\left(\frac{\theta}{\pi}, -\frac{\theta}{\pi}; \frac{1}{2} + i\Delta; z\right). \tag{5.87}$$

This expression yields the amplitude as $t \to \infty$

$$\begin{aligned}\psi_0(t)|_{t\to\infty} &= \psi_0(z)|_{z=1} = {}_2F_1(\frac{\theta}{\pi}, -\frac{\theta}{\pi}; \frac{1}{2} + i\Delta; 1) \\ &= \frac{\Gamma(\frac{1}{2}+i\Delta)\Gamma(\frac{1}{2}+i\Delta)}{\Gamma(\frac{1}{2}+i\Delta-\frac{\theta}{\pi})\Gamma(\frac{1}{2}+i\Delta+\frac{\theta}{\pi})}\end{aligned} \tag{5.88}$$

in terms of Γ-functions. The corresponding population reads

$$\rho_0(t)|_{t\to\infty} = \frac{\cosh(2\pi\Delta) + \cos(2\theta)}{2\cosh^2(\pi\Delta)}. \tag{5.89}$$

where (5.46) and the relations

$$\begin{aligned}\Gamma(x+1) &= x\Gamma(x) \\ |\Gamma(1-x+i\Delta)\Gamma(1+x+i\Delta)|^2 &= \frac{2\pi^2(x^2+\Delta^2)}{\cosh(2\pi\Delta)-\cos(2\pi x)}\end{aligned} \tag{5.90}$$

have been employed. For the transition probability, after trigonometric transformations one obtains

$$\rho_1(t)|_{t\to\infty} = \frac{\sin^2\theta}{\cosh^2(\pi\Delta)}. \tag{5.91}$$

In Fig. 5.6 we show the time-dependent transition probability $1 - |\psi_0(t)|^2$ corresponding to the amplitude (5.87), this amplitude, and the asymptotic probability transition (5.91).

We note one very specific property of the problem under consideration. Unlike other cases of time-dependence, for the interaction profile $V/\cosh\alpha t$ the transition probability is a product of two factors, one of which is a function of the detuning Δ only, whereas the other depends only on the coupling amplitude V. Moreover, the last enters the expression in the form of the so-called pulse area $\theta = \int_{-\infty}^{\infty} V/\cosh(\alpha t)\ dt$. Each of these two factors has a clear physical meaning. Indeed, for $\Delta_{0;1} = 0$ equations (5.29) take a particularly simple form, $i\dot{\psi}_0 = V(t)\psi_1$, $i\dot{\psi}_1 = V(t)\psi_0$, and have an exact solution:

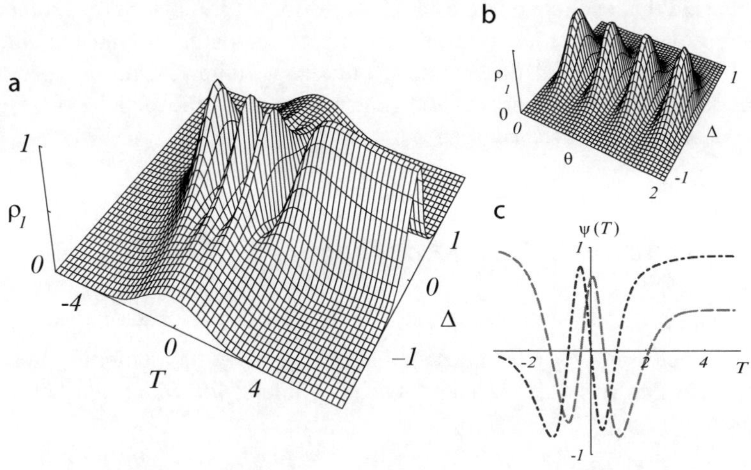

Fig. 5.6 Excitation of a two-level system by a pulse $V/\cosh^{-2}(2\alpha t)$. (**a**) Transition probability as a function of time $T = 2\alpha t$ and detuning $\Delta = \Delta_1/2\alpha$. One sees oscillations in resonance, and the adiabatic regime for large detunings. (**b**) The asymptotic probability as a function of the detuning and the pulse area $\theta = V/\alpha$. (**c**) Real (*dashed line*) and imaginary (*dot-dash line*) parts of the probability amplitude to remain in the state $|0\rangle$ as a function of time for fixed detuning $\Delta_1/2\alpha = 0.7$ and pulse area $\theta = 2.45$

$$\psi_0(t) = \cos[\int_{-\infty}^{t} V(t')\,dt'],$$

$$\psi_1(t) = i\sin[\int_{-\infty}^{t} V(t')\,dt'], \tag{5.92}$$

which yields $\psi_1(\infty) = i\sin\theta$, and hence the expression $\rho_1(\infty) = \sin^2\theta$ coinciding completely with (5.91) in resonance. The other factor allows for non-zero detunings, being just the line shape $|V_\omega|^2$ at $\omega = \Delta_1$ given by the Fourier transform of the $V/\cosh\alpha t$ dependence and normalized by the condition $V_{\omega=0} = 1$. Indeed, employing the Cauchy formula for residuals at the points $\alpha t = i(\pi/2 + n\pi)$, for $n = 0, 1 \ldots$, yields

$$\frac{1}{\pi i}\int_{-\infty}^{\infty}\frac{\exp(i\Delta_1 t)}{\cosh(\alpha t)}\,dt = 2\sum_{n=0}^{\infty}(-1)^n\exp(-\frac{\Delta_1}{\alpha}(\pi/2 + n\pi))$$

$$= \frac{2\exp(-\frac{\Delta_1\pi}{2\alpha})}{1+\exp(-\frac{\Delta_1\pi}{\alpha})} = \cosh^{-1}(\frac{\Delta_1\pi}{2\alpha}). \tag{5.93}$$

We emphasize that such a simple asymptotic form of (5.91) is more a coincidence than the result of separation between the effects of detuning and coupling that remain strongly entangled during the evolution at finite t. One can see this in Fig. 5.6a, where the difference of the time behavior of the population $\rho_1(t)$ for different detunings Δ is clearly seen.

5.2.5 *Exponentially Rising Coupling*

We consider one more example of exactly soluble problems for which the interaction $V(t)$ between the states rises exponentially, while the detuning Δ_1 remains constant, also known as the Demkov–Olson problem, for which the Schrödinger equation (5.29) reads

$$\begin{aligned} i\dot{\psi}_0 &= Ve^{\alpha t}\psi_1, \\ i\dot{\psi}_1 &= \Delta_1\psi_1 + Ve^{\alpha t}\psi_0. \end{aligned} \tag{5.94}$$

We set $\Delta_0 = 0$ as earlier, and introduce the variable $\theta = \int^t Ve^{\alpha x}dx = Ve^{\alpha t}/\alpha$ such that the system takes the form

$$\begin{aligned} i\frac{\partial\psi_0}{\partial\theta} &= \psi_1, \\ i\frac{\partial\psi_1}{\partial\theta} &= \frac{\Delta_1}{\alpha\theta}\psi_1 + \psi_0, \end{aligned} \tag{5.95}$$

where the initial condition is $\psi_0(\theta = 0) = 1$.

Now one can either employ the transformation (5.30) that yields the Bessel equation, or look for the solution in the form of a power series $\psi_0(\theta) = \sum_{n=0}^{\infty} c_{2n}\theta^{2n}$ and $\psi_1(\theta) = \sum_{n=0}^{\infty} c_{2n+1}\theta^{2n+1}$, as has been done for (5.37). The last method immediately allows one to incorporate the initial condition by setting $c_0 = 1$, and after making use of the recurrence relations

$$\begin{aligned} i2nc_{2n} &= c_{2n-1}, \\ i\,(2n-1)\,c_{2n-1} &= \frac{\Delta_1}{\alpha}c_{2n-1} + c_{2(n-1)}, \end{aligned} \tag{5.96}$$

that is, $c_{2n} = -c_{2(n-1)}/2n(2n-1+i\frac{\Delta_1}{\alpha})$, this approach results in

$$c_{2n} = \left(\frac{-1}{4}\right)^n \frac{1}{n!}\frac{\Gamma\left(\frac{1}{2}+\frac{i\Delta_1}{2\alpha}\right)}{\Gamma\left(n+\frac{1}{2}+\frac{i\Delta_1}{2\alpha}\right)}. \tag{5.97}$$

This means that

$$\psi_0(\theta) = \Gamma\left(\frac{1}{2}+\frac{i\Delta_1}{2\alpha}\right)\sum_{n=0}^{\infty}\frac{1}{n!}\frac{\left(-\theta^2/4\right)^n}{\Gamma\left(n+\frac{1}{2}+\frac{i\Delta_1}{2\alpha}\right)}$$
$$= \Gamma\left(\frac{1}{2}+i\Delta\right)\left(\frac{\theta}{2}\right)^{1/2-i\Delta} J_{i\Delta-1/2}(\theta), \qquad (5.98)$$

where a standard representation of the Bessel function $J_\nu(\theta)$ in the form of a power series has been employed, and where we make use of the notation (5.85), although now the parameter α corresponds to an exponential, and not the $1/\cosh x$ dependence. From the first equation of the set (5.95) and with allowance for the rule for the Bessel function derivatives one also obtains

$$\psi_1(\theta) = -i\Gamma\left(\frac{1}{2}+i\Delta\right)\left(\frac{\theta}{2}\right)^{1/2-i\Delta} J_{i\Delta+1/2}(\theta). \qquad (5.99)$$

Equation (5.98) with allowance for (5.46) gives the probabilities

$$\rho_0(\theta) = \left|\Gamma\left(\frac{1}{2}+i\Delta\right)\right|^2 \frac{\theta}{2}\left|J_{i\Delta-1/2}(\theta)\right|^2$$
$$= \frac{\pi\theta}{2}\cosh^{-1}(\pi\Delta)\left|J_{i\Delta-1/2}(\theta)\right|^2,$$
$$\rho_1(\theta) = \frac{\pi\theta}{2}\cosh^{-1}(\pi\Delta)\left|J_{i\Delta+1/2}(\theta)\right|^2. \qquad (5.100)$$

The population of the state $|0\rangle$ is shown in Fig. 5.7. It is also interesting to trace the populations of the adiabatic states, that is, the instantaneous eigenstates $|\pm\rangle$ of the Hamiltonian that can be found by analogy to (5.48) as

$$|\pm\rangle = \frac{\theta\,|1\rangle + \left(\Delta \pm \sqrt{\Delta^2+\theta^2}\right)|0\rangle}{\sqrt{\left(\Delta \pm \sqrt{\Delta^2+\theta^2}\right)^2+\theta^2}}. \qquad (5.101)$$

Let us assume $\Delta > 0$ and denote by Θ the angle in Hilbert space between the state vector $|+\rangle$ and the state $|0\rangle$, that is,

$$\cos\Theta = \frac{\left(\Delta + \sqrt{\Delta^2+\theta^2}\right)}{\sqrt{\left(\Delta + \sqrt{\Delta^2+\theta^2}\right)^2+\theta^2}},$$

and find the population

$$\rho_-(\theta) = \frac{\pi\theta \left| J_{i\Delta-1/2}(\theta)\cos\Theta - iJ_{i\Delta+1/2}(\theta)\sin\Theta \right|^2}{2\cosh(\pi\Delta)} \tag{5.102}$$

of the state $|-\rangle$ that determines the probability of a non-adiabatic transition. In

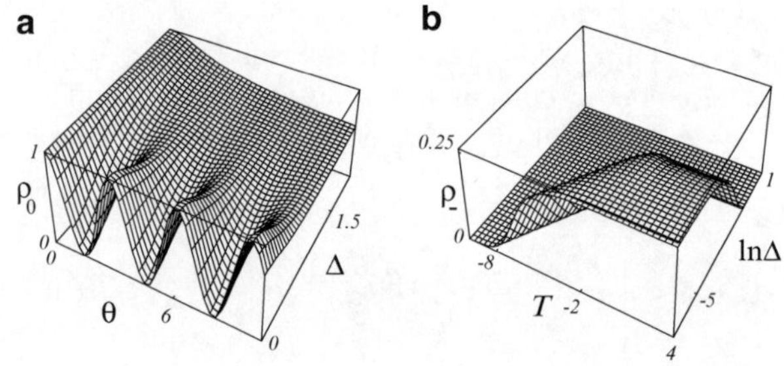

Fig. 5.7 Excitation of a two-level system by an exponentially increasing interaction $V/e^{\alpha t}$. (**a**) Transition probability as a function of the time parameter $\theta = Ve^{\alpha t}/\alpha$ and the detuning $\Delta = \Delta_1/2\alpha$. One sees oscillations for small detunings and the stationary regime with $\rho_{0;1} \to 1/2$. (**b**) The probability of the adiabatic transition as a function of $T = \alpha t$ and $\ln(\Delta_1/2\alpha)$. At small detunings one can identify a regime where the non-adiabatic transition probability tends to $1/4$

Fig. 5.7b we depict this population as a function of time and the detuning. One sees that for small detunings in the long-time regime it approaches the asymptotic value $1/4$, which can be seen directly from (5.102) with the help of the asymptotic expression

$$J_{i\Delta\pm1/2}(\theta) \simeq \sqrt{\frac{1}{\pi\theta}}\cosh\left(i\theta + \Delta\pi - (1\pm1)\frac{\pi i}{4}\right) \tag{5.103}$$

and with allowance for the fact that $\cos\Theta \simeq \sin\Theta \simeq 1/\sqrt{2}$ as $t \to \infty$ for large θ and interactions. The substitution yields

$$\rho_{na}(\theta \to \infty) = \frac{\exp\left(-2\left|\Delta\right|\pi\right)}{4\cosh(\pi\Delta)}, \tag{5.104}$$

where we have also taken into account that for negative Δ the probability of the non-adiabatic transition $\rho_{na}(\theta)$ is given by ρ_+.

We note one general property common for all two-level problems. For slow variation of the system $\alpha \to 0$ and the probability to change the adiabatic state given by (5.50), (5.91), and (5.104) is always small.

5.3 Semiclassical Analysis of Time-Dependent Systems

Considering the Landau–Zener problem on p. 207, we have seen that the populations of adiabatic states change slightly when the Hamiltonian varies slowly. This is a general situation, also valid in the case of a large system consisting of many quantum levels. For the description of the population dynamics, one may make use of a treatment similar to the semiclassical approach to the Schrödinger equation of a particle moving in an external potential, with the only difference that time t plays the role of the coordinate. A straightforward substitution of the semiclassical ansatz $\exp[-iS(t)]$ either into (5.36) or (5.37) results however in a complex-valued solution for $S(t)$, which means that there is no direct classical analog for the mechanical action in these representations. Moreover, the Hamilton–Jacobi equation, obtained for the classical action $S(t)$ after the substitution $\exp[-iS(t)]$ into the Schrödinger equation, turns out to be dependent on the choice of representation, while the allowance for the second derivative of $S(t)$ via perturbation theory does not improve the situation much. Therefore, in order to make use of the semiclassical analysis, one needs first to find a representation where the equation for the action would have an appropriate form that would demonstrate the vanishing transition probabilities among the adiabatic states.

5.3.1 Two-Level Systems and the Dykhne Formula

We start the analysis with the semiclassical consideration of the Landau–Zener problem and determine the reference system most suitable for the semiclassical consideration. To this end we employ the adiabatic representation by finding a time-dependent transformation $\hat{O}(t)$ diagonalizing the Hamiltonian (5.32) by the transformation $\widehat{H'}(t) = \hat{O}^{-1}(t)\hat{H}(t)\hat{O}(t)$, which corresponds to the rotation from our initial diabatic representation to the adiabatic basis. For the Landau–Zener problem, the unitary matrix performing the rotation has the form

$$\hat{O}(t) = \exp\left[i\hat{\sigma}_y \Theta(t)/2\right] \tag{5.105}$$

where $\Theta(t) = \pi/2 + \arctan\left(\alpha t/2V\right)$ is the angle of rotation around the y-axis that aligns the vector $(V, 0, \alpha t/2)$ with the z-axis, and the Pauli matrices $\hat{\sigma}_x, \hat{\sigma}_y, \hat{\sigma}_z$ are given explicitly by (5.33). The unitary matrix (5.105) transforms the Hamiltonian $\hat{H} = V\hat{\sigma}_x + \alpha t\hat{\sigma}_z/2$ to the diagonal form $\hat{H}' = \sqrt{(\alpha t/2)^2 + V^2}\hat{\sigma}_z$.

We note here that $\hat{O}(t)$ of (5.105) equals the unit matrix at $t = -\infty$, and at $t = \infty$ it reads

$$\hat{O}(\infty) = e^{i\hat{\sigma}_y\pi/2} = \begin{pmatrix} 0 & 1 \\ -1 & 0 \end{pmatrix}, \tag{5.106}$$

which means that not only the amplitudes of the states $|0\rangle$ and $|1\rangle$ have been interchanged after the diabatic states have passed through the resonance, but also that an additional phase factor $e^{i\pi} = -1$ appears as a result of this change in the relative position of the levels. This is a particular example of the general phenomenon known as topological or Berry phases, associated with rotation of the adiabatic energy eigenstates with respect to the original basis in Hilbert space.

By substituting $|\psi\rangle = \hat{O}(t)\,|\psi'\rangle$ into (5.31)

$$i\frac{\partial}{\partial t}\hat{O}(t)\,|\psi'\rangle = i\widehat{\frac{\partial O(t)}{\partial t}}\,|\psi'\rangle + i\hat{O}(t)\frac{\partial}{\partial t}\,|\psi'\rangle = \hat{H}(t)\hat{O}(t)\,|\psi'\rangle\,, \tag{5.107}$$

one obtains the Schrödinger equation for the state vector $|\psi'\rangle$ in the adiabatic representation

$$i\frac{\partial}{\partial t}\,|\psi'\rangle = \left[\hat{H}'(t) - \hat{C}(t)\right]|\psi'\rangle\,, \tag{5.108}$$

where $\hat{H}'(t) = \sqrt{V^2 + (\alpha t/2)^2}\hat{\sigma}_z$ is a diagonal matrix, and

$$\hat{C}(t) = i\hat{O}^{-1}(t)\dot{\hat{O}}(t) = -\hat{\sigma}_y\dot{\Theta}(t)/2.$$

In other words the Schrödinger equation has the explicit form

$$\begin{aligned} i\frac{\partial}{\partial t}\psi'_0 &= -\sqrt{V^2 + (\frac{\alpha t}{2})^2}\psi'_0 - \frac{iV\alpha}{4V^2 + (\alpha t)^2}\psi'_1 \\ i\frac{\partial}{\partial t}\psi'_1 &= \sqrt{V^2 + (\frac{\alpha t}{2})^2}\psi'_1 + \frac{iV\alpha}{4V^2 + (\alpha t)^2}\psi'_0. \end{aligned} \tag{5.109}$$

One sees that in spite of the fact that the matrix $\hat{O}(t)$ diagonalizes the Hamiltonian, the cross-terms between the adiabatic states persist due to the time-dependence of this transformation, and the rotation rate $\dot{\Theta}(t)$ of the adiabatic reference frame represents the probability amplitude of this process.

We now express ψ'_1 with the help of the first equation (5.109) and substitute it into the second one, which yields

$$\frac{\partial^2\psi'_0}{\partial t^2} + \frac{\alpha^2 t/2}{V^2 + \frac{\alpha^2 t^2}{4}}\frac{\partial\psi'_0}{\partial t} = \left[\frac{3i\alpha^2 t/4}{\sqrt{V^2 + \frac{\alpha^2 t^2}{4}}} - V^2 - \frac{\alpha^2 t^2}{4} - \frac{\alpha^2/16}{V^2 + \frac{\alpha^2 t^2}{4}}\right]\psi'_0, \tag{5.110}$$

take the amplitude in the form $\psi'_0 = \Psi_0\left(4V^2 + (\alpha t)^2\right)^{-1/2}$ in order to get rid of the coefficient in front of the first derivative, and arrive at

$$\frac{\partial^2\Psi_0}{\partial t^2} + \left[\frac{3i\alpha^2 t/2}{\sqrt{4V^2 + (\alpha t)^2}} + V^2 + \frac{(\alpha t)^2}{4} + \frac{5\alpha^2/4}{\left(4V^2 + (\alpha t)^2\right)^2}\right]\Psi_0 = 0. \tag{5.111}$$

The last term in square brackets can be omitted when the adiabaticity parameter α/V^2 is small, and we obtain the WKB solution

$$\psi_0' = \frac{C \exp\left[-i\int\left(\frac{3i\alpha^2 t/2}{\sqrt{4V^2+(\alpha t)^2}} + V^2 + \frac{(\alpha t)^2}{4}\right)^{1/2} dt\right]}{\left(\frac{3i\alpha^2 t/2}{\sqrt{4V^2+(\alpha t)^2}} + V^2 + \frac{(\alpha t)^2}{4}\right)^{1/4} (4V^2 + (\alpha t)^2)^{1/2}}, \tag{5.112}$$

where C is a constant.

In the adiabatic condition, the imaginary term is always small and the square root in the integrand can be developed as a Taylor series, which after integration yields

$$\psi_0' = \frac{C \exp\left[-i\int\sqrt{V^2 + \frac{(\alpha t)^2}{4}}dt + \int\frac{3\alpha^2 t/2}{4V^2+(\alpha t)^2}dt\right]}{\left(V^2 + \frac{(\alpha t)^2}{4}\right)^{1/4} (4V^2 + (\alpha t)^2)^{1/2}} = C' e^{-i\int\sqrt{V^2+(\alpha t)^2/4}dt}, \tag{5.113}$$

where to be consistent with the order of approximation we have omitted terms in the pre-exponential factors that are small and vanishing as $t \to \infty$ and have substituted the explicit expression $\int\frac{3\alpha^2 t/2}{4V^2+(\alpha t)^2}dt = (3/4)\ln[4V^2 + (\alpha t)^2]$. The constant C' equals unity for the initial conditions $\rho_0(t \to -\infty) \to 1$.

From the result (5.113) one sees that the adiabatic representation has a remarkable property—all of the first-order corrections over the parameter α/V^2 within the approximation given by the diagonal Hamiltonian $\hat{H}'(t)$ cancel each other, such that the latter gives much better precision than is suggested by the linear perturbation theory for other representations. The reason for this is the fact that the probability amplitudes oscillate rapidly as compared to the typical time-dependence of the non-diagonal matrix elements $iV\alpha(4V^2 + (\alpha\tau)^2)^{-1}$ responsible for the transition in (5.109). But how do we determine the accuracy of the approximation? Substituting (5.113) into the second equation of the set (5.109) one finds the other probability amplitude

$$\psi_1' = ie^{i\Phi(t)}\int^t \frac{iV\alpha}{4V^2 + (\alpha\tau)^2} e^{-2i\Phi(\tau)} d\tau. \tag{5.114}$$

Evaluation of the integrals $\Phi(t) = \int^t \sqrt{V^2 + (\frac{\alpha t}{4})^2}dt$ yields $\frac{t}{2}\sqrt{V^2 + (\frac{\alpha t}{4})^2} + \frac{V^2}{\alpha}\ln[\frac{t\alpha}{2V} + \sqrt{1 + (\frac{\alpha t}{2V})^2}]$, and hence (5.114) takes the form

$$\psi_1' = -e^{i\Phi(t)}\int^t \frac{V\alpha(\frac{\tau\alpha}{2V} + \sqrt{1 + (\frac{\alpha\tau}{2V})^2})^{-i2V^2/\alpha}}{4V^2 + (\alpha\tau)^2} e^{-i\tau\sqrt{V^2+(\alpha\tau)^2/4}} d\tau, \tag{5.115}$$

and results for $t \to \infty$, after the substitution $\tau = 2V\ \sinh[\chi]/\alpha$, in

$$|\psi_1'|^2 = \left| \int_{-\infty}^{\infty} \frac{\exp[\frac{V^2}{i\alpha}(2\chi + \sinh[2\chi])]}{2\cosh[\chi]} d\,\chi \right|^2 . \tag{5.116}$$

By analogy to (5.93), the integral in (5.116) has simple poles at the points $\chi = -\pi i/2 - ik\pi$, where $\cosh[\chi] = \sinh[2\chi] = 0$. We have to take positive k, corresponding to the vanishing residuals, in order to obtain the result

$$|\psi_1'|^2 = \left| \sum_{k=0}^{\infty} \frac{\pi \exp[-\frac{\pi V^2}{\alpha}(1+2k)]}{-i(-1)^k} \right|^2 = \left| \frac{i\pi \exp[-\frac{\pi V^2}{\alpha}]}{2(1+\exp[-\frac{2\pi V^2}{\alpha}])} \right|^2 \simeq \pi^2 e^{-\frac{2\pi V^2}{\alpha}}, \tag{5.117}$$

coinciding with the exact result of (5.44) for the adiabatic limit $\frac{2\pi V^2}{\alpha} \gg 1$ under consideration. Note that the exponentially small probability of the non-adiabatic transition has an evident physical explanation: the frequency spectrum of the time-dependent perturbation $V\alpha \left(4V^2 + (\alpha t)^2\right)^{-1}$, given by (5.109) and originating from the slow rotation of the adiabatic basis, simply has vanishing intensities for the harmonics corresponding to the transition energies between the adiabatic states, even at the moment of their closest approach.

This result can be generalized to an arbitrary time-dependent two-level Hamiltonian, provided the typical width of the frequency spectrum associated with the rotation of the adiabatic frame remains much smaller than the transition frequency between the adiabatic levels. For the Hamiltonian

$$\hat{H} = H_z(t)\hat{\sigma}_z + H_x(t)\hat{\sigma}_x \tag{5.118}$$

Equation (5.114) takes the form

$$\psi_1' = e^{i\Phi(t)} \int^t \frac{\dot{\Theta}(\tau)}{2} e^{-2i\Phi(\tau)} d\tau, \tag{5.119}$$

where $\Phi(t) = \int^t \sqrt{H_z^2(t) + H_x^2(t)}dt$ is the accumulated adiabatic phase, and

$$\dot{\Theta}(t) = \frac{[\dot{H}_z(t)H_z(t) - \dot{H}_z(t)H_x(t)]}{[H_z^2(t) + H_x^2(t)]}$$

is the derivative of the angle $\Theta(t) = \pi/2 - \arctan\left(H_z(t)/H_x(t)\right)$ in the transformation matrix $\exp\left[i\hat{\sigma}_y\Theta(t)/2\right]$ relating the diabatic and the adiabatic reference frames (5.105). The main contribution to the integral (5.119) comes from the domain around the points t_n in the complex plane t given by the relation $[H_z^2(t_n) + H_x^2(t_n)] = 0$ where the derivative $\dot{\Theta}(t)$ has poles, and the phase $\Phi(t)$

has a stationary point $\dot{\Phi}(t) = \sqrt{H_z^2(t) + H_x^2(t)} = 0$. However, the fact that $\dot{\Phi}(t)$ is not analytic at these points does not allow one to employ either the Cauchy formula or the stationary point approximation for $\Phi(t)$. Nevertheless, the Taylor expansion of the square root and the substitution $t_n - t = x^2$ result in

$$\psi_1'(t \to \infty) \simeq e^{-2i\Phi(t_n)+i\Phi(t)} \int e^{-iAx} \frac{dx}{2x} = \pi e^{-2i\Phi(t_n)+i\Phi(t)}, \qquad (5.120)$$

and apparently all the contributions of different points t_n in that half of the complex plane t where $-2i\Phi(t_n)$ has a negative real part have to be added. Note that the coefficient A originating from the third-order term in the Taylor expansion does not enter the result. The main contribution corresponds to t_n with the largest negative real part.

For the transition probabilities P_{ab} between the adiabatic states a and b the result (5.120) gives a short and elegant generalization known as the Dykhne expression:

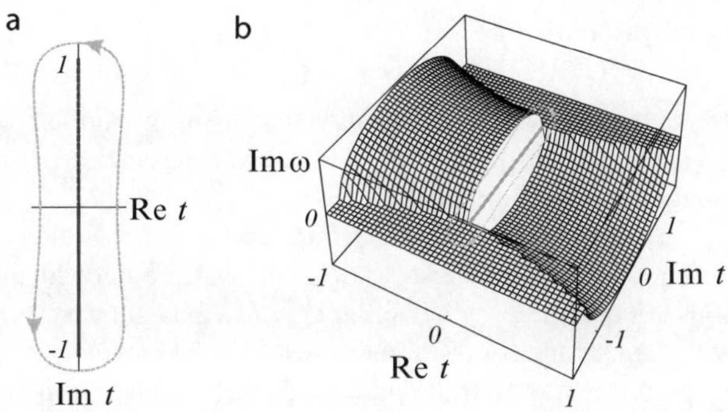

Fig. 5.8 (**a**) The integration contour C for the calculation of the non-adiabatic transition probability. (**b**) Imaginary part of the transition frequency as a function of the complex-valued time. The contour circumvents two complex conjugated branching points $t_n = i$ and $t_n^* = -i$, where the energies corresponding to the state a and the state b coincide, that is, $\omega_{a,b} = 0$. One has to take either the open contour from $t_n^* = -i$ to $t_n = i$ or take half of the integral over the loop shown

$$P_{ab} = |\psi_1'(t \to \infty)| \simeq \pi^2 \exp\left\{ i \int_{t_n}^{t_n^*} \omega_{ab}(t) dt \right\}, \qquad (5.121)$$

where the integral of the time-dependent energy difference $\omega_{ab}(t)$ between the adiabatic states a and b calculated in the limits suggested by the two closest complex conjugated roots t_n and t_n^* of the equation $\omega_{ab}(t) = 0$ takes a positive imaginary value. The other possibility is to integrate half of the frequency over a closed loop, shown in Fig. 5.8.

5.3.2 Multilevel Systems

The adiabatic reference system can also be introduced for a multilevel quantum system with a time-dependent Hamiltonian, if all the eigenstates are non-degenerate. Indeed, let $\hat{H}(t)$ be a time-dependent Hamiltonian entering the Schrödinger equation

$$i\frac{\partial}{\partial t}|\psi\rangle = \hat{H}(t)|\psi\rangle \tag{5.122}$$

for a large system of $n+1$ interacting non-degenerate levels corresponding to the states $|k\rangle\,; k = 0, 1, \ldots, n$, The state vector therefore reads $|\psi\rangle = \sum_k \psi_k |k\rangle$. For each time t one can find a set of eigenvalues $E_{k'}(t)$ of the Hamiltonian, and the corresponding eigenvectors $|k'\rangle$, which form an adiabatic basis set related to the initial basis via the expression $|k\rangle = \hat{O}(t)|k'\rangle$. Here $\hat{O}(t)$ is a unitary time-dependent operator of the transformation diagonalizing the Hamiltonian, as in (5.105) for the Landau–Zener problem. By analogy, substitution of $|\psi\rangle = \hat{O}(t)|\psi'\rangle$ into (5.105) yields (5.107) for the state vector $|\psi'\rangle$ in the adiabatic representation and results in the Schrödinger equation (5.109) with the diagonal Hamiltonian $\hat{H}'(t) = \hat{O}^{-1}(t)\hat{H}(t)\hat{O}(t) = \delta_{k'm'}E_{k'}(t)$ perturbed by a small and slowly changing operator $\hat{C}(t) = \hat{O}^{-1}(t)\dot{\hat{O}}(t)$ originating from a slow rotation of the adiabatic basis in Hilbert space. In the adiabatic representation, the last operator does not have diagonal matrix elements.

Note that by passing to the adiabatic reference system, we did not gain much from a formal point of view, since instead of an initial time-dependent Hamiltonian $\hat{H}(t)$, which does not commute with itself at $t' \neq t$, we obtain another one, $\hat{H}'(t) - \hat{C}(t)$, neither self-commuting for different times. The only advantage is that the off-diagonal elements of the new Hamiltonian are directly related to the rotation of the adiabatic basis, and vanish when the rotation rate approaches 0. When the latter is the case, the transformation $\hat{O}(t)$ diagonalizing the Hamiltonian also diagonalizes the evolution operator

$$\hat{U}_O(t) = \hat{O}(t)\exp\left[-i\int^t \hat{H}'(\tau)d\tau\right]\hat{O}^{-1}(t), \tag{5.123}$$

which implies that no transitions occur among the adiabatic states. On the contrary, in the general case of (5.22) the operator $\hat{U}(t)$ allows for such non- adiabatic transitions, and therefore is not diagonal in the $|k'\rangle$ representation. This difference originates from the finite $\hat{C}(t)$ not commuting with $\hat{H}'$ and results in a remainder $\hat{R}$ encountered in (5.22) which, as we have seen, is exponentially ($\sim e^{-\pi V^2/\alpha}$) small for the two-level Landau–Zener problem. However, in a multilevel system the large variety of possible paths leading from the initial to the final state may change the situation considerably by augmenting the number of possible transition channels.

Let us consider this situation in more detail, and calculate the remainder $\hat{R}$ responsible for the discrepancy between the set of adiabatic states $|k'\rangle$ and the set of eigenvectors of the evolution operator. To get the maximum benefit from the adiabaticity condition, we make use of the interaction representation, that is, we take the evolution operator in the form $\hat{U}(t) = e^{-i\hat{R}(t)}\hat{U}_O(t)$. The differential equation for $\hat{R}(t)$ follows from the equation for the evolution operator $\hat{U}_{int}(t) = e^{-i\hat{R}(t)}$ in the interaction representation

$$i\frac{\partial}{\partial t}\hat{U}_{int}(t) = \hat{H}''(t)\hat{U}_{int}(t) \tag{5.124}$$

with the Hamiltonian $\hat{H}''(t) = -e^{i\int \hat{H}'(t)dt}\hat{C}(t)e^{-i\int \hat{H}'(t)dt}$ which in the adiabatic basis contains only small and rapidly oscillating non-diagonal matrix elements. This allows us to find the equation for the coefficients in the from of the Baker–Campbell–Hausdorff formula (5.3). Indeed, from (5.124) one obtains the expression $\hat{U}(t+dt) = \exp\left[-i\hat{H}''(t)dt\right]\hat{U}(t)$ for an infinitesimally small variation of the evolution operator, which in agreement with (5.1) yields

$$e^{-i\hat{R}(t+dt)} = e^{-i\hat{H}''(t)dt}e^{-i\hat{R}(t)}. \tag{5.125}$$

We take the logarithm of both sides of this equation and expanding the first exponent in *a* Taylor series obtain

$$\begin{aligned} -i\hat{R}(t+dt) &= \ln[e^{-i\hat{R}(t)} - i\hat{H}''(t)e^{-i\hat{R}(t)}dt] \\ &= \ln[1-(1-e^{-i\hat{R}(t)} + i\hat{H}''(t)e^{-i\hat{R}(t)}dt)] \\ &= \sum_n \frac{(-1)^{n+1}}{n}\left[1-e^{-i\hat{R}(t)} + i\hat{H}''(t)dte^{-i\hat{R}(t)}\right]^n. \end{aligned} \tag{5.126}$$

A Taylor expansion of the logarithm over $\hat{H}''$ yields

$$\begin{aligned} -i\hat{R}(t+dt) = &\sum_{n=1} \tfrac{(-1)^{n+1}}{n}\left[1-e^{-i\hat{R}(t)}\right]^n \\ &+i\,dt\sum_{n,k=0}\tfrac{(-1)^{n+k}}{n+k+1}\left[1-e^{-i\hat{R}(t)}\right]^n \hat{H}''(t)e^{-i\hat{R}(t)}\left[1-e^{-i\hat{R}(t)}\right]^k, \end{aligned} \tag{5.127}$$

which after performing the summation takes the form

$$-i\hat{R}(t+dt) = -i\hat{R}(t) + i\,dt\sum_{k=0} a_k\Big\{[-i\hat{R}(t),\Big\}^k \hat{H}''(t)\Big\{]\Big\}^k, \tag{5.128}$$

where the coefficients a_k in the commutator series are given by the Taylor expansion $\sum a_k x^k = x(e^x - 1)^{-1}$. It worth mentioning that the coefficients a_k for $k > 0$ differ from zero only for even k, and, moreover, they are related to the even Bernoulli numbers $B_{2n} = a_{2n}(2n)!$. Here we have employed the integral representation $(n + k + 1)^{-1} = \int_0^1 z^{n+k} dz$, which enables one to rewrite the sum in the last line of (5.127) in the form

$$\begin{aligned}
&\int_0^1 \sum\nolimits_{n,k=0} (-z)^n \left[1 - e^{-i\hat{R}(t)}\right]^n \hat{H}''(t) e^{-i\hat{R}(t)} (-z)^k \left[1 - e^{-i\hat{R}(t)}\right]^k dz \\
&= \int_0^1 \left[1 + z\left(1 - e^{-i\hat{R}(t)}\right)\right]^{-1} \hat{H}''(t) e^{-i\hat{R}(t)} \left[1 + z\left(1 - e^{-i\hat{R}(t)}\right)\right]^{-1} dz.
\end{aligned} \tag{5.129}$$

For the matrix elements of (5.129) in the representation where $\hat{R}(t)$ is diagonal we have performed the integration

$$\begin{aligned}
&\int_0^1 \frac{1}{1+z\left(1-e^{-i\hat{R}_{ll}(t)}\right)} \hat{H}''_{lm}(t) \frac{e^{-i\hat{R}_{mm}(t)}}{1+z\left(1-e^{-i\hat{R}_{mm}(t)}\right)} dz \\
&= \frac{\log\left[e^{-i\hat{R}_{ll}(t)}\right] - \log\left[e^{-i\hat{R}_{mm}(t)}\right]}{e^{-i\hat{R}_{ll}(t)} - e^{-i\hat{R}_{mm}(t)}} e^{-i\hat{R}_{mm}(t)} \hat{H}''_{lm}(t) \\
&= \sum\nolimits_{k=0} a_k \left[i\hat{R}_{mm}(t) - i\hat{R}_{ll}(t)\right]^k \hat{H}''_{lm}(t),
\end{aligned} \tag{5.130}$$

which yields the matrix form of the commutator expansion (5.128).

The nonlinear differential equation (5.128)

$$\frac{\partial \hat{R}(t)}{\partial t} = -\sum_{k=0} a_k \left\{[-i\hat{R}(t),\right\}^k \hat{H}''(t) \left\{]\right\}^k, \tag{5.131}$$

for the matrix $\hat{R}(t)$ has, in the adiabatic case, an approximate solution based on the fact that $\hat{H}''(t)$ is small and rapidly changing. Note that assuming $\hat{H}''(t > t_A) = \hat{H}_B$ and $\hat{R}(t_A) = t_A \hat{H}_A$ and solving this equation iteratively one can also determine the coefficients $c_{k,s}$ in the Baker–Campbell–Hausdorff formula (5.3)–(5.4). Also note that for the high-order commutators $\left\{[\hat{A},\right\}^k \hat{B} \left\{]\right\}^k$ there exists an alternative notation—either $(-1)^k \mathrm{ad}^k(\hat{B}, \hat{A})$, or $(-1)^k \mathrm{ad}^k \hat{B}(\hat{A})$.

For small and rapidly changing $\hat{H}''(t)$, by taking only the first term on the right-hand side of (5.131) one obtains $\hat{R}(t) = \int_{-\infty}^t \hat{H}''(t)dt + \hat{o}(t)$ where the small higher-order correction $\hat{o}(t)$ can also be found by iterations if necessary. Neglecting the correction immediately results in

$$\hat{R}(\infty) = -\int_{-\infty}^{\infty} e^{i\int^t \hat{H}'(\tau)d\tau} \hat{O}^{-1}(t)\dot{\hat{O}}(t)\, e^{-i\int^t \hat{H}'(\tau)d\tau} dt \tag{5.132}$$

where we have substituted the explicit form of $\hat{H}''(t)$. The matrix elements

$$R_{lm}(\infty) = -\sum_k \int_{-\infty}^{\infty} e^{i\int^t \hat{H}'_{ll}(\tau)d\tau} \left[\hat{O}^{-1}(t)\right]_{lk} \dot{O}_{km}(t)\, e^{-i\int^t \hat{H}'_{mm}(\tau)d\tau} dt \quad (5.133)$$

found by an approximate calculation similar to (5.121) give the amplitudes of the direct transitions between the states l and m in the presence of a number of other states k, whereas the overall probabilities that include sequential transitions in this order of approximation are given by the expression

$$P_{lm} = \left| \left[e^{-i\hat{R}(\infty)}\right]_{lm} \right|^2 . \quad (5.134)$$

In the adiabatic regime each matrix element R_{kn} of $\hat{R}(\infty)$ is exponentially small. However, this does not necessarily mean that the matrix itself is small, since the rank $\sim N$ of the matrix and hence its norm $\sim \sqrt{N\,(R_{kn})^2}$ can be large for a large number N of levels in the system. Therefore, the Taylor expansion of the exponent in (5.134) cannot be automatically truncated at the first-order terms. The question arises whether Eq. (5.131) for $\hat{R}(t)$ can be restricted to the zero-order approximation in such a situation that has given the result (5.132)? The answer can be positive if the system is "complex enough", such that both the Hamiltonian $\hat{H}''(t)$ and its primitive $\int \hat{H}''(t)dt$ are generic, that is, each of them contains all or almost all of the $\sim N^2$ terms in their representations in the form of linear combinations of the operators $\hat{\mathcal{G}}_n$ generating the Lie group of the $N \times N$ matrix of evolution. Among the total number $N^2 \times N^2$ of generator pairs the number of non-commuting pairs is of the order of N^2, and hence a typical matrix element of the commutator of two generic matrices is of the order of the product of their typical matrix elements. Therefore the norms of commutators of small generic matrices are small values of a higher order, and thus the correction $\hat{o}(t) = \hat{R}(t) - \int_{-\infty}^{t} \hat{H}''(t)dt$ can indeed be neglected.

5.4 Time-Dependent Level–Band System

Earlier, in Chap. 3, we have seen that a single level interacting with a homogeneous band of levels is a very particular example of complex multilevel quantum systems that allow an exact analytic description. A number of problems for the time-dependent level–band systems with the Schrödinger equation

$$\begin{aligned} i\dot{\psi}_n &= \delta\, n\psi_n + V_{n0}(t)\psi_0 \\ i\dot{\psi}_0 &= \Delta(t)\psi_0 + \sum_{n=1}^{\infty} V_{0n}(t)\psi_n \end{aligned} \quad (5.135)$$

can also be solved analytically for certain particular cases of the time-dependence of the detuning $\Delta(t)$ and the couplings $V_{n0}(t)$. Here we consider just some of them, which manifest meaningful and qualitatively important features.

One peculiarity, common to all of the problems under consideration, has to be mentioned in advance: we assume that the size of the band X is very large, much larger as compared to any of the energy parameters of the system. However, we cannot simply set it to infinity since the logarithm of the ratio X/δ, that is, the logarithm of the typical number of levels in the system, remains one of the crucial parameters of the problem. It has a very moderate size even when the bandwidth X is enormous, and enters many of the characteristics of the system, such as the typical response time.

5.4.1 The Demkov–Osherov Problem

Let us consider now an example of complex multilevel quantum systems with time-dependent parameters that have an exact analytic solution. A system shown in Fig. 5.9 composed of a non-degenerate quantum state $|0\rangle$ with a time-dependent energy $E_0(t) = \alpha t$ interacting with an equidistant spectrum $E_n = \delta\, n$ via a constant coupling $V_{0n} = V$ was suggested by Demkov and Osherov as a model for chemical reactions. The Schrödinger equation (5.135) for such a system

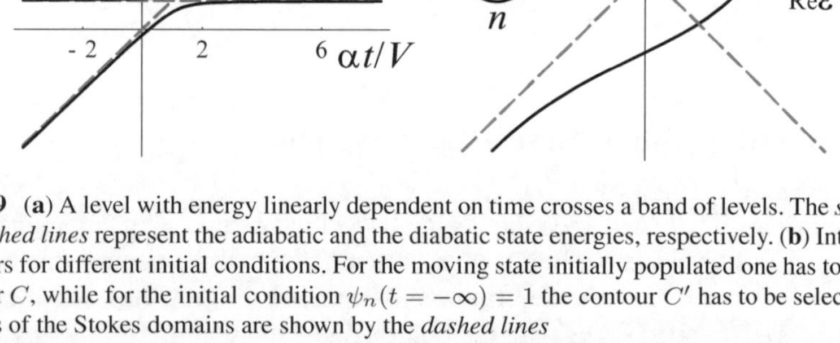

Fig. 5.9 (**a**) A level with energy linearly dependent on time crosses a band of levels. The *solid* and the *dashed lines* represent the adiabatic and the diabatic state energies, respectively. (**b**) Integration contours for different initial conditions. For the moving state initially populated one has to take the contour C, while for the initial condition $\psi_n(t = -\infty) = 1$ the contour C' has to be selected. The borders of the Stokes domains are shown by the *dashed lines*

$$
\begin{aligned}
i\dot{\psi}_n &= \delta n \psi_n + V\psi_0 \\
i\dot{\psi}_0 &= \alpha t \psi_0 + V\sum_{n=1}^{\infty}\psi_n
\end{aligned}
\qquad (5.136)
$$

is similar to (3.23) for the level–band system of Sect. 3.2 with the only difference that the energy of the initially populated state $|0\rangle$ depends linearly on time. The Laplace contour integral method (see p. 214) allows one to find an exact solution. We look for a solution in the form

$$\psi_n(t) = \int\limits_C \psi_n(\varepsilon) e^{-i\varepsilon t} d\varepsilon, \tag{5.137}$$

where the integration contour C is selected in such a way that $\psi_n(\varepsilon)$ assumes the same values at the ends of C for any $n = 0, 1, \ldots$ while the amplitudes themselves satisfy the initial conditions. This choice allows one to perform integration in parts, replacing the time derivatives $\partial/\partial t$ by the factors $-i\varepsilon$ and factors t by $-i\partial/\partial\varepsilon$, which results in a set of equations for the amplitudes entering the integrands

$$\begin{aligned}
\varepsilon\psi_n(\varepsilon) &= \delta\, n\psi_n(\varepsilon) + V\psi_0(\varepsilon) \\
\varepsilon\psi_0(\varepsilon) &= -i\alpha\frac{\partial\psi_0(\varepsilon)}{\partial\varepsilon} + V\sum_{n=1}^{\infty}\psi_n(\varepsilon).
\end{aligned} \tag{5.138}$$

We note here an important difference between (5.138) and equations resulting from the standard Fourier–Laplace method of solving systems of linear differential equations discussed earlier, in Chap. 3: this equation does not explicitly contain initial conditions, and therefore one has to look for a proper integration contour instead.

For $\psi_0(\varepsilon)$ (5.138) immediately yields a first-order differential equation,

$$i\alpha\frac{\partial\psi_0}{\partial\varepsilon} + \varepsilon\psi_0 - \sum_{n=1}^{\infty}\frac{V^2\psi_0}{\varepsilon - \delta\, n} = 0, \tag{5.139}$$

with the solution

$$\begin{aligned}
\psi_0 &= \exp\left\{\int\left(i\frac{\varepsilon}{\alpha} - \sum_{n=1}^{\infty}\frac{iV^2/\alpha}{\varepsilon - \delta\, n}\right)d\varepsilon\right\} \\
&= \exp\left\{i\frac{\varepsilon^2}{2\alpha} - \sum_{n=1}^{\infty}\frac{iV^2}{\alpha}\ln(\frac{\varepsilon - \delta\, n}{A})\right\},
\end{aligned} \tag{5.140}$$

where A is an integration constant. Substitution of this expression into (5.137) yields

$$\psi_0(t) = \int\limits_C \exp\left\{i\frac{\varepsilon^2}{2\alpha} - i\varepsilon t - \sum_{n=1}^{\infty}\frac{iV^2}{\alpha}\ln\left(\frac{\varepsilon - \delta\, n}{A}\right)\right\} d\varepsilon. \tag{5.141}$$

Now we have to find an integration contour that would correspond to the initial conditions imposed. The expression in the exponent has no poles, but only branching points at each $\varepsilon = \delta\, n$, such that closed loops are incapable of giving non-trivial solutions, while the possibility to find an open loop with a pair of initial and final points where the integrand (5.137) takes identical values for each n is quite questionable. The other possibility is to take contours at the ends of which all of the integrands vanish. This can be done either for $\varepsilon \to \pm i e^{\pm i\pi/4}\infty$ or for $\arg(\varepsilon - \delta\, n) \to -\infty$. These two possibilities, shown in Fig. 5.9, correspond to different initial conditions. If all of the population was initially in the moving level, the contour C going from $\varepsilon = \sqrt{i}\infty$ to $\varepsilon = -\sqrt{i}\infty$ along the Stokes lines can be selected. At $t \sim t_0 \to \infty$ where the functional dependence of the logarithmic terms is negligible, the contour passes through the saddle point $\varepsilon = \alpha t$ on the real axis, which yields

$$\psi_0(t) = \mathrm{const} \int\limits_C \exp\left\{i\frac{\varepsilon^2}{2\alpha} - i\varepsilon t\right\} d\varepsilon = \exp\left\{-i\frac{t^2\alpha}{2}\right\}. \tag{5.142}$$

The contour C' appropriate for the initial condition $\psi_n(t)|_{t\to-\infty} = 1$ is a spiral winding towards the point $\varepsilon = \delta\, n$, and the main contribution for large negative t comes from the vicinity of the saddle point $\varepsilon = \delta\, n - (V^2/\alpha + i)/t$

$$\begin{aligned} \psi_n(t) &= \int\limits_{C'} \tfrac{V}{\varepsilon - \delta\, n} \exp\left\{i\tfrac{\varepsilon^2}{2\alpha} - i\varepsilon t - \sum\nolimits_{n=1}^{\infty} \tfrac{iV^2}{\alpha} \ln(\tfrac{\varepsilon - \delta\, n}{A})\right\} d\varepsilon \\ &\simeq \exp\left\{-i\delta\, nt + \frac{iV^2}{\alpha}\ln(-t/A')\right\}, \end{aligned} \tag{5.143}$$

where A' is a constant different from A.

Let us consider the problem in the limit $\delta \to 0, V^2/\delta \to W$, and for $t \ll 1/\delta$, namely assuming that our time interval t, being long, still remains shorter than the Heisenberg recurrence time given by the inverse of the level spacing δ. Replacing the sum in (5.140) by the integral we arrive at

$$\psi_0 = \exp\left\{i\frac{\varepsilon^2}{2\alpha} - \frac{iW}{\alpha}\int \ln\left(\frac{\varepsilon - x}{A}\right) dx\right\}. \tag{5.144}$$

Evaluation of the integral in the finite limits $0 < x < X < \infty$ results in

$$\begin{aligned} \int\limits_0^X \ln(\frac{\varepsilon - x}{A}) dx &= \varepsilon \ln(\tfrac{-\varepsilon}{X - \varepsilon}) - X + X\ln\left[\tfrac{-X+\varepsilon}{A}\right] \\ &\simeq \varepsilon \ln(\tfrac{-\varepsilon}{X}) - \varepsilon + const = \varepsilon \log(\tfrac{-\varepsilon}{Xe}) + const, \end{aligned} \tag{5.145}$$

where we have ignored the small variable ε in the arguments of logarithms as compared to a large constant X in the first term on the right-hand side and retain only the correction linear in ε resulting from the third term; these are the only two non-constant parts in this extreme case. Equation (5.144) then takes the form

$$\psi_0 = const \int \exp\left\{ i\frac{\varepsilon^2}{2\alpha} - \frac{iW}{\alpha}\varepsilon \ln\left(\frac{-\varepsilon}{Xe}\right)\right\} d\varepsilon, \tag{5.146}$$

and for the time-dependent amplitudes results in

$$\psi_0 = \frac{1}{\sqrt{2\alpha\pi}} \int\limits_C \exp\left\{ i\frac{\varepsilon^2}{2\alpha} + i\left(\frac{W}{\alpha}\ln\frac{X}{W} - t\right)\varepsilon - \varepsilon\frac{iW}{\alpha}\ln\left(\frac{-\varepsilon}{We}\right)\right\} d\varepsilon, \tag{5.147}$$

where we have employed the relation $\log(-\varepsilon/X) = \ln(-\varepsilon/W) - \ln(X/W)$, and have chosen the constant corresponding to the initial condition $|\psi_0\,|_{t\to-\infty}| = 1$. By the replacement $\varepsilon \to yW$ and $t = (W/\alpha)\ln(X/W) + WT/\alpha$ we express the probability amplitude in terms of a contour integral,

$$\psi_0 = \sqrt{\frac{W^2}{2\alpha\pi}} \int\limits_C \exp\left\{ \frac{iW^2}{\alpha}\left[\frac{y^2}{2} - T\,y - y\ln(\frac{-y}{e})\right]\right\} dy, \tag{5.148}$$

which is a universal function of two parameters, T and W^2/α. In Fig. 5.10 we depict $|\psi_0|$ as a function of these parameters. One sees that with the elapse of time the population $|\psi_0|^2$, initially equal to 1, approaches 0 with a rate proportional to the parameter W^2/α. This process is not just an exponential relaxation—a tiny reminiscence of the coherence manifests itself in small oscillations of the real and imaginary parts of the amplitude. The asymptotic expression for the probability of the state $|0\rangle$ reads

$$\rho_0 \simeq e^{-2\pi W^2 T/\alpha}. \tag{5.149}$$

Equation (5.149) written in the ordinary time variable $t \sim TW/\alpha$ corresponds to the exponential decay $\rho_0 \sim e^{-2\pi V^2 t/\delta}$ identical to that in the case of a stationary level–band system, which is given by (3.34).

In Sect. 5.1 we have noticed that the complexity of the multilevel quantum system dynamics originates from the algebraic properties of the Hamiltonian and its time derivatives. If the operator algebra of their commutators is rich enough, the dynamics of the system does not have a simple description either in terms of explicit algebraic expressions nor in terms of integrals. The fact that the Demkov–Osherov problem under consideration possesses such an explicit solution implies that it corresponds to a very particular structure of the underlying operator relations. Inspection immediately shows that only the first derivative of the Hamiltonian corresponding to (5.136) is a non-zero matrix of rank 1, and is proportional to the

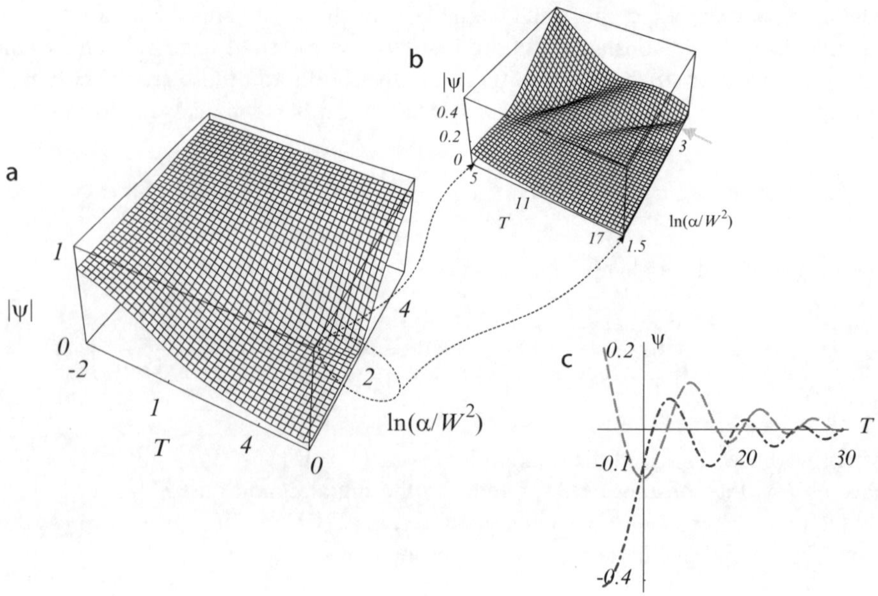

Fig. 5.10 Probability amplitude to remain in the initial state $|\psi|$ as a function of time $T = \alpha t/W - \ln(X/W)$ and the interaction parameter W^2/α for the linearly moving level coupled to a continuum. (**a**) The amplitudes tends from its initial value 1 to the final value 0 with the rate depending on the scaled coupling $\sqrt{W^2/\alpha}$. (**b**) Manifestation of a coherent transitional process that takes place at the moment that the level enters the continuum. (**c**) Real (*dashed line*) and imaginary (*dot-dashed line*) parts of the amplitude during the transitional process, for the value of W^2/α shown in (**b**) by the *arrow*

operator $\hat{P}_0$ of projection to the moving state $|0\rangle$. Nevertheless, the direct application of the idea of algebraic description is not of much use, since the ensemble of all possible commutations of the Hamiltonian $i\hat{H}(t)\,|_{t=0}$ and the projector $i\hat{P}_0$ spans all operator space. In other words, any Hermitian operator in the Hilbert space of quantum states can be given in the form of a linear combination of $i\hat{H}(0), i\hat{P}_0$, and their commutators of first and all higher orders, which form therefore a complete basis set. Hence from the viewpoint of the algebra of operators in the Hilbert space of quantum states, we have encountered a problem of the general type.

The possibility of finding the analytical solution comes from the fact that the relation $\partial\hat{H}(t)/\partial t\,|_{t=0} \sim \hat{P}_0$ results in a very important simplification of the problem. After the Fourier transformation $t \rightarrow \varepsilon$ the Schrödinger equation results in a linear first-order differential equation (5.139) in the space ε. In other words, the algebraic structure of the problem is very simple from the viewpoint of operators in the frequency space. Indeed, the commutator of the operator $\partial/\partial\varepsilon$ canonically conjugated with ε and any multiplicative operator $U(\varepsilon)$ results in another multiplicative operator $U'(\varepsilon) = [\partial/\partial\varepsilon, U(\varepsilon)]$ which is equivalent to $U(\varepsilon)$ as an operator. We note that any further complication of the problem, such as parabolic time-dependence of the detuning, or simultaneous linear time-dependence of two states, results, instead of (5.139), in a second-order differential equation for ψ_0, whose algebraic structure is much more involved, and which in the general case does not have an explicit analytic solution.

5.4.2 The Landau–Zener Transition at the Continuum Edge

When considering on p. 80 the role of spectrum edges, we noticed that a discrete quantum state exists near the end of the continuum interacting with a non-degenerate level $|0\rangle$. The positions of such states in the vicinity of each of two continuum edges are found from the solution of a transcendental equation illustrated in Fig. 3.6, and their population remains constant with the course of time. Therefore, even though weakly populated, these states bring the dominating contribution to the long-time asymptotic of the level population, while the contribution of the rest of the spectrum decays exponentially. For the level–band problem with time-dependent parameters such discrete eigenvalues of instantaneous Hamiltonians exist as well and although they move, their population also plays an important role in the long-time dynamics of the system.

5.4.2.1 The Single Moving Level

One can trace the origin of these discrete states for the Demkov–Osherov problem in the limit $\delta \to 0$, $V^2/\delta \to W$ shown in Fig. 5.11a. We concentrate here on the role of the lower edge, considering the problem only as long as the level remains far below the upper edge of the band. Even in the condition where the level $|0\rangle$ crosses the lower edge and penetrates deeper and deeper into the band, the adiabatic state $|ad\rangle$ corresponding at $t \to -\infty$ to the position $E_{ad}(t) \to \alpha t$ of the level $|0\rangle$ remains discrete, although it approaches the lower edge of the continuum, that is, $E_{ad}(t) \to 0$ at $t \to \infty$. Our aim is to find the time-dependent population of this adiabatic state.

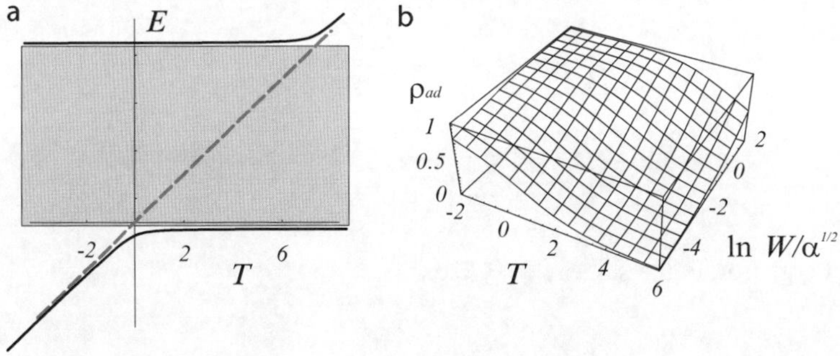

Fig. 5.11 Level crosses a continuous band (**a**). There are two discrete adiabatic states above and below the continuum. Population (**b**) of the lower adiabatic state as a function of the scaled time $T = \alpha t/V^2 g - \ln X/W$ and the adiabaticity parameter $W/\sqrt{\alpha}$

Let us first express the probability amplitude of the adiabatic state in terms of the amplitudes ψ_n of diabatic states $|n\rangle$ of (5.136) and then take the limit $\delta \to 0$. From the equation for the eigenvalues one finds immediately

$$E_{ad}(t)\tilde{\psi}_n = \delta\, n\tilde{\psi}_n + V\tilde{\psi}_0$$
$$E_{ad}(t)\tilde{\psi}_0 = \alpha t\tilde{\psi}_0 + V\sum_{n=1}^{\infty}\tilde{\psi}_n, \tag{5.150}$$

which yields algebraic equations for the projections $\langle ad \mid n\rangle = \tilde{\psi}_n$ and for the adiabatic energy

$$\tilde{\psi}_n = \frac{V}{E_{ad}(t)-\delta\, n}\tilde{\psi}_0$$
$$E_{ad}(t) = \alpha t + \sum_{n=1}^{\infty}\frac{V^2}{E_{ad}(t)-\delta\, n}, \tag{5.151}$$

respectively. The first equation allows one to express the normalized adiabatic state in the form

$$|ad\rangle = \frac{1}{\mathcal{N}}\left(|0\rangle + \sum_{n=1}^{\infty}\frac{V}{E_{ad}(t)-\delta\, n}|n\rangle\right) \tag{5.152}$$

where

$$\mathcal{N} = \sqrt{1+\sum_{n=1}^{\infty}\left(\frac{V}{E_{ad}(t)-\delta\, n}\right)^2}, \tag{5.153}$$

and therefore relates the amplitude $\psi_{ad} = \langle ad \mid st\rangle$ of the adiabatic state to the solutions $\psi_n(t)$ of (5.136) via the standard definition of the state vector $|st\rangle = \sum\psi_n(t)\,|n\rangle$,

$$\psi_{ad}(t) = \frac{1}{\mathcal{N}}\left(\langle 0| + \sum_{n=1}^{\infty}\frac{V}{E_{ad}(t)-\delta\, n}\langle n|\right)\sum_{n=0}^{\infty}\psi_n(t)\,|n\rangle$$
$$= \frac{1}{\mathcal{N}}\left(\psi_0 + \sum_{n=1}^{\infty}\frac{V\psi_n}{E_{ad}(t)-\delta\, n}\right). \tag{5.154}$$

The first equation in the set (5.136) yields

$$\psi_n(t) = -iV\int_{-\infty}^{t} e^{-i\delta n(t-t')}\psi_0(t')dt, \tag{5.155}$$

and hence

$$\psi_{ad}(t) = \frac{1}{\mathcal{N}}\left(\psi_0(t) - i\int\limits_{-\infty}^{t}\sum_{n=1}^{\infty}\frac{V^2 e^{-i\delta n(t-t')}\psi_0(t')}{E_{ad}(t) - \delta\, n}dt\right), \tag{5.156}$$

where the probability amplitude $\psi_0(t)$ to remain in the state $|0\rangle$ is given by (5.141).

Now we are in a position to consider the limit $\delta \to 0$ and replace summation by integration. This yields an explicit expression for the probability amplitude

$$\psi_{ad}(t) = \frac{\psi_0(t) - i\int\limits_{-\infty}^{t} W \int\limits_{0}^{\infty} \frac{\exp[-ix(t-t')]}{E_{ad}(t)-x} dx\psi_0(t')\, dt'}{\sqrt{1 - W/E_{ad}(t)}}, \tag{5.157}$$

where we have taken into account the normalization constant

$$\mathcal{N} = \sqrt{1 + \int\limits_{0}^{\infty}\frac{W\, dx}{\left(E_{ad}(t) - x\right)^2}} = \sqrt{1 - \frac{W}{E_{ad}(t)}}. \tag{5.158}$$

At the same extreme equation (5.151) reads

$$E_{ad}(t) = \alpha t + \int\limits_{0}^{X\to\infty}\frac{W\, dx}{E_{ad}(t) - x} = \alpha t + W\ln\left(\frac{E_{ad}(t)}{-X}\right), \tag{5.159}$$

where we should take precautions and cut the logarithmic divergence at the upper integration limit, by choosing a large but finite width X for the continuum band in the same way as earlier in (5.145)–(5.148). As then, the replacement $t \to (W/\alpha)\log(X/W) + WT/\alpha$, along with the substitutions $E_{ad}(t) = \mathcal{E}(T)W$, $W^2/\alpha = w$, allows one to get rid of the large constant X, and after a proper change of the integration variable x together with (5.148) yields

$$\begin{aligned}
\psi_{ad} &= \tfrac{1}{\sqrt{1-\mathcal{E}^{-1}}}\left(\psi_0 - i\int\limits_{-\infty}^{T} w \int\limits_{0}^{\infty} \tfrac{\exp[-ix(T-T')]}{w\mathcal{E}-x}\psi_0(T')\, dx\, dT'\right),\\
\mathcal{E} &= T + \ln\left[-\mathcal{E}\right],\\
\psi_0 &= \sqrt{\tfrac{w}{2\pi}}\int\limits_{C}\exp\left\{iw\left[\tfrac{y^2}{2} - T\, y - y\ln(\tfrac{-y}{e})\right]\right\}dy,
\end{aligned} \tag{5.160}$$

where we have to keep in mind that $\mathcal{E} = \mathcal{E}(T) < 0$, where the integration contour C is shown in Fig. 5.9b.

Substitution of the last equation of the set (5.160) into the second term in the parentheses of the first equation followed by integration over T' yields

$$-\sqrt{\frac{w^3}{2\pi}}\int\limits_C\int\limits_0^\infty \frac{\exp\left\{iw\left[\frac{y^2}{2}-T\,y-y\ln(\frac{-y}{e})\right]\right\}}{(w\mathcal{E}(T)-x)\,(ix-iw\,y)}\,dx\,dy. \tag{5.161}$$

We note that the position $\mathcal{E}(T)$ of the discrete adiabatic level underneath the band is related to time by the second equation (5.160), and therefore the replacement $T\to\mathcal{E}-\ln[-\mathcal{E}]$ can be performed in order to simplify the integrals. After integration of (5.161) over x and substitution of the result into the first equation (5.160) we arrive at

$$\psi_{ad}=\sqrt{\frac{w}{2\pi}}\int\limits_C\left(1+\frac{\ln\frac{-\mathcal{E}}{y}}{y-\mathcal{E}}\right)\frac{\exp\left\{iw\left[\frac{y^2}{2}-\mathcal{E}\,y-y\ln(\frac{y}{e\mathcal{E}})\right]\right\}}{\sqrt{\mathcal{E}^{-1}-1}}\,dy. \tag{5.162}$$

In the exponent one recognizes the primitive of $y-\mathcal{E}+\log(-\mathcal{E}/y)$ which is also the numerator of the prefactor, and therefore after integration by parts, the integral takes the form

$$\psi_{ad}=\sqrt{\frac{w}{2\pi}}\int\limits_C\frac{\exp\left\{iw\left[\frac{y^2}{2}-\mathcal{E}\,y-y\ln(\frac{y}{e\mathcal{E}})\right]\right\}}{(y-\mathcal{E})^2\sqrt{\mathcal{E}^{-1}-1}}\,dy, \tag{5.163}$$

convenient for numerical calculations.

In Fig. 5.11b we show the results of the numerical calculation of the probability $|\psi_{ad}|^2$ given by (5.163) as a function of two parameters T and w where for large couplings and slow motion, one sees that the population decay of the adiabatic state at $T\to\infty$ is slower than for the state ψ_0. We demonstrate this in Fig. 5.12, where the

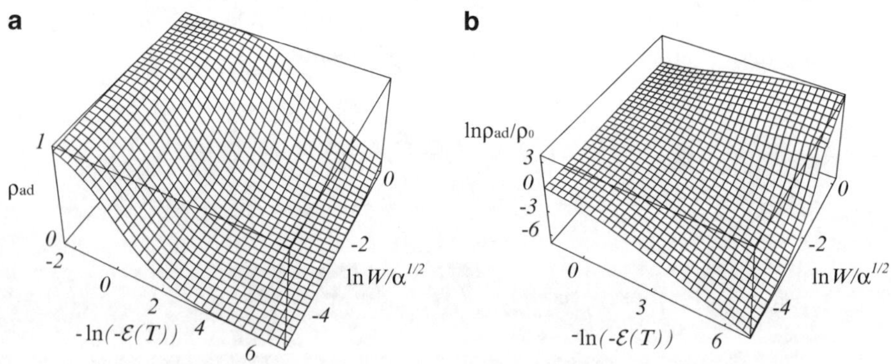

Fig. 5.12 Population (**a**) of the adiabatic state as a function of the adiabaticity parameter $W/\alpha^{1/2}$ and the position $\mathcal{E}(T)$ of the adiabatic level, given by (5.160) and its ratio (**b**) to the diabatic state population as a function of the same variables

adiabatic state population, $\rho_{ad} = |\psi_{ad}|^2$, and its ratio to the diabatic state population $\rho_0 = |\psi_0|^2$ are shown as a function of the parameter $W/\sqrt{\alpha}$ and the energy position $\mathcal{E}(T)$ of the adiabatic level, which serves as a nonlinear time-scale.

As we have mentioned in the previous section, the amplitude ψ_0 exponentially decreases in the long-time limit such that the population given by (5.149) decreases as $e^{-2\pi Wt}$, and in a certain regime its asymptotic contribution to the amplitude of the adiabatic state can become small as compared to the contribution of the ensemble of the band states, the states near the edge in particular. In order to find the asymptotic behavior of the adiabatic population, we evaluate the integral (5.163) approximately for large w and long times (that is, small $\mathcal{E}$). We transform the integration contour to a loop circumventing the branching point 0 and consider the sum of two parts of the integral as the single integral

$$\psi_{ad} = \int_{-\infty}^{0} (1 - e^{2\pi wy}) \frac{\exp\left\{iw\left[\frac{y^2}{2} - \mathcal{E}\, y - y\ln(\frac{y}{e\mathcal{E}})\right]\right\}}{(y-\mathcal{E})^2\sqrt{2\pi/w}\sqrt{\mathcal{E}^{-1}-1}}\, dy. \tag{5.164}$$

We now set $(1 - e^{2\pi wy}) \simeq 2\pi wy$ and keep only the leading term in the exponent, which yields

$$\begin{aligned} \psi_{ad} &= \frac{-\sqrt{2\pi w^3}}{\sqrt{\mathcal{E}^{-1}-1}} \int_{-\infty}^{0} y \frac{\exp\{iwy\ln(-\mathcal{E})\}}{(y-\mathcal{E})^2}\, dy \\ &\simeq \frac{-\ln(-w\mathcal{E}\ln(-\mathcal{E}))}{\sqrt{\mathcal{E}^{-1}-1}} \sqrt{2\pi w^3}. \end{aligned} \tag{5.165}$$

We note that at long times $T \simeq -\ln[-\mathcal{E}]$ and hence $\mathcal{E} \simeq -e^{-T}$, and therefore

$$\begin{aligned} |\psi_{ad}|^2 &\simeq 2\pi w^3 e^{-T}\left[\ln\left(we^{-T}\ln\left(e^{-T}\right)\right)\right]^2 \\ &\simeq 2\pi w^3 e^{-T} T^2. \end{aligned} \tag{5.166}$$

It is expedient to compare the population (5.166) of the adiabatic state with the population of the diabatic state $|0\rangle$ of (5.149). The comparison

$$\begin{aligned} \rho_{ad} &\simeq 2\pi w^3 e^{-T} T^2 \\ \rho_0 &\simeq e^{-2\pi wT}, \end{aligned} \tag{5.167}$$

shows clearly that for a slow motion ($w \to \infty$), the adiabatic contribution dominates. Both expressions (5.167) contain implicitly the width of the band, which determines, however, just the typical time T when the asymptotic regime is attained. Comparison of (5.149) with (5.166) shows that the adiabatic asymptotic regime certainly becomes more important at $T > \log(2\pi w^3)/\,(1 - 2\pi w)$.

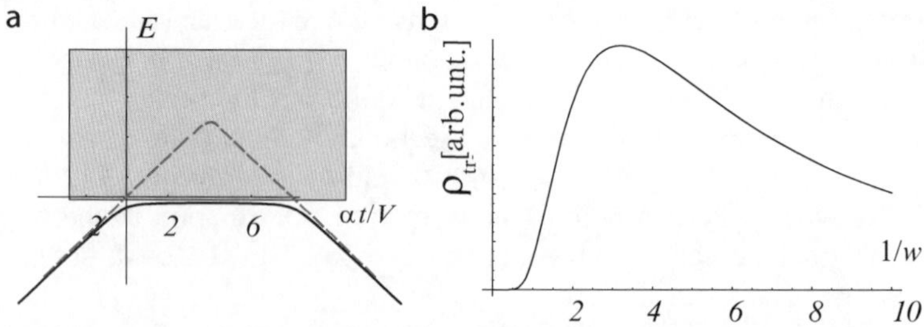

Fig. 5.13 Double passage of the continuum edge (**a**). The diabatic (*dashed line*) and the adiabatic (*solid line*) channels. Population (**b**) of the non-adiabatic transition as a function of the level velocity $v = \alpha/W^2$ for $W = 10^{-2}$. The interaction time is taken such that $\mathcal{E}(T_{max})/W = 1/e$

Note another important consequence. Imagine that the level $|0\rangle$ enters the band and remains for a while before returning back, as shown in Fig. 5.13a. Let us assume that during the return, the system has lost the phase coherence such that the probability to return back to the initial state after the double passage of the continuum edge is just the product of two equal probabilities for the forward and for the back transition, no matter whether we consider the adiabatic or the diabatic trajectory. As for the overall probability $\rho_{0\to 0}$ to return in the initial state $|0\rangle$ at $t \to \infty$ one has to take into account the fact that the adiabatic and the diabatic discrete states are not orthogonal, and therefore this probability has to be taken in the form

$$\rho_{0\to 0} \simeq \rho_0^2 + \rho_{ad}^2 - \frac{\rho_0\rho_{ad}}{|\langle 0|ad\rangle|^2} = \rho_0^2 + \rho_{ad}^2 - \frac{\rho_0\rho_{ad}}{|\mathcal{N}|^2}, \tag{5.168}$$

where the normalization factor $\mathcal{N}$ is given by (5.153). The irreversible transition probability $\rho_{tr} = 1 - \rho_{0\to 0}$ as a function of the velocity has a maximum when the adiabatic and the diabatic populations are of the same order of magnitude, which is illustrated in Fig. 5.13b.

The Exponentially Rising Interaction

Earlier, on p. 220 we considered the two-level system with an exponentially rising coupling. Here we generalize this consideration to the case of the level–band system. The Schrödinger equation (5.135) for this case reads

$$\begin{aligned} i\dot{\psi}_n &= n\delta\psi_n + Ve^{\alpha t}\psi_0 \\ i\dot{\psi}_0 &= \Delta_0\psi_0 + Ve^{\alpha t}\sum_{n=1}^{\infty}\psi_n, \end{aligned} \tag{5.169}$$

and with the notation $\theta = \sqrt{\alpha/2\delta}\int^t Ve^{\alpha x}dx = Ve^{\alpha t}/\sqrt{2\alpha\delta}$ and $\Delta = \Delta_0/2\alpha$, similar to that introduced earlier in (5.94) and (5.85), it adopts the form

$$
\begin{aligned}
i\frac{\partial\psi_n}{\partial\theta} &= \frac{\Delta + n\delta'}{\theta}\psi_n + \sqrt{2\delta'}\psi_0 \\
i\frac{\partial\psi_0}{\partial\theta} &= \sqrt{2\delta'}\sum_{n=1}^{\infty}\psi_n
\end{aligned}
\tag{5.170}
$$

resembling (5.95) for a two-level system. Here $\delta' = \delta/\alpha$ and the initial condition $\psi_0(\theta = 0) = 1$ is the same as earlier.

Now one can employ the same method and look for the solution in the form of a power series,

$$
\begin{aligned}
\psi_0(\theta) &= \sum_{k=0}^{\infty}(-i)^k\frac{c_{0,k}}{k!}\theta^{2k}, \\
\psi_n(\theta) &= \sum_{k=0}^{\infty}(-i)^k\frac{c_{n,k}}{k!}\theta^{2k+1},
\end{aligned}
\tag{5.171}
$$

as was done for (5.37). The recurrence relations for the coefficients

$$
\begin{aligned}
\left[i\,(2k+1) - (\Delta + \delta'\,n)\right]c_{n,k} &= \sqrt{2\delta'}c_{0,k} \\
2c_{0,k} &= \sqrt{2\delta'}\sum_{n=1}^{\infty}c_{n,k-1}
\end{aligned}
\tag{5.172}
$$

yield

$$
\begin{aligned}
c_{0,k} &= c_{0,k-1}\sum_{n=1}^{\infty}\frac{\delta'}{\left[i\,(2k-1) - (\Delta + \delta'\,n)\right]} \\
&= c_{0,k-1}\int_0^{X\to\infty}\frac{dx}{\left[i\,(2k-1) - \Delta - x\right]} = c_{0,k-1}\ln\frac{i\,(2k-1) - \Delta}{-X}
\end{aligned}
\tag{5.173}
$$

and result in

$$
c_{0,k} = \prod_{p=1}^{k}\ln\frac{2p - 1 + i\Delta}{iX}. \tag{5.174}
$$

such that

$$
\psi_0(\theta) = \sum_{k=0}^{\infty}\frac{(-i\theta^2)^k}{k!}\prod_{p=1}^{k}\ln\frac{2p - 1 + i\Delta}{iX}. \tag{5.175}
$$

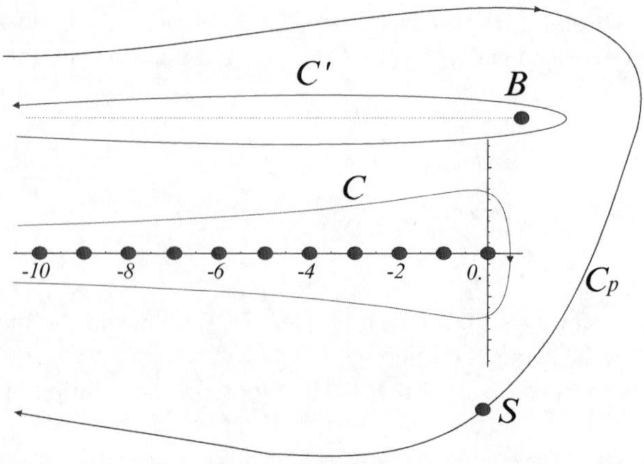

Fig. 5.14 Integration contour C circumventing the negative integer points in the complex plane can be deformed as $C \to C_p + C'$ such that it passes the saddle point S. The loop C' which goes around the branching (B) point $y = (1+i\Delta)/2$ allows for the dying oscillations of the populations, whereas the principal part C_p of the contour gives the main contribution

In the calculation of the sum (5.175), we make use of the fact that the bandwidth X is large. We employ the representation

$$c_{0,k} = \exp\left\{\sum_{p=1}^{k} \ln\left[\ln \frac{2p-1+i\Delta}{iX}\right]\right\} \tag{5.176}$$

and develop the double logarithm in the exponent in the Taylor series over a small parameter $1/\ln X$ allowing for the fact that the leading term $\ln[-\ln X]$ of this series is non-analytic at $1/\ln X \to 0$. Although in the first-order approximation the coefficients

$$c_{0,k} = k \exp \ln[-\ln X] - \frac{1}{\ln X}\sum_{p=1}^{k} \ln\left(\frac{2p-1}{i} + \Delta\right) \tag{5.177}$$

can be found explicitly,

$$\sum_{p=1}^{k} \ln\left(\frac{2p-1}{i} + \Delta\right) = \ln \prod_{p=1}^{k}\left(\frac{2p-1}{i} + \Delta\right) = \ln \frac{2^k}{i^k} \frac{\Gamma\left(k + \dfrac{1+i\Delta}{2}\right)}{\Gamma\left(\dfrac{1+i\Delta}{2}\right)}, \tag{5.178}$$

in order to be consistent within the approximation one has to take the asymptotic $k, \Delta \gg 1$ for the Γ-functions and substitute $\ln\Gamma(z) = (z-1/2)\ln z - z + \ln\sqrt{2\pi}$. Equation (5.175) then yields

$$\psi_0(\theta) = \sum_{k=0}^{\infty} \frac{(i\theta^2 \ln X)^k}{k!} \frac{\left(\dfrac{\Delta - i - 2ik}{2}\right)^{-(2k+i\Delta)/2\ln X}}{\left(\dfrac{\Delta - i}{2}\right)^{-i\Delta/2\ln X}}. \tag{5.179}$$

For small θ, Eqs. (5.175) and (5.179) are both equally good for the numerical evaluation of the probability amplitude of the initial state. However, for large θ, the series starts to explode and become inconvenient for numerical calculations. There is an alternative integral representation

$$\psi_0(\theta) = \int\limits_C \frac{(-i\theta^2 \ln X)^{-y}}{2\pi i} \Gamma(y) \frac{\left(\dfrac{\Delta - i}{2} + iy\right)^{(2y-i\Delta)/2\ln X}}{\left(\dfrac{\Delta - i}{2}\right)^{-i\Delta/2\ln X}} dy, \tag{5.180}$$

which relies on the fact that the Γ-function has simple poles with the residuals $(-1)^k/k!$ all at negative integer points $-k$, which the integration contour C shown in Fig. 5.14 has to circumvent.

For large θ one can make use of the asymptotic expression $\Gamma(y) = y^y e^{-y}/\sqrt{2\pi y}$ for the Γ-function

$$\psi_0(\theta) = \int\limits_C \frac{\left(\dfrac{\Delta - i}{2}\right)^{i\Delta/2\ln X}}{\sqrt{2\pi y}\left(\dfrac{\Delta - i}{2} + iy\right)^{(i\Delta-2y)/2\ln X}} \left(\frac{iy}{e\,\theta^2 \ln X}\right)^y dy, \tag{5.181}$$

and move the integration contour at the saddle point located near $y = -i\theta^2 \ln X$. For moderate θ^2 the integral can be evaluated numerically with the help of the saddle point method, which in this case requires numerical solution of the equation determining the position of the saddle point, and in the extreme case of large θ^2 (5.181) yields for the population

$$|\psi_0(\theta)|^2 \simeq \exp\left[-\frac{4\theta^2 + \Delta/\ln X}{2\theta^2 \ln X + \Delta} + \frac{\Delta}{\ln X} \arctan \frac{1}{\Delta}\right]. \tag{5.182}$$

In Fig. 5.15a we show the population $\rho_{tr} = 1 - |\psi_0(\theta)|^2$ transferred from the discrete level as a function of the integrated interaction $\theta^2 = \int^t dt' V^2(t')/\beta = \int^t dt' W(t')$ and the detuning $\Delta_0 = \Delta 2\alpha$. At detunings smaller than the growth rate α of the interaction, one observes dying oscillations, typical of the coherent damping discussed earlier in Chap. 4. It is worth mentioning that the oscillation frequency amounts to $\ln X$, which indicates the important role of the cooperative behavior which was previously discussed on p. 82. For detunings much larger than the

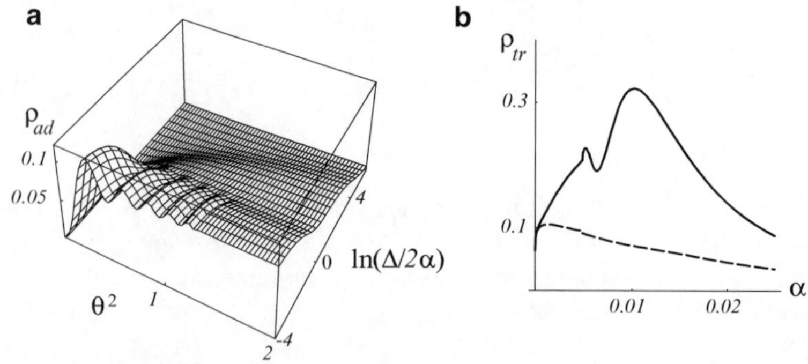

Fig. 5.15 A level near the continuum edge with an exponentially rising coupling $V = V_0 e^{\alpha t}$. Population of the level (**a**) as a function of the integrated interaction $\theta^2 = \int dt V_0^2 e^{2\alpha t}/\beta = We^{2\alpha t}/2\alpha$, and the detuning Δ_0 of the level from the lower edge of the band. (**b**) Probability of the transition as a function of the growth rate α for the case of phase decorrelated double interaction: the exponential increase followed by an exponential decrease of the same rate. Small detuning $\Delta_0 \sim W \sim 0.001$ (*solid line*) and large detuning $\Delta_0 \sim 20W$ (*dashed line*). The maximum of the *solid curve* corresponds to the velocity $\alpha \sim \pi W \ln(X/W)/2$, where the band width X is taken to be unity

growth rate, the transferred population remains zero as long as the interaction remains smaller than the detuning. Note an interesting phenomenon: even for small detunings the transition probability remains small, being of the order of $2/\ln X$. This somewhat strange effect becomes clear when we recall that the repulsion of the level from the band resulting by analogy to (5.159) in the adiabatic energy

$$E_{ad}(t) - \Delta_0 = W \ln\left(\frac{E_{ad}(t)}{-X}\right), \tag{5.183}$$

differs from the probability W of the resonant decay approximately by the factor $\ln X$, which means that even for small detunings, the adiabatic state remains separated from the bottom of the continuum by a gap broader than the width of the Lorentzian profile (3.36) of the resonant transition.

Chapter 6
Time-Dependent Complex Systems

We now turn to time-dependent multilevel systems, which are similar to the two-band models discussed in Chap. 4. The main difference between them and the time-dependent level–band problem discussed in Chap. 5 is in the topology of the transitions; for two-band models the transition topology is much more involved, and a transition from one level to another does not necessarily imply the immediate inverse transition. On the contrary, the next transition rather occurs to a different energy eigenstate of the unperturbed Hamiltonian. The algebraic structure of these problems is such that they cannot be reduced to simpler systems, and therefore no explicit solution is possible without taking advantage of the statistical models. Therefore the main technique employed for the description is based on the ensemble average over the matrix elements of the coupling, and on the topological selection rules for leading terms in each order of the perturbation series, followed by the exact analytic summation of these series, in complete analogy with the approach developed in Chap. 4. Note that in this chapter the consideration of multilevel systems is done for times short as compared to the Heisenberg return time. For times longer than or comparable with the Heisenberg time, the revival effects may result in a phenomenon known as dynamic localization, which will be considered later, in Chap. 8.

We start with the problem of a degenerate level crossing a band, resembling the Demkov–Osherov case. However, the presence of the many quantum states in the degenerate level drastically changes the structure of the solution, such that it can no longer be exactly solved by the Laplace contour integral method. The moving level becomes split into individual components and does not lose its population exponentially, but retrieves it when the new component passes the same resonance with a band state that has been populated while crossed by the former component.

We continue by considering perturbations of multilevel systems by a time-dependent random matrix. A generic time-dependent perturbation contains different harmonics, which provoke transitions among the eigenstates of the unperturbed Hamiltonian separated by energies equal to the energies of quanta corresponding to the harmonics. Therefore, under the action of a time-dependent perturbation, the

V.M. Akulin, *Dynamics of Complex Quantum Systems*, Theoretical and Mathematical Physics, DOI 10.1007/978-94-007-7205-2_6,

energy distribution can differ dramatically from the Lorentzian distribution typical of the time-independent perturbations switched on at a given moment of time. The ensemble averaged population distribution can be found exactly for a quantum system perturbed by a random matrix with an amplitude arbitrarily changing in time. We consider this problem prior to calculation of the responses of the systems to the random perturbations. Note that a non-trivial coherent response exists only in the case of the perturbations that belong to the Gaussian orthogonal ensemble.

We conclude the chapter by considering the special cases of harmonic time-dependent perturbations for the case of an abrupt and adiabatic switching. These results can be generalized to the bi-harmonic time-dependence and to the case of a two band system. It turns out that the energy distribution of the populations over the bands is sensitive to the photon statistics in the case when the time-dependent perturbation is considered as a quantized electromagnetic field.

6.1 Degenerate Level Crosses an Infinite Band

Consider a system consisting of a degenerate level with energy linearly increasing in time, which crosses an infinite band. In many respects this problem is similar to the Demkov–Osherov problem considered in Sect. 5.4.1. The two main differences are in the fact that the level is degenerate, as was the case in Sect. 4.4 for a similar time-independent system, and in that the band is infinite on both sides. We also assume that initially, at $t = 0$, only one sublevel of the degenerate level is populated, which from the physical point of view implies that the population has been injected into this sublevel, while the degenerate level was already in resonance with the band.

The corresponding Schrödinger equation then reads

$$
\begin{aligned}
i\dot{\psi}_n &= n\delta\psi_n + \sum_{m=1}^{N} V_{nm}\psi_m \\
i\dot{\psi}_m &= \alpha t\psi_m + \sum_{n=-\infty}^{\infty} V_{mn}\psi_n,
\end{aligned}
\tag{6.1}
$$

and implies that an N-fold degenerate level of energy αt moving linearly in time, enters across an infinite band of levels spaced by small energy intervals δ, which we consider vanishing $\delta \to 0$. The index n enumerates the levels of the band, and m denotes the sublevels of the degenerate level.

We take the position of the degenerate level at $t = 0$ as the energy reference point and attribute the detuning αt, linearly depending on time, to the levels n of the band. After such a replacement, Fourier–Laplace transformation of (6.1) yields

$$\varepsilon\psi_n(\varepsilon) = n\delta\psi_n(\varepsilon) - i\alpha\frac{\partial\psi_n(\varepsilon)}{\partial\varepsilon} + \sum_{m=1}^{N} V_{nm}\psi_m(\varepsilon)$$

$$\varepsilon\psi_m(\varepsilon) = i\delta_{m,1} + \sum_{n=1}^{\infty} V_{mn}\psi_n, \tag{6.2}$$

where $\delta_{m,m'}$ is the Kronecker δ. Equation (6.2) differs from (5.138) not only in the fact that we allow here for the sublevels m of the degenerate level and move the time-dependence of the degenerate level detuning to the states of the band, but also in the presence of the initial condition $i\delta_{m,1}$, which means that initially only the first sublevel of the degenerate level is populated.

Let us now compare (6.2) with (4.5) for the two-band system of Sect. 4.2.1. at $\Delta_m = 0$. The main discrepancy is in the fact that the detunings of the band states $\Delta_n = \delta\, n - i\alpha\frac{\partial}{\partial\varepsilon}$ include the derivative over the spectral variable ε, and therefore, after the ensemble average, the renormalized Green's functions $X_n(\varepsilon), X_m(\varepsilon)$ given by (4.20), (4.21) and corresponding to the static degenerate-level–band problem of (4.5) became, for the time-dependent system of (6.2) considered, integral operators $\hat{X}_n, \hat{X}_m$ over ε that read

$$\hat{X}_n = \frac{1}{\varepsilon-(n\delta-i\alpha\frac{\partial}{\partial\varepsilon})-\langle V^2\rangle N\hat{X}_m},$$

$$\hat{X}_m = \frac{1}{\varepsilon-\langle V^2\rangle\sum_n\hat{X}_n}. \tag{6.3}$$

In complete analogy with Sect. 4.4.1 we replace the sum by the integral and obtain again (4.71)–(4.73), that is,

$$\sum_n \hat{X}_n = i\pi g$$

$$\hat{X}_m = X(\varepsilon) = \frac{1}{\varepsilon+i\pi g\langle V^2\rangle}, \tag{6.4}$$

regardless of the presence of the derivative in the denominator of the first of (6.3). Here g, as earlier, is the density of the band states. This yields

$$\hat{X}_n(\varepsilon) = \frac{1}{\varepsilon-(n\delta-i\alpha\frac{\partial}{\partial\varepsilon})-\langle V^2\rangle N(\varepsilon+i\pi g\langle V^2\rangle)^{-1}}, \tag{6.5}$$

and by analogy for the complex conjugate Green's function

$$\hat{X}_n(\xi) = \frac{1}{\xi-(n\delta-i\alpha\frac{\partial}{\partial\xi})-\langle V^2\rangle N(\xi-i\pi g\langle V^2\rangle)^{-1}},$$

$$X(\xi) = \frac{1}{\xi-i\pi g\langle V^2\rangle}. \tag{6.6}$$

We also note that apart from the replacements $n \leftrightarrow m$ and $2 \leftrightarrow 0$ expressions (4.27) remain valid for the total population of the degenerate level $\rho_0(\varepsilon, \xi)$ and of the band $\rho_1(\varepsilon, \xi)$, and therefore after having evaluated the sum

$$\sum_n \hat{X}_n(\varepsilon)\hat{X}_n(\xi) = \frac{-2\pi i g}{\varepsilon - \xi + \langle V^2 \rangle N\left(X(\xi) - X(\varepsilon)\right) + i\alpha(\frac{\partial}{\partial\varepsilon} + \frac{\partial}{\partial\xi})} \tag{6.7}$$

by calculating the correspondent integral, one finds for the total population

$$\begin{aligned} \rho_1 &= \tfrac{-2\pi i V^2 g}{\varepsilon-\xi+\langle V^2\rangle N(X(\xi)-X(\varepsilon))-i\alpha(\frac{\partial}{\partial\varepsilon}+\frac{\partial}{\partial\xi})}\rho_0 \\ \rho_0 &= V^2 N X(\varepsilon) X(\xi)\rho_1 + X(\varepsilon)X(\xi). \end{aligned} \tag{6.8}$$

After substitution of $X(\xi)$ and $X(\varepsilon)$ from (6.4) and (6.6) this immediately results in

$$\left(\varepsilon - \xi + \frac{\mathcal{V}^2\left(\varepsilon - \xi\right)}{(\varepsilon + iW)(\xi - iW)} - i\alpha(\frac{\partial}{\partial\varepsilon} + \frac{\partial}{\partial\xi})\right)\rho_1 = \frac{-1}{\varepsilon + iW}\frac{2iW}{\xi - iW} \tag{6.9}$$

where $\langle V^2 \rangle N = \mathcal{V}^2$ is the squared cooperative coupling (see p. 82) and $\pi g \langle V^2 \rangle = W$ is the probability amplitude decay rate (3.34). After the replacement $\varepsilon - \xi \to 2vW, \varepsilon + \xi \to 2uW$ this yields the equation

$$\left(2v + \frac{\mathcal{V}^2 2v/W^2}{u^2 - (v+i)^2} + \frac{i\alpha}{W^2}\frac{\partial}{\partial u}\right)\rho_1 = \frac{-2i/W^2}{u^2 - (v+i)^2}. \tag{6.10}$$

By standard methods, one finds the solution of the first-order differential equation (6.10) in the form

$$\rho_1 = \frac{-2i}{\alpha}\left(\frac{u - v - i}{u + v + i}\right)^{-b}\int_{-\infty}^{0}\exp\left(\frac{-2iv\lambda}{\alpha/W^2}\right)\frac{(u+\lambda-v-i)^{b-1}}{(u+\lambda+v+i)^{b+1}}d\lambda, \tag{6.11}$$

where $b = -\mathcal{V}^2 v/\alpha(1 - iv)$, or in the original notation

$$\rho_1 = \frac{W}{\alpha}\left(\frac{\varepsilon + iW}{\xi - iW}\right)^b\int_{-\infty}^{0} e^{i\lambda(\xi-\varepsilon)/\alpha}\frac{(\xi+\lambda-iW)^{b-1}}{(\varepsilon+\lambda+iW)^{1+b}}d\lambda \tag{6.12}$$

where $b = \frac{-\mathcal{V}^2(\xi-\varepsilon)}{\alpha(2W+i(\xi-\varepsilon))}$. Another representation of the integral

$$\rho_1 = \frac{W}{\alpha}\frac{1}{(\varepsilon + iW)(\xi - iW)}\int_{-\infty}^{0} e^{i\lambda(\xi-\varepsilon)/\alpha}\frac{(1+\frac{\lambda}{\xi-iW})^{b-1}}{(1+\frac{\lambda}{\varepsilon+iW})^{1+b}}d\lambda \tag{6.13}$$

is more convenient for further calculations and the asymptotic analysis.

Let us first consider the limit $\alpha \to 0$. The main contribution to the integral comes from the domain of small λ, which allows one to approximate (6.13) and write

$$\rho_1 = \frac{W}{\alpha} \frac{1}{(\varepsilon+iW)(\xi-iW)} \int_{-\infty}^{0} e^{i\lambda(\xi-\varepsilon)/\alpha+\frac{\lambda}{\xi+iW}(b-1)-\frac{\lambda}{\varepsilon-iW}(b+1)} d\lambda$$

$$= \frac{W/\alpha}{(\varepsilon+iW)(\xi-iW)i(\xi-\varepsilon)/\alpha+(\varepsilon+iW)(b-1)-(\xi-iW)(b+1)}. \tag{6.14}$$

After substitution of b, replacements $(\varepsilon + \xi)/2 \to \eta$, $(\varepsilon - \xi) \to \zeta$ and integration over η this yields

$$\rho_1 = \frac{2\pi iW}{\sqrt{4\alpha^2 + (4V^2 - (2iW + \zeta)^2)\zeta^2}}, \tag{6.15}$$

which for $\alpha = 0$ after inverse Fourier transformation over ζ results in (4.80) for the time-independent system of a band and a degenerate level. This equation shows that the ratio of band population to the degenerate level population in this extreme is a constant, in accordance with (4.82).

Since the limit just considered implies $\alpha T \to 0$, that is, $\alpha/(\varepsilon - \xi) \to 0$, it cannot be employed in the physically interesting case of $\alpha \to 0$, $\alpha T = const$ corresponding to the adiabatic displacement of the degenerate level by a finite energy interval αT, which implies that $(\varepsilon - \xi) \sim \alpha$. In order to describe the dynamics of populations in this limit, one has to make the replacements $(\varepsilon+\xi)/2 \to \eta$, $(\varepsilon - \xi) \to \zeta$ and $-\mathcal{V}^2\zeta/\alpha(i\zeta - 2W) \to b$ in (6.13) and find

$$\rho_1 = \frac{W}{\alpha} \left(\frac{\eta+\zeta/2+iW}{\eta-\zeta/2-iW}\right)^b \int_{-\infty}^{0} e^{i\lambda\zeta/\alpha} \frac{(\eta-\zeta/2+\lambda-iW)^{b-1}}{(\eta+\zeta/2+\lambda+iW)^{1+b}} d\lambda. \tag{6.16}$$

Now we note that the integration to be performed over $d\eta$ from $-\infty$ to ∞ along the real axis is equivalent to the integration along a contour circumventing two singularities in the upper part of the complex plane at the points $\eta = \zeta/2-\lambda+iW$ and $\eta = \zeta/2 + iW$. This can be done explicitly after the replacement

$$\eta \to -\frac{(2iW + \zeta)(2iW + \zeta + \lambda + \lambda x)}{2(2iW + \zeta + \lambda - \lambda x)} \tag{6.17}$$

which reduces to an integral over dx coinciding with the integral representation of the hypergeometric function and yields

$$\rho_1 = \int_{-\infty}^{0} \frac{-2iW}{\alpha(\zeta+2iW)} e^{i\lambda\zeta/2\alpha} \left(\frac{\zeta-\lambda+2iW}{\zeta+\lambda+2iW}\right)^b {}_2F_1\left(b, -b, 1, \frac{\lambda^2}{\lambda^2-(\zeta+2iW)^2}\right) d\lambda. \tag{6.18}$$

In this exact expression we can neglect ζ compared to W in the long-time limit, since the crucial dependence on ζ already enters the indices $\pm b$ of the hypergeometric function, and after introducing the variables $w = W/\mathcal{V}$, $y = \lambda w$, $x = \zeta/2\alpha w$, $\Theta = \alpha T/\mathcal{V}$ we obtain the inverse Fourier transform

$$\rho_1(\Theta) = \frac{-1}{\pi}\int\limits_{-\infty}^{0} \exp\left(ixy - 2i\Theta x + \tfrac{2ix}{w}\arctan(\tfrac{y}{w})\right)$$
$$_2F_1\left(\frac{x}{w}, -\frac{x}{w}, 1, \tfrac{y^2}{w^2+y^2}\right) dxdy. \tag{6.19}$$

Now one has to take into account that the Fourier transformation of this hypergeometric function over the indices reads

$$\int\limits_{-\infty}^{\infty} {}_2F_1\left(u, -u, 1, z\right) \cos\left(uv\right) du = \frac{\cos\left(\frac{v}{2}\right)}{\sqrt{z - \sin^2\left(\frac{v}{2}\right)}}, \tag{6.20}$$

which can be found by a certain exponential replacement in the corresponding integral representation of $_2F_1$. The spectrum of the hypergeometric function ranges from $v = -2\arcsin\sqrt{z}$ to $v = 2\arcsin\sqrt{z}$, and therefore after the replacement $y \to y/w$ one arrives at

$$\rho_1(\Theta) = \int\limits_{-\infty}^{0} \frac{\cos\left(\frac{y}{2} - \Theta w + \arctan\frac{y}{w^2}\right)}{\sqrt{\sin^2\left(\arctan\frac{y}{w^2}\right) - \sin^2\left(\frac{y}{2} - \Theta w + \arctan\frac{y}{w^2}\right)}}\frac{dy}{\pi}$$
$$= \int\limits_{-\infty}^{0} \frac{\cos\left(\frac{y}{2} - \Theta w + \arctan\frac{y}{w^2}\right)}{\sqrt{\sin\left(\Theta w - \frac{y}{2} + 2\arctan\frac{y}{w^2}\right)\sin\left(\frac{y}{2} - \Theta w\right)}}\frac{dy}{\pi}, \tag{6.21}$$

where the arguments of sine and cosine must remain within the interval $(-\pi, \pi)$. Here we have also taken into account that $\sin^2(\arctan z) = z^2/(1+z^2)$.

In Fig. 6.1 we show the population $\rho_0(\Theta) = 1 - \rho_1(\Theta)$ of the degenerate level as a function of the displacement $\Delta = \alpha T$ and the coupling parameter $\mathcal{V}/W = 1/w$, calculated with the help of (6.21). Note that $1/w^2$ equates to the ratio of the number of states N comprising the degenerate level and the number Wg of the band levels in resonance, discussed on p. 82. One sees the main trend of the population behavior, namely an almost linear decrease of $\rho_0(T)$ toward zero with a slope proportional to the parameter w. This is a natural consequence of the fact that in the course of the motion of the degenerate level across the band, the degeneracy is raised by the interaction. Therefore the populations of the resulting components make transitions to the states of the band via the mechanism of adiabatic population transport. Each transition can occur in both directions: from a state of the level to a state of the band and vice versa, depending on which one of these two states corresponds to

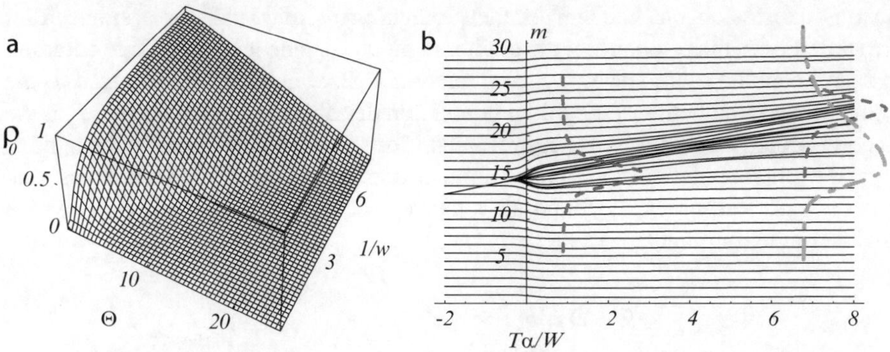

Fig. 6.1 Population (**a**) of the degenerate level as a function of the dimensionless energy shift $\Theta = \Delta W/V^2N$ and the scaled coupling $1/w = \sqrt{V^2N/W^2}$ in the limit $\alpha \to 0$, and $T\alpha \to \Delta$, for the system of an N-fold degenerate level interacting with a band. The eigenvalues of the system (**b**) are shown as a function of time. In order to distinguish among the eigenstates originating from the degenerate level and those originating from the states of the band we took a finite time for the interaction switch. The density of states that belong to the degenerate level are shown by dashed lines for two different times. The population of the adiabatic levels (*dash-dot line*) given by the initial distribution of the degenerate level's density of states propagated along the adiabatic curves after convolution with the final distribution of these states gives the population (**a**)

the populated adiabatic level before the crossing. The number of such transitions is proportional to the product of the band state density and the displacement, whereas the population of each component is inversely proportional to the degeneracy N of the level. One can see this phenomenon by tracing the time evolution of the eigenstates shown in Fig. 6.1b. The population of the adiabatic states does not change, whereas the population of the degenerate level at a given moment in time depends on the number of the initially populated adiabatic states that still belong to this manifold at that moment.

On a long-time scale, the band population probability approaches unity. Deviation from unity can be found from (6.21) when we denote $\frac{y}{2} - \Theta w + \arctan\frac{y}{w^2} = u$, take this quantity as the integration variable, and find approximately the derivative $\partial u/\partial y \simeq 1/2 + 1/(w^2 + 4\Theta^2)$. This yields

$$\rho_1(\Theta) = \int \frac{du}{\sqrt{\sin^2\left(\arctan\frac{2\Theta w}{w^2}\right) - u^2}} \frac{1}{\pi\frac{\partial u}{\partial y}}$$

$$\simeq \frac{1}{1 + 2/(w^2 + 4\Theta^2)} \simeq 1 - \frac{\mathcal{V}^2}{2(\alpha T)^2}, \tag{6.22}$$

that is

$$\rho_0(T) \simeq \frac{\mathcal{V}^2}{2(\alpha T)^2}. \tag{6.23}$$

In this expression one recognizes the population ρ_0 of (3.19) for a single level separated from the resonance with another populated single level by a large detuning αT and coupled to this level by an interaction $\mathcal{V}/2$, which can be attributed to the joint effect of the wings of the initial population distribution and the influence of the second-order perturbation of the level's states by the populated states of the band.

An explicit analytical solution for another important limit of high velocities, that is, $\alpha \to \infty$, can be easily found after the replacement $\lambda \to \alpha\lambda$ in (6.18) which yields, in the limit under consideration

$$\rho_1 = \int_{-\infty}^{0} \frac{-2iWe^{i\lambda\zeta/2}}{(\zeta + 2iW)} (-1)^b \; {}_2F_1(b, -b, 1, 1)\, d\lambda$$
$$= \frac{-i4W}{2\pi b(\zeta + 2iW)\zeta} \left(1 - e^{2i\pi b}\right), \tag{6.24}$$

where the relation

$$(-1)^b \; {}_2F_1(b, -b, 1, 1) = \frac{i}{2\pi b} \left(1 - e^{2i\pi b}\right) \tag{6.25}$$

has been employed. We have also to set $b \to 0$, and hence

$$\rho_1 = \frac{-4W}{(\zeta + 2iW)\zeta}, \tag{6.26}$$

which after inverse Fourier transformation gives a transparent expression

$$\rho_1(t) = 1 - e^{-2\pi g \langle V^2 \rangle t} \tag{6.27}$$

in complete analogy with (3.34). This means that, at such a high velocity of displacement of the degenerate level across the band, each component of the level encounters such a large number of band states that it decays to these states individually, as if they were a real continuum, while all of the interference phenomena such as recurrences and revivals are completely suppressed.

6.2 Perturbation Proportional to a Random Matrix

Thus far, we have considered particular forms of the time-dependencies of the interaction or the level positions that allow one to find the result analytically. Looking at this class of problems from the viewpoint of the algebraic properties of quantum systems discussed in Sect. 5.1, one can say that their algebraic structure is such that the representation

$$\hat{U}(t) = e^{\sum_{n=0}^{N_O} f_n(t)\hat{\mathcal{G}}_n} \tag{6.28}$$

of the evolution operator $\hat{U}(t)$ given in terms of the generators $\hat{\mathcal{G}}_n$ of the unitary group, by analogy to (5.5), has a very special relationship among the functions $f_n(t)$ that allow one to express the solution in terms of integrals or special functions, something that is completely impossible for a generic problem. In this section we turn to the problem of a perturbation with an arbitrary time-dependence, which does not have such a hidden symmetry. In order to achieve an analytical result for this case, we have to make some simplifying assumptions compensating for the complex character of the exact dynamics. We therefore assume that the Hamiltonian has the form

$$\hat{H}(t) = \hat{H}_0 + q(t)\,\hat{V} \tag{6.29}$$

of an unperturbed Hamiltonian $\hat{H}_0$ with a dense and uniform spectrum, plus a perturbation in the form of a time-independent random matrix $\hat{V}$ multiplied by an arbitrary time-dependent coefficient $q(t)$. In the representation of the eigenfunctions of $\hat{H}_0$ shown in Fig. 6.2a the corresponding Schrödinger equation reads

$$i\dot{\psi}_n = \Delta_n \psi_n + q(t) \sum_m V_{nm}\psi_m, \tag{6.30}$$

where Δ_n stand for the eigenstate energies.

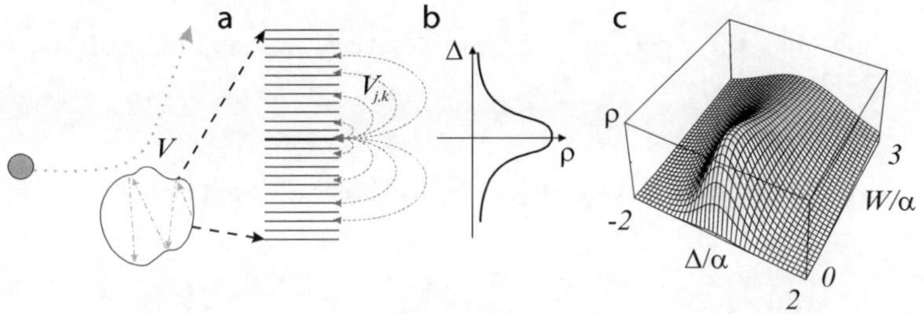

Fig. 6.2 Model of a collision. Dense spectrum (**a**) subjected to the action of a perturbation which has the structure $q(t)\hat{V}$ of a random matrix $\hat{V}$ multiplied by a time-dependent function $q(t)$. Typical population distribution after the collision (**b**). Population distribution (**c**) as a function of the interaction strength $W = \pi g V^2$ and the detuning Δ for a Gaussian $q(t) = e^{-\alpha^2 t^2}$ time-dependence

We now make use of the idea already employed in Sect. 5.1, and replace the Schrödinger equation (6.30) with time-dependent coefficients by an equation of a higher dimensionality with time-independent coefficients by analogy to (5.8). For the evolution operator of this system in a Fourier representation one obtains

$$\varepsilon \hat{U}(\varepsilon,\tau) - \left[\hat{H}_0 + q(\tau)\,\hat{V} - i\frac{\partial}{\partial \tau}\right]\hat{U}(\varepsilon,\tau) = \hat{I}, \tag{6.31}$$

that is

$$\hat{U}(\varepsilon,\tau) = \left(\varepsilon - \hat{H}_0 + i\frac{\partial}{\partial\tau} - q(\tau)\,\hat{V}\right)^{-1}. \tag{6.32}$$

In other words

$$\psi_n(t) = \int \frac{e^{-i\varepsilon t}}{2\pi} \sum_m \left(\frac{1}{\varepsilon - \hat{H}_0 + i\frac{\partial}{\partial\tau} - q(\tau)\,\hat{V}} \right)_{n,m} \Bigg|_{\tau=t} \psi_m(t=0). \tag{6.33}$$

This equation takes exactly the same form as (4.10) when we attribute the derivative $-i\frac{\partial}{\partial\tau}$ to the unperturbed Hamiltonian $\hat{H}_0$. Therefore all of the calculations performed in Sect. 4.1 such as the expansion of the evolution operator in a Taylor series over $\hat{V}$ followed by ensemble averaging can also be repeated for the problem under consideration. The only new requirement is to take into account the order of operators, since $-i\frac{\partial}{\partial\tau}$ and $q(\tau)$ do not commute.

By complete analogy to (4.21), the calculations yield the equation

$$\hat{X}_n(\varepsilon,\tau) = \frac{1}{\varepsilon - \Delta_n + i\frac{\partial}{\partial\tau} - q(\tau)\,\langle V^2\rangle \sum\limits_m \hat{X}_m(\varepsilon)\,q(\tau)}, \tag{6.34}$$

for the renormalized operators $\hat{X}_n(\varepsilon)$ although now they have to be considered also as operators in τ-space. However for the sum $\hat{Q}(\varepsilon) = \sum\limits_m \hat{X}_m(\varepsilon)$, under the condition that the spectrum of $\hat{H}_0$ is broad enough, we still obtain the solutions (4.32). This yields

$$\hat{X}_n(\varepsilon,\tau) = \frac{1}{\varepsilon - \Delta_n + i\frac{\partial}{\partial\tau} + i\pi g\,\langle V^2\rangle\, q^2(\tau)}. \tag{6.35}$$

6.2.1 Population Distribution

We now consider the population ρ_n of an arbitrary state $|n\rangle$. Since we have just one band, the perturbation series for the population is a sum of all the diagrams shown in Fig. 6.3 similar to Fig. 4.12, and ρ_n is given by (4.25) with the only difference that instead of (4.24) the factor Ξ reads

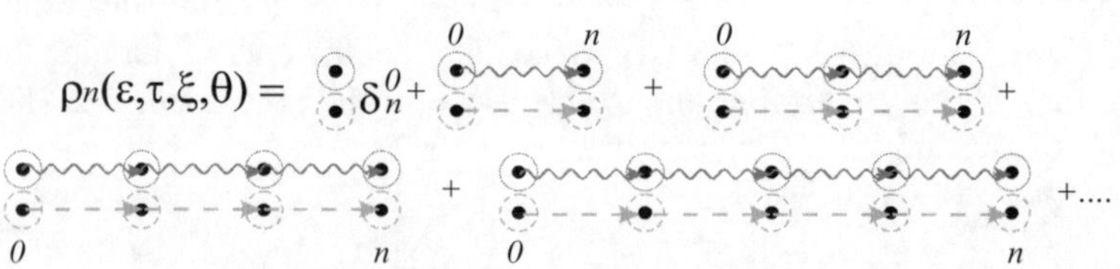

Fig. 6.3

$$\hat{\Xi}(\varepsilon,\xi,\tau,\theta) = \sum_n \hat{X}_n(\varepsilon,\tau)\hat{X}_n(\xi,\theta)V^2 q(\tau)q(\theta)$$
$$= \int_{-\infty}^{\infty} \frac{g\langle V^2\rangle}{\varepsilon-\Delta+i\frac{\partial}{\partial\tau}+i\pi g\, V^2\, q^2(\tau)} \frac{d\Delta}{\xi-\Delta-i\frac{\partial}{\partial\theta}-i\pi g\, V^2\, q^2(\theta)} q(\tau)q(\theta)$$
$$= \frac{2i\pi g V^2}{\varepsilon-\xi+i\frac{\partial}{\partial\theta}+i\frac{\partial}{\partial\tau}+i\pi g\, V^2\, (q^2(\tau)+\, q^2(\theta))} q(\tau)q(\theta). \tag{6.36}$$

Here the variable θ plays for $X_{n'}(\xi,\theta)$ the same role as τ does for $X_{n'}(\varepsilon,\tau)$, and the integral replacing the sum has been calculated by taking the residual in the pole of the upper part of the complex plane. The sum of the series (4.25) for the initial distribution $\rho_m(t=0)$ yields for the resulting population

$$\begin{aligned}\rho_n\,(\varepsilon,\xi,\tau,\theta) &= \hat{X}_n(\varepsilon,\tau)\hat{X}_n(\xi,\theta)\rho_n(t=0)\\ &+\hat{X}_n(\varepsilon,\tau)\hat{X}_n(\xi,\theta)V^2 q(\tau)q(\theta)\frac{1}{1-\hat{\Xi}(\varepsilon,\xi)}\sum_m \hat{X}_m(\varepsilon,\tau)\hat{X}_m(\xi,\theta)\rho_m(t=0).\end{aligned} \tag{6.37}$$

We note that the relation

$$\frac{1}{1-\hat{\Xi}(\varepsilon,\xi)} = \frac{\varepsilon-\xi+i\frac{\partial}{\partial\theta}+i\frac{\partial}{\partial\tau}+i\pi g\, V^2\, \left(q^2(\tau)+\, q^2(\theta)\right)}{\varepsilon-\xi+i\frac{\partial}{\partial\theta}+i\frac{\partial}{\partial\tau}+i\pi g\, V^2\, \left(q(\tau)-\, q(\theta)\right)^2} \tag{6.38}$$

follows from (6.36), and consider the initial condition $\rho_m(t=0)=\delta_m^0$, where $\Delta_0=0$. For $n\neq 0$ (6.37) takes the form

$$\begin{aligned}\rho_n\,(\varepsilon,\xi,\tau,\theta) &= \hat{X}_n(\varepsilon,\tau)\hat{X}_n(\xi,\theta)V^2 q(\tau)q(\theta)\hat{Z}(\xi,\theta,\varepsilon,\tau)\\ &\left[\varepsilon-\xi+i\tfrac{\partial}{\partial\theta}+i\tfrac{\partial}{\partial\tau}+i\pi g\, V^2\, \left(q^2(\tau)+\, q^2(\theta)\right)\right]\hat{X}_0(\varepsilon,\tau)\hat{X}_0(\xi,\theta),\end{aligned} \tag{6.39}$$

where

$$\hat{Z}(\xi,\theta,\varepsilon,\tau) = \frac{1}{\varepsilon-\xi+i\frac{\partial}{\partial\theta}+i\frac{\partial}{\partial\tau}+i\pi g\, V^2\, \left(q(\tau)-\, q(\theta)\right)^2}, \tag{6.40}$$

and the term $\hat{X}_n(\varepsilon,\tau)\hat{X}_n(\xi,\theta)\delta_0^n$ is omitted. We now note the consequence of (6.35)

$$\left[\hat{X}_m^{-1}(\varepsilon,\tau)-\hat{X}_m^{-1}(\xi,\theta)\right] = \left[\varepsilon-\xi+i\tfrac{\partial}{\partial\theta}+i\tfrac{\partial}{\partial\tau}+i\pi g\, V^2\, \left(q^2(\tau)+\, q^2(\theta)\right)\right], \tag{6.41}$$

which is valid for any m, including $m=0$. Therefore, the product of the right-hand side of (6.39) with $\hat{X}_0(\varepsilon,\tau)\hat{X}_0(\xi,\theta)$ yields $\left[\hat{X}_0(\xi,\theta)-\hat{X}_0(\varepsilon,\tau)\right]$, and hence substitution of this product into (6.39) results in

$$\rho_n\,(\varepsilon,\xi,\tau,\theta) = \hat{X}_n(\varepsilon,\tau)\hat{X}_n(\xi,\theta)V^2 q(\tau)q(\theta)\hat{Z}(\xi,\theta,\varepsilon,\tau)\left[\hat{X}_0(\xi,\theta)-\hat{X}_0(\varepsilon,\tau)\right]. \tag{6.42}$$

The relation

$$\left[\hat{Z}^{-1}(\xi,\theta,\varepsilon,\tau)+\hat{X}_m^{-1}(\varepsilon,\tau)-\hat{X}_m^{-1}(\xi,\theta)\right]=2\pi gV^2q(\tau)q(\theta) \tag{6.43}$$

allows one to make the last simplification for the general form of operators

$$\begin{aligned}2\pi g\rho_n\left(\varepsilon,\xi,\tau,\theta\right)=-i\hat{X}_n(\varepsilon,\tau)\hat{X}_n(\xi,\theta)&\\ \left[\hat{Z}^{-1}(\xi,\theta,\varepsilon,\tau)+\hat{X}_m^{-1}(\varepsilon,\tau)-\hat{X}_m^{-1}(\xi,\theta)\right]&\\ \hat{Z}(\xi,\theta,\varepsilon,\tau)\left[\hat{X}_0(\xi,\theta)-\hat{X}_0(\varepsilon,\tau)\right],&\end{aligned} \tag{6.44}$$

which results in

$$\begin{aligned}2\pi g\rho_n\left(\varepsilon,\xi,\tau,\theta\right)&=-i\hat{X}_n(\varepsilon,\tau)\hat{X}_n(\xi,\theta)\left[\hat{X}_0(\xi,\theta)-\hat{X}_0(\varepsilon,\tau)\right]\\ -i\left[\hat{X}_n(\xi,\theta)-\hat{X}_n(\varepsilon,\tau)\right]&\hat{Z}(\xi,\theta,\varepsilon,\tau)\left[\hat{X}_0(\xi,\theta)-\hat{X}_0(\varepsilon,\tau)\right].\end{aligned} \tag{6.45}$$

Further simplifications require the substitution of explicit expressions for $\hat{X}_n(\varepsilon,\tau)$, $\hat{X}_n(\xi,\theta)$, and $\hat{Z}(\xi,\theta,\varepsilon,\tau)$, which, being inverse to the first-order differential operators (6.35), (6.40), are integral operators of the form

$$\hat{X}_n(\varepsilon,\tau)\psi(\tau)=\int\limits_{-\infty}^{\infty}X_n(\varepsilon,\tau,\tau')\psi(\tau')d\tau'. \tag{6.46}$$

For the kernels $Z(\xi,\theta,\theta',\varepsilon,\tau,\tau')$, $X_n(\varepsilon,\tau,\tau')$, and $X_n(\xi,\theta,\theta')$ of these operators one finds the first-order differential equations

$$\begin{aligned}&\left[\varepsilon-\xi+i\pi g\,V^2\,\left(q(\tau)-\,q(\theta)\right)^2\right]Z(\xi,\theta,\theta',\varepsilon,\tau,\tau')\\ &\qquad+i\left(\frac{\partial}{\partial\theta}+\frac{\partial}{\partial\tau}\right)Z(\xi,\theta,\theta',\varepsilon,\tau,\tau')=\delta(\tau-\tau')\delta(\theta-\theta')\\ &\left(\varepsilon-\Delta_n+i\pi g\,\left\langle V^2\right\rangle\,q^2(\tau)\right)X_n(\varepsilon,\tau,\tau')+i\frac{\partial X_n(\varepsilon,\tau,\tau')}{\partial\tau}=\delta(\tau-\tau')\\ &\left(\xi-\Delta_n-i\pi g\,\left\langle V^2\right\rangle\,q^2(\theta)\right)X_n(\xi,\theta,\theta')-i\frac{\partial X_n(\xi,\theta,\theta')}{\partial\theta}=\delta(\theta-\theta').\end{aligned} \tag{6.47}$$

The solutions

$$X_n(\varepsilon,\tau,\tau')=-i\Theta(\tau-\tau')e^{i(\varepsilon-\Delta_n)(\tau-\tau')-\pi g\,V^2\int\limits_{\tau'}^{\tau}q^2(x)dx}$$

$$X_n(\xi,\theta,\theta')=i\Theta(\theta-\theta')e^{-i(\xi-\Delta_n)(\theta-\theta')-\pi g\,V^2\int\limits_{\theta'}^{\theta}q^2(x)dx}$$

$$Z(\xi,\theta,\theta',\varepsilon,\tau,\tau') = -i\Theta(\tau+\theta-\tau'-\theta')\delta(\tfrac{\tau-\theta-\tau'+\theta'}{2})\exp\Bigg[i\left(\varepsilon-\xi\right)$$

$$\times\frac{\tau+\theta-\tau'-\theta'}{2} - \pi g\,V^2 \int\limits_{\frac{\tau'+\theta'}{2}}^{\frac{\tau+\theta}{2}} \left(q(\tfrac{\tau-\theta}{2}+x) - q(x-\tfrac{\tau-\theta}{2})\right)^2 dx\Bigg] \quad (6.48)$$

substituted into (6.45), integrated over all intermediate indices, followed by the substitution $\tau \to t$, $\theta \to t$ after the inverse Fourier transformation yield for the spectral population density at $t > 0$

$$\rho_n\left(t\right) = \mathrm{Re}\int\limits_0^t \exp\left[i\Delta_n \upsilon - 2\pi g V^2\left(\int\limits_0^t q^2(x)dx - \int\limits_0^{t-\upsilon} q(x)q(\upsilon+x)dx\right)\right]\frac{d\upsilon}{\pi g}. \quad (6.49)$$

We note that the contribution comes only from the second term of the right-hand side of (6.45), since the first term is a $\delta(\Delta_n)$-like one.

We present more details of the calculations. One can obtain the result (6.49) by considering two of six terms of the expanded (6.45), namely $\hat{X}_n(\varepsilon,\tau)\hat{X}_n(\xi,\theta)\hat{X}_0(\xi,\theta)$ and $\hat{X}_n(\xi,\theta)\hat{Z}(\xi,\theta,\varepsilon,\tau)\hat{X}_0(\varepsilon,\tau)$. Two other terms given by the replacement $\varepsilon \rightleftharpoons \xi$ are just the complex conjugate of these terms, and the two remaining terms $\hat{X}_n(\xi,\theta)\hat{Z}(\xi,\theta,\varepsilon,\tau)\hat{X}_0(\xi,\theta)$ and its complex conjugate vanish, since all of the singularities of this expression are located at the lower (upper) part of the complex plane ξ (upper part of ε plane for the conjugate). Substitution of (6.48) yields

$$\begin{aligned}\hat{X}_n(\varepsilon,\tau)\hat{X}_n(\xi,\theta)\hat{X}_0(\xi,\theta) = {} & i\Theta(\tau-\tau')\int d\theta''\,\Theta(\theta-\theta'')\Theta(\theta''-\theta') \\ & e^{i(\varepsilon-\Delta_n)(\tau-\tau')-\pi g\,V^2\int_{\tau'}^{\tau} q^2(x)dx} \\ & e^{-i(\xi-\Delta_n)(\theta-\theta'')-\pi g\,V^2\int_{\theta''}^{\theta} q^2(x)dx} \\ & e^{-i(\xi-\Delta_n)(\theta''-\theta')-\pi g\,V^2\int_{\theta'}^{\theta''} q^2(x)dx}, \end{aligned} \quad (6.50)$$

which after performing the integration of the inverse Fourier transformation over ε and ξ results in

$$\frac{1}{4\pi^2}\int e^{-i(\varepsilon-\xi)t}\hat{X}_n(\varepsilon,\tau)\hat{X}_n(\xi,\theta)\hat{X}_0(\xi,\theta) =$$

$$\Theta(t)\delta(\theta')\delta(\tau')e^{-2\pi g\,V^2\int_0^t q^2(x)dx}\frac{1-e^{-i\Delta_n t}}{i\Delta_n}. \quad (6.51)$$

The real part of this expression is proportional to $\sin(\Delta_n t)/\Delta_n$, which in the long-time asymptotic gives the singular term $\delta(\Delta_n)$, unimportant for non-zero detunings.

The principal term contributing to (6.48) reads

$$
\begin{aligned}
&- \hat{X}_n(\xi,\theta)\hat{Z}(\xi,\theta,\varepsilon,\tau)\hat{X}_0(\varepsilon,\tau) = \Theta(\theta-\theta'')e^{-i(\xi-\Delta_n)(\theta-\theta'')-\pi g V^2 \int_{\theta''}^{\theta} q^2(x)dx} \\
&i\Theta(\tau+\theta''-\tau''-\theta')\delta(\tfrac{\tau-\theta''-\tau''+\theta'}{2})\Theta(\tau''-\tau')e^{i\varepsilon(\tau''-\tau')-\pi g V^2 \int_{\tau'}^{\tau''} q^2(x)dx} \\
&e^{i(\varepsilon-\xi)\frac{\tau+\theta''-\tau''-\theta'}{2}-\pi g V^2 \int_{\frac{\tau''+\theta'}{2}}^{\frac{\tau+\theta''}{2}} \left(q(\frac{\tau+\theta''}{2}+x)-q(\frac{\tau+\theta''}{2}-x)\right)^2 dx}, \qquad (6.52)
\end{aligned}
$$

whereas its inverse Fourier transform over ε and ξ gives the Dirac δ-functions $\delta(\frac{\tau-\theta''-\tau''+\theta'}{2})$ and $\delta(-t-\tau'+\frac{\tau+\theta''+\tau''-\theta'}{2})$ and yields

$$
\begin{aligned}
&- \int e^{-i(\varepsilon-\xi)t}\hat{X}_n(\xi,\theta)\hat{Z}(\xi,\theta,\varepsilon,\tau)\hat{X}_0(\varepsilon,\tau)\frac{d\varepsilon d\xi}{4\pi^2} = i\Theta(\theta''-\theta')\Theta(\theta-\theta'') \\
&\delta(t-\theta+\theta')e^{i\Delta_n(\theta-\theta'')-\pi g V^2 \int_{\theta''}^{\theta} q^2(x)dx}\delta(\frac{\tau-\theta''-\tau''+\theta'}{2})\delta(-t-\tau'+\tau) \\
&e^{-\pi g V^2 \int_{\frac{\tau''+\theta'}{2}}^{\frac{\tau+\theta''}{2}} \left[q(\frac{\tau-\theta''}{2}+x)-q(x-\frac{\tau-\theta''}{2})\right]^2 dx} \Theta(\tau''-\tau')e^{-\pi g V^2 \int_{\tau'}^{\tau''} q^2(x)dx}, \qquad (6.53)
\end{aligned}
$$

where we have to set $\theta=\tau=t$. This results in

$$
\begin{aligned}
&- \int e^{-i(\varepsilon-\xi)t}\hat{X}_n(\xi,\theta)\hat{Z}(\xi,\theta,\varepsilon,\tau)\hat{X}_0(\varepsilon,\tau)\tfrac{d\varepsilon d\xi}{4\pi^2} = -i\Theta(t-\tau'')\Theta(\tau'') \\
&\delta(\tau')\delta(\theta')e^{i\Delta_n\tau''}e^{2\pi g V^2 \int_0^{t-\tau''} q(x)\, q(x+\tau'')dx - 2\pi g V^2 \int_0^t q^2(x)dx}, \qquad (6.54)
\end{aligned}
$$

which after the substitution $\tau'' \to v$ and after adding the complex conjugate, yields (6.49).

For the interaction vanishing at $|t| \to \infty$ the integration limits $[0,t]$ have to be extended to $\pm\infty$

$$
\pi g \rho_n(t) = \mathrm{Re} \int_{-\infty}^{\infty} \exp\left\{ i\Delta_n v - 2\pi g V^2 \int_{-\infty}^{\infty} \left[q^2(x) - q(x)q(v+x)\right] dx \right\} dv. \qquad (6.55)
$$

Briefly, the result obtained means that the population distribution over the spectrum of a system with a uniform density of states perturbed by a time-dependent random matrix equals the Fourier transform $v \to \Delta_n$ of the exponent of the interaction autocorrelation. Figure 6.2c illustrates the situation for a particular example of a Gaussian time-dependence $q(x)$ resulting in Gaussian autocorrelation.

Note that the result (6.55) can be generalized to the case of time-dependent detunings $\Delta_n(t)$. To this end the combination $\Delta_n v$ has to be replaced by the integral $\int_{-\infty}^{\infty}[\Delta_n(t+v) - \Delta_n(t)]dt$, as follows from the exact solution of the first-order differential equations (6.47). The same generalization can be done for (6.49); in this case the replacing integral reads $\int_0^t \Delta_n(x)dx - \int_0^{t-v} \Delta_n(x)dx$.

6.2.2 *Response to Perturbation Proportional to a Random Matrix*

Thus far our main concern was the population distribution over states of time-dependent, complex quantum systems. However, it is not the only possible quantity that manifests universal behavior independent of spectrum details and interaction in the systems. Another example of such a quantity is interaction with external fields. We consider an external field $\mathcal{F}$ where the interaction Hamiltonian $\mathcal{F}\hat{x}$ is proportional to a generalized coordinate operator $\hat{x}$. If the time-dependent mean value $x = \langle\Psi(t)|\,\hat{x}\,|\Psi(t)\rangle$ of this operator does not vanish after the ensemble average discussed in Chap. 4, the complex system interacts with such a field, and at the macroscopic level such an interaction is responsible for the polarization of the media corresponding to $\mathcal{F}$. Electric polarization is a typical example, when $\mathcal{F}$ is simply the electric field strength, and $\hat{x}$ is the dipole moment operator.

Diagram Series Structure

The terms of perturbation series contributing to this interaction have a similar, although somewhat different structure as compared to the perturbation series for the population shown in Figs. 4.10–4.12. The main difference results from the necessity to find a counterpart for the matrix elements of $\hat{x}$ among the transition matrix elements $\hat{V}$ belonging either to the perturbation series for $|\Psi(t)\rangle$ or that for $\langle\Psi(t)|$, such that $\langle x_{km}V_{mk}\rangle \neq 0$. Also important is the difference between the statistical ensemble for the imaginary and for the real parts of the perturbation matrix elements, concisely written as two possibilities $\langle V_{km}V_{km}\rangle = 0$ and $\langle V_{km}V_{km}\rangle = \langle V_{km}V_{mk}\rangle = \langle V^2\rangle \neq 0$, respectively. For the Gaussian statistics of the matrix elements these cases correspond respectively to the Gaussian orthogonal ensemble (GOE) and the Gaussian unitary ensemble (GUE). In Fig. 6.4 we show the diagrams that, after the ensemble average, give a non-vanishing contribution for the mean matrix element for these two cases.

Thus we have encountered for the first time a quantity for which the results depend on the symmetry with respect to time inversion, a property of random ensembles, typically associated with the presence of a magnetic field. In the absence

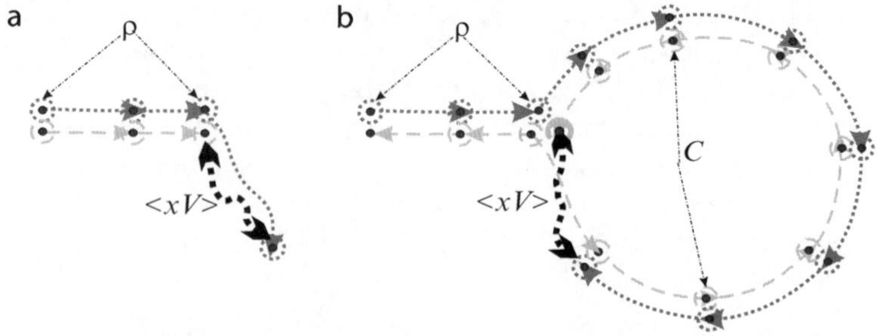

Fig. 6.4 Non-vanishing contribution of the perturbation series for the matrix element of a generalized coordinate operator $\hat{x}$. *Dotted lines* correspond to $|\Psi\rangle$ and dashed lines to $\langle\Psi|$. (**a**) Case of complex random couplings, typical of the unitary Gaussian ensemble. The main contribution results from the susceptibility $\langle \hat{x}\hat{X}(\varepsilon)\hat{V}\rangle + c.c.$ averaged over the population distribution ρ. One recognizes the parallel diagrams contributing to the series for the populations ρ, similar to that of Fig. 4.11. (**b**) Real coupling $\hat{V}$, typical of the orthogonal Gaussian ensemble. Apart from the diagrams for the population one sees part C, originating from the averages $\langle V_{km}V_{km}\rangle$. Terms of the same topology are responsible for the Cooper pairs in the theory of superconductivity

of magnetic fields, all of the interaction matrix elements are real (or can be selected as real in a certain basis), and therefore $\langle V_{km}V_{km}\rangle = \langle V_{km}V_{mk}\rangle$. Therefore, apart from the diagrams where the non-vanishing contribution results either from the transitions coupled to the antiparallel transition in the perturbation series for the same wavefunction or from the transitions coupled with the parallel transition in the perturbation series for the conjugated wavefunctions, we have to take into account the contributions $\hat{C}$, resulting from the coupling of the antiparallel transitions in the series for the different wavefunctions. We note that the diagrams of such a topology are important and well known in the theory of superconductivity, allowing for the formation of Cooper pairs, or "Cooperons". The relative number of these terms is, however, usually smaller by a factor $1/N$ of the inverse number of levels participating in the process, since topologically an extra self-crossing of the diagram is required for this purpose. That was the reason why these diagrams have not been taken into account for the populations. Nevertheless, for susceptibilities they may give the leading contributions.

Let us consider first the other case, $\langle V_{km}V_{km}\rangle = 0$, typical of the systems with no time-reversal symmetry. The principal contribution to the perturbation series for $\langle\Psi(t)|\,\hat{x}\,|\Psi(t)\rangle$ has the structure shown in Fig. 6.4a and reads

$$\begin{aligned} x\,(t) &= \langle\Psi(t)|\,\hat{x}\,|\Psi(t)\rangle \\ &= \operatorname{Re}\sum_{n,m}\int e^{i(\varepsilon-\xi)\,t}\hat{\rho}_n\,(\varepsilon,\xi,\tau,\theta)\,\langle x_{nm}V_{mn}\rangle\,\hat{X}_m(\varepsilon,\tau)\tfrac{d\varepsilon d\xi}{4\pi^2} \end{aligned} \tag{6.56}$$

where the operators $\hat{\rho}_n$ and $\hat{X}_m$ are given in (6.35), (6.42) and the effect is proportional to the correlation value $\langle x_{nm}V_{mn}\rangle$. When this average is independent of the index m, the summation over the uniform spectrum yields a purely imaginary value $\sum_m \hat{X}_m(\varepsilon,\tau) = \hat{Q}(\varepsilon)$, such that the result (6.56) vanishes. For the more general case of non-uniform density of states and the correlation $\langle x_{nm}V_{mn}\rangle$, the combination Re $\sum_m \langle x_{nm}V_{mn}\rangle \hat{X}_m(\varepsilon,\tau)$ may be interpreted as the linear response of the system in the state n, whereas summation over n implies taking the average of the linear response by convolution with the instantaneous population distribution $\hat{\rho}_n(t) = \int e^{i(\varepsilon-\xi)t}\hat{\rho}_n(\varepsilon,\xi,\tau,\theta)\frac{d\varepsilon d\xi}{4\pi^2}$.

We note that (6.56) is asymmetric with respect of the change ε to ξ, however this asymmetry disappears after taking the real part of the expression. Due to this, there is no need for special consideration of the diagram symmetric with respect to the interchange of $\langle\Psi(t)|$ and $|\Psi(t)\rangle$ to the one shown in Fig. 6.4b.

The Case of Inverted Ordering

We now turn to the case $\langle V_{km}V_{km}\rangle = (V')^2 \neq 0$, and consider the general case $V \neq V'$, assuming that a real V' is different from another real number V. The main contribution comes from the diagrams of the type shown in Fig. 6.4b, and reads

$$x(t) = \langle\Psi(t)|\,\hat{x}\,|\Psi(t)\rangle = \mathrm{Re}\sum_{n,m}\int e^{i(\varepsilon-\xi)t}\langle x_{mn}V_{mn}\rangle$$
$$\times\hat{C}_{mn}(\varepsilon,\xi,\tau,\theta)\,\hat{X}_n^{-1}(\varepsilon,\tau)\hat{\rho}_n(\varepsilon,\xi,\tau,\theta)\frac{d\varepsilon d\xi}{4\pi^2}, \tag{6.57}$$

where $\hat{C}_{mn}$ denotes an operator similar to $\hat{\rho}_n$ in (6.37) which corresponds to the series Fig. 6.5, and the order of operations is in accordance with that shown in

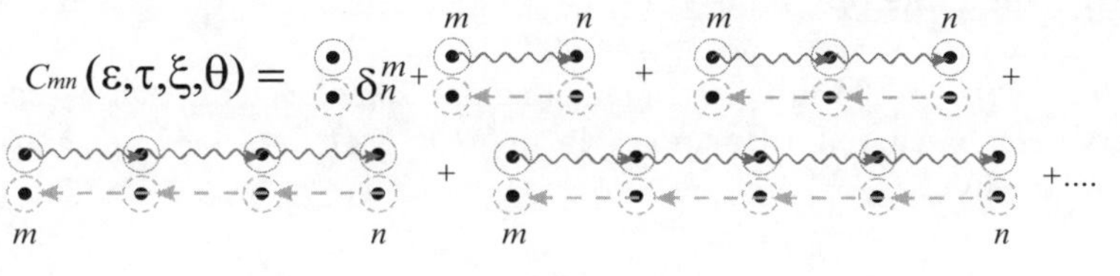

Fig. 6.5

Fig. 6.4(b). The operator $\hat{X}_n^{-1}(\varepsilon,\tau)$ compensates for the repetition of the operator $\hat{X}_n(\varepsilon,\tau)$ found both at the end of $\hat{\rho}_n$ and the beginning of $\hat{C}_{mn}$.

We note that the perturbation operators originating from the perturbation series for $\langle\Psi(t)|$ are ordered in a sequence opposite to those of $|\Psi(t)\rangle$ in the "Cooperon" loop $\hat{C}$ of the diagram in Fig. 6.4b. This may be accounted for when performing calculations for $\hat{C}$ by analogy to (6.37) for $\hat{\rho}$ without rewriting all of the cumbersome intermediate expressions. All that is necessary is to change the sign of the derivative over θ in all of the operators related to $\hat{C}$. Those operators with inverted ordering are indicated by primes and by analogy to (6.35)–(6.37) we can write

$$\hat{X}'_n(\xi,\theta) = \frac{1}{\xi - \Delta_n + i\frac{\partial}{\partial\theta} - i\pi g\,\langle V^2\rangle\, q^2(\theta)}, \tag{6.58}$$

$$\hat{\Xi}'(\varepsilon,\xi,\tau,\theta) = \frac{2i\pi g\,(V')^2}{\varepsilon - \xi - i\frac{\partial}{\partial\theta} + i\frac{\partial}{\partial\tau} + i\pi g\,V^2\,\left(q^2(\tau) + q^2(\theta)\right)} q(\tau)q(\theta), \tag{6.59}$$

$$\begin{aligned}\hat{C}_{mn}(\varepsilon,\xi,\tau,\theta) = {} & \hat{X}_n(\varepsilon,\tau)\hat{X}'_n(\xi,\theta)\delta_n^m + \hat{X}_m(\varepsilon,\tau)\hat{X}'_m(\xi,\theta)\,(V')^2 \\ & \times q(\tau)q(\theta)\tfrac{1}{1-\hat{\Xi}'(\varepsilon,\xi)}\hat{X}_n(\varepsilon,\tau)\hat{X}'_n(\xi,\theta).\end{aligned} \tag{6.60}$$

The reasoning behind changing the sign of the derivative becomes evident when we recall the physical image associated with the technique discussed on p. 255. The derivatives $\frac{\partial}{\partial\theta}$ and $\frac{\partial}{\partial\tau}$ appear when we replace the time-dependent system by the time-independent system with a dimensionality higher by unity, and assume that the system moves along this new degree of freedom with a constant velocity, while the Hamiltonian depends on this new coordinate. The opposite ordering therefore implies motion along the same coordinate in the opposite direction.

Equation (6.57) takes the form

$$x(t) = \mathrm{Re}\,\langle xV\rangle \sum_{n,m} \int e^{i(\varepsilon-\xi)t}\hat{C}_{mn}(\varepsilon,\xi,\tau,\theta)\,\hat{X}_n^{-1}(\varepsilon,\tau)\hat{\rho}_n(\varepsilon,\xi,\tau,\theta)\,\frac{d\varepsilon d\xi}{4\pi^2}, \tag{6.61}$$

where, again, the operator $\hat{X}_n^{-1}(\varepsilon,\tau)$ is needed to avoid double counting of the operator $\hat{X}_n(\varepsilon,\tau)$ originating from the beginning of the series for $\hat{C}_{mn}$ and that coming from the termination of the diagrams for $\hat{\rho}_n$.

Factorization of the Diagrams

Equations (6.58)–(6.61) still remain rather cumbersome operator expressions, ones not very convenient for calculations. In order to simplify, one can employ the operator identity

$$\begin{aligned}\hat{X}_m(\varepsilon,\tau)\hat{X}_m(\xi,\theta) &= \left[\hat{X}_m^{-1}(\xi,\theta) - \hat{X}_m^{-1}(\varepsilon,\tau)\right]^{-1}\left[\hat{X}_m(\varepsilon,\tau) - \hat{X}_m(\xi,\theta)\right] \\ &= \frac{-1}{\varepsilon - \xi + i\frac{\partial}{\partial\theta} + i\frac{\partial}{\partial\tau} + i\pi g\,V^2\,\left(q^2(\tau) + q^2(\theta)\right)} \\ &\quad \times \left[\hat{X}_m(\varepsilon,\tau) - \hat{X}_m(\xi,\theta)\right],\end{aligned} \tag{6.62}$$

which is easy to prove with the help of (6.41) and the evident relation

$$\frac{1}{\hat{X}_m(\xi,\theta)} - \frac{1}{\hat{X}_m(\varepsilon,\tau)} = \frac{\hat{X}_m(\varepsilon,\tau) - \hat{X}_m(\xi,\theta)}{\hat{X}_m(\varepsilon,\tau)\hat{X}_m(\xi,\theta)}, \tag{6.63}$$

and which allows one to replace the products $\hat{X}_m(\varepsilon,\tau)\hat{X}_m(\xi,\theta)$ of the mean Green's functions by their differences $\hat{X}_m(\varepsilon,\tau) - \hat{X}_m(\xi,\theta)$ subject to the action of an m-independent operator. From (6.36) one also sees that

$$\frac{2i\pi g}{\varepsilon - \xi + i\frac{\partial}{\partial\theta} + i\frac{\partial}{\partial\tau} + i\pi g\, V^2\ (q^2(\tau) +\ q^2(\theta))} = \sum_n \hat{X}_n(\varepsilon,\tau)\hat{X}_n(\xi,\theta), \quad (6.64)$$

and hence (6.62) reads

$$\hat{X}_m(\varepsilon,\tau)\hat{X}_m(\xi,\theta) = \frac{-i}{2\pi g}\sum_n \hat{X}_n(\varepsilon,\tau)\hat{X}_n(\xi,\theta)\left[\hat{X}_m(\varepsilon,\tau) - \hat{X}_m(\xi,\theta)\right]. \quad (6.65)$$

Equation (6.65) allows one to simplify the triple products of the mean Green's functions to products of independent binary products, such as

$$\begin{aligned}\sum_m \hat{X}_m(\varepsilon,\tau)\hat{X}_m(\xi,\theta)\hat{X}'_m(\xi,\theta) &= \frac{-i}{2\pi g}\sum_n \hat{X}_n(\varepsilon,\tau)\hat{X}_n(\xi,\theta) \\ &\quad\times\sum_m\left[\hat{X}_m(\varepsilon,\tau) - \hat{X}_m(\xi,\theta)\right]\hat{X}'_m(\xi,\theta) \\ &= \frac{-i}{2\pi g}\sum_n \hat{X}_n(\varepsilon,\tau)\hat{X}_n(\xi,\theta)\sum_m \hat{X}_m(\varepsilon,\tau)\hat{X}'_m(\xi,\theta),\end{aligned} \quad (6.66)$$

for instance, which gives a very convenient tool for diagram factorization. In the last equation we have taken into account the fact that

$$\sum_m \hat{X}_m(\xi,\theta)\hat{X}'_m(\xi,\theta) = 0, \quad (6.67)$$

since both $\hat{X}_m(\xi,\theta)$ and $\hat{X}'_m(\xi,\theta)$, as functions of Δ_m, have poles in the upper part of the complex plane. In Fig. 6.6 we illustrate the topological meaning of this transformation.

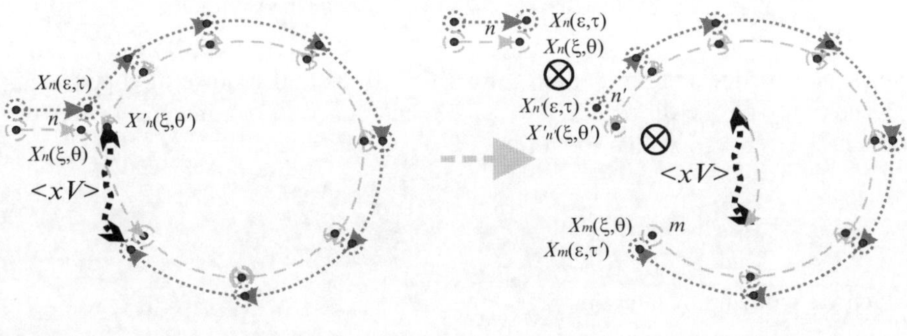

Fig. 6.6

Application of (6.66) to (6.61) yields

$$x\left(t\right)=\mathrm{Re}\,\frac{-i\langle xV\rangle}{2\pi g}\int e^{i(\varepsilon-\xi)t}\sum_{n',m}\hat{C}_{mn'}\left(\varepsilon,\xi,\tau,\theta\right)\sum_{n}\hat{\rho}_n\left(\varepsilon,\xi,\tau,\theta\right)\frac{d\varepsilon d\xi}{4\pi^2},\quad(6.68)$$

which means that the operators $\hat{C}$ and $\hat{\rho}$ do not retain common indices. We note that the interaction operator V coupled to the operator x by the ensemble average $\langle xV\rangle$ has been factored out from the integral. This means that we consider the time-dependent susceptibility of the system to the perturbation V, which yields the response only after being multiplied by the time-dependence of this perturbation. We will occasionally refer to this value as a "response", since it indeed gives a linear response to a small δ-like additional increase in the interaction.

The Total Population Operator

Considering (6.37), which includes the first term $\hat{X}_n(\varepsilon,\tau)\hat{X}_n(\xi,\theta)\delta_n^0$, that has been neglected during consideration of the population transfer, and with the allowance for (6.45) and (6.64) one finds

$$\sum_n\rho_n\left(\varepsilon,\xi,\tau,\theta\right)=\hat{Z}(\xi,\theta,\varepsilon,\tau)$$
$$\left[\varepsilon-\xi+i\tfrac{\partial}{\partial\theta}+i\tfrac{\partial}{\partial\tau}+i\pi g\,V^2\,\left(q^2(\tau)+\,q^2(\theta)\right)\right]\hat{X}_0(\varepsilon,\tau)\hat{X}_0(\xi,\theta),\quad(6.69)$$

which can be reduced to a simpler form

$$\sum_n\rho_n\left(\varepsilon,\xi,\tau,\theta\right)=\hat{Z}(\xi,\theta,\varepsilon,\tau)\left[\hat{X}_0(\xi,\theta)-\hat{X}_0(\varepsilon,\tau)\right],\qquad(6.70)$$

with the help of the expression

$$\left[\varepsilon-\xi+i\tfrac{\partial}{\partial\theta}+i\tfrac{\partial}{\partial\tau}+i\pi g\,V^2\,\left(q^2(\tau)+\,q^2(\theta)\right)\right]$$
$$\times\hat{X}_0(\varepsilon,\tau)\hat{X}_0(\xi,\theta)=\left[\hat{X}_0(\xi,\theta)-\hat{X}_0(\varepsilon,\tau)\right],\qquad(6.71)$$

which is a particular form of (6.62) for $m=0$. Note that since the operator $\hat{Z}$ depends only on the difference $\zeta=\varepsilon-\xi$, and hence (6.70), integrated over the variable $\eta=\left(\varepsilon+\xi\right)/2$, has the form

$$\sum_n\rho_n\left(\zeta,\tau,\theta\right)=\hat{Z}(\xi,\theta,\varepsilon,\tau)2\pi i,\qquad(6.72)$$

which we will employ later on.

Substitution of the explicit expressions (6.48) into (6.70) yields

$$\sum_n \rho_n(\varepsilon,\xi,\tau,\theta) = \delta(\tfrac{\tau-\theta-\tau'+\theta'}{2})e^{-\pi g\, V^2 \int\limits_{\tau''}^{\tau} q^2(x)dx - \pi g V^2 \int\limits_{\theta''}^{\theta} q^2(x)dx}$$
$$e^{i\varepsilon(\tau-\tau'')-i\xi(\theta-\theta'')+2\pi g V^2 \int\limits_{\frac{\tau'+\theta'}{2}}^{\frac{\tau+\theta}{2}} \left(q(\frac{\tau-\theta}{2}+x)\; q(x-\frac{\tau-\theta}{2})\right)dx} \qquad \Theta(\tau-\tau')\Theta(\theta-\theta')$$
$$[\delta(\tau'-\tau'')\Theta(\theta'-\theta'')+\delta(\theta'-\theta'')\Theta(\tau'-\tau'')]\,. \tag{6.73}$$

Note that the Fourier transform of the operator has for $\tau=\theta=t$, a δ-like kernel

$$\int e^{i(\varepsilon-\xi)t}\sum_n \hat{\rho}_n(\varepsilon,\xi,\tau,\theta)\,\frac{d\varepsilon d\xi}{4\pi^2}\bigg|_{\tau=\theta=t} \to \delta(\theta'')\delta(\tau'') \tag{6.74}$$

which is consistent with the normalization requirement

$$\int\limits_{-\infty}^{\infty} d\tau'' d\theta'' \int e^{i(\varepsilon-\xi)t}\sum_n \hat{\rho}_n(\varepsilon,\xi,\tau,\theta)\,\frac{d\varepsilon d\xi}{4\pi^2}\bigg|_{\tau=\theta=t} \qquad \rho(t=0,\tau'',\theta'')=1 \tag{6.75}$$

following the application of the operator (6.74) to the τ-independent initial condition $\rho(t=0,\tau,\theta)=1$.

Sum of the "Cooperon" Series

Substitution of the combination

$$\hat{\Xi}'(\varepsilon,\xi,\tau,\theta) = \sum_n \hat{X}_n(\varepsilon,\tau)\hat{X}'_n(\xi,\theta)\,(V')^2\, q(\tau)q(\theta) \tag{6.76}$$

entering (6.59) into (6.60) yields, after taking the sum over n' and m,

$$\sum_{n',m}\hat{C}_{n'm}(\varepsilon,\xi,\tau,\theta) = \sum_{n',m}\hat{X}_{n'}(\varepsilon,\tau)\hat{X}'_{n'}(\xi,\theta)\delta_n^m$$
$$+\sum_{n'}\hat{X}_{n'}(\varepsilon,\tau)\hat{X}'_{n'}(\xi,\theta)\,(V')^2\, q(\tau)q(\theta)\tfrac{1}{1-\hat{\Xi}'}\sum_m \hat{X}_m(\varepsilon,\tau)\hat{X}'_m(\xi,\theta)$$
$$=\left(1+\tfrac{\hat{\Xi}'}{1-\hat{\Xi}'}\right)\sum_m \hat{X}_m(\varepsilon,\tau)\hat{X}'_m(\xi,\theta) = \tfrac{1}{1-\hat{\Xi}'}\sum_m \hat{X}_m(\varepsilon,\tau)\hat{X}'_m(\xi,\theta). \tag{6.77}$$

With the allowance for (6.64) and (6.59) this results in

$$\hat{Z}'(\xi,\theta,\varepsilon,\tau) = \sum_{n',m}\hat{C}_{n'm}(\varepsilon,\xi,\tau,\theta)$$
$$= \frac{2i\pi g}{\varepsilon-\xi-i\frac{\partial}{\partial\theta}+i\frac{\partial}{\partial\tau}+i\pi g V^2\left((q(\tau)-q(\theta))^2+\lambda\, q(\tau)\, q(\theta)\right)}, \tag{6.78}$$

where $\lambda = 2\left(V^2 - (V')^2\right)/V^2$ is the relative difference between the mean squared matrix elements of the interaction and their mean moduli squared. For real matrix elements, λ vanishes.

By analogy to (6.47) one finds the kernel $Z'(\xi,\theta,\theta',\varepsilon,\tau,\tau')$ of this integral operator by solving the differential equation

$$\begin{aligned} &\left[\varepsilon - \xi + i\pi g\, V^2\ \left((q(\tau) - q(\theta))^2 + \lambda\, q(\tau)\, q(\theta)\right)\right] Z'(\xi,\theta,\theta',\varepsilon,\tau,\tau') \\ &+ i\left(\tfrac{\partial}{\partial\tau} - \tfrac{\partial}{\partial\theta}\right) Z'(\xi,\theta,\theta',\varepsilon,\tau,\tau') = 2i\pi g\delta(\tau-\tau')\delta(\theta-\theta'), \end{aligned} \tag{6.79}$$

which yields

$$\begin{aligned} &Z'(\xi,\theta,\theta',\varepsilon,\tau,\tau') = 2\pi g\Theta(\tau-\theta-\tau'+\theta')\delta\left(\tfrac{\tau+\theta-\tau'-\theta'}{2}\right) \\ &e^{i(\varepsilon-\xi)\frac{\tau-\theta-\tau'+\theta'}{2} - \pi g V^2 \int_{\frac{\tau'-\theta'}{2}}^{\frac{\tau-\theta}{2}} \left[\left(q(\frac{\tau+\theta}{2}+x) - q(\frac{\tau+\theta}{2}-x)\right)^2 + \lambda q(\frac{\tau+\theta}{2}+x) q(\frac{\tau+\theta}{2}-x)\right] dx} \end{aligned} \tag{6.80}$$

The Response

We are now in a position to determine the response. To this end we substitute (6.70), (6.78) into (6.68) and obtain

$$\begin{aligned} x(t) = \mathrm{Re}\, \tfrac{-i\langle xV\rangle}{2\pi g} \int & e^{i(\varepsilon-\xi)t} \hat{Z}'(\xi,\theta,\varepsilon,\tau) \\ & \hat{Z}(\xi,\theta,\varepsilon,\tau)\left[\hat{X}_0(\xi,\theta) - \hat{X}_0(\varepsilon,\tau)\right] \tfrac{d\varepsilon d\xi}{4\pi^2}. \end{aligned} \tag{6.81}$$

We note that both $\hat{Z}'(\xi,\theta,\varepsilon,\tau)$ and $\hat{Z}(\xi,\theta,\varepsilon,\tau)$ depend only on the variable ζ given by the difference $\varepsilon - \xi = \zeta$, whereas the functional dependence on the sum $\varepsilon + \xi = 2\eta$ enters only via the combination $\hat{S} = \hat{X}_0(\xi,\theta) - \hat{X}_0(\varepsilon,\tau)$, related to the initial condition. This immediately results in

$$x(t) = \mathrm{Re}\, \frac{-i\,\langle xV\rangle}{2\pi g} \int e^{i(\varepsilon-\xi)t} \hat{Z}'(\zeta,\theta,\tau)\hat{Z}(\zeta,\theta,\tau)\hat{S}(\zeta,\eta)\frac{d\eta d\zeta}{4\pi^2}. \tag{6.82}$$

From (6.35) one finds

$$\begin{aligned} \int_{-\infty}^{\infty} \hat{S}(\zeta,\eta)d\eta = & \int_{-\infty}^{\infty} \frac{d\eta}{\eta - \frac{\zeta}{2} - i\frac{\partial}{\partial\theta} - i\pi g\ \langle V^2\rangle\ q^2(\theta)} \\ & - \int_{-\infty}^{\infty} \frac{d\eta}{\eta - \frac{\zeta}{2} + i\frac{\partial}{\partial\tau} + i\pi g\ \langle V^2\rangle\ q^2(\tau)} = 2\pi i, \end{aligned} \tag{6.83}$$

which is easy to obtain by displacement of the integration contour in the lower part of the complex plane toward $-i\infty$ for the first integral, and in the upper part of the complex plane towards $i\infty$ for the second integral, where the corresponding integrands have no singularities because of the causality principle. The integrals therefore amount to the accumulated phases $\pm i\pi$ of the corresponding logarithms. This yields

$$\begin{aligned} x(t) &= \mathrm{Re}\,\frac{\langle xV\rangle}{4\pi^2 g}\int e^{i(\varepsilon-\xi)\,t}\hat{Z}'(\zeta,\theta,\tau)\hat{Z}(\zeta,\theta,\tau)d\zeta \\ &= \mathrm{Re}\int Z'(\xi,\theta'',\theta,\varepsilon,\tau,\tau'')Z(\xi,\theta'',\theta',\varepsilon,\tau'',\tau')|_{\tau=\theta=t} \\ &\quad e^{i(\varepsilon-\xi)t}\frac{\langle xV\rangle}{4\pi^2 g}d\zeta d\tau'' d\theta'' d\tau' d\theta'. \end{aligned} \tag{6.84}$$

where we have taken into account that the convolution of the kernels of $\hat{Z}'$ and $\hat{Z}$ has to be performed with an inverted order of θ-arguments in the operator $\hat{Z}'$, as illustrated in Fig. 6.7.

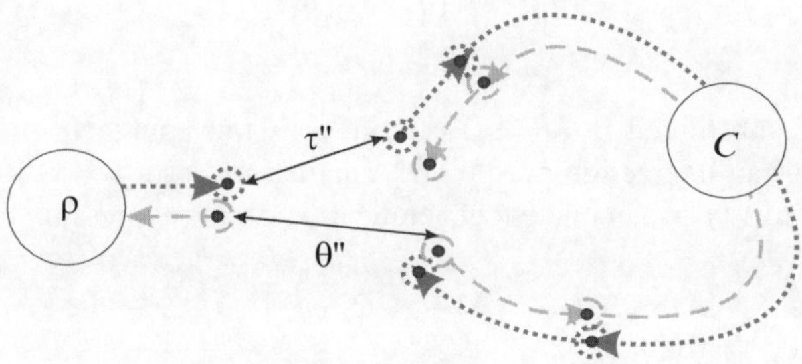

Fig. 6.7 The correspondence of the arguments τ and θ of the operators $\hat{Z}$ and $\hat{Z}'$ in (6.84) follows from the topology of the diagrams shown in Fig. 6.6

We now make use of (6.48) and (6.80) for the kernels of $\hat{Z}'$ and $\hat{Z}$ which can be written in the form

$$\begin{aligned} Z'(\xi,\theta,\theta',\varepsilon,\tau,\tau') &= -i\delta(\tfrac{\tau+\theta-\tau'-\theta'}{2})\Theta(\tau-\tau') \\ &\quad e^{i\zeta(\tau-\tau')-\pi g V^2\int_{\tau'}^{\tau}\left[(q(x)-\,q(\tau+\theta-x))^2+\lambda q(x)q(\tau+\theta-x)\right]dx}, \\ Z(\xi,\theta,\theta',\varepsilon,\tau,\tau') &= 2\pi g\delta(\tfrac{\tau-\theta-\tau'+\theta'}{2})\Theta(\tau-\tau') \\ &\quad e^{i\zeta(\tau-\tau')-\pi g V^2\int_{\tau'}^{\tau}(q(x)-\,q(x-\tau+\theta))^2dx}, \end{aligned} \tag{6.85}$$

and arrive at

$$x(t) = \mathrm{Re}\,\frac{\langle xV\rangle}{2\pi i}\int d\zeta d\tau'' d\theta'' d\tau' d\theta'\; e^{-i\zeta\, t}\Theta(\tau-\tau'')\delta(\tfrac{\tau+\theta''-\tau''-\theta}{2})$$
$$e^{i\zeta(\tau-\tau'')-\pi g\, V^2\int_{\tau''}^{\tau}\left[\left(q(x)-\ q(\tau+\theta''-x)\right)^2+\lambda q(x)q(\tau+\theta''-x)\right]dx}\Theta(\tau''-\tau')$$
$$\delta(\frac{\tau''-\theta''-\tau'+\theta'}{2})\; e^{i\zeta(\tau''-\tau')-\pi g\, V^2\int_{\tau'}^{\tau''}\left(q(x)-\ q(x-\tau'+\theta')\right)^2dx}\Big|_{\tau=\theta=t}. \tag{6.86}$$

Integration over $d\zeta$ yields one more δ-function, $2\pi\delta\left(\tau-\tau''-t\right)$, and after straightforward integration over $d\theta'' d\tau' d\theta'$ of the expression with three δ-functions one finds

$$x(t) = \mathrm{Re}\; 2i\,\langle xV\rangle\int_0^t d\tau'' e^{-\pi g\lambda\, V^2\int_0^{\tau''} q^2(x)dx}$$
$$\exp\left\{-\pi g\, V^2\int_{\tau''}^{t}\left(q(x)-\ q(t+\tau''-x)\right)^2dx\right\}, \tag{6.87}$$

where we have substituted $\tau=\theta=t$, as required in this approach.

Note that for an interaction profile $q(t)$ ranging not from $t=0$ to $t=\infty$, but from $t=-\infty$ to $t=\infty$, an evident generalization of this expression reads

$$x(t) = \mathrm{Re}\; 2i\,\langle xV\rangle\int_{-\infty}^{t} d\tau'' e^{-\pi g\lambda\, V^2\int\limits_{-\infty}^{\tau''} q^2(x)dx}$$
$$\exp\left\{-\pi g\, V^2\int_{\tau''}^{t}\left(q(x)-\ q(t+\tau''-x)\right)^2dx\right\}. \tag{6.88}$$

This short and elegant expression (6.88) has only one disadvantage—it vanishes for a generic quantum system, since all of the averages V^2, $\langle xV\rangle$, and λ are normally real. Therefore a generic complex quantum system with a uniform spectrum does not manifest either the linear response nor any of the higher responses of odd orders.

The situation can be different for the even-order responses, if the triple means $\langle x_{nm}V_{mk}V_{kn}\rangle = \left\langle xV^2\right\rangle$ differ from zero. This is, for example, the case for the interaction $\hat{V}=\mathcal{E}_x\hat{d}_x+\mathcal{B}_y\hat{\mu}_y$, proportional to the electric field $\mathcal{E}$ in the x-direction and the magnetic field $\mathcal{B}$ in the y-direction and the operator of the Hall current $\hat{x}=\hat{j}_z$ in the z-direction. In this case the average $\langle xV\rangle$ entering the perturbation series shown in Fig. 6.6 has to be replaced by the mean value

$$\langle xV\rangle \to \sum_k \left\langle x_{nm}V_{mk}\hat{X}'_k(\xi,\theta)V_{kn}\right\rangle$$
$$= \langle xV^2\rangle \sum_k \hat{X}'_k(\xi,\theta) = \langle xV^2\rangle\, i\pi g \qquad (6.89)$$

where we have employed (4.32) for $Q(\xi) = \sum_k \hat{X}_k(\xi)$. We illustrate the meaning of this replacement and subsequent factorization of the series in Fig. 6.8. This yields

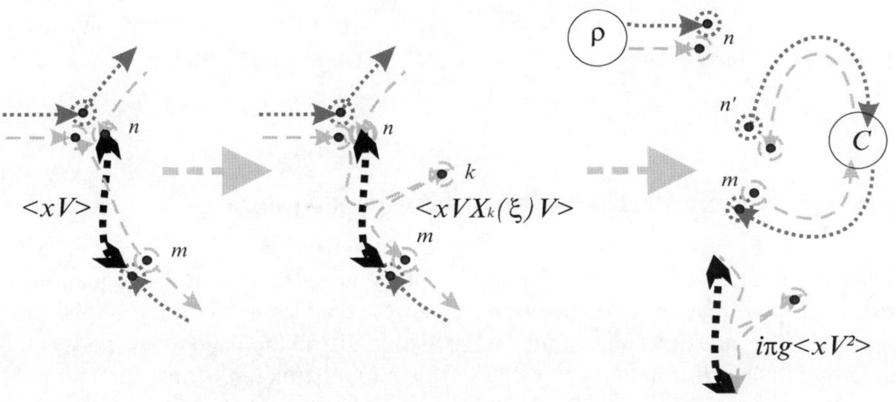

Fig. 6.8

$$x(t) = -\pi g \left\langle xV^2\right\rangle \int_{-\infty}^{t} d\tau'' e^{-\pi g\lambda\, V^2 \int_{-\infty}^{\tau''} q^2(x)dx}$$
$$\exp\left\{-\pi g\, V^2 \int_{\tau''}^{t} \left(q(x) -\ q(t+\tau''-x)\right)^2 dx\right\}, \qquad (6.90)$$

for the response, which in the dimensionless units $\pi g\ V^2 \to 1$, $\pi g \left\langle xV^2\right\rangle \to \mathcal{X}$ takes the simple form

$$x(t) = -\mathcal{X} \int_{-\infty}^{t} d\tau'' e^{-\lambda \int_{-\infty}^{\tau''} q^2(x)dx - \int_{\tau''}^{t} \left(q(x)-\ q(t+\tau''-x)\right)^2 dx}. \qquad (6.91)$$

It is worth mentioning that, for a complex system without central symmetry, the even-order response given by the average $\sum_k \left\langle V_{nm}x_{mk}\hat{X}'_k(\xi,\theta)V_{kn}\right\rangle$ can also contribute to the result.

In order to gain a first insight into the physics of the response of complex quantum systems, let us consider the simplest step-like time-dependence $q(t) = \Theta(t)$, which allows one to write (6.91) in the form

$$x(t) = -\mathcal{X} \int_0^t d\tau'' e^{-\lambda \int_0^{\tau''} q^2(x)dx} = \mathcal{X} \frac{1 - e^{-\lambda t}}{\lambda}. \tag{6.92}$$

This shows that the response approaches its stationary value $\mathcal{X}/\lambda$ with the rate λ proportional to the difference $V^2 - (V')^2$ given by the mean square $(\mathrm{Im}V)^2$ of the imaginary part of the coupling matrix elements. Note that the stationary value itself is inversely proportional to the imaginary part of the coupling, since $\mathcal{X} \sim \langle xV^2 \rangle \sim \langle x\, \mathrm{Re}V\ \mathrm{Im}V \rangle$.

6.3 Harmonic Perturbation of Complex Systems

One of the standard characteristics of a physical system is its behavior under harmonic perturbation. Consideration of Sect. 6.2 enables us to give an immediate answer to the question concerning the population distribution and the response of a system with a uniform density of states subject to a harmonic perturbation proportional to a random matrix. Analysis of this case will be followed by the consideration of several more sophisticated examples of quantum systems with interaction matrices of a block structure. The treatment of these cases is facilitated by the use of the quasienergy technique (Sect. 5.1), to be presented below in more detail.

6.3.1 Population Distribution Over a Uniform Spectrum

Equations (6.49) and (6.55) allow one to explicitly find the time-dependent population distribution and the response of complex quantum systems for the time-dependent interaction $\chi(t)\cos(\omega t + \varphi)\,\hat{V}$, proportional to a random matrix $\hat{V}$. To shorten the expressions that follow we set $\omega = 1,\ 2\pi g V^2 = W$ and first consider the population dynamics.

Population Dynamics for an Abrupt Switching

In the case $\varphi = 0$ for the step-like envelope $\chi(t) = \Theta(t)$ the population density reads

$$\pi g \rho_n(t) = \mathrm{Re} \int_0^t e^{i\Delta_n \upsilon - W\left(\int_0^t \cos^2(x)dx - \int_0^{t-\upsilon} \cos(x)\cos(\upsilon + x)dx\right)} d\upsilon. \tag{6.93}$$

We perform the integration in the exponent and arrive at

$$\pi g\rho_n(t) = \mathrm{Re}\int_0^t dy \exp\left\{i\Delta_n y - \tfrac{W}{4}(2t + \sin 2t)\right.$$
$$\left. + \tfrac{W}{4}(2(t-y)\cos y + \sin(2t-y) - \sin y)\right\}. \qquad (6.94)$$

For W small, as compared to the frequency, one can neglect the oscillating terms with constant amplitudes and obtain the integral

$$g\rho_n(t) = \mathrm{Re}\int_0^t dy \exp\left[i\Delta_n y - \frac{W}{2}t + \frac{W}{2}(t-y)\cos y\right], \qquad (6.95)$$

which for times exceeding the typical time $1/W$ of a single transition can be calculated approximately. We first note that for a large number of transitions $Wt \gg 1$, the main contribution can come only from the vicinities of the points where $\cos y \simeq 1$, otherwise the contribution is small. This immediately yields $y = 2\pi k + z$ with $z \ll 1$, the integration over z can be extended from $-\infty$ to ∞ for all k except of $k = 0$, and the population (6.95) takes the form

$$\pi g\rho_n(t) = \mathrm{Re}\int_0^\infty dz \exp\left[i\Delta_n z - \tfrac{Wt}{2}\tfrac{z^2}{2}\right]$$
$$+\mathrm{Re}\sum_{k=1}^{\infty}\int_{-\infty}^{\infty} dz \exp\left[i\Delta_n(2\pi k + z) - \tfrac{Wt}{2}\tfrac{z^2}{2} - W\pi k\right]. \quad (6.96)$$

The integral in the first term is half of the integral from $-\infty$ to ∞, and the rest of the terms form a geometric series which results in

$$\begin{aligned} g\rho_n(t) &= \tfrac{2}{\sqrt{\pi W t}} \exp\left[-\tfrac{\Delta_n^2}{Wt}\right] \mathrm{Re}\left[\tfrac{1}{2} + \tfrac{e^{2\pi i\Delta_n - W\pi}}{1 - e^{2\pi i \Delta_n - W\pi}}\right] \\ &= \tfrac{1}{\sqrt{\pi W t}} \exp\left[-\tfrac{\Delta_n^2}{Wt}\right] \mathrm{Im}\cot(\pi\Delta_n + iW/2) \\ &= \tfrac{1}{\sqrt{\pi W t}} \exp\left[-\tfrac{\Delta_n^2}{Wt}\right] \sum_{m=-\infty}^{\infty} \tfrac{W/2\pi}{(\Delta_n - m)^2 + W^2/4}, \end{aligned} \qquad (6.97)$$

where we have taken into account the singularities of $\cot(\pi\Delta_n + iW/2)$ which has simple poles at the points $\Delta_n = -iW/2\pi + m$, where m is an integer. One sees that the population is located in the $W/2\pi$-vicinities of the resonant frequencies $\Delta = m$, whereas the population distribution among these vicinities is a Gaussian of width proportional to the square root of the dimensionless time $\sqrt{Wt}$. In Fig. 6.9a we illustrate this dependence.

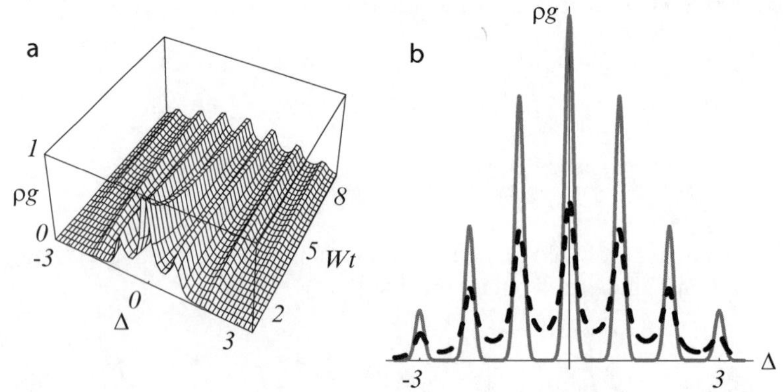

Fig. 6.9 A complex quantum system under harmonic perturbation. (**a**) Population distribution over the uniform spectrum as a function of time for $W/\omega = 0.3$. (**b**) Population distribution for the adiabatic envelope (6.98) for $\alpha/\omega^2 = 3 \times 10^{-3}$ (*solid line*) as compared to the distribution for an abrupt switch (*dash line*). The typical number of transitions $Wt = 5$ is taken for both cases. The transition rate W is the same as that in (**a**)

As seen in Fig. 6.9a, the population remains localized in the vicinities of the resonances. The width of all these populated vicinities is the same and is equal to the transition probability W. Moreover, consideration of the general case $q(t) = \cos(\omega t + \varphi)$ shows that this result does not depend on the initial phase φ as long as W is small as compared to the frequency, which demonstrates once again the evident fact that all dependencies with fronts shorter than a typical relaxation time $1/W$ act as an instantaneous switching. A natural question arises as to whether this width is formed during the development of the population dynamics, or is just a result of an abrupt switching of the interaction at $t = 0$.

Population Dynamics for an Adiabatic Switching

In order to answer this question let us consider the population distribution $\rho_n\,(t = \infty)$ following a quasiperiodic perturbation with a smooth Gaussian envelope $q(t) = e^{-\alpha t^2} \cos \omega t$. Equation (6.55) reads

$$\pi g \rho_n\,(\infty) = \mathrm{Re} \int_{-\infty}^{\infty} e^{i\Delta_n v - W \int_{-\infty}^{-\infty} \exp(-2\alpha x^2)\cos^2(x)dx}$$

$$e^{W \int_0^{-\infty} \exp(-\alpha x^2 - \alpha(v+x)^2)\cos(x)\cos(v+x)dx} dv, \qquad (6.98)$$

and for $\omega = 1$ yields after integration over x

$$\pi g \rho_n(\infty) = \mathrm{Re} \int_{-\infty}^{\infty} dv\, e^{i\Delta_n v} \exp\left[-W\sqrt{\tfrac{\pi}{8\alpha}}\left(1 - e^{-v^2\alpha/2}\cos v\right)\right]$$
$$\exp\left[-e^{-1/2\alpha}\sqrt{\tfrac{\pi}{8\alpha}}\, W\left(1 - e^{-v^2\alpha/2}\right)\right]. \tag{6.99}$$

The last factor in the integrand is exponentially close to unity for $\alpha \to 0$, and therefore can be neglected, while the exponent of the second factor can be expanded in a power series and gives

$$\pi g \rho_n(\infty) = \mathrm{Re} \int_{-\infty}^{\infty} dv \exp\left[i\Delta_n v - W\sqrt{\frac{\pi}{8\alpha}}\left(1 - \cos v + \frac{v^2\alpha}{2}\right)\right]. \tag{6.100}$$

Here we have also taken into account the fact that $W/\sqrt{\alpha}$ is a large factor and therefore the main contribution arises from the domains where $\cos v \simeq 1$.

The population distribution (6.100) is an inverse Fourier transformed product of two factors $e^{-Wv^2\sqrt{\pi\alpha/2}/4}$ and $e^{-W\sqrt{\pi/8\alpha}\,(1-\cos v)}$, each of which has an explicit inverse Fourier transform—a narrow Gaussian

$$\int_{-\infty}^{\infty} \frac{dv}{2\pi} e^{i\Delta_n v - v^2 W\sqrt{\pi\alpha/32}} = \left(\frac{2}{\alpha\pi^3 W^2}\right)^{1/4} e^{-\Delta_n^2\sqrt{2/\pi\alpha W^2}}, \tag{6.101}$$

and a series

$$\int_{-\infty}^{\infty} \frac{dv}{2\pi} e^{i\Delta_n v - W\sqrt{\pi/8\alpha}\,(1-\cos v)} = \sum_k \delta\left(\Delta_n - k\right) e^{-W\sqrt{\pi/8\alpha}} I_k\left(W\sqrt{\frac{\pi}{8\alpha}}\right) \tag{6.102}$$

with the coefficients given by the modified Bessel functions $I_k(x)$, respectively. The last equation can be obtained with the help of the Poisson summation formula (3.89) and the Schlafley integral representation for the Bessel functions.

The result is the convolution of (6.101) and (6.102)

$$g\rho_n(\infty) = \sum_k \left(\frac{2\pi}{\alpha W^2}\right)^{1/4} e^{-(\Delta_n - k)^2\sqrt{2/\pi\alpha W^2}} e^{-\sqrt{\pi W^2/8\alpha}} I_k\left(\sqrt{\frac{\pi W^2}{8\alpha}}\right). \tag{6.103}$$

This has a clear physical meaning. The narrow vicinities of the resonances $\Delta_n = k$ of width $\left(\pi\alpha W^2/2\right)^{1/4}$ given by the span of the Fourier transformation of the Gaussian envelope are populated according to the weights suggested by the solution

$e^{-\sqrt{\pi W^2/8\alpha}} I_k(\sqrt{\pi W^2/8\alpha})$ of the one-dimensional diffusion equation on a grid, while the step of the grid is given by the pumping frequency ($\omega = 1$) as we will see from (8.30) later on. The typical number of transitions $\langle |k| \rangle$ amounts to the probability W of a single transition multiplied by the typical interaction time $\sqrt{\frac{\pi}{2\alpha}}$. The product $W\sqrt{\frac{\pi}{2\alpha}}$ plays the same role as the combination Wt in the previous case. Implicit in this treatment is the condition $\left(\pi\alpha W^2/2\right)^{1/4} g \gg 1$ of sufficient spectral density, which means that the populated vicinity of each resonance contains many of quantum levels of the system. In Fig. 6.9b we compare the population distributions for the abrupt and the adiabatic switch of the perturbations.

Now we can give an answer to the question concerning the origin of the population distribution in the vicinities of the resonances. It has a coherent nature since the distribution is given by the Fourier transform of the perturbation envelope, being Lorentzian for an abrupt switching, while narrow and Gaussian for a long adiabatic Gaussian pulse of interaction. This implies that the population distribution near the resonance is formed as a result of a reversible process. Indeed, being a broad $\sim W$ Lorentzian for the middle part of the adiabatic interaction pulse, the distribution shrinks to a narrower by a factor of $\left(W^2/\alpha\right)^{1/4}$ Gaussian when the pulse is over. We can therefore expect coherent interference phenomena in the case $\omega \sim W$ when the vicinities of different resonances start to overlap. Later on we will consider such a phenomenon for a two-band quantum system.

6.3.2 *Response of the Uniform Spectrum to a Harmonic Perturbation*

We consider the susceptibility of the quantum system with a uniform spectrum to the periodic perturbation switched either abruptly or adiabatically. We employ (6.90) for a perturbation with frequency $\omega = 1$ and $2\pi g V^2 = W$, and ignore, as earlier, the time-dependence of $\langle xV^2 \rangle$.

Response for an Abrupt Switching

For the abrupt switch one has

$$x(t) = -\pi g \left\langle xV^2 \right\rangle \int_0^t d\tau'' e^{-\lambda\, W \int_0^{\tau''} \cos^2(x) dx}$$
$$\exp\left\{-W \int_{\tau''}^{t} \left(\cos(x) - \cos(t+\tau''-x)\right)^2 dx\right\} \qquad (6.104)$$

for the complex quantum system with a uniform spectrum on a periodic perturbation $\hat{V}\cos t$ starting at $t = 0$. Here we set $\omega = 1$. We concentrate here on the case of small W and small λ, which allows one to retain only the linearly increasing component of the first integral in the exponent and write

$$x(t) = -\pi g \left\langle xV^2 \right\rangle \int_0^t d\tau'' e^{-\lambda W \tau''/2}$$

$$\exp\left\{-2W \sin^2(\frac{t+\tau''}{2})(t-\tau'' - \sin(t-\tau''))\right\}. \quad (6.105)$$

Consider this expression in the limit of large t. We note that the last exponent is always less than unity, and therefore the contribution of the domain $\tau'' \gg 1/\lambda W$ is negligible, and also we neglect the term $\sin(t-\tau'')$ as compared to the large quantity $t-\tau''$ thus arriving at

$$x(t) = -\pi g \left\langle xV^2 \right\rangle \int_0^t d\tau'' e^{-\lambda W \tau''/2} e^{-2W(t-\tau'')\sin^2[(t+\tau'')/2]}, \quad (6.106)$$

where the main contribution to this integral comes from the domains $t+\tau'' \simeq 2k\pi$ around zeros of the argument of the second exponent. We therefore set $\tau'' = 2k\pi - \arg e^{it} + z$ and split the integration interval into a set of 2π-long intervals, with the total number of these intervals equal to the integer part of the ratio $t/2\pi$, that is, $K = (t - \arg e^{it})/2\pi$, and arrive at

$$x(t) = -\pi g \left\langle xV^2 \right\rangle \sum_{k=0}^{K} \exp\left[-\lambda W(k\pi)\right]$$

$$\int_{-\pi}^{\pi} dz \exp\left[-2W\left(t - 2k\pi + \arg e^{it} - z\right)\sin^2[z/2]\right], \quad (6.107)$$

where the small terms $\lambda W(\arg(\exp(it)) - z)$ have been ignored. We make use of the fact that for the all important $k \le 1/W\lambda$, the prefactor in front of $\sin^2[z/2]$ is large, of the order of Wt, and therefore we can expand the expression in a Taylor series around $z = 0$ and keep only the first non-vanishing quadratic term. We can also set both integration limits and the upper limit of the sum to infinity, that is,

$$x(t) = -\pi g \left\langle xV^2 \right\rangle \sum_{k=0}^{\infty} e^{-\lambda W k\pi} \int_{-\infty}^{\infty} dz \exp\left[-\frac{Wt}{2} z^2\right]$$

$$= \frac{-\pi g \left\langle xV^2 \right\rangle}{1 - e^{-\lambda W \pi}} \sqrt{\frac{2\pi}{Wt}} \simeq \frac{-g\left\langle xV^2 \right\rangle}{\lambda W} \sqrt{\frac{2\pi}{Wt}}. \quad (6.108)$$

We have thus encountered an interesting universal property of complex systems, namely that the second-order response to an abruptly switched harmonic perturbation decreases as the square root of time. This is remarkable in the fact that the periodic perturbation does not result in the oscillation of the permeability of the system. On the contrary, the spectrum of the permeability is located near zero frequencies and manifests there a non-analytic behavior $\sim 1/\sqrt{\omega}$, as one can find by performing the Fourier transformation of (6.108). It is also important to note that, as we have mentioned on p. 266, for calculation of the total response, the multiplication of the $1/\sqrt{t}$ dependence by the time-dependent part of $\langle xV^2 \rangle$ is required.

Response to an Adiabatic Switching

The long-lasting $1/\sqrt{t}$ character of the response suggests we consider the case of the adiabatic switching of the harmonic interaction. For the case $q(t) = e^{-\alpha t^2} \cos \omega t$, (6.90) for the response reads

$$x\,(t) = -\pi g \left\langle xV^2 \right\rangle \int\limits_{-\infty}^{t} d\tau'' e^{-\lambda W \int_{-\infty}^{\tau''} \exp(-2\alpha x^2) \cos^2 x \, dx}$$
$$\exp\left\{ -W \int_{\tau''}^{t} \left(e^{-\alpha x^2} \cos x - e^{-\alpha (t+\tau''-x)^2} \cos(t+\tau''-x) \right)^2 dx \right\}, \tag{6.109}$$

where the rate α of the front rising is a very small number, and the factor in front of the integral we treat as a constant for the moment. We replace $\cos^2 x$ entering the first integral in the exponent by its mean value $1/2$ assuming $\lambda\, W \ll 1$, and note that this integral is large $\sim \lambda\, W/\sqrt{\alpha}$ unless the integration limit τ'' is a large negative number $\tau'' \ll -1/\sqrt{\alpha}$. Since the second integral in the exponent is negative, the last inequality is also a necessary condition for the total integral not to vanish. Indeed, for $\tau'' \sim -1/\sqrt{\alpha}$ the result of the integration will be small because of the first integral in the exponent, which takes a large negative value. In turn, the second integral in the exponent becomes large when two oscillating terms of the integrand have different amplitudes. This means that the response is exponentially small, unless $t \sim -\tau'' \gg 1/\sqrt{\alpha}$. Therefore the sum $t + \tau''$ must be a small value, such that $\alpha(t+\tau'')^2 \ll 1$ and the limits of the second integral in the exponent therefore can be taken as infinite. After straightforward integration of the second term one obtains

$$x\,(t) = -\pi g \left\langle xV^2 \right\rangle \int\limits_{-\infty}^{t} d\tau'' e^{-\lambda\, W \int_{-\infty}^{\tau''} \exp(-2\alpha x^2)/2 \, dx}$$
$$\exp\left\{ -W \sqrt{\frac{\pi}{2\alpha}} \left(1 - e^{-\alpha (t+\tau'')^2/2} \cos(t+\tau'') \right) \right\}, \tag{6.110}$$

where we have also set $e^{-1/\alpha}$ to zero.

The second exponential factor dominates in (6.110), since it contains a large parameter $W\sqrt{\frac{\pi}{2\alpha}}$, and the main contributions to the integral arise in the vicinities of the points $\tau'' = -t + 2k\pi$ periodically arranged on the integration axis, where $\cos(t+\tau'') \to 1$. Each interval of length 2π gives the contribution $(\frac{\pi\alpha}{2})^{1/4}W^{-1/2}e^{\alpha(t+\tau'')^2/4}$, equal to the width of the corresponding vicinity, which one can find by expanding the second exponent in series near these points. Calculation of the integrals (6.110) for small α results in

$$x(t) = -\frac{\pi g\langle xV^2\rangle}{2\sqrt{W}}\left(\frac{\pi\alpha}{2}\right)^{1/4}\sum_{k=-\infty}^{\infty} e^{-\lambda W\int_{-\infty}^{-t+2k\pi}\exp(-2\alpha x^2)/2\,dx}$$
$$\exp\left\{\alpha(k\pi)^2 - W\sqrt{\frac{\pi}{2\alpha}}\left(1-e^{-2\alpha(k\pi)^2}\right)\right\}. \quad (6.111)$$

In order to be consistent within the approximation, one has to neglect the first term in the second exponent as compared to the second one and take the Taylor expansion of $1-e^{-2\alpha(k\pi)^2}$. This yields

$$x(t) = -\frac{\pi g\langle xV^2\rangle}{2\sqrt{W}}\left(\frac{\pi\alpha}{2}\right)^{1/4}$$
$$\sum_{k=-\infty}^{\infty} e^{-W\sqrt{2\alpha\pi}(k\pi)^2-\lambda W\int_{-\infty}^{-t+2k\pi}\exp(-2\alpha x^2)/2\,dx}. \quad (6.112)$$

For small α the sum can be replaced by an integral which results in

$$x(t) = -\frac{g\langle xV^2\rangle}{4\sqrt{2\alpha}\sqrt{W}}\left(\frac{\pi\alpha}{2}\right)^{1/4}\int_{-\infty}^{\infty} dy$$
$$\exp\left[-\frac{W\sqrt{\pi}}{4\sqrt{2\alpha}}\left((y+t\sqrt{2\alpha})^2 + \frac{\lambda}{\sqrt{\pi}}\int_{-\infty}^{y}\exp(-x^2)\,dx\right)\right], \quad (6.113)$$

which can be found by the saddle-point method.

For small λ the saddle point is located near $y = -t\sqrt{2\alpha}$, whereas the second derivative is close to $W\sqrt{\pi/8\alpha}$ and therefore

$$x(t) = -\frac{g\langle xV^2\rangle\sqrt{\pi}}{2\sqrt{2W}}\exp\left[-\frac{W\lambda}{4\sqrt{2\alpha}}\int_{-\infty}^{-t\sqrt{2\alpha}}\exp(-x^2)\,dx\right], \quad (6.114)$$

which in the notation of (6.91) and after the replacements $\pi g\langle xV^2\rangle \to \mathcal{X}$ for dimensionless permeability, $W\lambda\sqrt{\pi}/8\sqrt{2\alpha} \to A$ for the parameter representing a typical number of transitions during the adiabatic pulse, and $t\sqrt{2\alpha} \to T$ for dimensionless time takes the short form

$$x\,(t) = -\frac{\mathcal{X}}{2\sqrt{2\pi}} \exp\left\{A\,\mathrm{Erf}(T) - A\right\}, \tag{6.115}$$

where the Gaussian integral is replaced by the error function

$$\int\limits_{-\infty}^{-X} \exp(-x^2)\,dx = \frac{\sqrt{\pi}}{2}(1 - \mathrm{Erf}[X]), \tag{6.116}$$

that has already been encountered in (3.133).

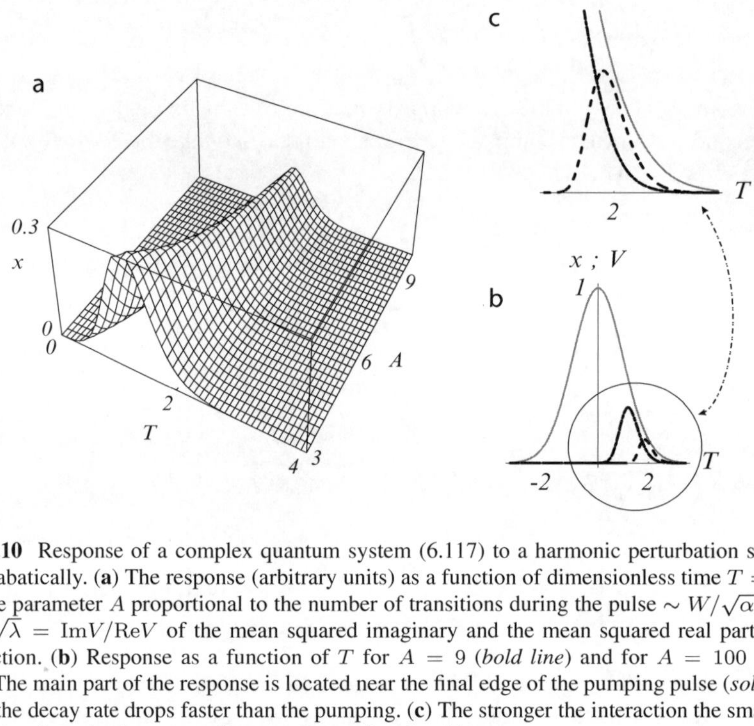

Fig. 6.10 Response of a complex quantum system (6.117) to a harmonic perturbation switched on adiabatically. (**a**) The response (arbitrary units) as a function of dimensionless time $T = t\sqrt{2\alpha}$ and the parameter A proportional to the number of transitions during the pulse $\sim W/\sqrt{\alpha}$ and the ratio $\sqrt{\lambda} = \mathrm{Im}V/\mathrm{Re}V$ of the mean squared imaginary and the mean squared real parts of the interaction. (**b**) Response as a function of T for $A = 9$ (*bold line*) and for $A = 100$ (*dashed line*). The main part of the response is located near the final edge of the pumping pulse (*solid line*), when the decay rate drops faster than the pumping. (**c**) The stronger the interaction the smaller the response, since the decay rate drops to the required level later

By adding $-T^2$ to the exponent one can also take into account the time-dependence of $\langle xV^2\rangle$. Discussing the response (6.92), we have already mentioned that $\lambda \sim \mathrm{Im}V^2$ and hence $A \sim \lambda$, whereas $\mathcal{X} \sim \mathrm{Im}V \sim \sqrt{A}$. For small λ, assuming the mean squared real part of the perturbation fixed, we can write

$$x\,(T) \sim \sqrt{A}\exp\left\{A\,\mathrm{Erf}(T) - A - T^2\right\}, \tag{6.117}$$

for the time-dependent response given by the product of the permeability and the square of the time-dependent Gaussian envelope of the interaction $q(t)$. In Fig. 6.10a we illustrate this dependence. The response is mainly located near the end of the pumping pulse, which is a natural consequence of the fact that the decay rate increases rapidly at the pulse front, thus resulting in a rapid decrease of the response, whereas at the end of the pulse the equally rapid decrease of the rate does not last a sufficiently long time to completely destroy the response. We note that the integral of $\exp\left\{A\,\mathrm{Erf}(T) - A - T^2\right\}$ over T tends for large A to the dependence $0.88/A$. With dimensional variables, this yields the asymptotic behavior

$$\int x(t)\,dt = -\frac{g\left\langle xV^2\right\rangle\sqrt{2\pi}0.88}{W^2\lambda} \sim \frac{x_t g}{gW}, \tag{6.118}$$

for the integrated total response, decreasing with the increase of the interaction strength and independent of the width α^{-1} of the pulse envelope.

The numeric factor in (6.118) relates to the chosen model of a Gaussian envelope, but the inverse dependence on the interaction intensity $\sim W^{-1}$ is a general property of the physical quantities given by the self-intersecting diagrams, as is the case in Fig. 6.4 for the response diagrams, since each self-intersection reduces the number of corresponding diagrams by a factor equal to the number of levels in resonance, discussed on p. 81. The order of magnitude estimations suggest that the integrated response is of the order of the typical matrix element x_t multiplied by the typical return time g and divided by the number gW of levels in the populated vicinities of the resonances.

6.4 Two-Frequency Excitation of Complex Systems

We now turn to complex quantum systems subject to the simultaneous action of two harmonics and consider the universal properties of nonlinear wave mixing in such systems. We start with the population distribution given by (6.55) for $q(x) = e^{-\alpha t^2}(\cos t + \cos bt)$ and then consider the corresponding response (6.87).

6.4.1 Population Dynamics for Bi-Harmonic Excitation

After a straightforward calculation one finds the integral

$$\int_{-\infty}^{\infty}\left[q^2(x) - q(x)q(v+x)\right]dx = \frac{1}{2}\sqrt{\frac{\pi}{2\alpha}}\left[2 - e^{-v^2\alpha/2}\left(\cos v + \cos bv\right)\right] \tag{6.119}$$

where the terms $\sim \exp[-1/\alpha]$ have been ignored, and equation (6.55) yields

$$\pi g\rho_n(t) = \mathrm{Re}\int_{-\infty}^{\infty} \exp\left\{i\Delta_n v - W\sqrt{\frac{\pi}{8\alpha}}\left[2 - e^{-v^2\alpha/2}(\cos v + \cos bv)\right]\right\} dv. \tag{6.120}$$

One recognizes here the Fourier transformation of the product of two similar expressions

$$\exp\left\{-W\sqrt{\frac{\pi}{8\alpha}}\left[1 - e^{-v^2\alpha/2}\cos v\right]\right\} \tag{6.121}$$

and

$$\exp\left\{-W\sqrt{\frac{\pi}{8\alpha}}\left[1 - e^{-v^2\alpha/2}\cos bv\right]\right\}, \tag{6.122}$$

each of which has exactly the same structure as the expression in (6.99), which we have encountered in considering the harmonic excitation of the complex system. The Fourier transform of a product $F[ab]$ yields the convolution $F[a] * F[b]$ of the Fourier transforms of the factors, and therefore with the help of (6.103)

$$\begin{aligned} g\rho_n(\infty) = {} & \textstyle\sum_k \left(\frac{2\pi}{\alpha W^2}\right)^{1/4} e^{-(\Delta_n - k)^2\sqrt{2/\pi\alpha W^2}} e^{-\sqrt{\pi W^2/8\alpha}} I_k\left(\sqrt{\frac{\pi W^2}{8\alpha}}\right) \\ & * \sum_m \left(\frac{2\pi}{\alpha W^2}\right)^{1/4} e^{-(\Delta_n - mb)^2\sqrt{2/\pi\alpha W^2}} e^{-\sqrt{\pi W^2/8\alpha}} I_m\left(\sqrt{\frac{\pi W^2}{8\alpha}}\right) \\ = {} & \sum_{km} \left(\frac{2\pi}{\alpha W^2}\right)^{1/4} e^{-(\Delta_n - k - bm)^2\sqrt{2/\pi\alpha W^2}} e^{-\sqrt{\pi W^2/2\alpha}} \\ & I_m\left(\sqrt{\frac{\pi W^2}{8\alpha}}\right) I_k\left(\sqrt{\frac{\pi W^2}{8\alpha}}\right). \end{aligned} \tag{6.123}$$

Equation (6.123) has a clear physical meaning. It shows that the population is located in narrow ($\sim \sqrt{W\sqrt{\alpha}}$) vicinities of the resonances $\Delta = k + bm$, whereas the total populations P_{mk} of different vicinities are distributed according the solution $e^{-w} I_m(w/2) I_k(w/2)$ of the diffusion equation on a two-dimensional grid (see (8.30)), given in terms of the modified Bessel functions and corresponding to the effective number of transitions $w = W\int q^2(t)dt = W\sqrt{\pi/2\alpha}$. For large T one finds the asymptotic relation

$$P_{mk} = e^{-w} I_m(w/2) I_k(w/2) \simeq e^{-(k^2+m^2)/w}/\pi w \tag{6.124}$$

and (6.123) for the population spectral density $\rho(\Delta_n) = g\rho_n(t = \infty)$ adopts the form

$$\rho(\Delta) = \sum_{km} \left(\frac{8\alpha\pi}{W^2}\right)^{1/4} e^{-(\Delta-k-bm)^2\sqrt{2/\pi\alpha W^2}} e^{-(k^2+m^2)\sqrt{2\alpha/\pi W^2}} \frac{1}{\pi^2 W}. \quad (6.125)$$

This expression gives the population distribution over the uniform energy spectrum, which results from the overlap of the vicinities of different resonances, located in the same energy positions. The way these vicinities are distributed along the energy axes depends on the ratio b of the pumping frequencies. The situation is similar to the Bragg scattering patterns as a function of different orientations of solids with rectangular lattices. In Fig. 6.11 we illustrate this dependence.

It turns out that the commensurability of two pumping frequencies plays an important role and determines the pattern of the population distribution. Indeed, when the frequency ratio $b = p/q$ is a rational number given by an irreducible fraction of two integer numbers p and q, then according to (6.123), the energy positions of different resonances $\Delta = k + bm = (kq + pm)/q$ are separated by a minimum distance $1/q$. The resulting distribution than turns out to be a set of narrow spikes of population separated by unpopulated intervals. The width of the spikes depends only on the spectral resolution $\sim \left(2/\pi\alpha W^2\right)^{-1/4}$ corresponding to

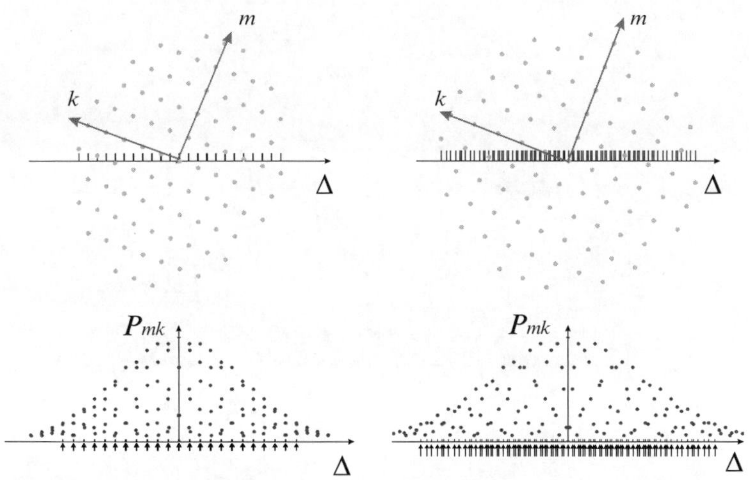

Fig. 6.11 Population distribution (6.123) of a complex quantum system subjected to the action of an adiabatically switched two-frequency field. The distribution results from the diffusion on the two-dimensional grid (*upper panel*) given by the resonant absorption or emission of quanta at frequencies $\omega_1 = 1$ (index k) and $\omega_2 = b$ (index m) of the field. The result depends on the diffusion-weighted $P_{m,k}$ projection of this grid on the one-dimensional energy axis (*lower panel*). For the commensurate frequencies $b = p/q$ (*left panel*) and for the non-commensurate frequencies $b \neq p/q$ the distributions are significantly different. The positions of the resonances are shown by *arrows*

the transition rate W and the pulse duration $\sim 1/\sqrt{\alpha}$, whereas the total population of each vicinity depends on the number of steps needed for arriving at this resonance from the initially populated state and the interaction pulse duration.

The situation is completely different for an irrational b. For asymptotically long pulse durations this results in a uniform distribution over the energy scale. However, this distribution is not attained at once but arises through a series of quasiperiodical distributions, each of which is given by the denominator of the best rational approximation to the irrational number b corresponding to the actual spectral resolution $\sim \left(2/\pi\alpha W^2\right)^{-1/4}$. In Fig. 6.12 we illustrate this situation for the example of $b = 1/3$ and $b = 1/\pi$. For the last case, the best approximations are

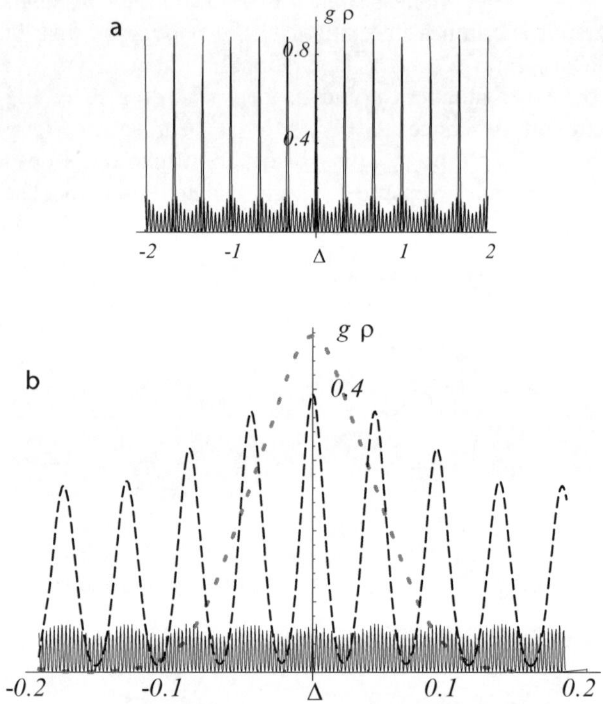

Fig. 6.12 Density of population [see (6.123) and Fig. 6.11] for different pulse durations $1/\sqrt{\alpha}$ and different frequency ratios $\omega_1/\omega_2 = b$. Transition rate $W = 0.1$ for all plots. (**a**) For commensurate frequencies $b = 1/3$, the population distribution is given by the spikes located periodically at the position of the inverse of the largest common divider of the oscillation periods, $\Delta_p = 1/3$. The non-commensurate ratio $b = 1/\pi$ yields a multiperiodic structure. (**b**) The genesis of multiperiodicity for irrational b. For different pulse durations the spectral resolution are different, and the period $1/q$ is given by the denominators of the best rational approximation $b \simeq p/q$ of the frequency ratio. For $\alpha = 10^{-4}$ one finds $q = 3$ (*dotted line*), for $\alpha = 10^{-7}$ $q = 22$ (*dashed line* scaled $\times 3$ relative to (**a**)) and $q = 355$ for $\alpha = 10^{-11}$ (*solid line* scaled $\times 9$), which corresponds the best rational approximations $\pi \simeq 3 \simeq 22/7 \simeq 355/113$, respectively

$b = 1/\pi \simeq 1/3 \simeq 7/22 \simeq 113/335 \simeq 99532/312689$, and one sees the periodic structures corresponding to three first denominators $1/q = 1/3\ ; 1/22\ ; 1/355$ that sequentially attains with increasing pulse duration.

6.4.2 Response to Bi-Harmonic Excitation

A complex periodic structure, originating in the number theory relating to the best rational approximation of irrational numbers also appears, although in a less pronounced way, in the spectrum for the universal time-dependent response of a complex quantum system with a uniform density of states, in the presence of a two-frequency perturbation. We shall consider the bi-harmonic time-dependence $q(x) = \chi(t)\,(\cos t + \cos bt)$ with an envelope $\chi(t)$. The particular choice of $\chi(t)$ does not play a crucial role and can be made in a way minimizing technical complications during the calculations by assuming that most of the time the envelope takes a constant value, say unity. The frequency spectrum of the response according to (6.87) reads

$$x\,(\Delta) \sim \int_{-\infty}^{\infty} dt\, \chi^2(t) \int_{-\infty}^{0} dy\, \exp\left[i\Delta t - \lambda\, W \int_{-\infty}^{y+t} q^2(x)dx \right.$$
$$\left. -W \int_{y}^{0} \left(q(x+t) - q(t+y-x)\right)^2 dx \right], \qquad (6.126)$$

where we have made the replacements $\tau'' \to t + y$, $x \to x + t$, and by introducing the factor $\chi^2(t)$ we have also taken into account the profile of the pumping field. Before taking the Fourier transform over time we note that the integrand can be represented as a product of two factors,

$$x\,(\Delta) \sim \int_{-\infty}^{\infty} dt\, e^{i\Delta i}\, \tilde{R}(t)\tilde{f}(t)$$
$$\tilde{f}(t) = \chi^2(t) \exp\left[-\lambda\, W \int_{-\infty}^{t} q^2(x)dx \right]$$
$$\tilde{R}(t) = \int_{-\infty}^{0} dy\, e^{-W \int_{y}^{0} \left[(q(x+t) - q(t+y-x))^2 + \lambda q^2(x+t)\right] dx}, \qquad (6.127)$$

which means that the Fourier transform of the response is a convolution of two profiles: the term

$$R(\Delta) = \int_{-\infty}^{\infty} \frac{dt}{2\pi} e^{i\Delta t} \int_{-\infty}^{0} dy\, e^{-W \int_{y}^{0} \left[(q(x+t) - q(t+y-x))^2 + \lambda q^2(x+t)\right] dx}, \qquad (6.128)$$

which allows for the frequency dependence of the response on a large scale and which has the structure of a number of δ-like spikes similar to (6.17), and the term

$$f(\Delta) = \int_{-\infty}^{\infty} \frac{dt}{2\pi} \chi^2(t) e^{i\Delta t - \lambda W \int_{-\infty}^{t} q^2(x)dx} \tag{6.129}$$

allowing for the profile (line shape) of each spike that mainly depends on the steepness of fronts and the shape of the envelope $\chi(t)$.

In the limit of small λ we can neglect the oscillating component of $q^2(x)$ and taking $\chi^2(t) = 1 + \tanh \alpha x$, for instance, obtain after the replacement $t \to T/\alpha$

$$\begin{aligned} f(\Delta) &= \int_{-\infty}^{\infty} \frac{dt}{2\pi} \left[1 + \tanh \alpha t)\right] e^{i\Delta t - \lambda W \int_{-\infty}^{t} [1+\tanh \alpha x)]dx} \\ &= \int_{-\infty}^{\infty} \frac{dT}{\pi\alpha} \frac{e^{2T}}{1+e^{2T}} \exp\left\{ i\frac{\Delta}{\alpha}T - \frac{\lambda W}{\alpha} \ln\left[1+e^{2T}\right] \right\}. \end{aligned} \tag{6.130}$$

The substitution $z = e^{2T}$ results in the line shape

$$f(\Delta) = \int_{0}^{\infty} \frac{z^{i\Delta/2\alpha} dz}{2\pi\alpha\,(1+z)^{1+\lambda W/\alpha}} = \frac{\Gamma\left(1+i\frac{\Delta}{2\alpha}\right)\Gamma\left(\frac{\lambda W}{\alpha} - i\frac{\Delta}{2\alpha}\right)}{2\pi\lambda W \Gamma\left(\frac{\lambda W}{\alpha}\right)} \tag{6.131}$$

whose absolute value has a bell-like shape of width $\sim \alpha$ centered at $\Delta = 0$ and with phase modulations antisymmetric with respect to this point.

Let us turn now to the term (6.128). The integrand in the exponent is always positive, and hence only $y \sim 1/\lambda W$ can be important. In this domain we can ignore the variation of the envelope and set $\chi(t) = 1$. Moreover, we assume $W \ll 1$ and ignore the oscillatory terms of constant amplitude, retaining only the leading terms with amplitudes proportional to y. This yields for the integral in the exponent

$$\begin{aligned} &\int_{y}^{0} \left[\left(q(x+t) - q(t+y-x)\right)^2 + \lambda q^2(x+t) \right] dx \\ &\simeq -y\left\{2 + \lambda - \cos\left[2t+y\right] - \cos\left[b(2t+y)\right]\right\}, \end{aligned} \tag{6.132}$$

and hence

$$\begin{aligned} R(\Delta) &= \int_{-\infty}^{\infty} \tfrac{dt}{2\pi} e^{i\Delta t} \int_{-\infty}^{0} dy\, e^{Wy\{2+\lambda-\cos[2t+y]-\cos[b(2t+y)]\}} \\ &= \int_{-\infty}^{0} dy e^{-i\Delta y/2 + Wy\lambda} \int_{-\infty}^{\infty} \tfrac{dt}{2\pi} e^{i\Delta t} e^{Wy\{2-\cos 2t - \cos 2bt\}}. \end{aligned} \tag{6.133}$$

In (6.120) we have already encountered the last integral, which can be given in terms of Bessel functions, and therefore we arrive at

$$R(\Delta) = \int_{-\infty}^{0} dy e^{Wy\lambda - i\Delta y/2}\, e^{2Wy} \sum_{m,n} I_m(-Wy) I_n(-Wy)\delta(\Delta - n - bm). \tag{6.134}$$

The structure of this expression resembles that for the population distribution (6.125), which is also a set of narrow (here δ-like) spikes weighted with an intensity distribution function $P_{n,m}$. Following convolution with the profile (6.131) this results in

$$R(\Delta) = \sum_{m,n} P_{n,m}(W,\lambda) f(\Delta - n - bm)$$

$$P_{n,m} = \int_{-\infty}^{0} dy e^{Wy\lambda - i(n-bm)y/2}\, e^{2Wy} I_m(-Wy) I_n(-Wy). \tag{6.135}$$

However, in contrast to the population distribution, the weighting function for the response remains essentially different from the Gaussian. When we make use of the Gaussian asymptotic expression for the Bessel functions (6.124) it takes the form

$$\begin{aligned} P_{n,m} &= \int_{-\infty}^{0} \frac{2dy}{\pi W y} \exp\left\{ Wy\lambda - \frac{i(n+bm)y}{2} + \frac{(k^2+m^2)2}{Wy} \right\} \\ &= -\frac{4}{W} K_0 \left[2\sqrt{\left(\frac{i(bm-n)}{W} + 2\lambda\right)(m^2+n^2)} \right], \end{aligned} \tag{6.136}$$

where $K_0(z)$ is the modified Bessel function.

Combination of (6.131) and (6.136) gives the spectral composition of the response

$$x(\Delta) = 2\pi g \left\langle x V^2 \right\rangle \sum_{m,n} K_0 \left[2\sqrt{\left(\frac{i(bm-n)}{W} + 2\lambda\right)(m^2+n^2)} \right] \frac{\Gamma\left(1 + i\frac{(\Delta - n - bm)}{2\alpha}\right) \Gamma\left(\frac{\lambda W}{\alpha} - i\frac{(\Delta - n - bm)}{2\alpha}\right)}{\pi \lambda W^2 \Gamma\left(\frac{\lambda W}{\alpha}\right)}. \tag{6.137}$$

The main difference between this function and the population distribution (6.125) is due to the narrow weighting function (6.136) given by $K_0(z)$ exponentially decreasing at large values of the argument z. Therefore the typical detuning $(bm - n)$ between the higher-order harmonics n and m should be smaller than $W/\left(m^2 + n^2\right)$, otherwise there is no substantial contribution to the sum (6.137). However a typical detuning scales as $1/n$ whereas for the best rational

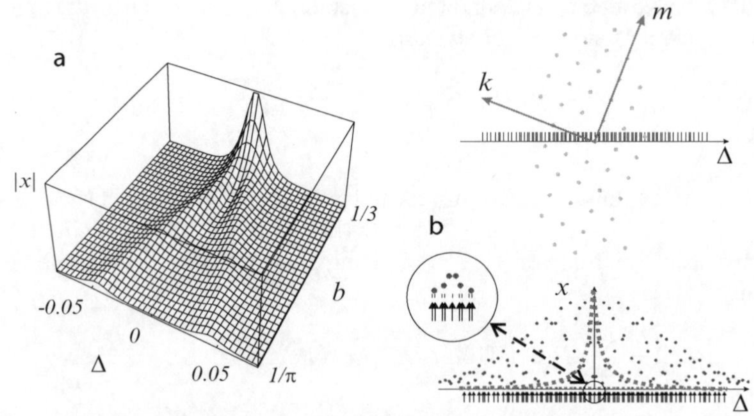

Fig. 6.13 Frequency dependence of the response (6.137) of a complex quantum system with a uniform density of states to a bi-harmonic excitation. The principal effect occurs around zero frequency. (**a**) Coalescence of two resonances for a frequency ratio b, assuming a rational value. (**b**) In contrast to the population distribution Fig. 6.11, the contribution of higher-order resonances is suppressed by the spectral profile (*dashed line*) decreasing exponentially for large detunings. Therefore only a few resonances near zero frequency can interfere as illustrated in the *inset*. The parameters of the plot are $W = 0.1$, $\lambda = 0.05$, $\alpha = 0.01$

approximation it can be estimated as $1/n^2$, assuming a uniform, random distribution of the higher-order resonances on the frequency axis. Therefore for a typical b, the minimum value of the combination $(bm-n)\left(m^2+n^2\right)$ for integer n and m ranging from $-N$ to N does not depend much on the size N of the interval, remaining of the order of unity. For example, in the case $b = 1/\pi$ for the first four best rational approximations this combination takes the values 0.15, -0.068, -0.0037, and -0.317. This means that for a generic ratio of frequencies at certain conditions one may observe the response consisting of several higher-order resonances at once, one for each best rational approximation, but it never forms a periodic pattern as was the case for the population distribution. Typically one can see the coalescence of only two resonances when b assumes a rational value, as shown in Fig. 6.13.

6.5 Two-Band System in a Periodic Field

The behavior of two-band systems under the action of a harmonic perturbation shown in Fig. 6.14a is almost identical to the dynamics of the single-band spectra of Sect. 6.3. The main distinction is in the fact that the even-order resonances correspond to the first band, while the odd-order resonances correspond to the second one. We therefore address a natural question specific for the two-band problem, about the time-dependence of the population difference of the two bands. We also consider here the role of the phase statistics of the pumping field and the regime of overlapping resonances—two problems that have not been considered earlier, although they could be formulated for a single band as well. Note that the two band model can be applied to a single-band system when we consider in the resonant approximation a high-frequency perturbation $V_h = Vq(t)\cos\omega_h t$ at a frequency ω_h comparable with the total width $\Gamma \sim \omega_h$ of the band. In the last case, the oscillation of the field amplitude $Vq(t)$ plays the role of the time-dependent perturbation coupling two resonant parts, as shown in Fig. 6.14b.

6.5.1 Dynamics of Total Band Populations

We address this problem with the help of the technique employed earlier in this chapter, and then propose a somewhat different point of view at the obtained results, which is based on the consideration of a quantized pumping field.

The Direct Consideration

Equation (4.28), valid for the total population of the bands in the case of time-independent perturbations, can also be employed for the harmonic pumping by simply replacing the static $\hat{X}$-operators (4.21) by similar operators for the time-dependent situation. For two identical bands the simplest forms have the expressions for the sum $\hat{\rho}_+ = \hat{\rho}_1 + \hat{\rho}_2$ and the difference $\hat{\rho}_- = \hat{\rho}_1 - \hat{\rho}_2$ of the populations

$$\hat{\rho}_+(\varepsilon,\xi,\tau,\theta) = \frac{1}{1-\hat{\Xi}(\varepsilon,\xi,\tau,\theta)}\hat{X}_0(\varepsilon,\tau)\hat{X}_0(\xi,\theta),$$
$$\hat{\rho}_-(\varepsilon,\xi,\tau,\theta) = \frac{1}{1+\hat{\Xi}(\varepsilon,\xi,\tau,\theta)}\hat{X}_0(\varepsilon,\tau)\hat{X}_0(\xi,\theta), \tag{6.138}$$

where $\hat{X}_0(\varepsilon,\tau)$ and $\hat{\Xi}(\varepsilon,\xi,\tau,\theta)$ are given by (6.35), (6.36).

We note that (6.38) can be generalized in order to include both signs in the denominator

$$\frac{1}{1\pm\hat{\Xi}(\varepsilon,\xi)} = \frac{1}{\varepsilon-\xi+i\frac{\partial}{\partial\theta}+i\frac{\partial}{\partial\tau}+i\pi g\,V^2\,\left(q(\tau)\pm\,q(\theta)\right)^2}$$
$$\left[\varepsilon-\xi+i\tfrac{\partial}{\partial\theta}+i\tfrac{\partial}{\partial\tau}+i\pi g\,V^2\,\left(q^2(\tau)+\,q^2(\theta)\right)\right], \tag{6.139}$$

which with the allowance for (6.71) results in the equation

$$\left[\varepsilon-\xi+i\frac{\partial}{\partial\theta}+i\frac{\partial}{\partial\tau}+i\pi g\,V^2\,\left(q(\tau)+\,q(\theta)\right)^2\right]$$
$$\hat{\rho}_-(\varepsilon,\xi,\tau,\theta) = \left[\hat{X}_0(\xi,\theta)-\hat{X}_0(\varepsilon,\tau)\right], \tag{6.140}$$

for the population difference.

After the standard replacements $\varepsilon - \xi \to \zeta$, $\varepsilon + \xi \to 2\eta$ and integration over η this yields

$$\left[\zeta + i\tfrac{\partial}{\partial\theta} + i\tfrac{\partial}{\partial\tau} + i\pi g\, V^2\, \left(q(\tau) + \, q(\theta)\right)^2\right] \hat{\rho}_-\left(\zeta, \tau, \theta\right) = 2\pi i. \tag{6.141}$$

The solution of this equation yields the population difference at a given time t, if after performing an inverse Fourier transformation one sets $\tau = \theta = t$. Of course, one can always act in this way and obtain the correct result, however we note that the same result can be obtained directly when we remember that the axillary variables τ and θ have been introduced in order to replace the time-dependent Schrödinger equation by a time-independent one with one more coordinate for the wavefunction ψ or two more coordinates for the density matrix ρ. Therefore, (6.141) is just a representation of the equation

$$\left[\frac{\partial}{\partial t} + 4\pi g\, V^2\, q^2(t)\right] \rho_-\left(t\right) = 0, \tag{6.142}$$

with the initial condition $\hat{\rho}_-\left(t = 0\right) = 1$ which is much easier to solve directly, without making use of such a technique. This yields

$$\rho_-\left(t\right) = e^{-4\pi g V^2 \int\limits_0^t q^2(x) dx}, \tag{6.143}$$

and for $q(t) = \cos[t + \varphi]$ results in

$$\rho_-\left(t\right) = \exp\left[-2\pi g V^2 (t + \cos(t + 2\varphi)\sin t)\right], \tag{6.144}$$

or in dimensional variables for frequency ω

$$\rho_-\left(t\right) = \exp\left[-2\pi g V^2 (t + \omega^{-1}\cos(\omega t + 2\varphi)\sin \omega t)\right]. \tag{6.145}$$

This function shown in Fig. 6.14d, demonstrates a decay with an oscillating rate.

Quantized Pumping Field

Thus far in our consideration, we have always assumed that the external field is classical in nature. The employed technique of introducing the additional variables τ and θ is identical to that known as the quasienergy method briefly discussed in Sect. 5.1, which is closely related to the quantum consideration of the pumping field. The main idea of the method is the following.

We substitute the anzats

$$\psi_n = \sum_k e^{-ik\omega t} \psi_{kn}(t) \tag{6.146}$$

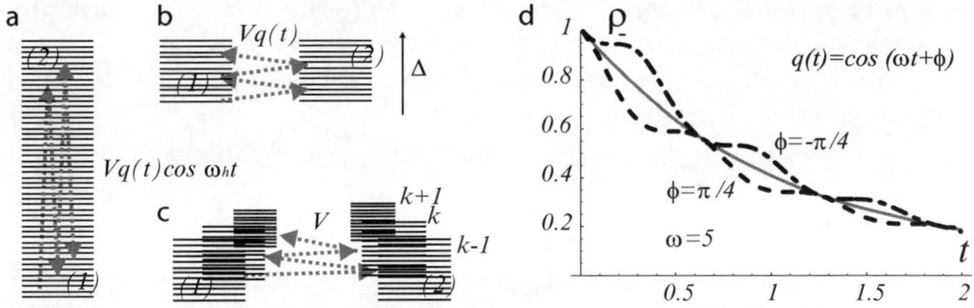

Fig. 6.14 One band in a bi-harmonic field with frequencies $\omega_h \pm \omega$ comparable with the band width (**a**) can be considered as a two-band problem (**b**) with the harmonic interaction at the beat frequency ω. Application of the quasienergy method (**c**) suggests the consideration of an equivalent multiband system with a time-independent coupling, where each harmonic of the population amplitude ψ_{kn}, (6.146), corresponds to an individual band. (**d**) The difference between the total band populations (6.145) for $W = 1$, $\omega = 5$, and the phase shifts $\phi = \pi/4$ (*dashed line*) or $\phi = -\pi/4$ (*dot dash line*) is a decaying function of time, with an oscillating decrement. For the field in the quantum state corresponding to a given number of photons (6.157), the dependence has no oscillations (*solid line*)

into the Schrödinger equation (6.30) with $q(t) = \cos(\omega t)$, and considering all of the coefficients corresponding to the harmonics $e^{-ik\omega t}$ arrive at the expression

$$i\dot{\psi}_{kn} = (\Delta_n - \omega k)\,\psi_{kn} + \frac{1}{2}\sum\nolimits_m V_{nm}\left(\psi_{k-1,m} + \psi_{k+1,m}\right), \qquad (6.147)$$

which is known as the quasienergy representation of (6.30) with periodic coefficients. This new equation does not contain the time-dependent coefficients and therefore can be solved by standard methods. Formally it can be considered as a multiband system as shown in Fig. 6.14c, where each quasienergy index k corresponds to an individual band of levels enumerated either by the index m or by index n, depending on the number of the band (1 or 2) to which it belongs initially.

But now we have encountered the problem of undefined initial conditions. Indeed, the requirement $\psi_0(t = 0) = 1$ with the allowance for (6.146) only states that $\sum_k \psi_{k0}(0) = 1$ and does not give any individual initial conditions for each amplitude $\psi_{kn}(t = 0)$. This problem has deep roots in the question of the quantum statistics of the pumping field, and its solution is different for the coherent states of the quantum field corresponding to the classical limit, and for the photon number state of the field which corresponds to a definite field energy.

Let us look at (6.147) from the point of view of a quantized field. Considering the two equivalent problems of a single band in the two-frequency field at $\omega_h \pm \omega$ and two bands in a harmonic field at the frequency ω we favor the latter approach and denote by ψ_{kn} the probability amplitude for the system to be at the level n of the first band and the quantized field to be in the state with $k \gg 1$ photons at the frequency ω. This corresponds to the state $|n\rangle \otimes |k\rangle$ of the compound system. By ψ_{km} we denote the analogous amplitudes for the second band, and therefore (6.147) now may be directly interpreted as the Schrödinger equation for the compound system. By introducing the operators $\hat{D}$ and $\hat{q}$ with the matrix elements $D_{kk'} = \omega k \delta_{k,k'}$ and $q_{kk'} = (\delta_{k,k'+1} + \delta_{k,k'-1})/2$ one obtains for this system the Fourier transformed evolution operator

$$\hat{U}(\varepsilon) = \left(\varepsilon - \hat{H}_0 + \hat{D} - \hat{q} \otimes \hat{V}\right)^{-1}, \tag{6.148}$$

which actually coincides with (6.32) if we note that the operators $\hat{D}$ and $\hat{q}$ are just a Fourier (in k) representation of $i\frac{\partial}{\partial \tau}$ and the harmonic time-dependent profile, respectively.

There is no need to repeat here all of the derivation, since by analogy one can write equations (6.138) and (6.140) for the total populations of the bands in a form similar to (6.72)

$$\begin{aligned}
\hat{Z}_{-}^{-1}(\zeta)\hat{\rho}_{+}(\zeta) &= 2\pi i, \\
\hat{Z}_{+}^{-1}(\zeta)\hat{\rho}_{-}(\zeta) &= 2\pi i, \\
\hat{Z}_{\mp}^{-1}(\zeta) &= \zeta - \hat{D} + \hat{D}' + i\pi g\, V^2 \left(\hat{q} \pm \hat{q}'\right)^2,
\end{aligned} \tag{6.149}$$

where the operators $\hat{D}'$ and $\hat{q}'$ act on the density operators $\hat{\rho}_{\pm}$ from right to left. The density matrices $\hat{\rho}_{\pm}$ do not depend on the indices enumerating the sublevels m of the system, but they do depend on the photon variables k. Note that two first equations (6.149) after the Fourier transformation in ζ takes the form of an equation for commutators and anticommutators

$$\begin{aligned}
i\frac{\partial}{\partial t}\hat{\rho}_{+} &= \left[\hat{D}, \hat{\rho}_{+}\right] - i\pi g\, V^2 \left[\hat{q}, \left[\hat{q}, \hat{\rho}_{+}\right]\right], \\
i\frac{\partial}{\partial t}\hat{\rho}_{-} &= \left[\hat{D}, \hat{\rho}_{-}\right] - i\pi g\, V^2 \left\{\hat{q}, \left\{\hat{q}, \hat{\rho}_{-}\right\}\right\}.
\end{aligned} \tag{6.150}$$

The traces of $\hat{\rho}_{+}$ and $\hat{\rho}_{-}$ will give the sum and the difference of the total band populations respectively. We concentrate on the second trace given by the second equation (6.150), since for the first one the result is evidently unity. After the trace-preserving transformation $\hat{\rho}_{-} \to e^{-i\hat{D}t}\hat{\rho}_{-}e^{i\hat{D}t}$ one obtains

$$\frac{\partial}{\partial t}\hat{\rho}_{-} = -\pi g\, V^2 \left\{e^{-i\hat{D}t}\hat{q}e^{i\hat{D}t}, \left\{e^{-i\hat{D}t}\hat{q}e^{i\hat{D}t}, \hat{\rho}_{-}\right\}\right\}, \tag{6.151}$$

which in the explicit matrix notation reads

$$\begin{aligned}
-\frac{4}{\pi g\, V^2}\frac{\partial}{\partial t}\rho_{k,k'} &= \left(4\rho_{k,k'} + 2\rho_{k+1,k'+1} + 2\rho_{k-1,k'-1}\right) \\
&\quad + e^{i2\omega t}\left(\rho_{k+2,k'} + 2\rho_{k+1,k'-1} + \rho_{k,k'-2}\right) \\
&\quad + e^{-i2\omega t}\left(\rho_{k-2,k'} + 2\rho_{k-1,k'+1} + \rho_{k,k'+2}\right),
\end{aligned} \tag{6.152}$$

and we have omitted here the index "–". Let us now introduce $\rho_s = \sum_k \rho_{k,k+s}$ and note that $\rho_{s=0}$ gives the band population difference. Equation (6.152) yields for this quantity

$$-\frac{1}{\pi g\, V^2}\frac{\partial}{\partial t}\rho_s = 2\rho_s + e^{i2\omega t}\rho_{s-2} + e^{-i2\omega t}\rho_{s+2}. \tag{6.153}$$

We make use of the generation function $\rho(\beta, t) = \sum_s \rho_s(t)e^{is\beta}$ for which one immediately finds

$$-\frac{1}{2\pi g\, V^2}\frac{\partial}{\partial t}\rho(\beta, t) = (1 + \cos\left[2\omega t + 2\beta\right])\rho(\beta, t), \tag{6.154}$$

that is

$$\rho(\beta, t) = C(\beta)\exp\left\{-2\pi g\, V^2[t + \frac{1}{2\omega}\sin(2\omega t + 2\beta)]\right\}, \tag{6.155}$$

and hence

$$\rho_s(t) = \int_0^{2\pi} d\beta \tfrac{C(\beta)}{2\pi}\exp\left\{-2\pi g\, V^2[t + \tfrac{1}{2\omega}\sin(2\omega t + 2\beta)] - is\beta\right\}. \tag{6.156}$$

For the case $C(\beta) = 2\pi\delta(\beta)$ (6.156) coincides with the result (6.145) obtained for a classical field of phase $\varphi = 0$. This result also follows from the consistent quantum consideration of the problem: for a quantized field in a coherent state with a large mean number of photons the initial photonic density matrix $\sim \exp\left(-i(k - k')\varphi\right)$ results in $\rho_s(t = 0) \sim \exp\left(-is\varphi\right)$, and hence according to (6.155) yields $C(\beta) = 2\pi\delta(\beta - \varphi)$, which indeed gives $C(\beta) = 2\pi\delta(\beta)$ for the field phase $\varphi = 0$.

For the field initially in the state with a given number of photons n_0, the photonic density matrix has the form $\delta_{k,n_0}\delta_{k',n_0}$ and therefore $C(\beta) = \rho(\beta, t = 0) = 1$. The population difference can therefore be given in terms of a modified Bessel function

$$\begin{aligned}\rho_-(t) = \rho_{s=0}(t) &= \frac{e^{-2\pi g\, V^2 t}}{2\pi}\int_0^{2\pi} d\beta \exp\left[-\tfrac{\pi g\, V^2}{\omega}\sin\left(2\omega t + 2\beta\right)\right] \\ &= e^{-\pi g\, V^2 t} I_0(\tfrac{\pi g\, V^2}{\omega}\sin\omega t),\end{aligned} \tag{6.157}$$

which is different from the absorption rate in a coherent field (6.145). However, this expression can be obtained from (6.145) by averaging it over the field phase φ. We have thus obtained an interesting result, which shows that the complex systems absorbs radiation with the rate depending on the quantum field statistics.

6.5.2 Population Distribution Over the Bands

We now consider the distribution of the populations over the bands. Actually, this problem has already been solved for the sum ρ_{n+} of the band populations, for which the derivation coincides with that for (6.49) and (6.55) while for the population difference ρ_{n-} the result is identical apart from the change of sign in the cross-term of the correlation function, such that the general expression reads

$$\pi g \rho_{n\pm}(t) = e^{-W\int_0^t q^2(x)dx} \int_0^t \cos(\Delta_n y) \exp\left[\pm W \int_0^{t-y} q(x)q(y+x)dx\right] dy. \tag{6.158}$$

The change of sign originates from the fact that for ρ_{n-}, instead of the operator $\hat{Z}$ given by (6.40), one has to make use of the operator $\hat{Z}_+$ given by (6.149).

As we have already seen, for small W/ω and an abrupt switch of the interaction, the population is distributed over the narrow vicinities of the resonances with Lorentzian profiles according to (6.97). This is also evidently the case for the two-band system, as long as the resonances do not overlap—we have just to take into account that the even resonances correspond to the first and the odd resonances to the second bands. We therefore concentrate here on the other extreme case, $W \gg \omega$, which we referred to on p. 276 restricting ourselves to the case of a coherent field. Taking as earlier the frequency units ($\omega = 1$) and performing the integration in the exponent of (6.158) one obtains

$$\pi g \rho_{n\pm}(t) = e^{-W[t+\cos(t+2\varphi)\sin t]/2}$$

$$\int_0^t \cos(\Delta_n y) \exp\left[\pm \tfrac{W}{2}\left((t-y)\cos y + \cos(t+2\varphi)\sin(t-y)\right)\right] dy. \tag{6.159}$$

In Fig. 6.15 we show the population distribution over the first and the second bands as a function of time.

For the asymptotically long times, the population difference ρ_{n-} becomes small, since the maximum index of the exponent in the integrand for the case of negative sign cannot compensate for the exponentially decreasing factor in front of the integral. The population distributions of the first and the second bands are therefore identical for long times, and in order to find them one can consider their sum ρ_{n+}. The main contribution comes from small $y \sim 1/\sqrt{W}$ such that one can cast the expression in the exponent of the integrand in the form of a Taylor series and obtain

$$\pi g \rho_{n\pm}(t) = \int_0^\infty \cos(\Delta_n y)$$

$$\exp\left[\frac{-W}{2}\left(t\frac{y^2}{2} + y\cos(t+2\varphi)\cos t\right)\right] dy, \tag{6.160}$$

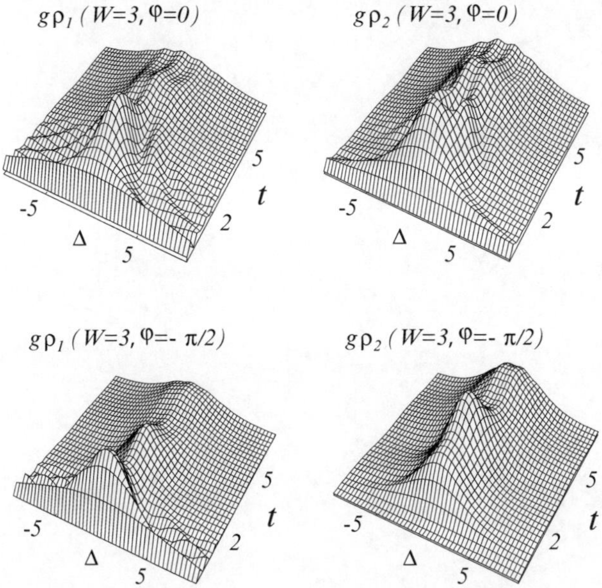

Fig. 6.15 Density of the population distribution (arb. units) of the first band $g\rho_1 = (\rho_+ + \rho_-)/2$ and of the second band $g\rho_2 = (\rho_+ - \rho_-)/2$, as functions of time. For short times, one sees that the profiles resembling those shown in Figs. 4.16 and 4.21 for the level–band system and for the degenerate level–band system respectively, although in the asymptotic of long times, these profiles behave differently, showing a diffusive broadening

where the integration interval has been extended to infinity. Evaluation of the integral for $t \gg W$ (with the allowance for $\omega = 1$) results in the distribution profile

$$\pi g\rho_+ (\Delta, t) \simeq \sqrt{\tfrac{\pi}{tW}} e^{-\Delta^2/Wt} \left\{ \cos\left[\tfrac{\Delta}{t} \cos t \cos(t + 2\varphi)\right] \right.$$
$$\left. + \operatorname{Erfi}\left(\tfrac{\Delta}{\sqrt{Wt}}\right) \sin\left[\tfrac{\Delta}{t} \cos t \cos(t + 2\varphi)\right] \right\}, \qquad (6.161)$$

where $\operatorname{Erfi}(x)$ is the error function. The Gaussian dependence implies that $\Delta \sim \sqrt{Wt}$, and therefore $\Delta/t \ll 1$, which means that in order to be consistent within the approximation we have to replace this expression by an even simpler Gaussian dependence

$$\pi g\rho_+ (\Delta, t) \simeq \sqrt{\frac{\pi}{tW}} e^{-\Delta^2/Wt}. \qquad (6.162)$$

In other words the population distribution is broadening in the long-time limit, according to the diffusion law.

Chapter 7
The Dynamics of One-Dimensional Relay-Type Systems

Thus far we have mainly considered an extreme of quantum complexity, where each pair of quantum states has been coupled by a matrix element, which is statistically equivalent to any matrix element coupling another pair of levels. Although this was not strictly speaking the case for two-band systems where states of the same bands are not coupled, the resulting impact of this fact on their behavior is so tiny that we could attribute them to the same class of problems. The structure of the couplings among different quantum states is closely related to the spatial dimensionality of the system, and the situation where all of the coupling matrix elements have the same order of magnitude is often referred to as the 0-dimensional case. In this chapter we consider a different situation, where quantum systems are arranged, satisfying certain "selection rules". We begin with the consideration of multilevel systems where not all of the states are coupled to each other, but where this coupling is organized in some way. The first natural example is a system of isolated levels, coupled only with their closest neighbors. After presenting several examples of exactly soluble problems, we consider systems conforming to the approximative WKB description, and then turn to the systems with randomly distributed couplings and energy level positions that manifest the phenomenon known as Anderson Localization. This will give us an opportunity to introduce a number of mathematical methods useful for the analysis of complex systems.

7.1 Exactly Soluble Relays of Isolated Levels

Consider a relay of isolated levels, that is, a set of quantum states $|n\rangle$ of energies E_n coupled with their closest neighbors by the interaction matrix elements $V_{n,m} = V_n\delta_{n,m+1} + V_n^*\delta_{m+1,n}$, where $\delta_{m,n}$ is the Kronecker delta. The Schrödinger equation for the probability amplitudes ψ_n corresponding to the states $|n\rangle$ reads

$$i\frac{\partial\psi_n}{\partial t} = E_n\psi_n + V_n\psi_{n-1} + V_{n+1}^*\psi_{n+1}. \tag{7.1}$$

V.M. Akulin, *Dynamics of Complex Quantum Systems*, Theoretical and Mathematical Physics, DOI 10.1007/978-94-007-7205-2_7,

The corresponding equation for the eigenvalues

$$0 = \Delta_n \psi_n + V_n \psi_{n-1} + V^*_{n+1} \psi_{n+1} \tag{7.2}$$

with $\Delta_n = E_n - E$ has the form of a three-term recurrence relation, typical of orthogonal polynomials or some other special functions. Each of these special functions corresponds to an exactly soluble problem for the particular case of Δ_n and V_n. We consider a number of such problems.

7.1.1 Uniform Coupling and Linear Detuning

The simplest case $\Delta_n = \alpha n - E$, $V_n = V^*_n = V$ conforms to the recurrence relations for the Bessel functions

$$0 = (\alpha n - E) J_{n-\frac{E}{\alpha}}\left(\frac{-V}{2\alpha}\right) + V J_{n-1-\frac{E}{\alpha}}\left(\frac{-V}{2\alpha}\right) + V J_{n+1-\frac{E}{\alpha}}\left(\frac{-V}{2\alpha}\right), \tag{7.3}$$

and therefore the energy eigenstates can indeed be given in terms of the Bessel functions

$$\psi_n(E) = A J_{n-\frac{E}{\alpha}}\left(\frac{-V}{2\alpha}\right) + B J_{-n+\frac{E}{\alpha}}\left(\frac{V}{2\alpha}\right) \tag{7.4}$$

where A and B are constants.

Dealing with the time-dependent problem for the initial condition $\psi_0(t = 0) = 1$, it is, however, easier to make use of the generating function $\Psi(t, \varphi) = \sum_n \psi_n(t) e^{i\varphi n}$ for which, for the chosen dependencies of Δ_n and V_n after substitution into (7.1), one finds the equation

$$\begin{aligned} i\frac{\partial \Psi(t,\varphi)}{\partial t} &= \sum_n \alpha n \psi_n(t) e^{i\varphi n} + (V e^{i\varphi} + V^* e^{-i\varphi}) \Psi(t,\varphi) \\ &= -i\alpha \frac{\partial \Psi(t,\varphi)}{\partial \varphi} + (V e^{i\varphi} + V^* e^{-i\varphi}) \Psi(t,\varphi), \end{aligned} \tag{7.5}$$

which can also be extended to the case of a time-dependent coupling $V(t)$ and results in

$$i\frac{\partial \ln[\Psi(t,\varphi)]}{\partial t} + i\alpha \frac{\partial \ln[\Psi(t,\varphi)]}{\partial \varphi} = V(t) e^{i\varphi} + V^*(t) e^{-i\varphi}. \tag{7.6}$$

We now introduce the new variables $x = (\alpha t + \varphi)/2$ and $y = (\alpha t - \varphi)/2$ which yields

$$\ln[\Psi(t,\varphi)] = \int V(\tfrac{x+y}{\alpha}) e^{i(x-y)} + V^*(\tfrac{x+y}{\alpha}) e^{i(y-x)} \frac{dx}{i\alpha}. \tag{7.7}$$

With the allowance for the initial condition $\Psi(t=0,\varphi)=1$ one therefore finds

$$\ln\left[\Psi(t,\varphi)\right] = \int\limits_{\varphi/2}^{(\alpha t+\varphi)/2} V(\tfrac{x+y}{\alpha})e^{i(x-y)} + V^*(\tfrac{x+y}{\alpha})e^{i(y-x)}\tfrac{dx}{i\alpha}$$

$$= \int\limits_{t/2}^{t} V(z)e^{i(\alpha z-\alpha t+\varphi)} + V^*(z)e^{-i(\alpha z-\alpha t+\varphi)}\tfrac{dz}{i}, \qquad (7.8)$$

where the replacement $z \to (x+y)/\alpha$ has been made.

From the generating function one immediately finds the amplitudes

$$\psi_n(t) = \int\limits_0^{2\pi}\frac{d\varphi}{2\pi}\exp\left\{-i\left[e^{i\varphi}\int\limits_{t/2}^{t} V(z)e^{i(\alpha z-\alpha t)}dz + \text{c.c.}\right] - in\varphi\right\}, \qquad (7.9)$$

which, with the help of the integral representation for the Bessel functions

$$J_n(X) = \frac{1}{2\pi}\int\limits_{-\pi}^{\pi} e^{-iX\sin\vartheta+in\vartheta}d\vartheta \qquad (7.10)$$

yields

$$\psi_n(t) = i^n e^{in\arg Z} J_n(2\,|Z|), \qquad (7.11)$$

where $Z = \int_{t/2}^{t} V(z)e^{i(\alpha z-\alpha t)}dz$.

We note that the problem can also be solved for any dependence $\alpha(t)$ which results in (7.11) with $Z = \int_{t/2}^{t} V(z)\exp\left[i\int_t^z \alpha(\theta)d\theta\right]dz$. The reason why this can be solved for arbitrary dependences $V(t)$ and $\alpha(t)$ lies in the algebraic structure of the Hamiltonian, which only contains the operators $\hat{a}_\pm \to (\delta_{n,m+1} \pm \delta_{m+1,n})/2$ and $\hat{a}_d \to n\delta_{n,m}$ forming a closed subalgebra in the infinite-dimensional Hilbert space of the entire system. These operators satisfy the commutation relations $[\hat{a}_+,\hat{a}_-]=0$, $[\hat{a}_d,\hat{a}_\pm]=\hat{a}_\mp$ which is easy to see after Fourier transformation over the indices (momentum representation $n\to\varphi$), since $\hat{a}_d \to -i\partial/\partial\varphi$; $\hat{a}_+ \to \cos\varphi$, and $\hat{a}_- \to i\sin\varphi$. That is why the representation (5.5) for the evolution operator of this infinite-dimensional system contains only these three generators, demonstrating once again the fact that an exactly soluble problem usually has a hidden operator structure symmetry.

In Fig. 7.1a we show the population distribution corresponding to the amplitudes (7.11) for different values of the parameter Z. Note that the states with $n < V$, apart from periodic oscillations, are populated almost uniformly, with a certain augmentation for $n \simeq V$, whereas for $n > V$ the population decreases exponentially.

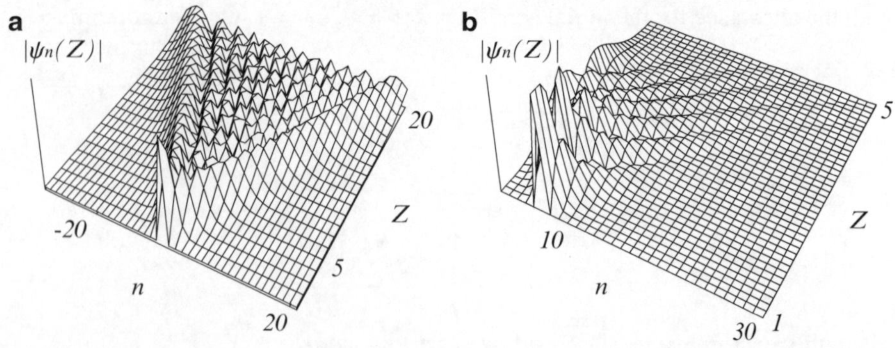

Fig. 7.1 Population distribution in systems with closest neighbor interaction. (**a**) Linear detuning and constant coupling. The absolute value of the amplitude is given by the Bessel functions (7.11). (**b**) Harmonic oscillator. The absolute value of the amplitude is given by the Laguerre polynomials (7.20). The dependence is shown for the initially populated state $n' = 6$

7.1.2 The Harmonic Oscillator in an Arbitrary Time-Dependent Field

The next important case of (7.1) is $E_n = \alpha n$, $V_n = V\sqrt{n}$ which after the replacement $\psi_n = e^{-i\alpha n t}\tilde{\psi}_n$ results in

$$i\frac{\partial}{\partial t}\tilde{\psi}_n = Ve^{i\alpha t}\sqrt{n}\tilde{\psi}_{n-1} + V^* e^{-i\alpha t}\sqrt{n+1}\tilde{\psi}_{n+1}, \tag{7.12}$$

which it is expedient to rewrite with the help of the operators $\hat{a}$ and $\hat{a}\dagger$ of the harmonic oscillator

$$i\frac{\partial}{\partial t}\left|\tilde{\psi}\right\rangle = Ve^{i\alpha t}\hat{a}\left|\tilde{\psi}\right\rangle + V^* e^{-i\alpha t}\hat{a}\dagger\left|\tilde{\psi}\right\rangle. \tag{7.13}$$

This form of the equation will help us to find a solution even for an arbitrary time-dependence of $V(t)$ and $\alpha(t)$ when (7.13) reads

$$i\frac{\partial}{\partial t}\left|\tilde{\psi}\right\rangle = \left[V(t)e^{i\int \alpha(t)dt}\hat{a} + V^*(t)e^{-i\int \alpha(t)dt}\hat{a}^\dagger\right]\left|\tilde{\psi}\right\rangle. \tag{7.14}$$

We introduce the quantity

$$Z(t) = -i\int V(t)e^{i\int \alpha(t)dt}dt \tag{7.15}$$

and consider a transformation $\hat{U}(t) = \exp\left[Z(t)\hat{a} + Z^*(t)\hat{a}\dagger\right]$, which according to the Baker–Campbell–Hausdorff formula (5.3) and with allowance for the commutation relation $[\hat{a}, \hat{a}\dagger] = 1$ can be written in the form

$$\hat{U}(t) = \exp\left[Z(t)\hat{a} + Z^*(t)\hat{a}\dagger\right]$$
$$= \exp\left[Z(t)\hat{a}\right]\exp\left[Z^*(t)\hat{a}\dagger\right]\exp\left[\tfrac{1}{2}Z(t)Z^*(t)\right]. \qquad (7.16)$$

For $\left|\tilde{\psi}\right\rangle = \hat{U}(t)\left|\psi'\right\rangle$ after straightforward calculations, (7.14) yields

$$\frac{\partial}{\partial t}\left|\psi'\right\rangle = \left[Z(t)\frac{\partial Z^*(t)}{2\partial t} - Z^*(t)\frac{\partial Z(t)}{2\partial t}\right]\left|\psi'\right\rangle, \qquad (7.17)$$

where in the course of calculations the expansion

$$e^{\hat{X}}\hat{Y}e^{-\hat{X}} = \hat{Y} + \left[\hat{X},\hat{Y}\right] + \frac{1}{2!}\left[\hat{X},\left[\hat{X},\hat{Y}\right]\right] + \ldots, \qquad (7.18)$$

has been employed.

Equation (7.17) describes an unimportant common phase shift, and hence the operator $\hat{U}(t)$ of (7.16), depending on two real functions $\mathrm{Re}Z(t)$ and $\mathrm{Im}Z(t)$, gives the evolution of the system. However, if the initial state of the system corresponds to a single number state, that is, $\psi_n(t=0) = \delta_{n,n'}$, and if we are interested only in the population distribution and not in the relative phases of different probability amplitudes, the result depends only on a single function $|Z(t)|$ and reads

$$|\psi_n|^2 = \left|\langle n|\exp\left[|Z(t)|\left(\hat{a} - \hat{a}^\dagger\right)\right]|n'\rangle\right|^2, \qquad (7.19)$$

which can be obtained by the transformation

$$\exp\left(-i\Phi(t)\hat{a}^\dagger\hat{a}\right)\hat{U}(t)\exp\left(i\Phi(t)\hat{a}^\dagger\hat{a}\right)$$

corresponding to rotation of the operator (7.16) by an angle $\Phi(t) = \pi/2 + \arg Z(t)$ in the phase space. In the combination $\left(\hat{a} - \hat{a}^\dagger\right)/\sqrt{2}$ one recognizes the momentum operator $i\hat{p} = \partial/\partial x$, and therefore $\exp\left[|Z(t)|\left(\hat{a} - \hat{a}^\dagger\right)\right]$ is just an operator of finite shift (5.13) of the coordinate by the value $|Z(t)|\sqrt{2}$. The easiest way to determine the matrix elements (7.19) of this operator is simply to perform calculations in the coordinate representation, where the wave functions on the oscillator energy eigenstates $\varphi(x)$ are given by Hermite polynomials $H_n(x)e^{-x^2/2}/\sqrt{2^n n!}$. This yields

$$|\psi_n|^2 = \frac{e^{-|Z(t)|^2}}{2^{n+n'}n!n'!}\left|\int H_{n'}\left(x + \tfrac{|Z(t)|}{\sqrt{2}}\right)H_n\left(x - \tfrac{|Z(t)|}{\sqrt{2}}\right)e^{-x^2}dx\right|^2$$
$$= \pi e^{-|Z(t)|^2}|Z(t)|^{2(n-n')}\tfrac{n'!}{n!}\left[L_{n'}^{n-n'}\left(|Z(t)|^2\right)\right]^2, \qquad (7.20)$$

where $n > n'$ and $L_k^m(x)$ is a Laguerre polynomial. For $n > n'$ these two indices have to be interchanged.

In Fig. 7.1b we show the population distribution corresponding to the amplitudes (7.20) for different values of the parameter Z. The existence of an explicit solution again originates in the structure of the operator algebra corresponding to the problem. The Hamiltonian for the harmonic oscillator, interacting with an external field, can be given in terms of three operators—the unperturbed Hamiltonian $\hat{H}_0 = \hat{a}^\dagger\hat{a} = (\hat{P}^2 + \hat{Q}^2)/2$, the momentum $\hat{P} = i(\hat{a}^\dagger - \hat{a})/\sqrt{2}$ and the coordinate $\hat{Q} = (\hat{a} + \hat{a}^\dagger)/\sqrt{2}$ operators, which satisfy the canonical commutation relations $\left[\hat{H}_0\hat{Q}\right] = i\hat{P}$, $\left[\hat{H}_0\hat{P}\right] = -i\hat{Q}$, and $\left[\hat{P}\hat{Q}\right] = -i$.

7.1.3 Raman Pumping of a Harmonic Oscillator

The case $E_n = 2\alpha n$, $V_n = V\sqrt{2n(2n+1)}$, where V and α are functions of time, also corresponds to a harmonic oscillator, although subjected to a different pumping. In Eq. (7.1) for such parameters, one recognizes the Schrödinger equation

$$i\frac{\partial}{\partial t}\left|\tilde{\psi}\right\rangle = \left[\alpha\,\hat{a}\hat{a}^\dagger + V\,\hat{a}^2 + V^*\,\hat{a}^{\dagger 2}\right]\left|\tilde{\psi}\right\rangle. \tag{7.21}$$

for the linear combinations $\left|\tilde{\psi}\right\rangle = \sum \tilde{\psi}_n\left|2n\right\rangle$ of even numbered states of the harmonic oscillator, written with the help of the creation $\hat{a}\dagger$ and annihilation $\hat{a}$ operators squared. The odd functions also satisfy the same equation. By analogy to the previous two cases, these squares $\hat{a}^2$ and $\hat{a}^{\dagger 2}$, together with the unperturbed Hamiltonian $\hat{a}\dagger\hat{a}$, also form a closed $su(1,1)$ subalgebra in the entire Hilbert space of the system, satisfying the commutation relations $\left[\hat{a}^\dagger\hat{a},\hat{a}^2\right] = -2\hat{a}^2$, $\left[\hat{a}^\dagger\hat{a},\hat{a}^{\dagger 2}\right] = 2\hat{a}^{\dagger 2}$, and $\left[\hat{a}^2,\hat{a}^{\dagger 2}\right] = 2\hat{a}\hat{a}^\dagger$. Therefore one can try to find an explicit solution for this problem as well.

In the coordinate and momentum representation (7.21) reads

$$i\frac{\partial}{\partial t}\left|\tilde{\psi}\right\rangle = \left[\alpha\,\frac{\hat{Q}^2+\hat{P}^2}{2} + \mathrm{Re}V\left(\hat{Q}^2-\hat{P}^2\right) - \mathrm{Im}V\left(\hat{P}\hat{Q}+\hat{Q}\hat{P}\right)\right]\left|\tilde{\psi}\right\rangle, \tag{7.22}$$

and we note that the operators $\hat{\mu}_+ = (\hat{Q}^2+\hat{P}^2)/2$, $\hat{\mu}_- = (\hat{Q}^2-\hat{P}^2)/2$, and $\hat{\mu}_s = (\hat{P}\hat{Q}+\hat{Q}\hat{P})/2$ satisfy the commutation relation resembling that for the spin operators σ_x, σ_y, and σ_z,

$$[\hat{\mu}_s,\hat{\mu}_\pm] = -2i\hat{\mu}_\mp;\;\; [\hat{\mu}_+,\hat{\mu}_-] = -2i\hat{\mu}_s \tag{7.23}$$

which, however, have an important difference in sign, resulting in a non-compact $SU(1,1)$ Lie group which apart from rotation in the oscillator phase space $\exp(-i\phi\hat{\mu}_+)$, also contains a squeezing operation $\exp(-iZ\hat{\mu}_s)$.

Let us assume V real (which can always be done by proper choice of the representation) and perform two sequential transformations $|\psi'\rangle =$

$\exp\left(-iZ\hat{\mu}_s\right)\exp\left(-i\phi\hat{\mu}_+\right)\left|\tilde{\psi}\right\rangle$ with time-dependent Z and ϕ. The first transformation mixes up the operators $\hat{\mu}_-$ and $\hat{\mu}_s$

$$\begin{aligned}\exp\left(i\phi\hat{\mu}_+\right)\hat{\mu}_-\exp\left(-i\phi\hat{\mu}_+\right) &= \hat{\mu}_-\cos 2\phi - \hat{\mu}_s\sin 2\phi \\ \exp\left(i\phi\hat{\mu}_+\right)\hat{\mu}_s\exp\left(-i\phi\hat{\mu}_+\right) &= \hat{\mu}_s\cos 2\phi + \hat{\mu}_-\sin 2\phi \end{aligned} \tag{7.24}$$

in the same way as the transformation $\exp\left(-i\phi\hat{\sigma}_x\right)$ mixes up the σ_y, and σ_z operators. The second transformation

$$\begin{aligned}\exp\left(iZ\hat{\mu}_s\right)\hat{\mu}_+\exp\left(-iZ\hat{\mu}_s\right) &= \hat{\mu}_+\cosh 2Z + \hat{\mu}_-\sinh 2Z \\ \exp\left(iZ\hat{\mu}_s\right)\hat{\mu}_-\exp\left(-iZ\hat{\mu}_s\right) &= \hat{\mu}_-\cosh 2Z + \hat{\mu}_+\sinh 2Z \end{aligned} \tag{7.25}$$

corresponds to squeezing of the phase space in the coordinate direction and stretching it in the momentum direction:

$$\begin{aligned}\exp\left(iZ\hat{\mu}_s\right)\hat{Q}\exp\left(-iZ\hat{\mu}_s\right) &= \hat{Q}e^{-Z} \\ \exp\left(iZ\hat{\mu}_s\right)\hat{P}\exp\left(-iZ\hat{\mu}_s\right) &= \hat{P}e^{Z}. \end{aligned} \tag{7.26}$$

In this representation we find

$$\begin{aligned} i\frac{\partial}{\partial t}\left|\psi'\right\rangle = {} & \left[\left(\alpha+\dot{\phi}\right)\cosh 2Z + 2V\cos\phi\sinh 2Z\right]\hat{\mu}_+\left|\psi'\right\rangle \\ & + \left[V\cos\phi\cosh 2Z + \tfrac{1}{2}\left(\alpha+\dot{\phi}\right)\sinh 2Z\right]2\hat{\mu}_-\left|\psi'\right\rangle \\ & - \left(V\sin\phi - \dot{Z}\right)2\hat{\mu}_s\left|\psi'\right\rangle. \end{aligned} \tag{7.27}$$

We now require the coefficients in front of $\hat{\mu}_-$ and $\hat{\mu}_s$ to vanish, and determine $Z(t)$ and $\phi(t)$ by solving the set of two coupled nonlinear equations

$$\begin{aligned}\dot{\phi} &= -\alpha - 2V\cos\phi\coth 2Z \\ \dot{Z} &= V\sin\phi \end{aligned} \tag{7.28}$$

which depend on the particular time-dependence of $V(t)$ and $\alpha(t)$. The remaining terms in (7.27) result in a transformation $\exp\left(-i\theta\hat{\mu}_+\right)$ where the phase reads $\theta = \int\left[\left(\alpha+\dot{\phi}\right)\cosh 2Z + 2V\cos\phi\sinh 2Z\right]dt$, and hence each number state $|n\rangle$ acquires an individual phase shift $n\theta$. The resulting distribution is given by multiplication of the initial condition by the resulting evolution matrix

$$\left|\psi_n\right|^2 = \left|\sum_{n'}\langle n|\exp\left(-i\theta\hat{\mu}_+\right)\exp\left(-iZ\hat{\mu}_s\right)\exp\left(-i\phi\hat{\mu}_+\right)|n'\rangle\,\psi_{n'}\right|^2, \tag{7.29}$$

and for a single initially populated level n', the value of $Z(t)$ is the only parameter governing the distribution at time t.

One again finds the population distribution (7.29), considering the overlap integrals of the properly normalized squeezed initial coordinate wavefunction $\exp(-iZ\hat{\mu}_s)\, H_{n'}(x)e^{-x^2/2} = H_{n'}(xe^{-Z})e^{-x^2e^{-2Z}/2}$ and the final wavefunction

$$|\psi_n|^2 = \frac{e^{-Z}}{2^{n+n'}n!n'!}\left|\int H_n(x)e^{-x^2/2}H_{n'}(xe^{-Z})e^{-x^2e^{-2Z}/2}dx\right|^2. \tag{7.30}$$

One can say that the symmetry corresponding to the algebraic operator relations (7.23) manifests itself in the coordinate representation—the wavefunction of the oscillator remains the same throughout the evolution apart from the fact that the coordinate itself experiences a scaling by a time-dependent complex number. In particular, this means that Gaussian wavefunctions remain Gaussian although the complex width parameter of the Gaussian distribution changes with the course of time. Such wavefunctions are known as squeezed states of the harmonic oscillator.

7.1.4 The Harmonic Oscillator in the Simultaneous Presence of Dipole and Raman Pumping

We note that for the harmonic oscillator, the more general problem of simultaneous excitation via both dipole and Raman interaction has an explicit analytical solution. For this case, the closed subalgebra also includes the operators $\hat{P}\hat{Q}$ and $\hat{Q}^2 - \hat{P}^2$, or the operators $\hat{a}^{\dagger 2}$ and $\hat{a}^2$ in the representation of second quantization. Instead of (7.2), this problem gives a five-term recurrent relation, which being considered with the help of generating functions results in awkward intermediate expressions. That is why we do not present this approach here. However, because of the important role played by harmonic oscillator models in quantum physics, we give here an alternative description.

Consider the Hamiltonian

$$\hat{H} = a(t)\hat{P}^2 + b(t)\hat{Q}^2 + c(t)\hat{P}\hat{Q} + f(t)\hat{P} + h(t)\hat{Q}, \tag{7.31}$$

and perform several sequential transformations in order to find a representation where this expression has the most simple form. We first consider the shift transformation $\hat{U}_1(t) = \exp\left[i\left(A(t)\hat{P} + B(t)\hat{Q}\right)\right]$ and find the time-dependent shifts of the coordinate $A(t)$ and of the momentum $-B(t)$ such that the transformed Hamiltonian

$$\hat{H}_1 = \hat{U}_1(t)\hat{H}\hat{U}_1^{-1}(t) \tag{7.32}$$

does not contain terms linear in $\hat{P}$ and $\hat{Q}$. To this end we note that

$$\begin{aligned}\hat{U}_1(t)\hat{P}\hat{U}_1^{-1}(t) &= \hat{P} - B(t)\\ \hat{U}_1(t)\hat{Q}\hat{U}_1^{-1}(t) &= \hat{Q} + A(t)\end{aligned} \tag{7.33}$$

as follows from the commutation relation and (7.18), and substitute (7.33) into (7.31) and (7.32). This yields

$$\begin{aligned}\hat{H}_1 = a(t)\hat{P}^2 + b(t)\hat{Q}^2 + c(t)\hat{P}\hat{Q} + [c(t)A(t) - 2a(t)B(t) + f(t)]\,\hat{P}\\ + [2b(t)A(t) - c(t)B(t) + h(t)]\,\hat{Q} + b(t)A(t)^2\\ -c(t)B(t)A(t) + a(t)B(t)^2 - f(t)B(t) + h(t)A(t),\end{aligned} \tag{7.34}$$

and for

$$A(t) = \frac{2a(t)h(t) - c(t)f(t)}{c^2(t) - 4a(t)b(t)};\ B(t) = \frac{2b(t)f(t) - c(t)h(t)}{c^2(t) - 4a(t)b(t)}, \tag{7.35}$$

the terms linear in $\hat{P}$ and $\hat{Q}$ disappear, whereas the free terms do not contain an operator part and therefore can be ignored.

We have therefore reduced the problem to the former one, and hence the total evolution can be given in the form

$$\begin{aligned}\hat{U}(t) = \exp\left[-\frac{\theta(t)}{2i}\left(\hat{P}^2 + \hat{Q}^2\right)\right] \exp\left[-iZ(t)\hat{P}\hat{Q}\right]\\ \exp\left[-i\frac{\phi(t)}{2}\left(\hat{P}^2 + \hat{Q}^2\right)\right] \exp\left[-iA(t)\hat{P} - iB(t)\hat{Q}\right],\end{aligned} \tag{7.36}$$

as follows from (7.29) whereas five time-dependent coefficients $A(t)$, $B(t)$, $\phi(t)$, $Z(t)$, $\theta(t)$ are related to the arbitrary functions $a(t)$, $b(t)$, $c(t)$, $f(t)$, and $h(t)$ by (7.28), (7.35). For an initial state, localized on just a single number state $|n\rangle$, the population distribution depends only on three parameters, since in this case the operators with $\hat{P}^2 + \hat{Q}^2$ in the exponents give only an irrelevant phase dependence, whereas the important part reads

$$\hat{U}(t) = \exp\left[-iZ(t)\hat{P}\hat{Q}\right] \exp\left[-iA'(t)\hat{P} - iB'(t)\hat{Q}\right], \tag{7.37}$$

with

$$\begin{aligned}A'(t) &= A\cos\phi(t) + B\sin\phi(t)\\ B'(t) &= B\cos\phi(t) - A\sin\phi(t).\end{aligned} \tag{7.38}$$

We note that the last exponent in (7.37) corresponds to the case of dipole excitation (7.20) whereas the first exponent is related to the Raman pumping (7.30).

7.2 General Case of an Exactly Soluble Relay

We now consider the general case of (7.1) with time-independent coefficients, and try to find the conditions where the solution can be expressed explicitly in terms of known functions. The matrix form of the Schrödinger equation

$$i\begin{Vmatrix}\dot{\psi}_1\\ \dot{\psi}_2\\ \dot{\psi}_3\\ .\\ \dot{\psi}_{N-2}\\ \dot{\psi}_{N-1}\\ \dot{\psi}_N\end{Vmatrix}=\begin{Vmatrix}E_1 & V_1^* & 0 & . & 0 & 0 & 0\\ V_1 & E_2 & V_2^* & . & 0 & 0 & 0\\ 0 & V_2 & E_3 & . & 0 & 0 & 0\\ . & . & . & . & . & . & .\\ 0 & 0 & 0 & . & E_{N-2} & V_{N-2}^* & 0\\ 0 & 0 & 0 & . & V_{N-2} & E_{N-1} & V_{N-1}^*\\ 0 & 0 & 0 & . & 0 & V_{N-1} & E_N\end{Vmatrix}\begin{Vmatrix}\psi_1\\ \psi_2\\ \psi_3\\ .\\ \psi_{N-2}\\ \psi_{N-1}\\ \psi_N\end{Vmatrix}, \tag{7.39}$$

allows one to write, for the Fourier transforms

$$\begin{Vmatrix}\psi_1\\ \psi_2\\ .\\ \psi_{N-1}\\ \psi_N\end{Vmatrix}=i\begin{Vmatrix}E_1-\varepsilon & V_1^* & . & 0 & 0\\ V_1 & E_2-\varepsilon & . & 0 & 0\\ . & . & . & . & .\\ 0 & 0 & . & E_{N-1}-\varepsilon & V_{N-1}^*\\ 0 & 0 & . & V_{N-1} & E_N-\varepsilon\end{Vmatrix}^{-1}\begin{Vmatrix}\psi_1'\\ \psi_2'\\ .\\ \psi_{N-1}'\\ \psi_N'\end{Vmatrix} \tag{7.40}$$

where primes denote initial conditions at $t = 0$. In accordance with the standard rules of linear algebra, the inverse of the three-diagonal matrix $\hat{H} - \varepsilon$ can be found with the help of the diagonal minors, M_k^+ corresponding to the $k \times k$ upper-left corner and M_n^- corresponding to the $(N - n) \times (N - n)$ lower-right corner of the matrix $\hat{H} - \varepsilon$,and for the matrix elements with $k < n$ one finds

$$\left(\hat{H}-\varepsilon\right)^{-1}_{kn}=\frac{M_k^+M_n^-}{\mathrm{Det}\left|\hat{H}-\varepsilon\right|}\prod_{l=k}^{n-1}V_l^* \tag{7.41}$$

that for the matrix element with $k = 4$ and $n = N - 3$ in particular means

$$\left(\hat{H}-\varepsilon\right)^{-1}_{4,N-3}=\frac{\begin{vmatrix}E_1 & V_1^* & 0\\ V_1 & E_2 & V_2^*\\ 0 & V_2 & E_3\end{vmatrix}\begin{vmatrix}E_{N-2} & V_{N-2}^* & 0\\ V_{N-2} & E_{N-1} & V_{N-1}^*\\ 0 & V_{N-1} & E_N\end{vmatrix}}{\mathrm{Det}\left|\hat{H}-\varepsilon\right|}V_4^*\ldots V_{N-4}^*. \tag{7.42}$$

The minors M_n^- and M_k^+ satisfy the three-term recurrence relations

$$M_n^- = (E_{n+1}-\varepsilon)\,M_{n+1}^- - |V_{n+1}|^2\,M_{n+2}^-,$$
$$M_k^+ = (E_{k-1}-\varepsilon)\,M_{k-1}^+ - |V_{k-1}|^2\,M_{k-2}^+. \tag{7.43}$$

which coincide with the Fourier transformed initial Schrödinger equation. Indeed, after the replacement $\psi_n \to \phi_n \prod_{k=0}^{n-1} V_k^*$, Eq. (7.39) yields

$$(E_n - \varepsilon)\, \phi_n = \phi_{n-1} + |V_n|^2\, \phi_{n+1} \tag{7.44}$$

where we have ignored for the moment the initial conditions in the form of δ-functions, discussed earlier in the context of (3.9). We also note that the relation

$$M_{n+1}^+ M_n^- - |V_n|^2\, M_n^+ M_{n+1}^- = \text{Det}\left|\hat{H} - \varepsilon\right| \tag{7.45}$$

is valid for any n.

7.2.1 Separability, Conditions for the Existence of a Polynomial Solution

Let us consider the conditions where the recurrence relations (7.43), (7.44) have an explicit algebraic solution, representable by an analytic expression. We concentrate only on the particular case where E_{n+1} is a polynomial in n and $|V_{n+1}|^2 = F_n/D_n$ is a ratio of two polynomials. The minors in this case will also be ratios of polynomials, as well as the combinations $u_n^- = M_n^-/M_{n+1}^-$ and $u_k^+ = M_k^+/M_{k-1}^+$ satisfying the equations

$$\begin{aligned} u_n^- &= (E_{n+1} - \varepsilon) - \frac{|V_{n+1}|^2}{u_{n+1}^-}, \\ u_k^+ &= (E_{k-1} - \varepsilon) - \frac{|V_{k-1}|^2}{u_{k-1}^+}. \end{aligned} \tag{7.46}$$

These two equations are equivalent. Therefore, we take just the second one and substitute $u_k^+ = P_k/Q_k$ in the form of a ratio of a polynomial P_k of order p and a polynomial Q_k of order q. This yields

$$D_k P_k^+ P_{k+1}^+ = D_k\,(E_k - \varepsilon)\, P_k^+ Q_{k+1}^+ - F_k Q_k^+ Q_{k+1}^+. \tag{7.47}$$

Now one can ask if a solution of (7.47) exists for p and q independent of k. In this case, the minors can be found in the form of an explicit algebraic expression. Otherwise no general algebraic expression exists for the minors simpler than their definitions. The $p + q + 2$ coefficients of the polynomials P_k and Q_k play the role of independent variables in (7.47), whereas the coincidence of the coefficients corresponding to the same powers of k on the left- and on the right-hand sides results in a set of at least $2p + d + 1$ equations, where d is the order of the polynomial D_k, as suggested by the polynomial order of left-hand side of (7.47). A solution exists for $p+q+2 \geq 2p+d+1$, that is, for $q-p \geq d-1$. Moreover, the last term on the right-hand side suggests that the number of equations is not smaller than $2q+f+1$, where f is the order of F_k, and therefore $d-1 \leq q-p \leq 1-f$. Therefore $q-p$ can only assume the values $0, \pm 1$. One more condition $1 \geq d+\epsilon$ emerges from the polynomial order of the first term on the right-hand side, where ϵ is the polynomial order of E_k. The latter yields two possibilities: $d=1; \epsilon=0$, or $d=0; \epsilon=1$, which implies $f \leq 1$ for the first and $f \leq 2$ for the second cases.

The case $d=0; \epsilon=1; f=0$ coincides with (7.3), and the case $d=0; \epsilon=1; f=1$ corresponds to the harmonic oscillator (7.12). Hence we are left with only two more possibilities

$$\begin{aligned} d=1; \quad \epsilon=0; \quad f=1; \quad (p=q)\,, \\ d=0; \quad \epsilon=1; \quad f=2; \quad (p=q+1)\,, \end{aligned} \tag{7.48}$$

where $d=1$, $\epsilon=0$, $f=0$ is a particular case of the first possibility. The second possibility includes Raman pumping (7.21) as a particular case. Equation (7.43) for minors now read

$$\begin{aligned} (an+b)\left(M_{n-1}^{-}+\varepsilon M_n^{-}\right)+M_{n+1}^{-}=0, \\ (ak+b)\left(M_{k+1}^{+}+\varepsilon M_k^{+}\right)+M_{k-1}^{+}=0, \end{aligned} \tag{7.49}$$

for the decreasing coupling $|V_n|^{-2}=an+b$; and strict resonance $E_n=0$, and

$$\begin{aligned} M_{n-1}^{-}=(\alpha n-\varepsilon)\,M_n^{-}-a\,(n-b)\,(n-c)\,M_{n+1}^{-}, \\ M_{k+1}^{+}=(\alpha k-\varepsilon)\,M_k^{+}-a\,(k-b)\,(k-c)\,M_{k-1}^{+}, \end{aligned} \tag{7.50}$$

for the increasing coupling $|V_n|^2 = a\,(n-b)\,(n-c)$ and an arbitrary linear detuning $E_n = \alpha n$. These couplings are shown in Fig. 7.2. The corresponding equations (7.44) read

$$\begin{aligned} (an+b)\,(\varepsilon\phi_n+\phi_{n-1})+\phi_{n+1}=0, \\ (\alpha n-\varepsilon)\,\phi_n-\phi_{n-1}-a\,(n-b)\,(n-c)\,\phi_{n+1}=0. \end{aligned} \tag{7.51}$$

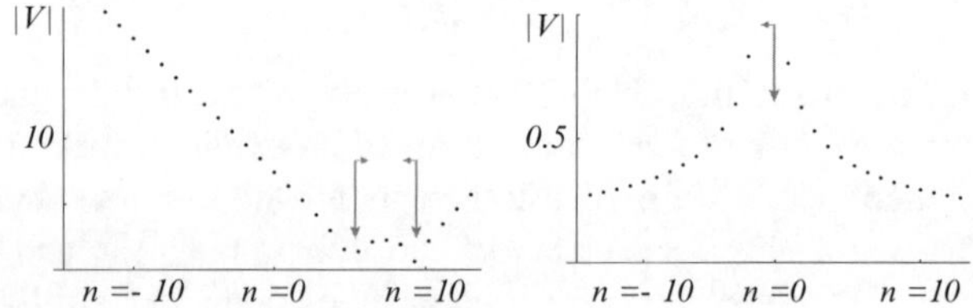

Fig. 7.2 Two types of dependences of the coupling strength on the level number corresponding to exactly soluble problems (Sect. 7.2.2, *left* and Sect. 7.2.3, *right*). *Arrows* show the domains of the imaginary matrix elements

For the case of increasing coupling, it is convenient to make the replacement $\phi_n \to \phi_n \Gamma(1-c)/\Gamma(n-c)$. Equations (7.51), with allowance for the initial conditions $\psi_n(0)$ at $t=0$, ignored earlier in (7.44), take the form

$$(an+b)\left(\varepsilon\phi_n + \phi_{n-1}\right) + \phi_{n+1} = i\,(an+b)\,\psi_n(0)\prod_{k=0}^{n-1}\left(V_k^*\right)^{-1},$$
$$(\varepsilon - \alpha n)\,\phi_n + (n-1-c)\,\phi_{n-1} + a\,(n-b)\,\phi_{n+1} = i\psi_n(0)\prod_{k=0}^{n-1}\frac{(k-c)}{V_k^*}. \tag{7.52}$$

Equation (7.52) is linear in n, and can be solved as earlier (7.5), by introducing the generating functions $\mathcal{M}(\varphi) = \sum_n \phi_n e^{i\varphi n}$. The generating function method suggests the replacement $n \to -i\partial/\partial\varphi$, and we arrive at the first-order differential equations

$$ia\frac{\partial\left(e^{i\varphi}+\varepsilon\right)\mathcal{M}(\varphi)}{\theta\varphi} = \left(be^{i\varphi} + b\varepsilon + e^{-i\varphi}\right)\mathcal{M}(\varphi) + q(\varphi), \tag{7.53}$$

$$i\frac{\partial\left(\alpha - ae^{-i\varphi} - e^{i\varphi}\right)\mathcal{M}(\varphi)}{\theta\varphi} = \left[(1+c)\,e^{i\varphi} - \varepsilon + abe^{-i\varphi}\right]\mathcal{M}(\varphi) + q(\varphi), \tag{7.54}$$

with the inhomogeneous terms $q(\varphi)$ originating from the initial conditions.

Note that (7.53)–(7.54) are the first-order linear differential equations, and that they always have explicit solutions in the form of integrals. The reason why the original problem reduces to an exactly soluble one originates in the condition (7.47), which implies a sort of translational symmetry, which follows from the requirement of the independence of the polynomial orders of P_k and Q_k entering (7.46) for $u_k^+ = P_k/Q_k$ on the index k. Moreover, it points to the fact that the Schrödinger equation, with the detunings and interactions restricted by the condition (7.48), describe evolution in a subspace of the entire Hilbert space of the multilevel system, whereas the corresponding evolution operators belong to a certain subgroup. For certain particular cases of increasing coupling, this is the subgroup $SU(2)$ of rotations generated by the angular momentum operators $L_\pm, L_z$, whereas the Hamiltonian can be written as a sum $aL_+ + a^*L_- + \alpha L_z$ of these operators.

The homogeneous parts of (7.53)–(7.54) have the form

$$i\frac{\partial f(\varphi)\mathcal{M}(\varphi)}{\theta\varphi} = g(\varphi)\mathcal{M}(\varphi), \tag{7.55}$$

where $f(x)$ and $g(x)$ are arbitrary functions. Therefore their solution has the form

$$\mathcal{M}(\varphi) = \frac{1}{f(\varphi)}\exp\left\{-i\int^{\varphi}\frac{g(x)}{f(x)}dx\right\}, \tag{7.56}$$

which results in a general solution of the homogeneous equation

$$\phi_n = C \int_0^{2\pi} \frac{e^{-i\varphi n}}{f(\varphi)} \exp\left\{ -i \int^{\varphi} \frac{g(x)}{f(x)} dx \right\} d\varphi. \tag{7.57}$$

The complete equations

$$i \frac{\partial f(\varphi)\mathcal{M}(\varphi)}{\theta\varphi} = g(\varphi)\mathcal{M}(\varphi) + q(\varphi), \tag{7.58}$$

have the solution

$$\mathcal{M}(\varphi) = \frac{1}{if(\varphi)} \int^{\varphi} q(y) \exp\left\{ i \int_{\varphi}^{y} \frac{g(x)}{f(x)} dx \right\} dy. \tag{7.59}$$

and hence

$$\phi_n = \int_0^{2\pi} \frac{e^{-i\varphi n}}{if(\varphi)} \left[\int^{\varphi} q(y) \exp\left\{ i \int_{\varphi}^{y} \frac{g(x)}{f(x)} dx \right\} dy \right] d\varphi \tag{7.60}$$

is the solution of the original problem. Straightforward calculations of the integrals (7.60) are rather awkward. We therefore separately consider the case of increasing coupling, for which another approach is more convenient, and then we concentrate on the case of decreasing coupling.

7.2.2 Relay with $|V_n| = a\sqrt{(n+1)(n-N)}$, Equivalent to a Two-Level System

This special case corresponds to a N-dimensional representation of $SU(2)$. In fact, the operators $\hat{L}_+$, $\hat{L}_-$, and $\hat{L}_z$, given by the relations

$$\begin{aligned}
\left(\hat{L}_+\right)_{nm} &= \delta_{n,m-1}\sqrt{\frac{(n+1)(N-n)}{2}} \\
\left(\hat{L}_-\right)_{nm} &= \left(\hat{L}_+\right)_{mn} \\
\left(\hat{L}_z\right)_{nm} &= \delta_{n,m}(n - N/2)
\end{aligned}$$

satisfy the $SU(2)$ commutation relations $\left[\hat{L}_+, \hat{L}_-\right] = -\hat{L}_z$, $\left[\hat{L}_z, \hat{L}_\pm\right] = -\hat{L}_\pm$, typical of the Pauly matrices for two-level systems. This can be easily understood,

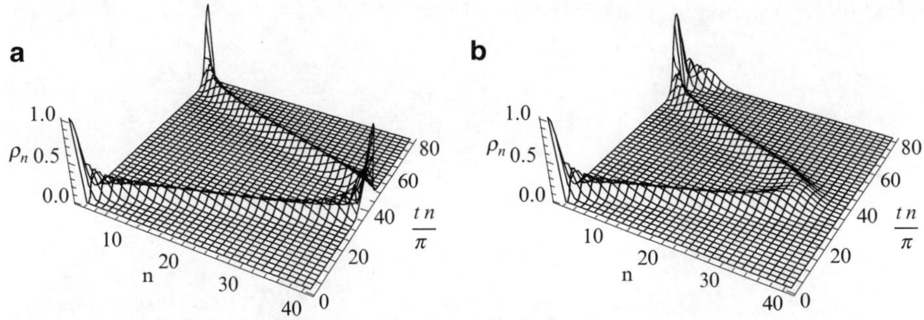

Fig. 7.3 (**a**) Population dynamics in a sequence of resonant levels with the nearest neighbor coupling $\sqrt{(n+1)(N-n)}$. The process is periodic, and at the half of the period, the population of the first level in the chain gets completely transfered to the last level. (**b**) The population dynamics in a detuned chain of levels

since these matrices correspond to the spherical vector components of the spin $N/2$ operators. As a consequence, the recurrent relation (7.40) is nothing but the equation

$$\left(a\hat{L}_x + \alpha\hat{L}_z\right)|\psi\rangle = \varepsilon\,|\psi\rangle\,,$$

describing rotation along the axis pointed in the direction $(a, 0, \alpha)$ in the three-dimensional (x, y, z) space in the Fourier representation.

General solution for this case is a particular case of the problem considered in the following section, and therefore it is explicitly given later on, while here we note that in an important case of the exact resonance, $\alpha = 0$, the population initially localized at the level $n = 0$ gets completely transfered to the level $n = N$ at the time moment $t = \frac{\pi}{a}$, as shown in Fig. 7.3. For a non resonant relay of levels this is no longer the case, although the motion remains periodic, and the population returns back to the initial state completely after the period.

7.2.3 *The Increasing Coupling* $|V_n| = \sqrt{a\,(n-b)\,(n-c)}$

Equation (7.54), after inverse Fourier transformation, reads

$$i\left(\alpha - ae^{-i\varphi} - e^{i\varphi}\right)\frac{\partial\mathcal{M}(\varphi)}{\partial\varphi} + i\frac{\partial\mathcal{M}(\varphi)}{\partial t} = \left[ce^{i\varphi} + a(b-1)e^{-i\varphi}\right]\mathcal{M}(\varphi), \tag{7.61}$$

where the factor a can be set to one, by introducing $\sqrt{a}$ as the frequency unit. As any of the first-order linear differential equations, it has a general solution

$$\mathcal{M}(\varphi,t) = e^{i\varphi(2+b)}\left(e^{i\varphi} - x_-\right)^{A_2-A_1}\left(e^{i\varphi} - x_+\right)^{A_2+A_1}\mathcal{F}\left(\frac{e^{i\varphi}-x_-}{e^{i\varphi}-x_+}e^{t\sqrt{4a+\alpha^2}}\right), \tag{7.62}$$

found by the method of characteristics, where

$$x_{\pm} = \frac{i\alpha}{2} \pm \frac{i\sqrt{4+\alpha^2}}{2};\ A_1 = i\frac{1+b+c}{2\sqrt{4-\alpha^2}}\alpha;\ A_2 = \frac{c-3-b}{2}, \tag{7.63}$$

and $\mathcal{F}(x)$ is an arbitrary function.

The choice of $\mathcal{F}(x)$ is suggested by the initial condition $\mathcal{M}(\varphi, 0) = \sum \phi_n(t = 0)e^{i\varphi n}$

$$\mathcal{F}(\frac{e^{i\varphi} - x_-}{e^{i\varphi} - x_+}) = e^{-i\varphi(2+b)} \left(\frac{e^{i\varphi} - x_-}{e^{i\varphi} - x_+}\right)^{A_1} \frac{\mathcal{M}(\varphi, 0)}{\left[(e^{i\varphi} - x_-)(e^{i\varphi} - x_+)\right]^{A_2}}, \tag{7.64}$$

which after the substitution $e^{i\varphi} = (yx_- - x_+)/(y-1)$ yields

$$\mathcal{F}(y) = \frac{y^{-A_2-A_1}(y-1)^{c-1}}{(yx_- - x_+)^{b+2}(x_+ - x_-)^{2A_2}} \mathcal{M}\left(-i\ln\left(\frac{yx_- - x_+}{y-1}\right), 0\right). \tag{7.65}$$

Since any initial condition $\phi_n(t=0) = \delta_{n,n_0}$ after the replacements $b \to b+n_0$, $c \to c+n_0$ results in $\phi_n(t=0) = \delta_{n,0}$, we can consider only $\mathcal{M}(\varphi, 0) = 1$ and obtain

$$\phi_k(t) = \int_0^{2\pi} \frac{d\varphi}{2\pi} \frac{e^{(A_2-A_1)T - i\varphi(k-b-2)}}{(x_+ - x_-)^{2A_2}} \frac{\left(e^{i\varphi}(e^T - 1) - x_+e^T + x_-\right)^{c-1}}{\left(e^{i\varphi}(e^T x_- - x_+) + 1 - e^T\right)^{b+2}}, \tag{7.66}$$

where $T = t\sqrt{4a + \alpha^2}$. The other form of the same expression

$$\phi_k(t) = \int_0^{2\pi} d\varphi \frac{e^{-i\varphi k}}{2\pi} \frac{e^{(A_2-A_1)T}}{(x_+ - x_-)^{2A_2}} \frac{\left(x_- - x_+e^T\right)^{c-1}}{\left(e^T x_- - x_+\right)^{b+2}}$$
$$\left(\frac{e^{i\varphi}(e^T - 1)}{-x_+e^T + x_-} + 1\right)^{c-1} \left(1 + \frac{1 - e^T}{e^{i\varphi}(e^T x_- - x_+)}\right)^{-b-2}, \tag{7.67}$$

is more convenient for casting in the form of a power series over $\left(e^T - 1\right)$.

This results in

$$\psi_k(t) = \int_0^{2\pi} d\varphi \frac{e^{-i\varphi k}}{2\pi} \frac{e^{(A_2-A_1)T}}{(x_+ - x_-)^{2A_2}} \frac{\left(x_- - x_+e^T\right)^{c-1}}{\left(e^T x_- - x_+\right)^{b+2}}$$
$$\sum_{m=0}^{\infty} \frac{\Gamma(c)e^{im\varphi}}{\Gamma(c-m)m!} \left(\frac{e^T - 1}{x_- - x_+e^T}\right)^m$$
$$\sum_{n=0}^{\infty} \frac{\Gamma(-1-b)e^{-in\varphi}}{\Gamma(-1-b-n)n!} \left(\frac{1 - e^T}{e^T x_- - x_+}\right)^n, \tag{7.68}$$

where we have assumed that $b < 0, c > 0$. Integration yields $m = k + n$ and we are left with the known sum

$$\sum_{m=0}^{\infty} \frac{\Gamma(c)}{\Gamma(c-k-n)(-k-n)!} \frac{\Gamma(-1-b)}{\Gamma(-1-b-n)n!} x^n = \frac{\Gamma(c)}{k!\Gamma(c-k)} \, {}_2F_1\left(2+b, 1+k-c; k+1; x\right), \tag{7.69}$$

given in terms of hypergeometric function ${}_2F_1(\alpha, \beta; \gamma; z)$. Substitution of this sum yields for the probability amplitudes

$$\psi_k(t) = \frac{i^k e^{(A_2-A_1)T}}{(x_+ - x_-)^{2A_2}} \sqrt{\frac{\Gamma(c)\Gamma(k-b)}{\Gamma(c-k)\Gamma(-b)}} \frac{\left(x_- - x_+ e^T\right)^{c-1-k}}{\left(e^T x_- - x_+\right)^{b+2}} \frac{\left(e^T - 1\right)^k}{k!}$$
$${}_2F_1\left(2+b, 1+k-c; k+1; \frac{e^T - 1}{x_- - x_+ e^T} \frac{1 - e^T}{e^T x_- - x_+}\right). \tag{7.70}$$

Note that the transformation formulas for the hypergeometric function can give an expression, symmetric with respect to b and c, such that the imposed conditions $c > b$ can be lifted. Otherwise one has to make the replacements $b \rightleftharpoons c$, $x_- \rightleftharpoons x_+$. In the opposite case $\alpha < 1$, the population distribution remains localized near the initially populated state. Note that in order to get the correct result of the form (7.70) in the domain where $|V_n| \sim \sqrt{(n-b)(n-c)}$ assumes imaginary values, one has to make the replacements $\alpha \to i\alpha$ and $t \to it$. Also note that for the particular case of integer b and c, the problem describes the dynamics of an angular momentum L, with $2L + 1 = |b - c|$, in a constant magnetic field, and therefore corresponds to a $2L + 1$-dimensional representation of the SU_2 group.

In Fig. 7.4 we show two different regimes of the population dynamics corresponding to small detunings $\alpha < 1$ and large detunings $\alpha > 1$. In the first regime, the population initially localized at the level $k = 0$ moves toward infinite values of k in the direction of increasing couplings V_n. We note that many different wavepackets are formed during the time evolution, and these packets follow different trajectories.

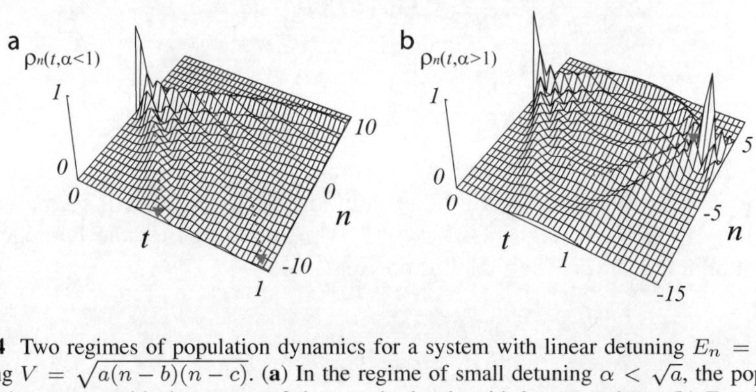

Fig. 7.4 Two regimes of population dynamics for a system with linear detuning $E_n = \alpha n$ and coupling $V = \sqrt{a(n-b)(n-c)}$. (**a**) In the regime of small detuning $\alpha < \sqrt{a}$, the population distribution moves with the course of time to the levels with large numbers. (**b**) For the large detunings $\alpha > \sqrt{a}$ the distribution oscillates around the initially populated state. *Arrows* show the directions of propagation of different wavepackets comprising the population distribution

7.2.4 Decreasing Coupling $|V_n| = 1/\sqrt{an+b}$

Consider the case of decreasing coupling. Calculating the integrals entering the exponents of (7.57), (7.59) one finds

$$\int^{\varphi} \frac{be^{ix} + b\varepsilon + e^{-ix}}{e^{ix} + \varepsilon} dx = b\varphi + \frac{ie^{-i\varphi}}{\varepsilon} - \frac{i}{\varepsilon^2} \ln\left(1 + \varepsilon e^{-i\varphi}\right), \tag{7.71}$$

and hence (7.57) yields a solution

$$\begin{aligned} \phi_k &= A \int_0^{2\pi} \frac{e^{-i\varphi(k+b)}}{(1+\varepsilon e^{-i\varphi})^{1+1/\varepsilon^2}} \exp\left\{\frac{e^{-i\varphi}}{\varepsilon}\right\} de^{i\varphi} \\ &= \varepsilon^k A' \int_C \frac{y^{(k+b)}}{(1+y)^{1+1/\varepsilon^2}} \exp\left\{\frac{y}{\varepsilon^2}\right\} de^{i\varphi} \end{aligned} \tag{7.72}$$

for the homogeneous equation (7.52). Here A and A' are constants, and the replacement $e^{i\varphi} \to y/\varepsilon$ has been performed in the last line. We have also set $a = 1$ by employing the scaled time $t \to t\sqrt{a}$ and frequency $\varepsilon \to \varepsilon/\sqrt{a}$.

Note that the integration contour C must not necessarily be a circle of radius ε around the point $y = 0$, as suggested by the replacement. On the contrary, any contour, such that the integrand takes identical values at his ends, can be chosen in the spirit of the Laplace contour integral method (5.137). In this case, one can show by integration by parts that (7.72) satisfies the homogeneous recurrence relation (7.52). For $k + b > 0$, among the possibilities is the contour of the type shown in Fig. 3.23b which goes around the points $y = 0$ and $y = -1$ and yields the confluent hypergeometric function

$$\begin{aligned} \chi_k = b \frac{\varepsilon^{k-1} e^{-1/\varepsilon^2} \Gamma(\frac{1}{\varepsilon^2} - b)}{\Gamma(1 - b - k + \frac{1}{\varepsilon^2})} \, {}_1F_1\left(-\frac{1}{\varepsilon^2}, 1 + b - \frac{1}{\varepsilon^2}; \frac{1}{\varepsilon^2}\right) \\ {}_1F_1\left(-k - b, 1 - k - b + \frac{1}{\varepsilon^2}; \frac{1}{\varepsilon^2}\right), \end{aligned} \tag{7.73}$$

where a convenient normalization has been done.

For $k + b < 0$ one finds the other solution by noting that the combination $\tilde{\chi}_k(b, \varepsilon) = (-i)^k \chi_{-k}\,(1 - b, i\varepsilon)\Gamma(k + b)/\Gamma(b)$ also satisfies the homogeneous recurrent relation (7.52). The substitution yields

$$\begin{aligned} \chi'_k = \frac{(1-b)}{(-\varepsilon)^{k-1}} e^{-1/\varepsilon^2} \frac{\Gamma\left(b - 1 - \frac{1}{\varepsilon^2}\right)}{\Gamma\left(b + k - \frac{1}{\varepsilon^2}\right)} \, {}_1F_1\left(\frac{1}{\varepsilon^2}, 2 - b + \frac{1}{\varepsilon^2}; \frac{1}{\varepsilon^2}\right) \\ \frac{\Gamma(k+b)}{\Gamma(b)} \, {}_1F_1\left(1 - \frac{1}{\varepsilon^2}, k + b - \frac{1}{\varepsilon^2}; \frac{1}{\varepsilon^2}\right), \end{aligned} \tag{7.74}$$

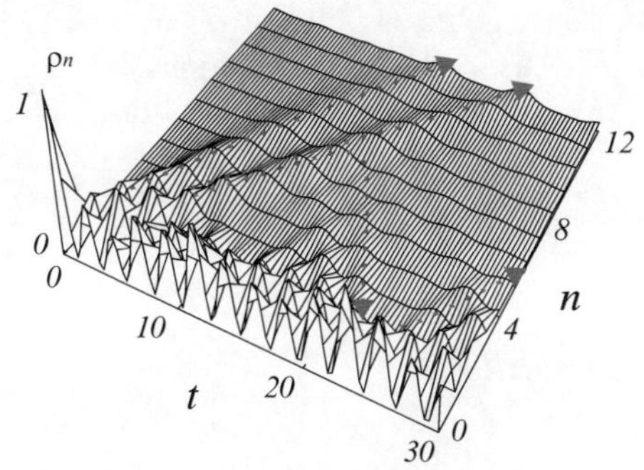

Fig. 7.5 Two regimes of population dynamics for a resonant system with a decreasing coupling $V = 1/\sqrt{a(n-b)}$. One sees that the interference of different non-commensurate harmonics results in involved population dynamics: a part of the population remains at the levels neighboring the initially populated state, whereas the rest of the population leaves toward states with large numbers n. The last part of the population distribution forms wavepackets, gradually losing their population with the course of propagation toward large n with a constant velocity. The *arrows* show the dynamics of the main wavepackets formed as a result of the interference

where the relation ${}_1F_1(\alpha,\gamma;x) = e^x\ {}_1F_1(\gamma-\alpha,\gamma;-x)$ has been employed. The normalization of χ_k and χ'_k has been chosen such that the combination

$$\phi_k = \chi_k\,\Theta(k) + \chi'_k\,\Theta(-k) \tag{7.75}$$

satisfies (7.52) with $\psi_0\,(t=0) = 1$. This enables us to avoid cumbersome calculations of the integrals suggested by the explicit expression (7.60). Time-dependent amplitudes have to be found by numerical calculation of the inverse Fourier transformation of (7.75).

It is worth mentioning that the k dependence of the solutions (7.73)–(7.74) coincides with the k dependence of the minors (7.49), and hence the determinant $\mathrm{Det}\left|\hat{H}-\varepsilon\right|$ in (7.41) behaves as $\chi'_{k\to\infty}$. For $|\varepsilon| < 1$ it vanishes at the points $\varepsilon = 1/\sqrt{b+n}$ of the complex ε-plane, where n is an integer. This gives the frequency spectrum of the system. Note that $1/\sqrt{b+n}$ coincides by accident with the Rabi frequencies of two-level systems composed of neighboring levels. In Fig. 7.5 one sees the interplay of these non commensurable eigenfrequencies resulting in a partial transfer of the population to the states with large numbers. The corresponding wavepackets, however, lose their population in the course of propagation at a constant velocity.

7.3 Smooth Variation of the Parameters

As we have seen, the exactly soluble problems correspond to certain subgroups of the full group of unitary transformations in the Hilbert space of the multilevel systems. This implies that the corresponding Hamiltonians in (7.1) consist of several additive parts that form an algebra of relatively small size, that is, commutators of these parts are given by their linear combinations. This is not always the case for systems with couplings V_n and energies E_n arbitrarily changing with the level number n. However, if these quantities change slowly, the commutators are small and can be ignored.

7.3.1 WKB Approximation

In this extreme, one can also consider the level number as a continuous coordinate, and write the Schrödinger equation (7.1) in the form

$$i\frac{\partial\psi(n)}{\partial t} = E(n)\psi(n)+V(n)\exp\left(-\tfrac{\partial}{\partial n}\right)\psi(n)+V^*(n+1)\exp\left(\tfrac{\partial}{\partial n}\right)\psi(n) \quad (7.76)$$

where the finite shift operator (5.12) has been employed. The generating function $\Psi(t,p) = \sum_n \psi_n(t)e^{ipn}$ of (7.5) may now be interpreted as the probability amplitude in the momentum representation, while the finite shift operator in this representation is multiplicative in nature: $\exp\left(\frac{\partial}{\partial n}\right) \to e^{ip}$.

Equation (7.76) can be interpreted as the Schrödinger equation of a particle with a non-quadratic relation between energy and momentum, moving in direction n. The momentum operator $\hat{p} = -i\frac{\partial}{\partial n}$ commutes neither with the analog of the potential energy $E(n)$ nor with the analog of the inverse mass $V(n)$, and the commutators given by the derivatives of these functions, generally nonlinear, do not commute with $\hat{p}$ either. However, when we consider the dimensionless mechanical action $S(n,t)$ given by the WKB substitution $\psi(n,t) = e^{-iS(n,t)}$ for slowly changing $E(n)$ and $V(n)$, the higher-order derivatives turn out to be small, as compared to the powers of the first derivative. In other words, higher-order commutators of the action with the momentum operator can be ignored, which yields

$$\frac{\partial S(n,t)}{\partial t} = E(n) + V(n)\exp\left(i\frac{\partial S(n,t)}{\partial n}\right) + V^*(n)\exp\left(\frac{\partial S(n,t)}{i\,\partial n}\right), \quad (7.77)$$

where the difference between $V^*(n+1)$ and $V^*(n)$ has also been ignored within the order of the approximation. In this equation, one recognizes the Hamilton–Jacobi equation for a classical particle with the Hamiltonian

$$\mathcal{H}(n,p) = E(n) + 2V(n)\cos p, \quad (7.78)$$

where $2V(n)\cos p$ plays the role of the kinetic energy. Let us first assume that the classical total energy of the system is ε. Then one immediately sees two regimes: (a) classically forbidden $|E(n) - \varepsilon| > |2V(n)|$, where the potential energy exceeds the kinetic energy, and (b) classically allowed, where $|E(n) - \varepsilon| < |2V(n)|$.

Substitution of $S(n,t) = \varepsilon t + S_m(n)$ into (7.78) yields the equation

$$\frac{\partial S_m(n)}{\partial n} = \arccos \frac{\varepsilon - E(n)}{2V(n)} \tag{7.79}$$

for the reduced action $S_m(n)$ which determines the eigenfunction $\psi(n,\varepsilon) = e^{-iS_m(n)}$ for the eigenstate of energy ε. This allows us to find $S_m(n)$ explicitly, and we obtain the WKB approximate result

$$\psi(n,t) = \int f(\varepsilon) \exp\left[-i\varepsilon t - i\int^{n} \arccos \frac{\varepsilon - E(x)}{2V(x)} dx\right] d\varepsilon \tag{7.80}$$

which has a clear physical meaning: the wavefunction is a linear superposition of the eigenfunctions of different energies ε with arbitrary amplitudes $f(\varepsilon)$. The latter have to be found from the initial conditions. Note that in the particular case $E(n) = const$; $V(n) = const$, considered on p. 298, the amplitude (7.80) gives an exact solution of the problem.

7.3.2 Position and Width of the Ehrenfest Wavepacket

Let us analyze some physical consequences of (7.80). Let us assume that initially the wave packet is localized near the point $x = x_0$ at zero energy. At long times, the main contribution to the integral over energies ε comes from the vicinity of the point $\varepsilon = 0$. We therefore can develop the classical action $S(n,t)$

$$S(n,t) = \varepsilon t + \int^{n} \arccos \frac{\varepsilon - E(x)}{2V(x)} dx = 0 \tag{7.81}$$

in a Taylor series near this point. This yields

$$S(n,t) = \int^{n} \arccos \frac{-E(x)}{2V(x)} dx$$
$$+\varepsilon\left(t - \int^{n} \frac{dx}{[4V(x)^2 - E(x)^2]^{1/2}}\right) + \frac{\varepsilon^2}{4}\int^{n} \frac{E(x)\, dx}{[4V(x)^2 - E(x)^2]^{3/2}}. \tag{7.82}$$

A relatively smooth function $f(\varepsilon)$ can be factored out of the integral (7.80) and replaced by its value at zero, and the integration with the allowance for (7.82) yields

$$\psi(n,t) = \frac{f(0)\sqrt{2\pi}}{\sqrt{\int^{n} \frac{E(x)\,dx}{[4V(x)^2 - E(x)^2]^{3/2}}}} \exp\left[-i\int^{n} \arccos\frac{-E(x)}{2V(x)}dx\right]$$
$$\exp\left[-\left(t + \int^{n} \frac{dx}{[4V(x)^2 - E(x)^2]^{1/2}}\right)^2 \Big/ \int^{n} \frac{E(x)\,dx}{[4V(x)^2 - E(x)^2]^{3/2}}\right]. \tag{7.83}$$

This means that a wavepacket passes the point n at time

$$t = \int^{n} \frac{dx}{[4V(x)^2 - E(x)^2]^{1/2}}, \tag{7.84}$$

at velocity

$$\frac{\partial n}{\partial t} = [4V(n)^2 - E(n)^2]^{1/2}, \tag{7.85}$$

and the passage takes time

$$\delta t = \int^{n} \frac{E(x)\,dx}{[4V(x)^2 - E(x)^2]^{3/2}}. \tag{7.86}$$

Note that the width δt is given by the integral diverging at the classical turning points where $4V(x)^2 = E(x)^2$, and the main contribution comes from the vicinity of this point. Therefore it depends strongly on the parameters of the last, classically available state, and sometimes it is necessary to replace the integrals by sums at the integer points $x = n$ near the turning points. Another important feature of the wavepacket duration follows from the fact that the integral (7.86) converges at the upper limit if the dependence $V(n)$ rises faster than $n^{1/3}$ in the classically accessible region $n \to \infty$. Therefore δt does not depend on the level number for large n.

We also note that if the combination $4V(n)^2 - E(n)^2$ is positive in the limit of large n and tends to infinity faster than n^2, the integral (7.84) also converges as $n \to \infty$, and therefore the center of the classical wavepacket reaches infinity at a finite time. In this regime, a level population with a Gaussian time-dependence does not imply a Gaussian distribution over the levels at a given time. On the contrary, this distribution has an essentially non-Gaussian character, being strongly asymmetric, since the higher the number n is, the higher is the local velocity of the populations. This asymmetry can already be seen for $V(x) \sim x^a$ for $1/2 < a < 1$, as shown in Fig. 7.6b.

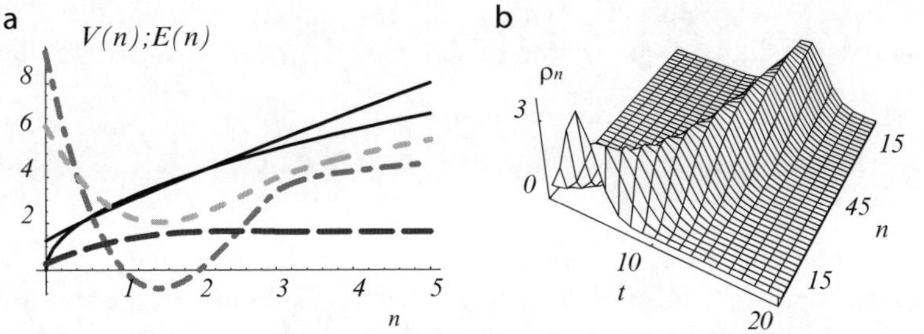

Fig. 7.6 Semiclassical analysis of the population distributions. (**a**) The energy dependence $E(n)$ increases at $n \to \infty$ slower, as compared to the coupling $V(n)$ (*solid lines*). The arrival time $t(n)$ (*dashed line*), the power index $S_m(n)$ of the tunneling probability $e^{-\mathrm{Im}S_m(n)}$ (*long dash line*), and the width $\delta t(n)$ (*dash-dot line*) are not shown to scale. (**b**) Population as a function of the level number n and time t for $V(n) = 0.42 + 0.53n^{0.9}$ and $E(n) = n^{1/2}$ as an example

7.3.3 The Tunneling Probability

The wavepacket cannot propagate in the classically forbidden region where $|E(n) - \varepsilon| > |2V(n)|$. The integrand of (7.79) assumes the imaginary values and the integral

$$\mathrm{Im}(S_m) = i \int\limits_{n_b}^{n_a} \arccos \frac{\varepsilon - E(x)}{2V(x)} dx. \tag{7.87}$$

between the points $n_{a,b}$ where $|E(n_{a,b}) - \varepsilon| = |2V(n_{a,b})|$, gives the probability amplitude $e^{-\mathrm{Im}(S_m)}$ for tunneling through the classically forbidden region. A more instructing form of this expression can be gained if we replace the arccosine function by the corresponding logarithm and the integral by a sum

$$\begin{aligned} e^{-\mathrm{Im}(S_m)} &= \exp\left[-i \int\limits_{n_b}^{n_a} \arccos \tfrac{\varepsilon - E(x)}{2V(x)} dx\right] \\ &= \exp\left\{\sum_{n=n_b}^{n_a} \ln\left[\left|\tfrac{\varepsilon - E(n)}{2V(n)}\right| - \sqrt{\left(\tfrac{\varepsilon - E(n)}{2V(n)}\right)^2 - 1}\right]\right\} \\ &= \prod_{n=n_b}^{n_a} \left|\frac{|\varepsilon - E(x)| - \sqrt{(\varepsilon - E(x))^2 - 4V(x)^2}}{2V(x)}\right| \end{aligned} \tag{7.88}$$

where $n_{a,b}$ now represents the first and the last integer points in the classically forbidden region. Expansion over the small ratio $2V(x)/(\varepsilon - E(x))$ gives the result

$$e^{-\mathrm{Im}(S_m)} = \prod_{n=n_b}^{n_a} \left| \frac{V(n)}{\varepsilon - E(n)} \right| \tag{7.89}$$

which coincides with the first non-vanishing order of perturbation theory. Indeed, as we have seen in (3.17), the amplitude ψ_1 of a level $|1\rangle$ strongly detuned by an energy $\Delta = \varepsilon - E(1)$ from the energy ε of the initially populated state $|0\rangle$ amounts at most to a fraction $V(1)/(\varepsilon - E(1))$ of the amplitude ψ_0. The level $|1\rangle$ serves, in turn, as a starting state for the subsequent set of two levels $|1\rangle$ and $|2\rangle$. Therefore, by analogy, $\psi_2 \sim V(1)V(2)/(\varepsilon - E(1))(\varepsilon - E(2))$, and repeating the same reasonings for all strongly detuned levels of the classically forbidden region, we arrive at (7.89).

7.3.4 Applicability of the WKB Approximation

We note that, contrary to the WKB analysis of the Schrödinger equation with a quadratic dependence of the kinetic energy on the momentum, the zero-order WKB approximation cannot be easily improved for the discrete level systems considered. The main reason for this is that a consistent equation for the corrections to the mechanical action (7.79) has the same structure as the original equation (7.76) and contains all-order derivatives. However, from WKB analysis of the regular Schrödinger equation, we know that this correction is important mainly near turning points where the integrals for action correction diverge. The turning points seldom coincide with the integer points, important for the ensembles of discrete levels. Therefore, cut-off or replacement of integrals diverging near the turning points by the corresponding sums may compensate to some extent for the impossibility of improving the zero-order approximation.

The other limitation of the WKB analysis has to do with the fact that, for the validity of the approach, the duration $\delta t(n)$ (7.86) of the wavepacket at a given state n should be much shorter compared to the arrival time $t(n)$ (7.84) to this level. This is not the case when the coupling increases faster than $\sim n$, when both quantities assume finite values. Neither is it the case for decreasing couplings, when the width increases faster with n than the arrival time. Therefore, the applicability of this approach is restricted to the classically accessible domains where $V(n) \sim n^a$ at $n \to \infty$ with $0 < a < 1$. It is worth mentioning that in the limit $a = 0$, that is, for constant coupling, the expression (7.80) is valid and, moreover, it yields solutions of the type (7.10), but the saddle-point approximation of the integrals is incorrect. The population distribution does not have the structure of a Gaussian wavepacket, although the front of this distribution indeed propagates with velocity (7.85).

7.4 Relay with Disordered Parameters

Efficient transport of population from a quantum level to another one, separated by a large interval in the relay-like sequence of coupled states, is possible only if both levels are equally represented in the large number of different eigenfunctions of the system. Formally this statement can be given in terms of the resolvent $1/(\varepsilon - \hat{H})$ (4.10) which is another name for Green's operator. In the basis of eigenfunctions $|e_k\rangle$ the Green's operator reads

$$\frac{1}{\varepsilon - \hat{H}} = \sum_k \frac{|e_k\rangle\langle e_k|}{\varepsilon - \varepsilon_k} \tag{7.90}$$

where ε_k denotes energies of the eigenstates, and the amplitude of the transition from the level $n_0 = 0$ to the level n adopts the form

$$\psi_n(t) = \int d\varepsilon\, e^{-i\varepsilon t} \sum_k \frac{\psi_n(\varepsilon_k)\psi_0^*(\varepsilon_k)}{\varepsilon - \varepsilon_k} \tag{7.91}$$

where $\psi_n(E_k) = \langle n | e_k \rangle$ is the amplitude of the level n in the eigenstate k. Comparison of (7.91) and (7.41) for $V_n = 1$ shows a useful correspondence:

$$\begin{aligned} \psi_n(\varepsilon_k)\psi_0^*(\varepsilon_k) &= \frac{M_n^+(\varepsilon = \varepsilon_k)M_0^-(\varepsilon = \varepsilon_k)}{\left(\partial \mathrm{Det}\left|\hat{H} - \varepsilon\right| / \partial\varepsilon\right)_{\varepsilon=\varepsilon_k}} \\ &= \frac{M_0^+(\varepsilon_k)M_0^-(\varepsilon_k)}{\left(\partial \mathrm{Det}\left|\hat{H} - \varepsilon\right| / \partial\varepsilon\right)_{\varepsilon=\varepsilon_k}} \prod_1^{l=n} u_l^+(\varepsilon_k), \end{aligned} \tag{7.92}$$

between the eigenfunctions $\psi_n(\varepsilon_k)$ and the minors $M_n^+(\varepsilon_k)$ of the matrix $\hat{H} - \varepsilon$, where now, in contrast to (7.39), the system is infinite on both sides, and $u_k^+ = M_k^+/M_{k-1}^+$ as earlier.

From (7.41) one immediately finds the expression

$$\begin{aligned} \rho_n(t) &= \int_{-\infty-i\nu}^{\infty-i\nu} \frac{d\xi}{2\pi i} \int_{-\infty+i\nu}^{\infty+i\nu} \frac{d\varepsilon}{2\pi i} e^{-i(\varepsilon-\xi)t} \left(\hat{H} - \varepsilon\right)^{-1}_{kn} \left(\hat{H} - \xi\right)^{-1}_{kn} \\ &= \int_{-\infty-i\nu}^{\infty-i\nu} \frac{d\xi}{2\pi i} \int_{-\infty+i\nu}^{\infty+i\nu} \frac{d\varepsilon}{2\pi i} e^{-i(\varepsilon-\xi)t} \frac{M_k^+(\varepsilon)M_n^-(\varepsilon)}{\mathrm{Det}\left|\hat{H}-\varepsilon\right|} \frac{M_k^+(\xi)M_n^-(\xi)}{\mathrm{Det}\left|\hat{H}-\xi\right|} \end{aligned} \tag{7.93}$$

for the populations $\rho_n(t) = |\psi_n(t)|^2$, which with the help of (7.90) can also be given in terms of the eigenfunctions $\psi_n(\varepsilon_k)$ of the Hamiltonian

$$\rho_n(t) = \int\int d\xi d\varepsilon\, e^{-i(\varepsilon-\xi)t} \sum_{k,k'} \frac{\psi_n(\varepsilon_k)\psi_n^*(\varepsilon_{k'})\psi_0^*(\varepsilon_k)\psi_0(\varepsilon_{k'})}{4\pi^2(\varepsilon-\varepsilon_k)(\xi-\varepsilon_{k'})}$$
$$= \int d\zeta\, e^{-i\zeta t} \sum_{k,k'} \frac{\psi_n(\varepsilon_k)\psi_n^*(\varepsilon_{k'})\psi_0^*(\varepsilon_k)\psi_0(\varepsilon_{k'})}{2\pi i(\zeta+\varepsilon_k-\varepsilon_{k'})}, \tag{7.94}$$

where, as earlier in (3.113), we have introduced the variables $\eta = (\varepsilon+\xi)/2$ and $\zeta = \varepsilon - \xi$ and have performed integration over $d\eta$. Now one sees that, when the product $\psi_n(\varepsilon_k)\psi_0^*(\varepsilon_k)$ of two amplitudes corresponding to the same energy ε_k of the eigenstate k in the numerator is small, the transition probability is small as well.

It turns out that for the vast majority of the system described by the three-term recurrence equation (7.2), the probability of transition from an initially populated state to another state at some distance Δn vanishes at large distances. Indeed, a typical realization of a relay-like system of interacting levels corresponds to the energies E_k irregularly changing with the level number n. Numerical analysis shows that the eigenfunctions of disordered systems are exponentially localized. This phenomenon is known as Anderson localization. More precisely, for almost all one-dimensional quantum systems, the amplitudes of eigenstates exponentially decrease when $|n| \to \infty$, whereas all of the exactly solved problems considered earlier in this section are exceptions, resulting from a special underlying group symmetry. In this subsection we give a semiqualitative description of the localization phenomenon, useful for an intuitive understanding of the dynamics of multilevel systems in the opposite limit of total absence of such a symmetry. The main tool of the analysis is the ensemble average which will be performed with a model distribution function of the states energies. The exactly soluble case, known as the Lloyd model, allows one to find the ensemble averaged resolvents for the Cauchy distribution of the level energies. More rigorous analysis based on the resolvents for populations requires a more sophisticated field theory technique whose basic elements are given in the next subsection.

It is worth mentioning that the Hamiltonian $\hat{H}_0 = E_n\delta_{n,n'}$ and the coupling $\hat{V} = \delta_{n+1,n'} + \delta_{n-1,n'}$ do not commute. Moreover, for the case of random E_n the higher-order commutators remain linearly independent. However, these commutators are not all of the same order of magnitude, since with the increasing order of commutation r, their sizes decrease like $r^{-1/2}$, demonstrating the dependence typical of a random process. That is the algebraic reason for the existence of the localization phenomenon considered in this section.

7.4.1 Ensemble Averaged Amplitudes and Corresponding Populations

We start with a quantity most convenient for calculation—the mean probability amplitudes $\langle\psi_n(t)\rangle$ for the system to be in a given level n. Following P. Lloyd, we assume that in an ensemble of systems (7.2), the level energies E_n have independent, Lorentzian (Cauchy) distributions

$$\tilde{g}(E_n) = \frac{\gamma/\pi}{(E_n - \Omega)^2 + \gamma^2}, \tag{7.95}$$

where $|V_n|^2 = 1$. This situation is shown in Fig. 7.7a.

The resolvent $1/(\varepsilon - \hat{H})$ of (7.2)

$$0 = (E_n - \varepsilon)\,\psi_n + \psi_{n-1} + \psi_{n+1}$$

given by the explicit expression (7.41) is a regular function in the lower part of complex planes of all E_n, since the Fourier variable ε is shifted to the upper part

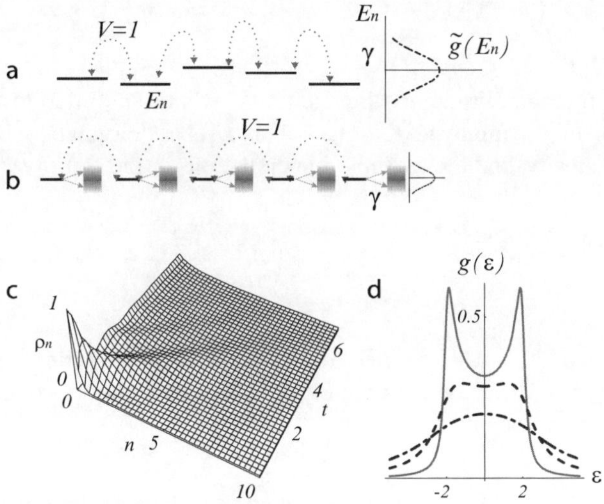

Fig. 7.7 Anderson localization in one dimension and its simplistic description by the mean Green's function. (**a**) A set of levels with the nearest neighbor coupling and disordered level positions given by the distribution function $\tilde{g}(E_n)$ of width γ suggested by (7.95). (**b**) Naive description assuming that the Lorentzian distribution of the level positions is equivalent to level decay to continua with a rate γ. (**c**) Distribution of the populations (7.100) suggested by the ensemble averaged resolvent. (**d**) Density of states of the disordered chain given by (7.103) for $\gamma = 0.2$ (*solid line*), $\gamma = 1$ (*dashed line*), $\gamma = 2.5$ (*dash-dot line*). For $\gamma = 0$ the density of states tends to infinity at the edges of the spectrum $\epsilon \to \pm 2$

of the complex plane by a vanishing value $\nu \to 0$, as shown in Fig. 3.2. Therefore calculation of the average with the distributions (7.95) can simply be done by taking the residuals at the points $E_n = \Omega - i\gamma$ which means the replacements $E_n \to \Omega - i\gamma$ in the resolvent. But such an average resolvent corresponds to the Schrödinger equation

$$i\frac{\partial \psi_n}{\partial t} = (\Omega - i\gamma)\,\psi_n + \psi_{n-1} + \psi_{n+1} \tag{7.96}$$

which after the replacement $\psi_n(t) \to \psi_n(t)e^{-\gamma t}$ coincides with (7.1) for $E_n = \Omega$, $V_n = 1$. With the help of the solution (7.11), for (7.96) we immediately find

$$\psi_n(t) = i^n e^{-\gamma t - i\Omega t} J_n(2\,|t|). \tag{7.97}$$

The ensemble averaged equation (7.96) coincides with the Schrödinger equation of a cascade of resonant levels, each of which is coupled to an independent uniform continuum, by analogy to the case of a single level coupled to a continuum considered in Sect. 3.2.2 as shown in Fig. 7.7b. Let us calculate the harmonics of the total "fluence"

$$\Phi_n(\varepsilon) = \gamma \int_0^\infty e^{i\varepsilon t}\psi_n(t)dt = i^n \left[4 + (\gamma + i\Omega - i\varepsilon)^2\right]^{-(n+1)/2} \tag{7.98}$$

of the population amplitude to the continua, where in order to perform the integration we have employed the Hankel integral representation of the Bessel function. We can also find the asymptotic populations of the continua

$$\tilde{\rho}_n = 2\gamma \int_0^\infty |\psi_n(t)|^2\, dt = \frac{\Gamma(n+\frac{1}{2})}{\gamma^{2n} n! \sqrt{\pi}}\, {}_2F_1\left(n + \frac{1}{2}, n + \frac{1}{2}; 1 + 2n; \frac{-4}{\gamma^2}\right) \tag{7.99}$$

that coincide with $|\Phi_n(\varepsilon)|^2$ integrated over the spectrum ε. In order to gain a deeper insight into the population dynamics in Fig. 7.7c we depict the sum of the population of the levels and corresponding continua

$$\rho_n(t) = |\psi_n(t)|^2 + 2\gamma \int_0^t |\psi_n(t)|^2\, dt \tag{7.100}$$

as a function of time, which shows how the asymptotic distribution (7.99) is attained. One sees that the population initially localized at level $n = 0$ spreads over the neighboring states with the course of time, until it reaches a steady state distribution, exponentially decreasing for $|n| \to \infty$.

7.4.2 *Ensemble Averaged Spectrum*

We can also find the ensemble averaged spectrum density of the eigenstates. To this end we note that the energy spectrum of a quantum system can be found from the resolvent with the help of the expression

$$g(\varepsilon) = \frac{-1}{\pi} \lim_{\nu\to 0} \mathrm{Im}\left(\mathrm{Tr}\frac{1}{\varepsilon - \hat{H} + i\nu}\right)$$
$$= \frac{-1}{\pi} \lim_{\nu\to 0} \mathrm{Im}\left(\sum_k \frac{\langle e_k\, | e_k\rangle}{\varepsilon - \varepsilon_k + i\nu}\right) = \sum_k \delta(\varepsilon - \varepsilon_k). \qquad (7.101)$$

The simplest way to calculate the average spectrum of a disordered chain is to express the Hamiltonian with the help of the momentum operator $p = -i\frac{\partial}{\partial n}$ which we have already employed when considering the classical limit (7.78)

$$\hat{H} = E_n + 2\cos\hat{p}, \qquad (7.102)$$

and obtain after integration over the energies E_n with the distributions (7.95)

$$g(\varepsilon) = \frac{-1}{\pi}\mathrm{Im}\left(\mathrm{Tr}\frac{1}{\varepsilon - \Omega + i\gamma - 2\cos\hat{p}}\right)$$
$$= \frac{-1}{\pi}\mathrm{Im}\int_{-\pi}^{\pi}\frac{dp}{\varepsilon - \Omega + i\gamma - 2\cos p}$$
$$= \mathrm{Re}\frac{2}{\sqrt{4 - (\varepsilon - \Omega + i\gamma)^2}}, \qquad (7.103)$$

where in the last line we have taken advantage of the invariance of the trace operator and have employed the momentum representation. One sees that this spectrum in Fig. 7.7d. It is given by the inverse of the combination $\sqrt{4 - (\varepsilon - \Omega + i\gamma)^2}$, the same as that which determines the amplitude fluence distribution (7.98).

7.4.3 *Distribution of the Amplitude Ratios*

Strictly speaking, the population distribution (7.100) calculated with the help of the ensemble averaged amplitudes has nothing to do with the ensemble averaged populations in the relay-like system with random positions of the levels. It just gives an "eye guiding line" and some parameters that might be of use for a consistent consideration, which is technically much more involved. Let us consider this phenomenon more carefully with the help of a different technique. From (7.2)

$$0 = (E_n - \varepsilon)\,\psi_n(\varepsilon) + \psi_{n-1}(\varepsilon) + \psi_{n+1}(\varepsilon)$$

and from the distribution of the level positions (7.95) we find distribution functions $g_n^{\pm}(u_n^{\pm})$ of the ratios

$$u_{n+1}^{+}(\varepsilon) = \psi_{n+1}(\varepsilon)/\psi_n(\varepsilon) \tag{7.104}$$

and

$$u_{n-1}^{-}(\varepsilon) = \psi_{n-1}(\varepsilon)/\psi_n(\varepsilon) \tag{7.105}$$

that satisfy (7.46)

$$\begin{aligned} u_{n+1}^{+} &= (\varepsilon - E_n) - \frac{1}{u_n^{+}}, \\ u_{n-1}^{-} &= (\varepsilon - E_n) - \frac{1}{u_n^{-}}. \end{aligned}$$

Note that we can also have in mind the ratios of minors of (7.92), as in the original form of (7.46).

We assume the distribution (7.95) for the level energies $\tilde{g}(E_n)$, and ask for the distributions $g_n^{\pm}(u_n^{\pm})$ of the ratios $u_n^{\pm}$. Due to the physical meaning of the probability distribution, the functions $g_n^{\pm}(x)$ have to be real and positive at any complex value of the argument. For the sake of simplicity, let us consider only real x. With the help of (7.46) we arrive at the recurrence equation

$$\begin{aligned} g_{n+1}^{+}(u_{n+1}^{+}) &= \int_{-\infty}^{\infty}\int_{-\infty}^{\infty} \delta\left[u_{n+1}^{+} - \varepsilon + E_n + \tfrac{1}{u_n^{+}}\right] g_n^{+}(u_n^{+})\tilde{g}(E_n)\, dE_n du_n^{+} \\ g_{n+1}^{+}(x) &= \int_{-\infty}^{\infty} \frac{\gamma/\pi}{(x - \varepsilon + 1/y + \Omega)^2 + \gamma^2}\, g_n^{+}(y) dy, \end{aligned} \tag{7.106}$$

where in the last line we have replaced the integration variable u_n^{+} by y and u_{n+1}^{+} by x. A similar equation holds for $g_n^{-}(u_n^{-})$.

The functional equation (7.106) may have many different complicated solutions. We will consider the simplest solution which has the structure

$$g_n^{+}(y) = \frac{\gamma_n/\pi}{(y - \omega_n)^2 + \gamma_n^2}. \tag{7.107}$$

In other words, we assume that distribution functions for different n have the same functional structure, and differ only by the positions ω_n and width γ_n of the single spike. Substitution of (7.107) into (7.106), followed by the integration around the poles in the upper part of the complex plane, yields

$$g^{\pm}_{n+1}(y) = \frac{\gamma_{n+1}/\pi}{(y-\omega_{n+1})^2 + \gamma^2_{n+1}}, \tag{7.108}$$

where

$$\begin{aligned} \gamma_{n+1} &= \frac{\gamma_n}{\omega_n^2 + \gamma_n^2} + \gamma, \\ \omega_{n+1} &= \frac{-\omega_n}{\omega_n^2 + \gamma_n^2} + \Omega + \varepsilon. \end{aligned} \tag{7.109}$$

The last equations can be written in a more compact form

$$z_{n+1} = \frac{-1}{z_n} + Z, \tag{7.110}$$

when we introduce complex numbers $z_n = \omega_n + i\gamma_n$ and $Z = \Omega - \varepsilon + i\gamma$. For $g^-_n(y)$ one obtains a similar relation

$$z_{n-1} = \frac{-1}{z_n} + Z. \tag{7.111}$$

Equations (7.110), (7.111) map the complex plane z to itself. There is a clear geometric meaning, as shown in Fig. 7.8. The transformation $-1/z$ corresponds

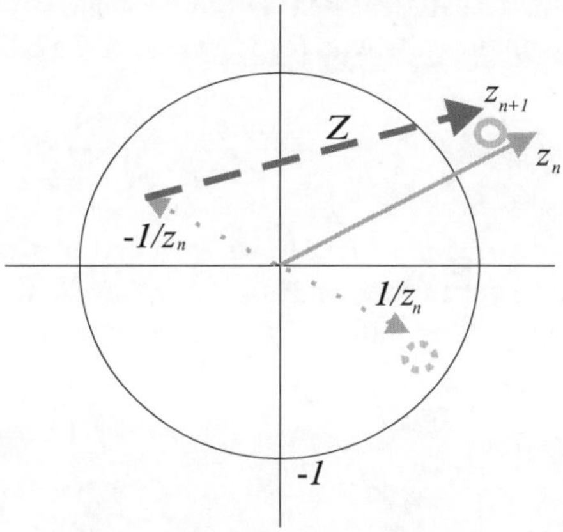

Fig. 7.8 Transformation of the complex plane given by (7.110). A complex number z_n (shown by the *solid vector*) after taking the inverse value $1/z_n$, change of sign (*dotted vectors*), and after displacement by the complex number Z (*dashed vector*), becomes a complex number z_{n+1}. The stable stationary point of this transformation (center of the *small solid circle*) is located outside of the unit circle, whereas the inverse value corresponds to the unstable stationary point (center of the *dotted circle*)

to inversion of a complex point z with respect to the unit circle, followed by the sign inversion of the real part, whereas addition of the complex number Z means displacement of this point. This mapping has two stationary points, given by the solution of the algebraic equation

$$z = \frac{-1}{z} + Z, \tag{7.112}$$

that corresponds to the situation where displacement of the point as a result of inversion and reflection with respect to the imaginary axis exactly equals $-Z$. The solution of (7.16) suggests

$$z_{\pm} = \frac{Z}{2} \pm \sqrt{\frac{Z^2}{4} - 1}. \tag{7.113}$$

The stationary point z_+ with $|z_+| > 1$ is stable for (7.110), whereas $|z_-| < 1$ corresponds to an unstable stationary point z_-. This follows from the fact that inversion to the inner part of a circle is a compressing mapping. For (7.111), the situation is the opposite. Note that the indices $\pm$ of $z_\pm$ do not always correspond to the sign of the square root in (7.113), which for complex Z may depend on the choice of the square root branch, and hence the condition $|z_+| > 1$ should be checked.

The presence of stationary points means that, after several iterations, any initial Lorentzian distribution of the ratios (7.104), (7.105) results in a Lorentzian distribution of width $\mathrm{Im}\ z_\pm$ around the points $\mathrm{Re}\ z_\pm$. Moreover, the linearity of (7.106) implies that any distribution that can be represented as a sum of Lorentzians adopts this form. Therefore, any analytical distribution that has an arbitrary number of simple poles, placed symmetrically with respect to the real axis, becomes Lorentzian

$$g_n^{\pm}(u_n^{\pm}) = \frac{\mathrm{Im}\ z_+/\pi}{(u_n^{\pm} - \mathrm{Re}\ z_+)^2 + \mathrm{Im}\ z_+^2}. \tag{7.114}$$

With the help of the distribution (7.107) we can find a mean value which will play an important role in our further analysis. We calculate the mean $\log(u_n^{\pm})$ and concentrate on its real part. Straightforward integration yields

$$\begin{aligned} \langle \ln |u_n^{\pm}(\varepsilon)| \rangle &= \frac{1}{\pi} \int_{-\infty}^{\infty} du_n^{\pm} \frac{\ln |u_n^{\pm}(\varepsilon)|\, \mathrm{Im}\ z_+}{\left(u_n^{\pm} - \mathrm{Re}\ z_+\right)^2 + (\mathrm{Im}\ z_+)^2} \\ &= \ln |z_+(\varepsilon)| = \ln \left| \tfrac{Z}{2} \pm \sqrt{\tfrac{Z}{4}^2 - 1} \right| \end{aligned} \tag{7.115}$$

where the sign of the square root should be chosen such that $|z_+(\varepsilon)| > 1$. Substitution of $Z = 2 \cosh X$ yields another form of this expression

$$\ln|z_+(\varepsilon)| = |\text{Re}X| = \left|\text{Re}(\text{arc}\cosh\tfrac{Z}{2})\right|$$
$$= \text{arc}\cosh\left[\tfrac{1}{2}\left|1-\tfrac{Z}{2}\right| + \tfrac{1}{2}\left|1+\tfrac{Z}{2}\right|\right]. \tag{7.116}$$

Note that the mean values (7.116) of both $\log u_n^+$ and $\log u_n^-$ are positive, although the product $u_n^+ u_n^-$ is equal to unity for any specific realization of the system, as follows from the definitions (7.104)–(7.105). There is no contradiction in this fact, since (7.2) for a generic ε has two solutions—one exponentially increasing for $n \to \infty$ and the other exponentially increasing for $n \to -\infty$. Only when ε equals an eigenvalues ε_k of a specific realization of the random system, do these two solutions match at some point n, but only for this specific realization. At a fixed ε, this does not happen for a typical realization in the ensemble, and that is why $\langle\ln|u_n^+|\rangle + \langle\ln|u_n^-|\rangle \neq \langle\ln|u_n^+ u_n^-|\rangle$ but tend to the largest stationary values.

Equation (7.115) gives the idea of the typical behavior of the eigenstate amplitudes $\psi_n(\varepsilon_k)$ as a function of the level number n. These functions are typically growing with the growth rate $\log|z_+(\varepsilon_k)|$ when the number n comes from $\pm\infty$ to a certain finite number $n_{\max}(\varepsilon_k)$ where, for a particular realization of the random system, these functions match the three-term recurrence relation. This exponential behavior is typical of all eigenfunctions and of any particular realization of the Hamiltonian, whereas the precise value of the energy ε_k and specific position $n_{\max}(\varepsilon_k)$ are sensitive to the particular realization.

In Fig. 7.9a we show three typical examples of the eigenvector amplitudes, corresponding to different energies of the eigenstates. One clearly sees the

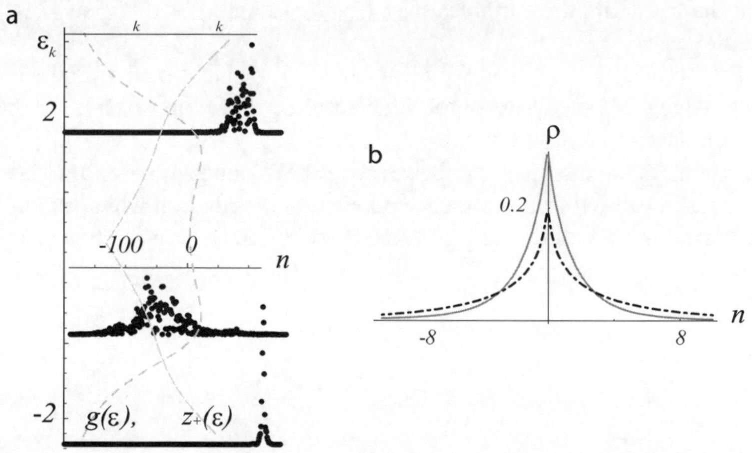

Fig. 7.9 Anderson localization in 1D systems. (**a**) Moduli of eigenvectors (*dots*, not to scale) as functions of the level number n are depicted for three values of eigenenergies ϵ_k for $\gamma = 0.8$. The values of ϵ_k serve as reference levels of the corresponding plots. The mean state density $g(\epsilon_k)$ (*dashed line*) of (7.103) and mean ratios of neighboring amplitudes $z_+(\epsilon_k)$ (*long dash line*) of (7.116) are also shown, but not to scale. (**b**) "Naive" distribution of the populations (7.100) (*solid line*) compared to the population distribution (7.124) (*dash-dot line*) for $\gamma = 0.15$

exponential decrease of the amplitudes. The ensemble averaged density of states and the mean ratios of the neighboring amplitudes are also shown as a function of energy.

7.4.4 Distribution of the Populations for Long Times

Let us now consider the population distribution (7.94) which has the form of a product of two mutually complex conjugate probability amplitudes. After the ensemble average, the sums over the eigenenergies ε_k and $\varepsilon_{k'}$ can be replaced by integrals over their distributions $g(\varepsilon_k, \varepsilon_{k'})$ which can always be represented as a sum $g(\varepsilon_k, \varepsilon_{k'}) = g(\varepsilon_k)g(\varepsilon_{k'}) + g_2(\varepsilon_k, \varepsilon_{k'})$ of the independent binary distribution $g(\varepsilon_k)g(\varepsilon_{k'})$, given by (7.103), and a correlation term $g_2(\varepsilon_k, \varepsilon_{k'})$. This yields

$$
\begin{aligned}
\rho_n(t) = & \int\int e^{-i\varepsilon t} g(\varepsilon_k) \frac{\psi_n(\varepsilon_k)\psi_0^*(\varepsilon_k)}{2\pi i(\varepsilon - \varepsilon_k)} d\varepsilon_k d\varepsilon \\
& \times \int\int e^{i\xi t} g(\varepsilon_{k'}) \frac{\psi_n^*(\varepsilon_{k'})\psi_0(\varepsilon_{k'})}{-2\pi i(\xi - \varepsilon_{k'})} d\varepsilon_{k'} d\xi \\
& + \int\int\int g_2(\varepsilon_k, \varepsilon_{k'})\, e^{-i\zeta t} \frac{\psi_n(\varepsilon_k)\psi_n^*(\varepsilon_{k'})\psi_0^*(\varepsilon_k)\psi_0(\varepsilon_{k'})}{2\pi i(\zeta + \varepsilon_k - \varepsilon_{k'})} d\zeta d\varepsilon_k d\varepsilon_{k'}
\end{aligned}
\tag{7.117}
$$

where in the last term, the variables $\eta = (\varepsilon + \xi)/2$ and $\zeta = \varepsilon - \xi$ of (7.94) have been employed.

The first term of (7.117) has the clear physical meaning of the product of two ensemble averaged Green's operators that result in the term $|\psi_n(t)|^2$ of (7.100), whereas the last term allows for the correction $\tilde{\rho}_n(t)$ which in (7.100) has been replaced by the heuristic term $2\gamma \int_0^t |\psi_n(t)|^2 dt$. We concentrate on this last term, since it is responsible for the stationary population distribution which is attained in the long-time limit. With the help of (7.104) and (7.105) we write this term in the form

$$
\tilde{\rho}_n(t) = \int\int\int \frac{|\psi_0(\varepsilon_k)|^2 |\psi_0(\varepsilon_{k'})|^2}{2\pi i(\zeta + \varepsilon_k - \varepsilon_{k'})} \frac{e^{-i\zeta t} g_2(\varepsilon_k, \varepsilon_{k'})\, d\zeta d\varepsilon_k d\varepsilon_{k'}}{\prod_{m=1}^{n-1} u_m^-(\varepsilon_k) \left(u_m^-(\varepsilon_{k'})\right)^*}, \tag{7.118}
$$

which shows that, as $t \to \infty$, the dominating time-independent contribution comes from $k = k'$. This yields

$$
\tilde{\rho}_n|_{t\to\infty} = \int |\psi_0(\varepsilon_k)|^4\, g_2(\varepsilon_k, \varepsilon_k) \prod_{m=1}^{n-1} \left|u_m^-(\varepsilon_k)\right|^{-2} d\varepsilon_k, \tag{7.119}
$$

whereas the time-dependent part reads

$$\tilde{\rho}_n(t) - \tilde{\rho}_n|_{t\to\infty} = \int g_n^{(2)}(\zeta)e^{-i\zeta t}d\zeta, \tag{7.120}$$

where the function

$$g_n^{(2)}(\zeta) = \int dx \frac{\left|\psi_0(x+\frac{\zeta}{2})\right|^2 \left|\psi_0(x-\frac{\zeta}{2})\right|^2 g_2\left(x+\frac{\zeta}{2}, x-\frac{\zeta}{2}\right)}{\prod\limits_{m=1}^{n-1} u_m^-(x+\frac{\zeta}{2})\left(u_m^-(x-\frac{\zeta}{2})\right)^*} \tag{7.121}$$

has the physical meaning of the correlation of the energy distance between eigenstates, weighted by the correlations of the probability amplitudes to be in the initial and the final levels corresponding to these states. Note that the joint distribution $g_2(x,x')$ of two eigenvalues is discontinuous at $x = x'$, assuming at this point the dependence (7.103) of a single eigenvalue distribution $g(x)$. Therefore, from the integral (7.120) we have to exclude the point $\zeta = 0$ which is included in the part (7.119) with $g_2(\varepsilon_k, \varepsilon_k) = g(\varepsilon_k)N$, where N is the total number of levels.

Consider first the stationary distribution (7.119). According to (7.115)

$$\prod_{m=1}^{n-1} \left|u_m^-(\varepsilon_k)\right|^{-2} = e^{-2\sum_{m=1}^{n-1}\ln\left|u_m^-(\varepsilon_k)\right|} \simeq \left|z_+(\varepsilon_k)\right|^{-2|n|} \tag{7.122}$$

since the sum of many randomly distributed logarithms can be replaced by their mean value multiplied by the total number of terms. Therefore

$$\tilde{\rho}_n|_{t\to\infty} \simeq \int |\psi_0(x)|^4 \, Ng(x)\, |z_+(x)|^{-2|n|}\, dx, \tag{7.123}$$

where the procedure of taking the average of $|\psi_0(x)|^4$ absorbs the deviations of sums of logarithms in (7.122) from their mean values $\left\langle |\psi_0(x)|^4 \right\rangle = \left\langle |\psi_0(x)|^4 \prod_{m=1}^{n-1} |z_+(x)/u_m^-(\varepsilon_k)|^2 \right\rangle$ and should be done carefully. To this end, let us take into account the normalization $\sum_n \tilde{\rho}_n|_{t\to\infty} = 1$ of the population distribution and establish with the help of this condition a relation between the average $\left\langle |\psi_0(x)|^2 \right\rangle$ and $|z_+(x)|$. If $\left\langle |\psi_0(x)|^2 \right\rangle$ entering the product $\left\langle |\psi_0(x)|^2 \right\rangle |z_+(x)|^{-2n}$ absorbs the deviations $\sum_{m=1}^{n-1} \ln |u_m^-(\varepsilon_k)/z_+(x)|^2$, then the product equals the mean population $\tilde{\rho}_n = \left\langle |\psi_0(x)|^2 \prod_{m=1}^{n-1} |u_m^-(\varepsilon_k)|^{-2} \right\rangle$ of the level n corresponding to a given eigenenergy x. Therefore the sum $\sum_n \left\langle |\psi_0(x)|^2 \right\rangle |z_+(x)|^{-2n}$ equals 1, whereas the average of the remaining factor $|\psi_0(x)|^2$ in (7.123) is uniform and equals the inverse total number N of levels in the system, thus normalizing the distribution function of the eigenvalues $Ng(x)$ to unity. This yields the stationary ensemble averaged population distribution

$$\tilde{\rho}_n|_{t\to\infty} \simeq \int \frac{g(x)\,|z_+(x)|^{-2|n|}}{\sum_{k=-\infty}^{\infty} |z_+(x)|^{-2|k|}} dx = \int \frac{1-|z_+(x)|^{-2}}{1+|z_+(x)|^{-2}} g(x)\,|z_+(x)|^{-2|n|}\,dx \tag{7.124}$$

which is systematically slightly broader than our first "naive" guess (7.99), as if the distribution (7.124) results from a dynamical process which takes into account partial recurrences from the continua shown in Fig. 7.7b. In Fig. 7.9b we depict both the "naive" guess and the distribution (7.124) calculated with the help of (7.103) and (7.113).

7.4.5 Dynamics of the Asymptotic Populations

We consider next the dynamics of populations at long t. According to (7.120), the population approaches its stationary distribution (7.124) following a time-dependence given by the Fourier transformation of the correlator (7.121). It can also be given in the form

$$g_n^{(2)}(\zeta) \simeq \int dx \frac{\left|\psi_0(x+\frac{\zeta}{2})\right|^2 \left|\psi_0(x-\frac{\zeta}{2})\right|^2 g_2\left(x+\frac{\zeta}{2}, x-\frac{\zeta}{2}\right)}{|z_+(x)|^{2n-2}}, \tag{7.125}$$

where, as earlier, we have replaced the amplitude ratios by their mean values, and have ignored the difference between $u_m^-(x+\frac{\zeta}{2})$ and $u_m^-(x-\frac{\zeta}{2})$ in the limit $\zeta \to 0$. Indeed, in the limit of long times, the most important domain corresponds to $\zeta \sim 0$. Therefore the main contribution arises from close eigenenergies in the system. The integrand (7.125) can be interpreted as the spectral correlation g_2 weighted by the joint probability

$$w = \left|\psi_0(x+\frac{\zeta}{2})\right|^2 \left|\psi_n(x-\frac{\zeta}{2})\right|^2$$

to be at the levels $|0\rangle$ and $|n\rangle$ corresponding to energy eigenstates $\left|x+\frac{\zeta}{2}\right\rangle$ and $\left|x-\frac{\zeta}{2}\right\rangle$, respectively. In a large, disordered system, the probability $g_2(x+\frac{\zeta}{2}, x-\frac{\zeta}{2})$ to have two close eigenvalues, $x+\zeta/2$ and $x-\zeta/2$, is high. However, typically these states are localized in different domains of n, and therefore the weight factor w is exponentially small for most of x. On the other hand, the eigenstates localized in the same domain of n and having large w, typically correspond to distinct eigenenergies. The probability of finding close eigenvalues increases linearly with the increasing size of the considered domain of n, whereas their contribution to the integral (7.125) drops exponentially because of the decreasing weight factors w. The interplay between these two tendencies forms the asymptotic spectral dependence $g_n^{(2)}(\zeta)$ as $\zeta \to 0$.

Let us discuss a possible form of this asymptotic dependence. On the axis of n, consider a long interval of N successive levels symmetrically placed around the point $n = 0$. The density of states in the energy domain x amounts to $Ng(x)$ and the mean spacing between the neighboring energy eigenstates localized in this domain *is* $\delta \sim 1/Ng(x)$. Smaller spacings in larger intervals result from the eigenstates localized in the domain with $n > N/2$. Therefore, in order to have the mean level spacing smaller than ζ, one has to consider levels at long distance with $n > 1/2g(x)\,|\zeta|$. But the overall contribution of these levels to the integral (7.125) is exponentially small, and can be estimated as a double sum,

$$C(x,\zeta) = \sum_{k,k'>1/2g(x)\zeta} A^2 e^{-\ln|z_+(x)|(k+k')} = \frac{A^2 e^{-2\ln|z_+(x)|/g(x)\zeta}}{\left(1-|z_+(x)|^{-2}\right)^2}$$
$$= \left(1+|z_+(x)|^{-2}\right)^2 \exp\left[-\frac{\ln|z_+(x)|}{g(x)\,|\zeta|}\right], \tag{7.126}$$

where the normalization $A = \left(1-|z_+(x)|^{-2}\right)/\left(1+|z_+(x)|^{-2}\right)$, similar to that in (7.124), has been taken into account.

The derivative $\partial/\partial\zeta$ of the total contribution $C(x,\zeta)$ of all resonances closer than ζ gives the weighted correlation $wg_n^{(2)}(x,\zeta)$ for a given x and hence

$$g_n^{(2)}(\zeta) \simeq \int dx \frac{\left(1+|z_+(x)|^{-2}\right)^2}{|z_+(x)|^{2n+2}} \frac{\ln|z_+(x)|}{g(x)\zeta^2} \exp\left[-\frac{\ln|z_+(x)|}{g(x)\,|\zeta|}\right]. \tag{7.127}$$

Therefore, for the time-dependence (7.120), by performing the inverse Fourier transformation with the help of (7.103) and (7.113) one obtains

$$\tilde{\rho}_n(t) - \tilde{\rho}_n|_{t\to\infty} = \int dx \frac{\left(1+|z_+(x)|^{-2}\right)^2}{|z_+(x)|^{2n+2}} \frac{\ln|z_+(x)|}{g(x)} \int_0^\infty \exp\left[-\frac{\ln|z_+(x)|}{g(x)|\zeta|}\right] \cos(\zeta t)\, \frac{d\zeta}{\zeta^2}, \tag{7.128}$$

and to be consistent within the order of approximation, after exact representation of the last integral

$$\tilde{\rho}_n(t) - \tilde{\rho}_n|_{t\to\infty} = \int dx \frac{\left(1+|z_+(x)|^{-2}\right)^2}{|z_+(x)|^{2n+2}} 2\sqrt{t\frac{\log|z_+(x)|}{g(x)}}\,\mathrm{Re}K_1\left[2\sqrt{it\frac{\log|z_+(x)|}{g(x)}}\right], \tag{7.129}$$

in terms of the Bessel function $K_1(x)$, one has to take the asymptotic of long times which implies $t/ng(x=0) \gg 1$.

Equation (7.129) yields the dying oscillations at a frequency decreasing in time as $t^{-1/2}$ and of an amplitude changing as $t^{-1/4}\exp\left[-\mathrm{const}\sqrt{t}\right]$. In order to obtain the last dependence we have taken into account that the main contribution to the integral (7.129) over x at $t \to \infty$ comes from the band center, that is, from the domain $x \sim 0$ where $|z_+(x)| - |z_+(0)| \propto |x|$, as one sees in Fig. 7.9a. The localization length $L = 1/2\log|z_+(0)|$ assumes in this spectral region the longest possible value. In Sect. 3.4, we have already seen that strongly non-homogeneous bands may result in non-exponential decay laws. The $e^{-\sqrt{t}}$ dependence obtained here indicates once again the fact that the evolution of a disordered system may differ considerably from the standard exponential relaxation, typical of continua. The common feature of such systems is an infinitely dense but not continuous spectrum, where the eigenfunctions and related matrix elements of physical operators are discontinuous at every point and may change considerably, by orders of magnitudes, with infinitesimal changes of the energy.

7.5 Field Theory Method for Disordered Systems

Considering in Sect. 7.2.1, the Anderson localization for a one-dimensional disordered relay of levels, we have employed the Cauchy distribution (7.95) of the level energies. This particular type of ensemble has enabled us to determine explicitly the distribution (7.106) of the neighboring amplitude ratios, which has been further employed for the analysis of the population dynamics. The analysis, however, relied on qualitative reasonings rather than on straightforward calculations of the physical quantities of interest. Though this approach helps to develop an intuitive vision of the problem, one needs another, based on straightforward calculations that would leave no doubts about the results obtained. It turns out that this goal is achieved with the help of a Gaussian distribution of the level energies

$$\tilde{g}(E_n) = \frac{1}{\sqrt{\pi}\gamma}\exp\frac{-\left(E_n - \Omega\right)^2}{\gamma^2}. \tag{7.130}$$

The required technique for this calculation is well known in field theory. In this subsection we discuss it, trying to keep the narration close (as much as possible) to the manner of presentation of the rest of the text.

7.5.1 *Tunneling Transparency and Classical Bosonic Fields*

We concentrate first on the simplest problem, and consider the tunneling transparency of a disordered chain. Equation (7.41) yields the probability

$$\rho_N(t) = \int\limits_{-\infty-i\nu}^{\infty-i\nu}\frac{d\xi}{2\pi i}\int\limits_{-\infty+i\nu}^{\infty+i\nu}\frac{d\varepsilon}{2\pi i}\frac{\exp\left[-it(\varepsilon-\xi)\right]}{\mathrm{Det}\left|\hat{H}-\varepsilon\right|\mathrm{Det}\left|\hat{H}-\xi\right|} \tag{7.131}$$

to find a particle at the last level N of the chain at a time t, provided that at $t = 0$ it has been at the first level. The Hamiltonian

$$\hat{H} = \left\| \begin{matrix} E_1 & 1 & 0 & . & 0 & 0 & 0 \\ 1 & E_2 & 1 & . & 0 & 0 & 0 \\ 0 & 1 & E_3 & . & 0 & 0 & 0 \\ . & . & . & . & . & . & . \\ 0 & 0 & 0 & . & E_{N-2} & 1 & 0 \\ 0 & 0 & 0 & . & 1 & E_{N-1} & 1 \\ 0 & 0 & 0 & . & 0 & 1 & E_N \end{matrix} \right\| \tag{7.132}$$

contains level energies E_n distributed according to (7.130). An explicit expression for the multidimensional Gaussian integral

$$\prod_{n=1}^{N} \left\{ \int_{-\infty}^{\infty} \frac{dx_n}{\sqrt{\pi}} \right\} \exp\left[-i \sum_{n,m=1}^{N} x_n a_{nm} x_m \right] = \frac{1}{\sqrt{\mathrm{Det}\, |a_{nm}|}}, \tag{7.133}$$

valid for any matrix a_{nm} with eigenvalues in the lower part of the complex plane, suggests a practical way to evaluate the ensemble averaged population (7.131). Indeed, taking into account the signs of the imaginary parts of ε and ξ for $a_{nm} = H_{nm} - \delta_{n,m}\varepsilon$ (or ξ), one finds

$$\begin{aligned} \rho_N(t) = & \int_{-\infty-i\nu}^{\infty-i\nu} \frac{d\xi}{2\pi} \int_{-\infty+i\nu}^{\infty+i\nu} \frac{d\varepsilon}{2\pi} \prod_{n=1}^{N} \left\{ \int_{-\infty}^{\infty} \int_{-\infty}^{\infty} \int_{-\infty}^{\infty} \int_{-\infty}^{\infty} \frac{dx_n d\overline{x}_n dy_n d\overline{y}_n}{\pi^2} \right\} \\ & \exp\left[-it(\varepsilon - \xi) - 2i \sum_{n=1}^{N-1} (x_n x_{n+1} + \overline{x}_n \overline{x}_{n+1} - y_n y_{n+1} - \overline{y}_n \overline{y}_{n+1}) \right. \\ & \left. -i \sum_{n=1}^{N} \left((E_n - \varepsilon)(x_n^2 + \overline{x}_n^2) - (E_n - \xi)(y_n^2 + \overline{y}_n^2) \right) \right], \end{aligned} \tag{7.134}$$

which after the ensemble average with (7.130) yields

$$\begin{aligned} \rho_N(t) = & \int_{-\infty-i\nu}^{\infty-i\nu} \frac{d\xi}{2\pi} \int_{-\infty+i\nu}^{\infty+i\nu} \frac{d\varepsilon}{2\pi} \prod_{n=1}^{N} \left\{ \int_{-\infty}^{\infty} \int_{-\infty}^{\infty} \int_{-\infty}^{\infty} \int_{-\infty}^{\infty} \frac{dx_n d\overline{x}_n dy_n d\overline{y}_n}{\pi^2} \right\} \\ & \times \exp\Big[-it(\varepsilon - \xi) - 2i \sum_{n=1}^{N-1} (x_n x_{n+1} + y_n y_{n+1} - \overline{x}_n \overline{x}_{n+1} - \overline{y}_n \overline{y}_{n+1}) \\ & + \sum_{n=1}^{N} \Big(i\varepsilon(x_n^2 + \overline{x}_n^2) - i\xi(y_n^2 + \overline{y}_n^2) - \frac{\gamma^2}{4}(x_n^2 + \overline{x}_n^2 - y_n^2 - \overline{y}_n^2)^2 \Big) \Big]. \end{aligned} \tag{7.135}$$

These integrals are known in field theory as functional integrals on a grid and can be viewed as integrals over all trajectories $\overrightarrow{R}(n)$, given by all of the possible dependencies of the four-dimensional radius vector $\overrightarrow{R} = (x, \overline{x}, y, \overline{y})$ on the fictitious "time" n that take discrete, integer, values. Note that one can also consider this four-dimensional real space as two-dimensional complex space of the variables $z = x + i\overline{x}$ and $z' = y + i\overline{y}$ for which instead of (7.133) one has

$$\prod_{n=1}^{N} \left\{ \int_{-\infty}^{\infty} \frac{dz_n^* dz_n}{\pi} \right\} \exp\left[-i \sum_{n,m=1}^{N} z_n a_{nm} z_m^* \right] = \frac{1}{\mathrm{Det}\, |a_{nm}|}, \qquad (7.136)$$

which implies integration over all complex planes of each variable. One can also consider the complex plane z as the phase space of a classical oscillator, where the square of the absolute value $|z_n|^2$ and the phase $\mathrm{Im} \ln z_n$ play the roles of action and phase, respectively.

An Analogy Between the Feynman Path Integral for a Boson and an Inverse Determinant

Prior to a detailed consideration of the integral (7.135), let us dwell on a simpler integral (7.133) for the case of $a_{nm} = H_{nm} - \delta_{n,m}\varepsilon$ and H_{nm} of (7.132) with $E_n \equiv 0$, that is, on the many-fold integral

$$\begin{aligned} I_N(\varepsilon) &= \prod_{n=1}^{N} \left\{ \int_{-\infty}^{\infty} \tfrac{dx_n}{\sqrt{\pi}} \right\} \exp\left[-2i \sum_{n=1}^{N-1} x_n x_{n+1} + i \sum_{n=1}^{N} \varepsilon x_n^2 \right] \\ &= \prod_{n=1}^{N} \left\{ \int_{-\infty}^{\infty} \tfrac{dx_n}{\sqrt{\pi}} \right\} \exp\left[i \sum_{n=0}^{N} (x_n - x_{n+1})^2 + (\varepsilon - 2) x_n^2 \right], \end{aligned} \qquad (7.137)$$

where in the last line we have assumed $x_0 = x_{N+1} = 0$.

We trace an analogy between this expression and the Feynman path integral for a quantum particle. Indeed, considering n as a continuous "time" variable and replacing finite differences by the first derivatives and the sums by the integrals, we obtain

$$I_N \simeq \int \exp\left\{ i \int_0^{N+1} \left[\left(\frac{\partial x(n)}{\partial n} \right)^2 + (\varepsilon - 2) x^2(n) \right] dn \right\} \mathcal{D}\, x(n), \qquad (7.138)$$

with

$$x(N+1) = x(0) = 0. \qquad (7.139)$$

This is a standard path integral representation for the probability amplitude of a quantum particle with a classical Lagrangian

$$L\left(x, \frac{\partial x}{\partial n}\right) = \left(\frac{\partial x}{\partial n}\right)^2 + (\varepsilon - 2)x^2 \tag{7.140}$$

to remain in the initial position $x = 0$ after a "time" $N + 1$. This Lagrangian corresponds to a particle of mass $m = 2$ moving in a parabolic potential $(2 - \varepsilon)x^2$, whereas (7.138) simply represents the integral of $\exp(-iS)$ for the action $S = \int L(x, \frac{\partial x}{\partial n})dn$ over all the trajectories starting at $x = 0$ at $n = 0$ and ending at the same point at $n = N + 1$. The corresponding Hamiltonian

$$\hat{H} = -\frac{1}{4}\left(\frac{\partial}{\partial x}\right)^2 + (2 - \varepsilon)x^2 \tag{7.141}$$

coincides with the Hamiltonian of a one-dimensional quantum harmonic oscillator of a "frequency" $\sqrt{4 - 2\varepsilon}$. Note that the "time"-dependent Hamiltonian

$$\hat{H} = \frac{1}{4}\hat{p}^2 + (2 - \varepsilon)x^2 \sum_{k=0}^{\infty} \delta(n - k) \tag{7.142}$$

corresponds to the original integral (7.137) with the finite difference in the action sum. It employs the trajectories on a one-dimensional grid of integer numbers, as shown in Fig. 7.10. The parabolic dependence of the Hamiltonian (7.142) on the coordinate and momentum, suggests the assumption that the dynamics of the system is closely related to the dynamics of a harmonic oscillator.

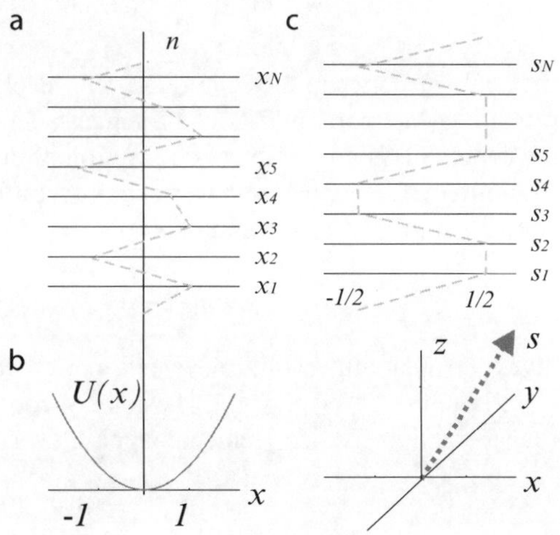

Fig. 7.10 The transmission probability amplitude for a set of N levels, coupled by relay-like interaction, can be given in terms of Feynman trajectories on a discrete grid (**a**) and the corresponding potential curve (**b**) for a particle of mass $m = 2$ which has probability amplitude to be in the initial state $x = 0$ at the fictitious "time" $n = N$. Calculation of the population also requires integration over classical trajectories of spin 1/2 (**c**)

Bosonic Field, a Quantum Oscillator and Mapping Over a Period

Earlier we have mentioned that the possibility to find an exact solution of (7.22) for a harmonic oscillator subject to an arbitrary Raman field results from an underlying algebraic symmetry, summarized in (7.23). The same symmetry also exists for a harmonic oscillator with a time-dependent frequency, where the Hamiltonian (7.142) can be rewritten in terms of bosonic creation and annihilation operators $\hat{a}^\dagger = c\hat{x} + if\hat{p}$ and $\hat{a} = c^*\hat{x} - if^*\hat{p}$:

$$\hat{H} = -\frac{1}{16f^2}\left(\hat{a} - \hat{a}^\dagger\right)^2 + \frac{(2-\varepsilon)}{4c^2}\left(\hat{a} + \hat{a}^\dagger\right)^2 \sum_{k=0}^{\infty} \delta(n-k), \tag{7.143}$$

in a form similar to the Hamiltonian in (7.21). A particular choice of the constants c and f, constrained by the condition $\left[\hat{a}, \hat{a}^\dagger\right] = 1$, is a matter of convenience. The symmetry manifests itself in the independence of the oscillation period on the oscillator amplitude, and is responsible for another important feature: the action S corresponding to the Lagrangian (7.140) identically equals zero for all classical trajectories.

Though from the viewpoint of calculation, the path integral representation (7.137) with the allowance for (7.139) adds nothing new to the exact expression (7.133),

$$I_N(\varepsilon) = \prod_{n=1}^{N}\left\{\int_{-\infty}^{\infty} \frac{dx_n}{\sqrt{\pi}}\right\} \exp\left\{i\sum_{n=0}^{N}\left[(x_n - x_{n+1})^2 + (\varepsilon - 2)x_n^2\right]\right\}$$

$$= \mathrm{Det}^{-1/2}\begin{Vmatrix} -\varepsilon & 1 & . & 0 \\ 1 & -\varepsilon & . & 0 \\ . & . & . & . \\ 0 & 0 & . & -\varepsilon \end{Vmatrix} = \frac{1}{\sqrt{\cos\left[N \arccos(-\varepsilon)\right]}}, \tag{7.144}$$

the analogy between the relay-like multilevel system and a free particle subject to periodic δ-like parabolic "kicks", suggests an additional intuitive insight, useful for the choice of an approximate description of disordered chains.

Note in this context that since integration with the kernel $e^{i(x_n - x_{n+1})^2}$ is equivalent to the transformation $\exp\left\{-i\frac{1}{4}\hat{p}^2\right\}$, and hence the operator

$$\hat{U} = \exp\left\{-i(2-\varepsilon)x^2\right\}\exp\left\{-i\tfrac{1}{4}\hat{p}^2\right\} \tag{7.145}$$

gives the quantum evolution over a period for the particle, corresponding to the Hamiltonian (7.143). For the Heisenberg coordinate and momentum operators, the transformation (7.145) yields a mapping over a period

$$\hat{U}\hat{p}\hat{U}^{-1} = \hat{p} - 2(2-\varepsilon)\hat{x}$$
$$\hat{U}\hat{x}\hat{U}^{-1} = (-1+\varepsilon)\hat{x} + \frac{1}{2}\hat{p}, \tag{7.146}$$

which can also be written in the matrix form

$$\begin{pmatrix} \hat{p} \\ \hat{x} \end{pmatrix}_{n+1} = \begin{pmatrix} 1 & -2(2-\varepsilon) \\ \frac{1}{2} & -1+\varepsilon \end{pmatrix} \begin{pmatrix} \hat{p} \\ \hat{x} \end{pmatrix}_{n}. \tag{7.147}$$

The transposed matrix gives the change of the coefficients r and s

$$\begin{pmatrix} r \\ s \end{pmatrix}_{n+1} = \begin{pmatrix} 1 & \frac{1}{2} \\ -2(2-\varepsilon) & -1+\varepsilon \end{pmatrix} \begin{pmatrix} r \\ s \end{pmatrix}_{n}. \tag{7.148}$$

for a linear combination $r\hat{p} + s\hat{x}$, as a result of the mapping. This 2×2 matrix has two eigenvalues $\lambda_{1,2} = \frac{\varepsilon}{2} \pm \sqrt{\frac{\varepsilon^2}{4} - 1}$, such that $|\lambda_{1,2}| = 1$ for $|\varepsilon| < 2$, and the determinant of the matrix is equal to unity. For $\varepsilon = 2\sin\phi$; $\pi/2 < \phi < 3\pi/2$ one has $\lambda_{1,2} = \exp\left[\mp i(\phi - \pi/2)\right]$. The corresponding eigenvectors

$$\hat{a} = \frac{1}{\sqrt{-\cos\phi}} \left[\hat{p}\frac{e^{i(\phi+\pi/2)/2}}{2} + 2\hat{x}\sin\left(\frac{\phi}{2} - \frac{\pi}{4}\right)\right]$$
$$\hat{a}^{\dagger} = \frac{1}{\sqrt{-\cos\phi}} \left[\hat{p}\frac{e^{-i(\phi+\pi/2)/2}}{2} + 2\hat{x}\sin\left(\frac{\phi}{2} - \frac{\pi}{4}\right)\right] \tag{7.149}$$

of the mapping (7.147) suggest a convenient choice of the bosonic creation and annihilation operators encountered in (7.143).

Mapping of $\hat{a}$ and $\hat{a}^{\dagger}$ is given by a simple multiplication $\hat{a} = -i\exp(i\phi)\,\hat{a}$, $\hat{a}^{\dagger} = i\exp(-i\phi)\,\hat{a}^{\dagger}$, whereas the state $|0\rangle$ corresponds for a positive ϕ to the wavefunction

$$\langle x\,|0\rangle = \left|\frac{2\cos\phi}{\pi}\right|^{1/4} e^{ix^2(1-i\exp(-i\phi))}, \tag{7.150}$$

which satisfies the relation $\hat{a}\,|0\rangle = 0$, thus being the ground state of the effective Hamiltonian $i\ln\hat{U}$ of (7.145). In terms of the operators $\hat{a}$ and $\hat{a}^{\dagger}$ the operator $\hat{U}$ of evolution over a period (7.145) adopts the form

$$\hat{U} = \exp\left\{-i(\phi - \frac{\pi}{2})\left(\hat{a}\dagger\hat{a} + \frac{1}{2}\right)\right\}. \tag{7.151}$$

For $\varepsilon = 2\sin\phi$; $\pi/2 < \phi < 3\pi/2$ one obtains the complex conjugate amplitudes (7.150) and inverse evolution. We note an explicit expression

$$\hat{x} = \frac{1}{2\sqrt{-\cos\phi}}\left(e^{i\tau}\hat{a}^{\dagger} + e^{-i\tau}\hat{a}\right),$$
$$\tau = \phi/2 + \pi/4 \tag{7.152}$$

for the coordinate operator in terms of the creation and the annihilation operators (7.149), where the coefficients in front of $\hat{a}^\dagger$ and $\hat{a}$ are just mutually conjugated phase factors.

The Tunneling Transparency as Propagation of Four Bosons

The ensemble averaged transparency (7.135) can also be considered as a path integral for a "particle", by analogy to (7.138), although this particle now moves in a four-dimensional space and discrete time, having negative values of mass for motion in the directions $\overline{x}$ and $\overline{y}$, whereas the problem itself does not have "oscillator" symmetry. The corresponding Hamiltonian

$$\hat{H} = \frac{\hat{p}_x^2}{4} + \frac{\hat{p}_{\overline{x}}^2}{4} - \frac{\hat{p}_y^2}{4} - \frac{\hat{p}_{\overline{y}}^2}{4} + \frac{1}{4}\left[i\gamma^2(x^2+\overline{x}^2-y^2-\overline{y}^2)^2\right.$$
$$\left.-4(2-\xi)(y^2+\overline{y}^2)+4(2-\varepsilon)(x^2+\overline{x}^2)\right]\sum_{k=0}^{\infty}\delta(n-k) \quad (7.153)$$

no longer has the same symmetry as (7.142), containing a bi-quadratic imaginary part of the "particle's" potential energy. This imaginary part results in an irreversible "decay", which implies that the rate of transition from the first to the last (N-th) level of the chain decreases exponentially with the "time" N. The probability amplitude for the "particle" to be at the origin $x = y = \overline{x} = \overline{y} = 0$ at "time" N, in the presence of the decay, gives, after inverse Fourier transformation, the population of the last level of the disordered chain.

Since at $\gamma = 0$, (7.153) is a sum of four independent Hamiltonians (7.142), for $\gamma \neq 0$, we can consider the problem in the interaction representation by just replacing coordinates in the interaction term $i\gamma^2(x^2+\overline{x}^2-y^2-\overline{y}^2)^2/4$ by the unperturbed Heisenberg operators of these coordinates, that is,

$$\hat{H}_{int} = \sum_{k=0}^{\infty}\frac{i\gamma^2}{4}(x_H^2+\overline{x}_H^2-y_H^2-\overline{y}_H^2)^2\delta(n-k). \quad (7.154)$$

According to (7.152) and the mapping relations on p. 339 for the creation and annihilation operators, the Heisenberg operators of coordinates have the form

$$\hat{x}_H = \frac{1}{2\sqrt{-\cos\phi}}\left(e^{i\tau}\left(-ie^{i\phi}\right)^n\hat{a}^\dagger + e^{-i\tau}\left(ie^{-i\phi}\right)^n\hat{a}\right),$$
$$\hat{\overline{x}}_H = \frac{1}{2\sqrt{-\cos\phi}}\left(e^{i\tau}\left(-ie^{i\phi}\right)^n\hat{b}^\dagger + e^{-i\tau}\left(ie^{-i\phi}\right)^n\hat{b}\right),$$
$$\hat{y}_H = \frac{1}{2\sqrt{-\cos\varphi}}\left(e^{i\tau'}\left(ie^{-i\varphi}\right)^n\hat{\overline{a}}^\dagger + e^{-i\tau'}\left(-ie^{i\varphi}\right)^n\hat{\overline{a}}\right),$$
$$\hat{\overline{y}}_H = \frac{1}{2\sqrt{-\cos\varphi}}\left(e^{i\tau'}\left(ie^{-i\varphi}\right)^n\hat{\overline{b}}^\dagger + e^{-i\tau'}\left(-ie^{i\varphi}\right)^n\hat{\overline{b}}\right), \quad (7.155)$$

with $\xi = 2\sin\varphi$; $-\pi/2 < \varphi < \pi/2$, which results from the mapping over n periods of all of the bosonic operators $\hat{a}$, $\hat{a}^\dagger$, $\hat{b}$, $\hat{b}^\dagger$, $\hat{\overline{a}}$, $\hat{\overline{a}}^\dagger$, $\hat{\overline{b}}$, and $\hat{\overline{b}}^\dagger$, corresponding to all four dimensions. The phase factor $e^{i\tau'}$, implicitly given by (7.149), coincides with $e^{i\tau}$ after replacing ϕ by φ and complex conjugation.

We concentrate on the case of $\gamma \ll 1$, since the opposite extreme corresponds to the case when the neighboring levels in the cascade are strongly off resonance and therefore their population is small, according to (7.89). In this limit we can consider an average of the Hamiltonian (7.154) over a large number of periods

$$\hat{H}_{int} = \frac{i\gamma^2}{4}\left\langle (x_H^2 + \overline{x}_H^2 - y_H^2 - \overline{y}_H^2)^2 \right\rangle_n \tag{7.156}$$

neglect all the terms oscillating with "time", that is, the terms dependent on ϕn or φn, and concentrate only on the smallest eigenvalue of the remaining Hamiltonian, which corresponds to the vacuum state $|0\rangle$ of all four oscillators. One determines this value by making use of the explicit forms

$$\begin{aligned}
\hat{x}_H^2 &= \tfrac{1}{-4\cos\phi}\left(e^{2i\tau}\left(-ie^{i\phi}\right)^{2n}\hat{a}^{\dagger 2} + e^{-2i\tau}\left(ie^{-i\phi}\right)^{2n}\hat{a}^2 + \{\hat{a}^\dagger\hat{a}\}\right), \\
\left\langle \hat{x}_H^2 \right\rangle_n &= \tfrac{1}{-4\cos\phi}\{\hat{a}^\dagger\hat{a}\} \stackrel{\langle 0|..|0\rangle}{\to} \tfrac{1}{-4\cos\phi} \\
\left\langle \hat{\overline{x}}_H^4 \right\rangle_n &= \tfrac{1}{16\cos^2\phi}\left(\{\hat{a}^{\dagger 2}\hat{a}^2\} + \{\hat{a}^\dagger\hat{a}\}^2\right) \stackrel{\langle 0|..|0\rangle}{\to} \tfrac{3}{16\cos^2\phi},
\end{aligned} \tag{7.157}$$

of the Heisenberg coordinate operators and their averages in the ground state. Substituting into (7.156) the explicit mean values (7.157) and similar expressions for the other averaged operators, we find the smallest eigenvalue $i\gamma^2(\cos^{-2}\phi + \cos^{-2}\varphi - \cos^{-1}\varphi\cos^{-1}\phi)/8 \sim i\gamma^2/8\cos\phi\cos\varphi$ which gives the slowest rate of "decay" with the course of "time" n. It therefore determines the asymptotic behavior at the limit of long chain $N \gg 1$.

The ground state dominates in the expression for the evolution operator in the interaction representation

$$\hat{U}_{int}(n) \simeq |0\rangle\, e^{-\gamma^2 n/8\cos\phi\cos\varphi}\,\langle 0|\,, \tag{7.158}$$

and hence for the original problem we find an expression

$$\rho_N(t) = \int_{-\infty-i\nu}^{\infty-i\nu}\frac{d\xi}{2\pi}\int_{-\infty+i\nu}^{\infty+i\nu}\frac{d\varepsilon}{2\pi}e^{-it(\varepsilon-\xi)-\gamma^2 N/8\cos\phi\cos\varphi}\left|\frac{2\cos\varphi}{\pi}\frac{2\cos\phi}{\pi}\right|^{1/2}$$
$$\mathrm{Det}^{-1}\left\|\begin{matrix} -\varepsilon & 1 & . & 0 \\ 1 & -\varepsilon & . & 0 \\ . & . & . & . \\ 0 & 0 & . & -\varepsilon' \end{matrix}\right\| \mathrm{Det}^{-1}\left\|\begin{matrix} -\xi & 1 & . & 0 \\ 1 & -\xi & . & 0 \\ . & . & . & . \\ 0 & 0 & . & -\xi' \end{matrix}\right\|, \tag{7.159}$$

where one recognizes the exponent corresponding to the evolution operator (7.158) in the interaction representation, the factors $\frac{2\cos\varphi}{\pi}$ and $\frac{2\cos\phi}{\pi}$ that according to (7.150) allow for the projection of the states $\langle 0|$ to the point $x_0 = y_0 = \overline{x}_0 = \overline{y}_0 = 0$ which is the origin of the trajectories, and the determinants that describe the free evolution in agreement with (7.144). The last diagonal terms $\varepsilon' = 1+\varepsilon-i\cos\phi$ and $\xi' = 1+\xi+i\cos\varphi$ in these determinants come from the projection of the evolution operators for unperturbed motion to the slowest decaying state $|0\rangle$, instead of setting $x_{N+1} = y_{N+1} = \overline{x}_{N+1} = \overline{y}_{N+1} = 0$ which modifies the integral (7.137). Note that $\varepsilon = 2\sin\phi$; $\xi = 2\sin\varphi$ as earlier.

Equation (7.159) has a clear physical meaning: the population of the last level of a disordered chain of N levels has almost the same time behavior as that of the infinite system of resonant levels depicted in Fig. 7.1a for $V = 1$, becoming different from zero once the front of the population distribution $|\psi_n(t)|^2 = J_n^2(2t)$ of (7.11) approaches the level $n = N$ at the time $t \sim N/2$. The disorder just results in attenuation of this distribution by the exponentially small factor $\sim e^{-\gamma^2 N/8}$ by analogy to the case of (7.97).

For the case of small γ, let us consider the long-time limit $\varepsilon \to \xi$ of the tunneling probability of a specific domain of spectrum at a given energy $\eta = (\varepsilon+\xi)/2$ corresponding to a certain value of the density of states $g(\eta)$. For the case of the Lorentzian distribution (7.95) of the level energies, the density $g(\eta)$ coincides with $g(\varepsilon)$, given by (7.103) and shown in Fig. 7.7d. The mean density of states for the Gaussian distribution (7.130) is clearly identical to (7.103) for small γ, and can be written in the form $g(\eta) = 1/\cos\left(\arcsin(\eta/2)\right)$. Therefore, in terms of the state density, the smallest eigenvalue of $\hat{H}_{int}$ is $i\gamma^2 g^2(\eta)/8$, which takes the maximum value at the edges of the spectrum $\eta \to \pm 2$ where $g(\eta) \to \infty$. The maximum transparency comes from the center of the spectrum $\eta \to 0$, which is the spectrum domain responsible for the overall transparency $\sim e^{-\gamma^2 N/8}$. This is consistent with the dependence of the localization length on the energy depicted in Fig. 7.9a which clearly shows that the localization length assumes the minimum values at the spectrum edges.

7.5.2 An Analogy with the Liouville Equation

The different signs of the x and y variables in (7.153), and the imaginary interaction (7.154), suggest the employment of the density matrix formalism and the Liouville equation instead of the Hamiltonian formalism and the Schrödinger equation, as a more suitable technique. Indeed, let us consider the combination

$$\tilde{\rho}_n(x_n, \overline{x}_n; y_n, \overline{y}_n) = \prod_{k=1}^{n-1} \left\{ \int_{-\infty}^{\infty}\int_{-\infty}^{\infty}\int_{-\infty}^{\infty}\int_{-\infty}^{\infty} \frac{dx_k d\overline{x}_k dy_k d\overline{y}_k}{\pi^2} \right\}$$
$$\exp\left[-2i\sum_{k=1}^{n-1}(x_k x_{k+1} + \overline{x}_k\overline{x}_{k+1} - y_k y_{k+1} - \overline{y}_k\overline{y}_{k+1})\right.$$

$$\sum_{k=1}^{n}\left(i\eta(x_k^2+\overline{x}_k^2-y_k^2-\overline{y}_k^2)+i\zeta(x_k^2+\overline{x}_k^2+y_k^2+\overline{y}_k^2)\right)$$
$$\left.-\tfrac{\gamma^2}{4}\sum_{k=1}^{n}(x_k^2+\overline{x}_k^2-y_k^2-\overline{y}_k^2)^2\right], \tag{7.160}$$

entering (7.135), where we take the part of the integrand dependent only on n first sets of variables and perform integrations over all variables except the last four: $x_n, \overline{x}_n; y_n$, and $\overline{y}_n$. We identify this combination with the "density matrix", evolving in the course of fictitious "time" n, which we mark by the tilde sign in order to distinguish between this fictitious object and the real density matrix. For this object one finds the recurrence relation

$$\tilde{\rho}_{n+1}(x,\overline{x};y,\overline{y})=\hat{U}(x)\hat{U}(\overline{x})\tilde{\rho}_n(x,\overline{x};y,\overline{y})\hat{U}^*(y)\hat{U}^*(\overline{y})$$
$$\exp\left[i\zeta(x^2+\overline{x}^2+y^2+\overline{y}^2)-\tfrac{\gamma^2}{4}(x^2+\overline{x}^2-y^2-\overline{y}^2)^2\right] \tag{7.161}$$

with

$$\hat{U}(x)=\exp\left\{-i(2-\eta)x^2\right\}\exp\left\{\frac{-i}{4}\hat{p}_x^2\right\} \tag{7.162}$$

by analogy to (7.145).

In the interaction representation

$$\tilde{\rho}_n(x,\overline{x};y,\overline{y})=\left(\hat{U}(x)\hat{U}(\overline{x})\right)^n\tilde{\rho}_n^{(int)}(x,\overline{x};y,\overline{y})\left(\hat{U}^*(y)\hat{U}^*(\overline{y})\right)^n$$

and for small γ and ζ one finds by a Taylor expansion in γ^2

$$\tilde{\rho}_{n+1}^{(int)}(x,\overline{x};y,\overline{y})-\tilde{\rho}_n^{(int)}(x,\overline{x};y,\overline{y})=-\tfrac{\gamma^2}{2}\hat{Z}_H^2\tilde{\rho}_n^{(int)}(x,\overline{x};y,\overline{y})\widehat{Z'}_H^2$$
$$+\left(i\zeta\hat{Z}_H^2-\tfrac{\gamma^2}{4}\hat{Z}_H^4\right)\tilde{\rho}_n^{(int)}(x,\overline{x};y,\overline{y})+\tilde{\rho}_n^{(int)}(x,\overline{x};y,\overline{y})\left(i\zeta\widehat{Z'}_H^2-\tfrac{\gamma^2}{4}\widehat{Z'}_H^4\right) \tag{7.163}$$

where $\hat{Z}_H^2=(\hat{x}_H^2+\hat{\overline{x}}_H^2)$, $\widehat{Z'}_H^2=(\hat{y}_H^2+\hat{\overline{y}}_H^2)$ are the left and right Heisenberg operators, respectively. Equations (7.155) and (7.157) specify these operators and their powers in terms of the bosonic creation and annihilation operators (7.149), where ε or ξ should be replaced by η. To be consistent within the approximation of small γ and ζ, we should also replace the finite difference with the "time" derivative $\partial/\partial n$ and take averages of the operators over many periods. This yields

$$\frac{\partial\tilde{\rho}_n^{(int)}}{\partial n}=i\zeta\left\{\hat{H}_o,\tilde{\rho}_n^{(int)}\right\}-\tfrac{\gamma^2}{4}\left[\hat{H}_o,\left[\hat{H}_o,\tilde{\rho}_n^{(int)}\right]\right]$$
$$-\tfrac{\gamma^2}{4}\left[\hat{A}_2^\dagger,\left[\hat{A}_2,\tilde{\rho}_n^{(int)}\right]\right]-\tfrac{\gamma^2}{4}\left[\hat{A}_2,\left[\hat{A}_2^\dagger,\tilde{\rho}_n^{(int)}\right]\right], \tag{7.164}$$

where $[\,,]$ and $\{,\}$ denote commutator and anticommutator, respectively, and

$$\hat{H}_o = \omega(\hat{a}^{\dagger}\hat{a} + \hat{b}^{\dagger}\hat{b} + 1);\; \hat{A}_2^{\dagger} = \omega(\hat{a}^{\dagger 2} + \hat{b}^{\dagger 2});$$
$$\hat{A}_2 = \omega(\hat{a}^2 + \hat{b}^2);\; \omega = 1/4\cos(\arcsin\eta/2)\,. \tag{7.165}$$

The operators b and $b^{\dagger}$ correspond here to the coordinates with bars.

Equation (7.164) resembles the Liouville equation, where the relaxation term governing the evolution of the density matrix in the interaction representation originates from the ensemble averaged disorder γ^2. It is an analog of the relaxation operator for the regular density matrix of an open quantum system, originating from random time-dependent external interventions into the coherent dynamics, which we will consider later on. Note that after substitution of the explicit expressions for the commutators and the anticommutator, the relaxation operator contains three different components $\hat{R}_{ll}, \hat{R}_{rr}$, and $\hat{R}_{lr}$ that act on $\tilde{\rho}_n^{(int)}$ from the left-hand side, from the right-hand side, and from both sides, respectively. These relaxation operators are linear, although their action on the density matrix cannot always be written as a matrix multiplication as was discussed in the context of (5.58). We therefore employ the notations with double "hats" $\hat{\hat{R}}_{ll}, \hat{\hat{R}}_{rr}$, and $\hat{\hat{R}}_{lr}$ on top of these operators in the situation where this circumstance is important.

For large n, the dominating contribution comes from the eigenvector (in this case of the Liouville equation it is the eigen "density" matrix) which corresponds to the eigenvalue of the Liouville operator with the smallest imaginary part. The "density" matrix satisfying (7.164) is a diagonal matrix in the oscillator quantum numbers representation $\sum_{mk} |mk\rangle\, \tilde{\rho}_n^{(mk)}\, \langle km|$. The main contribution to the slowest decaying term comes from the vacuum state $m = k = 0$ for which the Liouville equation reads

$$\frac{\partial \tilde{\rho}_n^{(0,0)}}{\partial n} = i2\zeta\omega\tilde{\rho}_n^{(0,0)} \;-2(\gamma\omega)^2\tilde{\rho}_n^{(0,0)}. \tag{7.166}$$

The dominating contribution therefore reads

$$\tilde{\rho}_n^{(int)}(x,\overline{x};y,\overline{y}) = \langle x,\overline{x}|0\rangle\, \tilde{\rho}_n^{(0,0)}\, \langle 0|y,\overline{y}\rangle\,,$$

where the coordinate representation of the wavefunctions is given explicitly in (7.150). We have thus reproduced the result of the previous subsection (p. 342) and have obtained for $\zeta \to 0$ the static transparency decreasing as $\sim e^{-\gamma^2 N/8}$ in the center of the band ($\eta = 0$) where the "frequency" assumes, according to (7.165), the minimum value $\omega = 1/4$.

7.5.3 Classical Fermionic Fields for the Population Dynamics

We now consider the evolution of the population $\rho_n(t)$ in an infinite relay of levels of energies E_n, randomly distributed around the mean position $\Omega = 0$ with Gaussian statistics (7.130). We calculate $\rho_n(t)$ for the case where all of the population of the chain has been initially localized at level 0. In other words, the initial condition $\psi_0(t=0) = 1$ holds for the Schrödinger equation (7.1), where the index n assumes both positive and negative integer values. The constant coupling V_n is taken as the energy unit. From (7.93) one immediately finds the population

$$\rho_n(t) = \int_{-\infty-i\nu}^{\infty-i\nu} \int_{-\infty+i\nu}^{\infty+i\nu} \frac{d\xi\, d\varepsilon}{4\pi^2} e^{-i(\varepsilon-\xi)t} \frac{M_0^+(\varepsilon)M_n^-(\varepsilon)}{\mathrm{Det}\left|\hat{H}-\varepsilon\right|} \frac{M_0^+(\xi)M_n^-(\xi)}{\mathrm{Det}\left|\hat{H}-\xi\right|}. \quad (7.167)$$

Though this expression resembles (7.131) for the tunneling transparency, it contains an important additional factor in the numerator of the integrand given by the product of four minors of the matrices $(\hat{H}-\varepsilon)_{n,n'}$ and $(\hat{H}-\xi)_{n,n'}$. As earlier, one can represent the determinants in the denominator with the help of the functional integrals (7.133). We are now going to introduce a technique which allows one to represent the factors in the numerator also in the form of functional integrals, and thereby to perform the Gaussian ensemble average by analogy to (7.134) and (7.135).

Analogy Between Minors and an Ensemble of Spins $1/2$

As we have already seen in the previous section, the functional integral (7.137) for the inverse square root of the determinant can be interpreted in terms of a one-dimensional particle, with a Hamiltonian (7.142) moving in a parabolic potential, periodically changing in time. The integral equals the probability amplitude for the particle to remain at the origin after $N+1$ periods, whereas the evolution after one period is given by (7.145). We can build a similar construction for the minors starting from the recurrence relation (7.43). However, now the representing particle does not correspond to an infinite-dimensional Hilbert space of a harmonic oscillator, but to just two-dimensional Hilbert space of a spin $1/2$. Indeed, by introducing "spinors 1/2" given as $\{M_n^+, iM_{n-1}^+\}$ one can rewrite (7.43) for $V_n \equiv 1$ in the matrix form

$$\begin{Bmatrix} M_{n+1}^+ \\ iM_n^+ \end{Bmatrix} = \begin{pmatrix} E_n-\varepsilon & i \\ i & 0 \end{pmatrix} \begin{Bmatrix} M_n^+ \\ iM_{n-1}^+ \end{Bmatrix} \quad (7.168)$$

where the 2×2 matrix describes the evolution over the period by analogy to (7.145).

The upper diagonal matrix element ($s = 1/2$) of the product of these matrices corresponding to the evolution over N periods yields the entire determinant Det $\left|\hat{H} - \varepsilon\right|$, whereas that for $n + 1$ periods yields M_n^+. In other words, projection to the upper state of the "spin" plays the same role as the projection to the state $|x = 0\rangle$ of the oscillator. Note that the operator of the evolution over a period

$$\begin{pmatrix} E_n - \varepsilon & i \\ i & 0 \end{pmatrix} = (E_n - \varepsilon)\hat{\sigma}^+\hat{\sigma} + i\hat{\sigma}^+ + i\hat{\sigma} \tag{7.169}$$

can be given in terms of the spin raising and lowering operators $\hat{\sigma}^+$ and $\hat{\sigma}^-$. One can introduce "trajectories" (see Fig. 7.10c) for the spin 1/2 that evidently are discrete, not only in "time" n but also in "space", since the spin assumes only two values $s_\pm = \pm 1/2$. The trajectories are just sequences $\{(s_\pm)_n\}$ of s_+ or s_- for each n, and each trajectory corresponds unambiguously to one of 2^N terms in the explicit representation of the three-diagonal determinant Det $\left|\hat{H} - \varepsilon\right|$. We assume the standard vector notation

$$|s_+\rangle = \begin{pmatrix} 1 \\ 0 \end{pmatrix}; \quad |s_-\rangle = \begin{pmatrix} 0 \\ 1 \end{pmatrix}$$
$$\langle s_+| = \begin{pmatrix} 1 & 0 \end{pmatrix}; \quad \langle s_-| = \begin{pmatrix} 0 & 1 \end{pmatrix} \tag{7.170}$$

for the spin eigenstates.

The correspondence between the terms of the determinant and the spin trajectories can be established as follows. We develop the determinant over the elements of the last column. The diagonal matrix element then corresponds to s_+ and has the factor $(E_1 - \varepsilon)$ in front of the minor M_1^+, whereas the off-diagonal matrix element corresponds to s_- and gives the unity factor in front of the minor $\left(M_2^+\right)_{ap}$. The latter results from appending the minor M_2^+ by one line and one column, containing zeros everywhere apart from unity on the diagonal position. The same development applied to M_1^+ or $\left(M_2^+\right)_{ap}$ provides us with the second element of the trajectory, and the procedure can be repeated up to the end, thus yielding the representation of the determinant as a sum of different products of unities and the factors $(E_n - \varepsilon)$, whereas each of the products corresponds to a different trajectory.

We illustrate this for the simplest example of a 2×2 determinant. On the one hand, straightforward calculation via expansion over the elements of the last column yields

$$\begin{aligned} \text{Det} \begin{vmatrix} E_2 - \varepsilon & 1 \\ 1 & E_1 - \varepsilon \end{vmatrix} &= M_1^+ (E_1 - \varepsilon) - \left(M_2^+\right)_{ap} 1 \\ &= (E_2 - \varepsilon)(E_1 - \varepsilon) - 1, \end{aligned} \tag{7.171}$$

where the 1×1 minors are $M_1^+ = (E_2 - \varepsilon)$ and $\left(M_2^+\right)_{ap} = (1)$. On the other hand, with the help of (7.168) one obtains

$$\begin{aligned}
\mathrm{Det}\left|\hat{H} - \varepsilon\right| &= \langle s_+| \begin{pmatrix} E_2 - \varepsilon & i \\ i & 0 \end{pmatrix} \begin{pmatrix} E_1 - \varepsilon & i \\ i & 0 \end{pmatrix} |s_+\rangle \\
&= \langle s_+| \begin{pmatrix} E_2 - \varepsilon & i \\ i & 0 \end{pmatrix} |s_+\rangle \langle s_+| \begin{pmatrix} E_1 - \varepsilon & i \\ i & 0 \end{pmatrix} |s_+\rangle \\
&\quad + \langle s_+| \begin{pmatrix} E_2 - \varepsilon & i \\ i & 0 \end{pmatrix} |s_-\rangle \langle s_-| \begin{pmatrix} E_1 - \varepsilon & i \\ i & 0 \end{pmatrix} |s_+\rangle , \quad (7.172)
\end{aligned}$$

where in agreement with (7.169, 7.170), the last two terms show the contribution $(E_2 - \varepsilon)\ {}_2\langle s_+|\,\hat{\sigma}^+\hat{\sigma}\,|s_+\rangle_2\, (E_1 - \varepsilon)\ {}_1\langle s_+|\,\hat{\sigma}^+\hat{\sigma}\,|s_+\rangle_1$ of the trajectory $(s_+)_2\,(s_+)_1$ and the contribution ${}_1\langle s_+|\,\hat{\sigma}^+\,|s_-\rangle_2\ (-1)\ {}_2\langle s_-|\,\hat{\sigma}\,|s_+\rangle_1$ of the trajectory $(s_-)_2\,(s_+)_1$. The indices n of $|s_\pm\rangle_n$ enumerate columns of the determinant, whereas that of ${}_n\langle s_\pm|$ enumerate the lines. Note that for a higher-order determinant, the trajectory $(s_r)_N\,(s_u)_c \cdots (s_s)_b\,(s_v)_a\,(s_w)_1$ implies that the operators (7.169) are sequentially placed in between ${}_a\langle s_v|$ and $|s_w\rangle_1$, in between ${}_b\langle s_s|$ and $|s_v\rangle_a$, etc. until one finally reaches the last operator in between ${}_N\langle s_r|$ and $|s_u\rangle_c$, at the end of the trajectory. The running indices $a = 1, 2$; $b = 1, 2$; $c = 1, 2$ mark the trajectory.

Classical Anticommuting Variables

The main problem in describing a quantum system, with the help of a functional integral over classical trajectories, is finding a proper "action" functional $\exp[S]$ and the corresponding classical integration variables. For the particular problem under consideration, the functional integration should be equivalent to summation over the "spin" trajectories. Each "spin" trajectory $\{(s_\pm)_n\}$ corresponds to a spin eigenstate of an ensemble of N spins $1/2$, explicitly given by the tensor product $\prod_{n=1}^{N} \otimes |s_\pm\rangle_n$ of N individual spin states (either s_+ or s_-) enumerated by the index n, such that the trajectory sum can also be considered as a trace over these eigenstates. However, the functional integrals convenient for the description of spin ensembles do not employ spins as the coordinates of classical trajectories, but rely on a different construction.

The fact that the problem is formulated in terms of the ensemble of spins of $1/2$ suggests the use of the anticommuting fermionic operators. In the context of tunneling transparency, we have seen the correspondence between a quantum harmonic oscillator described by the bosonic operators $\hat{a}^\dagger$ and $\hat{a}$ with the commutation relation $\left[\hat{a}, \hat{a}^\dagger\right] = 1$ and the functional integral over the regular (commuting) classical coordinates $x_n, \overline{x}_n$. By analogy, classical anticommuting coordinates χ_n and $\overline{\chi}_n$ with the anticommutators $\{\chi_n, \overline{\chi}_m\} = 0$, $\{\chi_n, \chi_m\} = 0$, and $\{\overline{\chi}_n, \overline{\chi}_m\} = 0$ are related to the quantum spin operators $\hat{\sigma}^+$ and $\hat{\sigma}$ of (7.169) satisfying the anticommutation relation $\{\hat{\sigma}^+, \hat{\sigma}\} = 1$. These classical anticommuting coordinates are known as nilpotent, graded, or Grassmann variables. The anticommutation relations imply that $\chi_n^2 = 0$ and $\overline{\chi}_n^2 = 0$. Moreover, they also imply that each anticommuting coordinate enters, at most once, to any non-vanishing product of the coordinates, and in particular $\chi_n \overline{\chi}_n \chi_n \overline{\chi}_n = 0$.

A certain physical meaning can be attributed to the variables χ_n and $\overline{\chi}_n$. The variable $\overline{\chi}_n$ can be interpreted as the creation of a "particle" at the position n, whereas χ_n denotes either annihilation of the existing "particle" or creation of an "antiparticle" at the same position. The combination $\overline{\chi}_n \chi_n$ may thus be interpreted as the number of particles (0 or 1) at the site n, whereas the combination $\overline{\chi}_{n+1} \chi_n$ indicates creation of a "particle–antiparticle" couple at the sites n and $(n+1)$, respectively. Each "particle" or antiparticle can be created and annihilated only once, as follows from the anticommutation relations for the fermionic operators.

Since each "particle" has "spin" $1/2$, creation or annihilation of a couple of "particles" changes the total "spin" by unity. In the case of the nearest neighbor, relay-like interaction, resulting in three-term recurrence relations for the minors, and in accordance with (7.137) following from these relations, the total "spin" always remains $1/2$, whereas creation or annihilation of a couple just corresponds to "spin flip". We note that the "pair" corresponding to spin 1, and its creation and annihilation operators given by products of two Grassmann coordinates (bi-Grassmann), are bosonic operators commuting among themselves. The operators of "pairs" also commute with single "particle" operators. Therefore one can manipulate with these operators as regular numbers.

Now each term of the determinant, which corresponds to a certain choice of the matrix elements in the sequential development of the minors over the elements of the last column, can also be attributed to a "spin" trajectory, and each of the "spin" trajectories corresponds in turn to a certain sequence of creation and annihilation of "particles" and "pairs". Let us illustrate this for the simplest example of (7.171)–(7.172):

$$\begin{aligned}
&\mathrm{Det}\begin{vmatrix} E_2-\varepsilon & 1 \\ 1 & E_1-\varepsilon \end{vmatrix} = (E_2-\varepsilon)(E_1-\varepsilon) - 1; \\
&(E_2-\varepsilon)(E_1-\varepsilon) \to (s_+)_2 (s_+)_1 \to \overline{\chi}_2 \chi_2 \overline{\chi}_1 \chi_1; \\
&-1 \to (s_-)_2 (s_+)_1 \to \overline{\chi}_1 \chi_2 \overline{\chi}_2 \chi_1.
\end{aligned} \tag{7.173}$$

The first term of the determinant resulting from the choice of the matrix element $(E_1 - \varepsilon)$ in the last column corresponds to the trajectory $(s_+)_2 (s_+)_1$ and to the creation and annihilation of "particle" 1 followed by the creation and annihilation of "particle" 2, whereas the second term resulting from the choice of the matrix element 1 in the last column corresponds to the creation of a "pair" consisting of a "particle" at site 1 and an "antiparticle" at site 2 followed by their annihilation. Our next step is to find an algebraic way to reveal this correspondence.

Functional Integrals over Anticommuting Variables

Consider a bilinear form

$$\overleftarrow{\chi}(\hat{H}-\varepsilon)\overrightarrow{\chi} \equiv (\overline{\chi}_2\,\overline{\chi}_1)\begin{pmatrix} E_2-\varepsilon & 1 \\ 1 & E_1-\varepsilon \end{pmatrix}\begin{pmatrix} \chi_2 \\ \chi_1 \end{pmatrix}$$
$$= (E_2-\varepsilon)\,\overline{\chi}_2\chi_2 + (E_1-\varepsilon)\,\overline{\chi}_1\chi_1 + \overline{\chi}_2\chi_1 + \overline{\chi}_1\chi_2 \equiv -S \tag{7.174}$$

and calculate the exponent $\exp[S]$ of this quantity, casting it in a Taylor series

$$\exp[S] = 1 - (E_2-\varepsilon)\,\overline{\chi}_2\chi_2 - (E_1-\varepsilon)\,\overline{\chi}_1\chi_1 - \overline{\chi}_2\chi_1 - \overline{\chi}_1\chi_2$$
$$+ (E_2-\varepsilon)\,\overline{\chi}_2\chi_2\,(E_1-\varepsilon)\,\overline{\chi}_1\chi_1 + \overline{\chi}_2\chi_1\overline{\chi}_1\chi_2 \tag{7.175}$$

where we retain only the terms linear in each coordinate, and allow for the fact that all binary products of the variables commute among themselves. Let us concentrate only on the two last terms of (7.175) containing all of the fermionic coordinates, and with the help of the anticommutation relations, make the order of the variables of the last term identical to those of the preceding term. This yields

$$\exp[S] = \ldots + (E_2-\varepsilon)\,(E_1-\varepsilon)\,\overline{\chi}_2\chi_2\overline{\chi}_1\chi_1 - \overline{\chi}_2\chi_2\overline{\chi}_1\chi_1, \tag{7.176}$$

where in the numerical factor in front of the coordinate product $\overline{\chi}_2\chi_2\overline{\chi}_1\chi_1$ one recognizes the determinant (7.173).

The functional S can serve as the action of the path integrals over anticommuting variables, provided this integration is defined as a procedure ensuring the selection of all of the terms in the Taylor expansion that only contain products of all of the coordinates. In other words, a "particle" should be created and annihilated at each site n. This goal is met by following the definitions of the integrals over the Grassmannian variables,

$$\int d\chi_n = 0, \quad \int \chi_n d\chi_n = 1, \tag{7.177}$$

where the differentials of the fermionic variables also anticommute. Equation (7.177) together with the anticommutation relations

$$\{d\chi_n d\chi_m\} = \{d\chi_n d\overline{\chi}_m\} = 0,$$
$$\{\chi_n, \overline{\chi}_m\} = 0; \ \{\chi_n, \chi_m\} = 0; \ \{\overline{\chi}_n, \overline{\chi}_m\} = 0 \tag{7.178}$$

can serve as an axiomatic definition of the Grassmann variables.

We therefore have

$$\int \exp[S]\ d\chi_1 d\overline{\chi}_1 d\chi_2 d\overline{\chi}_2 = (E_2-\varepsilon)\,(E_1-\varepsilon) - 1, \tag{7.179}$$

whereas the functional integral

$$\int \exp\left[-\sum_{n,m=1}^{N} \overline{\chi}_n a_{nm} \chi_m\right] \prod_{n=1}^{N} \{\, d\chi_n d\overline{\chi}_n\} = \mathrm{Det}\, |a_{nm}|\,, \tag{7.180}$$

is the generalization of (7.179) for the case of an arbitrary order N of the determinant, and is an analog of (7.136) for the regular variable. This expression is valid for the general matrix a_{nm} even when the three-term recurrence relations (7.43) and (7.168) do not hold. Note that for a minor M_{kl} resulting from the elimination of the k-th line and l-th column from the matrix a_{nm} one has the integral representation

$$M_{kl} = \int \overline{\chi}_k \chi_l \exp\left[-\sum_{n,m=1}^{N} \overline{\chi}_n a_{nm} \chi_m\right] \prod_{n=1}^{N} \{\, d\chi_n d\overline{\chi}_n\}\,. \tag{7.181}$$

Two coordinates in the pre-exponential factor eliminate all of the terms containing the matrix elements of the k-th line and l-th column from the Taylor expansion of the integrand.

7.6 Population Dynamics in a Disordered Chain

With the help of Grassmann notation, we now consider the population of an infinite relay of levels with disordered energies given by (7.167). We note that for the relay-like systems with $V_n = 1$ one has $M_0^+ M_n^- = M_{0n}$ and hence

$$\begin{aligned}\rho_n(t) = & \int_{-\infty-i\nu}^{\infty-i\nu} \int_{-\infty+i\nu}^{\infty+i\nu} \frac{d\xi\; d\varepsilon}{4\pi^2}\, e^{-i(\varepsilon-\xi)t} \prod_{k=-\infty}^{\infty} \left\{\int d\chi_k d\overline{\chi}_k d\chi'_k d\overline{\chi}'_k\right\} \\ & \overline{\chi}'_0 \chi'_n \overline{\chi}_0 \chi_n \exp\left[-\sum_{k,m=-\infty}^{\infty} \left(\overline{\chi}_k a_{km} \chi_m - \overline{\chi}'_k a'_{km} \chi'_m\right)\right] \\ & \prod_{k=-\infty}^{\infty} \left\{\int_{-\infty}^{\infty} \frac{dz_k^* dz_k dz_k'^* dz'_k}{4\pi^2}\right\} \exp\left[-i \sum_{k,m=-\infty}^{\infty} \left(z_k a_{km} z_m^* - z'_k a'_{km} z_m'^*\right)\right]\end{aligned} \tag{7.182}$$

where $a_{km} = (E_k - \varepsilon)\delta_{k,m} + \delta_{k,m+1} + \delta_{k,m-1}$ and $a'_{km} = (E_k - \xi)\delta_{k,m} + \delta_{k,m+1} + \delta_{k,m-1}$. The integrals over the variables z^*, z, z'^*, and z' are just another form of the integral (7.134), given in the notation of (7.136), whereas the integrals over the anticommuting variables of the type (7.181), take into account the minors in the numerators of (7.167).

By analogy to (7.135), for the Gaussian distribution function (7.130) centered around $\Omega = 0$, one finds the ensemble average of (7.182),

$$\rho_n(t) = \int_{-\infty-i\nu}^{\infty-i\nu} \int_{-\infty+i\nu}^{\infty+i\nu} \frac{d\xi\, d\varepsilon}{4\pi^2} \prod_{k=-\infty}^{\infty} \left\{ \int_{-\infty}^{\infty} \frac{dz_k^* dz_k dz_k'^* dz_k'}{4\pi^2} \right\}$$

$$\prod_{k=-\infty}^{\infty} \left\{ \int d\chi_k d\overline{\chi}_k d\chi_k' d\overline{\chi}_k' \right\} \overline{\chi}_0' \chi_n' \overline{\chi}_0 \chi_n \, e^{-i(\varepsilon-\xi)t+S};$$

$$S = \sum_{k=-\infty}^{\infty} \left[\varepsilon \overline{\chi}_k \chi_k - \xi \overline{\chi}_k' \chi_k' + i\varepsilon \left| z_k \right|^2 - i\xi \left| z_k' \right|^2 + \overline{\chi}_k' \chi_{k\pm1}' - \overline{\chi}_k \chi_{k\pm1} \right.$$

$$\left. + i z_k' z_{k\pm1}'^* - i z_k z_{k\pm1}^* - \frac{\gamma^2}{4} \left(i \overline{\chi}_k' \chi_k' - i \overline{\chi}_k \chi_k + \left| z_k \right|^2 - \left| z_k' \right|^2 \right)^2 \right], \tag{7.183}$$

where the subscript $k \pm 1$ denotes the sum of the terms with indices $k-1$ and $k+1$.

7.6.1 *Population Dynamics and Propagating Fictitious Particles*

Considering the inverse square root of a three-diagonal determinant in the context of tunneling transparency as a function of its order n, we have seen that for $\gamma = 0$, the functional integral (7.137) representing this determinant can be interpreted as the probability amplitude of a periodically driven harmonic oscillator evolving in a fictitious "time" n. The evolution over a period (7.151) written with the help of the creation and the annihilation operators (7.149) suggests the employment of the oscillator energy eigenfunctions, where the most important role is played by the ground state (7.150). We have expressed the tunneling transparency in terms of four oscillators composing a "density matrix" (7.160) and considering $\gamma^2 \neq 0$ as a perturbation. Now we employ the same idea for the anticommuting variables.

First we separate the "fermionic" part of (7.183) and introduce for $\gamma = 0$ the "fermionic density matrix" $\overline{\rho}_n(\overline{\chi}_k, \chi_k, \overline{\chi}_k', \chi_k')$

$$\overline{\rho}_n(\overline{\chi}_n, \chi_n, \overline{\chi}_n', \chi_n') = \exp\left(\varepsilon \overline{\chi}_n \chi_n - \xi \overline{\chi}_n' \chi_n'\right) \prod_{k=-\infty}^{n-1} \left\{ \int d\chi_k d\overline{\chi}_k d\chi_k' d\overline{\chi}_k' \right\}$$

$$\exp\left[\sum_{k=-\infty}^{n-1} \varepsilon \overline{\chi}_k \chi_k - \xi \overline{\chi}_k' \chi_k' + \overline{\chi}_k' \chi_{k+1}' - \overline{\chi}_k \chi_{k+1} + \overline{\chi}_{k+1}' \chi_k' - \overline{\chi}_{k+1} \chi_k \right], \tag{7.184}$$

by finishing the Grassmann integration at the "time moment" $n-1$. In order to distinguish between the real density matrix and the fictitious "bosonic" and "fermionic density matrices" (7.184), the latter is indicated by a bar. The "fermionic density matrix" is a polynomial consisting of 16 terms where each of four coordinates $(\overline{\chi}_n, \chi_n, \overline{\chi}'_n, \chi'_n)$ occurs either in the first or in zero power, that is,

$$\begin{aligned}
&\overline{\rho}_n(\overline{\chi}, \chi, \overline{\chi}', \chi') = \overline{\rho}_{0000}(n) + \chi'\overline{\rho}_{0001}(n) \\
&+\overline{\chi}'\overline{\rho}_{0010}(n) + \overline{\chi}'\chi'\overline{\rho}_{0011}(n) + \chi\overline{\rho}_{0100}(n) \\
&+\chi\chi'\overline{\rho}_{0101}(n) + \chi\overline{\chi}'\overline{\rho}_{0110}(n) + \chi\overline{\chi}'\chi'\overline{\rho}_{0111}(n) \\
&+\overline{\chi}\overline{\rho}_{1000}(n) + \overline{\chi}\chi'\overline{\rho}_{1001}(n) \\
&+\overline{\chi}\overline{\chi}'\overline{\rho}_{1010}(n) + \overline{\chi}\overline{\chi}'\chi'\overline{\rho}_{1011}(n) + \overline{\chi}\chi\overline{\rho}_{1100}(n) \\
&+\overline{\chi}\chi\chi'\overline{\rho}_{1101}(n) + \overline{\chi}\chi\overline{\chi}'\overline{\rho}_{1110}(n) + \overline{\chi}\chi\overline{\chi}'\chi'\overline{\rho}_{1111}(n).
\end{aligned} \tag{7.185}$$

With the help of the leftmost line vector $\left(1, \chi, \overline{\chi}, \overline{\chi}\chi,\right)$ and the corresponding rightmost column vector this can also be written as a bi-linear form

$$\begin{aligned}
&\overline{\rho}_n(\overline{\chi}, \chi, \overline{\chi}', \chi') = \left(1, \chi, \overline{\chi}, \overline{\chi}\chi,\right) \\
&\times \begin{pmatrix} \overline{\rho}_{0000} & \overline{\rho}_{0001} & \overline{\rho}_{0010} & \overline{\rho}_{0011} \\ \overline{\rho}_{0100} & \overline{\rho}_{0101} & \overline{\rho}_{0110} & \overline{\rho}_{0111} \\ \overline{\rho}_{1000} & \overline{\rho}_{1001} & \overline{\rho}_{1010} & \overline{\rho}_{1011} \\ \overline{\rho}_{1100} & \overline{\rho}_{1101} & \overline{\rho}_{1110} & \overline{\rho}_{1111} \end{pmatrix} \begin{pmatrix} 1 \\ \chi' \\ \overline{\chi}' \\ \overline{\chi}'\chi' \end{pmatrix}
\end{aligned} \tag{7.186}$$

where the coefficients forming a 4×4 matrix and depending on n encode all of the fermionic information. Later, we will consider the evolution of an entangled multiplet consisting of both "fermions" and "bosons", for which all matrix elements are functions of bosonic variables $(x, \overline{x}; y, \overline{y})$, or in short (z, z'). We attribute variables with primes to "antiparticles" ; z' to "antibosons", and $\chi'$and $\overline{\chi}'$ to "antifermions". We assume that "bosonic" coordinates are implicit in the representation (7.186).

We determine the recurrence relation for this matrix propagating it in "time" to $n+1$. The result of such a propagation is known: it should be equivalent to the transformation given by a 2×2 matrix in (7.168) for $E_n = 0$, which has to be applied to each of the participating fermions, that is, separately for each index $(0, 1)$, of the matrix elements. We consider this in more detail and show how this transformation appears in the matrix notation (7.186). We perform integration over all but the last four anticommuting variables and find the recurrence relations

$$\begin{aligned}
\overline{\rho}_n(\overline{\chi}, \chi, \overline{\chi}', \chi') = \int d\chi_1 d\overline{\chi}_1 d\chi'_1 d\overline{\chi}'_1 \overline{\rho}_{n-1}(\overline{\chi}_1, \chi_1, \overline{\chi}'_1, \chi'_1) \\
\exp\left(\varepsilon\overline{\chi}\chi - \xi\overline{\chi}'\chi' + \overline{\chi}'\chi'_1 - \overline{\chi}\chi_1 + \overline{\chi}'_1\chi' - \overline{\chi}_1\chi\right)
\end{aligned} \tag{7.187}$$

for the functions of the Grassmann variables.

Let us first consider the result of integration over one variable, say $d\overline{\chi}_1'$, which involves only the last column vector of the quadratic form (7.186). We take into account the anticommutation relations (7.178) among the differentials $d\chi_1 d\overline{\chi}_1 d\chi_1' d\overline{\chi}_1' = -d\overline{\chi}_1' d\chi_1' d\chi_1 d\overline{\chi}_1$, allow for the relation $\exp(\overline{\chi}_1'\chi') = (1+\overline{\chi}_1'\chi')$, and perform the integration following the rules (7.177). For the last vector of (7.186) this yields

$$\begin{pmatrix} 1 \\ \chi_1' \\ \overline{\chi}_1' \\ \overline{\chi}_1'\chi_1' \end{pmatrix} \overset{d\overline{\chi}_1'}{\rightarrow} \begin{pmatrix} -\int (1+\overline{\chi}_1'\chi')\, d\overline{\chi}_1' \\ -\int \chi_1' (1+\overline{\chi}_1'\chi')\, d\overline{\chi}_1' \\ -\int \overline{\chi}_1' (1+\overline{\chi}_1'\chi')\, d\overline{\chi}_1' \\ -\int \overline{\chi}_1'\chi_1' (1+\overline{\chi}_1'\chi')\, d\overline{\chi}_1' \end{pmatrix} = \begin{pmatrix} \chi' \\ \chi_1'\chi' \\ -1 \\ \chi_1' \end{pmatrix}. \tag{7.188}$$

Continuing by integration over $d\chi_1'$ allowing for the relation $\exp(\overline{\chi}'\chi_1') = (1+\overline{\chi}'\chi_1')$ we obtain

$$\begin{pmatrix} \chi' \\ \chi_1'\chi' \\ -1 \\ \chi_1' \end{pmatrix} \overset{d\chi_1'}{\rightarrow} \begin{pmatrix} \int \chi' (1+\overline{\chi}'\chi_1')\, d\chi_1' \\ \int \chi_1'\chi' (1+\overline{\chi}'\chi_1')\, d\chi_1' \\ -\int (1+\overline{\chi}'\chi_1')\, d\chi_1' \\ \int \chi_1' (1+\overline{\chi}'\chi_1')\, d\chi_1' \end{pmatrix} = \begin{pmatrix} -\overline{\chi}'\chi' \\ -\chi' \\ -\overline{\chi}' \\ 1 \end{pmatrix}. \tag{7.189}$$

In other words, replacement of the variables with subscript 1 by variables without the subscript, is equivalent to multiplication of the vector column by a transformation matrix

$$\begin{pmatrix} 1 \\ \chi_1' \\ \overline{\chi}_1' \\ \overline{\chi}_1'\chi_1' \end{pmatrix} \overset{d\overline{\chi}_1' d\chi_1'}{\rightarrow} \begin{pmatrix} 0 & 0 & 0 & -1 \\ 0 & -1 & 0 & 0 \\ 0 & 0 & -1 & 0 \\ 1 & 0 & 0 & 0 \end{pmatrix} \begin{pmatrix} 1 \\ \chi' \\ \overline{\chi}' \\ \overline{\chi}'\chi' \end{pmatrix}. \tag{7.190}$$

Let us allow now for multiplication by $\exp(-\xi\overline{\chi}'\chi') = 1 - \xi\overline{\chi}'\chi'$ which yields

$$\begin{pmatrix} 1 \\ \chi' \\ \overline{\chi}' \\ \overline{\chi}'\chi' \end{pmatrix} \overset{\exp(-\xi\overline{\chi}'\chi')}{\rightarrow} \begin{pmatrix} 1-\xi\overline{\chi}'\chi' \\ \chi' \\ \overline{\chi}' \\ \overline{\chi}'\chi' \end{pmatrix}, \tag{7.191}$$

and therefore modifies (7.190)

$$\begin{pmatrix} 1 \\ \chi' \\ \overline{\chi}' \\ \overline{\chi}'\chi' \end{pmatrix}_1 \overset{d\overline{\chi}_1' d\chi_1' \times \exp(-\xi\overline{\chi}'\chi')}{\rightarrow} \begin{pmatrix} -\xi & 0 & 0 & -1 \\ 0 & -1 & 0 & 0 \\ 0 & 0 & -1 & 0 \\ 1 & 0 & 0 & 0 \end{pmatrix} \begin{pmatrix} 1 \\ \chi' \\ \overline{\chi}' \\ \overline{\chi}'\chi' \end{pmatrix}. \tag{7.192}$$

We can calculate the transformation of the line vectors analogously and obtain

$$(1, \chi, \overline{\chi}, \overline{\chi}\chi)_1 \to (1\ \chi, \overline{\chi}, \overline{\chi}\chi)\begin{pmatrix} 0 & 0\,0\,1 \\ 0 & 1\,0\,0 \\ 0 & 0\,1\,0 \\ -1 & 0\,0\,\varepsilon \end{pmatrix}. \tag{7.193}$$

In other words, the matrix $\hat{\overline{\rho}}$ of the coefficients $\overline{\rho}_{ijkl}$ of (7.186) experiences the transformation $\hat{U}_R\hat{\overline{\rho}}\hat{U}_A$, where $\hat{U}_A$ and $\hat{U}_R$ are the matrices entering (7.192) and (7.193), respectively.

Note that the fermionic part of the line and the column vectors that contain only linear terms $\chi', \overline{\chi}'$ or $\chi, \overline{\chi}$ is not affected by the transformation over the period and does not change, apart from an unimportant variation of the signs. This allows us to concentrate only on the first and the last lines and columns of the matrices, discard the second and the third line and column and write

$$\begin{pmatrix} 1 \\ \overline{\chi}'\chi' \end{pmatrix}_1 \to \begin{pmatrix} -\xi, & -1 \\ 1, & 0 \end{pmatrix}\begin{pmatrix} 1 \\ \overline{\chi}'\chi' \end{pmatrix}. \tag{7.194}$$

$$(1, \overline{\chi}\chi)_1 \to (1, \overline{\chi}\chi)\begin{pmatrix} 0, & 1 \\ -1, & \varepsilon \end{pmatrix}, \tag{7.195}$$

which is equivalent to (7.168). In other words, combinations of "particles" containing an odd number of "fermions" or "antifermions" are not affected by the "time" evolution. Neither are they affected by the disorder. This circumstance is very important to note, since the operators $\overline{\chi}_0'\overline{\chi}_0$ and $\chi_n\chi_n'$ in front of the exponent in the integrand (7.182) do change the number of "fermions" from even to odd. This occurs at the "time moments" corresponding to the starting level 0 and the final level n in our original problem of the population dynamics, and means that during this "interval of time", the propagation of the "fermions" is disentangled from that of the "bosons", where the latter propagate independently according to (7.166).

After discarding from (7.186) the lines and the columns corresponding to odd numbers of fermions and antifermions, the "density matrix" for fictitious particles adopts the form

$$\overline{\overline{\rho}}_n(\overline{\chi}, \chi, x, \overline{x}; \overline{\chi}', \chi', y, \overline{y}) = (1, \overline{\chi}\chi)\,\hat{\rho}_n(z, z')\begin{pmatrix} 1 \\ \overline{\chi}'\chi' \end{pmatrix};$$

$$\hat{\rho}_n(z, z') = \begin{pmatrix} \tilde{\overline{\rho}}_{0000}(z, z', n), & \tilde{\overline{\rho}}_{0011}(z, z', n) \\ \tilde{\overline{\rho}}_{1100}(z, z', n), & \tilde{\overline{\rho}}_{1111}(z, z', n) \end{pmatrix} \tag{7.196}$$

where the bosonic variables $(x, \overline{x}; y, \overline{y})$ are written explicitly as complex numbers (z, z'), as in (7.136). At $\gamma = 0$ according to (7.162), (7.194), and (7.195), the evolution of the "density matrix" over the period reads

$$\hat{\rho}_{n+1}(z,z') = \hat{U}_l \hat{u}_L \hat{\rho}_n(z,z') \hat{u}_R \hat{U}_r \tag{7.197}$$

where

$$\hat{u}_R = \begin{pmatrix} -\xi & -1 \\ 1 & 0 \end{pmatrix};\ \hat{u}_L = \begin{pmatrix} 0 & 1 \\ -1 & \varepsilon \end{pmatrix},$$

$$\hat{U}_r = \exp\left\{ i\frac{1}{4} |\hat{p}|^2_{z'} \right\} \exp\left\{ i(2-\xi)\, |z'|^2 \right\}$$

$$\hat{U}_l = \exp\left\{ -i(2-\varepsilon)\, |z|^2 \right\} \exp\left\{ -i\frac{1}{4} |\hat{p}|^2_{z} \right\}. \tag{7.198}$$

7.6.2 *Mapping Over a Period and the Ensemble Average*

For $\gamma \neq 0$, as a result of an ensemble averaging of the expression

$$\rho_n(t) = \int\limits_{-\infty - i\nu}^{\infty - i\nu} \int\limits_{-\infty + i\nu}^{\infty + i\nu} \frac{d\xi\, d\varepsilon}{4\pi^2} \prod_{k=-\infty}^{\infty} \left\{ \int\limits_{-\infty}^{\infty} \frac{dz_k^* dz_k dz_k'^* dz_k'}{4\pi^2} \right\} \tag{7.199}$$

$$\prod_{k=-\infty}^{\infty} \left\{ \int d\chi_k d\overline{\chi}_k d\chi_k' d\overline{\chi}_k' \right\} \overline{\chi}_0' \chi_n' \overline{\chi}_0 \chi_n \, e^{-i(\varepsilon - \xi)t + S};$$

$$S = \sum_{k=-\infty}^{\infty} \Big[(\varepsilon - \gamma_k)\overline{\chi}_k \chi_k - (\xi - \gamma_k)\overline{\chi}_k' \chi_k' + i(\varepsilon - \gamma_k)\, |z_k|^2$$

$$-i(\xi - \gamma_k)\, |z_k'|^2 + i z_k' z_{k\pm 1}'^* - i z_k z_{k\pm 1}^* + \overline{\chi}_k' \chi_{k\pm 1}' - \overline{\chi}_k \chi_{k\pm 1} \Big],$$

with the distribution function (7.130) for the energies E_n, emerges equation (7.183). Here, as earlier, $z_k = x_k + i\overline{x}_k$, $z_k' = y_k + i\overline{y}_k$. The representation for level populations given by this equation implies calculation of non-Gaussian integrals, which not always is an easy task. An alternative suggests to work with the Gaussian integrals Eq. (7.199) by considering the recurrence relations for fictitious "density matrices" $\overline{\rho}_n$ (7.186) for each individual realization of the disorder

$$\overline{\rho}_n(\overline{\chi}_n, \chi_n, x_n, \overline{x}_n; \overline{\chi}_n', \chi_n', y_n, \overline{y}_n) = \prod_{k=-\infty}^{n-1} \left\{ \int\limits_{-\infty}^{\infty} \frac{dx_k dy_k d\overline{x}_k d\overline{y}_k}{\pi^2} \right\}$$

$$\times \prod_{k=-\infty}^{n-1} \left\{ \int d\chi_k d\overline{\chi}_k d\chi_k' d\overline{\chi}_k' \right\} \exp \sum_{k=-\infty}^{n} \Big[2i x_k x_{k-1} + 2i \overline{x}_k \overline{x}_{k-1} - 2i y_k y_{k-1}$$

$$- 2i \overline{y}_k \overline{y}_{k-1} + \overline{\chi}_k' \chi_{k-1}' - \overline{\chi}_k \chi_{k-1} + \overline{\chi}_{k-1}' \chi_k' - \overline{\chi}_{k-1} \chi_k + (\varepsilon - E_k) \overline{\chi}_k \chi_k$$

$$- (\xi - E_k)\overline{\chi}_k' \chi_k' + i(\varepsilon - E_k)\left(x_k^2 + \overline{x}_k^2\right) - i(\xi - E_k)\left(y_k^2 + \overline{y}_k^2\right) \Big], \tag{7.200}$$

where the integrations are performed over all variables with indices less than n. Since the result of such a multistep Gaussian integration is a Gaussian function, one can assume that the "density matrix" has the form

$$\overline{\rho}_{n-1} = \exp\left[iM_{11}^{(n-1)}\left|z_{n-1}\right|^2 - iM_{22}^{(n-1)}\left|z'_{n-1}\right|^2 - M_{33}^{(n-1)}\overline{\chi}_{n-1}\chi_{n-1} + M_{44}^{(n-1)}\overline{\chi}'_{n-1}\chi'_{n-1}\right], \tag{7.201}$$

and after substitution of (7.201) into (7.200), followed by integration, we find a Gaussian function again for the next n

$$\overline{\rho}_n = \exp\left[iM_{11}^{(n)}\left|z_n\right|^2 - iM_{22}^{(n)}\left|z'_n\right|^2 - M_{33}^{(n)}\overline{\chi}_n\chi_n + M_{44}^{(n)}\overline{\chi}'_n\chi'_n\right], \tag{7.202}$$

where the diagonal matrices $\hat{M}^{(n)}$ and $\hat{M}^{(n-1)}$ are defined as

$$\hat{M}^{(n)} = \begin{pmatrix} M_{11}^{(n)} & 0 & 0 & 0 \\ 0 & M_{22}^{(n)} & 0 & 0 \\ 0 & 0 & M_{33}^{(n)} & 0 \\ 0 & 0 & 0 & M_{44}^{(n)} \end{pmatrix}$$

and satisfy the recurrence relation

$$\hat{M}^{(n)} = -\left(\hat{M}^{(n-1)}\right)^{-1} + \begin{pmatrix} \varepsilon - E_n & 0 & 0 & 0 \\ 0 & \xi - E_n & 0 & 0 \\ 0 & 0 & \varepsilon - E_n & 0 \\ 0 & 0 & 0 & \xi - E_n \end{pmatrix}. \tag{7.203}$$

This is a direct analog of mapping (7.110). One sees that in this construction, $M_{11}^{(n)}$ always coincides with $M_{33}^{(n)}$, whereas $M_{22}^{(n)}$ coincides with $M_{44}^{(n)}$. We have already encountered such a situation considering the tunneling transparency.

Successive application of the mapping yields the matrix $\hat{M}^{(n)}$ as a function of the initial condition $\hat{M}^{(0)}$ and all of the string of intermediate energies E_k with $0 < k < n$. Averaging over the ensemble of random energies E_k therefore generates an ensemble of random matrices $\hat{M}^{(n)}$ with a certain distribution function $g(M_{11}^{(n)}, M_{22}^{(n)}, \varepsilon, \xi)$. For an infinitely long chain of levels, this distribution function loses the memory of the initial condition $\hat{M}^{(0)}$ and becomes "stationary", not depending on the level number n, that is, on the fictitious "time". Hence for the ensemble average "density matrix" one finds

$$\overline{\rho}_n(\overline{\chi}, \chi, z; \overline{\chi}', \chi', z') = \int dM_{11}dM_{22}\, g(M_{11}, M_{22}, \varepsilon, \xi) \exp\left[iM_{11}\left|z\right|^2 - iM_{22}\left|z'\right|^2 - M_{11}\overline{\chi}\chi + M_{22}\overline{\chi}'\chi'\right], \tag{7.204}$$

where the "stationary" distribution function $g(M_{11}^{(n)}, M_{22}^{(n)}, \varepsilon, \xi)$ of matrices satisfies the equation

$$g(M_{11}, M_{22}, \varepsilon, \xi) = \int dE_n \frac{\tilde{g}(E_n)}{(\varepsilon - E_n - M_{11})^2} \frac{1}{(\xi - E_n - M_{22}^2)^2}$$
$$g\left(\frac{1}{\varepsilon - E_n - M_{11}}, \frac{1}{\xi - E_n - M_{22}}, \varepsilon, \xi\right), \qquad (7.205)$$

which follows from the recurrence relation (7.203) and the random energy distribution (7.130),

$$\tilde{g}(E_n) = \frac{1}{\sqrt{\pi}\gamma} \exp \frac{-E_n^2}{\gamma^2}.$$

Integration over $dM_{11}dM_{22}$ can be referred to as integration over a matrix $d\hat{M}$, which for the particular case under consideration is reduced to integration over only two diagonal matrix elements. For the real numbers ε, ξ, E_n the matrix elements M_{11} and M_{22} are real, but even a vanishing imaginary part of ε and ξ makes these matrix elements complex when mapping (7.203) is performed an infinite number of times. In the latter case the symbol $dM_{11}dM_{22}$ implies independent integration over the real and the imaginary parts of M_{11} and M_{22}, for which one always has two constraints: $\mathrm{Im}M_{11} > 0 > \mathrm{Im}M_{22}$. The distribution $g(M_{11}, M_{22}, \varepsilon, \xi)$ in this case is a real function of the complex variables M_{11}, M_{22}.

Note that the representation

$$\tilde{\rho}_n(z, z') = \int dM_{11}dM_{22}\, \overline{g}_n(M_{11}, M_{22}, \varepsilon, \xi) \times$$
$$\exp\left[iM_{11}\left|z\right|^2 - iM_{22}\left|z'\right|^2\right], \qquad (7.206)$$

similar to (7.204) can be given for the purely "bosonic" density matrix (7.161) although with a slightly different recurrence relation

$$\overline{g}_{n+1}(M_{11}, M_{22}, \varepsilon, \xi) = \int dE_n \frac{\tilde{g}(E_n)}{\varepsilon - E_n - M_{11}}$$
$$\frac{1}{\xi - E_n - M_{22}}\, \overline{g}_n\left(\frac{1}{\varepsilon - E_n - M_{11}}, \frac{1}{\xi - E_n - M_{22}}, \varepsilon, \xi\right), \qquad (7.207)$$

for the distribution function. The difference originates in the fact that integration over bosonic variables z, z' yields denominators $M_{11}'\, M_{22}'$ that are not compensated by the same factors in numerators resulting from Grassmann integrations. Due to this difference, the normalization of the distribution function (7.207) is not conserved, but exponentially decreases over the course of iterations, where the function itself does not have a non-zero asymptotic profile. The decrease rate is given by the modulus of the smallest eigenvalue $e^{\lambda(\varepsilon,\xi)}$ of the integral operator (7.207), and we denote the corresponding eigenfunction by $\overline{g}_{\min}(M'_{11}, M'_{22}, \varepsilon, \xi)$.

Equation (7.205) implies that a stationary distribution $g(M_{11}, M_{22}, \varepsilon, \xi)$ is attained as a result of the interplay between the compressing mapping (7.203) at $E_n = 0$ and the random walk resulting from the irregular energy shifts E_n. Mapping in the complex planes M_{11} and M_{22} tends to compress the distribution to the stable stationary points shown for (7.110) in Fig. 7.8. For $\mathrm{Im}(\varepsilon - \xi) < 0$ these points are located outside of the unit circles $|M_{11}| = 1$ and $|M_{22}| = 1$ at a distance $|\mathrm{Im}\zeta| / \sqrt{4-\eta^2}$, which can be found for small ζ by straightforward calculation of the positions of the stationary points. Here $\eta = (\varepsilon + \xi)/2$ and $\zeta = \varepsilon - \xi$, as usual. The compression factor $1 - 2|\mathrm{Im}\zeta| / \sqrt{4-\eta^2}$ narrows the distribution around the stationary points, whereas the random values of E_n result in a diffusive broadening of the distribution. Relative broadening of a Gaussian distribution $\exp[-a|M_{ii} - M_{st}|^2]$ as a result of one additional random diffusion step of length γ amounts to $\gamma^2 a + 1$, and therefore is completely compensated by the compression if $\gamma^2 a = 2|\mathrm{Im}\zeta| / \sqrt{4-\eta^2}$. Hence, for $|\mathrm{Im}\zeta| / \gamma^2 \gg 1$ the stationary distribution is a Gaussian

$$g(M_{11}, M_{22}, \varepsilon, \xi) = \exp\left[-2|\mathrm{Im}\zeta| \frac{\left|M_{11} - e^{i \arccos \eta/2}\right|^2}{\gamma^2 \sqrt{4-\eta^2}}\right]$$
$$\exp\left[-2|\mathrm{Im}\zeta| \frac{\left|M_{22} - e^{-i \arccos \eta/2}\right|^2}{\gamma^2 \sqrt{4-\eta^2}}\right] \frac{\gamma^4 (4-\eta^2)}{4\pi^2 |\mathrm{Im}\zeta|^2}, \tag{7.208}$$

where we have determined the positions of the stationary points

$$M_{st} = \exp\left(\pm i \arccos \eta/2\right)$$

in the approximation $|\zeta| \ll \eta$ and have allowed for the normalization factor.

The inverse Fourier transformation is performed, however, along the contours, where the imaginary parts of both ε and ξ are small as compared to any physical parameter. Therefore the other extreme $|\mathrm{Im}\zeta| / \gamma^2 \ll 1$ is important. In this limit, the distribution tends to its asymptotic form

$$g(M_{11}, M_{22}, \varepsilon, \xi) = \frac{\sqrt{(4-\varepsilon^2)(4-\xi^2)}}{\pi^2 \left(2 + 2M_{11}^2 - 2M_{11}\varepsilon\right)\left(2 + 2M_{22}^2 - 2M_{22}\xi\right)}, \tag{7.209}$$

which is a solution of (7.205) for $\gamma \to 0$, $(\mathrm{Im}\zeta)/\gamma \to 0$ in complete agreement with the distribution (7.107), (7.113). In Fig. 7.11 we show the results of a large number of sequential mappings (7.203) performed numerically, with randomly chosen sequences of E_n for both the $|\mathrm{Im}\zeta| / \gamma^2 \gg 1$ and $|\mathrm{Im}\zeta| / \gamma^2 \ll 1$ cases.

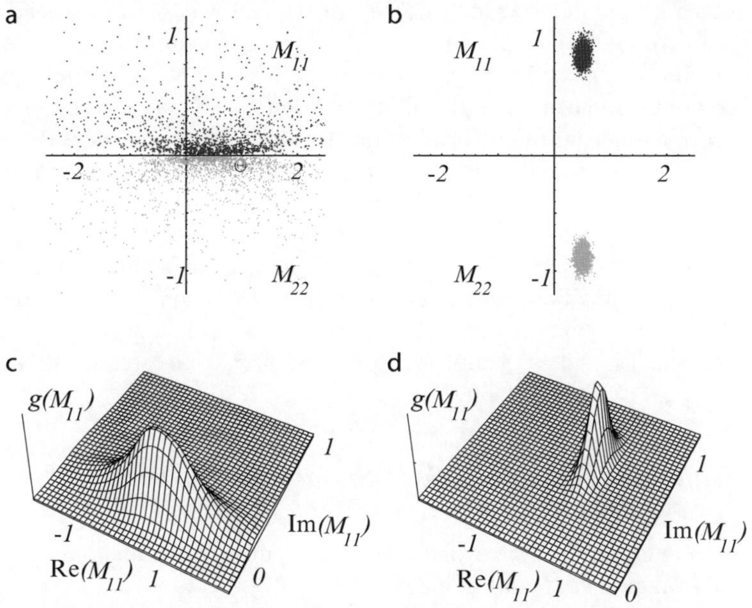

Fig. 7.11 Distribution of the points M_{11} and M_{22} in the upper and lower parts of the complex plane after a large number of mappings (7.203) with randomly distributed E_n. (**a**) A Lorentzian stationary distribution is attained close to the real axis for the case $\mathrm{Im}\zeta/\gamma^2 \ll 1$, whereas in the opposite limit (**b**) one sees a narrow Gaussian distribution around the attracting stable points. The corresponding distribution functions $g(M_{11})$ are shown in (**c**) and (**d**) respectively

One sees that the values of M_{11} and M_{22} obtained for different realizations of the disorder from the same initial conditions are scattered in the upper and lower parts of the complex plane, respectively. Two regimes of small and large ratio $|\mathrm{Im}\zeta|/\gamma^2$ corresponding to distributions (7.208) and (7.209), respectively, are clearly seen.

For the "bosonic density matrix" (7.207) the distribution is similar. For the case $|\mathrm{Im}\zeta|/\gamma^2 \ll 1$ one finds

$$\overline{g}_n(\hat{M},\varepsilon,\xi) = \overline{g}_{\min}(\hat{M},\varepsilon,\xi)e^{-n\lambda(\varepsilon,\xi)},$$
$$\overline{g}_{\min}(\hat{M},\varepsilon,\xi) = \frac{4}{\pi^2\left(\varepsilon+i\sqrt{4-\varepsilon^2}-2M_{11}\right)\left(\xi-i\sqrt{4-\xi^2}-2M_{22}\right)}, \tag{7.210}$$
$$\exp\left[\lambda(\varepsilon,\xi)\right] = \left(\frac{\varepsilon}{2}+i\sqrt{1-\frac{\varepsilon^2}{4}}\right)\left(\frac{\xi}{2}-i\sqrt{1-\frac{\xi^2}{4}}\right) = e^{i\arccos\varepsilon/2-i\arccos\xi/2},$$

which satisfy the recurrence equation (7.207) at $\gamma = 0$. We note that substitution of (7.209), (7.210) into (7.204), (7.206), respectively, yields the "density matrices", which do not change as a result of the transformations (7.162), (7.198), thus remaining constants in the interaction representation.

Allowance for the finite width γ of the distribution of E_n can be considered as a perturbation of (7.209), (7.210). For the distribution function $\overline{g}_{\min}$ of (7.210) this yields a small ($\sim \gamma^2$) correction, and also results in a correction for the eigenvalue parameter λ of the same order of magnitude. The latter is of major importance, since this very correction leads to the localization effect and thereby changes the form of the asymptotic population distribution in the long-time limit. Corrections resulting from the difference between the identity operator applied to $\overline{g}_{\min}$ and a convolution with a sharp symmetric distribution (7.130) are proportional to the second derivative of this function, which should be projected back to $\overline{g}_{\min}$, according to the standard prescription of perturbation theory. Performing the calculations for $\zeta \ll 1$ one finds $\lambda = i\zeta/\sqrt{4-\eta^2} - \gamma^2/2(4-\eta^2)$ which coincides with the index of the exponent suggested by the Liouville equation (7.166) in the interaction representation.

7.6.3 *Time Dependence of the Population Distribution*

Now we can interpret the time evolution of the population distribution $\rho_n(t)$ of the relay (7.183) in terms of "density matrices" (7.204), (7.206) for fictitious particles evolving in fictitious time n. We denote the running value of this fictitious time by the index k. From $k = -\infty$ until the moment $k = 0$, the "n-evolution" corresponds to the joint propagation of bosons and bi-fermions, and is given by their stationary "density matrix" (7.204). Note that in this case the matrix contains only the part $\tilde{\overline{\rho}}_n$ given by (7.196) of the entire density matrix (7.186). At the point $k = 0$, the Grassmann operators $\overline{\chi}_0'\overline{\chi}_0$ entering the pre-exponential factor in the integrand of (7.183) annihilate one of the fermions, and as a result, the remaining fermion becomes decoupled from the bosons, propagating without change as we have seen in the context of (7.194). Note in this case that only the matrix element $\overline{\rho}_{1010}$ of (7.186) is different from zero and

$$\tilde{\overline{\rho}}_k(\overline{\chi}, \chi, z; \overline{\chi}', \chi', z') = \tilde{\rho}_k(z, z')\overline{\chi}_k'\overline{\chi}_k.$$

The bosons, in turn, start to experience decay, and the dominating contribution to their density matrix $\tilde{\rho}_k(z, z')$ (7.206) comes from the distribution given by the eigenfunction $\tilde{g}_{\min}(M_{11}', M_{22}', \varepsilon, \xi)$ with the smallest eigenvalue $e^{\lambda(\varepsilon,\xi)}$ corresponding to the slowest decay. At the point $k = n$, the Grassmann operators $\chi_n'\chi_n$ restore the fermion, and the further propagation again corresponds to the stationary "density matrix" (7.204). We illustrate this in Fig. 7.12.

The population $\rho_n(t)$ can therefore be written as overlap integrals,

$$\rho_n(t) = \int_{-\infty-i\nu}^{\infty-i\nu} \int_{-\infty+i\nu}^{\infty+i\nu} \frac{d\xi\, d\varepsilon}{4\pi^2}\, e^{-i(\varepsilon-\xi)t} e^{n\lambda(\varepsilon,\xi)} \int \frac{dz^* dz dz'^* dz'}{4\pi^2} d\chi d\overline{\chi} d\chi' d\overline{\chi}'$$
$$\overline{\rho}(\overline{\chi}, \chi, z; \overline{\chi}', \chi', z')\overline{\chi}'\overline{\chi}\tilde{\rho}(z, z')\chi'\chi \qquad (7.211)$$

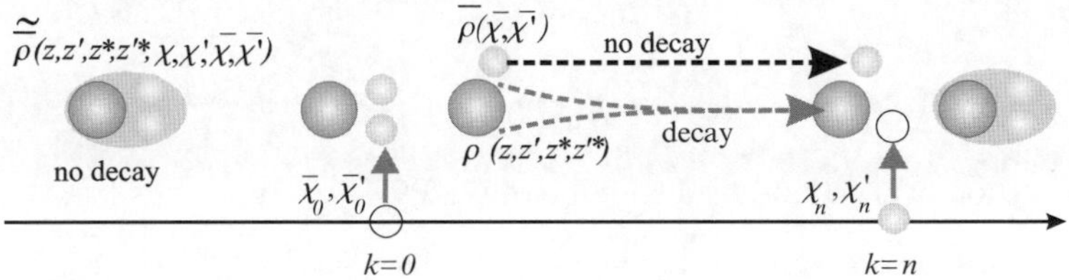

Fig. 7.12 Representation of the localization with the help of the Liouville equation for a decaying compound. The dependence of the population on n can be interpreted as a Liouvillian evolution of a compound comprising "bosons" and "fermions" along a fictitious time axis n. The stationary compound being perturbed at the point $k = 0$ becomes decomposed to "fermions" propagating without decay and "bosons" exponentially decaying with the course of time. In order to determine the population ρ_n at the point n one has to find an overlap of the decaying density matrix at the point $k = n$ with the density matrix of the stationary compound

of the "density matrices", Eqs. (7.204, 7.206), of fictitious particles, where we have taken into account that these "density matrices" do not change with n, apart from attenuation by a factor $e^{n\lambda(\varepsilon,\xi)}$. After substitution of the right-hand sides of these equations and after performing integrations over "fermionic" and "bosonic" coordinates, one obtains another representation of the overlap integral,

$$\rho_n(t) = \int_{-\infty-i\nu}^{\infty-i\nu} \int_{-\infty+i\nu}^{\infty+i\nu} \frac{d\xi\, d\varepsilon}{4\pi^2}\, e^{-i(\varepsilon-\xi)t-n\lambda(\varepsilon,\xi)}$$

$$\int dM_{11} dM_{22}\; g(M_{11}, M_{22}, \varepsilon, \xi)\overline{g}_{\min}(M_{11}, M_{22}, \varepsilon, \xi), \quad (7.212)$$

now in terms of the distribution functions (7.205) and (7.207). Note that substituting of the distribution functions (7.209)–(7.210), which are exact in the limit $\gamma \to 0$, into this equation, followed by integration over the matrix elements dM_{11}, dM_{22} , dM'_{11}, and dM'_{22} along the real axis and over $d\xi\; d\varepsilon$, one finds

$$\rho_n(t) = |J_n(2t)|^2\,, \quad (7.213)$$

in complete agreement with (7.11) for the relay of resonant levels. Evidently in this limit, one cannot see any trace of the disorder.

In order to take into account the effect of the disorder, one therefore has to allow for the deviation of the distribution functions entering (7.205) and (7.207) from their asymptotic forms given by (7.209) and (7.210). Let us assume that for both g and $\overline{g}_{\min}$ this deviation can be expressed in the form of a convolution of the asymptotic distribution with a narrow function of width vanishing in the limit $\gamma \to 0$. Although each of the functions g and $\overline{g}_{\min}$ may have an individual narrow counterpart, the population distribution is governed by the convolution g_γ of these two counterparts. This modifies (7.212)

$$\rho_n(t) = \int_{-\infty-i\nu}^{\infty-i\nu} \int_{-\infty+i\nu}^{\infty+i\nu} \frac{d\xi\, d\varepsilon}{4\pi^2} e^{-i(\varepsilon-\xi)t-n\lambda(\varepsilon,\xi)} \int dM_{11} dM_{22}\, du\, g_\gamma(u)$$
$$g(M_{11}+u, M_{22}+u, \varepsilon, \xi)\overline{g}_{\min}(M_{11}-u, M_{22}-u, \varepsilon, \xi) \quad (7.214)$$

and integration over $\hat{M}$, after substitution of (7.209)–(7.210), results in

$$\rho_n(t) = \int_{-\infty-i\nu}^{\infty-i\nu} \int_{-\infty+i\nu}^{\infty+i\nu} \frac{d\xi\, d\varepsilon}{4\pi^2} e^{-i(\varepsilon-\xi)t-n\lambda(\varepsilon,\xi)} \int du \frac{g_\gamma(u)}{\left(-i\sqrt{4-\varepsilon^2}+u\right)\left(i\sqrt{4-\xi^2}+u\right)}, \quad (7.215)$$

where $\lambda \simeq i(\arccos \varepsilon/2 - \arccos \zeta/2) - \gamma^2/2(4-\eta^2)$. Note that replacement of the distribution $g_\gamma(u)$ by the Dirac δ-function, followed by integration over ε and ξ, immediately yields a result similar to that of (7.159).

After the replacements $\varepsilon \to \cos\phi$ and $\xi \to \cos\varphi$ we take into account the contributions of the poles at the points $\phi = \arcsin \frac{u}{2i}$, $\phi = \pi - \arcsin \frac{u}{2i}$, $\varphi = -\arcsin \frac{u}{2i}$, and $\varphi = \pi + \arcsin \frac{u}{2i}$ to the integrals over ϕ and φ and find

$$\rho_n(t) = \int du \frac{g_\gamma(u)u^2}{\cos^2 \arcsin u/2i} \exp\left[-2in \arcsin \frac{u}{2i} - \frac{n\gamma^2}{u^2}\right]$$
$$-\mathrm{Re} \int du \frac{g_\gamma(u)u^2}{\cos^2 \arcsin u/2i} \exp\left[-4it \cos \arcsin \frac{u}{2i} - in\pi - \frac{n\gamma^2}{u^2}\right]. \quad (7.216)$$

In the limit $n \to \infty$, $t \to \infty$, this integral has the asymptotic form resulting from the contribution of the saddle points $u = (2\gamma^2)^{1/3}$, $u = \left(\frac{n\gamma^2}{it}\right)^{1/3}$

$$\rho_n(t) \simeq \frac{A}{\sqrt{n}} \exp\left[-3(\gamma/2)^{2/3} n\right] - B\, \frac{n^{5/6}}{t^{4/3}} \left|g_\gamma\left[\left(\frac{n\gamma^2}{it}\right)^{1/3}\right]\right|$$
$$\exp\left[-\frac{3^{3/2}\gamma^{2/3}}{2} n^{1/3} t^{2/3}\right] \cos\left[\frac{3\gamma^{2/3}}{2} n^{1/3} t^{2/3} - n\pi - \frac{2}{3}\pi + \theta_\gamma\right], \quad (7.217)$$

where

$$A \simeq \sqrt{\frac{4\pi}{3}}\, g_\gamma(2^{1/3}\gamma^{2/3}) 2^{5/6} \gamma^{5/3};$$
$$B = \sqrt{\frac{2\pi}{3}} \gamma^{5/3}; \quad \theta_\gamma = \arg g_\gamma \left[\left(\frac{n\gamma^2}{it}\right)^{1/3}\right]. \quad (7.218)$$

This means that an exponentially localized distribution of localization length $\sim \gamma^{-2/3}/3$ is attained, while the deviation from this distribution has the asymptotic time-dependence $\exp\left(-\text{const}\, n^{1/3} t^{2/3}\right)$. We note that the particular form of the distribution function g_γ does not affect the shape of the asymptotic population distribution, but only influences the pre-exponential factor of the time-dependent deviation from this distribution, which vanishes with the course of time.

Chapter 8
Composite Complex Quantum Systems

We now consider level grouping in multilevel systems other than the one-dimensional relay of isolated levels discussed in the previous chapter. The population dynamics of such systems can display various manifestations that are typical of the systems already considered, such as exponential decay at short times, quantum recurrences and revivals, cooperative behavior, and quantum localization. Such systems can also demonstrate diffusive behavior. Typically, a problem of this type does not correspond to any hidden symmetry resulting in specific commutation relations among different parts of the Hamiltonian that comprise a closed subalgebra, and the evolution occurs in the entire Hilbert space, being constrained only by energy conservation. We first consider several problems for the relay of multilevel bands, for which diffusive evolution is strongly affected by correlations between the energy position and the coupling of the levels, and then address a general question concerning quantum dynamics of the systems that in the classical limit manifest random walk behavior. We relate the quantum localization phenomenon to the dimensionality of those random walks responsible for classical returns and to the quantum recurrences. At a particular example we also consider the phenomenon of dynamic localization, typical of complex quantum systems perturbed by a periodic interaction. This phenomenon manifests itself in the localization of the populated states in a finite energy domain, that is, in the fact that due to the quantum interference phenomenon, a complex quantum system does not absorb energy from the time-dependent external field in spite of the fact that the resonance conditions are satisfied. The case of weak dynamic localization when a complex system absorbs energy logarithmically in time will also be addressed.

8.1 Relay of Multilevel Bands

Consider a number of statistically equivalent, or identical spectral bands arranged in a one-dimensional array where each of the bands is coupled to each of its two neighbors via an interaction. The unperturbed Hamiltonian of the n-th band is given

V.M. Akulin, *Dynamics of Complex Quantum Systems*, Theoretical and Mathematical Physics, DOI 10.1007/978-94-007-7205-2_8,

by its eigenvalues $E_{n,m}$ enumerated by the index m, while the coupling matrices are $V_{n,m}^{n+1,m'}$ and $V_{n,m}^{n-1,m'}$. The Schrödinger equation for the system reads

$$i\dot{\psi}_{n,k} = E_{n,k}\psi_{n,k} + \sum_{m}\left(V_{n,k}^{n+1,m}\psi_{n+1,m} + V_{n,k}^{n-1,m}\psi_{n-1,m}\right), \tag{8.1}$$

and serves as a model for the physical problems corresponding to (2.4) and (2.56). In this section we consider three particular cases of (8.1). The first case relates to the situation where the total width of each of the bands is negligibly small as compared to the cooperative coupling, that is, $\max\left(E_{n,k} - E_{n,k'}\right) \ll N\left\langle V^2\right\rangle$ where N is the total number of levels in each band and $\left\langle V^2\right\rangle$ is the mean squared interaction, as earlier. During a period of time shorter than $1/\max\left(E_{n,k} - E_{n,k'}\right)$ we can neglect the difference in level energies and consider the system as resonant. In the opposite extreme of broad bands, the total number N of levels in each band plays an unimportant role, while the dynamics of the system is governed by the density g of the band states. For time intervals shorter than the quantum recurrence time, the band populations satisfy the discrete diffusion equation or a master equation in the case of statistically non-equivalent bands.

Earlier, in Sect. 4.5, we have considered the influence of correlations among the coupling amplitudes and energy positions of the band levels on the population

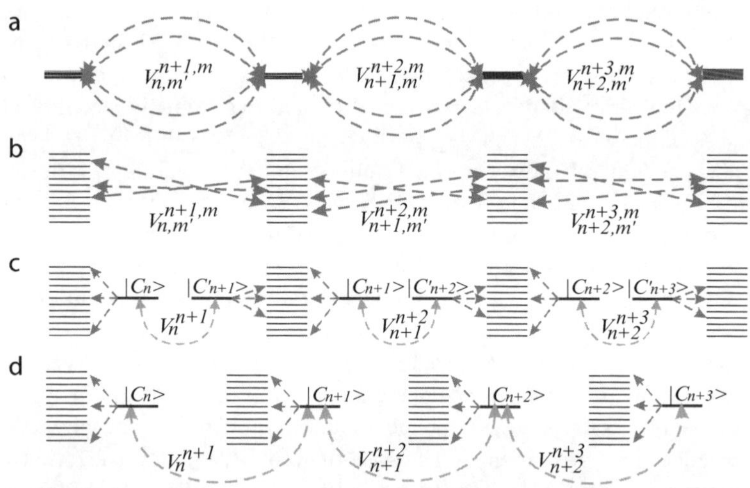

Fig. 8.1 Relay of bands. (**a**) The bands can be considered as degenerate levels when the cooperative coupling $\langle V^2\rangle N \gg \max|\Delta_{n,m} - \Delta_{n,m'}|$. (**b**) For the broad bands the population remains in the vicinities of resonances, whose total populations satisfy the master equation. (**c**) Correlation among the matrix elements $\langle V_{n,m}^{n+1,k} V_{n,m'}^{n+1,k'}\rangle \neq 0$ occurs when the rank of the interaction matrix is considerably smaller than its order. For rank 1 the correlation can be taken into account by introducing eigenstates of the coupling matrix, corresponding to non-vanishing eigenvalues. For the non-Hermitian matrix $V_{n,m}^{n+1,k}$ the right eigenvector $|C_n\rangle$ may not coincide with the left one $|C_n'\rangle$. (**d**) When the right and the left eigenstates coincide the system resembles a relay of decaying levels

dynamics of two-band systems. The same influence exists in multiband problems, where the correlations are responsible for the initial stage of the evolution, which resembles the ballistic evolution of a classical particle preceding its diffusion in a disordered medium. As for the two-band systems, the correlations can manifest themselves in the ranks of the coupling matrices $V_{n,k}^{n\pm 1,m}$, which can be considerably smaller than the order of the matrices. In this case one can choose the eigenstates of the coupling matrices corresponding to non-zero eigenvalues as basis states, appending this incomplete basis by orthogonal combination of the eigenstates of unperturbed band Hamiltonians, in the manner shown in Fig. 4.24b) for two-band systems. Four particular cases considered in this section are sketched in Fig. 8.1.

For the initial condition we always take $\psi_{n,k}(t=0) = \delta_n^0 \delta_k^0$, since for any other Kronecker δ-like initial condition the consideration is similar, whereas in general, initial conditions can be given as a linear combination of the δ-like ones. We choose the energy position $E_{0,0}$ of the level $n = 0$, $k = 0$ as the energy reference point and denote $E_{n,k} - E_{0,0} = \Delta_{n,k}$. Fourier transformation in time of (8.1) therefore adopts the form

$$\varepsilon\psi_{n,k}(\varepsilon) = \Delta_{n,k}\psi_{n,k} + \sum_m \left[V_{n,k}^{n+1,m}\psi_{n+1,m}(\varepsilon) + V_{n,k}^{n-1,m}\psi_{n-1,m}(\varepsilon)\right] + i\delta_n^0\delta_k^0, \tag{8.2}$$

where, as earlier, we denote by the same symbol $\psi_{n,k}$ the time-dependent amplitude $\psi_{n,k}(t)$ and its Fourier transform $\psi_{n,k}(\varepsilon)$, since the argument clearly specifies which one of these two objects we have in mind. This equation is a generalization of (4.5) for the case of many bands.

Green's function, corresponding to (8.2), can be represented as a perturbation series (4.13) for the resolvent, in complete analogy with the two-band systems with random matrix elements, and can be depicted as diagrams Fig. 4.4, which after the ensemble average for the case $\left\langle V_{n,k}^{n+1,m} V_{n,k'}^{n+1,m'} \right\rangle = \left\langle V^2 \right\rangle \delta_m^{m'} \delta_k^{k'}$, adopt a tree-like structure (Fig. 4.7). This structure, resulting in the renormalized factors $X_n(\varepsilon)$ satisfying the equations

$$\begin{aligned} X_{n,k}(\varepsilon) &= \frac{1}{\varepsilon - \Delta_{n,k} - \sum_k \langle V^2\rangle X_{n+1,k}(\varepsilon) - \sum_k \langle V^2\rangle X_{n-1,k}(\varepsilon)} \\ X_{n,k}(\xi) &= \frac{1}{\xi - \Delta_{n,k} - \sum_k \langle V^2\rangle X_{n+1,k}(\xi) - \sum_k \langle V^2\rangle X_{n-1,k}(\xi)} \end{aligned} \tag{8.3}$$

by analogy to (4.20), (4.21), is shown in Fig. 4.8. This is also valid for the perturbation series for ensemble averaged populations, shown in Figs. 4.10 and 4.11. The only difference is in the fact that now instead of two alternating factors $X_m(\varepsilon)X_m(\xi)$ and $X_n(\varepsilon)X_n(\xi)$, we have a sequence of factors $X_{n,k}(\varepsilon)X_{n,k}(\xi)$ where the running indices n for two neighboring factors are different by ± 1, as shown in Fig. 8.2.

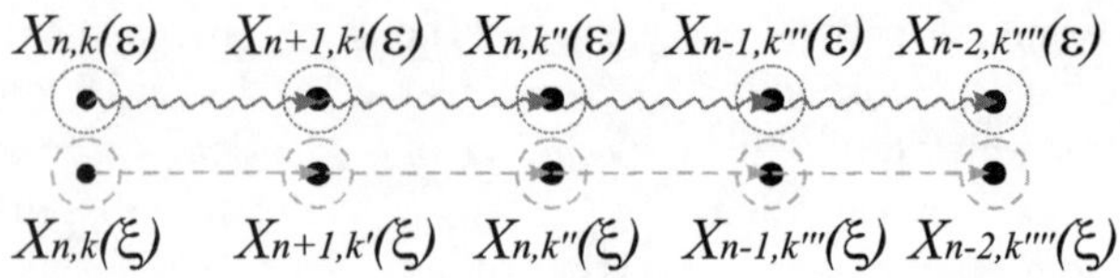

Fig. 8.2

By analogy to (4.27) and Fig. 4.14 one can write the equation for the population

$$\begin{aligned}\rho_n(\varepsilon,\xi) = {} & \rho_{n-1}(\varepsilon,\xi)\textstyle\sum_k \langle V^2\rangle X_{n,k}(\varepsilon)X_{n,k}(\xi) \\ & +\rho_{n+1}(\varepsilon,\xi)\textstyle\sum_k \langle V^2\rangle X_{n,k}(\varepsilon)X_{n,k}(\xi) \\ & +\delta_n^0\delta_k^0 X_{0,0}(\varepsilon)X_{0,0}(\xi),\end{aligned} \quad (8.4)$$

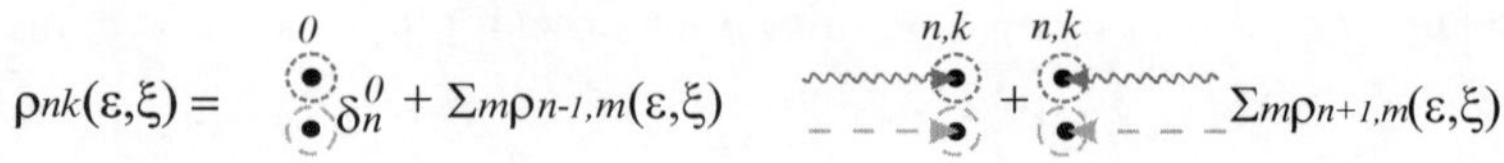

Fig. 8.3

which corresponds to the diagram shown in Fig. 8.3.

8.1.1 Degenerate Bands with Random Coupling

For an infinite relay of degenerate bands shown in Fig. 8.1a the factors $X_{n,k}(\varepsilon)$ and $X_{n,k}(\xi)$ in (8.3) do not depend on the index k thus satisfying the equations

$$\begin{aligned}X_n(\varepsilon) &= \left[\varepsilon - \langle V^2\rangle N X_{n+1}(\varepsilon) - \langle V^2\rangle N X_{n-1}(\varepsilon)\right]^{-1} \\ X_n(\xi) &= \left[\xi - \langle V^2\rangle N X_{n+1}(\xi) - \langle V^2\rangle N X_{n-1}(\xi)\right]^{-1}.\end{aligned} \quad (8.5)$$

For the case of statistically equivalent bands, by analogy to (4.55) they have n-independent solutions

$$\begin{aligned}X_{n,k}(\varepsilon) = X(\varepsilon) &= \tfrac{2}{\varepsilon - \sqrt{\varepsilon^2 - 8\langle V^2\rangle N}}, \\ X_{n,k}(\xi) = X(\xi) &= \tfrac{2}{\xi + \sqrt{\xi^2 - 8\langle V^2\rangle N}},\end{aligned} \quad (8.6)$$

different from (4.55) only by a factor of 2 in front of the number of states N in the degenerate bands, which allows for the presence of two neighbors for each band.

There are many ways to solve (8.4), with the factors $X_{n,k}(\varepsilon)$ and $X_{n,k}(\xi)$ given by (8.6), which do not depend on indices. We choose one, based on the calculation of the trajectory sum for the diagrams of the type shown in Fig. 8.2. For this purpose, we note that the total number $T_{l,n}$ of trajectories of length l that start at the band number 0 and finish at the band number n, is given by the binomial coefficient

$$T_{l,n} = C_l^{(n+l)/2} = \frac{l!}{(\frac{l+n}{2})!(\frac{l-n}{2})!}, \tag{8.7}$$

as one can conclude from the sketch shown in Fig. 8.4a. Each trajectory of

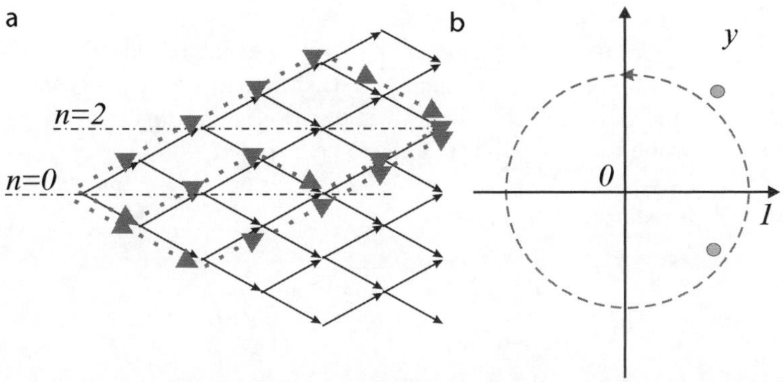

Fig. 8.4 (**a**) A number of random walk trajectories of length l (*dotted arrows*) that result in the displacement n, given by the number of choices of $k+n$ steps upward and k steps downward from the total number of $l = n + 2k$ steps. (**b**) Integration contours (*dashed lines*) for calculation of the generating function $\sum_{n=0}^{\infty} T_{l,n} x^n$ corresponding to these numbers. Poles of the integrand (8.11) are shown by *circles*

length l brings a contribution $[X(\varepsilon)X(\xi)]^{l+1} (\langle V^2 \rangle N)^l$ to the trajectory sum for populations, and therefore the net sum of all trajectories reads

$$\rho_n(\varepsilon,\xi) = \sum_{l=n}^{\infty} T_{l,n} [X(\varepsilon)X(\xi)]^{l+1} (\langle V^2 \rangle N)^l . \tag{8.8}$$

The sum in (8.8) can be found with the help of the generating function

$$\Phi_n(x) = \sum_{l=n}^{\infty} \frac{l!}{(\frac{l+n}{2})!(\frac{l-n}{2})!} x^l = \sum_{l=0}^{\infty} \frac{l!}{(\frac{l+n}{2})!(\frac{l-n}{2})!} x^l, \tag{8.9}$$

which can be extended to the integer $l < n$ since in this case, the second factorial in the denominator takes an infinite value. We calculate the numbers $T_{l,n}$ of (8.7) with the help of the integral representation

$$T_{l,n} = C_l^{(n+l)/2} = \frac{1}{l+1} \frac{1}{B\left(\frac{l+n}{2}+1, \frac{l-n}{2}+1\right)} = \frac{2^l}{2\pi} \int_{-\pi}^{\pi} dt \, \cos(nt) \cos^l t \tag{8.10}$$

for the B-function, which is valid for l and n of the same parity. Substitution of (8.10) into (8.9), followed by summation of the geometric series yields

$$\begin{aligned}\Phi_n(x) &= \sum_{l=0}^{\infty} \tfrac{2^l}{2\pi} \int_0^{2\pi} dt \cos(nt) \cos^l t\, x^l \\ &= \frac{1}{2\pi} \int_0^{2\pi} dt\, \tfrac{\cos(nt)}{1-2x\cos t} = \frac{(2x)^n}{2\left(1+\sqrt{1-4x^2}\right)^n \sqrt{1-4x^2}},\end{aligned} \tag{8.11}$$

where in the last line after the replacement $e^{it} \to y$, the integration contour becomes a circle of unit radius around the point $y = 0$ shown in Fig. 8.4b, separating the singularities of the integrand. For $n \geq 0$ integration along this contour gives the result directly, while for $n \leq 0$ the sign has to be changed.

Substituting $\varepsilon \to \sqrt{8\langle V^2\rangle N} \cos\phi$, $\xi \to \sqrt{8\langle V^2\rangle N} \cos\varphi$ into (8.6) we find the explicit expressions for the factors

$$\begin{aligned}1/X(\varepsilon) &= \sqrt{2\langle V^2\rangle N} e^{-i\phi}, \\ 1/X(\xi) &= \sqrt{2\langle V^2\rangle N} e^{i\varphi},\end{aligned} \tag{8.12}$$

and making the replacement

$$x \to X(\varepsilon)X(\xi)\left(\langle V^2\rangle N\right) = \frac{1}{2} e^{i\phi - i\varphi} \tag{8.13}$$

in (8.11) we obtain

$$\rho_n(\phi, \varphi) = \frac{e^{i(n+1)(\phi-\varphi)}}{2\langle V^2\rangle N \left(1+\sqrt{1-e^{2i(\phi-\varphi)}}\right)^n \sqrt{1-e^{2i(\phi-\varphi)}}} \tag{8.14}$$

for the populations (8.8). The inverse Fourier transform of $\rho_n(\varepsilon, \xi)$ in the coordinates (ϕ, φ) therefore reads

$$\rho_n(\tau) = \frac{1}{\pi^2} \int\limits_{C_2 C_1} d\phi d\varphi\, \frac{\sin\phi \sin\varphi\, e^{i(n+1)(\phi-\varphi) - i\tau(\cos\phi - \cos\varphi)}}{\left(1+\sqrt{1-e^{2i(\phi-\varphi)}}\right)^n \sqrt{1-e^{2i(\phi-\varphi)}}} \tag{8.15}$$

where $\tau = t\sqrt{8\langle V^2\rangle N}$, and the integration contours for ϕ and φ coincide with the contours C_1 and C_2, respectively, shown in Fig. 4.17b.

For further simplification we introduce the variables

$$\begin{aligned}v &= \tfrac{\phi+\varphi}{2} \\ u &= \varphi - \phi.\end{aligned} \tag{8.16}$$

Due to the periodicity of the integrand (8.15) the integration contours for u and v are the same intervals $(-\pi, \pi)$, and merely shifted by $-i0$ for u. Note that this shift ensures the convergence of the sum (8.9) for the generating function. We therefore arrive at

$$\rho_n(\tau) = \int_{-\pi}^{\pi} \int_{-\pi-i0}^{\pi-i0} dudv \, \frac{[\cos u - \cos 2v] \, e^{i2\tau \sin(u/2) \sin v} e^{-i(n+1)u}}{2\pi^2 \left(1 + \sqrt{1 - e^{-2iu}}\right)^n \sqrt{1 - e^{-2iu}}}$$
$$= \int_{-\pi-i0}^{\pi-i0} du \, \frac{\cos u \, J_0(2\tau \sin \frac{u}{2}) - J_2(2\tau \sin \frac{u}{2})}{\pi \left(1 + \sqrt{1 - e^{-2iu}}\right)^n \sqrt{1 - e^{-2iu}}} e^{-i(n+1)u} \tag{8.17}$$

where in the last line the integral representation (4.60) of the Bessel functions $J_k(x)$ has been employed.

In the asymptotic of large n and large τ the main contribution comes from the domains $u \sim 1/n^2$ and $u - \pi \sim 1/n^2$ where the denominator is not an exponentially small value. Replacing $\left(1 + \sqrt{1 - e^{-2iu}}\right)^n$ in these domains by its approximate value $\exp n\sqrt{2iu}$ we find

$$\rho_n(\tau) = \int_{-\infty}^{\infty} du \, \frac{J_0(\tau u) - J_2(\tau u)}{\pi \sqrt{2iu}} e^{-n\sqrt{2iu}}$$
$$+ \int_{-\infty}^{\infty} du \, \frac{J_0(2\tau) + J_2(2\tau)}{\pi \sqrt{2iu}} e^{-in\pi} e^{-n\sqrt{2iu}} \tag{8.18}$$

where the first and the second integral correspond to the first and the second domain, respectively, and the integration limits for these exponentially converging integrals are extended to infinity. For large n and τ the second integral scales as $1/n\sqrt{\tau}$ and can be ignored, as compared to the first integral, which scales as $1/\sqrt{\tau}$. Replacing positive u by y^2/τ and negative u by $-y^2/\tau$ in the first integral we arrive at

$$\rho_n(\tau) \simeq \frac{2\sqrt{2}}{\pi\sqrt{\tau}} \int_0^{\infty} dy \, \left[J_0(y^2) - J_2(y^2)\right] e^{-Yy} \cos(Yy + \pi/4) \tag{8.19}$$

where $Y = n/\sqrt{\tau}$. The integral on the right-hand side of (8.19) is a universal function $\sqrt{\pi}\Phi(Y)/4$ of the parameter Y, which for large Y is not very much different from a Gaussian function $\exp\left(-Y^2/2\right)$, so that the population distribution (8.19) resembles a diffusive one.

In Fig. 8.5a we show the total population of the first few degenerate levels calculated with the help of the exact expression (8.17). One sees dying oscillations resembling those shown in Fig. 4.19c for two degenerate levels. Over the course of time, the population distribution becomes broader and its width increases as the

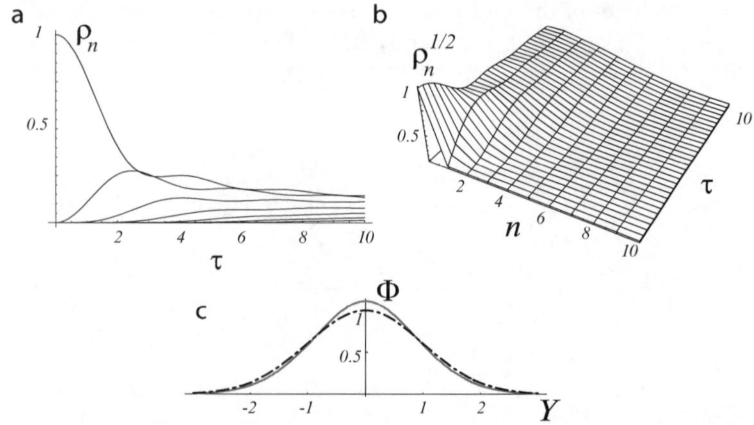

Fig. 8.5 Population distribution in an infinite relay of degenerate levels coupled by random matrices. Initially all the population is in level $n = 0$. The time t is scaled as $\tau = \sqrt{8N\langle V^2\rangle}t$. (**a**) Time dependence of the population for the first several levels. (For negative n the populations are identical.) (**b**) Dependence of population on time and index n. (Nonlinear scale $\sqrt{\rho_n}$ for the population.) (**c**) Universal asymptotic profile $\Phi(Y)$ of the population distribution (*solid line*) depends only on the parameter $Y^2 = n^2/\tau$, the same as that which enters the diffusion profile, $\exp(-Y^2/2)$ (*dash-dot line*)

square root of time, as one can see in Fig. 8.5b, whereas the asymptotic profile $\Phi(Y)$ of the distribution is shown in Fig. 8.5c.

The validity of the approach is limited in time by the assumption of the tree-like topological structure of the diagrams shown in Fig. 4.7. Indeed, for a one-dimensional random walk the mean width of the diffusive population distribution, and hence the mean displacement of the random trajectory is of the order of the square root of the random steps. In the context of the problem under consideration, this means that at the time t, a typical length of the trajectory is of the order of $\tau = t\sqrt{8\langle V^2\rangle N}$ and a typical width of the distribution is $n_t \sim \sqrt{\tau} \sim t^{1/2}(8\langle V^2\rangle N)^{1/4}$. This means that the trajectory returns $\tau/n_t \sim t^{1/2}(8\langle V^2\rangle N)^{1/4}$ times to the initially populated level, and this number of returns should remain much smaller as compared to the total degeneracy N of the level. Therefore, at $t \sim N^{3/2}(8\langle V^2\rangle)^{-1/2}$, the system no longer conforms to the model employed. At this time, the population recurrences to energy eigenstates become important, and as a consequence one may encounter a multiband analog of the Localization effect of Sect. 7.6, which we will consider in Sect. 8.2 in the context of a generic random walk. Note that when the multiband structure describes a complex system subject to a periodic excitation whereas the band number corresponds to the quasienergy index as shown in Fig. 6.14, one encounters a similar situation known as "dynamical localization", when the process of energy absorption stops as a result of quantum interference, which will be considered later on in this chapter.

8.1.2 Non-degenerate Bands with Random Coupling

We now consider the case of non-degenerate bands, as shown in Fig. 8.1b. The three-term recurrence relation (8.4) instead of the operators $X(\varepsilon)$ and $X(\xi)$ of (8.6) now contains the operators (4.33) for non-degenerate bands, which are slightly modified by the presence of two neighbors of each band and read

$$X_{n,k}(\varepsilon) = \frac{1}{\varepsilon - \Delta_{n,k} + i\pi \left\langle V_{n+1}^2 \right\rangle g_{n+1} + i\pi \left\langle V_n^2 \right\rangle g_{n-1}},$$
$$X_{n,k}(\xi) = \frac{1}{\xi - \Delta_{n,k} - i\pi \left\langle V_{n+1}^2 \right\rangle g_{n+1} - i\pi \left\langle V_n^2 \right\rangle g_{n-1}}, \tag{8.20}$$

in the general case of bands of different density of states g_n and different mean square interactions $\left\langle V_n^2 \right\rangle$ coupling the band n with the band $n-1$. Substitution of (8.20) into (8.4), after summation over k, yields

$$\rho_n(\varepsilon,\xi) = \frac{2\pi i\, \rho_{n-1}(\varepsilon,\xi)\left\langle V_n^2 \right\rangle g_n}{\varepsilon-\xi+2i\pi\left\langle V_{n+1}^2 \right\rangle g_{n+1}+2i\pi\left\langle V_n^2 \right\rangle g_{n-1}}$$
$$+\frac{2\pi i\, \rho_{n+1}(\varepsilon,\xi)\left\langle V_{n+1}^2 \right\rangle g_n}{\varepsilon-\xi+2i\pi\left\langle V_{n+1}^2 \right\rangle g_{n+1}+2i\pi\left\langle V_n^2 \right\rangle g_{n-1}} + \delta_n^0 X_{0,0}(\varepsilon) X_{0,0}(\xi). \tag{8.21}$$

Introducing the variables $\zeta = \varepsilon - \xi$, $\eta = (\varepsilon + \xi)/2$ and integrating this equation over $d\eta$ one finds

$$\rho_n(\zeta) = \frac{2i\pi\rho_{n-1}(\zeta)\left\langle V_n^2 \right\rangle g_n}{\zeta+2i\pi\left\langle V_{n+1}^2 \right\rangle g_{n+1}+2i\pi\left\langle V_n^2 \right\rangle g_{n-1}}$$
$$+\frac{2i\pi\rho_{n+1}(\zeta)\left\langle V_{n+1}^2 \right\rangle g_n}{\zeta+2i\pi\left\langle V_{n+1}^2 \right\rangle g_{n+1}+2i\pi\left\langle V_n^2 \right\rangle g_{n-1}} + \frac{2i\pi\delta_n^0}{\zeta+2i\pi\left\langle V_1^2 \right\rangle g_1+2i\pi\left\langle V_0^2 \right\rangle g_{-1}}. \tag{8.22}$$

for $\rho_n(\zeta) = \int \rho_n(\zeta,\eta)\, d\eta/2\pi$. This immediately yields

$$\zeta\rho_n(\zeta) = 2i\pi \left\langle V_n^2 \right\rangle \left[\rho_{n-1}(\zeta) g_n - \rho_n(\zeta) g_{n-1}\right]$$
$$+2i\pi \left\langle V_{n+1}^2 \right\rangle \left[\rho_{n+1}(\zeta) g_n - \rho_n(\zeta) g_{n+1}\right] + \delta_n^0 \delta_k^0, \tag{8.23}$$

and after inverse Fourier transformation over ζ results in the master equation

$$\dot\rho_n(t) = 2\pi \left\langle V_n^2 \right\rangle \left[\rho_{n-1}(t) g_n - \rho_n(t) g_{n-1}\right]$$
$$+2\pi \left\langle V_{n+1}^2 \right\rangle \left[\rho_{n+1}(t) g_n - \rho_n(t) g_{n+1}\right] + \delta_n^0 \delta(t). \tag{8.24}$$

where the last term containing the Dirac δ-function allows for the initial condition $\rho_n(t=0) = \delta_n^0$. One can interpret this term as a unit population instantaneously injected into the band $n = 0$ at time $t = 0$.

For the population per quantum state $\tilde{\rho}_n(t) = \rho_n(t)/g_n$, (8.24) takes a more symmetric form

$$\dot{\tilde{\rho}}_n(t) = D_n^- [\tilde{\rho}_{n-1}(t) - \tilde{\rho}_n(t)] + D_n^+ [\tilde{\rho}_{n+1}(t) - \tilde{\rho}_n(t)] + \frac{\delta_n^0 \delta(t)}{g_0}, \quad (8.25)$$

where $D_n^+ = 2\pi g_{n+1} \langle V_{n+1}^2 \rangle$ and $D_n^- = 2\pi g_{n-1} \langle V_n^2 \rangle$ denote the transition probability to the neighboring bands with the larger and smaller n, respectively. An attempt to solve this equation analytically leads to the same problems that have already been discussed in the context of the Schrödinger equation (7.1) for the relay of isolated levels—there are only a small number of particular cases of dependences D_n^- and D_n^+ where the problem possesses an exact analytic solution. However, since the eigenfunctions of (8.25) decay exponentially in time, the approximate WKB calculations describing the time evolution of the "wavepackets" turn out to be rather accurate, and applicable to most cases.

Constant Transition Rates

For the case of n-independent coefficients D_n^- and D_n^+, solution of (8.25) can be found with the help the generating function

$$\Phi(x,t) = \sum_{n=-\infty}^{\infty} e^{inx} \tilde{\rho}_n(t). \quad (8.26)$$

One finds the differential equation

$$\dot{\Phi}(x,t) = \left[D^- \left(e^{ix} - 1\right) + D^+ \left(e^{-ix} - 1\right)\right] \Phi(x,t) + \frac{\delta(t)}{g_0} \quad (8.27)$$

for this function, which immediately yields

$$\Phi(x,t) = \frac{1}{g_0} \exp\left[tD^- \left(e^{ix} - 1\right) + tD^+ \left(e^{-ix} - 1\right)\right]. \quad (8.28)$$

From the generating function (8.28) one finds

$$\tilde{\rho}_n(t) = \frac{1}{2\pi} \int_{-\pi}^{\pi} dx\, e^{-inx} \Phi(x,t) = \frac{e^{-t\left(D^- + D^+\right)}}{2\pi i g_0} \int_C \frac{dy}{y^{n+1}} \exp\left[tD^- y + tD^+ \frac{1}{y}\right]$$
$$= e^{-t\left(D^- + D^+\right)} \frac{I_n(2t\sqrt{D^- D^+})}{g_0 (D^+/D^-)^{n/2}}, \quad (8.29)$$

where the replacement $e^{ix} \to y$ has been done. The contour C is a loop around $y = 0$,and $I_n(x)$ is a modified Bessel function. Note that the requirements $D_n^- = \text{const}$ and $D_n^+ = \text{const}$ imply that $g_n = \text{const}$ and $\langle V_n^2 \rangle = \text{const}$, and consequently one finds $D_n^- = D_n^+ = D$, which yields for the total populations

$$\rho_n(t) = e^{-2Dt} I_n(2Dt). \tag{8.30}$$

Let us find the asymptotic form of this equation for large t when the main contribution to the integral (8.29) comes from the vicinity of the saddle point $y = 1$, that is, $x = 0$. By developing the argument of the exponent in (8.28) over x, up to second order and by extending the integration limits for x to $\pm\infty$, one finds the distribution

$$\begin{aligned} \rho_n(t) &= e^{-2Dt} I_n(2Dt) \\ &\simeq \frac{1}{2\pi} \int_{-\infty}^{\infty} dx \exp\left[-tDx^2 - inx\right] = \frac{1}{\sqrt{2\pi Dt}} \exp\left[-\frac{n^2}{4Dt}\right] \end{aligned} \tag{8.31}$$

coinciding with the classical diffusion expression. We note that at $t > 1/D$ for the population distribution $\rho_n(\Delta, t)$ over the band one finds the Lorentzian profile (4.39) with $W = 2D$ by considering the poles of $\rho_{n,k}(\zeta, \eta)$ on the real axis of ζ by analogy to (4.38), which yields

$$\rho_n(\Delta, t) \simeq \rho_n(t) \frac{2D}{\pi(4D^2 + \Delta^2)}. \tag{8.32}$$

In the context of (8.32), it is expedient to discuss the applicability of the approach based on the master equation (8.25) for asymptotically long times. As for the case of degenerate levels, the validity of the approach is limited in time by the assumption of the tree-like structure of diagrams, which implies that self-intersections of the diagrams are unlikely. This assumption fails when the typical number of returns of the random trajectory $\sqrt{Dt}$ becomes of the order of the total number gD of band levels in the populated domain of width D, suggested by the distribution (8.31). This occurs for $t \sim g^2 D$, when the distribution over the band has a width gD coinciding with the number of populated levels in each band.

Linearly Increasing Transition Rates

Equation (8.25) can also be solved exactly for linearly changing coefficients $D_n^+ = 2\pi g_{n+1} \langle V_{n+1}^2 \rangle$ and $D_n^- = 2\pi g_{n-1} \langle V_n^2 \rangle$. This variation can result from a linear dependence of the state density, $g_n = g(1 + \alpha n)$, on the band number n or from the dependence of the mean square coupling, $\langle V_n^2 \rangle = V^2(1 + \alpha n)$, on n. A slight difference between these two cases manifests itself in a difference between the coefficients D_n^-: the coefficient for the first case, $D_{n+1}^- = 2\pi g \langle V^2 \rangle (1 + \alpha n)$, coincides with the coefficient D_n^- for the second case, whereas the coefficients D_n^+ are identical. Equation (8.25) takes the form

$$\dot{\tilde{\rho}}_n(\tau) = (1 - \alpha' + \alpha n)\left[\tilde{\rho}_{n-1}(\tau) - \tilde{\rho}_n(\tau)\right] \\ +(1 + \alpha + \alpha n)\left[\tilde{\rho}_{n+1}(\tau) - \tilde{\rho}_n(\tau)\right] + \frac{\delta_n^0 \delta(\tau)}{g} \tag{8.33}$$

where $\tau = 2\pi g \left\langle V^2 \right\rangle t$, and either $\alpha' = \alpha$ for the first case, or $\alpha' = 0$ for the second case.

For the generating function (8.26) one finds the equation

$$\dot{\Phi}(x,\tau) = \left[(1-\alpha')\left(e^{ix}-1\right) + (1+\alpha)\left(e^{-ix}-1\right)\right]\Phi(x,\tau) \\ -i\alpha\frac{\partial}{\partial x}\left\{\left[\left(e^{ix}-1\right) + \left(e^{-ix}-1\right)\right]\Phi(x,\tau)\right\} + \frac{\delta(\tau)}{g}. \tag{8.34}$$

where linear dependence on n corresponds to the derivative $-i\partial/\partial x$. The solution of this first-order linear differential equation in two dimensions can be found by the standard method of characteristics. This reads

$$\Phi(x,\tau) = \frac{1}{g}e^{-ix/\alpha}(e^{ix} + \alpha\tau - e^{ix}\alpha\tau)^{1/\alpha}(1 + \alpha\tau - e^{ix}\alpha\tau)^{-1+\alpha'/\alpha-1/\alpha} \tag{8.35}$$

and yields the population

$$\tilde{\rho}_n(\tau) = \frac{1}{2\pi g}\int_0^{2\pi} dx e^{-inx-ix/\alpha}(e^{ix}+\alpha\tau-e^{ix}\alpha\tau)^{1/\alpha}(1+\alpha\tau-e^{ix}\alpha\tau)^{-1+\alpha'/\alpha-1/\alpha}. \tag{8.36}$$

By making the replacement $e^{ix} \to y$ and by deformation of the integration contour one can write this integral in the form

$$\tilde{\rho}_n(\tau) = \frac{-i}{2\pi g}\int_{C_0} dy\, y^{-n-1/\alpha}(y + \alpha\tau - y\alpha\tau)^{1/\alpha}(1 + \alpha\tau - y\alpha\tau)^{-1+\alpha'/\alpha-1/\alpha}, \tag{8.37}$$

where the integration contour C_0 circumvents the zeros of two last factors of the integrand. This three-center integral resembles the integral representation of the hypergeometric function ${}_2F_1(a, b; c; X)$, and indeed, after the replacement $y \to (\alpha^2 t^2 - z)/(\alpha^2 t^2 - \alpha t)$ the integral over dz takes the form of this integral representation and yields

$$\tilde{\rho}_n(\tau) = \frac{\Gamma(n-1/\alpha)(-\alpha\tau)^n\, {}_2F_1(n-(1-\alpha')/\alpha, n-1/\alpha; n+1; \alpha^2\tau^2)}{g\,(1+\alpha\tau)^{1+2/\alpha-\alpha'/\alpha-n}\Gamma(-1/\alpha)\, n!}, \tag{8.38}$$

which for large n and τ has the asymptotic form

$$\tilde{\rho}_n(\tau) \simeq \frac{1}{2\pi g}\int\limits_{-\infty}^{\infty} dx e^{-inx-ix/\alpha}(1-ix\alpha\tau)^{-1} \simeq \frac{e^{-n/\alpha\tau}}{\alpha\tau g} \tag{8.39}$$

in the case $\alpha' = \alpha$. For $\alpha' = 0$ the asymptotic behavior differs by the pre-exponential factor.

In Fig. 8.6 we illustrate the dependence of the population per state on the band number. One sees that the population distribution is limited in the domain of n, where the coefficients D_n^+ and D_n^- are positive. For the case of linearly dependent coupling, the distribution tends to zero when n approaches the band for which one of the coefficients D equals zero. For the linearly dependent state density at this point, the distribution function is discontinuous. Both functions manifest asymptotic behavior; see (8.39).

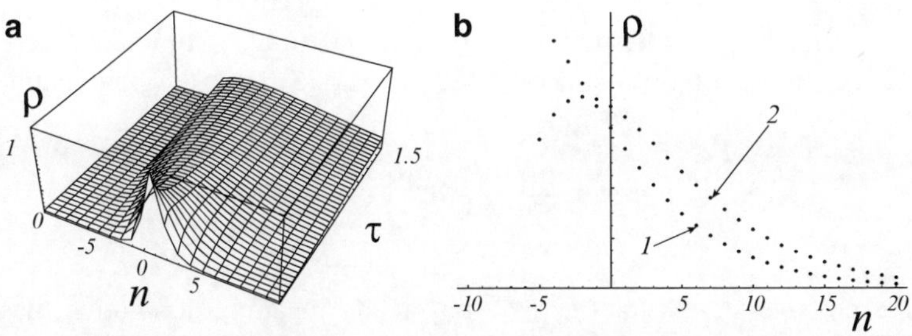

Fig. 8.6 Dependence of the population per state on the band number for the linearly changing transition probability. (**a**) Dependence of the population on the band number n and on the scaled time $\tau = 2\pi g\langle V^2\rangle t$ in the case of constant state density and linearly increasing coupling ($\alpha = 0.2$, $\alpha' = 0$). (**b**) Population distributions for the linearly increasing state density $g_n = g(1+0.2n)$ and for linearly increasing mean square coupling $\langle V_n^2\rangle = \langle V^2\rangle(1+0.2n)$ are shown by *dotted lines* 1 and 2, respectively

WKB Approximation

For smooth dependencies of the kinetic coefficients D_n^+ and D_n^- on the band number, one can employ the WKB approximation by analogy to that considered on p. 316. With the help of the finite shift operators (5.15) one can write the continuous analog of (8.25) in the form

$$\dot{\tilde{\rho}}(n,t) = \left\{D^-(n)\left[\exp(-\tfrac{\partial}{\partial n}) - 1\right] + D^+(n)\left[\exp(\tfrac{\partial}{\partial n}) - 1\right]\right\}\tilde{\rho}(n,t), \tag{8.40}$$

which just requires specification of the initial condition. Smooth dependence of the positive coefficients results in a smooth distribution function, for which one can expand exponents into a Taylor series up to second order and obtain

$$\dot{\tilde{\rho}}(n,t) = \left[D^+(n) - D^-(n)\right] \frac{\partial \tilde{\rho}(n,t)}{\partial n} + \left[D^+(n) + D^-(n)\right] \frac{\partial^2 \tilde{\rho}(n,t)}{2\partial n^2}. \quad (8.41)$$

The WKB ansatz $\tilde{\rho}(n,t) = \int d\lambda\, g(\lambda) \exp\left[-\lambda t + S(n,\lambda)\right]$ yields in the same order of approximation

$$-\lambda = \Pi(n) \frac{\partial S(n,\lambda)}{\partial n} + D(n) \left[\frac{\partial S(n,\lambda)}{\partial n}\right]^2 \quad (8.42)$$

where

$$\Pi(n) = \left[D^+(n) - D^-(n)\right],$$
$$D(n) = \frac{1}{2}\left[D^+(n) + D^-(n)\right], \quad (8.43)$$

whereas the function $g(\lambda)$ has to be found from the initial condition at $t = 0$.

Solution of the quadratic equation (8.42) allows one to find the exponential factor

$$S(n,\lambda) = \int_0^n dn' \left[\frac{-\Pi(n')}{2D(n')} \pm \frac{\sqrt{\Pi^2(n') + 4\lambda D(n')}}{2D(n')}\right], \quad (8.44)$$

where the positive sign, which gives a physically meaningful solution, has to be chosen. For long times only small λ are important, and therefore one can make use of the Taylor expansion of the right-hand side of (8.44), that is,

$$S(n,\lambda) = \int_0^n dn' \left[\frac{\lambda}{\Pi(n')} - \frac{\lambda^2 D(n')}{\Pi^3(n')}\right]. \quad (8.45)$$

Therefore the WKB ansatz takes the form

$$\tilde{\rho}(n,t) \simeq \int d\lambda\, g(0) \exp\left[-\lambda\left(t - \int_0^n \frac{dn'}{\Pi(n')}\right) - \lambda^2 \int_0^n dn' \frac{D(n')}{\Pi^3(n')}\right]. \quad (8.46)$$

Integration over $d\lambda$ yields

$$\tilde{\rho}(n,t) \simeq \sqrt{2\pi \int_0^n \frac{D(n')dn'}{\Pi^3(n')}}\, g(0) \exp\left[-\frac{\left(t - \int_0^n \frac{dn'}{\Pi(n')}\right)^2}{4\int_0^n dn' \frac{D(n')}{\Pi^3(n')}}\right]. \quad (8.47)$$

If the dependencies $\Pi(n)$ and $D(n')$ are such that for large n

$$\int_0^n \frac{dn'}{\Pi(n')} \gg \left(\int_0^n dn' \frac{D(n')}{\Pi^3(n')}\right)^{1/2}, \tag{8.48}$$

then at a given time moment t, expression (8.47) corresponds to a distribution, localized around the band number n_0 for which

$$\int_0^{n_0} \frac{dn'}{\Pi(n')} = t \tag{8.49}$$

while the distribution width is

$$\Delta n \sim \Pi(n_0)\left(\int_0^{n_0} dn' \frac{D(n')}{\Pi^3(n')}\right)^{1/2}. \tag{8.50}$$

Note that if the relation (8.48) is not valid, the width of the distribution Δn exceeds the displacement of the center of the distribution n_0 and the truncation of the Taylor expansion (8.45) at the second-order terms might not be justified. In this case (8.47) might not be very accurate. In the latter case one has to perform an exact integration over $dn'd\lambda$ employing (8.44).

8.1.3 Correlated Coupling

Now we consider the case of (8.1) when the coupling matrices $\hat{V}_n^{n+1}$ with the matrix elements $V_{n,k}^{n+1,m}$ connecting each level of a band with the levels of the neighboring band have low ranks $R \ll N$. Here N denotes a typical number of levels in each band. In other words, the matrix elements $\hat{V}_n^{n+1} = \sum_{j=1}^R \left|C_n^{(j)}\right\rangle \mathcal{V}^{(j)} \left\langle C_{n+1}^{(j)}\right|$ are strongly correlated, since each of the matrices is a sum of tensor products for a small number R of eigenvectors. Earlier, on p. 166 we considered such a situation for two-band systems. Now, considering multiband problems, we concentrate on the extreme, when the rank of the matrix is one, and therefore the matrix elements have a factorized form $V_{n,k}^{n+1,m} = C'_{n,k}\mathcal{V}_n C_{n+1,m}$ determined by the single non-zero eigenvalue $\mathcal{V}_n$ and the corresponding left $C'_{n,k}$ and right $C_{n+1,m}$ eigenvectors normalized to unity, for each band number n. We assume that the components $C'_{n,k}$ and $C_{n+1,m}$ of the vectors are real, although a straightforward generalization of the complex values of these coefficients can also be performed. We note that the eigenvalue $\mathcal{V}_n$ may be physically interpreted as the cooperative matrix element of the transition between band n and band $n+1$. A typical matrix element V, coupling two levels of neighboring bands, is of the order of $\mathcal{V}/N$.

Substitution of this particular form of $V_{n,k}^{n+1,m}$ into (8.1) after Fourier transformation results in

$$(\varepsilon - E_{n,k})\,\psi_{n,k}(\varepsilon) = i\psi_{n,k}(t=0) + C'_{n,k}\mathcal{V}_n \sum_m C_{n+1,m}\psi_{n+1,m}(\varepsilon)$$
$$+C_{n,k}\mathcal{V}_{n-1}\sum_m C'_{n-1,m}\psi_{n-1,m}(\varepsilon), \tag{8.51}$$

where the term $i\psi_{n,k}(t = 0)$ on the right-hand side corresponds to the initial condition. For the combinations

$$\Phi_n(\varepsilon) = \sum_k C_{n,k}\psi_{n,k}(\varepsilon)$$
$$\overline{\Phi}_n(\varepsilon) = \sum_k C'_{n,k}\psi_{n,k}(\varepsilon) \tag{8.52}$$

(8.51) yields

$$\Phi_n = \mathcal{V}_n\Phi_{n+1}\sum_k \frac{C_{n,k}C'_{n,k}}{\varepsilon - E_{n,k}} + \mathcal{V}_{n-1}\overline{\Phi}_{n-1}\sum_k \frac{C_{n,k}C_{n,k}}{\varepsilon - E_{n,k}} + iU_n,$$
$$\overline{\Phi}_n = \mathcal{V}_n\Phi_{n+1}\sum_k \frac{C'_{n,k}C'_{n,k}}{\varepsilon - E_{n,k}} + \mathcal{V}_{n-1}\overline{\Phi}_{n-1}\sum_k \frac{C'_{n,k}C_{n,k}}{\varepsilon - E_{n,k}} + i\overline{U}_n, \tag{8.53}$$

where

$$U_n(\varepsilon) = \sum_k \frac{C_{n,k}\psi_{n,k}(t=0)}{\varepsilon - E_{n,k}},$$
$$\overline{U}_n(\varepsilon) = \sum_k \frac{C'_{n,k}\psi_{n,k}(t=0)}{\varepsilon - E_{n,k}}. \tag{8.54}$$

In Fig. 8.1c, d one sees two different cases of the system with factorized interaction. The case of Fig. 8.1c corresponds to the situation where the left $\left|C_n^{(j)}\right\rangle$ and the conjugate of the right eigenstate $\left|C_{n+1}^{(j)}\right\rangle$ of the coupling matrices coincide, and therefore the interaction couples these "bright" states of neighboring bands directly. In this case the population propagates along this relay of isolated "bright" levels, decaying to the other "dark" states of the bands, by analogy to the situation discussed on p. 176 for the two-band system. In the case of Fig. 8.1d the left and right eigenvectors for the coupling are different. This means that the state which is "bright" for the transition $n-1 \to n$ does not coincide with the "bright" state for the transition $n \to n+1$. The matrix element of the resolvent coupling these two states determines the efficiency of the transition, by analogy to the quantity W_1^0 in (4.114) introduced for the problem of two states interacting via the band shown in Fig. 4.22.

For statistically identical bands, and for times shorter as compared to the typical density of states of one band, the sums

$$W_0^1 = \mathcal{V}_n \sum_k \frac{C_{n,k} C'_{n,k}}{\varepsilon - E_{n,k}}, \qquad W_0^0 = \mathcal{V}_{n-1} \sum_k \frac{C_{n,k} C_{n,k}}{\varepsilon - E_{n,k}},$$

$$W_1^1 = \mathcal{V}_n \sum_k \frac{C'_{n,k} C'_{n,k}}{\varepsilon - E_{n,k}}, \qquad W_1^0 = \mathcal{V}_{n-1} \sum_k \frac{C'_{n,k} C_{n,k}}{\varepsilon - E_{n,k}}, \tag{8.55}$$

do not depend on the index n, $W_0^1 = W_1^0 = W$ and therefore (8.53) takes the form

$$\begin{aligned} \Phi_n &= \Phi_{n+1} W + \overline{\Phi}_{n-1} W_0^0 + iU_n, \\ \overline{\Phi}_n &= \Phi_{n+1} W_1^1 + \overline{\Phi}_{n-1} W + i\overline{U}_n, \end{aligned} \tag{8.56}$$

and results in the equation

$$\begin{aligned} \Phi(x) &= e^{-ix}\Phi(x)W + e^{ix}\overline{\Phi}(x)W_0^0 + iU(x), \\ \overline{\Phi}(x) &= e^{-ix}\Phi(x)W_1^1 + e^{ix}\overline{\Phi}(x)W + i\overline{U}(x), \end{aligned} \tag{8.57}$$

for the generating functions

$$\begin{aligned} \Phi(x) &= \textstyle\sum_n e^{inx}\Phi_n, & \overline{\Phi}(x) &= \textstyle\sum_n e^{inx}\overline{\Phi}_n, \\ U(x) &= \textstyle\sum_n e^{inx}U_n, & \overline{U}(x) &= \textstyle\sum_n e^{inx}\overline{U}_n. \end{aligned} \tag{8.58}$$

Solution of (8.58) yields for the functions (8.52)

$$\Phi_n = \frac{i}{2\pi} \int_{-\pi}^{\pi} \frac{(1 - e^{ix}W)U(x) - e^{-ix}W_1^1\overline{U}(x)}{1 - e^{ix}W - e^{-ix}W} e^{-inx}\,dx$$

$$\overline{\Phi}_n = \frac{i}{2\pi} \int_{-\pi}^{\pi} \frac{\left(1 - e^{-ix}W\right)\overline{U}(x) - e^{ix}W_0^0 U(x)}{1 - e^{ix}W - e^{-ix}W} e^{-inx}\,dx, \tag{8.59}$$

where the argument ε of W is implicit.

If all of the population is initially in the band $n = 0$, then the generating function of the initial conditions $U(x)$ does not depend on the variable x. In this case the integrands (8.59) only have singularities at the points where $\cos x = 1/2W$ and the integration yields

$$\Phi_n = U\delta_n^0 \pm \frac{UW - W_1^1\overline{U}}{\sqrt{1 - 4W^2}} e^{\mp i(n+1)\arccos(1/2W)}$$

$$\overline{\Phi}_n = \overline{U}\delta_n^0 \pm \frac{\overline{U}W - W_0^0 U}{\sqrt{1 - 4W^2}} e^{\mp i(n-1)\arccos(1/2W)} \tag{8.60}$$

where the choice of the sign $+$ or $-$ depends on the sign of the imaginary part of $\arccos 1/2W$ and the sign of the number n in such a way that the result converges for $|n| \to \infty$. Substitution of (8.60) into (8.51) gives the probability amplitude

$$\psi_{n,k}(\varepsilon) = \frac{C'_{n,k}\mathcal{V}\left(UW - W_1^1\overline{U}\right)}{(\varepsilon - E_{n,k})\sqrt{1-4W^2}} e^{\mp i(n+2)\arccos(1/2W)} + \frac{C_{n,k}\mathcal{V}\left(\overline{U}W - W_0^0 U\right)}{(\varepsilon - E_{n,k})\sqrt{1-4W^2}} e^{\mp i(n-2)\arccos(1/2W)} \quad (8.61)$$

for the initially non-populated state.

For further calculations one needs to make a particular choice of the real coefficients $C_{n,k}$ and $C'_{n,k}$. We take the simplest model and assume the distribution

$$(C_{n,k})^2 = (C'_{n,k})^2 = \frac{N}{\pi(k^2+N^2)}, \qquad C_{n,k}C'_{n,k} = \frac{\mu N}{\pi(k^2+N^2)} \quad (8.62)$$

for the squares and the product of these coefficients by analogy to (4.138). Here the factor μ entering the last equation allows for the fact that the left and the right eigenvectors of the coupling matrix can be different, although they have an identical distribution for the absolute values. The difference manifests itself in the statistics of the coefficients, and we assume that the product $C'_{n,k}C_{n,k}$ averaged over a narrow strip of energy, near the energy position $E_{n,k}$, differs only by an energy-independent factor μ from the product $C_{n,k}C_{n,k}$ averaged over the same strip. We have already encountered such a situation in (4.114), considering the transition between two isolated levels which occurs via a band. Equation (8.62) also implies that the bands are infinitely broad, although the levels located at a large distance from the energy position $E_{n,k} \simeq 0$ are weakly coupled to the levels of neighboring bands, since they are weakly represented in the eigenvectors $|C_n\rangle$ of the coupling matrices.

Substitution of (8.62) into (8.55) yields

$$W_0^0 = W_1^1 = \mathcal{V}\sum_k \frac{1}{\varepsilon - k\delta}\frac{N}{\pi(k^2+N^2)} = \frac{\mathcal{V}}{\varepsilon + iN\delta}, \qquad W = \mathcal{V}\sum_k \frac{1}{\varepsilon - k\delta}\frac{\mu N}{\pi(k^2+N^2)} = \frac{\mathcal{V}\mu}{\varepsilon + iN\delta}, \quad (8.63)$$

for $\mathcal{V}_n$ independent of n, and for $E_{n,k} = \delta k$, where δ is a typical energy spacing among neighboring levels in the same band. As we have already seen in Chap. 3, the detailed structure of the spectrum is not important for time intervals $t \ll 1/\delta$ shorter then the recurrence period, and this spectrum can be considered as an equidistant one, whereas the sums can be replaced by integrals.

The main effects associated with correlated matrix elements are seen already for the simplest case of the initial condition

$$\psi_{n,k}(t=0) = C_{n,k}, \tag{8.64}$$

which means that the state $|C_n\rangle$ corresponding to the eigenvector of the coupling matrix is populated at $t = 0$. Therefore comparing (8.54) with the allowance for (8.64) and (8.55) we find

$$U(\varepsilon) = \sum_k \frac{C_{n,k}C_{n,k}}{\varepsilon - E_{n,k}} = \frac{W}{\mathcal{V}},$$
$$\overline{U}(\varepsilon) = \sum_k \frac{C'_{n,k}C_{n,k}}{\varepsilon - E_{n,k}} = \frac{\mu W}{\mathcal{V}}, \tag{8.65}$$

and substitution of (8.63, 8.65) into (8.61) yields

$$\psi_{n,k}(\varepsilon) = \frac{(1-\mu^2)(\mathcal{V}\mu)^2}{(iN-k)\,\delta}\sqrt{\frac{N}{\pi(k^2+N^2)\left[(\varepsilon+iN\delta)^2 - 4(\mathcal{V}\mu)^2\right]}}$$
$$\left(\frac{1}{\varepsilon - k\delta} - \frac{1}{\varepsilon - iN\delta}\right)\exp\left[\mp i(n-2)\arccos\left(\frac{\varepsilon+iN\delta}{2\mathcal{V}\mu}\right)\right]. \tag{8.66}$$

This expression, as a function of ε, has four singularities: two poles at the points $\varepsilon = k\delta$ and $\varepsilon = -iN\delta$, and two branching points at $\varepsilon = -iN\delta \pm 2\mathcal{V}\mu$. Only the first point is located on the real axis and therefore is important at times $t > 1/N\delta$, whereas the other points yield exponentially decreasing contributions responsible for transitional dynamics resembling that of (7.97) for the relay of isolated levels. The contribution of the pole at the real axis yields the stationary distribution

$$\psi_{n,k}(t\to\infty) = \frac{\pi i(\mu^2-1)}{2\mathcal{G}\,(iN+k)}$$
$$\sqrt{\frac{N}{\pi(k^2+N^2)}}\frac{\exp\left[\mp i(n-2)\arccos(\mathcal{G}\,(k+iN))\right]}{\sqrt{\mathcal{G}^2\,(k+iN)^2 - 1}}. \tag{8.67}$$

which depends only on one dimensionless parameter $\mathcal{G} = \delta/2\mathcal{V}\mu$ and results in the expression

$$\rho_{n,k} = \left(\frac{(1-\mu^2)}{2\mathcal{G}}\right)^2 \frac{N\pi}{(N^2+k^2)^2}$$
$$\frac{\exp\left[-2(n-2)\left|\mathrm{Re}\arccos(\mathcal{G}\,(k+iN))\right|\right]}{\sqrt{\left[\mathcal{G}^2\,(k-iN)^2 - 1\right]\left[\mathcal{G}^2\,(k+iN)^2 - 1\right]}} \tag{8.68}$$

for the population distribution.

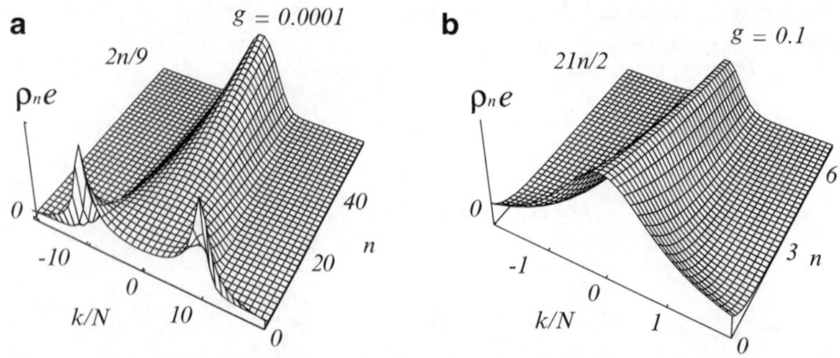

Fig. 8.7 Population distribution (not to scale) over the bands for the case of correlated coupling of band levels. The typical number of interacting levels in each band is $N = 1{,}000$. Exponential factors have been introduced in order to illustrate the energy distribution of the population for large n. (**a**) For strong interaction $\mathcal{G}N \ll 1$ one sees the population peaks detuned from the center, resembling that shown in Fig. 4.25. (**b**) Weak interaction. Population distribution over the energy remains Lorentzian, whereas the total population distribution amongst the bands is localized practically only on one, the initially populated band

From (8.68), it is clear that at $k = 0$ in the center of the distribution over the bands, the population as a function of n manifests an exponential behavior that is $\rho_{n,0} \sim \exp\left[-2n\left|\mathrm{Re}\beta\right|\right]$, where $\beta = \arccos\left(i\mathcal{G}N\right)$. The parameter $\mathcal{G}N = N\delta/2\mathcal{V}\mu$ may be physically interpreted as the ratio of the effective band width $N\delta$ and the cooperative matrix element of the interaction $\mathcal{V}$. We have already encountered a similar parameter $1/\sqrt{w} = \sqrt{N\delta_0 M\delta_1}/\mathcal{V}$ that is responsible for the population distribution width in the two-band system shown in Fig. 4.25. When this parameter is large one finds $\left|\mathrm{Re}\beta\right| \simeq \log\left(\mathcal{G}N\right)$ and the population distribution over the bands behaves as $\rho_{n,0} \sim \left(\mathcal{G}N\right)^{-2n}$, decreasing rapidly with the increase of n. As we have seen on p. 377, a typical matrix element V which couples two levels of neighboring bands is of order $\mathcal{V}/N$, and therefore the parameter $\mathcal{G}N$ is of the order of the ratio of the typical detuning $\delta/2$ between the two closest levels of the neighboring bands and their coupling V. That is why the dependence $\rho_{n,0} \sim \left(\mathcal{G}N\right)^{-2n}$ coincides with the perturbation theory limit (7.89) for the relay of weakly coupled isolated levels. When the interaction V exceeds the typical distance δ between the neighboring levels of a band, the population of bands with large n becomes larger. In Fig. 8.7 we show the dependence (8.68) as a function of the band number n and the ratio k/N for two different cases (a) $\mathcal{G}N \ll 1$ and (b) $\mathcal{G}N \gg 1$. In order to illustrate the population distribution over the bands we scale the band population by an exponential factor. The most interesting feature of the distribution is the spikes of the population near the position $k\delta \sim \mathcal{V}$ corresponding to the cooperative "Rabi frequency" $\mathcal{V}$.

8.2 Random Walks and Coherent Behavior

While considering the behavior of the level–band systems in Chap. 3, we have seen that for time intervals exceeding the inverse energy distance between neighboring levels, the recurrences and revivals bring the population partially back to the initial state. This phenomenon results in an incomplete decay of the levels. For infinite systems such as relays, the density of states approaches infinity, thus making the estimate of the revival time infinite. However, from the results discussed in Chap. 7, we know that even infinite relays with irregularly positioned energy levels manifest the localization phenomenon, which implies that an initially populated level does not experience complete decay. This means that in the general case, the estimate of the time when the revivals and recurrences begin to play the dominating role is governed by other parameters, and therefore must be improved. We have already encountered an example of incomplete decay to an infinitely dense spectrum when considering, in Sect. 3.4, a band with extremely irregular behavior of the matrix elements, coupling states of the band to the isolated level. The incomplete decay may occur when the density of states with vanishing coupling dominates in the total state density. In this section we will trace more attentively the relation between the random walks in an infinite complex multilevel system and the quantum recurrences resulting either in non-exponential or even incomplete decay of the initially populated state. Prior to this, we consider manifestation of the returns of a random walk on the spectral properties of a quantum system.

8.2.1 Level Decay to a Band of Random Walks

Let us consider two coupled levels, one of which interacts with a band. Such a system has already been considered in the context of the Fano problem shown in Fig. 3.8(a). However, now we assume that only one of these two levels is coupled to a band, as was the case in Fig. 2.24b and that the transition amplitudes coupling this level with the energy eigenstates of the band do not depend smoothly on the state energy, resembling the case shown in Fig. 2.5c, and corresponding to the case considered in p. 113 and to (3.111). The Schrödinger equation for the level scheme has the same structure

$$\begin{aligned} i\dot{\psi}_0 &= \varDelta_0\psi_0 + V_{01}\psi_1 \\ i\dot{\psi}_1 &= \varDelta_1\psi_1 + V_{10}\psi_0 + \sum_n V_{1n}\psi_n \\ i\dot{\psi}_n &= \varDelta_n\psi_n + V_{n1}\psi_1, \end{aligned} \tag{8.69}$$

as (3.48) for the Fano problem, although the coupling strength V_{1n} now changes strongly, by orders of magnitude, for the neighboring levels, and therefore cannot be factored out from the sum. Here ψ_0, ψ_1, and ψ_n are amplitudes of the states $|0\rangle$, $|1\rangle$ and $|n\rangle$, respectively, and Δ_n is the detuning of the state $|n\rangle$ in the band from the state $|1\rangle$.

In this section we do not specify the matrix elements of the coupling V_{1n}, but make use of the phenomenological diffusion model. To this end we perform Fourier transformation of (8.69), express ψ_n in terms of ψ_1 with the help of the third equation, and substitute the result into the second equation. This yields

$$\begin{aligned} \varepsilon\psi_0(\varepsilon) &= V_{01}\psi_1(\varepsilon) + \Delta_0\psi_0(\varepsilon) + i, \\ \varepsilon\psi_1(\varepsilon) &= \Delta_1\psi_1 + V_{10}\psi_0(\varepsilon) + \sum_n \tfrac{V_{1n}V_{n1}}{\varepsilon-\Delta_n+i0}\psi_1(\varepsilon), \end{aligned} \tag{8.70}$$

and hence

$$\psi_0(\varepsilon) = \frac{i}{\varepsilon - \Delta_0 - V_{01}V_{10}\left(\varepsilon - \Delta_1 - \sum_n \frac{V_{1n}V_{n1}}{\varepsilon-\Delta_n+i0}\right)^{-1}}. \tag{8.71}$$

The sum

$$G_b(\varepsilon) = \sum_n \frac{V_{1n}V_{n1}}{\varepsilon - \Delta_n + i0} \tag{8.72}$$

in (8.71) has a clear physical meaning: up to the factor $\langle V^2\rangle$ it coincides with the Fourier transform of the probability amplitude to return back to the state $|b\rangle = \sum V_{1n}/V\,|n\rangle$ at time t, once the system leaves this state for the band at time $t=0$. Here $V = \langle V^2\rangle^{1/2}$.

In this section we take the Fourier transform $G_b(t)$ of $G_b(\varepsilon)$, proportional to the square root of the probability

$$\rho_b(t) = \frac{\exp\left\{-a^2/4Dt\right\}}{(2\pi Dt)^{-d/2}} \tag{8.73}$$

to find the system in the $|b\rangle$ state, assuming that this probability corresponds to the classical diffusion in a d-dimensional space. This yields

$$G_b(\varepsilon) = -i\langle V^2\rangle e^{i\theta}\int_0^\infty (2\pi Dt)^{-d/4}\exp\left\{i\varepsilon t - a^2/8Dt\right\}dt, \tag{8.74}$$

where $e^{i\theta}$ is an arbitrary phase factor, and hence

$$\psi_0(\varepsilon) = \frac{i}{\varepsilon - \Delta_0 - |V_{01}|^2\left[\varepsilon - \Delta_1 + iV^2e^{i\theta}\int\limits_0^\infty (2\pi Dt)^{-d/4}e^{i\varepsilon t - a^2/8Dt}dt\right]^{-1}}. \tag{8.75}$$

Here a finite parameter a ensures the convergence of the integral (8.74) and may be interpreted as the typical size of a domain occupied by the state $|b\rangle$ in the space where the diffusion occurs. We assume that only this domain is coupled directly to the level $|1\rangle$ and take its size to be small. Note that for $a \to 0$ the integrand in (8.75) behaves as $\varepsilon^{d/4-1}$ for $d \neq 4$ and as $\ln(\varepsilon)$ for $d = 4$, and (8.75) takes the form

$$\psi_0(\varepsilon) = \frac{i}{\varepsilon - \Delta_0 - |V_{01}|^2 \left[\varepsilon - \Delta_1 - e^{i\theta} V^2 \varepsilon^{d/4-1} (2\pi i D)^{-d/4} \Gamma(1 - \frac{d}{4})\right]^{-1}}. \tag{8.76}$$

Equation (8.76) has two simple poles in the complex plane corresponding to the exponentially decreasing contributions to the population $\rho_0(t) = |\psi_0(t)|^2$ and a branching point at $\varepsilon = 0$ which yields a power dependence dominating in the limit of long times. We concentrate on the last contribution which is the most important one and consider two different cases $0 < d < 4$ and $4 < d < 8$ corresponding to the different asymptotic behavior of the Green's functions at long times. For $d > 4$ one finds that $G_b(\varepsilon) \to 0$ at $\varepsilon \to 0$, and the Taylor expansion over $G_b(\varepsilon)$ yields the first non-vanishing contribution,

$$\psi_0(\varepsilon) = \frac{i |V_{01}|^2 e^{i\theta} V^2}{\left(\Delta_0 \Delta_1 - |V_{01}|^2\right)^2} \varepsilon^{d/4-1} (2\pi i D)^{-d/4} \Gamma\left(1 - \frac{d}{4}\right), \tag{8.77}$$

provided $\Delta_0 \Delta_1 \neq V_{01} V_{10}$. For $\Delta_0 \Delta_1 = V_{01} V_{10}$ the expansion reads

$$\psi_0(\varepsilon) = \frac{-i \Delta_1 e^{-i\theta}}{\Delta_0 \varepsilon^{d/4-1} V^2 (2\pi i D)^{-d/4} \Gamma(1 - \frac{d}{4})}. \tag{8.78}$$

For $d < 4$ the Green's function $G_b(\varepsilon) \to \infty$ for $\varepsilon \to 0$, and for $\Delta_0 \neq 0$ the Taylor expansion over $1/G_b(\varepsilon)$ results in

$$\psi_0(\varepsilon) = \frac{\varepsilon^{1-d/4}}{i \Delta_0^2} \frac{e^{-i\theta} |V_{01}|^2}{V^2 (2\pi i D)^{-d/4} \Gamma(1 - \frac{d}{4})}, \tag{8.79}$$

whereas for $\Delta_0 = 0$ it yields

$$\psi_0(\varepsilon) = \frac{e^{i\theta} \varepsilon^{d/4-1} V^2 (2\pi i D)^{-d/4} \Gamma(1 - \frac{d}{4})}{|V_{01}|^2}. \tag{8.80}$$

For $d > 4$ after inverse Fourier transformation (8.77) and (8.78) yield

$$\rho_0(t) \simeq \frac{V^4 |V_{01}|^4 \pi^2}{(2\pi D t)^{d/2} \left(\Delta_0 \Delta_1 - |V_{01}|^2\right)^4} \tag{8.81}$$

and

$$\rho_0(t) \simeq \frac{(2\pi Dt)^{d/2}\,\Delta_1^2(1-d/4)^2\sin^2(\pi d/4)}{V^4t^4\Delta_0^2}, \tag{8.82}$$

respectively. This means that the population of the state $|0\rangle$ manifests the time behavior suggested by the diffusion. However, at resonance, when the detuning of the ground state shifted by the interaction with the state$|1\rangle$ vanishes, that is, $\Delta_0 - |V_{01}|^2/\Delta_1 = 0$, the decay gets slower by the factor $\sim t^{d-4}$.

For $d < 4$, Eqs. (8.79) and (8.80) result in

$$\rho_0(t) \simeq \frac{(2\pi Dt)^{d/2}\,|V_{01}|^4\,(1-d/4)^2\sin^2(\pi d/4)}{V^4t^4\Delta_0^4}, \tag{8.83}$$

and

$$\rho_0(t) \simeq \frac{\pi^2V^4}{(2\pi Dt)^{d/2}\,|V_{01}|^4}. \tag{8.84}$$

In this regime, the behavior is similar—the slowest decay occurs in the resonance $\Delta_0 = 0$, but corresponds now to diffusive behavior. The off-resonant decay is faster by the factor $\sim t^{d-4}$. We note that the presence of the state $|1\rangle$ no longer results in a shift of the resonance between the band and the state $|0\rangle$, since the state $|1\rangle$ is strongly broadened due to the interaction with the continuum.

The crucial role of returns of the population amplitudes (or the population amplitude wavepackets) in the course of a random walk becomes much more clear in another particular case related to the level–band problem. Let us assume that a probability amplitude is injected into the state $|0\rangle$ with the rate $\Pi(\varepsilon)$, similar to the case shown in Fig. 3.3 for a level–band system. Equation (8.76) then adopts the form

$$\psi_0(\varepsilon) = \frac{i\Pi(\varepsilon)}{\varepsilon - \Delta_0 - |V_{01}|^2\left[\varepsilon - \Delta_1 - e^{i\theta}V^2\varepsilon^{d/4-1}(2\pi iD)^{-d/4}\,\Gamma(1-\frac{d}{4})\right]^{-1}}. \tag{8.85}$$

From (8.85) one finds the Fourier transform $\rho(t)$

$$\begin{aligned}\rho_0(t) &= \frac{1}{2\pi}\int_{-\infty}^{\infty}\psi_0(\varepsilon)\psi_0^*(\xi)e^{-i(\varepsilon-\xi)t}d\varepsilon d\xi \\ &= \frac{1}{2\pi}\int\frac{\Pi(\varepsilon)\Pi(\xi)}{\left[\varepsilon-\Delta_0-\frac{|V_{01}|^2}{\varepsilon-\Delta_1-G_b(\varepsilon)}\right]\left[\xi-\Delta_0-\frac{|V_{01}|^2}{\xi-\Delta_1-G_b^*(\xi)}\right]}e^{-i(\varepsilon-\xi)t}d\varepsilon d\xi\end{aligned} \tag{8.86}$$

of the probability to be at the state $|0\rangle$ at time t. Assuming the injected population amplitude $\Pi(t)$ to be a random function, with a δ-like time autocorrelation that results in $\langle \Pi(\varepsilon)\Pi(\xi)\rangle \sim \delta(\varepsilon - \xi)$, one finds a steady state population

$$\rho_0 \sim \int \frac{d\varepsilon}{\left[\varepsilon - \Delta_0 - \frac{|V_{01}|^2}{\varepsilon - \Delta_1 - G_b(\varepsilon)}\right]\left[\varepsilon - \Delta_0 - \frac{|V_{01}|^2}{\varepsilon - \Delta_1 - G_b^*(\varepsilon)}\right]}. \tag{8.87}$$

If we assume that the amplitude injection is saturated, such that the population of the state $|0\rangle$ remains of the order of its maximum possible value $\rho_0 \simeq 1$ then the flux Φ of the population to the continuum corresponding to the random walk is given by the inverse value of the integral (8.87).

$$\Phi \sim 1/\int \frac{d\varepsilon}{\left[\varepsilon - \Delta_0 - \frac{|V_{01}|^2}{\varepsilon - \Delta_1 - G_b(\varepsilon)}\right]\left[\varepsilon - \Delta_0 - \frac{|V_{01}|^2}{\varepsilon - \Delta_1 - G_b^*(\varepsilon)}\right]}. \tag{8.88}$$

In Fig. 8.8 we show this flux Φ as a function of the detunings Δ_0 and Δ_1, and as a function of Δ_0 and V_{01} for different dimensions d of the random walk. Diffusive returns of the population amplitude play an important role inhibiting the decay at small detuning Δ_0 and result in a depth in the center of the line at $d < 4$. In other words, interference of the population decay with these recurrences strongly affects the line shapes of the resonances. For $d > 4$ returns of the population amplitude do not occur, and the line shape resembles a Lorentzian profile.

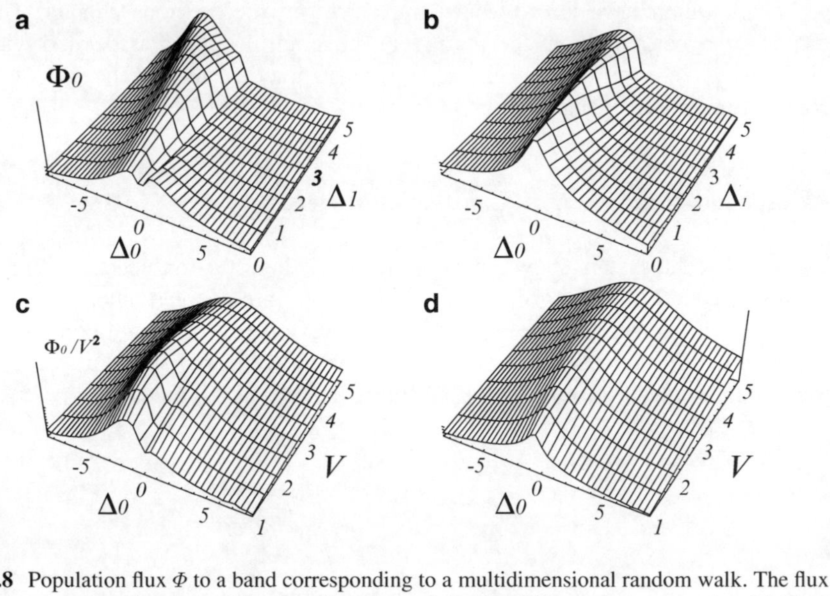

Fig. 8.8 Population flux Φ to a band corresponding to a multidimensional random walk. The flux Φ for $V_{01} = 1.3$ as a function of the detunings Δ_0 and Δ_1, for different dimensions $d = 3$ (**a**) and $d = 5$ (**b**). The flux Φ for $\Delta_1 = 1.3$ as a function of Δ_0 and V_{01} for $d = 3$ (**c**) and $d = 5$ (**d**)

8.2.2 *Interference of Random Returns at Long Times: General Consideration*

We now consider the behavior of quantum systems for asymptotically long times. For a complex system, this behavior might be strongly affected by the important role played by self-intersections in the diagrams. Under certain conditions, the main contribution to the perturbation series give the diagrams that return many times at the same quantum levels, and allow for the self-interference of the wavepackets. The most important characteristic which determines the behavior of a complex quantum system is therefore closely related to a property of the corresponding classical random walk, given by the probability for a randomly moving particle to return back to the points which it has previously visited. In the quantum limit, this corresponds to multiple self-intersections of the trajectories, which results in strong interference of many essentially different quantum paths corresponding to the same graph.

The direct approach can be based on the combinatorial calculation of the numbers of different self-intersecting trajectories and the numbers of different ways to pass along these trajectories. The latter is well known in mathematics as the number of Euler trails on the oriented graphs. An explicit formula exists for such a number, which turns out to be a function of the self-intersecting diagram topology, and the number of paths entering each self-intersection. However the calculations based on this intuitively clear idea are usually rather cumbersome, and therefore we demonstrate this approach in the next section, for a simpler example. Here we suggest a less instructive consideration based on the integrals over Grassmann variables, which have already been introduced in the previous chapter. Though it hides to a certain extent the intuitively meaningful images associated with the diagrams of interfering trajectories, it is most practical technically, giving more compact intermediate expressions.

Level–Band–Random Walk Continuum Model

We consider the system shown in Fig. 8.9a consisting of a non-degenerate level $|0\rangle$ directly coupled to a band of levels $|n\rangle$. The latter are coupled among themselves by an indirect interaction which occurs via a very dense or even a continuous set of states $|\kappa\rangle$ that represents a random walk with certain statistical properties. The Schrödinger equation for the system initially in the state $|0\rangle$ reads

$$\begin{aligned} i\dot{\psi}_0 &= V_{0n}\psi_n + i\delta(t) \\ i\dot{\psi}_n &= \Delta_n\psi_n + V_{n0}\psi_0 + \sum_{\kappa} V_{n\kappa}\psi_\kappa \\ i\dot{\psi}_\kappa &= \Delta_\kappa\psi_\kappa + \textstyle\sum_n V_{\kappa n}\psi_n, \end{aligned} \tag{8.89}$$

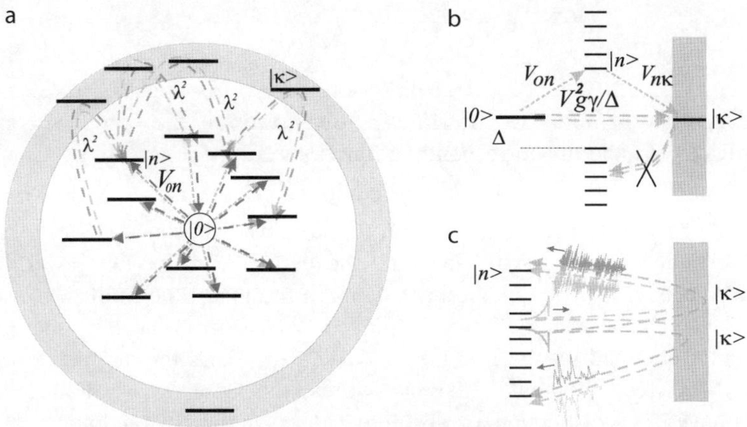

Fig. 8.9 The type of random walk affects the level decay at long times. (**a**) A non-degenerate level $|0\rangle$ is coupled directly by interactions V_{0n} of a random strength to a discrete set of quantum states randomly distributed in energy that are coupled among themselves via the states $|\kappa\rangle$ of an irregular continuum. The Fourier transforms of the decay rate $G(\varepsilon)$ to the continuum and the correlation function $\lambda^2(\varepsilon,\xi)$ of the return amplitudes are the main characteristics of the process. (**b**) The unlikely ($\sim \exp(-\Delta g)$), long decaying realizations of the band, with a large gap Δ in the resonance position, determine the long-time behavior for a non-returnable random walk. (**c**) Two types of returnable random walks. The exponential wavepackets leaving the band states acquire, after returning from the continuum, either zero or a finite autocorrelation time

and adopts the form

$$\begin{aligned}\varepsilon\psi_0 &= V_{0n}\psi_n + i\\ \varepsilon\psi_n &= \Delta_n\psi_n + V_{n0}\psi_0 + \sum_\kappa V_{n\kappa}\psi_\kappa\\ \varepsilon\psi_\kappa &= \Delta_\kappa\psi_\kappa + \textstyle\sum_n V_{\kappa n}\psi_n \end{aligned} \tag{8.90}$$

after the Fourier transformation. We assume a uniform distribution Δ_n of the level positions in the entire interval from $-\infty$ to ∞ and a Gaussian distribution of $e^{-|V^2|}$ complex valued couplings V_{0n} with unit variance. All energy variables are therefore scaled by the mean square coupling $V = \langle V^2\rangle^{1/2}$ whereas the mean density of states g is understood as a large dimensionless parameter gV, representing the expectation number of resonances, which have been considered in Sect. 3.4.4.

Introducing in complete analogy with (8.72) the function

$$G_{nn'}(\varepsilon) = \sum_\kappa \frac{V_{n\kappa}V_{\kappa n'}}{\varepsilon - \Delta_\kappa} \tag{8.91}$$

we exclude from the consideration the levels of the band representing the random walk and the Schrödinger equation adopts the form

$$\varepsilon\psi_0 = V_{0n}\psi_n + i$$
$$\varepsilon\psi_n = \Delta_n\psi_n + V_{n0}\psi_0 + \sum_{n'} G_{nn'}(\varepsilon)\psi_{n'}. \tag{8.92}$$

A similar equation holds for the Fourier transforms of the complex conjugate amplitudes $\psi_n(\xi)$ and the corresponding functions $G_{nn'}(\xi)$.

The moments $\langle G^k_{nn'}(\varepsilon)\rangle$ and the correlations $\left\langle G^k_{nn'}(\varepsilon)G^{k'}_{mm'}(\xi)\right\rangle$ of the functions (8.71) contain information about all of the statistical properties of the random walk. Later on we will discuss in detail the particular properties governing the asymptotic behavior of the quantum systems in the long-time limit. Here we just specify several general hypotheses employed. We assume that the random walk is completely defined by the first $G(\varepsilon) = \langle G_{nn'}(\varepsilon)\rangle$ and the second $\lambda^2(\varepsilon,\xi) = \langle G_{nn'}(\varepsilon)G_{mm'}(\xi)\rangle$ moments. In other words, the Green's functions are well-defined, have zero dispersion in absolute values, and vary randomly in phase. Therefore the non-vanishing higher moments are explicitly given by the two first moments via the relations

$$\left\langle G^k_{nn'}(\varepsilon)G^{k'}_{mm'}(\xi)\right\rangle = \delta_{kk'}\delta_{nm}\delta_{n'm'}\left[\lambda^2(\varepsilon,\xi)\right]^k$$
$$\langle G^k_{nn'}(\varepsilon)\rangle = \delta_{nn'}\left[G(\varepsilon)\right]^k. \tag{8.93}$$

Effective Hamiltonians and Averaging Over the Random Walk

Once the amplitudes of the states $|\kappa\rangle$ are excluded, the dynamics of the remaining system consisting of the level $|0\rangle$ and the band of the states $|n\rangle$ satisfies the Schrödinger equation (8.92) with the effective Hamiltonian

$$\hat{H}(\varepsilon) = \begin{pmatrix} 0 & V_{01} & V_{02} & \dots & V_{0N} \\ V_{10} & \Delta_1 + G_{11}(\varepsilon) & G_{12}(\varepsilon) & \dots & G_{1N}(\varepsilon) \\ V_{20} & G_{21}(\varepsilon) & \Delta_2 + G_{22}(\varepsilon) & \dots & G_{2N}(\varepsilon) \\ \dots & \dots & \dots & \dots & \dots \\ V_{0N} & G_{N1}(\varepsilon) & G_{N2}(\varepsilon) & \dots & \Delta_N + G_{NN}(\varepsilon) \end{pmatrix}. \tag{8.94}$$

The Fourier transform of the population $\rho_0(\varepsilon,\xi)$ of the state $|0\rangle$ reads

$$\rho_0 = \frac{\begin{vmatrix} \Delta_{1\varepsilon} & G_{12}(\varepsilon) & \dots & G_{1N}(\varepsilon) \\ G_{21}(\varepsilon) & \Delta_{2\varepsilon} & \dots & G_{2N}(\varepsilon) \\ \dots & \dots & \dots & \dots \\ G_{N1}(\varepsilon) & G_{N2}(\varepsilon) & \dots & \Delta_{N\varepsilon} \end{vmatrix} \begin{vmatrix} \Delta_{1\xi} & G_{12}(\xi) & \dots & G_{1N}(\xi) \\ G_{21}(\xi) & \Delta_{2\xi} & \dots & G_{2N}(\xi) \\ \dots & \dots & \dots & \dots \\ G_{N1}(\xi) & G_{N2}(\xi) & \dots & \Delta_{N\xi} \end{vmatrix}}{\begin{vmatrix} -\varepsilon & V_{01} & \dots & V_{0N} \\ V_{10} & \Delta_{1\varepsilon} & \dots & G_{1N}(\varepsilon) \\ \dots & \dots & \dots & \dots \\ V_{0N} & G_{N1}(\varepsilon) & \dots & \Delta_{N\varepsilon} \end{vmatrix} \begin{vmatrix} -\xi & V_{01} & \dots & V_{0N} \\ V_{10} & \Delta_{1\xi} & \dots & G_{1N}(\xi) \\ \dots & \dots & \dots & \dots \\ V_{0N} & G_{N1}(\xi) & \dots & \Delta_{N\xi} \end{vmatrix}} \tag{8.95}$$

as directly follows from the solution of (8.92). Here the replacements $\Delta_n + G_{nn}(\xi) - \xi \to \Delta_{n\xi}$ and $\Delta_n + G_{nn}(\varepsilon) - \varepsilon \to \Delta_{n\varepsilon}$ are performed to shorten the notation.

We have to find the average of the population over all possible realizations of the couplings V_{0n} and the positions Δ_n of the levels which we assume to be statistically independent, that is, following Poissonian statistics. In the representation (8.95) this is impossible, since $\rho_0(\varepsilon, \xi)$ depends on complex combinations of V_{0n} and Δ_n. A possible approach should rely on factorization of the expression for the population in such a way that each cofactor corresponds to a different $|n\rangle$ and is distributed in the same way. To this end, the ratios of the determinants can be represented as Gaussian integrals over complex numbers x and the Grassmann variables χ with the help of (7.133), (7.136), and (7.180). This yields

$$\rho_0(\varepsilon, \xi) = \int e^{-\sum_{n,n'=1}^{N}\left[\hat{H}'_{nn'}(\varepsilon)\overline{\chi}_n\chi_{n'} + \hat{H}'_{nn'}(\xi)\overline{\chi}'_n\chi'_{n'}\right]} \prod_{n=1}^{N} d\overline{\chi}_n d\chi_{n'} d\overline{\chi}'_n d\chi'_{n'}$$
$$\times \int e^{-\sum_{n,n'=0}^{N}\left[\hat{H}'_{nn'}(\varepsilon)\overline{x}_n x_{n'} + \hat{H}'_{nn'}(\xi)\overline{x}'_n x'_{n'}\right]} \prod_{n=0}^{N} d\overline{x}_n dx_n d\overline{x}'_n dx'_n, \quad (8.96)$$

where we denote $\hat{H}_{nn'}(\varepsilon) - \varepsilon = \hat{H}'_{nn'}(\varepsilon)$ and $\hat{H}_{nn'}(\xi) - \xi = \hat{H}'_{nn'}(\xi)$ to shorten the notation. For the same reason, the normalization $1/\pi$ of Gaussian integrals over $d\overline{x}_n dx_n$ is implicit.

We first make use of the statistical properties of the random walk. We cast the right-hand side of (8.96),

$$\begin{aligned}&\int \prod_{n=0}^{N} d\overline{x}_n dx_{n'} d\overline{x}'_n dx'_{n'} \prod_{n=1}^{N} d\overline{\chi}_n d\chi_{n'} d\overline{\chi}'_n d\chi'_{n'} \exp\Big\{i\varepsilon\overline{x}_0 x_0 - i\xi\overline{x}'_0 x'_0 \\ &-i\sum_{n=1}^{N}\Delta_{n\varepsilon}\left(\overline{\chi}_n\chi_n + \overline{x}_n x_n\right) + i\sum_{n=1}^{N}\Delta_{n\xi}\left(\overline{\chi}'_n\chi'_n + \overline{x}'_n x'_n\right) \\ &+i\sum_{n=1}^{N} V_{0n}\left(\overline{x}_0 x_n - \overline{x}'_n x'_0\right) - i\sum_{n=1}^{N} V^*_{0n}\left(\overline{x}'_0 x'_n - \overline{x}_n x_0\right) \\ &+\sum_{n,n'=1}^{N}\left[iG_{n'n}(\varepsilon)\left(\overline{\chi}_{n'}\chi_n + \overline{x}_{n'}x_n\right) - iG_{n'n}(\xi)\left(\overline{\chi}'_{n'}\chi'_n + \overline{x}'_{n'}x'_n\right)\right]\Big\}\end{aligned} \quad (8.97)$$

in a Taylor series over G, make use of (8.93), performing the average over the random walks, and employ the integral representation for the factor $1/k!$ typical of Taylor expansion and write

$$\frac{\langle G^k_{nn'}(\varepsilon)G^k_{mm'}(\xi)\rangle}{k!k!} = \delta_{nm}\delta_{n'm'}\frac{1}{k!}\int_C \frac{e^{-\tau}}{2\pi i\tau}\left[\frac{\lambda^2(\varepsilon,\xi)}{-\tau}\right]^k d\tau, \quad (8.98)$$

where the contour C circumvents the point $\tau = 0$ in the positive direction starting and ending at $\tau = \infty$. Summing the Taylor series back we arrive at

$$\begin{aligned}\langle\rho_0(\varepsilon,\xi)\rangle = \int \tfrac{e^{-\tau}}{2\pi i\tau}d\tau \prod_{n=0}^{N} d\overline{x}_n dx_{n'} d\overline{x}'_n dx'_{n'} \prod_{n=1}^{N} d\overline{\chi}_n d\chi_{n'} d\overline{\chi}'_n d\chi'_{n'} \\ \exp\Big\{-i\sum_{n=1}^{N}\Delta_{n\varepsilon}\left(\overline{\chi}_n\chi_n+\overline{x}_n x_n\right)+i\sum_{n=1}^{N}\Delta_{n\xi}\left(\overline{\chi}'_n\chi'_n+\overline{x}'_n x'_n\right) \\ +i\sum_{n=1}^{N}V_{0n}\left(\overline{x}_0 x_n-\overline{x}'_n x'_0\right)-i\sum_{n=1}^{N}V^*_{0n}\left(\overline{x}'_0 x'_n-\overline{x}_n x_0\right)+i\varepsilon\overline{x}_0 x_0 \\ -i\xi\overline{x}'_n x'_n-\tfrac{\lambda^2}{\tau}\sum_{n,n'=1}^{N}\left(\overline{\chi}_{n'}\chi_n+\overline{x}_{n'}x_n\right)\left(\overline{\chi}'_{n'}\chi'_n+\overline{x}'_{n'}x'_n\right)\Big\} \quad (8.99)\end{aligned}$$

which now has to be averaged only over the ensemble of the band levels.

Factorization of the Population

Now our aim is to represent (8.99) in factorized form. To this end, in the exponent of (8.99), we multiply factors proportional to λ^2 and regroup them differently

$$\begin{aligned}\exp\left[-\tfrac{\lambda^2}{\tau}\sum_{n,n'=1}^{N}\left(\overline{\chi}_{n'}\chi_n+\overline{x}_{n'}x_n\right)\left(\overline{\chi}'_{n'}\chi'_n+\overline{x}'_{n'}x'_n\right)\right] \\ = \exp\left[-\tfrac{\lambda^2}{\tau}\left(S_{\overline{\chi}\overline{\chi}'}S_{\chi'\chi}+\sigma_{\overline{\chi}\overline{x}}\sigma_{x\chi'}+\sigma_{x'\overline{\chi}}\sigma_{\chi\overline{x}'}+S_{\overline{x}x'}S_{x\overline{x}'}\right)\right] \quad (8.100)\end{aligned}$$

with the help of the combinations

$$\begin{aligned}\sum_{n=1}^{N}\overline{x}_n x'_n = S_{\overline{x}x'};\ \sum_{n=1}^{N}x_n\overline{x}'_n = S_{x\overline{x}'};\ \sum_{n=1}^{N}\overline{\chi}_n\overline{\chi}'_n = S_{\overline{\chi}\overline{\chi}'};\ \sum_{n=1}^{N}\chi'_n\chi_n = S_{\chi'\chi}; \\ \sum_{n=1}^{N}\overline{\chi}'_n\overline{x}_n = \sigma_{\overline{\chi}\overline{x}};\ \sum_{n=1}^{N}x_n\chi'_n = \sigma_{x\chi'};\ \sum_{n=1}^{N}x'_n\overline{\chi}_n = \sigma_{x'\overline{\chi}};\ \sum_{n=1}^{N}\chi_n\overline{x}'_n = \sigma_{\chi\overline{x}'}\end{aligned} \quad (8.101)$$

that do not contain double sums. The combinations S contain complex numbers and bi-Grassmann variables, which from the viewpoint of the commutation relations are equivalent to complex numbers, while the combinations σ are equivalent to Grassmann variables. We note the following relations for Gaussian integrals over the complex conjugate variables $z = x + iy$ and $\overline{z} = x - iy$

$$\begin{aligned}\int_{-\infty}^{\infty}\tfrac{dzd\overline{z}}{\pi}\exp\left(iaz+ib\overline{z}-z\overline{z}\right) = \int_{-\infty}^{\infty}\tfrac{dxdy}{\pi}e^{i(a+b)x+(b-a)y-x^2-y^2} \\ = e^{(b-a)^2/4-(a+b)^2/4} = e^{-ab} \quad (8.102)\end{aligned}$$

are valid for any numbers a and b. Therefore the factorization of the binary combinations of the sums S in the exponent can be performed with the help of the integrals over two pairs of complex conjugated numbers—the pair f and $\overline{f}$ for $S_{\overline{x}x'}S_{x\overline{x}'}$ and the pair w and $\overline{w}$ for $S_{\overline{\chi}\overline{\chi}'}S_{\chi'\chi}$:

$$\exp\left[-\frac{\lambda^2}{\tau}\left(S_{\overline{\chi}\overline{\chi}'}S_{\chi'\chi}+S_{\overline{x}x'}S_{x\overline{x}'}\right)\right]=\int dwd\overline{w}dfd\overline{f}$$
$$\exp\left[\frac{i\lambda}{\sqrt{\tau}}\left(S_{\overline{\chi}\overline{\chi}'}w+\overline{w}S_{\chi'\chi}+S_{\overline{x}x'}f+\overline{f}S_{x\overline{x}'}\right)-w\overline{w}-f\overline{f}\right] \tag{8.103}$$

where the normalization factors $1/\pi$ for $dwd\overline{w}$ and for $dfd\overline{f}$ are implicit. For Grassmann combinations σ one finds an analog of the relation (8.102),

$$\int\exp[-\overline{\alpha}\alpha-\overline{\sigma}\alpha-\overline{\alpha}\sigma]d\overline{\alpha}d\alpha=\exp\overline{\sigma}\sigma. \tag{8.104}$$

With the help of (8.102) and (8.104) one obtains only a linear combination of sums (8.101) in the exponent, and therefore the population adopts the desired factorized form

$$\langle\rho_0(\varepsilon,\xi)\rangle=\int\frac{d\tau}{2\pi i\tau}dwd\overline{w}dfd\overline{f}d\overline{\alpha}d\alpha d\overline{\beta}d\beta\; e^{-\tau-w\overline{w}-f\overline{f}-\overline{\alpha}\alpha-\overline{\beta}\beta}$$
$$\left\{\prod_{n=0}^{N}\int d\overline{x}_n dx_n d\overline{x}'_n dx'_n d\overline{\chi}_n d\chi_n d\overline{\chi}'_n d\chi'_n\right\}e^{i\varepsilon\overline{x}_0x_0-i\xi\overline{x}'_0x'_0}\prod_{n=1}^{N}e^{L_n}, \tag{8.105}$$

where

$$\begin{aligned}L_n=\;&iV_{0n}(\overline{x}_0x_n-\overline{x}'_nx'_0)-iV^*_{0n}(\overline{x}'_0x'_n-\overline{x}_nx_0)\\&-i\Delta_{n\varepsilon}(\overline{\chi}_n\chi_n+\overline{x}_nx_n)+i\Delta_{n\xi}(\overline{\chi}'_n\chi'_n+\overline{x}'_nx'_n)\\&-\frac{\lambda}{\sqrt{\tau}}(x'_n\overline{\chi}_n\alpha+\overline{\alpha}\chi_n\overline{x}'_n+\overline{\chi}'_n\overline{x}_n\beta+\overline{\beta}x_n\chi'_n).\\&+\frac{i\lambda}{\sqrt{\tau}}(\overline{\chi}_n\overline{\chi}'_nw+\overline{w}\chi'_n\chi_n+\overline{x}_nx'_nf+\overline{f}x_n\overline{x}'_n).\end{aligned} \tag{8.106}$$

The expression for the ensemble averaged population therefore takes the form of a ninefold integral

$$\langle\rho_0(\varepsilon,\xi)\rangle=\int\frac{e^{-\tau}}{2\pi i\tau}d\tau dwd\overline{w}dfd\overline{f}d\overline{\alpha}d\alpha d\overline{\beta}d\beta\exp\left[-w\overline{w}-f\overline{f}-\overline{\alpha}\alpha-\overline{\beta}\beta\right]$$
$$\int d\overline{x}_0dx_0d\overline{x}'_0dx'_0\; e^{i\varepsilon\overline{x}_0x_0-i\xi\overline{x}'_0x'_0}\prod_{n=1}^{N}\left\langle\Psi(\Delta_n,x_0,\ldots,\overline{\beta})\right\rangle. \tag{8.107}$$

which includes the product of identical and independently distributed factors

$$\Psi(\Delta_n,x_0,\ldots,\overline{\beta})=\int d\overline{x}dxd\overline{x}'dx'd\overline{\chi}d\chi d\overline{\chi}'d\chi'\exp\left(L_n\right) \tag{8.108}$$

averaged over the distribution of statistically independent positions and couplings of the band levels. The index n of the integration variables x and χ is omitted since all the $|\kappa\rangle$ levels are statistically equivalent, and the averages of the integrals (8.108) for $n \neq 0$ are identical. The marker n of the statistically independent level position Δ_n can be omitted as well.

Average Over the Ensemble of the Band Levels

Our next step is to perform the ensemble average of $\Psi(\Delta, x_0, \ldots, \overline{\beta})$ over the positions Δ and the couplings V of the levels $|n\rangle$. For the distribution $g(V, \Delta)$ of these quantities we choose the uniform Poissonian dependence on Δ in the entire interval from $-\infty$ to ∞ and the Gaussian distribution $e^{-|V^2|}$ of the complex valued V. The average over the ensemble with the coupling distribution $e^{-|V^2|}$ is straightforward—according to (8.102) the average of the V-dependent part immediately yields

$$\begin{aligned} &\int e^{-|V^2|} \exp\left[iV\left(\overline{x}_0 x - \overline{x}' x_0'\right) - iV^*\left(\overline{x}_0' x' - \overline{x} x_0\right)\right] dV dV^* \\ &= \exp\left(-\overline{x}_0 \overline{x}_0' x x' + x_0' \overline{x}_0' \overline{x}' x' + x_0 \overline{x}_0 x \overline{x} - x_0' x_0 \overline{x}' \overline{x}\right). \end{aligned} \tag{8.109}$$

However, the average over the ensemble of uniformly distributed Δ is a more sophisticated procedure. Indeed, for large Δ the integral over the variables x and χ yields a value close to unity as one can see neglecting in (8.106) the terms of L_n, independent of Δ and performing the integration

$$\int d\overline{x} dx d\overline{x}' dx' d\overline{\chi} d\chi d\overline{\chi}' d\chi' e^{-i\Delta_\varepsilon(\overline{\chi}\chi + \overline{x}x) + i\Delta_\xi\left(\overline{\chi}'\chi' + \overline{x}'x'\right)} = \frac{-i\Delta_\varepsilon i\Delta_\xi}{-i\Delta_\varepsilon i\Delta_\xi} = 1. \tag{8.110}$$

Therefore the ensemble average of $\Psi(\Delta, x_0, \ldots, \overline{\beta})$ is a value infinitesimally close to 1. However, the product of an infinite number of such factors can be arbitrary. We have already encountered such a situation in Chap. 3, considering the decay of the level to a dense and irregular band. The proper way to overcome the problem has been demonstrated in (3.119). Employing the relation $\prod \Psi(\Delta_n) \simeq \exp\{\sum[\Psi(\Delta_n) - 1]\}$ one can put the small values $\Psi - 1$ into the exponent, and perform the integration of Ψ over Δ there, closing the integration contour C at infinity where the integrand equals zero, that is, $\int_{-\infty}^{\infty}[\Psi(\Delta) - 1]\, d\Delta \to \int_C \Psi(\Delta) d\Delta$. For the population (8.107) this yields

$$\begin{aligned} \langle \rho_0(\varepsilon, \xi)\rangle = &\int \tfrac{e^{-\tau}}{\tau} d\tau dw d\overline{w} df d\overline{f} d\overline{\alpha} d\alpha d\overline{\beta} d\beta \; e^{-w\overline{w} - f\overline{f} - \overline{\alpha}\alpha - \overline{\beta}\beta} \\ &\int d\overline{x}_0 dx_0 d\overline{x}_0' dx_0' \; e^{i\varepsilon \overline{x}_0 x_0 - i\xi \overline{x}_0' x_0'} \exp\left(\int_C \mathrm{g}\Psi dV dV^* d\Delta\right) \end{aligned} \tag{8.111}$$

where $\mathbf{g} = Ng(V, \Delta)$ stands for the total density of levels with given detuning and coupling and the contour C has to be closed either in the upper or in the lower part of the complex plane Δ where the integrand vanishes.

In order to find an explicit expression for the averaged Ψ we now perform the integration

$$\Psi_G = \int e^{L_G} d\overline{\chi} d\chi d\overline{\chi}' d\chi'$$

retaining in (8.106) only the terms L_G of L_n that contain Grassmann variables. Expanding e^{L_G} in a Taylor series over each of the Grassmann variables up to first order and combining the terms proportional to $\overline{\chi}\chi\overline{\chi}'\chi'$, which according to the definition (7.177) is the only non-vanishing term remaining after the integration, we find

$$\begin{aligned}\Psi_G &= \int d\overline{\chi} d\chi d\overline{\chi}' d\chi' \exp\Big[-i\Delta_\varepsilon \overline{\chi}\chi + i\Delta_\xi \overline{\chi}'\chi' \\ &\quad + \lambda\tau^{-1/2} \left(i\overline{\chi}\overline{\chi}' w + i\overline{w}\chi'\chi - x'\overline{\chi}\alpha - \overline{\alpha}\chi\overline{x}' - \overline{\chi}'\overline{x}\beta - \overline{\beta}x\chi' \right) \Big] \\ &= \left(i\overline{x}' x' \alpha\overline{\alpha}\Delta_\xi - i\beta\overline{\beta}\overline{x}x\Delta_\varepsilon + \tfrac{i\lambda}{\sqrt{\tau}} w\overline{\alpha}\overline{\beta}x\overline{x}' + \tfrac{i\lambda}{\sqrt{\tau}} x'\overline{x}\beta\alpha\overline{w} \right) \tfrac{\lambda^2}{\tau} \\ &\quad + \overline{x}x\overline{x}'x'\beta\overline{\beta}\alpha\overline{\alpha}\tfrac{\lambda^4}{\tau^2} + \Delta_\xi\Delta_\varepsilon - \tfrac{\lambda^2}{\tau} w\overline{w}. \end{aligned} \tag{8.112}$$

The next step is the calculation of the integral

$$\Psi(\Delta) = \int e^{L'} \Psi_G \, d\overline{x} dx d\overline{x}' dx' \tag{8.113}$$

where the remaining terms

$$\begin{aligned} L' &= -i\Delta_\varepsilon \overline{x}x + i\Delta_\xi \overline{x}' x' - \overline{x}_0 \overline{x}'_0 x x' + x'_0 \overline{x}'_0 \overline{x}' x' \\ &\quad + x_0 \overline{x}_0 x\overline{x} - x'_0 x_0 \overline{x}'\overline{x} + \tfrac{i\lambda}{\sqrt{\tau}} \overline{x} x' f + \tfrac{i\lambda}{\sqrt{\tau}} \overline{f} x \overline{x}' \end{aligned}$$

of L_n are taken into account with the allowance for (8.109) for the average over the couplings. With the help of the relations

$$\begin{gathered} \overline{x}x\overline{x}'x'\tfrac{\lambda^2}{\tau} e^{L'} = -\tfrac{\partial^2 e^{L'}}{\partial f \partial \overline{f}};\ i\overline{x}'x' e^{L'} = \tfrac{\partial e^{L'}}{\partial \Delta_\xi};\ -i\overline{x}x e^{L'} = \tfrac{\partial e^{L'}}{\partial \Delta_\varepsilon} \\ \tfrac{i\lambda}{\sqrt{\tau}} x\overline{x}' e^{L'} = \tfrac{\partial e^{L'}}{\partial \overline{f}};\ \tfrac{i\lambda}{\sqrt{\tau}} x'\overline{x} e^{L'} = \tfrac{\partial e^{L'}}{\partial f} \end{gathered}$$

one can write $\Psi(\Delta)$ in the form

$$\begin{aligned}\Psi(\Delta) &= \Big[\Big(\beta\overline{\beta}\tfrac{\Delta_\varepsilon \partial}{\partial \Delta_\varepsilon} + \alpha\overline{\alpha}\tfrac{\Delta_\xi \partial}{\partial \Delta_\xi} + \beta\alpha\overline{w}\tfrac{\partial}{\partial f} + w\overline{\alpha}\overline{\beta}\tfrac{\partial}{\partial \overline{f}} \\ &\quad - \tfrac{\partial^2}{\partial f \partial \overline{f}} \beta\overline{\beta}\alpha\overline{\alpha} - w\overline{w} \Big) \tfrac{\lambda^2}{\tau} + \Delta_\xi \Delta_\varepsilon \Big] \Xi, \end{aligned} \tag{8.114}$$

where the Gaussian integral

$$\Xi = \int e^{L'} d\overline{x} dx d\overline{x}' dx'$$

can be found explicitly by calculating the determinant of the quadratic form $\partial^2 L'/\partial x \partial x'$, in agreement with (7.133). This yields

$$\Xi = \left[\left(i\Delta_\xi r - i\Delta_\varepsilon r' - \Delta_\varepsilon \Delta_\xi - \frac{\lambda^2}{\tau} f\overline{f}\right)^2 + 4rr'\frac{\lambda^2}{\tau} f\overline{f}\right]^{-1/2}, \quad (8.115)$$

where the variables r and r' stand for the combinations $x_0\overline{x}_0$ and $x_0'\overline{x}_0'$ respectively.

Now we perform the last remaining integration over the level position

$$\int g\Psi d\Delta = \int g d\Delta \Big[\tfrac{\lambda^2}{\tau}\Big(w\overline{\alpha}\overline{\beta}\tfrac{\partial}{\partial \overline{f}} + \beta\alpha\overline{w}\tfrac{\partial}{\partial f} - \beta\overline{\beta}\alpha\overline{\alpha}\tfrac{\partial^2}{\partial f \partial \overline{f}} - w\overline{w} + \beta\overline{\beta}\tfrac{\Delta_\varepsilon \partial}{\partial \Delta_\varepsilon} + \alpha\overline{\alpha}\tfrac{\Delta_\xi \partial}{\partial \Delta_\xi}\Big) + \Delta_\xi \Delta_\varepsilon\Big]\Xi. \quad (8.116)$$

where $g = N\int g(V, \Delta) d^2V$ now denotes the uniform total density of the band levels, and find

$$\int_C g\Psi d\Delta = \int_C \Xi \Delta_\xi \Delta_\varepsilon d\Delta + \tfrac{\lambda^2}{\tau}\beta\overline{\beta}\int_C \tfrac{\Delta_\varepsilon \partial \Xi}{\partial \Delta_\varepsilon} d\Delta + \tfrac{\lambda^2}{\tau}\alpha\overline{\alpha}\int_C \tfrac{\Delta_\xi \partial \Xi}{\partial \Delta_\xi} d\Delta + \tfrac{\lambda^2}{\tau}\Big(\beta\overline{\beta} + \alpha\overline{\alpha} + w\overline{\alpha}\overline{\beta}\tfrac{\partial}{\partial \overline{f}} + \beta\alpha\overline{w}\tfrac{\partial}{\partial f} - \beta\overline{\beta}\alpha\overline{\alpha}\tfrac{\partial^2}{\partial f \partial \overline{f}} - w\overline{w}\Big)\int_C g\Xi d\Delta. \quad (8.117)$$

The integral $\Lambda = \int_C g\Xi d\Delta$ can be given explicitly in terms of the hypergeometric function

$$\Lambda = \int_C \frac{g dy}{\sqrt{-(a - iy^2)^2 - b^2}} = \sqrt{\frac{1}{a}}\pi g\, {}_2F_1\left(\frac{1}{4}, \frac{3}{4}; 1, \frac{b^2}{a^2}\right) \quad (8.118)$$

with

$$a = \tfrac{(r-r')^2}{4} + \tfrac{\lambda^2}{\tau} f\overline{f} + i\tfrac{r+r'}{2}(\Delta_\varepsilon - \Delta_\xi) - \tfrac{1}{4}(\Delta_\xi - \Delta_\varepsilon)^2$$
$$b = \sqrt{\tfrac{4rr'\lambda^2 f\overline{f}}{-\tau}} \quad (8.119)$$

whereas the remaining integrals

$$\Lambda_\varepsilon = \int \frac{\Delta_\varepsilon \partial}{\partial \Delta_\varepsilon} \Xi d\Delta = \frac{r'-r}{2i}\frac{\partial \Lambda}{\partial X} + \frac{X}{2}\frac{\partial \Lambda}{\partial X} - \frac{1}{2}\Lambda$$

$$\Lambda_\xi = \int \frac{\Delta_\xi \partial}{\partial \Delta_\xi} \Xi d\Delta = \frac{r-r'}{2i}\frac{\partial \Lambda}{\partial X} + \frac{X}{2}\frac{\partial \Lambda}{\partial X} - \frac{1}{2}\Lambda$$

$$\int d\Delta \Xi \Delta_\xi \Delta_\varepsilon = \Phi - \frac{(r-r')^2}{4}\Lambda - \frac{X^2}{4}\Lambda \tag{8.120}$$

can also be expressed in terms of the hypergeometric functions:

$$\Phi = \int_C \frac{gy^2 dy}{\sqrt{-(a-iy^2)^2 - b^2}} = -\sqrt{a}\pi g \, {}_2F_1\left(-\frac{1}{4}, \frac{1}{4}; 1, \frac{b^2}{a^2}\right) \tag{8.121}$$

where $X = \Delta_\varepsilon - \Delta_\xi = F(\varepsilon) - F(\xi) + \xi - \varepsilon$. Note that the hypergeometric functions have branching points at $b^2/a^2 = 1$, and the proper choice of the branches of Λ and Φ depends on the sign of the imaginary part of $a - b$, that is, on the sign of ζ. Equation (8.117) finally adopts the form

$$\begin{aligned} \int g\Psi d\Delta = {} & \beta\overline{\beta}\tfrac{\lambda^2}{\tau}\Lambda_\xi + \alpha\overline{\alpha}\tfrac{\lambda^2}{\tau}\Lambda_\varepsilon - \tfrac{\lambda^2}{\tau}\tfrac{\partial^2 \Lambda}{\partial f \partial \overline{f}}\beta\overline{\beta}\alpha\overline{\alpha} + w\overline{\alpha}\beta\tfrac{\lambda^2 \partial \Lambda}{\tau \partial \overline{f}} \\ & + \beta\alpha\overline{w}\tfrac{\lambda^2 \partial \Lambda}{\tau \partial f} - w\overline{w}\tfrac{\lambda^2 \Lambda}{\tau} + \Phi - \tfrac{(r-r')^2}{4}\Lambda - \tfrac{X^2}{4}\Lambda. \end{aligned} \tag{8.122}$$

We now substitute (8.122) into (8.111) and perform the exact integrations over the variables $d\overline{\alpha}d\alpha d\overline{\beta}d\beta$ starting with the Grassmann variables. Expanding in a Taylor series and retaining only the terms linear in $\overline{\alpha}, \alpha, \overline{\beta}, \beta$ we arrive at

$$\begin{aligned} \langle \rho_0(\varepsilon, \xi) \rangle = {} & \int \tfrac{e^{-\tau}}{\tau} d\tau dw d\overline{w} df d\overline{f} \int dr dr' \\ & \left[\left(1 - \tfrac{\lambda^2}{\tau}\Lambda_\xi\right)\left(1 - \tfrac{\lambda^2}{\tau}\Lambda_\varepsilon\right) - \tfrac{\lambda^2}{\tau}\tfrac{\partial^2 \Lambda}{\partial f \partial \overline{f}} + \overline{w}w\tfrac{\partial \Lambda \lambda^2}{\tau \partial \overline{f}}\tfrac{\partial \Lambda \lambda^2}{\tau \partial f}\right] \\ & \exp\left(i\varepsilon r - i\xi r' - w\overline{w}\tfrac{\Lambda \lambda^2}{\tau} - w\overline{w} - f\overline{f} + \Phi - \tfrac{(r-r')^2}{4}\Lambda - \tfrac{X^2}{4}\Lambda\right), \end{aligned} \tag{8.123}$$

where we have taken into account that

$$\frac{1}{\pi^2}\int_{-\infty}^{\infty} \ldots d\overline{x}_0 dx_0 d\overline{x}'_0 dx'_0 \to \int_0^{\infty} \ldots dr dr'.$$

After integration over $dwd\overline{w}$ this yields the final general and exact expression

$$\begin{aligned}\langle\rho_0(\varepsilon,\xi)\rangle = &\int \tfrac{e^{-\tau}}{\tau} d\tau df d\overline{f} \int dr dr' \Big[\tfrac{-\tau}{\Lambda\lambda^2+\tau}\tfrac{\lambda^2}{\tau}\tfrac{\partial^2\Lambda}{\partial f\partial\overline{f}} \\ &+\tfrac{\tau}{\Lambda\lambda^2+\tau}\left(1-\tfrac{\lambda^2}{\tau}\Lambda_\xi\right)\left(1-\tfrac{\lambda^2}{\tau}\Lambda_\varepsilon\right)+\left(\tfrac{\tau}{\Lambda\lambda^2+\tau}\right)^2\tfrac{\partial\Lambda\lambda^2}{\tau\partial\overline{f}}\tfrac{\partial\Lambda\lambda^2}{\tau\partial f}\Big] \\ &\exp\left(i\varepsilon r - i\xi r' - f\overline{f} + \Phi - \tfrac{(r-r')^2}{4}\Lambda - \tfrac{X^2}{4}\Lambda\right)\end{aligned} \tag{8.124}$$

for the ensemble averaged population of the state $|0\rangle$. Further analysis should rely on the particular choice of the functions $F(\varepsilon)$ and $\lambda^2(\varepsilon,\xi)$ that depend on the particular type of random walk.

8.2.3 Three Types of Random Walk and the Asymptotic Decay

As we have already mentioned, the analysis of population dynamics in the long-time limit, resulting from quantum interference of different self-intersecting trajectories, depends on particular properties of the random walk. The quantities $G(\varepsilon)$ and $\lambda^2(\varepsilon,\xi)$ responsible for the specific behavior of the random walk and entering (8.122) and (8.124) have a clear physical meaning for the system of band levels $|n\rangle$ and the states $|\kappa\rangle$ of the continuum that does not include the state $|0\rangle$. The quantity $G(\varepsilon)$ has been already introduced in Sect.3.2.1 for the level–band problem, and it allows for the level probability amplitude decay rate to the continuum. Now we consider $G(\varepsilon)$ as a characteristic of the band levels interacting with the continuum responsible for the random walk. The quantity $\lambda^2(\varepsilon,\xi)$ represents the Fourier transform of the time correlation of the probability amplitudes to be in a given state of the band under the condition that at the initial time moment $t=0$, the population was localized in another level of the band. After being integrated over the variable $\eta=(\varepsilon+\xi)/2$, this quantity depends only on the difference $\zeta=\varepsilon-\xi$ and represents the Fourier transform of the probability to be in the given state.

We consider two rather general particular cases of returnable and non-returnable random walks. For a returnable random walk, the Fourier transform $\lambda^2(\varepsilon,\xi)$ diverges at the limit $\varepsilon\to\xi$ according to the power law $(\varepsilon-\xi)^{-\nu}$ where the power index ν also describes the increase of the expectation number of returns $\mathcal{N}\sim t^\nu$ at time t. For a non-returnable random walk, $\lambda\to 0$ as $\varepsilon\to\xi$. For the first case we also identify two different particular cases: a random walk with a finite correlation time τ_r and a δ-correlated random walk. The correlation time of the random walk is a typical residence time τ_r, during which the probability amplitude of the quantum particle remains at a state of the band, once it has returned to the state from the continuum. This time is associated with dephasing of the probability amplitudes, whereas the absolute value of the amplitude changes smoothly. The width of the quantity $\lambda^2(\zeta,\eta)$ as a function of η is inversely proportional to τ_r. The order of magnitude of τ_r can be found from the second derivative

$$\frac{\partial^2\lambda^2(\eta)}{2\,\partial\eta^2} = -\frac{\lambda^2(\eta)}{\tau_r^2} \tag{8.125}$$

at the point $\eta = 0$. It also relates to the width of the continuum seen by each state of the band. The function $\lambda^2(\zeta,\eta)$ does not depend on η at all for a δ-correlated random walk, when the particle wavepacket rapidly comes and leaves the band state, keeping the mean population unchanged during some short time interval.

Non-returnable Random Walk

For $\lambda \to 0$, the only important singularity of the integrand (8.124) as a function of τ is the simple pole at $\tau = 0$, and for this case the integration over $d\tau$ yields

$$\begin{gathered}\langle\rho_0(\varepsilon,\xi)\rangle = \int df d\overline{f} \int dr dr' \\ \exp\left(i\varepsilon r - i\xi r' - f\overline{f} - \sqrt{a}\pi g - \tfrac{(r-r')^2}{4\sqrt{a}}\pi g - \tfrac{X^2}{4\sqrt{a}}\pi g\right).\end{gathered} \tag{8.126}$$

with

$$a = \frac{(r-r')^2}{4} + i\frac{r+r'}{2}X - \frac{X^2}{4}$$

where we have employed the fact that hypergeometric functions tend to one at zero argument. For the long-time limit, when $\varepsilon \sim \xi \sim 1/t$ the parameter $X = F(\varepsilon) - F(\xi) + \xi - \varepsilon \simeq -2i\gamma_c$ can be considered as independent of ε and ξ, and it is given by the decay rate γ_c of the amplitudes of the band states decaying to the continuum. In this case, the inverse Fourier transformation over ε and ξ immediately gives the Dirac δ-functions $\delta(r-t)$ and $\delta(r'-t)$, that is, $a = 2\gamma t$, and we arrive at

$$\langle\rho_0(t)\rangle = \int df d\overline{f} \exp\left(-f\overline{f} - \sqrt{a}\pi g - \frac{\gamma^2}{4\sqrt{a}}\pi g\right) \simeq \exp\left(-\pi g\sqrt{\gamma_c t}\right), \tag{8.127}$$

or $\langle\rho_0(t)\rangle \simeq \exp\left(-\pi g\sqrt{\langle V^2\rangle\,\gamma_c t}\right)$ for dimensional variables.

The exponent of the square root time-dependence emerges from the ensemble average over the randomly distributed energies of the levels in the band. Namely, it results from the unlikely long-leaving realizations of the system when the level $|0\rangle$ is separated from the closest level $|n\rangle$ of the band by a big energy gap Δ, as shown in Fig. 8.9b. The probability of such a realization for the independently distributed Δ_n is exponentially small $\sim \exp(-g\Delta)$. However, the contribution of this realization is dominating for large t since it is long leaving, with a decay rate inversely proportional to $\varDelta$. Indeed, the composite matrix element $V_{0n}V_{n\kappa}/\varDelta$ of the transition from the level $|0\rangle$ to a state $|\kappa\rangle$ of the continuum via a level $|n\rangle$ of the band detuned from the state $|0\rangle$ by the energy difference $\varDelta_n$, gives a contribution $(V_{0n}V_{n\kappa}/\varDelta)^2 g_\kappa = (V_{0n}/\varDelta_n)^2\gamma_c$ to the decay rate. The contribution of all levels with the detuning larger than a gap size $\varDelta$ is given by the integral $\int_\varDelta^\infty (V_{0n}/\varDelta_n)^2\gamma_c g d\varDelta_n$, that is, a quantity $\gamma_c g V^2/\varDelta$ which is small for a large gap size and which results in a decay $\exp(-t\gamma_c g V^2/\varDelta)$ of $\rho_0(t)$. Averaging over the gap size yields

$$\langle\rho_0(t)\rangle = \int d\varDelta \exp\left(-t\gamma_c g V^2/\varDelta - g\varDelta\right) \sim \exp\left(-\pi g\sqrt{V^2\gamma_c t}\right), \qquad (8.128)$$

where we have ignored a factor in front of the exponent, slowly dependent on time. In the units where $V = 1$ this expression coincides with (8.127).

Returnable Random Walk

Now let us consider the technically more difficult case of a returnable random walk. For $\lambda \neq 0$ the point $\tau = 0$ gives no contribution to the integral (8.124), since the hypergeometric function entering the term Φ in the exponent tends at this point to infinity. The contribution comes now from the singularity at the point $\tau = -\varLambda\lambda^2$. In order to avoid cumbersome calculations of the derivatives of hypergeometric functions associated with the second-order pole we make the replacement of the integration variables $f \to -\tau f$, which moves the variable τ out of the arguments of $\varLambda$ and Φ. Such a replacement does not affect the convergence of the integrals over $df d\overline{f}$, provided the term $\varLambda\lambda^2$ has a positive real part. If this is not the case, the replacement should be taken with the opposite sign. The integral (8.124) takes the form

$$\begin{aligned}\langle\rho_0(\varepsilon,\xi)\rangle = \textstyle\int e^{-\tau}d\tau df d\overline{f} \int dr dr' \Big[&\tfrac{1}{\varLambda\lambda^2+\tau}\tfrac{\partial^2\varLambda\lambda^2}{\tau\partial f\partial\overline{f}} \\ &-\tfrac{\tau}{\varLambda\lambda^2+\tau}\left(1-\tfrac{\lambda^2}{\tau}\varLambda_\xi\right)\left(1-\tfrac{\lambda^2}{\tau}\varLambda_\varepsilon\right) - \left(\tfrac{1}{\varLambda\lambda^2+\tau}\right)^2\tfrac{\partial\varLambda\lambda^2}{\partial f}\tfrac{\partial\varLambda\lambda^2}{\tau\partial\overline{f}}\Big] \\ &\exp\left(i\varepsilon r - i\xi r' + \tau f\overline{f} + \Phi - \tfrac{(r-r')^2}{4}\varLambda - \tfrac{X^2}{4}\varLambda\right),\end{aligned} \qquad (8.129)$$

where the arguments of Φ and $\varLambda$ read

$$\begin{aligned} a &= \tfrac{(r-r')^2}{4} - \lambda^2 f\overline{f} + i\tfrac{r+r'}{2}X - \tfrac{X^2}{4} \\ b &= \sqrt{4rr'\lambda^2 f\overline{f}}. \end{aligned} \qquad (8.130)$$

The integration over $d\tau$ yields

$$\langle\rho_0(\varepsilon,\xi)\rangle = \int df d\overline{f} \int dr dr' \Big[-\tfrac{\partial^2 \Lambda}{\Lambda \partial f \partial \overline{f}} + \Lambda\lambda^2 \left(1+\tfrac{\Lambda_\xi}{\Lambda}\right)\left(1+\tfrac{\Lambda_\varepsilon}{\Lambda}\right)$$
$$+\tfrac{\partial \Lambda}{\Lambda \partial \overline{f}}\tfrac{\partial \Lambda}{\Lambda \partial f}\left(1-\Lambda\lambda^2+f\overline{f}\Lambda\lambda^2\right)\Big]$$
$$\times \exp\left(i\varepsilon r - i\xi r' + \Phi - \tfrac{(r-r')^2}{4}\Lambda - \tfrac{X^2}{4}\Lambda + \Lambda\lambda^2 - \Lambda\lambda^2 f\overline{f}\right). \tag{8.131}$$

No further significant simplifications can be performed in the general form of (8.131), even by introducing a single integration variable u instead of the product $f\overline{f}$ of two variables. However, one can pursue the simplification by making the replacement $f\overline{f} \to u$ in the long-time limit, where the return probability λ^2 is asymptotically large and hence the asymptotic form

$$\Lambda\left(r,r',u,\lambda^2\right) \to \frac{\Lambda}{2u^{1/4}\lambda^{1/2}\sqrt{rr'}};\ \Lambda = \pi g\sqrt{i\pi}\Gamma\left(\frac{1}{4}\right)^{-2}$$

$$\Phi\left(r,r',u,\lambda^2\right) \to -2\Phi u^{1/4}\lambda^{1/2}\sqrt{rr'};\ \Phi = \pi g\sqrt{i^3\pi}\Gamma\left(\frac{3}{4}\right)^{-2} \tag{8.132}$$

can be employed for the functions $\Phi\left(r,r',u,\lambda^2\right)$ and $\Lambda\left(r,r',u,\lambda^2\right)$. Note that the functions are now independent of X. Hereafter the symbols Φ and Λ without arguments denote the large constants given by (8.132) in terms of Γ-functions. Substitution of (8.132) into (8.131) with allowance for (8.120) yields

$$\langle\rho_0(\varepsilon,\xi)\rangle = \int du \int dr dr'\ \tfrac{\Lambda\lambda^{3/2}(3+u)}{32\ u^{5/4}\sqrt{rr'}} e^{i\varepsilon r - i\xi r'}$$
$$\exp\left[-2\Phi u^{1/4}\lambda^{1/2}\sqrt{rr'} - \Lambda\tfrac{(r-r')^2+X^2-4\lambda^2+4\lambda^2 u}{8u^{1/4}\lambda^{1/2}\sqrt{rr'}}\right]. \tag{8.133}$$

We neglect the finite value X^2 compared to the large function λ^2 and employ the hyperbolic coordinates R,θ,ϕ introduced by the relations

$$r = R\,e^{\theta+\phi}\frac{\Lambda}{2\Phi};\ r' = R\,e^{-\theta-\phi}\frac{\Lambda}{2\Phi};\ u = \frac{e^{4\phi}}{\lambda^2}\left(\frac{\Lambda}{2\Phi}\right)^2 \tag{8.134}$$

with the Jacobian $-e^{6\phi}R\Lambda^4/2\lambda^2\Phi^4$. This results in

$$\langle\rho_0(\zeta,\eta)\rangle = \int d\phi(e^{4\phi}+12\overline{\lambda}^2)\int dR \exp\left[\left(-\sqrt{2}e^{\phi}R - \tfrac{4e^{-2\phi}\overline{\lambda}^2-e^{2\phi}}{\sqrt{2}R}\right)\tfrac{\overline{g}U}{2}\right]$$
$$\tfrac{i(\overline{g}U)^3}{16}\int d\theta \exp\left[\left(ie^{\phi}2\eta\sinh[\theta] - \sqrt{2}\zeta e^{\phi}\cosh[\theta] - \sinh[\theta]^2\right)\tfrac{R\overline{g}U}{2}\right], \tag{8.135}$$

where $\overline{g} = \sqrt{\Lambda\Phi} = \sqrt{\pi}g/\sqrt{2}$, $U = \Phi\Lambda^{-1}$, $\lambda = 2\overline{\lambda}U$, $\zeta = i\sqrt{2}\overline{g}(\varepsilon - \xi)$, and $\eta = \overline{g}(\varepsilon + \xi)/2$. By the last two substitutions we have scaled the time variable $t \to t/\overline{g}$, which means that time is now measured in the band recurrence periods (Heisenberg times) discussed in Sect. 3.3.1. We have also rotated the variable $\varepsilon - \xi$ in the complex plane such that the integration contour of the inverse Fourier transformation will circumvent the positive part of the real axis. The Fourier transformation over this variable thus becomes the Laplace transformation. The expression is multiplied by the Jacobian $i\sqrt{2}\overline{g}^2$ of the last two replacements as well.

Since the parameter $\overline{g}$ is large, the integration over $d\theta$ in (8.135) can be performed by the saddle point method. Expanding the function in the exponent near the point $\theta = 0$, one finds the first integral over the hyperbolic angle θ

$$\frac{\sqrt{2}}{\sqrt{R}}\left(\frac{\pi\sqrt{2}}{U\overline{g}}\right)^{1/2}\left(1 + ie^{\phi}\zeta\right)^{-1/2}\exp\left[\frac{-RU}{\sqrt{2}}\overline{g}\left(e^{\phi}\zeta + \frac{e^{2\phi}\eta^2}{1+\zeta e^{\phi}}\right)\right]. \tag{8.136}$$

The second integral over the hyperbolic radius R now has the structure

$$\int\limits_0^\infty \frac{dR}{\sqrt{R}}\exp\left(-AR - \frac{1}{R}B\right) = \sqrt{\frac{\pi}{A}}e^{-2\sqrt{AB}}$$

and yields

$$\langle\rho_0(\zeta,\eta)\rangle = \frac{2^{1/4}\pi iU^2}{8}\int d\phi \frac{e^{-\phi}(e^{4\phi}+12\overline{\lambda}^2)}{\sqrt{\eta^2+\zeta^2+1+2\zeta\cosh\phi}}$$
$$\exp\left[-\frac{U\overline{g}\left(4\overline{\lambda}^2-e^{4\phi}\right)^{1/2}}{(1+\zeta e^{\phi})^{1/2}}\left(\eta^2+\zeta^2+1+2\zeta\cosh\phi\right)^{1/2}\right] \tag{8.137}$$

One can neglect the small quantity ζ everywhere in (8.137) except in those terms where it is multiplied by exponentially large factors. We also note that for small ζ, two different contributions to the integral come from two distinct domains of large positive and large negative ϕ, where only one of the two exponents entering $\cosh\phi$ can be retained. This suggests the substitution $x = \zeta e^{\phi}$ for large positive ϕ and the substitution $x = \zeta e^{-\phi}$ for large negative ϕ with x ranging from 0 to ∞, so that the integral adopts the form

$$\langle\rho_0(\zeta,\eta)\rangle = \frac{2^{1/4}\pi iU^2}{8}\int\limits_0^\infty dx\frac{(x/\zeta^2+12\overline{\lambda}^2\zeta^2/x^2)}{\zeta\sqrt{\eta^2+1+x}}$$
$$\times\exp\left[-\frac{U\overline{g}\left(4\overline{\lambda}^2-\zeta^{-4}x\right)^{1/2}}{(1+x)^{1/2}}\left(\eta^2+1+x\right)^{1/2}\right]$$
$$+\int\limits_0^\infty dx\frac{3\overline{\lambda}^2 2^{1/4}\pi iU^2}{2\zeta\sqrt{\eta^2+1+x}}.\exp\left[-2U\overline{g}\left(\overline{\lambda}^2\right)^{1/2}\left(\eta^2+1+x\right)^{1/2}\right]. \tag{8.138}$$

Inspection shows that the first integral contributes only at short times and can be ignored. The second integral can be evaluated exactly and yields the probability to be in the state $|0\rangle$ for the returnable random walk

$$\langle \rho_0(\zeta,\eta)\rangle = \frac{3\sqrt{\pi}\lambda(\zeta,\eta)}{2^{1/4}\zeta g V} \exp\left[-\sqrt{2\pi}(\eta^2 + V^4 g^2)^{1/2}\lambda(\zeta,\eta)\right]. \tag{8.139}$$

In this final result for the general case of the returnable random walk, we have restored the original notation $\lambda^2(\zeta,\eta) = 4\overline{\lambda}^2 U^2$ for the Fourier transform of the random walk correlation function and the density of the band states $\overline{g} = \sqrt{\pi}g/\sqrt{2}$. The mean squared coupling $V = \langle V^2\rangle^{1/2}$ is also included.

The Role of the Correlation Time for a Returnable Walk

We now analyze the role of the correlation time τ_r. For a δ-correlated random walk, the Fourier transform $\lambda^2(\zeta,\eta)$ does not depend on η. Therefore the main contribution to the inverse Fourier transformation over $d\eta$ comes from the domain $\eta = 0$ where the exponent in (8.139) has the only saddle point. Evaluation of the integral results in

$$\langle \rho_0(\zeta)\rangle = \frac{3\pi^{3/4}\sqrt{\lambda(\zeta)}}{\zeta\sqrt{g}} \exp\left[-\sqrt{2\pi}V^2 g\lambda(\zeta)\right], \tag{8.140}$$

and for a random walk with the power law dependence $\lambda(\zeta) = \kappa\zeta^{-\alpha}$ at the limit of small ζ this gives the time-dependence

$$\begin{aligned}\langle \rho_0(t)\rangle &= \int \frac{3\pi^{3/4}\sqrt{\lambda(\zeta)}}{2\pi\zeta\sqrt{g}} e^{-\zeta t - \sqrt{2\pi}V^2 g\lambda(\zeta)} d\zeta \\ &\sim \frac{t^{1/(1+\alpha)}}{g^{(3+2\alpha)/(1+\alpha)}} \exp\left[-\text{const}\, t^{\alpha/(1+\alpha)}\right].\end{aligned} \tag{8.141}$$

For the case of a finite τ_r we employ the model $\lambda^2(\zeta,\eta) = \lambda^2(\zeta)/(\eta^2 + \gamma^2)$ where $\gamma = 1/\tau_r$ determines the width associated with the dephasing. After the replacement

$$\sqrt{2\pi}\sqrt{\frac{\eta^2 + V^4 g^2}{(\eta^2 + \gamma^2)}}\lambda(\zeta) \to Z \tag{8.142}$$

Equation (8.139) yields

$$\langle \rho_0(\zeta)\rangle = \int_0^\infty \frac{-i3\, 2^{1/4}\pi\lambda^2(\zeta)Z\sqrt{\gamma^2 - V^4 g^2}e^{-Z}}{\zeta g V\left(Z^2 + 2\pi\lambda^2(\zeta)\right)\sqrt{Z^2\gamma^2 + 2\pi\lambda^2(\zeta)V^4 g^2}} dZ. \tag{8.143}$$

Here we have employed the fact that the integration contour for η can be closed in the upper part of the complex plane (which was impossible for the δ-correlating random walk), whereas the lower and upper integration limits for Z correspond to the points $\eta = iV^2 g$ and $\eta = i\gamma$, respectively. The analytical property of $\lambda^2(\zeta, \eta)$ that makes such a transformation of the contour possible has an important consequence—contrary to (8.140) the integral (8.143) contains a term which corresponds to a population, non-vanishing in the long-time limit. Indeed, for small values of the parameter $V^4 g^2 \gamma^{-2}$, by neglecting Z^2 relative to the large function $2\pi\lambda^2(\zeta)$ we evaluate the integral exactly in terms of the Bessel functions, which we, however, give only in the asymptotic form for large $\lambda^2(\zeta)$

$$\langle \rho_0(\zeta) \rangle = \frac{3\,2^{-3/4}}{i\zeta g V} \int_0^\infty \frac{Z e^{-Z}}{\sqrt{Z^2 + 2\pi\lambda^2(\zeta) V^4 g^2 \gamma^{-2}}} dZ \tag{8.144}$$

$$\frac{3\,2^{3/4}}{i\zeta g V} + \frac{3i}{\zeta}\sqrt{\frac{\lambda(\zeta)}{\sqrt{\pi}\gamma g}} \sin\left(\frac{\lambda(\zeta) V^2 g}{\gamma}\right) \ln\left(\frac{\sqrt{2\pi}\lambda(\zeta) V^2 g}{\gamma}\right)$$

in order to be consistent within the approximation.

The asymptotic regime corresponding to (8.144) is attained only for very long times, when the number of returns $\sim \lambda^2(1/t)$ exceeds the number of levels $(\gamma g)^2$ in the spectral band of width $\gamma = 1/\tau_r$ corresponding to the correlation time. Therefore it is never attained for the δ-correlated random walk with $\tau_r = 0$.

The inverse Laplace transformation of (8.144) yields

$$\langle \rho_0(t) \rangle = \frac{3\,2^{3/4}}{gV} + \frac{\text{const}\; t^{1/(1+\alpha)}}{(g/\gamma)^{(3+2\alpha)/(1+\alpha)}} \exp\left[-\text{const}\;(g/\gamma)^{-1/(1+\alpha)} t^{\alpha/(1+\alpha)}\right] \tag{8.145}$$

where the last term is estimated for $\lambda(\zeta) = \kappa\zeta^{-\alpha}$ by analogy to (8.141). One sees that the asymptotic population is given by the inverse of the expectation value of resonances gV for the level–band system and does not contain any trace of the correlation time, which only determines the typical time when the constant asymptotic population of the state $|0\rangle$ is reached.

Summarizing the results obtained, we can say that a level does not decay completely to a band corresponding to a returnable random walk with a finite time of the correlation of the probability amplitudes. It keeps a residual population given by the inverse expectation number of resonances gV. For a non-returnable random walk, the decay manifests a universal time-dependence $\exp(-\sqrt{t})$, whereas for a δ-correlated returnable random walk, the decay law $\exp\left[-t^{\alpha/(1+\alpha)}\right]$ is governed by the power index α of the Fourier transformed return probability $\lambda(\zeta) = \kappa\zeta^{-\alpha}$.

8.3 Dynamical Localization in Complex Systems

Dynamical localization is a phenomenon, which occur in multilevel quantum systems subject to external time-dependent perturbations. The localization manifests itself in the regime where the interference of recurrences emerging from this discreteness of the energy spectrum is important, in contrast to the regime discussed earlier, in Chap. 6, where the dynamics of complex quantum system with time-dependent parameters was considered for a time domain shorter than the Heisenberg time, and where the recurrences were ignored. Note that when the external perturbation can be associated with an external field, say an electromagnetic field, and when amplitude of each harmonic of this field can be treated as an independent quantum variable, one can consider the multilevel system with time-dependent Hamiltonian as a composed quantum object with a time-independent Hamiltonian comprising both the multilevel system and the field modes. Being viewed in such a way, dynamical localization becomes an analog to the Anderson localization of Sect. 7.6 extended on the case of a more general combined system similar to that considered in Sect. 8.2.2. The same analogy can be drown when the external fields are not quantized, bus simply are treated with the help of the quasienergy technique presented on p. 202.

General formulation of the problem implies consideration of the quantum evolution induced by the time-dependent Hamiltonian of the form

$$\hat{H}(t) = \hat{H}_0 + \hat{V}(t),$$

which alternatively can be represented as

$$\hat{H}(t) = \hat{H}_0 + \sum_n \mathrm{f}_n(t)\hat{V}_n. \tag{8.146}$$

The matrix $\hat{H}_0$ and the perturbations $\hat{V}_n$ governing quantum evolution of the multilevel system belong to a certain algebra, and this very algebra, rather than a particular choice of the Hamiltonian itself, determines the character of the population dynamics at the long time scale. Here we are going to invoke one of the simplest models of quantum system recurrences—interfering revivals, suggested by the equidistant spectrum of $\hat{H}_0$. We will also employ one of the simplest versions of the perturbation, given by the factorized matrix $\hat{V}_{uf}$ that has all matrix element equal to unity, and which is multiplied by an arbitrary function $\mathrm{f}(t)$. Even with such a specific choice of the operators, the resulting algebra turns out to be the full algebra of the multilevel system, and therefore the results of the consideration have good chances to be universal at the asymptotic of long times, apart of the exceptional cases of zero measure. The basic questions we address are: (i) what are the functional properties of $\mathrm{f}(t)$ that govern the asymptotic population distribution over the eigenstates of $\hat{H}_0$, and (ii) how does this distribution relates to the distribution of (6.49) obtained for the perturbation by a random matrix at the time scale short as compared to the Heisenberg time? In the quasienergy representation, the simplest choice $\mathrm{f}(t) = \mathrm{f}\cos t$ yields (8.51), which now has to be considered in a time domain much longer than the density of states $1/\delta$.

8.3.1 Algebraic Conditions for a Generic Model

Considering in Sect. 5.1 the general algebraic properties of the time-dependent quantum systems at the example of the two-valued Hamiltonian, we have seen that in virtue of the Baker–Campbell–Hausdorff relation the evolution operator can be cast in the form (5.5), such that logarithm of this operator is a linear combination

$$\log \hat{U}(t) = \sum_{n=0}^{N_o} f_n(\mathrm{f}_n(t))\hat{\mathcal{G}}_n \tag{8.147}$$

of all generators $\hat{\mathcal{G}}_n$ of the relevant full dynamic group. The coefficients $f_n(\mathrm{f}_n(t))$ are determined by the particular realization of the time-dependent Hamiltonian (8.146) given by the choice of the functions $\mathrm{f}_n(t)$. For generic $\mathrm{f}_n(t)$ and $\hat{V}_n$, these coefficients are also generic, but for a specific choice, some of them may vanish or correlate to a certain extent. The reason for this might be hidden both in the algebraic properties of the operators $\hat{H}_0$ and $\hat{V}_n$ and in the functional properties of the functions $\mathrm{f}_n(t)$. The first option emerges from the commutation relations of the operators $i\hat{H}_0$ and $i\hat{V}_n$ and their higher-order commutators. For an N-level system, these commutators may either span the entire $su(N)$ algebra such that the system experience generic $SU(N)$ evolution for generic $\mathrm{f}_n(t)$, or remain restricted in a subalgebra of $su(N)$ such that the evolution also remains restricted to the relevant subgroup, whereas coefficients $f_n(\mathrm{f}_n(t))$ in front of the generators other that the generators of this subgroup are zeros. The second option corresponds to the situation where the commutators do span the $su(N)$ algebra, but the coefficients $f_n(\mathrm{f}_n(t))$ in front of certain generators vanish due to the functional properties of $\mathrm{f}_n(t)$.

Magnus Series

For the evolution operator corresponding to the Hamiltonian (8.146), one can make use of the expression

$$\hat{U} = e^{-i\hat{R}(t)},$$

with $\hat{R}(t)$ satisfying the nonlinear matrix differential equation (5.131)

$$-i\frac{\partial \hat{R}(t)}{\partial t} = i\sum_{k=0} a_k\Big\{\big[-i\hat{R}(t),\Big\}^k \hat{H}(t)\Big\{\big]\Big\}^k,$$

and the coefficients a_k given by the Taylor expansion $\sum a_k x^k = x(e^x - 1)^{-1}$, which we have already obtained considering corrections to the adiabatic process in multilevel systems. It is worthwhile to also present this equation in an elegant implicit Hausdorf form,

$$\sum_{n=0}^{\infty} \frac{(-1)^n}{(n+1)!}\{[\}^n i\frac{\partial \hat{R}(t)}{\partial t}\Big\{, i\hat{R}(t)]\Big\}^n = -i\hat{H}(t).$$

Solving (5.131) by iterations, one obtains the so-called Magnus series

$$\int_0^t dt_1 \frac{\hat{H}(t_1)}{-i} + \frac{1}{2}\int_0^t dt_1 \int_0^{t_1} dt_2 \frac{\Big[\hat{H}(t_1), \hat{H}(t_2)\Big]}{(-i)^2}$$

$$+\frac{1}{6}\int_0^t dt_1 \int_0^{t_1} dt_2 \int_0^{t_2} dt_3 \frac{\Big[\hat{H}(t_1), \Big[\hat{H}(t_2), \hat{H}(t_3)\Big]\Big] + \Big[\hat{H}(t_3), \Big[\hat{H}(t_2), \hat{H}(t_1)\Big]\Big]}{(-i)^3}$$

$$+\frac{1}{12}\int_0^t dt_1 \int_0^{t_1} dt_2 \int_0^{t_2} dt_3 \int_0^{t_3} dt_4 \frac{\Big[\hat{H}(t_1), \Big[\hat{H}(t_2), \Big[\hat{H}(t_3), \hat{H}(t_4)\Big]\Big]\Big] + \text{permut.}}{(-i)^4}$$

$$+\ldots$$

for the operator $-i\hat{R}(t)$, where coefficients in front of the integrals have no explicit expression but can be recursively related to each other and to the coefficients a_k, as was the case for the Baker–Campbell–Hausdorf relation. For high orders, it is easier to draw each term of the series as shown in Fig. 8.10 than explicitly write it down.

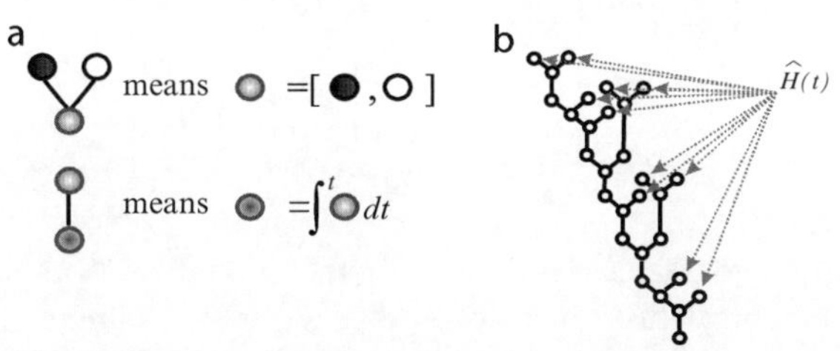

Fig. 8.10 Diagrams for Magnus series. (**a**) Rules; two circles connected to one denotes commutation, and the vertical line denotes time integration. (**b**) An example of the diagram for a term of the Magnus series. By convention, the left branch of the tree is longer that the right branch at each vertex

The important fact is that even in the case where only few functions $\mathrm{f}_n(t)$ in (8.146) differ from zero, or even just one of the coefficients does, the commutation relations entering Magnus series usually generate a complete set of coefficients $f_n(\mathrm{f}_n(t))$ in front of all the generators of the dynamical Lie group (8.147). We now illustrate this by a particular example of the unperturbed Hamiltonian $\hat{H}_0$ and the time-dependent perturbation $\mathrm{f}(t)\hat{V}$. We believe that for a given symmetry group, at the long time scale, dynamics of any multilevel quantum system becomes universal, while all the qualitative characteristics of the behavior depend on the functional properties of the dependency $\mathrm{f}(t)$ exclusively, and not on the particular realization of the operators $\hat{H}_0$ and $\hat{V}$.

Algebraic Structure for the Model of Interacting Revivals

For the particular choice

$$\hat{H}(t) = \hat{H}_0 + \mathrm{f}(t)\hat{V}$$

with two operators $\hat{H}_0$ and $\hat{V}$ given by the $N \times N$ matrices

$$\hat{H}_0 = \begin{pmatrix} \cdots & \cdots & \cdots & \cdots & \cdots & \cdots & \cdots \\ \cdots & 2 & 0 & 0 & 0 & 0 & \cdots \\ \cdots & 0 & 1 & 0 & 0 & 0 & \cdots \\ \cdots & 0 & 0 & 0 & 0 & 0 & \cdots \\ \cdots & 0 & 0 & 0 & -1 & 0 & \cdots \\ \cdots & 0 & 0 & 0 & 0 & -2 & \cdots \\ \cdots & \cdots & \cdots & \cdots & \cdots & \cdots & \cdots \end{pmatrix} \tag{8.148}$$

and

$$\hat{V} - \hat{1} = \begin{pmatrix} \cdots & \cdots & \cdots & \cdots & \cdots & \cdots & \cdots \\ \cdots & 0 & 1 & 1 & 1 & 1 & \cdots \\ \cdots & 1 & 0 & 1 & 1 & 1 & \cdots \\ \cdots & 1 & 1 & 0 & 1 & 1 & \cdots \\ \cdots & 1 & 1 & 1 & 0 & 1 & \cdots \\ \cdots & 1 & 1 & 1 & 1 & 0 & \cdots \\ \cdots & \cdots & \cdots & \cdots & \cdots & \cdots & \cdots \end{pmatrix}, \tag{8.149}$$

one finds that these two traceless matrices together with their all-order commutators span the $su(N)$ algebra. This can easily be checked at the example of $N = 3$, where a vector

$$\hat{\mathbf{v}}_C = \begin{pmatrix} \hat{H}_0 \\ \hat{V} - 1 \\ \left[\hat{H}_0, \hat{V}\right] \\ \left[\hat{H}_0, \left[\hat{H}_0, \hat{V}\right]\right] \\ \left[\left[\hat{H}_0, \hat{V}\right], \left[\hat{H}_0, \left[\hat{H}_0, \hat{V}\right]\right]\right] \\ \left[\hat{H}_0, \left[\left[\hat{H}_0, \hat{V}\right], \left[\hat{H}_0, \left[\hat{H}_0, \hat{V}\right]\right]\right]\right] \\ \left[\hat{V}, \left[\hat{H}_0, \left[\left[\hat{H}_0, \hat{V}\right], \left[\hat{H}_0, \left[\hat{H}_0, \hat{V}\right]\right]\right]\right]\right] \\ \left[\left[\left[\hat{H}_0, \hat{V}\right], \left[\hat{H}_0, \left[\hat{H}_0, \hat{V}\right]\right]\right], \left[\hat{H}_0, \left[\hat{H}_0, \hat{V}\right]\right]\right] \end{pmatrix} \tag{8.150}$$

composed of the commutators is related to the vector $\hat{\mathbf{G}} = \left\{\hat{\mathcal{G}}_n\right\}$ composed of the generators of the group $SU(3)$

$$\hat{\mathcal{G}}_1 = \begin{pmatrix} 1 & 0 & 0 \\ 0 & 0 & 0 \\ 0 & 0 & -1 \end{pmatrix}, \hat{\mathcal{G}}_2 = \begin{pmatrix} 1 & 0 & 0 \\ 0 & -2 & 0 \\ 0 & 0 & 1 \end{pmatrix}, \hat{\mathcal{G}}_3 = \begin{pmatrix} 0 & 0 & 1 \\ 0 & 0 & 0 \\ 0 & 0 & 0 \end{pmatrix}, \hat{\mathcal{G}}_4 = \begin{pmatrix} 0 & 1 & 0 \\ 0 & 0 & 0 \\ 0 & 0 & 0 \end{pmatrix},$$

$$\hat{\mathcal{G}}_5 = \begin{pmatrix} 0 & 0 & 0 \\ 0 & 0 & 1 \\ 0 & 0 & 0 \end{pmatrix}, \hat{\mathcal{G}}_6 = \begin{pmatrix} 0 & 0 & 0 \\ 0 & 0 & 0 \\ 1 & 0 & 0 \end{pmatrix}, \hat{\mathcal{G}}_7 = \begin{pmatrix} 0 & 0 & 0 \\ 1 & 0 & 0 \\ 0 & 0 & 0 \end{pmatrix}, \hat{\mathcal{G}}_8 = \begin{pmatrix} 0 & 0 & 0 \\ 0 & 0 & 0 \\ 0 & 1 & 0 \end{pmatrix}, \tag{8.151}$$

via the explicit matrix relation

$$\hat{\mathbf{G}} = \begin{pmatrix} \frac{1}{2} & 0 & 0 & 0 & 0 & 0 & 0 & 0 \\ 0 & \frac{1}{6} & 0 & \frac{-1}{12} & 0 & 0 & \frac{-1}{72} & 0 \\ 0 & \frac{-1}{6} & \frac{7}{24} & \frac{1}{6} & 0 & 0 & 0 & \frac{-1}{144} \\ \frac{-3}{4} & \frac{1}{3} & \frac{-13}{24} & \frac{-1}{12} & \frac{1}{24} & \frac{-1}{24} & 0 & \frac{1}{144} \\ \frac{3}{4} & \frac{1}{3} & \frac{-13}{24} & \frac{-1}{12} & \frac{-1}{24} & \frac{1}{24} & 0 & \frac{1}{144} \\ 0 & \frac{-1}{6} & \frac{-7}{24} & \frac{1}{6} & 0 & 0 & 0 & \frac{1}{144} \\ \frac{-3}{4} & \frac{1}{3} & \frac{13}{24} & \frac{-1}{12} & \frac{1}{24} & \frac{1}{24} & 0 & \frac{-1}{144} \\ \frac{3}{4} & \frac{1}{3} & \frac{13}{24} & \frac{-1}{12} & \frac{-1}{24} & \frac{-1}{24} & 0 & \frac{-1}{144} \end{pmatrix} \hat{\mathbf{v}}_C. \tag{8.152}$$

Therefore in the expansion (8.147) none of the generators is excluded due to the algebraic restrictions. In other words, no algebraic restriction on the coefficients in front of the generators of the dynamic Lie group exists, such that these coefficients are governed by the functional properties of the function $\mathrm{f}(t)$ exclusively. These properties will be of the main interest in the following section.

8.3.2 Dynamics Governed by the Interfering Revivals

Consider now the Schrödinger equation

$$i\dot{\psi}_n = n\psi_n + \mathrm{f}(t)\sum_n \psi_n \tag{8.153}$$

for the system with Hamiltonian $\hat{H}_0 + \mathrm{f}(t)\hat{V}_{uf}$ where the operators are given by (8.148, 8.149). We note that the energy distance among the neighboring levels in the uniform band corresponding to $\hat{H}_0$ is set to unity, such that the revival time amounts to 2π. For the initial condition $\psi_n(t=0)=1$ one finds the relation

$$\psi_n(t) = \delta_{n,0} - i\int_0^t d\tau e^{-in(t-\tau)}\mathrm{f}(\tau)\sum_n \psi_n(\tau), \tag{8.154}$$

which results in the equation

$$Q(t) = 1 - i\int_0^t d\tau G(t-\tau)\mathrm{f}(\tau)Q(\tau), \tag{8.155}$$

for the function $Q(t) = \sum_n \psi_n(t)$, where the kernel of the integral equation reads

$$G(z) = \sum_n e^{-inz} = 2\pi\sum_k \delta(z - 2\pi k). \tag{8.156}$$

The kernel corresponds to a periodic sequence of δ-like revivals over the period 2π, which is typical of the infinitely broad band of equidistant levels, as we know from Chap. 3. For a broad finite band, the kernel is a periodic sequence of short spikes of a finite width.

Equation (8.155) with the kernel (8.156) results in the algebraic expression

$$Q(t) = 1 - i\pi\mathrm{f}(t)Q(t) + 2i\pi\sum_{k=1}^{\mathrm{ip}\{\frac{t}{2\pi}\}} \mathrm{f}(t-2\pi k)Q(t-2\pi k) \tag{8.157}$$

for the function $Q(t)$ with the arguments shifted by the multiples of the revivals time, where $\mathrm{ip}\left\{\frac{t}{2\pi}\right\}$ stands for the integer part of $\frac{t}{2\pi}$. Here we have taken into account the fact that in the limit where the band width tends to infinity $\int_{t-\varepsilon}^{t} d\tau\delta(t-\tau) = 1/2$ for $t > \varepsilon > 0$. By $Q_k(\theta)$ and $\mathrm{f}_k(\theta)$ we now denote the function $Q(t)$ and $\mathrm{f}(t)$, respectively, in the time interval between $2\pi k$ and $2\pi\,(k-1)$ with $\theta = t - 2\pi\,(k-1)$ and reduce (8.157) to the form

$$Q_n(\theta) = 1 - i\pi\mathrm{f}_n(\theta)Q_n(\theta) + 2i\pi\sum_{k=1}^{n-1}\mathrm{f}_k(\theta)Q_k(\theta). \tag{8.158}$$

The sum on the right-hand side of (8.158) can be viewed as multiplication of the vector $\{Q_k(\theta)\}$ by a triangular matrix $\hat{\mathrm{f}}(\theta)$ with the matrix elements

$$\left(\hat{\mathrm{f}}(\theta)\right)_{n,m<n} = \mathrm{f}_{n-m}(\theta) \equiv \mathrm{f}(t = \theta + 2\pi\,(n - m - 1)),$$

which can be explicitly inverted. This yields the solution

$$Q_n(\theta) = \frac{1}{1 - i\pi \mathrm{f}_n(\theta)} \prod_{k=1}^{n} \frac{1 - i\pi \mathrm{f}_k(\theta)}{1 + i\pi \mathrm{f}_k(\theta)}. \tag{8.159}$$

For the amplitude $\psi_m(t)$ of the m-th level at the time moment $t = 2\pi r$, in accordance with (8.154), this solution results in

$$\psi_m(2\pi r) = -i \int_0^{2\pi} d\theta e^{-im\theta} \sum_{n=1}^{r} Q_n(\theta)$$

$$= i \int_0^{2\pi} d\theta e^{-im\theta} \sum_{n=0}^{r-1} \frac{-\mathrm{f}(\theta + 2\pi n)}{1 - i\pi \mathrm{f}(\theta + 2\pi n)} \prod_{k=0}^{n} \frac{1 - i\pi \mathrm{f}(\theta + 2\pi k)}{1 + i\pi \mathrm{f}(\theta + 2\pi k)} \tag{8.160}$$

$$= -\int_0^{2\pi} \frac{d\theta}{2\pi} e^{-im\theta} \sum_{n=0}^{r-1} \left(\prod_{k=0}^{n} \frac{1 - i\pi \mathrm{f}(\theta + 2\pi k)}{1 + i\pi \mathrm{f}(\theta + 2\pi k)} - \prod_{k=0}^{n-1} \frac{1 - i\pi \mathrm{f}(\theta + 2\pi k)}{1 + i\pi \mathrm{f}(\theta + 2\pi k)} \right)$$

$$= -\int_0^{2\pi} \frac{d\theta}{2\pi} e^{-im\theta} \left(\prod_{k=0}^{r-1} \frac{1 - i\pi \mathrm{f}(\theta + 2\pi k)}{1 + i\pi \mathrm{f}(\theta + 2\pi k)} - \frac{1 + i\pi \mathrm{f}(\theta)}{1 - i\pi \mathrm{f}(\theta)} \right). \tag{8.161}$$

The main term in the integrand of (8.161) is given by the Fourier transform of the product of a number of phase factors

$$\frac{1 - i\pi \mathrm{f}(\theta + 2\pi k)}{1 + i\pi \mathrm{f}(\theta + 2\pi k)}. \tag{8.162}$$

For large m, the main contribution comes from large r, that is, the long products that may rapidly depend on time θ and therefore may contain rapidly oscillating harmonics in their Fourier series decompositions. Note that for the chosen energy units, the typical size of the function $\mathrm{f}(t)$ is of the order of the ratio $V(t)/\delta$ of the typical perturbation matrix element and the typical level spacing, discussed on p. 81. This ratio must be large to satisfy the condition of the quasicontinuum existence when the number of levels in resonance given by the square of this quantity is large. This means that the main contribution of each of the factors (8.162) to the overall phase change comes from the domains where the corresponding function $\mathrm{f}(\theta + 2\pi k)$ crosses zero, and where, generally speaking, the phase change is most rapid. Outside of these domains, where the function is large, the relevant phase factors are close to -1. When in the vicinity of a point θ two functions, say $\mathrm{f}(\theta + 2\pi k)$ and $\mathrm{f}(\theta + 2\pi k')$, go zero having the same signs of the derivatives, such a coincidence yields two factors in the product of (8.160) that rapidly change their phases in the same direction, which, evidently, increases the overall rate of the phase change in this very moment of time θ. If the signs of the derivatives are opposite, the phase change decelerates.

We have thus identified a qualitative functional characteristic of $\mathrm{f}(t)$, which is responsible for broadening of the population distribution over the energy scale—this is the coincidence statistics of zeros of the function $\mathrm{f}(t)$ folded with the revival period 2π, which should be taken with the allowance for the derivatives $\mathrm{f}'(t)$ in these points. Two particular cases can immediately be suggested for the analysis: case (i), a periodic $\mathrm{f}(t)$, with a period $2\pi\frac{p}{q}$ which is a rational fraction of the revival period, and case (ii), a function $\mathrm{f}(t)$ randomly changing for every next revival period. Inspection reveals at least two more particular cases: case (iii), an irrational ratio of the function period and the revival period, and case (iv), many incommensurable irrational harmonics of $\mathrm{f}(t)$.

8.3.3 Resonance Excitation Diffusion and Localization as Regimes of the Revival Interference

We now consider four different regimes of the excitation listed above, which correspond to four asymptotically distinct types of the behavior of the population distribution over the energy scale. It turns out that these regimes are governed by rather subtle spectral properties of the perturbation, such as the commensurability of the harmonics in the Fourier decomposition of this signal.

Integer and Rational Frequency Spectrum of $\mathrm{f}(t)$

The simplest case corresponds to the situation where the spectrum of $\mathrm{f}(t)$ has integer frequencies. Each product entering the sum on the right-hand side of (8.160) is therefore simply the $(n+1)$-th power of the factor (8.162) with $k = 0$. Summation of the geometric series results in

$$\psi_m(2\pi r) = \frac{1}{2\pi}\int_0^{2\pi} d\theta e^{-im\theta}\left(\left(\frac{1 - i\pi\mathrm{f}(\theta)}{1 + i\pi\mathrm{f}(\theta)}\right)^{r-1} - \frac{1 + i\pi\mathrm{f}(\theta)}{1 - i\pi\mathrm{f}(\theta)}\right), \tag{8.163}$$

where for the long-time asymptotic $r \to \infty$, and the main contribution comes from the rapidly changing term $\left(\frac{1-i\pi\mathrm{f}(\theta)}{1+i\pi\mathrm{f}(\theta)}\right)^r$. In the situation where the typical size of the function $\mathrm{f}(\theta)$ is large, we can separately consider the contribution of each root

$\mathrm{f}(\theta_n) = 0$, by a Taylor expansion $\mathrm{f}(\theta_n) \simeq \mathrm{f}'_n(\theta - \theta_n)$ in the vicinity of this root and by making use of the stationary phase approximation for the integral over $x = (\theta - \theta_n)$. For the rapid part of the integrand (8.163)

$$\psi_m(2\pi r) \simeq \sum_{\text{roots} \quad n} \frac{e^{-im\theta_n}}{2\pi} \int\limits_{-\infty}^{\infty} dx \exp\left(-imx + (r-1)\log\frac{1 - i\pi \mathrm{f}'_n x}{1 + i\pi \mathrm{f}'_n x}\right),$$

evaluation by the stationary phase method yields

$$\psi_m(2\pi r) \simeq \sum_{\text{roots} \quad n} \frac{\exp\left(i(r-1)\Phi_n\right)}{i\sqrt{2\pi(r-1)}z^{3/4}\left(2\pi \mathrm{f}'_n + z\right)^{1/4}}, \tag{8.164}$$

where $z = \frac{m}{r-1}$ and Φ_n is a phase factor, which for positive f'_n implies negative z and reads

$$\Phi_n = \frac{\sqrt{z}\sqrt{2\pi \mathrm{f}_n + z}}{i\pi \mathrm{f}_n} - z\theta_n + \arg\frac{\sqrt{z} - \sqrt{2\pi \mathrm{f}_n + z}}{\sqrt{z} + \sqrt{2\pi \mathrm{f}_n + z}}.$$

For the negative f_n and positive z the phase factor is similar.

From the structure of the amplitude (8.164), where the time-dependence enters via the parameter exclusively, it becomes clear that in the long-time limit, the population distribution along the energy scale gets broader and it width linearly scales with the number of the revivals periods r given by the integer part of the dimensional time $t\delta/2\pi$. Zeros of $\mathrm{f}(t)$ with the positive derivatives account for the distribution propagation towards negative energies with the velocities of f'_n, while the negative ones account for the energy increase. One can interpret the result as an excitation of the equidistant spectrum of the band by a resonant field and its harmonics, which are just parametrized by the positions of zeros of $\mathrm{f}(t)$ and by the slopes of the function in these points. The resulting population distribution over the energy scale linearly broadens with time, and the velocity of this broadening is given by the typical slope $\overline{|\mathrm{f}'_n|}$.

When the field frequency spectrum is given not by integers, but by multiples of a rational number $\frac{p}{q}$, there exists a common period $2\pi q$ of the revivals and the perturbation, such that one still can perform summation of the geometric series and write an expression for the amplitudes

$$\psi_m(2\pi r)|_{r=vq+1} = \frac{1}{2\pi}\int\limits_0^{2\pi} d\theta e^{-im\theta}\left\{\left(\prod_{k=0}^{q-1}\frac{1-i\pi \mathrm{f}(\theta+2\pi k)}{1+i\pi \mathrm{f}(\theta+2\pi k)}\right)^{pv} - \frac{1+i\pi \mathrm{f}(\theta)}{1-i\pi \mathrm{f}(\theta)}\right\} \tag{8.165}$$

similar to (8.163). Again, for the long-time limit and large m, the main contribution comes from the most rapidly oscillating term—the r-th power of the product representing the net contributions during the common period $2\pi q$, while the summation over q revivals results in the phase $\sum_{k=0}^{q-1} pv \log \frac{1-i\pi \mathrm{f}(\theta+2\pi k)}{1+i\pi \mathrm{f}(\theta+2\pi k)}$ linearly increases with the number $r \sim \frac{pv}{q}$ of periods. The contribution of the second term in the parentheses is unimportant. We find a situation similar to the case of integer frequencies given by (8.164)—the energy distribution width increases linearly in time. This case corresponds to parametric resonance between the band frequency and the frequency of the field.

Random Function $\mathrm{f}(t)$

The situation is quite different in the case where $\mathrm{f}(\theta) \neq \mathrm{f}(\theta + 2\pi k)$ for all k, that is, in general, the function $\mathrm{f}(t)$ is not a periodic but a random one, which means that in the time interval $(0, 2\pi)$ it varies in a way different from that in the interval $(2\pi, 4\pi)$, and in a yet different way in each next time interval $(2\pi k, 2\pi (k+1))$. In other words, time-dependences of the perturbation during different revival periods are completely decorrelated and their actions may compensate each other on average. Equation (8.160) written in the form

$$\psi_m(2\pi n) = -\int_0^{2\pi} \frac{d\theta}{2\pi} \left(e^{-im\theta + i\Phi_n(\theta)} - e^{-im\theta} \frac{1 + i\pi \mathrm{f}(\theta)}{1 - i\pi \mathrm{f}(\theta)} \right) \tag{8.166}$$

shows that the phase factors

$$i\Phi_r(\theta) = \sum_{k=0}^{n-1} \log \frac{1 - i\pi \mathrm{f}(\theta + 2\pi k)}{1 + i\pi \mathrm{f}(\theta + 2\pi k)}$$

are no longer linear functions of the number of the revival periods, since the sum of n random functions is not n times but just $\sqrt{n}$ times larger than a typical term of this sum. In other words, for a given moment θ, the contributions of the terms with increasing phases are compensated on average by those with decreasing ones, and only statistical fluctuations allow for the non-vanishing result. This means that the width of the population distribution over the energy scale grows in the diffusive way—as a square root of time.

To show this, let us invoke a model of the random function

$$\mathrm{f}(\theta) = \frac{1}{i\pi} \frac{\prod_{r=1}^{N} (1 + i\pi \mathrm{f}_{k,r}(\theta - \theta_{k,r})) - \prod_{r=1}^{N} (1 - i\pi \mathrm{f}_{k,r}(\theta - \theta_{k,r}))}{\prod_{r=1}^{N} (1 + i\pi \mathrm{f}_{k,r}(\theta - \theta_{k,r})) + \prod_{r=1}^{N} (1 - i\pi \mathrm{f}_{k,r}(\theta - \theta_{k,r}))}, \tag{8.167}$$

determined by the random parameters $\theta_{k,r}$ and $\mathrm{f}_{k,r}$, which yields

$$\log\frac{1-i\pi\mathrm{f}(\theta+2\pi k)}{1+i\pi\mathrm{f}(\theta+2\pi k)}=\sum_{r=1}^{N}\log\frac{1-i\pi\mathrm{f}_{k,r}(\theta-\theta_{k,r})}{1+i\pi\mathrm{f}_{k,r}(\theta-\theta_{k,r})}$$

for the phase factors. In other words, within each period $(2\pi k, 2\pi\,(k+1))$, the model function (8.167) is characterized by $2N$ parameters—the positions $\theta_{k,r}$ of the roots $\mathrm{f}(\theta_{k,r}) = 0$ and the derivatives $\mathrm{f}_{k,r} = \mathrm{f}'(\theta_{k,r})$ in these points, with $r = 1, \ldots, N$. This choice of a model allows one to express the probability amplitudes (8.160) in the form

$$\psi_m(2\pi n)=\int_0^{2\pi}\frac{d\theta}{2\pi}e^{-im\theta}\left(\prod_{r=1,k=0}^{N,n-1}\frac{1-i\pi\mathrm{f}_{k,r}(\theta-\theta_{k,r})}{1+i\pi\mathrm{f}_{k,r}(\theta-\theta_{k,r})}-\ldots\right)\tag{8.168}$$

convenient for the analysis. Here by dots we denote unimportant slow terms.

Randomness of $\mathrm{f}(\theta)$ implies averaging over an ensemble of such functions. Here, the population

$$\rho_m(2\pi n)=\int_0^{2\pi}\frac{d\vartheta d\theta}{4\pi^2}e^{im\vartheta-im\theta+\sum_{r=1,k=0}^{N,n-1}\log\frac{1-i\pi\mathrm{f}_{k,r}(\theta-\theta_{k,r})}{1+i\pi\mathrm{f}_{k,r}(\theta-\theta_{k,r})}\frac{1+i\pi\mathrm{f}_{k,r}(\vartheta-\theta_{k,r})}{1-i\pi\mathrm{f}_{k,r}(\vartheta-\theta_{k,r})}}\tag{8.169}$$

of the level m after n periods given by $|\psi_m(2\pi s)|^2$ of (8.168) will be averaged over independent uniform distribution of the roots positions $\theta_{k,r}$ within the interval $(0, 2\pi)$ and the uniform distribution of the derivatives $\mathrm{f}_{k,r}$ within a band $(-\mathrm{F}, \mathrm{F})$, while the width 2F of this band serves as the parameter of the model characterizing typical front duration of the phase changes. Each term of the sum in the exponent is a statistically independent factor for the amplitude, which can be calculated separately, and yields

$$\int_{-\mathrm{F}}^{\mathrm{F}}\frac{dy}{2\mathrm{F}}\int_0^{2\pi}\frac{dx}{2\pi}\frac{1-i\pi y(\theta-x)}{1+i\pi y(\theta-x)}\frac{1+i\pi y(\vartheta-x)}{1-i\pi y(\vartheta-x)}.$$

In the limit of large 2F the integration limits for x can be extended to $\pm\infty$ and give

$$\int_{-\mathrm{F}}^{\mathrm{F}}\frac{dy}{2\mathrm{F}}\left(1-\frac{4\pi(\vartheta-\theta)}{\pi y(\vartheta-\theta)+2i}\mathrm{sign}\,(y)\right),$$

while the remaining integral results in

$$1 - \frac{2}{\mathrm{F}} \log \left[1 + \left(\frac{\mathrm{F}\pi\zeta}{2} \right)^2 \right],$$

where $\zeta = \vartheta - \theta$. After substitution to (8.169) and integration over $d\frac{\vartheta+\theta}{2}$ the last expression gives the population

$$\rho_m(2\pi n) = \int_0^{2\pi} \frac{d\zeta}{4\pi^2} \exp\left\{ im\zeta + Nn \log\left(1 - \frac{2}{\mathrm{F}} \log\left[1 + \left(\frac{\mathrm{F}\pi\zeta}{2} \right)^2 \right] \right) \right\},$$

which after evaluation of the integral by the saddle-point method yields the diffusion profile

$$\rho_m(2\pi n) = \frac{1}{2\pi\sqrt{2\pi Nn\mathrm{F}}} \exp\left\{ -\frac{m^2}{8\pi^2 \mathrm{F}\, Nn} \right\}, \tag{8.170}$$

in complete agreement with the intuitive expectations.

Irrational Frequency

We have thus identified two regimes, the linear increase of the distribution width, resulting from the constructive interference of the revivals, and the diffusive regime under the action of a random force where the interference is suppressed by the randomness. Is it possible to identify a regime where interference of the revivals is destructive? The answer to this question is affirmative, and the simplest example of such a case is a harmonic excitation f $\sin[\omega t]$ with a period incommensurate with that of revivals. Equation (8.161) for this case reads

$$\psi_m(2\pi r) = -\int_0^{2\pi} \frac{d\theta}{2\pi} \exp\left\{ -im\theta + \sum_{k=0}^{r-1} \log \frac{1 - i\pi \mathrm{f} \sin[\omega(\theta + 2\pi k)]}{1 + i\pi \mathrm{f} \sin[\omega(\theta + 2\pi k)]} \right\} + \ldots, \tag{8.171}$$

where ω is an irrational number. During each revival period, that is, for each k, the phase factor may turn to zero several times at the points $\theta_{p,k} = \pi p/\omega - 2\pi k$, and the derivatives in these points may take just two values $\pm 2i\pi\omega \mathrm{f}$.

Though for long times, due to the incommensurably of the field period and the revival period, the roots $\theta_{p,k}$ form a rather dense set of points in the interval $(0, 2\pi)$, this neither implies that the root density is uniform nor that the signs of derivatives are randomly distributed. Existence of correlations in the positions of the roots and in the signs of derivatives may and does result in the strong interference phenomena.

In order to reveal this interference we cast the sum of logarithms entering (8.171) to the Taylor series,

$$2\mathrm{Im}\sum_{k=0,m=0,n=0}^{r-1,\infty,\infty}\frac{(-1)^{n+1}\left(\frac{\pi\mathrm{f}}{2}\right)^{m+n}(m+n-1)!}{n!m!}e^{i(m-n)\omega(\theta+2\pi k)},$$

perform summation over $r-1$ revival periods k of the kernel $G(t)$

$$2\mathrm{Im}\sum_{m,n=0}^{\infty}\frac{(-1)^{n+1}\left(\frac{\pi\mathrm{f}}{2}\right)^{m+n}(m+n-1)!}{n!m!}\frac{\sin\left[(n-m)\pi r\omega\right]}{\sin\left[(n-m)\pi\omega\right]}e^{i(m-n)\omega(\theta+\pi(r-1))},$$

make the replacements $n\to m+s$ for $n>m$, or $m\to n+s$ otherwise, getting

$$2\mathrm{Im}\sum_{n,s=0}^{\infty}\frac{(-1)^{n+1}\left(\frac{\pi\mathrm{f}}{2}\right)^{s+2n}(s+2n-1)!}{n!\,(n+s)!}\frac{\sin\left[s\pi r\omega\right]}{\sin\left[s\pi\omega\right]}e^{is\omega(\theta+\pi(r-1))}+$$

$$2\mathrm{Im}\sum_{m,s=0}^{\infty}\frac{(-1)^{m+1}\left(\frac{\pi\mathrm{f}}{2}\right)^{s+2m}(s+2m-1)!}{n!\,(n+s)!}\frac{\sin\left[s\pi r\omega\right]}{\sin\left[s\pi\omega\right]}\left(-e^{i\omega(\theta+\pi(r-1))}\right)^{s},$$

perform summation over n (or m) noting that

$$\sum\frac{x^m(2m+s-1)!}{m!(m+s)!}=\frac{2^s}{(1+\sqrt{1-4x})^s s},$$

and finally obtain the expression

$$-4\sum_{s=2n+1}\left[\frac{\pi\mathrm{f}}{1+\sqrt{1+\pi^2\mathrm{f}^2}}\right]^s\sin\left[s\omega(\theta+\pi\,(r-1))\right]\frac{\sin\left[s\pi r\omega\right]}{s\sin\left[s\pi\omega\right]}\tag{8.172}$$

for the averaged sum of logarithms (8.171). Here summation is performed over odd natural numbers. In the summand, one recognizes the combination $\frac{\sin[s\pi r\omega]}{\sin[s\pi\omega]}$ typical of the r-beam interference, which amounts to r for the resonant frequencies corresponding to zero denominators, and which is rapidly drops due to the destructive interference, when the frequency is detuned from resonance.

The summand of (8.172) contains three factors, the first one exponentially decreasing with the increasing summation index and vanishing at typical sizes $s\sim\pi\mathrm{f}$, the second factor corresponding to oscillations with a constant amplitude, and the third, interference induced factor $\frac{\sin[s\pi r\omega]}{s\sin[s\pi\omega]}$, which is the most important for our analysis. For an irrational ω, the argument of the sine in the denominator never coincides with a multiple of π, and hence the denominator never assumes zero value, such that a finite difference depending on the parameter ε defined via the relation

$s\pi\omega = k\pi + s\pi\varepsilon$ always exists. Given an integer s, the minimum difference $s\pi\varepsilon_s$ is determined by the best rational approximation $\omega = \frac{k}{s} + \varepsilon_s$ of the irrational frequency by a rational number with the denominator s.

From the explicit expression for the derivative of the phase

$$\Phi_r'(\theta) = -4\omega \sum_{s=2n+1} \left[\frac{\pi \mathrm{f}}{1+\sqrt{1+\pi^2\mathrm{f}^2}}\right]^s \cos\left[s\omega(\theta + \pi(r-1))\right] \frac{\sin\left[s\pi r\omega\right]}{\sin\left[s\pi\omega\right]}$$

one finds

$$\Phi_r'(\theta) = \frac{4\omega}{(-1)^{r+1}} \sum_{s=2n+1} \left[\frac{\pi \mathrm{f}}{1+\sqrt{1+\pi^2\mathrm{f}^2}}\right]^s \frac{\cos\left[s\omega(\theta + \pi(r-1))\right]\sin\left[rs\pi\varepsilon_s\right]}{\sin\left[s\pi\varepsilon_s\right]}. \tag{8.173}$$

Being exponentially decreasing as $\exp\left[-s/\pi\mathrm{f}\right]$, the first factor of the summand suggests to consider at most $s \sim 2\pi\mathrm{f}$ first terms of the infinite sum, while the second factor has a numerator less than unity in absolute value, while the denominator is not less than the minimum distance $\varepsilon_{s<2\pi\mathrm{f}}$ between the irrational number and its best rational approximation with a denominator inferior to $2\pi\mathrm{f}$. Apart from a zero measure set of so-called Liouville numbers, for a typical real ω, this distance $\varepsilon_{s<q} \sim 1/q^2$ in virtue of the so-called Dirichlet's theorem, while, according to Hurwitz's theorem, $\sqrt{5}$ can be taken for the proportionality coefficient. One therefore can estimate the upper bound of the phase derivative as

$$|\Phi_{r\to\infty}'| \leq \sum_s 4\sqrt{5}\omega e^{-\frac{s}{\pi\mathrm{f}}} \frac{s}{\pi} \simeq \frac{4\sqrt{5}\omega}{\pi} (\pi\mathrm{f})^2. \tag{8.174}$$

The stationary phase method of evaluation of the integral (8.170) suggests to consider this quantity as an estimate of the maximum possible width of the population distribution over the energy scale, and this distribution attains after $r \sim 1/s\varepsilon_{s<2\pi\mathrm{f}} \sim \pi\mathrm{f}$ revival periods. One can interpret this result as a linear increase of the population distribution till the moment of time $t \sim r \sim \pi\mathrm{f}$ where the destructive interference of the revivals becomes important. This happens when couplings between the system and the field harmonics at the frequencies of parametric resonances no longer can overcome the detunings of these resonances given by the best rational approximations of the irrational frequency.

Equation (8.174) means that excitation with a typical period incommensurate with the revival period remains localized near its initial energy position. Exceptions of this rule are two sets of frequencies of zero measure: the rational frequencies considered above, for which the distribution width linearly increases in time, and the dense set of frequencies corresponding to the so-called Liouville numbers, which we are going to present and consider now.

Concept of the Liouville numbers relates to the Number Theory. One defines these numbers via their representation

$$\omega = \sum \omega_n b^{-n} \tag{8.175}$$

over an integer base b with the integer coefficients $\omega_n \in \{0, 1, \ldots, b-1\}$. The string of ordered coefficients $\{\omega_n\}$ contains infinite number of long and gradually getting longer sequences of zeros $\omega_{n \in Sj} = 0$. By $S_j = \{k_{2j-1}, k_{2j-1}+1, \ldots, k_{2j}-1\}$ one denotes here j-th set of sequential integers numerating the orders n in the sum (8.175) for which all the coefficients ω_n must vanish. Each set S_j has a length $L_{2j-1} = k_{2j} - k_{2j-1}$ that not only monotonously increase with increasing j but does this not slower than the lengths $L_{2j} = k_{2j+1} - k_{2j}$ of the complimentary intervals $\overline{S}_j = \{k_{2j}, \ldots, k_{2j+1}-1\}$ where $\omega_{n \in \overline{S}_j}$ not required to be zero. A string

$$\omega = 1.\underbrace{101}_{\overline{S}_1}\underbrace{0000}_{S_1}\underbrace{11011}_{\overline{S}_2}\underbrace{0000000}_{S_2}\underbrace{110101}_{\overline{S}_3}\underbrace{00000000}_{S_3}\ldots \tag{8.176}$$

where the sequences S_j are shown explicitly gives an example of Liouville number in the binary representation $b = 2$. Given a base b, each Liouville number can be defined by a set of the register positions $\{k_m\}$

$$\omega = \omega\left(\{k_m\}\right) \tag{8.177}$$

where the integers k_{2j-1} and k_{2j} denote the beginnings of the sets S_j and $\overline{S}_j$, respectively, while L_m can be considered as an increasing function $L_m = F(k_m)$ of k_m. The character of $F(x)$ increase at large arguments specifies a class of the Liouville numbers one deals with. The Liouville numbers in general have very good rational approximations,—just in virtue of the assumption that $F(x)$ monotonously increases, for a denominator $s_j = b^{k_{2j-1}}$, the accuracy $\varepsilon_{s_j} = b^{-k_{2j}}$ of the Liouville number rational approximation remains better than any power g approximation $\varepsilon_{s_j} < s_j^{-g}$. But when $F(k)$ grows fast enough, the approximation may approach the exponential accuracy.

For the latter case, smallness of the denominator $\sin\left[s_j \pi \omega\right] \simeq s_j \pi \varepsilon_{s_j}$ in (8.173) for j-th best approximation can compensate for the exponential smallness of the numerator $\exp\left[-s_j/\pi \mathrm{f}\right]$, and than the high s_j-order parametric resonances do occur in the system since the composite transition matrix element $\sim e^{-s_j/\pi \mathrm{f}}$ exceeds the resonance mismatch $s_j \pi \varepsilon_{s_j}$. Moreover, the high order resonances turns to be as important as the low order resonances although they manifest themselves at much longer time scales. The leading contribution to the population distribution width

$$W\left(t \sim \frac{1}{2n\varepsilon_{s_j}}\right) \sim \sum_{q=1}^{j} e^{-\frac{s_j}{\pi \mathrm{f}}} \frac{1}{\sin\left[s_j \pi \varepsilon_{s_j}\right]} \tag{8.178}$$

suggested by the derivative (8.173) where the summation goes now over the best rational approximation denominators s_j exclusively, and the relations $s_j = b^{k_{2j-1}}$, $\varepsilon_{s_j} = b^{-k_{2j}}$, $\sin\left[s_j \pi \varepsilon_{s_j}\right] \sim s_j \varepsilon_{s_j}$, yield

$$W \sim \sum_{q=1}^{j} b^{k_{2j}-k_{2j-1}} e^{-b^{k_{2j-1}}/\pi \mathrm{f}}. \tag{8.179}$$

for the time moment

$$t \sim \frac{1}{s_j \varepsilon_{s_j}} = b^{k_{2j}-k_{2j-1}},$$

which defines the length $L_{2j-1} = k_{2j} - k_{2j-1}$ of the longest interval S_j and thereby the corresponding denominator s_j of the largest j-th rational approximation to be taken into account.

One can alternatively parametrized the integers k_{2j-1} that define a Liouville number by the time parameters

$$t_m = b^{k_m - k_{m-1}},$$

such that $L_m = \frac{\ln t_m}{\ln b}$, and $s_j = \prod_m^{2j-1} t_m$, which allows one to explicitly represent the augmentation ΔW_{2j} of the width (8.179) that occurs at time $t \sim \frac{1}{s_j \varepsilon_{s_j}} = t_{2j}$ by the expression

$$\Delta W_{2j} = t_{2j} \exp\left[-\frac{1}{\pi \mathrm{f}} \prod_{m=1}^{2j-1} t_m\right], \tag{8.180}$$

independent on the base b. One can see that provided each next time parameter is of the order of the exponent of the product of all former time parameters, that is when

$$\ln t_n \sim \frac{1}{\pi \mathrm{f}} \prod_{m=1}^{n} t_m,$$

the width of the distribution may increase by an amount of the order of unity and hence it tends to infinity with the time, although, this increase is not graduate but goes in steps. Each of the steps occurs when the time $t \simeq t_{2j}$ becomes long enough, such that the ratio

$$\frac{\sin\left[s\pi r\omega\right]}{\sin\left[s\pi\omega\right]} = \frac{\sin\left[rs\pi\varepsilon_s\right]}{\sin\left[s\pi\varepsilon_s\right]}$$

in (8.173) approaches an exponentially large maximum value $1/s\pi\varepsilon_s$ corresponding to the next best rational approximation ε_s of the irrational frequency given by the Liouville number.

One can look at this situation from another point of view. Imagine that at a given strength of the harmonic perturbation, one has identified a multi-quantum resonance of a high order s_j, which as a consequence is very narrow having a width $\delta\omega_{2j-1}$ given by the exponentially small $\sim \exp\left(-s_j/\pi\mathrm{f}\right)$ composite matrix element of the multiphoton transition. Still within this narrow width $\delta\omega_{2j-1}$, one can find another resonance of even higher order, and hence even much narrower, $\delta\omega_{2j+1} \ll \delta\omega_{2j-1}$. By tuning the frequency to this second resonance, one will simultaneously have two resonances that correspond to two distinct Rabi periods. The process can be continued by identifying a third resonance in the vicinity of the second one, than a forth one etc., thus gradually approaching an irrational frequency corresponding to a Liouville number and generating a string of the corresponding Rabi half-periods $\{t_{2j-1}\}$. Since the width $\delta\omega_{2j-1}$ drops exponentially with the order s_j of the resonance, each next s_j exponentially increases with the resonance number j, while the Rabi half-periods t_{2j-1} increase as double exponents as the consequence. and the width ΔW_{2j-1} of the population distribution given by (8.180).

Now one can address the question: how to chose the sequence $\{t_m\}$ in order to obtain any desired asymptotic dependence $W(t)$? This situation corresponds to a general intermediate case placed in between the limiting case of parametric resonance with a linear growth of the width and the limiting case of a generic boarded dynamically localized distribution. Assuming a power-law dependence $\Delta W = t^\alpha$ with $\alpha < 1$, from (8.180) one finds the recurrent relation

$$\ln t_n = \frac{1}{(1-\alpha)\,\pi\mathrm{f}} \prod_{m=1}^{n-1} t_m$$

for the time parameters solving this problem. Due to the rapid growing of ΔW_{2j-1} with the index j, the width W_{2j-1} of the population distribution is mainly given by the last increment ΔW_{2j-1}., and hence the recurrent relation

$$\ln \frac{t_n}{W(t_n)} = \frac{1}{\pi\mathrm{f}} \prod_{m=1}^{n-1} t_m$$

solves the problem for the general case. The fact that the left hand side of the last equation must be positive, implies that in the asymptotic regime W grows with time slower than linearly.

8.3.4 A Number of Incommensurate Irrational Frequencies

In the case of excitation by a multi-harmonic external field

$$\mathrm{f}(t) = \mathrm{f}\left(\sin\omega t + \sin\nu t + \ldots + \sin\mu t\right)$$

with a number M of irrational incommensurable frequencies ω, $\nu\ldots$ and μ, equation (8.166) written for the probability amplitudes at the time moment corresponding to k-th period has the phase term

$$\Phi\left(\theta, r\right) = -i\sum_{k=0}^{r-1}\log\frac{1 - i\pi\mathrm{f}\left[\sin\omega(\theta+2\pi k) + \ldots + \sin\mu(\theta+2\pi k)\right]}{1 + i\pi\mathrm{f}\left[\sin\omega(\theta+2\pi k) + \ldots + \sin\mu(\theta+2\pi k)\right]},$$

which now governs the excitation dynamics to be analyzed in the asymptotic of large r. We cast each of the r terms of the sum to multiple Fourier series by noting that

$$-i\log\frac{1 - i\pi\mathrm{f}\left[\sin x + \sin y + \ldots + \sin z\right]}{1 + i\pi\mathrm{f}\left[\sin x + \sin y \ldots + \sin z\right]} = \sum_{n,l\ldots,m} c_{n,l\ldots,m}e^{inx+ily+\ldots+imz},$$

where the coefficients

$$c_{n,l\ldots,m} = \int\frac{dx}{\left(2\pi\right)^{1-M}}\frac{e^{-|x|}}{2x}J_n\left(\pi\mathrm{f}x\right)J_l\left(\pi\mathrm{f}x\right)\ldots J_m\left(\pi\mathrm{f}x\right) \tag{8.181}$$

are found with the help of the Fourier representation

$$-i\log\frac{1-iX}{1+iX} = 2\arctan X = \frac{1}{2\pi}\int\frac{ie^{-|x|}}{2x}e^{iXx}dx$$

and the definition of the Bessel function

$$2\pi J_n\left(Y\right) = \int_0^{2\pi}e^{iY\sin y - iny}dy,$$

we then change the order of summations

$$\Phi\left(\theta, r\right) = \sum_{n,l\ldots,m} c_{n,l\ldots,m}\sum_{k=0}^{r-1}e^{i(n\omega+l\nu+\ldots+m\mu)(\theta+2\pi k)},$$

and arrive at

$$\Phi(\theta, r) = \sum_{n,l\ldots,m} c_{n,l\ldots,m} e^{i(\theta+\pi(r-1))(n\omega+l\nu+\ldots+m\mu)} \frac{\sin[\pi r(l\nu + n\omega + \ldots + m\mu)]}{\sin[\pi(l\nu + n\omega + \ldots + m\mu)]}. \tag{8.182}$$

The Fourier coefficients $c_{n,l\ldots,m}$ for the case $M \gg 1$ can estimated as follows. The Bessel functions in (8.181) reach their maximum absolute values for the arguments of the order of the indices. For smaller arguments they rapidly drop to zero, and for larger arguments they oscillate with the amplitude decreasing like a square root of the argument. Therefore, the main contribution to the integral (8.181) comes from the domain

$$x \sim \frac{\max(n, l, \ldots, m)}{\pi \mathrm{f}},$$

of a length of the order of the oscillation period $1/\pi\mathrm{f}$, where the integrand is of the order of

$$\frac{e^{-|x|}}{2x} \frac{(2\pi)^M}{(\pi \mathrm{f} x)^{(M-1)/2}},$$

thus resulting in

$$c_{n,l\ldots,m} \sim \frac{e^{-\max(n,l,\ldots,m)/|\pi\mathrm{f}|}}{4\pi} \left(\frac{2\pi}{\sqrt{\max(n, l, \ldots, m)}} \right)^{M+1},$$

Now analyzing the sum (8.182), one sees that on one hand side, the Fourier coefficients $c_{n,l\ldots,m}$ are exponentially decreasing with the size of the maximum index, but on the other hand side, the interference factors

$$\frac{\sin[\pi r(l\nu + n\omega + \ldots + m\mu)]}{\sin[\pi(l\nu + n\omega + \ldots + m\mu)]}$$

may became rather large when these indices tend to infinity such that the best rational approximation of irrational numbers may become very small. In contrast to (8.173), the approximation concerns not just one but a sum of M rational numbers multiplied by integer coefficients. Indeed, given $l, n, \ldots$ and m, the fractional part $\varepsilon_{n,l\ldots,m}$ of the irrational number

$$l\nu + n\omega + \ldots + m\mu = k + \varepsilon_{n,l\ldots,m}$$

determines the maximum size of the interference factor

$$\left|\frac{\sin\left[\pi r(l\nu + n\omega + \ldots + m\mu)\right]}{\sin\left[\pi(l\nu + n\omega + \ldots + m\mu)\right]}\right| \leq \frac{1}{\pi\left|\varepsilon_{n,l\ldots,m}\right|}, \tag{8.183}$$

which takes the maximum value at $r \simeq \left|\varepsilon_{n,l\ldots,m}\right|^{-1}/2$.

Now one has to analyze the behavior of the inverse of the fractional part $\varepsilon_{n,l\ldots,m}$ of an irrational number $l\nu + n\omega + \ldots + m\mu$ as a function of the maximum value of the indices. This quantity is a very irregular function of the indices $l, n, \ldots$ and m : it takes very large values for some specific points $\{l_s, n_s, \ldots m_s\}$ that sparsely distributed over the M-dimensional manifold of integers, while for other points on this "hyperplane" $l\nu + n\omega + \ldots + m\mu$ the inverse of $\varepsilon_{n,l\ldots,m}$ remains of the order of unity. Therefore, summation in (8.182) can be taken only over these specific indices, because only their contribution may result in the energy distribution of the width

$$\sqrt{\left|m^2\right|} \sim \Phi'_\theta(\theta, r) \sim \sum_{n,l \in \cup\{l_s,n_s,\ldots m_s\}} c_{n,l\ldots,m} e^{i(\theta+\pi(r-1))(n\omega+l\nu+\ldots+m\mu)}$$
$$(l\nu + n\omega + \ldots + m\mu)\frac{\sin\left[\pi r(l\nu + n\omega + \ldots + m\mu)\right]}{\sin\left[\pi(l\nu + n\omega + \ldots + m\mu)\right]} \tag{8.184}$$

growing in the course of time $t \sim r$. Again, as it was the case of the excitation at a single frequency given by a Liouville number, this broadening augments not gradually but by steps, whereas each next step s manifests itself at a time moment $t \sim r \simeq \left|\varepsilon_{l_s,n_s,\ldots m_s}\right|^{-1}/2$ resulting from the contribution of a specific point $\{l_s, n_s, \ldots m_s\}$ that correspond to the best integer approximation of $l\nu + n\omega + \ldots + m\mu$, yet improving with the rising s and thus resulting in decrease of the denominator $\sin\left[\pi(l\nu + n\omega + \ldots + m\mu)\right]$. These very points bring the factor $\left(\pi\varepsilon_{n,l\ldots,m}\right)^{-1}$ large enough to compensate for the exponentially small $c_{n_s,l_s\ldots,m_s}$. In other words, each specific point $\{l_s, n_s, \ldots m_s\}$ of the hyperplane gives contribution at a specific time scale $r \simeq \left|\varepsilon_{l_s,n_s,\ldots m_s}\right|^{-1}/2$, while the delay among the sequential broadening events increases exponentially in time.

Now the question is whether in course of time the width of the population distribution increases unlimitedly, as it was in the case of a random f(t), or it remains localized, as it was in the case of a generic harmonic excitation? In order to answer this question one could formulate the best rational approximation problem in physical terms by figuring out what is the smallest detuning of the combinational frequency $l\nu + n\omega + \ldots + m\mu$ for a maximum order $K \leq M \max(n, l, \ldots, m)$ of the resonance. The discrepancy $\varepsilon_K = \min \varepsilon_{n,l\ldots,m<K/M}$ of the best integer approximation of $l\nu + n\omega + \ldots + m\mu$ given $l, n, \ldots, m < K/M$ corresponds to the minimum detuning of the resonance of the order inferior to K. An unlimitedly increasing width of the population distribution given by the condition

$$\sum_{n,l\ldots m\in\cup\{l_s,n_s,\ldots m_s\}} c_{n_s,l_s\ldots,m_s} \frac{l_s\nu + n_s\omega + \ldots + m_s\mu}{\pi\varepsilon_{l_s,n_s,\ldots m_s}} \to \infty \tag{8.185}$$

implies ε_K decreasing with K faster than

$$c_{n,l\ldots,m<K/K} \sim e^{-K/M|\pi\mathrm{f}|} \left(K/M\right)^{-M/2}.$$

This should never be the case for generic real frequencies $\nu, \omega \ldots, \mu$., since the Number Theory states that $\varepsilon_K = \min \varepsilon_{n,l\ldots,m<K/M}$ scales as a power $(K/M)^{-M}$ of the maximum integer size $\max(n, l, \ldots, m) \sim K/M$, and hence

$$\begin{aligned}\sqrt{\overline{|m^2|}} &\sim \sum_K \frac{M}{\pi} e^{-\frac{K}{M|\pi\mathrm{f}|}} \left(\frac{K}{M}\right)^{\frac{M-1}{2}} \\ &\sim (\pi\mathrm{f})^{\frac{M+1}{2}} \frac{M^2}{\pi} \Gamma\left(\frac{M+3}{2}\right) < \infty,\end{aligned} \tag{8.186}$$

where $\Gamma(x)$ is the Euler gamma-function. This expression means that though the asymptotic distribution width $\sqrt{\overline{|m^2|}}$ remains finite, it scales exponentially with the number M of distinct frequencies in the Fourier decomposition of the driving field.

Chapter 9
Spectra of Complex Systems

We start this chapter by considering spectral manifestations typical of a quantum system perturbed by a random matrix, when both the structure of the unperturbed Hamiltonian and the mean squared coupling by the random matrix play an important role. The spectral transformation which occurs as a result of such perturbation explains many aspects of the time evolution of complex quantum systems. This version of the random matrix model does not take into account the properties of the system associated with the dimensionality of the space, since coupling among the unperturbed quantum states is assumes statistically equivalent, no matter whether or not the coupled states correspond to the wave functions localized in the same or quite different regions in space. Therefore such a model is often referred to as the 0-dimensional random matrix model. We then turn to the case of a system, which has very specific arrangement in space, representing a particle moving along a fractal, and show spectral manifestations of the fractality, and the transformations of the spectra as a result of perturbation by a random matrix. The last issue considered in this chapter is the spectra of the so-called Euclidean random matrices, where each of the quantum state is associated with a point randomly placed in a few-dimensional space while the matrix elements coupling pairs of these states are functions of the Euclidean distances among the points of the pairs.

9.1 Manifestation of Quantum Complexity in the State Density

Thus far we were mainly focused on the dynamics of populations in complex quantum systems, and the time evolution of probability amplitudes was considered only as an auxiliary tool in this context. However, some important physical properties such as the density of states and the linear susceptibility of quantum objects at a given frequency are primarily related to the dynamics of the probability amplitudes.

V.M. Akulin, *Dynamics of Complex Quantum Systems*, Theoretical and Mathematical Physics, DOI 10.1007/978-94-007-7205-2_9,

Generally speaking, all of the physical properties that can be expressed as a linear functional of the Green's operator of a system, are associated with the dynamical and spectral properties of the probability amplitudes. Moreover, all results obtained for the ensemble averaged amplitudes $\langle\psi(\varepsilon)\rangle$ corresponding to the initial condition $\psi(t=0)=1$ are valid for the Green's functions $G_{0n}(E)=\psi_n(E)$, since in fact these two objects are identical. In this section, we consider a general rule which governs the transformation of the spectral density and other physical characteristics of a large number of interacting quantum states, linear in $G(E)$. The Fourier transforms of such spectral characteristics give the corresponding time-dependent response functions. The universal features in these spectral dependences, typical of complex multilevel quantum systems, give rise to some universality in the responses, which under certain conditions can manifest a universal behavior of physically different response functions at asymptotically long time.

We start by addressing the question: what happens to the density of states of an arbitrary multilevel quantum system when we couple the levels by random interaction? The rigorous analysis of this problem relies on involved calculations, but the final result which governs the reconstruction of the spectra in the limit of a large number of quantum states mixed up by a random perturbation has a very simple, although non-trivial form. It is closely related with the procedure of renormalization that has been employed in Sect. 4.3.1, in the context of the analysis of population dynamics for time intervals shorter than the density of states. We can also illustrate this result by simple heuristic speculation, based on second-order perturbation theory. Several examples of the transformation of the state density and the absorption line profiles follow.

The rigorous analysis, for the case when the interaction mixes up just a small number of states $gV\sim 1$, is a more challenging task. One has to take into account the role of recurrences and revivals, which manifests itself in the dominating contribution of the trajectories with multiple self-intersections, similar to the trajectories considered in the previous section in the context of the incomplete level decay, and population localization. In this section we give a description based directly on the physically more meaningful calculation of the self-intersecting trajectories and the corresponding diagrams, instead of the more compact approach of Sect. 8.2 involving the Grassmann integration. We try to show the parallels between these two approaches when it is possible.

9.1.1 Spectrum Transformation Induced by Random Perturbation

In the second-order perturbation theory, a weak coupling $\hat{V}$ shifts the energies Δ_l of levels such that they acquire the new values

$$\tilde{\Delta}_l=\Delta_l+\sum_m\frac{|V_{l,m}|^2}{\Delta_l-\Delta_m}. \tag{9.1}$$

This implies that the physical properties associated with a given level undergo a shift to a different spectral region. Let us take, for example, the linear susceptibility $f_l(E) = \mu_l^2(E - \Delta_l - i\gamma_l)^{-1}$corresponding to the level l. Here γ_l is the decay rate and μ_l the dipole moment of the transition to this level from the ground state. The real part of the susceptibility $\mathrm{Re} f_l(E; \Delta_l)$ gives the Raman scattering amplitude, and the imaginary part $\mathrm{Im} f_l(E; \Delta_l)$ is the absorption cross-section. For the perturbed spectra in these expressions we have to replace Δ_l by $\tilde{\Delta}_l$.

But what happens when the interaction is not small? Evidently, the level shifts become larger. Besides the shifts, the strong interaction mixes up the quantum states, so that the "individual properties" of each of the levels become distributed among a number of energy eigenstates. It turns out that for a random perturbation, this spreading of the "individual properties" does not manifest itself in the resulting spectra: after the ensemble average, only the level shifts remain important. These shifts may, however, contain imaginary components, thus moving the state responsible for a certain contribution to the quantity of interest to the complex plane of energies. In other words, a physical value $\Phi(E)$, which had a given magnitude at the energy E, now has the same magnitude at the energy $\tilde{E}$. Here the function $\Phi(E)$ is supposed to be proportional to the Green's function, that is, it corresponds to a linear property of the system.

A Simple Rule for Large Complex Systems

The main features of the transformation of the density of states follows from a procedure almost identical to the renormalization discussed in Sect. 4.1.3. It only has to be generalized to an arbitrary density of the unperturbed states. This gives the relation between E and $\tilde{E}$, that is between the domain of the unperturbed energy spectrum associated with the contribution to the particular physical property and the new energy position of the state responsible for this contribution. We note that as a result of such a transformation, the new energy position could, and usually does, acquire an imaginary part. In other words, the quantum states contributing in a specific way to the physical property of the system, are not necessarily energy eigenstates. In the limit of a large number of levels, this energy is given by the transcendental algebraic equation resembling (9.1), which reads

$$E = \tilde{E} + \sum_m \frac{\langle V^2 \rangle}{\tilde{E} - \Delta_m - i\gamma_m}, \tag{9.2}$$

where $\langle V^2 \rangle$ is the mean square of the random interaction. In the other notations, the non-uniform transformation of the energy scale (9.2) is given by the equation

$$E = \tilde{E} + \langle V^2 \rangle \,\mathrm{Tr} G(\tilde{E}), \tag{9.3}$$

where $\hat{G}_0(\tilde{E})$ is the unperturbed Green's function.

Equation (9.3) implicitly determines the dependence of new energies on the old ones, that is, the function $\tilde{E}(E)$, and thus enables us to calculate the new shapes of any spectral dependence $\tilde{\Phi}(E)$, provided we knew the old profiles $\Phi(E)$. This relation reads

$$\tilde{\Phi}(E) = \Phi\left[\tilde{E}(E)\right]. \tag{9.4}$$

Note that the transformation of the spectra, given by (9.3) and (9.4), is not as trivial as it may look at first glance: plural roots of (9.3) result in the multivalued dependence of $\tilde{E}(E)$. In this case we may obtain a superposition of profiles corresponding to different roots. For the roots of (9.3) to be taken into account, we use the following selection rule. We find the contour in the complex plane $\tilde{E}$ going from $-\infty$ to ∞, on which the imaginary part of the right-hand side of (9.3) equals zero. We take only the roots of (9.3) located on this contour, and numerate them according to the sequence in which they are ordered, starting from $-\infty$. The even roots give positive and the odd, negative contributions. Note that the roots of the transcendental equation (9.3) obtained in this way are not necessarily real: they usually have negative imaginary parts.

For the derivation of (9.3), there is no need to perform all of the calculations based on the ensemble averaging of the all-order perturbation series of the Green's functions in the general case, since all of the physically important ingredients of such a calculation have already been illustrated in Sect. 4.1.3 for the particular example of the Green's function projected to the unperturbed energy eigenstate $X_m(E) = \langle m| \langle G(E)\rangle |m\rangle$. Equation (4.21) can be directly generalized to the average Green's functions in any representation, which reads

$$\langle G(E)\rangle = \sum_m \frac{|m\rangle \langle m|}{E - \Delta_m - \langle V^2\rangle \operatorname{Tr} \langle G(E)\rangle}. \tag{9.5}$$

We simply notice that substitution of the renormalized energy $\tilde{E} = E - \langle V^2\rangle \operatorname{Tr} \langle G(E)\rangle$ into the definition of the Green's function

$$G(E) = \sum_m \frac{|m\rangle \langle m|}{E - \Delta_m}, \tag{9.6}$$

given in the representation of the unperturbed energy eigenstates $|m\rangle$, immediately gives $\langle G(E)\rangle = G(\tilde{E})$, while from the invariance of the trace operation one obtains $\operatorname{Tr} \langle G(E)\rangle = \operatorname{Tr} G(\tilde{E})$, which results in (9.3).

Example of Two Merging Multiplets

Let us illustrate this rule by simple examples. For a degenerate multiplet containing N components we find

$$\tilde{E} - E + \frac{N\langle V^2\rangle}{\tilde{E}} = 0. \tag{9.7}$$

This equation has two roots,

$$\tilde{E}_{1,2} = \frac{E}{2} \pm \sqrt{\left(\frac{E}{2}\right)^2 - N\langle V^2\rangle}. \tag{9.8}$$

The sign selection rule suggests minus, and the new Green's function therefore reads

$$\langle G(E)\rangle = \frac{N}{\tilde{E}} = \frac{N}{\frac{E}{2} - \sqrt{\left(\frac{E}{2}\right)^2 - N\langle V^2\rangle}}. \tag{9.9}$$

For $(\frac{E}{2})^2 < N\langle V^2\rangle$ the imaginary part of the Green's function yields the Wigner semicircular state density

$$g(E) = \mathrm{Im}G(E) = (\pi\langle V^2\rangle)^{-1}\sqrt{E^2 - 4N\langle V^2\rangle} \tag{9.10}$$

shown in Fig. 4.19.

One can also see the coalescence of two Wigner semicircles. We take N_a levels, at the energy $\Delta_a = \Delta$ and N_b levels of the energy $\Delta_b = -\Delta$. The trace of the imaginary part of the Green's function $\mathrm{Im}G(E) = \mathrm{Im}[N_a(E-\Delta-i0)^{-1} + N_b(E+\Delta - i0)^{-1}]$ gives the unperturbed density of states. For the transformed energy $\tilde{E}$ one finds a cubic equation

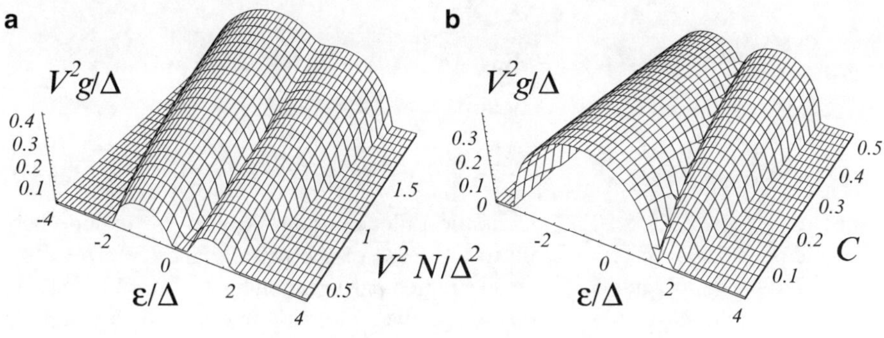

Fig. 9.1 Density of states $g(E)$ calculated with the help of (9.12) for various parameters V^2N/Δ^2 (**a**) and $C = N_a/(N_a + N_b)$ (**b**). This is an analog of the semicircular Wigner distribution (Fig. 4.19) for the case of two multiply degenerate levels coupled by a random matrix

$$\tilde{E} - E + V^2 G(\tilde{E}) \equiv \tilde{E} - E + \frac{V^2 N_a}{\tilde{E} - \Delta - i0} + \frac{V^2 N_b}{\tilde{E} + \Delta - i0} = 0 \tag{9.11}$$

with the roots $\tilde{E}_{1,2,3}(E)$ given by the Cardano formula. The selection rule yields the root with the negative imaginary part. We note that the variable E is real, and therefore, according to (9.5), the density of states $g(E)$ is given by $g(E) \equiv \mathrm{Im}\mathrm{Tr}\left\langle \hat{G}(E)\right\rangle = \mathrm{Im}G[\tilde{E}(E)]$. After substitution of $G[\tilde{E}(E)]$ from (9.11) this reads

$$g(E) = -\mathrm{Im}\ \tilde{E}_3(E). \tag{9.12}$$

In Fig. 9.1 we depict the density of states for various concentrations $N_a/(N_a+N_b)$.

Examples of Multidimensional Motion

For a free particle moving in d-dimensional space, with kinetic energy proportional to the momentum squared p^2, the density of states as a function of energy E reads $g(E) = d(2\pi)^{-d/2}E^{d/2-1}/\Gamma(d/2)$, and therefore the trace of the Green's function in the eigenstate representation reads

$$\mathrm{Tr}G(E) = \frac{d}{(2\pi)^{d/2}\Gamma(d/2)}\int\limits_0^\infty \frac{x^{d/2-1}e^{-\nu x}}{E-x-i0}dx \tag{9.13}$$

where a small cut-off constant ν is introduced in order to avoid the divergency of the upper limit. For $d = 1, 2, 3$ (9.13) yields

$$\mathrm{Tr}G(E) = \frac{i}{\sqrt{2\,(E-i0)}};$$

$$\mathrm{Tr}G(E) = \frac{1}{\pi}\left[\ln\left(-E+i0\right) + \ln\nu + \mathbf{C}\right];$$

$$\mathrm{Tr}G(E) = \frac{3i}{\sqrt{2}\pi}\sqrt{E-i0} - 3\sqrt{\frac{\pi}{2\pi^3\nu}}, \tag{9.14}$$

respectively, where $\mathbf{C} \simeq 0.577\ldots$ is the Euler constant. We do not include here the dimensional parameters related to the mass of the particle and the size of the system, but we simply define the mean squared perturbation by the rate $W = \langle V^2\rangle g$ of relaxation that moves the particle with a given momentum out of the momentum state. This implies that the perturbation amplitude is strong and rapidly changing in space, such that this rate is independent of the momentum.

We will not take into account the energy-independent terms of (9.14) that just shift the energy scale without affecting the profile of the density of states, but concentrate only on the energy-dependent ones that have to be substituted into (9.3). This yields

$$E = \tilde{E} - \frac{i}{\sqrt{2\left(\tilde{E}-i0\right)}}W;$$

$$E = \tilde{E} + \frac{W}{\pi}\left[\ln\left(-\tilde{E} + i0\right)\right];$$

$$E = \tilde{E} - \frac{3iW}{\sqrt{2}\pi}\sqrt{\tilde{E} - i0}, \tag{9.15}$$

where the first and the last equations can be solved analytically, whereas the second one requires a numerical solution. Equation (9.15) has a certain symmetry with respect to scaling and shift operations, such that their solutions result in the universal functional dependencies of the density of states, whereas the energy axis experiences shift and scaling, which may also be accompanied by a scaling of the density axis. Indeed, representing E and $\tilde{E}$ in the first equation as the quantities X_1 and $\tilde{X}_1$, respectively, that are both multiplied by the factor $W^{2/3}$, one gets rid of the parameter W. A similar representation of E and $\tilde{E}$ in the second equation containing the factor W/π, followed by the shift of the argument $X_2 \rightarrow X_2 - \ln W/\pi$ gives a similar result. The right-hand side of the third equation is quadratic with respect to $\sqrt{\tilde{E}}$. Therefore, by a simple shift of the argument E by the value $-9W^2/8\pi^2$ followed by taking the roots and squares one finds that the combination $-\mathrm{Im}(\tilde{E})/\pi W$ remains the same square root function, simply shifted along the energy axis. We therefore arrive at

$$\tilde{g}_1(E) = \frac{-1}{\pi W^{1/3}}\mathrm{Im}\tilde{X}_1(\frac{E}{W^{2/3}})$$

$$\tilde{g}_2(E) = -\frac{1}{\pi}\mathrm{Im}\tilde{X}_2(\frac{E\pi}{W} - \ln\frac{W}{\pi})$$

$$\tilde{g}_3(E) = \frac{3}{\sqrt{2}\pi^2}\sqrt{E - \frac{9W^2}{8\pi^2}}, \tag{9.16}$$

where $\tilde{X}_1(X)$ and $\tilde{X}_2(X)$ are universal functions given by the solutions of the algebraic and transcendental equations

$$\begin{aligned} \left(\tilde{X}_1 - X\right)^2 2\tilde{X}_1 + 1 = 0, \\ X = \tilde{X}_2 + \left[\ln\left(-\tilde{X}_2 + i0\right)\right], \end{aligned} \tag{9.17}$$

respectively. The first equation implies a standard solution for the cubic equation, while we have already encountered the second equation in (5.160) in the context of a band interacting with the moving level. There we were interested in the real solution of the latter equation that determines the position of an adiabatic state near the continuum edge, whereas now we are concentrating on the imaginary solution describing the transformation of the state density. In Fig. 9.2 we show the profiles of the imaginary parts of these functions.

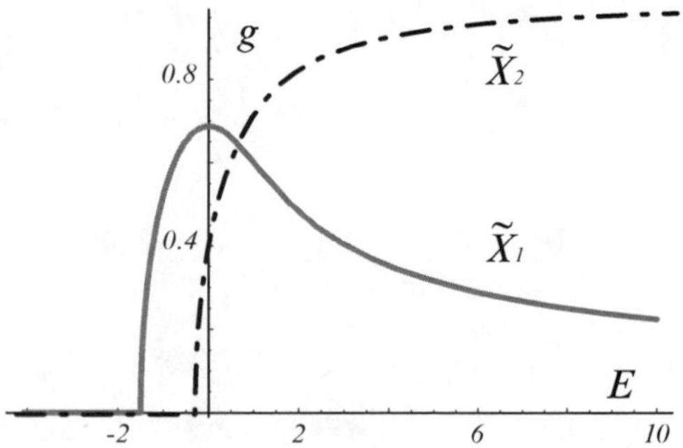

Fig. 9.2 Universal profiles of the state density $g(E)$ perturbed by a random matrix as a function of the scaled energy E, given by the solutions $\tilde{X}_1(X)$ (*solid line*) and $\pi^{-1}\tilde{X}_2(\pi X)$ (*dash-dot line*) of (9.17) for one- and two-dimensional motion, respectively

Examples of Absorption Lines

The density of states is not the only characteristic of a quantum system that is linear in the Green's function. The other important physical quantity is the transition probability. The natural example of this is the photon absorption cross-section, corresponding to dipole transitions from the ground state of a quantum system to a set of highly excited levels, coupled among themselves by a random perturbation. When the perturbation mixes up the dipole accessible odd states with even states, forbidden for the dipole transitions, the profile $F(E)$ of the absorption line experiences a significant transformation. We illustrate such a transformation for two merging multiplets that undergo transformation given by (9.11). We assume that only one of these multiplets, say a, is accessible from the ground state $|0\rangle$ and that the other multiplet is "dark". Therefore the profile

$$F(E) = \langle 0\,|\mu_{0a}G(E)\mu_{a0}|\,0\rangle \tag{9.18}$$

given for $\langle V^2\rangle = 0$ in terms of the unperturbed Green's function (9.6) and the matrix elements μ_{0a} of the dipole moments (taken here as identical) adopts the form

$$\langle F(E)\rangle = \frac{N_a\,|\mu_{a0}|^2}{\tilde{E}(E) - \Delta - i0}, \tag{9.19}$$

as suggested by the transformation (9.5) of the energy scale in the presence of the perturbation.

In Fig. 9.3 we show the evolution of $\langle F(E)\rangle$ for an increasing, random perturbation $\langle V^2\rangle$. For small perturbations the dipole activity is localized on the manifold a and gives the "triangular" shaped profile typical of the multiplets with a Wigner semicircular shape. When the nearest edges of the manifolds a and b get closer, part of the dipole activity moves to the multiplet b, while the superposition with a non-uniform state density at the edges creates a two-hump profile, shown in Fig. 9.3b. For strong coupling, when the multiplets merge, the line acquires an almost Lorentzian shape which returns to the "triangular" one only for very strong couplings, when both multiplets can be treated as one.

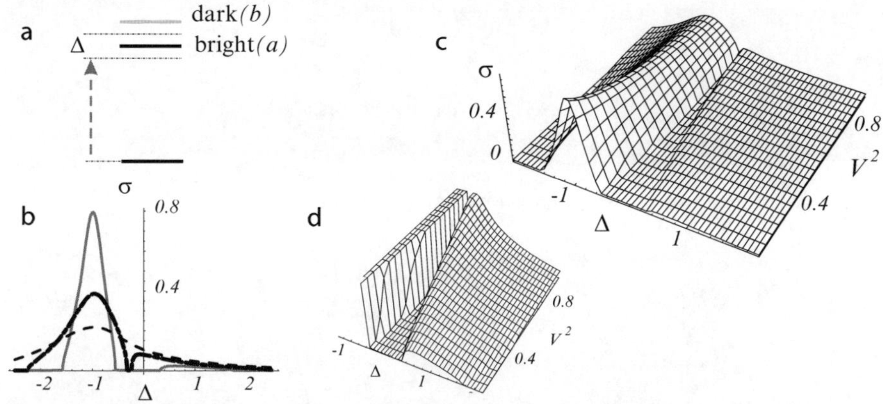

Fig. 9.3 Absorption contour of a "bright" multiplet merging with a "dark" one as a result of random coupling of the multiplet levels, calculated with the help of (9.19). (**a**) The level scheme for transitions to coupled bright and dark multiplets. (**b**) The absorption profiles $\sigma(\Delta, V)$ for small, medium, and strong random coupling. (**c**) The same as a function of the mean squared coupling. (**d**) The same near the point where the multiplets merge

9.1.2 Universal Features in Time Dependencies

When considering in Sect. 4.3 the population dynamics of two degenerate levels, we encountered a non exponential behavior of the population difference in the long-time limit. The non-exponential time-dependence also manifests itself in the frequency domain as discontinuities of derivatives of the corresponding spectra. The sharp edges of the Wigner semicircular distribution entering the imaginary part of $Q \sim \tilde{E}$ in (4.54) as the origin of the oscillating dependencies given by Bessel functions in (4.61) for the population difference is one of the examples. We can illustrate this by taking the Fourier transform of the Green's function in (9.9) and calculating the response function

$$R(t) \equiv \int_{\infty}^{-\infty} \frac{dE}{2\pi} \tilde{G}(E) e^{i\varepsilon t} = \int_{\infty}^{-\infty} \frac{dE}{2\pi} \frac{e^{iEt}}{E - \sqrt{E^2 - 4N\langle V^2 \rangle}} = \frac{J_1(t\sqrt{N\langle V^2 \rangle})}{t\sqrt{N\langle V^2 \rangle}}. \tag{9.20}$$

The square of the response function $R(t)$ gives the time-dependent probability of the transition. We recall the asymptotic behavior of the Bessel function $J_1(x) \sim x^{-1/2}\sin(x - \pi/4)$ and see that this probability decreases as t^{-3}. In this respect, the Wigner semicircular spectrum behaves like a six-dimensional random walk

We can also find some typical features of the optical response function $R(t)$ corresponding to the profile

$$\langle F(E) \rangle = \langle 0 \left| \mu_{0a} \left\langle G(E) \right\rangle \mu_{a0} \right| 0 \rangle \tag{9.21}$$

in the general case. Calculating the integral

$$R(t) = \int_{\infty}^{-\infty} \frac{dE}{2\pi} \left\langle 0 \left| \mu_{0a} G[\tilde{E}(E)] \mu_{a0} \right| 0 \right\rangle e^{iEt} \tag{9.22}$$

of the inverse Fourier transformation by parts one finds

$$R(t) = -\int_{\infty}^{-\infty} \frac{e^{iEt}}{2\pi i t} \frac{\partial \left\langle 0 \left| \mu_{0a} G[\tilde{E}(E)] \mu_{a0} \right| 0 \right\rangle}{\partial E} dE. \tag{9.23}$$

A spectrum edge located at a point $E = E_{ed}$ where $G[\tilde{E}(E)] \sim (E - E_{ed})^{1/2}$ gives the contribution

$$R(t) \sim \frac{\exp\left[iE_{ed}t\right]}{t^{3/2}} \tag{9.24}$$

to the integral (9.23), and hence the overall response in the long-time limit decreases as $t^{-3/2}$ and results from the interference of responses (9.24) of all spectrum edges. This yields a multifrequency interference pattern in the case of a spectrum consisting of many spectral bands. However, when for a certain value of the random coupling two closest edges of neighboring bands merge at a point $E = E_{cusp}$, as shown in Fig. 9.1, the density of state forms a cusp with the profile $\sim |E - E_{\text{cusp}}|^{5/3}$ and the corresponding contribution to the response

$$R(t) \sim \frac{\exp\left[iE_{\text{cusp}}t\right]}{t^{8/3}} \tag{9.25}$$

decays faster as compared to the edge contributions.

9.2 The Effect of Quantum Recurrences on the State Density Profiles

Thus far we have been considering the situation when the typical strength V of the random perturbation is much larger than the typical spacing $1/g$ between the neighboring levels of the unperturbed system and we were concentrating on the transformation of the spectral density on a scale much larger than $1/g$. In other words, the time uncertainty corresponding to our spectral resolution was much shorter than the typical Heisenberg return time $\sim g$. Now we turn to the case when the perturbation is of the order of the mean spacing $Vg \sim 1$ and the time uncertainty of the resolution has to be larger than the return time. The influence of quantum returns therefore becomes dominant.

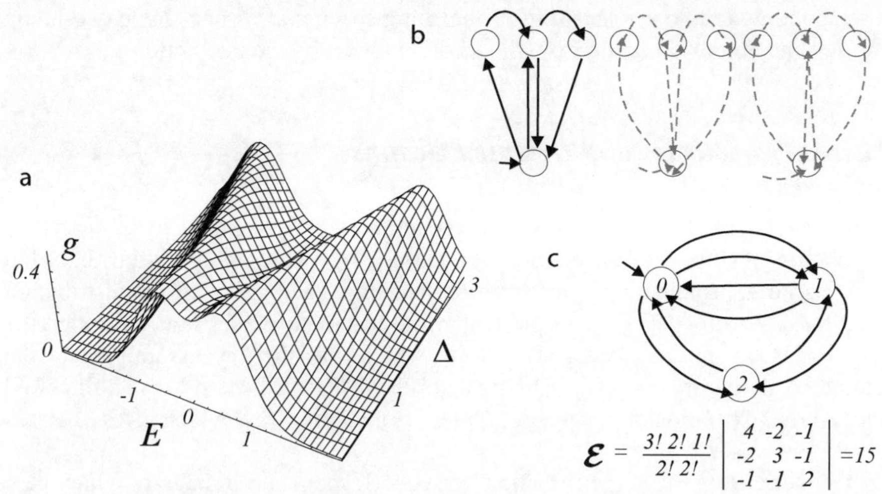

Fig. 9.4 The density of quantum states at a resolution scale smaller than the mean level spacing is affected by the interference of recurrences. For the simplest, two-level "complex" quantum system (**a**) it results in a two-hump profile (9.26). (**b**) Consistent description of larger systems requires that one takes into account that different trajectories (*dash lines*) give identical contributions when they correspond to the same oriented graph (*solid lines*). (**c**) Example of an orgraph with parallel transitions. For this orgraph $K_0 = 4$; $K_1 = 3$; $K_2 = 2$; $p_{0,1} = p_{1,0} = 2$; $p_{0,2} = p_{2,0} = p_{1,2} = p_{2,1} = 1$. The number of different Euler paths is 15 according to (9.31) and, indeed, the orgraph corresponds to the 15 trajectories: 010121030; 010102120; 010120210; 010201210; 010210120; 010212010; 012010210; 012021010; 012101020; 012102010; 020101210; 020121010; 021010120; 021012010; 021201010

In Fig. 9.4a we illustrate this situation by showing the average density of states $g(\varepsilon, \Delta)$ in the most simple case of a "complex" quantum system consisting of two levels with an energy spacing 2Δ, which is perturbed by a Hermitian 2×2 random matrix V_{km}. Averaging of the state density

$$\delta\left(\varepsilon - \tfrac{V_{11}+V_{22}}{2} + \sqrt{(\tfrac{V_{11}-V_{22}}{2} + \Delta)^2 + V_{12}V_{21}}\right)$$
$$+\delta\left(\varepsilon - \tfrac{V_{11}+V_{22}}{2} - \sqrt{(\tfrac{V_{11}-V_{22}}{2} + \Delta)^2 + V_{12}V_{21}}\right)$$

with the Wigner perturbation distribution function $g \sim e^{-V_{11}^2 - V_{22}^2 - 2V_{12}V_{21}}$ reads

$$\frac{(\varepsilon+\Delta)e^{2\Delta\varepsilon} - (\varepsilon-\Delta)}{2\sqrt{2\pi}\Delta} \exp\left[-\frac{(\varepsilon+\Delta)^2}{2}\right], \tag{9.26}$$

as one can find by performing the direct integration over $dV_{11}\, dV_{22}\, d\,|V_{12}|\, d \arg V_{12}$. Here ε stands for energy which is scaled in the mean perturbation $\langle V^2 \rangle^{1/2}$ as well as all other energy units in this subsection. One sees the two Gaussian profiles corresponding to two levels that for a vanishing spacing Δ form a single two-hump profile. The humps originate from the level repulsion by the interaction V_{12}.

9.2.1 Trajectories and Oriented Graphs

In order to obtain the state density in the general case, we begin the exact analytical calculation of the averaged perturbation series (4.2) which is quite unwieldy but contains no approximation apart from the ensemble average with the distribution function $g \sim \exp(-\mathrm{Tr}\hat{V}^2)$. In particular, the calculation does not imply that the diagrams have only a tree-like structure, which was the key assumption in the regime of short times, $t < g$. The final result of these calculations is a threefold integral representation of the average Green's function $\langle G(\varepsilon)\rangle$ which gives (9.26) as a particular case.

A generic term of the perturbation series corresponds to a diagram which may return many times to the same level. It has the structure

$$S(\{p_{n,m}\}) = \prod_{n,m} [(V_{nm})^{p_{nm}}] \prod_{n} \left[\frac{1}{(\varepsilon - \Delta_n)^{K_n}}\right], \tag{9.27}$$

where $K_n = \sum_m p_{mn}$ is the number of times the level $|n\rangle$ has been visited, and the numbers $p_{mn} = 0, 1, 2\ldots$ show how many times the matrix element V_{nm} is found on the trajectory. Each trajectory $0 \to \ldots \to n \to m \to \ldots \to 0$ starts and ends at the state $|0\rangle$, the only initially populated level. For the state $|0\rangle$ the power index is $K_0 = \sum_m p_{m0} + 1$. When needed, one obtains the trace of the Green's function, by taking for the state $|0\rangle$, all levels of the system and summing up the resulting contributions. The Green's function is the sum of the amplitudes S of all trajectories

$$G(\varepsilon) = \sum_{\mathtt{traject.}} S(\{p_{nm}\}), \tag{9.28}$$

where many of these trajectories may have the same set $\{p_{n,m}\}$ of indices and therefore give identical contributions.

Let us represent the states n of the system by points in a plane, and depict a self-intersecting trajectory by connecting these points by the lines representing the coupling V_{nm} which has been encountered on the trajectory. We obtain an oriented graph (orgraph), that is, the totality $\{p_{n,m}\}$ of lines connecting each pair n, m of the points, and p_{nm} gives the number of different lines going from the point n to the point m. The set of numbers $\{p_{mn}\}$ uniquely defines the orgraph.

There is no one-to-one correspondence between trajectories and orgraphs: any trajectory of the sum (9.28) corresponds only to one orgraph, but this orgraph may also correspond to some other trajectories. One can say that each of these trajectories goes round the orgraph in its own way, as shown in Fig. 9.4b. All of the trajectories corresponding to the same orgraph have identical amplitudes S and hence they interfere constructively. Therefore the sum (9.28) of the terms (9.27) can be represented in the form

$$G(\varepsilon) = \sum_{\text{orgraph}\{p_{n,m}\}} \prod_{n} \left[\frac{\prod_{m} (V_{nm})^{p_{n,m}}}{(\varepsilon - \Delta_n)^{K_n}} \right] \mathcal{E}(\{p_{mn}\}), \tag{9.29}$$

where the factor $\mathcal{E}(\{p_{mn}\})$ allows for a number of different trajectories on the orgraph $\{p_{nm}\}$.

9.2.2 The Number of Euler Paths and Sum Over Carriers

We now make use of the combinatorial expression for $\mathcal{E}(\{p_{m,n}\})$ entering (9.29) which is known explicitly for $p_{n,m} = 0, 1$. It is the so-called number of Euler paths, given as

$$\mathcal{E} = \prod_{n} [(K_n - 1)!] \operatorname{minor} \left| \delta_{n,m} \left(\sum_{m} p_{nm} \right) - p_{nm} \right|, \tag{9.30}$$

where the symbol "minor" stands for any of the M^2 identical $(M - 1)$-th order minors corresponding to the M-th order matrix $\delta_{nm} \sum_{m} p_{nm} - p_{nm}$. The determinant $|\delta_{nm} \sum_{m} p_{nm} - p_{nm}|$ of this matrix equals zero. Instead of calculating minors, one can calculate the determinant where unity is added to the diagonal element corresponding to the level $|0\rangle$. In this way, one can replace the sum of p_{mn} by the value of K_n. We illustrate this in the orgraphs by adding an entering arrow to the point corresponding to $|0\rangle$. One obtains the generalization of (9.30) for the case $p_{n,m} = 0, 1, 2, \ldots$ dividing it by the number $\prod_{n,m} p_{n,m}!$ of the transposition of all identical lines in the orgraph

$$\mathcal{E}(\{p_{n,m}\}) = \prod_{n} \left[\frac{(K_n - 1)!}{\prod_{m} p_{nm}!} \right] |\delta_{n,m} K_n - p_{n,m}| \,. \tag{9.31}$$

The summation over orgraphs (9.29) is not equivalent to the summation over all $p_{n,m} = 0, 1, \ldots$, because a trajectory may not visit all of the states of the system. As a consequence, the order of the matrix $\delta_{n,m} K_n - p_{nm}$ in (9.31) can be different for different orgraphs. We therefore perform the summation in two steps. Firstly we select the carrier $C(\{n\})$ of the orgraph, that is, a group $\{n\}$ of N levels, each of which must be visited by the corresponding trajectory at least once. The carrier always includes the state $|0\rangle$. Then we sum the contributions of all orgraphs with identical carriers, that is, we perform the summation over all $p_{n,m}$ running from one to infinity. In this sum we keep the size $N \times N$ of the matrix $p_{m,n}$ unchanged. When some of the numbers $K_n = \sum_m p_{n,m}$ equal zero, the determinant in (9.31) has a zero line and hence the corresponding $\mathcal{E}$ vanishes. Hence the imposed condition, that each trajectory must visit all the states of the carrier, holds automatically. Secondly we sum up the contributions of all possible selections of the carriers with different N and arrive at

$$G(\varepsilon) = \sum_{C(\{n\})} \sum_{p_{n,m}=0}^{\infty} \prod_{n} \left[\frac{\prod_{m} (V_{nm})^{p_{n,m}}}{(\varepsilon - \Delta_n)^{K_n}} \right] \mathcal{E}(\{p_{mn}\}). \tag{9.32}$$

Note that if we want to keep the natural enumeration of the states included in a carrier $(n, m = 1, 2, \ldots)$, we have to attribute to them new numbers according to the sequence of the initial numeration. The other possibility is to retain the original enumeration, which does not run over all natural numbers. It is convenient to choose the set-theory notation that gives identical expressions for both types of enumeration. We write $l, m \in \{n\}$ if the states l and m are included in the set $\{n\}$, and $l, m \in \{n\}/\{q\}$ if at the same time these states are not included in the set $\{q\}$. All states in the sets are ordered according to the original enumeration.

9.2.3 Factorization of $\mathcal{E}$ and the Ensemble Average

Our aim now is to factorize the number of Euler paths and represent the sum of products $\sum_{p_{nm}} \prod_{nm}$ as the product of sums $\prod_{nm} \sum_{p_{nm}}$. Following that, we perform the summation for each pair n, m independently. We employ the relation

$$\frac{(K_n - 1)!}{(\varepsilon - \Delta_n)^{K_n}} = \int_{C_t} \frac{dt_n}{t_n} t_n^{K_n} \exp\{-(\varepsilon - \Delta_n) t_n\}, \tag{9.33}$$

were C_t is the integration contour that starts at $t = 0$ and goes to infinity along the direction ensuring convergence of the integrals, and thereby in the integrand we obtain the product of the factors $t_n^{p_{mn}}$. Separation of the determinants can be performed with the help of Grassmann integration. Here we employ another approach, which to a certain extent illustrates the meaning of the graded anticommuting variables. We note that $\frac{\partial}{\partial \alpha} \exp(p_{lm} \alpha) \to p_{lm}$ for $\alpha \to 0$ and write

$$\begin{aligned} &\left| \delta_{lm} \sum_{k \in \{n\}} p_{km} + \delta_{l0} \delta_{m0} - p_{lm} \right| \\ &= \lim_{\alpha \to 0} \left| \delta_{lm} \sum_{k \in \{n\}} \frac{\partial}{\partial \alpha_{km}} + \delta_{l0} \delta_{m0} - \frac{\partial}{\partial \alpha_{lm}} \right| \exp \sum_{l,m} p_{lm} \alpha_{lm}, \end{aligned} \tag{9.34}$$

where $\delta_{l0}\delta_{m0}$ allows for the fact that $K_0 = \sum_m p_{0m} + 1$.

Let us make parallels between the technique based on the differential operator-determinant and the technique based on graded variables. Interchange of two neighboring columns in a determinant results in a change of its sign. This property corresponds to the anticommutation relations of the Grassmann variables $\chi\chi' = -\chi'\chi$, whereas the fact that elements of each line or column enter into each term of the determinant only once along with the requirement $\alpha \to 0$ corresponds to the vanishing of all powers of these variables. The fact that the argument should depend on the variables α_{lm} for Grassmann variables χ corresponds to the condition $\int d\chi = 0$.

We substitute (9.33), (9.34), and (9.31) into (9.32) and arrive at

$$G(\varepsilon) = \lim_{\alpha\to 0} \sum_{C(\{n\})} \left| \delta_{lm} \sum_{k\in\{n\}} \frac{\partial}{\partial\alpha_{km}} + \delta_{l0}\delta_{m0} - \frac{\partial}{\partial\alpha_{lm}} \right|$$
$$\prod_{l\in\{n\}} \int_{C_t} \frac{dt_l}{t_l} t_l^{\delta_{0l}} \exp\{-(\varepsilon - \Delta_l)t_l \prod_{m\in\{n\}} \left[\sum_{p_{n,m}=0}^{\infty} \frac{[t_l \exp(\alpha_{lm})V_{l,m}]^{p_{lm}}}{p_{l,m}!} \right]. \tag{9.35}$$

Here by the power δ_{l0} we take into account that all carriers $C(\{n\})$ include the state $n = 0$. In the summation over $p_{l,m} = 0, 1, \ldots$ we recognize the Taylor expansion of the exponents, and hence

$$G(\varepsilon) = \lim_{\alpha\to 0} \sum_{C(\{n\})} \left| \delta_{lm} \sum_{k\in\{n\}} \frac{\partial}{\partial\alpha_{km}} + \delta_{l0}\delta_{m0} - \frac{\partial}{\partial\alpha_{lm}} \right|$$
$$\prod_{l\in\{n\}} \int_{C_t} \frac{dt_l}{t_l} t_l^{\delta_{0l}} \exp\{-(\varepsilon - \Delta_l)t_l \prod_{m\in\{n\}} [\exp(t_l V_{l,m} \exp\alpha_{lm})]. \tag{9.36}$$

Now we perform the average over the Gaussian distribution of the matrix elements V_{nm}. Application of (8.102) to each matrix element V_{lm} and its complex conjugate counterpart V_{ml} yields

$$\left\langle \prod_{l,m\in\{n\}} \exp(t_l V_{l,m} \exp\alpha_{lm}) \right\rangle = \exp\left[\sum_{l,m\in\{n\}} t_l t_m \exp\left(\alpha_{ml} + \alpha_{lm}\right) \right]. \tag{9.37}$$

We now move the differential operator-determinant under the integral (9.36) and substitute (9.37) such that the integrand of the many-fold integral over $\prod dt_l$ adopts the form

$$\mathcal{Y} = \lim_{\alpha\to 0} \left| \delta_{lm} \sum_{k\in\{n\}} \frac{\partial}{\partial\alpha_{km}} + \delta_{l0}\delta_{m0} - \frac{\partial}{\partial\alpha_{lm}} \right|$$
$$\exp\left[\sum_{l,m\in\{n\}} t_l t_m e^{\alpha_{ml}+\alpha_{lm}} \right] \prod_{l\in\{n\}} \left\{ t_l^{\delta_{l,1}-1} \exp[-(\varepsilon - \Delta_l)t_l] \right\}. \tag{9.38}$$

Now we calculate the derivatives, and take the limit $\alpha = 0$. The determinant in (9.38) is a differential operator. It comprises only the first-order derivatives over each particular $\alpha_{l,m}$, since all of the terms generated by a determinant include only the matrix elements corresponding to different lines. Therefore in the last sum we can expand the exponents

$$t_l t_m e^{\alpha_{ml}+\alpha_{lm}} = t_l t_m + t_l t_m \alpha_{ml} + t_l t_m \alpha_{ml}\alpha_{lm},$$

move the zero-order terms in front of the determinant, and keep only the terms linear in α. Bilinear terms are not important in the limit $\alpha \to 0$. Indeed, when we take a bilinear term $\alpha_{ml}\alpha_{lm}$ and consider the structure of the minor of the determinant comprising derivatives over both these variables, we find that it has the structure

$$\begin{vmatrix} \frac{\partial}{\partial \alpha_{ml}} & -\frac{\partial}{\partial \alpha_{ml}} \\ -\frac{\partial}{\partial \alpha_{lm}} & \frac{\partial}{\partial \alpha_{lm}} \end{vmatrix} = 0,$$

vanishing identically. The integrand of the averaged equation (9.36) given by (9.38) adopts the form

$$\mathcal{Y} = \left| \delta_{lm} t_m \sum_{k\in\{n\}} t_k + \delta_{l0}\delta_{m0} - t_l t_m \right| \exp\left[\sum_{l,m\in\{n\}} t_l t_m \right] \prod_{l\in\{n\}} \left\{ t_l^{\delta_{l,1}-1} \exp[-(\varepsilon - \Delta_l)t_l] \right\}. \tag{9.39}$$

The determinant can be found explicitly, and for the $N \times N$ matrix reads $\left(\sum_{l\in\{n\}} t_l \right)^{N-2} \prod_{l\in\{n\}} t_l$. We can represent this in completely factorized form with the help of the integrals

$$\left(\sum_{l\in\{n\}} t_l \right)^{-1} = \int_0^\infty \exp\left\{ -u \sum_{l\in\{n\}} t_l \right\} du,$$
$$\left(\sum_{l\in\{n\}} t_l \right)^{N-1} = \frac{1}{2\pi i} \int_{C_0} \frac{dx}{x} \int_0^\infty dz e^{-z} \left(\frac{z}{x} \right)^{N-1} \exp\left\{ -x \sum_{l\in\{n\}} t_l \right\}, \tag{9.40}$$

where the contour C_0 circumvents the point $x = 0$ and allows one to select the $(N-1)$-th power of the sum according the Cauchy formula for derivatives of analytical functions, and the integral over dz compensates for the factorial in the Cauchy formula. For the factorization of the sums in the exponent, we employ a simpler version

$$\exp\left(\sum_{l\in\{n\}} t_l \sum_{m\in\{n\}} t_m\right) = \frac{1}{\sqrt{\pi}} \int_{-\infty}^{\infty} \exp\left(-y^2 - 2y \sum_{m\in\{n\}} t_m\right) dy, \quad (9.41)$$

of the standard factorization procedure (8.102). Substituting (9.40) and (9.41) into (9.39) for the integrand we obtain (9.36) in the form

$$\langle G(\varepsilon)\rangle = \sum_{C(\{n\})} \frac{1}{2\pi i\sqrt{\pi}} \int_{C_0} \frac{dx}{x} \int_0^{\infty} dz \int_{-\infty}^{\infty} dy \int_0^{\infty} du \left(\frac{z}{x}\right)^{N-1} e^{-z-y^2}$$
$$\prod_{l\in\{n\}} \int_{C_t} t_l^{\delta_{0l}} \exp\left[-(u + x + 2y + \varepsilon - \Delta_l)t_l\right] dt_l. \quad (9.42)$$

The last integration is standard and yields $(u + x + 2y + \varepsilon - \Delta_l)^{-1-\delta_{0l}}$ since the integration contour C_t can be directed from $t_l = 0$ to $-i\infty$ employing an infinitesimal imaginary part of the variable ε.

Now we perform the summation over all of the carriers that contain the state $n = 0$. All other states of the system can either be included or not included in the carrier $C(\{n\})$. We attribute the factor unity to the states which are not included in the carriers, and extend our product to all states of the system. Then the summation over all possible carriers reduces to the multiplication of a binomial, and we arrive at

$$\langle G(\varepsilon)\rangle = \frac{1}{2\pi i\sqrt{\pi}} \int_{C_0} \frac{dx}{x} \int_0^{\infty} dz \int_{-\infty}^{\infty} dy\, e^{-z-y^2} \int_0^{\infty} du$$
$$(u + x + 2y + \varepsilon - \Delta_0)^{-2} \prod_{l=1}^{N} \left[1 + \tfrac{z}{x}(u + x + 2y + \varepsilon - \Delta_l)^{-1}\right]. \quad (9.43)$$

Equation (9.43) corresponds by construction to the matrix element $G(\varepsilon)_{0,0}$ of the Green's operator, since we have considered only the trajectories that start at the level $|0\rangle$. In order to obtain the trace of the Green's operator we note that the factor in (9.43) corresponding to the starting level $|0\rangle$ is the derivative $\partial/\partial u$ of the factor identical to that of the other levels, and therefore after integration over du the trace reads

$$\mathrm{Tr}\,\langle G(\varepsilon)\rangle = \frac{1}{2\pi i\sqrt{\pi}} \int_{C_0} \frac{dx}{x} \int_0^{\infty} dz \int_{-\infty}^{\infty} dy\, e^{-z-y^2}$$
$$\frac{x}{z}\left\{\prod_{l=0}^{N} \left[1 + \tfrac{z}{x}(x + 2y + \varepsilon - \Delta_l)^{-1}\right] - 1\right\}, \quad (9.44)$$

where the integration constant -1 is chosen in order to exclude the non-physical constant term of the product, corresponding to the carrier containing no level.

Further simplifications have no physical background, being purely technical. We employ an integral representation similar to (9.41), replacing e^{-y^2} by the integral of $\pi^{-1/2}\int e^{-q^2-2iyq}dq$, change the variables $2y+x+\varepsilon \to 2y$ and $x \to zx$, simultaneously also changing the order of the integration when needed, perform integration over dz, and arrive at

$$\mathrm{Tr}\,\langle G(\varepsilon)\rangle = \frac{1}{2\pi^2 i}\int_{-\infty}^{\infty} dq \int_{C_0} \frac{dx}{x} \int_{-\infty}^{\infty} dy\, e^{-q^2-2iyq+iq(xz+\varepsilon)}$$
$$\frac{1}{1-iqx}x\left\{\prod_{l=0}^{N}\left[1+\tfrac{1}{x}(2y-\Delta_l)^{-1}\right]-1\right\}. \tag{9.45}$$

The integration contour C_0 is a loop separating the high-order poles at the point $x=0$ from the first-order pole at $x=1/iq$. We move this contour to infinity, where the integral vanishes and calculate the residual at the point $x=1/iq$. This yields

$$\mathrm{Tr}\,\langle G(\varepsilon)\rangle = \frac{1}{\pi}\int_{-\infty}^{\infty} dq \int_{-\infty}^{\infty} dy\, e^{-q^2-2iyq+iq\varepsilon}$$
$$\frac{1}{iq}\left\{\prod_{l=0}^{N}\left[1+iq(2y-\Delta_l)^{-1}\right]-1\right\}, \tag{9.46}$$

which is the final result for the general case. The limit of a large number of states follows from (9.46) after some algebraic manipulations, when we assume that each factor of the product is close to unity and set

$$\prod\nolimits_{l=0}^{N}\left[1+iq(2y-\Delta_l)^{-1}\right] = \exp\left[iq\sum\nolimits_{l=0}^{N}(2y-\Delta_l)^{-1}\right]$$

by analogy to (3.119) and (8.111).

The other interesting limit is that of identical levels, that is, $\Delta_l = 0$ for all l. In this case one can expand the binomial $(1+iq(2y)^{-1})^N$ and integrate each term independently allowing for the integral representation

$$H_n(x) = \frac{2^n}{\sqrt{\pi}}\int_{-\infty}^{\infty}(x+i\tilde{q})^n\, e^{-\tilde{q}^2}\, d\tilde{q}$$

for the Hermite polynomials, which yields

$$\langle G(\varepsilon)\rangle = \frac{e^{-\varepsilon^2/4}}{2\sqrt{\pi}}\sum_{k=0}^{N-1}\frac{N!H_{2k}(\varepsilon/2)}{2^k(k+1)!k!(N-k-1)!}. \tag{9.47}$$

In Fig. 9.5 we show the density of states for the case of $N=1,2,\ldots 8$ along with the Wigner semicircular profile corresponding to the asymptotic of $N\to\infty$.

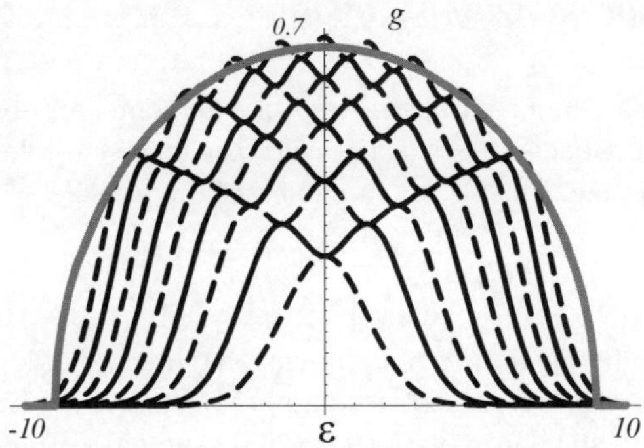

Fig. 9.5 The density of quantum states for multiplets perturbed by a random matrix. The even number of states $N = 2, 4, 6, 8$ are shown by *solid lines*, while the *dashed lines* show the profiles for the odd N. The asymptotic semicircular dependence, scaled to the parameters of $N = 10$, is also shown. The two-hump dependence of $N = 2$ coincides with that of Fig. 9.3

We conclude by summarizing the main results of this subsection. We sum up explicitly the all-order perturbation series for an arbitrary quantum system perturbed by a random matrix from a Gaussian unitary ensemble. We take into account all self-intersections of the trajectories exactly, with the help of the explicit expression for the number of Euler paths. We obtain the exact expression for the transformed spectra in the form of a twofold integral. A simple transformation rule, which governs the "macroscopic" transformation of spectra, follows from the exact expression in the extreme of a large number of states.

9.3 The Density of Quantum States of Fractals

The description of the interference phenomena inhibiting the level's decay and affecting the state density at high resolution was based on the analysis of self-intersections of the transition trajectories going over quantum states of a complex quantum system. Now we focus on another case of interference, when the orgraphs of the diagrams do not self-intersect but correspond to a tree-like topological structure. However, each branch of the "tree" may be passed an infinite number of times by the trajectory. One encounters a physical realization of this situation by considering a quantum particle moving along a ramified structure like that shown in Fig. 2.20. In contrast with the problems considered earlier in the energy representation, for this problem the coordinate representation is much more natural, whereas the Green's function of the particle is given by the Feynman path integral in its classical formulation, that is, as a sum over all possible trajectories, weighted by a phase factor suggested by the classical action along each trajectory.

9.3.1 The Renormalization Equation

In Fig. 9.6 we compare and contrast the tree-like diagrams for both cases, assuming that each vertex is connected to only two vertexes placed at a larger distance from the beginning of the tree. The Green's function of the particle

$$G_{nm}(p) = \exp\left(-ipL_{mn}\right) \tag{9.48}$$

moving with momentum p along a one-dimensional interval of length L_{mn} connecting nodes n and m of the ramified structure, corresponds to the line of the graph which coincides with standard notations in solid state physics and field theory. The vertexes that were denoting the Green's functions $(\varepsilon - \Delta_k)^{-1}$ of the unperturbed complex system correspond now, on the contrary, to the interaction matrix

$$\hat{S} = \begin{pmatrix} S_{bb} & S_{br} & S_{bl} \\ S_{rb} & S_{rr} & S_{rl} \\ S_{lb} & S_{rr} & S_{rr} \end{pmatrix}, \tag{9.49}$$

which describes the scattering of the particle at a node of the ramified structure. We assume that the scattering amplitudes are identical for all vertexes, depending only on the direction. The subscripts r and l denote the right and left direction,

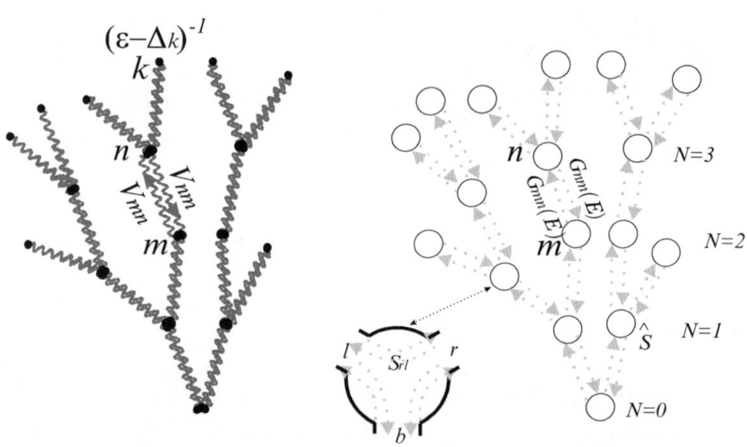

Fig. 9.6 Tree-like diagrams for the probability amplitude at times shorter than the recurrence time (*left*) and the diagrams for a particle moving along a ramified structure (*right*) have the same topological structure. However *line* and *vortex symbols* have opposite meanings. The free propagation of the complex system corresponds to the Green's function $(\varepsilon - \Delta_k)^{-1}$ shown by the vertex point, which corresponds to the *line* $G_{n,m} = \exp(-ipL_{n,m})$ for the propagating particle. The coupling $V_{n,m}$ corresponds in turn to the vertex $\hat{S}$ describing scattering of the particle in the node of the fractal. The length of the intervals $L_{n,m}$ may decrease with increasing number N of steps separating a vertex from the root

respectively, while b stands for the direction towards the root of the tree. We also assume that the distance L_{mn} can decrease by a factor a with each additional step in the number of intervals N separating the interval from the root. For simplicity, we take all intervals corresponding to the same N to be identical, that is, $L_{mn} = L_N$.

Let us derive an equation for the Green's function (9.48) which would be an analog of the renormalization equations (4.21) or (9.5) for complex systems. The derivation is illustrated in Fig. 9.7a. By X_N we denote the renormalized Green's

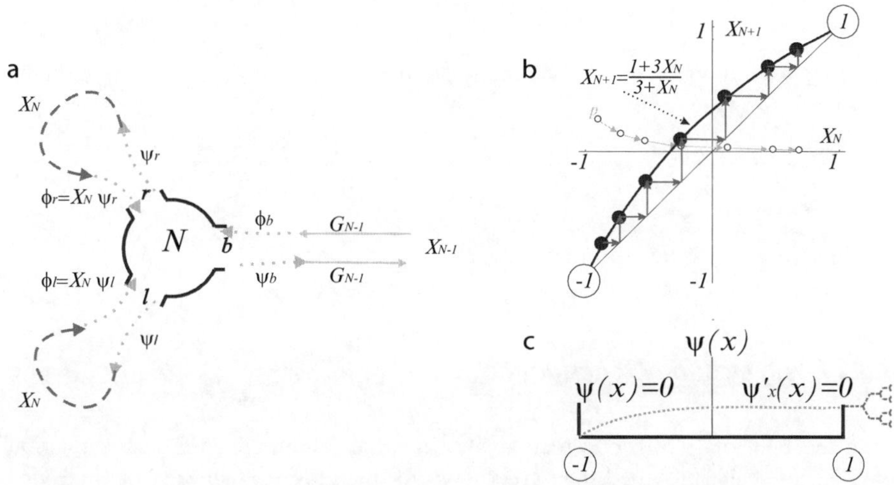

Fig. 9.7 Recurrence relation for the Green's functions. (**a**) The relations among the scattering amplitudes at a node N have to be complemented by the propagation conditions given by the phase factors X_N. (**b**) The trajectory $X(p/a^N)$ of mapping (9.52) shows the value of the asymptotic dependence after each consecutive momentum scaling. Starting from the vicinity of the unstable stationary point with $X = -1$ which corresponds to the boundary condition $\psi(x_N) = 0$ for the wavefunction in the coordinate representation, the trajectory goes to the stable stationary point that corresponds to the condition $\psi'_x(x_N) = 0$ of the constant flux (**c**)

function for a particle leaving a node N for the neighboring vertexes $N+1$ and coming back. This implies the relation $\phi_r = X_N\,\psi_r$ between the amplitude of the returning wave ϕ_r and the amplitude ψ_r of the wave leaving a node N along the right channel. The same condition for the incoming ϕ_l and outgoing ψ_l waves holds for the left channel. The renormalized Green's function X_{N-1} is a product of three factors, two functions G_{N-1} describing propagation before and after the passage of the vertex N and the ratio ψ_b/ϕ_b allowing for the scattering process in this node. By the relation $\boldsymbol{\psi} = \hat{S}\boldsymbol{\phi}$,the scattering matrix (9.49) relates the vector $\boldsymbol{\phi} = (\phi_b, \phi_r, \phi_l)$ of the incoming amplitudes for the vortex N with the vector $\boldsymbol{\psi} = (\psi_b, \psi_r, \psi_l)$ of the outgoing amplitudes.

For further analysis, we take the simplest model of the unitary scattering matrix $\hat{S}$, completely symmetric with respect to the channels interchange, which also satisfies the long-wave requirement, giving the $-\pi$ phase shift for the amplitudes

$\phi_b/\psi_b = -1$ when the same is valid for two other pairs $\phi_r/\psi_r = -1$ and $\phi_l/\psi_l = -1$. The latter condition means that if both channels r and l are closed, channel b sees the vertex as an infinite, vertical, completely reflecting wall. In other words, no quantum defect is associated with such a vertex. The only matrix satisfying these properties gives the following system of equations:

$$\hat{S}\phi = \begin{pmatrix} -1/3 & 2/3 & 2/3 \\ 2/3 & -1/3 & 2/3 \\ 2/3 & 2/3 & -1/3 \end{pmatrix} \begin{pmatrix} \phi_b \\ X_N\,\psi_r \\ X_N\,\psi_l \end{pmatrix} = \begin{pmatrix} \psi_b \\ \psi_r \\ \psi_l \end{pmatrix}$$

and after excluding ψ_r and ψ_l via the relation $X_{N-1} = G_{N-1}^2 \psi_b/\phi_b$ it yields the renormalization condition

$$X_{N-1}(L_{N-1}p) = G_{N-1}^2 \frac{1 - 3X_N(L_N\,p)}{X_N(L_N\,p) - 3} \tag{9.50}$$

relating two successive renormalized Green's functions X.

9.3.2 Asymptotic Dependence

With the help of (9.50), one recursively finds the renormalized Green's function on a larger scale, provided they are known for the closest smaller scale, and vice versa. In the long wave limit, when $G_{N-1}^2 = \exp(-ipL_{N-1}) \simeq 1$ this equation is a mapping which does not contain any parameters depending on the scale size, apart from the arguments of X. Each branch of the tree-like diagram has the same topology. If we assume that the size of the branches scales by a factor $1/a$ when we pass from the branch $N-1$ to the next branch N, the corresponding functions $X_{N-1}(L_{N-1}p)$ and $X_N(L_N\,p)$ should be identical, differing only by the scale factor of the argument, that is, $X_{N-1}(L_{N-1}p) = X(L_{N-1}p)$ and $X_N(L_N\,p) = X(L_N\,p) = X_{N-1}(L_{N-1}p/a)$. In other words, for small p (9.50) reads

$$X(p) = \frac{1 - 3X(p/a)}{X(p/a) - 3}; \quad \text{or} \quad X(p/a) = \frac{1 + 3X(p)}{X(p) + 3}. \tag{9.51}$$

This implies that starting from the value of the asymptotic dependence $X(p)$ for a given argument p and performing the mapping

$$X_N = \frac{1 + 3X_{N-1}}{X_{N-1} + 3}, \quad p = p/a \tag{9.52}$$

one can obtain the values of this function at all points $p = p/a^N$ approaching the point $p = 0$, as shown in Fig. 9.7b.

The mapping (9.52) has two stationary points at $X = \pm 1$, as one can see in Fig. 9.7b. The unstable stationary point $X = -1$ corresponds to the reflection of the wavefunction from the node as from a rigid potential wall, where the probability amplitude $\psi_b + \phi_b$ must tend to 0. Therefore the reflected wave ψ_b is shifted in phase by π with respect to the incident wave ϕ_b. The stable stationary point $X = 1$ corresponds to the probability amplitude flux going through the node to the ensuing branches of the fractal structure and returning. The total length of the branches can be very large, even infinite, and hence the totality of the branches occupies a vast domain of the phase space and is therefore capable of absorbing a big fraction of the population. However, the length of each particular trajectory going through the node to the end of the ramified structure decreases, and often even vanishes at $N \to \infty$, such that the total accumulated phase of the returning wave does not have enough distance to change. This implies that in this case, the wavefunction at the node satisfies the boundary condition with zero derivative $p\psi_b - p\,\phi_b = 0$, and the reflection occurs with zero phase shift, as illustrated in Fig. 9.7c. The latter condition changes the spectrum of the system drastically, since now the Bohr quantification rule permits momenta values close to zero. Let us call "the spectrum of fractals" the part of the spectrum at small p resulting from the large total length of the ramified structure and consider it in more detail.

The asymptotic behavior of $X_{as}(p)$ for $p \to 0$ can be found exactly. Comparing the mapping (9.52) with the trigonometric identity

$$\tanh\left(z + \ln\sqrt{2}\right) = \frac{1 + 3\tanh z}{3\tanh z + 3}$$

we find that $X_{as}(p) = \tanh\left[-\alpha \ln(Lp)\right]$ satisfies (9.51) exactly, provided that $\alpha \ln(a) = \ln\sqrt{2}$. Therefore

$$X_{as}(p) = \frac{1 - (Lp)^{2\alpha}}{1 + (Lp)^{2\alpha}}, \tag{9.53}$$

which indeed tends to -1 for $p \to 0$. Starting with $X_{as}(p)$ for a small argument p/a^{N_0} and performing N_0 iterations back to the larger scales, one finds the exact dependence $X(p)$ for all p with any predetermined accuracy, depending on the number N_0 of iterations. Even for a few iterations, the result perfectly reproduces all of the important features of the system, whereas $N_0 = 10$ is sufficient for practical reasons for $a > 1.5$ and gives an accuracy better than 1%. Note that the asymptotic form (9.53) remains valid even in the case when the scaling factor a depends on the interaction number N, with the only difference that we have to substitute the average of $\ln(a)$ over all of the iterations performed into the definition $\alpha \ln(a) = \ln\sqrt{2}$ of the index α. Note that α coincides with the fractal dimensionality of the tree-like structure.

9.3.3 Density of States

In order to find the state density profile for the fractal structures, we note that at a given energy two different momenta with the same absolute values $|p|$ and opposite signs contribute to the state density. The corresponding Green's functions interfere in the coordinate representation. Therefore, instead of generalization of the expression $g(E) = -\mathrm{Im}[\mathrm{Tr}G(E)/\pi]$ in the case of the double degeneracy of the energy levels, it is more convenient to make use of the known Fabri–Pérot formula

$$g(|p|) = \frac{1}{2\pi}\mathrm{Re}\frac{1+X(p)}{1-X(p)}, \tag{9.54}$$

which for $X(p) = R\exp[iLp]$ gives the state density of the rectangular potential well, with the reflection coefficient R at the edges. Actually, apart from a constant factor, this formula coincides with the expression for the transmission profile of the Fabri–Pérot resonator for the particle wave. In Fig. 9.8a we show the state density as a function of $|p|$, calculated for different scaling factors a with the help of (9.54). For $a > 1$ but relatively small, the position of the first resonance goes down with respect to the position of the first state in the potential well, which is consistent with the increase of the space domain occupied by the particle. The presence of the fractal manifests itself in the broadening of the profile. For larger a the resonance returns to the Fabri–Pérot position, and goes even higher, being repelled by the states appearing at the domain $|p| \to 0$. The appearance of these state close to zero energy is the most important manifestation of the fractal structure, and implies that the population is getting distributed over all of the domain occupied by the fractal.

The asymptotic dependence $g(|p|) \sim (Lp)^{-2\alpha}$ directly follows from the substitution of (9.53) into (9.54). In Fig. 9.8b we show this domain of momenta with a higher resolution, where the power dependencies different for different a are clearly seen. We note that the relation between the energy and momentum can be different for different cases, changing from $E \sim p^2$ for free particles to $E \sim p$ for the conducting electrons near the Fermi surface in metals. In the latter case the energy dependence of the state density for small E reads $g(E) \sim E^{-\ln 2/\ln a}$. For this case, an important qualitative conclusion can be drawn—for $a > 2$ the total number of low-laying states is finite, since the integral $\int_0 g(E)dE$ converges at the lower limit. This is consistent with the fact that the total length of the fractal branches is finite in this case, and is given by the sum of the geometric series $L/(1-2/a)$. For $a < 2$ the total length of the branches is infinite, and the integral diverges. For a free particle with $E \sim p^2$ the density of states at small energies behaves as $g(E) \sim E^{-\alpha-1/2}$, and therefore the combination $2(1-\alpha)$ can be interpreted as a spatial dimensionality of the fractal, by comparison with the expression $g(E) \sim E^{(d-2)/2}$ for the regular d-dimensional space.

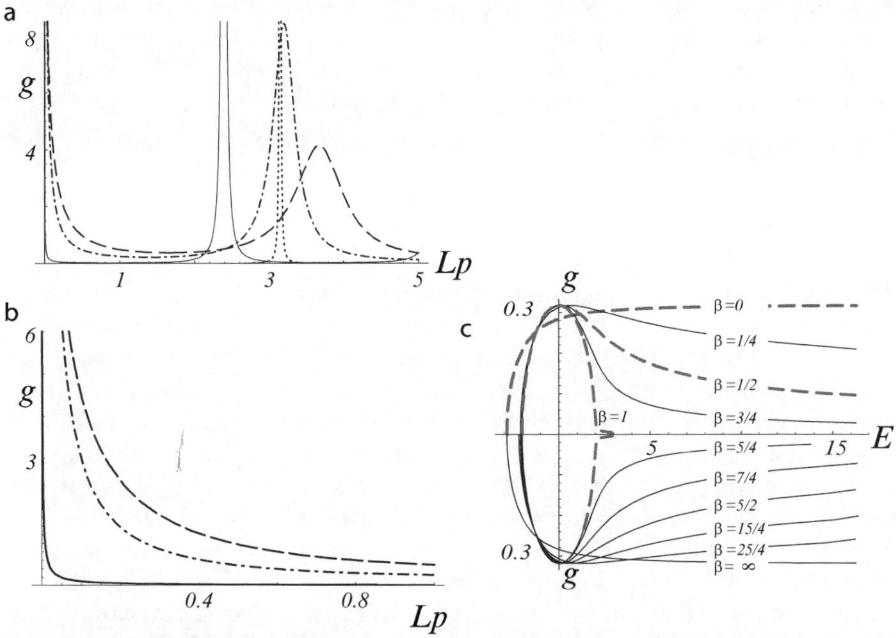

Fig. 9.8 Spectra of fractal structures. (**a**) Density g of states as a function of dimensionless momentum Lp. Fabri–Pérot resonances (*dotted line*) corresponding to the isolated first branch get shifted towards lower Lp when additional branches with a relatively small scaling factor ($a = 1.7$, *solid line*) are added. For a larger scaling factor ($a = 2.9$, *dot-dash line* and $a = 3.9$, *dashed line*) the resonances are broadened and shifted due to the presence of states at small Lp. (**b**) The states at small Lp typical of fractals for the same parameters. The state density $g(Lp)$ follows power laws with different indices for different scaling factors. (**c**) Fractal state density transformed by a random perturbation for different fractal dimensionality found from the solution of (9.55) with $\beta = 2\alpha$. *Dashed lines* show the particular cases corresponding to the Wigner semicircle and to one- and two-dimensional free particles, depicted in Figs. 4.19, 9.5, and 9.2, respectively. (The state density g is positive in both upward and downward directions.)

9.3.4 Randomly Perturbed Fractals

We now address the question: how general are our results obtained for the specific scattering at the nodes and for the strict scaling law of the branch sizes? One possibility to answer the question is to consider deviations of branch lengths from the mean value L/a^N and scattering matrices from the mean matrix S of (9.49) as random perturbations. This assumption allows us to employ the transformation (9.3), (9.4) that can be written in the dimensionless variables

$$E = \tilde{E} - \left(-\tilde{E} + i0\right)^{-2\alpha}, \tag{9.55}$$

by analogy to (9.15) for a free particle moving in regular space, whereas the mean squared random perturbation just scales the energies. This yields universal shapes of the state densities for each value of the fractal dimension shown in Fig. 9.8c. Note that this form is identical for the reciprocal dimensions.

9.4 Spectra of Euclidean Random Matrices

Thus far, considering Hamiltonians given by random matrices, we have been paying attention at the statistics of matrix elements, spectral densities, or parametric resonances. The consideration was mainly performed in the framework of the so-called 0-dimensional model where the special structure of the random perturbation has been ignored. Of course, some aspects of the spatial structure can be associated with the multiband relay-like systems or with the model of random walk, and, at certain extend, can be related with polychromatic excitation of a system, whereas each of the harmonics of the excitation field plays role of a spatial dimension. Still, in most of the cases a spatial dimensionality was associated with unperturbed Hamiltonians, as earlier in this section it was the case for free motion in the space of an integer or fractal dimensions. In this section we consider another type of random matrices, for which the role of the spatial dimension is crucial. These matrices can be interpreted as Hamiltonians of binary coupling of "quantum states" that are randomly distributed in multidimensional Euclidean space, while our aim is to trace relations between the range and the spatial dimension of this binary interaction from one hand side and the spectral properties of such matrices from the other.

9.4.1 Euclidean Random Matrices, Heuristic Consideration

In this section we introduce the notion of Euclidean random matrix and perform some preliminary heuristic consideration of their spectra. Consider a set on points uniformly scattered over d-dimensional Euclidean space, such that the coordinates of n-th point are given by a random d-dimensional vector

$$\mathbf{r}_n = \left\{ \underbrace{x_n, y_n, \ldots, u_n}_{d} \right\}. \tag{9.56}$$

Let each unit volume of this space contains N of such points. The matrix with elements

$$V_{mk} = V\left(\mathbf{r}_m - \mathbf{r}_k\right) \tag{9.57}$$

is called the Euclidean random matrix. One can say that each point m in the space is associated with a quantum state $|m\rangle$ localized at this point, and the matrix element V_{mk} of the Hamiltonian $\hat{V}$ accounts for the amplitude of transition from the point m to the point k. We will call these states "Euclidean states" and consider density of the eigenvalues of such a matrix averaged over random positions of the Euclidean states.

Short- and Long-Range Interactions

One can identified two distinct limiting cases suggested by the analogy to (3.110) for the level–band system:

$$\int Vg(V)dV \to \sum_k |V(\mathbf{r}_m - \mathbf{r}_k)| < \text{a number} \tag{9.58}$$

or

$$\int Vg(V)dV \to \sum_k |V(\mathbf{r}_m - \mathbf{r}_k)| \to \infty, \tag{9.59}$$

corresponding to the cases of short- and long-range interactions, respectively. Continuous analogies of these conditions are formulated in terms of the integral

$$\int Vg(V)dV \to \int_{r_{\min}}^{r_{\max}} |V(\mathbf{r})|\, d^d\mathbf{r}, \tag{9.60}$$

which diverges at the upper limit for the long-range interactions and converges for the short-range ones. At the same time, for the short-range interactions the integral may diverge at the lower limit. There exists a special critical case, where the interaction $|V(\mathbf{r})| \sim 1/|\mathbf{r}|^d$, which corresponds to the integral (9.60) logarithmically diverging in both upper and lower limits.

In the Fourier representation

$$V(\mathbf{k}) = \int d^d\mathbf{r}\, V(\mathbf{r})\, e^{-i(\mathbf{rk})}$$

$$V(\mathbf{r}) = \int \frac{d^d\mathbf{k}}{(2\pi)^d} V(\mathbf{k})\, e^{i(\mathbf{rk})},$$

in the limit $k \to 0$, for the long-range interactions $V(\mathbf{k}) \to \infty$, for the short-range ones $V(\mathbf{k}) \to 0$, and for the critical regime $V(\mathbf{k}) \to \text{const}$. Note that for the case of long-range interactions, some eigenvalue in the spectrum of Euclidean random matrices are sensitive to the shape of the domain uniformly seeded by the Euclidean states, while for the short-range interactions one encounters the so-called statistical limit, where the spectrum is governed exclusively by the density N. In the latter case, one can derive the so-called virial expansion, describing a physical observable of the distributed systems as a power series in the density.

Continuous Media Model

As a starting point of the analysis, one can implement the continuous media model, where the probability amplitude (9.57) of transition from one point to another depends on the distance between these points. In the momentum representation one can write

$$g(\varepsilon) = \frac{1}{\pi}\mathrm{Im}\int d^d\mathbf{k}\frac{p}{\varepsilon - pV(\mathbf{k}) - i0}, \tag{9.61}$$

which suggests that the interaction matrix element $V(\mathbf{k})$ multiplied by the density p of the Euclidean states serves as a model of the Hamiltonian of the system in the continuous limit. Note that tracing over the system is different from the tracing over the continuum by the density factor p. The integration can be performed explicitly for the power-law interaction $V(\mathbf{r}) \sim r^{-\alpha}$, and the proportionality coefficient determined by the relation $V(\mathbf{k}) = k^{\alpha-d}$ can be written in units of scaled the wave vector. From the standard expression

$$\frac{1}{\pi}\mathrm{Im}\frac{1}{\varepsilon - pV(\mathbf{k}) - i0} = \delta[\varepsilon - pV(\mathbf{k})],$$

after the replacement $d^d\mathbf{k} \to 2\pi^{d/2}\Gamma^{-1}\left(\frac{d}{2}\right)k^{d-1}dk$ valid for the direction independent interaction, we arrive at

$$\begin{aligned} g(\varepsilon) &= \frac{2\pi^{d/2}p}{\Gamma\left(\frac{d}{2}\right)}\int \delta\left[\varepsilon - pk^{\alpha-d}\right]k^{d-1}dk \\ &= \frac{2\pi^{d/2}}{\Gamma\left(\frac{d}{2}\right)}\frac{(\varepsilon/p)^{2d-\alpha}}{(\alpha-d)}, \end{aligned} \tag{9.62}$$

where $\Gamma(x)$ stands for the Gamma-function. Note that the Euclidean random matrices have zero trace, and therefore the dependence (9.62) has to be shifted in energy to satisfy the zero trace condition, $\int \varepsilon g(\varepsilon)\, d\varepsilon = 0$.

Still the fact discreteness of the Euclidean states brings about several important modifications of this continuous media profile, and the reasons for such modifications will be discussed in the following sections.

Eigenvalues Corresponding to the Cooperative Behavior

In Sect. 4.5.2 we have seen that dynamics of two band systems has some specific features in the case, where the coupling matrix manifests strong correlations among the matrix elements. In such a case the spectrum of V_{mk} has one eigenvalue much larger than the others, which results in the cooperative behavior of different levels that bring constructively interfering contributions in the perturbation series for the transition amplitudes. The same situation is typical of Euclidean random matrices V_{mk} corresponding to the long-range interactions, where the main contribution to the probability to leave an Euclidean state comes from the transitions corresponding to long distances. In other words, when the integral over the distance r of the square matrix element $|V(\mathbf{r})|^2$ corresponding to transitions at the distance $\mathbf{r}$ multiplied by the density of possible final Euclidean states $\sim |\mathbf{r}|^{d-1}$ diverges at the upper limit, there exists one eigenvalue separated from the other eigenvalues by a big gap. This eigenvalue corresponds to a cooperative eigenvector given by the symmetric linear combination of all Euclidean states $\sum_m |m\rangle$. One can check that for a power-law interaction $V(\mathbf{r}) \sim r^{-\alpha}$ such a cooperative state exists when $2\alpha \leq d$. In Fig. 9.9 we illustrate this situation.

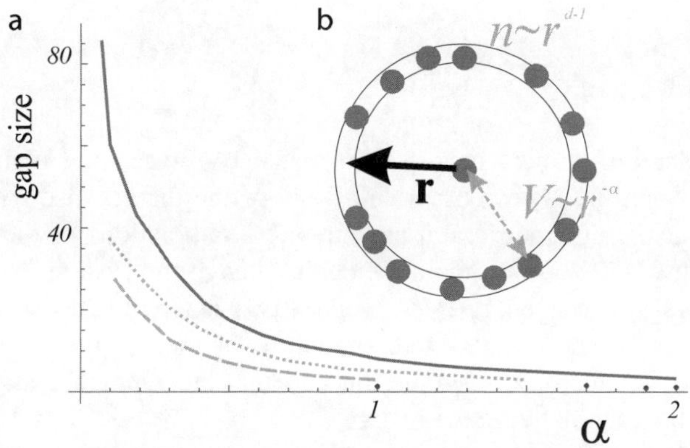

Fig. 9.9 (**a**) Size of the gap in units of the bandwidth as function of the power law $V(\mathbf{r}) \sim r^{-\alpha}$ index α for $d = 4$ *solid line*, $d = 3$ *dotted line*, and $d = 2$ *dashed line*. (**b**) Interaction $V(\mathbf{r})$ couples a state to other $n \sim r^{d-1}$ states at a distance r, such that the transition probability $|V(r)|^2 g(r) \sim r^{d-1-2\alpha}$ diverges at large distances at $\alpha > d/2$

Eigenvalues Corresponding to Dimers

There exists one more limiting case, the case of dimer fluctuations, which can easily be interpreted. In the situation where cooperative behavior does not take place, the largest in absolute values eigenenergies originate from the pairs of Euclidean states placed so closely each to the other that the distance between them are much shorter than the average one. Let us isolate a pair of points n and m with the Euclidean inter-state distance r_{nm} and find two energy eigenstates—the symmetric $(|n\rangle + |m\rangle)/\sqrt{2}$ and the antisymmetric $(|n\rangle - |m\rangle)/\sqrt{2}$ combinations of the Euclidean states that have energies $\varepsilon = -V(\mathbf{r}_{nm})$ and $\varepsilon = V(\mathbf{r}_{nm})$, respectively. For the power-law interactions $V(\mathbf{r}) \sim r^{-\alpha}$ in d-dimensional space, where the probability $P(\mathbf{r}_{nm})$ of Euclidean inter-state distance $|\mathbf{r}_{nm}|$ scales as $N|\mathbf{r}_{nm}|^{d-1}$, one immediately finds the large energy asymptotic dependence of the state density

$$g(\varepsilon) = \frac{2\pi^{d/2}N}{\Gamma\left(\frac{d}{2}\right)} \int \delta\left(|\varepsilon| - r^{-\alpha}\right) r^{d-1} dr = \frac{2\pi^{d/2}N}{\alpha\Gamma\left(\frac{d}{2}\right)} |\varepsilon|^{-(d+\alpha)/\alpha}, \tag{9.63}$$

where $\Gamma(x)$ stands for the Gamma-function.

In this regime dimers are relatively rare configurations and they weakly interact each with the other, as is implied by the condition of the absence of the collective behavior.

Interaction of Dimers

Two almost identical dimers at a certain distance b exceeding considerably the distance a between the states comprising each of the dimers also can interact, in such a way that the symmetric and asymmetric combinations of the dimer states experience a shift and do not remain the eigenstates of the system. Two symmetric states of dimers at lower energies as well as two antisymmetric states at higher energies repel each other, however the strengths of the couplings leading to this repulsion are different for the symmetric and the antisymmetric states. One can easily see this by diagonalizing the matrix

$$\begin{pmatrix} 0 & V(a) & V(b) & V(b) \\ V(a) & 0 & V(b) & V(b) \\ V(b) & V(b) & 0 & V(a) \\ V(b) & V(b) & V(a) & 0 \end{pmatrix},$$

which describes couplings of four Euclidean states comprising to two pairs with a same small distances a between the states of each pair and a large distance b between the pairs. Two eigenvalues $-V(a)$ of this matrix correspond to the degenerate antisymmetric dimer states, and two other eigenvalues $V(a) \pm 2V(b)$ describe splitting of the symmetric combinations.

At this simple example, one can see the reason why spectra of Euclidean random matrix can be asymmetric with respect of the zero energy, although the average positive energy equals the average negative energy, as should be the case for the traceless matrices.

Large Cluster Contributions

In the regime of the short-range interaction, assume that K Euclidean states are inside of a small sphere of a radius a. The largest eigenvalue of such a cluster can be estimated by the cooperative matrix element $V \sim V(a)K$, and the contribution of this cluster matrix element to the spectral density of the entire matrix can be written as

$$g_K(\varepsilon) = \int^{\infty} da \left(\frac{\pi^{d/2} N}{\Gamma\left(\frac{d}{2}+1\right)} a^d \right)^K \delta\left[\varepsilon - V(a)K\right],$$

whereas $\pi^{d/2} N \Gamma^{-1}(d/2+1)\, a^d$ stands for the probability to have one Euclidean state within a sphere of the radius a. Assuming K large and performing integration over the cluster size one gets

$$\int g_K(\varepsilon)\, dK = \int^{\infty} \frac{da}{V(a)} \left(\frac{\pi^{d/2} N}{\Gamma\left(\frac{d}{2}+1\right)} a^d \right)^{\varepsilon/V(a)},$$

and evaluating the integral by the saddle-point method for $V(\mathbf{r}) = r^{-\alpha}$ arrives at

$$\int g_K(\varepsilon)\, dK = \int^{\infty} a^{\alpha} da \left(\frac{\pi^{d/2} N}{\Gamma\left(\frac{d}{2}+1\right)} a^d \right)^{\varepsilon a^{\alpha}}$$
$$\simeq \left[\frac{\Gamma\left(\frac{d}{2}\right) d}{2N e^{d/\alpha}} \right]^{\frac{\alpha+2}{2d}} \frac{\sqrt{2} e^{-|\varepsilon| b}}{\pi^{\alpha/4} \sqrt{\alpha d\, |\varepsilon|}},$$

were $b = d\Gamma^{\alpha/d}\left(\frac{d}{2}+1\right)/N^{\alpha/d} \alpha e \pi^{\alpha/2}$. This quantity scales as $e^{-|\varepsilon| b}/\sqrt{|\varepsilon|}$, that is, at large energies it drops faster than the dimers contribution. Note that in this expression the most relevant cluster size scales as $K \simeq \varepsilon b \alpha / d$.

Numerical Results for the Power-Law Interactions

In Fig. 9.10 we present the state densities of Euclidean random matrices, obtained numerically for the interaction $V(\mathbf{r}) \sim r^{-\alpha}$ by random casting of $N = 600$ Euclidian states into a unit rectangular box in d = 2-, 3-, and 4-dimensional spaces. The results averaged over 500 casts are plotted as a function of the parameter $v = \alpha/d - 1$. One sees that for the long-range interactions $v < 0$, the density

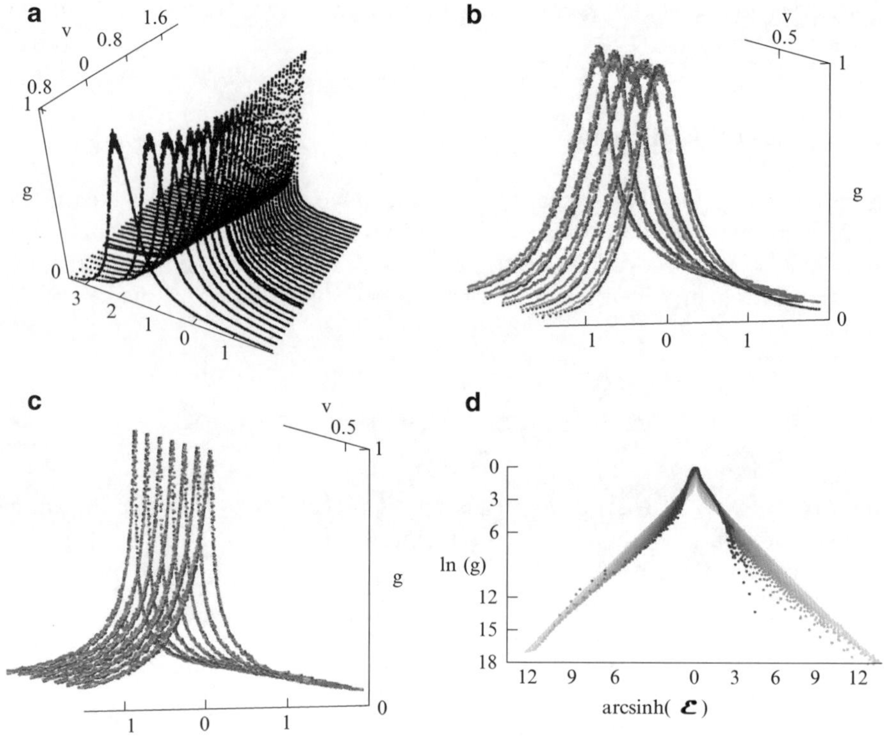

Fig. 9.10 Numerical simulation of the spectra of Euclidean random matrices obtained for the power-lax interaction $V \sim r^{-\alpha}$ and presented as function of the parameter $v = \alpha/d - 1$. (**a**) Results of the simulations with energies scaled by the factor $N^{3v/2}$. (**b**) Profiles scaled and displaced to have coinciding maximum for the long-range interactions. (**c**) Same for the short-range interactions. (**d**) Logarithmic scale, one sees difference in the long tails of the profiles due to the contribution of dimers, Eq. (9.63)

of states is displaced from the zero energy position, which is the consequence of the state repulsion from the cooperative eigenstate.

For the short-range interactions $v > 0$, the state densities are independent from the dimension d, while for the case of the short-range interaction, this dependence enters only via the energy scaling factor and practically does not affect the shape of the profile. Moreover, the shape of the profiles varies gradually with v although they manifest strong dependence of the energy scaling factors and the energy shift on this parameter. Dependence $\sim |\varepsilon|^{-v/(v+1)}$ on v typical of the contribution (9.63) of dimers manifests itself only in the far tails of these profiles.

9.4.2 *Spectral Density Averaged Over Random Positions of Euclidean States*

We now perform a more rigorous derivation of the state density of the Euclidean random matrices averaged over the positions over the Euclidean states, and suggest several models that allow one to gain a deeper insight into the mechanisms of these spectra formation. In the general case, the spectral state density of a system with Hamiltonian $\hat{V}$ is given by the imaginary part of the Green's function

$$g(\varepsilon) = \frac{1}{\pi}\mathrm{Im}\mathrm{Tr}G(\varepsilon) = \frac{1}{\pi}\mathrm{Im}\mathrm{Tr}\frac{1}{\varepsilon - \hat{V} - i0}, \tag{9.64}$$

which does not depend on the representation. In particular, for continuous media where the probability amplitude (9.57) of transition from one Euclidean state to another depends on the distance between these points, one can make use of the momentum representation and write

$$g(\varepsilon) = \frac{1}{\pi}\mathrm{Im}\int d^d\mathbf{k}\frac{1}{\varepsilon - V(\mathbf{k}) - i0}. \tag{9.65}$$

Naively, one may think that the same spectral density can be written for the Euclidean states interacting according to (9.57), which at the first glance may look similar to the continuous media case. However, this assumption is valid only partially, since the spectral state density may experience strong influence of the Euclidean state density fluctuations associated with the random seeding.

In order to emphasize the difference, one can write the spectral state density in the form

$$g(\varepsilon) = \frac{1}{\pi}\mathrm{Im}\mathrm{Tr}\frac{1}{\varepsilon - \hat{P}\hat{V}\hat{P} - i0},$$

which employees the operator $\hat{P}$ projecting the continuous media operator $\hat{V}$ onto the set of Euclidean points. In the momentum representation, the projection operator is a complicated function rapidly oscillating in the domain where the momenta are large as compared to the inverse of typical distance among these points. This does not allow one to directly employ (9.65) for the spectral state density, and the correct expression

$$g(\varepsilon) = \left\langle \frac{1}{\pi}\mathrm{Im}\mathrm{Tr}\frac{1}{\varepsilon - \hat{P}\hat{V}\hat{P} - i0} \right\rangle_{\hat{P}} \tag{9.66}$$

implies calculation of the spectrum for a given realization of seeding, followed by averaging over all possible seedings emerging from random positioning of each Euclidean state.

A Model of Sparsely Seeded Lattice

Let us present a model which allows one to practically perform the aforesaid averaging over random projections. Consider a large regular, say rectangular, lattice with the interaction (9.57) among the sites. The number M^d of the lattice sites is by many orders of magnitude larger than the number N of the Euclidean states, which may occupy the lattice sites with a small probability $p = N/M^d$ per randomly chosen site. Let us add an infinitesimally small imaginary term $-i0$ to the energy ε, cast the Green's operator in (9.65) in Taylor series, and re-sum this series in the following way:

$$
\begin{aligned}
\hat{G}(\varepsilon) &= \hat{P}\frac{1}{\varepsilon - \hat{P}\hat{V}\hat{P}}\hat{P} = \frac{\hat{P}}{\varepsilon} + \frac{\hat{P}}{\varepsilon}\hat{V}\frac{\hat{P}}{\varepsilon} + \frac{\hat{P}}{\varepsilon}\hat{V}\frac{\hat{P}}{\varepsilon}\hat{V}\frac{\hat{P}}{\varepsilon} + \ldots \\
&= \frac{\hat{P}}{\varepsilon} + \frac{\hat{P}}{\varepsilon}\frac{1}{\hat{V}^{-1}}\frac{\hat{P}}{\varepsilon} + \frac{\hat{P}}{\varepsilon}\frac{1}{\hat{V}^{-1}}\frac{\hat{P}}{\varepsilon}\frac{1}{\hat{V}^{-1}}\frac{\hat{P}}{\varepsilon} + \ldots \\
&= \hat{V}^{-1}\left(\frac{1}{\hat{V}^{-1}}\frac{\hat{P}}{\varepsilon} + \frac{1}{\hat{V}^{-1}}\frac{\hat{P}}{\varepsilon}\frac{1}{\hat{V}^{-1}}\frac{\hat{P}}{\varepsilon} + \ldots\right) \\
&= \hat{V}^{-1}\frac{\hat{V}\frac{\hat{P}}{\varepsilon}}{1 - \hat{V}\frac{\hat{P}}{\varepsilon}} = \hat{V}^{-1}\left(\frac{1}{\hat{V}^{-1} - \frac{\hat{P}}{\varepsilon}}\hat{V}^{-1} - 1\right).
\end{aligned} \tag{9.67}
$$

Now the tracing operation is performed over all sites of the lattice. For a lattice interaction matrix that has a property $\mathrm{Im}\mathrm{Tr}\hat{V}^{-1} = 0$, this yields the state density of the form

$$
g(\varepsilon) = \frac{1}{\pi}\mathrm{Im}\mathrm{Tr}\left[\hat{V}^{-2}\left\langle\frac{1}{\hat{V}^{-1} - \frac{\hat{P}}{\varepsilon}}\right\rangle_{\hat{P}}\right]. \tag{9.68}
$$

For the inverse matrix entering (9.68), the Kramer formula suggests

$$
\left(\frac{1}{\hat{V}^{-1} - \frac{\hat{P}}{\varepsilon}}\right)_{ij} = \frac{\mathrm{Minor}_{ij}\left[\hat{V}^{-1} - \frac{\hat{P}}{\varepsilon}\right]}{\det\left[\hat{V}^{-1} - \frac{\hat{P}}{\varepsilon}\right]},
$$

and one can employ the integral representations (7.136, 7.181) for the determinant and the minor thus obtaining

$$
\left(\frac{1}{\hat{V}^{-1} - \frac{\hat{P}}{\varepsilon}}\right)_{ij} = \prod_n\left\{\int\limits_{-\infty}^{\infty}\frac{dz_n^* dz_n}{\pi}\int d\chi_n d\overline{\chi}_n\right\}\overline{\chi}_i\chi_j
$$
$$
e^{-i\sum\limits_{n,m}\left(z_n\hat{V}_{nm}^{-1}z_m^* - \frac{\hat{P}}{\varepsilon}z_n z_n^*\right) - \left(\overline{\chi}_n\hat{V}_{nm}^{-1}\chi_m - \frac{\hat{P}}{\varepsilon}\delta_{nm}\overline{\chi}_n\chi_n\right)}.
$$

where $\hat{V}_{nm}^{-1}$ stands for the matrix element of the inverse matrix. Now one can notice that for the terms containing the projection operators, the summation over all sits n and m of the lattice reduces to summation over only the sites occupied by the Euclidean states, and hence

$$\left(\frac{1}{\hat{V}^{-1}-\frac{\hat{P}}{\varepsilon}}\right)_{ij} = \prod_n \left\{ \int_{-\infty}^{\infty} \frac{dz_n^* dz_n}{\pi} \int d\chi_n d\overline{\chi}_n \right\} \overline{\chi}_i \chi_j$$
$$\exp\left[- \sum_{\text{all } n,m} \left(iz_n \hat{V}_{nm}^{-1} z_m^* + \overline{\chi}_n \hat{V}_{nm}^{-1} \chi_m \right) \right]$$
$$\exp\left[\sum_{\text{seeded } n} \frac{iz_n z_n^* + \overline{\chi}_n \chi_n}{\varepsilon} \right]. \tag{9.69}$$

Consider separately the last factor in (9.69), and perform the averaging over site occupations, which is an analog to the summation over graph carriers performed earlier in (9.43). Each site n can either be occupied with the small probability p, and in such a case it brings about the factor $\exp\left[\frac{iz_n z_n^* + \overline{\chi}_n \chi_n}{\varepsilon}\right]$, or be empty with the probability $1-p$, bringing about the unit factor. Therefore

$$\left\langle \exp\left[\sum_{\text{seeded } n} \frac{iz_n z_n^* + \overline{\chi}_n \chi_n}{\varepsilon} \right] \right\rangle_{\hat{P}} = \prod_{\text{all } n} \left(1 - p + p e^{\frac{iz_n z_n^* + \overline{\chi}_n \chi_n}{\varepsilon}} \right)$$
$$\simeq \exp\left[p \sum_{\text{all } n} \left(e^{\frac{iz_n z_n^* + \overline{\chi}_n \chi_n}{\varepsilon}} - 1 \right) \right].$$

Substituting this expression to (9.69) one obtains an integral representation for the Green's function,

$$g(\varepsilon) = \frac{1}{\pi} \mathrm{Im} \prod_n \left\{ \int_{-\infty}^{\infty} \frac{dz_n^* dz_n}{\pi} \int d\chi_n d\overline{\chi}_n \right\} \sum_{\text{all } n,m} \overline{\chi}_n \hat{V}_{nm}^{-2} \chi_m$$
$$\exp\left[- \sum_{\text{all } n,m} \left(iz_n \hat{V}_{nm}^{-1} z_m^* + \overline{\chi}_n \hat{V}_{nm}^{-1} \chi_m \right) \right]$$
$$\exp\left[p \sum_{\text{all } n} \left(\exp \frac{iz_n z_n^* + \overline{\chi}_n \chi_n}{\varepsilon} - 1 \right) \right]. \tag{9.70}$$

This exact representation for the state density cannot be further simplified without making some physically reasonable approximations.

We now concentrate on the last factor in the integrand and cast the exponent in the summand to Taylor series up to the second order

$$F_n = \exp \frac{iz_n z_n^* + \overline{\chi}_n \chi_n}{\varepsilon} - 1 \simeq \frac{iz_n z_n^* + \overline{\chi}_n \chi_n}{\varepsilon} + \left(\frac{iz_n z_n^* + \overline{\chi}_n \chi_n}{\varepsilon} \right)^2 /2, \tag{9.71}$$

and by introducing auxiliary Hubbard–Stratanovich variables u_n obtain

$$\exp\left[pF_n\right] \simeq \exp p \frac{iz_n z_n^* + \overline{\chi}_n \chi_n}{\varepsilon} \exp \frac{p}{2} \left(\frac{iz_n z_n^* + \overline{\chi}_n \chi_n}{\varepsilon} \right)^2 \tag{9.72}$$

$$= e^{p \frac{iz_n z_n^* + \overline{\chi}_n \chi_n}{\varepsilon}} \int\limits_{-\infty}^{\infty} \frac{du_n}{\sqrt{2\pi}} \exp\left[\sqrt{p} u_n \frac{iz_n z_n^* + \overline{\chi}_n \chi_n}{\varepsilon} - \frac{u_n^2}{2} \right].$$

Substitution of this expression into (9.70) results in

$$g(\varepsilon) = \frac{1}{\pi} \mathrm{Im} \prod_n \left\{ \int\limits_{-\infty}^{\infty} \frac{du_n dz_n^* dz_n}{\sqrt{2\pi}\pi} e^{-\frac{u_n^2}{2}} \int d\chi_n d\overline{\chi}_n \right\} \sum_{\text{all } n,m} \overline{\chi}_n \hat{V}_{nm}^{-2} \chi_m$$

$$\exp\left[- \sum_{\text{all } n,m} iz_n \left(\hat{V}_{nm}^{-1} - \frac{\delta_{mn}}{\varepsilon} \left(p + \sqrt{p} u_n\right) \right) z_m^* \right]$$

$$\exp\left[- \sum_{\text{all } n,m} \overline{\chi}_n \left(\hat{V}_{nm}^{-1} - \frac{\delta_{mn}}{\varepsilon} \left(p + \sqrt{p} u_n\right) \right) \chi_m \right], \tag{9.73}$$

which implies

$$g(\varepsilon) = \prod_n \left\{ \int\limits_{-\infty}^{\infty} \frac{du_n}{\sqrt{2\pi}} \exp \frac{-u_n^2}{2} \right\} \frac{1}{\pi} \mathrm{Im}\mathrm{Tr} \left[\hat{V}^{-2} \frac{1}{\hat{V}^{-1} - \frac{p}{\varepsilon} \left(1 + \frac{\hat{U}}{\sqrt{p}}\right)} \right], \tag{9.74}$$

where $\left(\hat{U}\right)_{mn} = \delta_{mn} u_n$ and where the trace is taken over all sites of the lattice. Note that the requirement $\mathrm{Im}\mathrm{Tr}\hat{V}^{-1} = 0$ allows to rewrite this expression in the form

$$g(\varepsilon) = \prod_n \left\{ \int\limits_{-\infty}^{\infty} \frac{du_n}{\sqrt{2\pi}} \exp \frac{-u_n^2}{2} \right\} \frac{1}{\pi} \mathrm{Im}\mathrm{Tr} \left[\frac{p\left(1 + \frac{\hat{U}}{\sqrt{p}}\right)}{\varepsilon - p\hat{V}\left(1 + \frac{\hat{U}}{\sqrt{p}}\right)} \right]. \tag{9.75}$$

Now one can see the difference between the continuous media model of (9.64) and the Euclidean random matrix model on the lattice given by (9.74)—apart from the evident scaling of the intersection by the density factor p, in the latter case the fluctuations of the Euclidean states density in space manifests themselves in the random Gaussian fluctuations $\sqrt{p}u_n/\varepsilon$ of the inverse Hamiltonian in each point of the lattice. Note that by simple setting the lattice step to zero, one obtains the continuous media expression for the density of states of Euclidean random matrices, while (9.75) can be interpreted as expression for the state density of a propagating particle with a mass randomly varying in space. Also note that the Gaussian distribution $P_G(u) = e^{-u^2/2}/\sqrt{2\pi}$ for u_n following from the Taylor expansion (9.71) can be replaced by another distribution,

$$P(u) = \frac{\sqrt{p}}{2\pi}\int dx \exp\left[p\left(e^x - 1 - x\right) - i\sqrt{p}ux\right],$$

which does not imply the Taylor expansion.

A Toy Model

Imagine that the averaging over energies u_n in (9.74) is performed not with a Gaussian, but with a Lorentzian distribution

$$\int_{-\infty}^{\infty} \frac{du_n}{\sqrt{2\pi}} \exp\frac{-u_n^2}{2} \ldots \to \int_{-\infty}^{\infty} \frac{du_n}{\pi}\frac{\gamma}{u_n^2 + \gamma^2} \cdots \tag{9.76}$$

of a width γ. This allows one to explicitly calculate the integrals by simple replacement $u_n \to \pm i\gamma$, that is, $\hat{U} \to \pm i\gamma$, and from (9.74) obtain

$$g(\varepsilon) = \frac{1}{\pi}\mathrm{Im}\mathrm{Tr}\left[\hat{V}^{-2}\frac{1}{\hat{V}^{-1} - \frac{p}{\varepsilon} \pm i\frac{\gamma\sqrt{p}}{\varepsilon}}\right], \tag{9.77}$$

where the choice of the sign depends on the sign of ε and is dictated by the requirement of the positive state density. In d-dimensional space and for $V \sim r^{-\alpha}$ corresponding to $V(\mathbf{k}) = k^{\alpha-d}$ this yields

$$g(\varepsilon) = \frac{2\pi^{d/2-1}}{\Gamma\left(\frac{d}{2}\right)k_0^d}\mathrm{Im}\int_{k_0}^{k_c} \frac{\left(k^{\alpha-d} - D\right)^{-2} k^{d-1}dk}{\left(k^{\alpha-d} - D\right)^{-1} - \frac{p}{\varepsilon}\left(1 \pm i\gamma p^{-1/2}\right)}, \tag{9.78}$$

where

$$D = \frac{\int\limits_0^{k_c} k^{\alpha-1} dk}{\int\limits_0^{k_c} k^{d-1} dk} = \frac{d}{\alpha} k_c^{\alpha-d} \tag{9.79}$$

is a displacement insuring $\mathrm{Tr}\hat{V} = 0$. Here k_c is the upper cut-off wavenumber, related to the number of lattice sites M^d via the equality

$$\int\limits_0^{k_c} \frac{2\pi^{d/2}}{\Gamma\left(\frac{d}{2}\right) k_0^d} k^{d-1} dk = \frac{2\pi^{d/2} \left(k_c/k_0\right)^d}{d\Gamma\left(\frac{d}{2}\right)} = M^d, \tag{9.80}$$

and k_0 is the lower cut-off wavenumber given by the inverse size of the system. Note that the integration (9.78) performed for a finite p may give a finite constant state density

$$g_C = \frac{2\pi^{d/2-1}}{\Gamma\left(\frac{d}{2}\right)} \mathrm{Im} \int\limits_{k_0}^{k_c} \frac{k^{d-1} dk}{k^{\alpha-d} - D} = \frac{2\pi^{d/2-1} D^{\alpha/(\alpha-d)}}{\Gamma\left(\frac{d}{2}\right)(\alpha - d)} \tag{9.81}$$

in the limit $p \to 0$, which means that with this integration method one finds $\mathrm{Im}\mathrm{Tr}\hat{V}^{-1} \neq 0$, in contrast to the assumption done after (9.67). Therefore the constant g_C has to be subtracted from the expression (9.78) in accordance with the exact expression (9.67) for the state density.

For the short-range interactions $\alpha > d$, the straightforward integration in (9.78) yields an expression for the state density

$$g\left(\varepsilon\right) = \frac{N}{\pi} \mathrm{Im} \frac{-(Q+1)}{\varepsilon + Q + 1} {}_2F_1\left(1, \frac{1}{v}; \frac{v+1}{v}; \frac{(Q+1)(v+1)}{Q+\varepsilon+1}\right) \tag{9.82}$$

in terms of a hypergeometric function, where $v = \alpha/d - 1$, pD is chosen as the energy unit, and $Q = \pm i\gamma p^{-1/2}$, with the sign depending on the sign of ε. For the long-range interactions $v < 0$, the straightforward integration in (9.78) yields

$$g\left(\varepsilon\right) = \frac{-N/\pi}{v-1} \mathrm{Im}\left[{}_2F_1\left(1, \frac{v-1}{v}; \frac{2v-1}{v}; \frac{\varepsilon}{Q+1} + \frac{1}{v+1}\right)\right]. \tag{9.83}$$

In Fig. 9.11 we illustrate these dependence for different values of the parameter v and a certain (small) value of the parameter $\gamma p^{-1/2}$, which gives a reasonable agreement with the results of numerical work. This means that for the long-range interaction the density of states does not deviate much from the prediction of the continuous media model (9.62), with the profile shifted by the value of (9.79) and

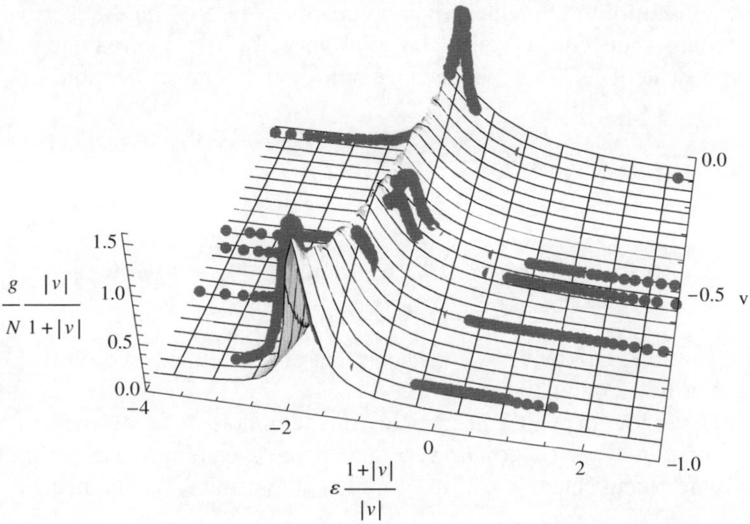

Fig. 9.11 Density of states of Euclidean random matrices suggested by the toy model for the long-range interactions. The width parameter is taken to be $\gamma p = -0.07v$. Points show the density of states of Fig. 9.10 obtained numerically for the case $d = 4$

where the discreteness of the Euclidean states manifests itself in a small imaginary shift of the energy resembling the Landau damping. In contrast, the prediction (9.82) of the toy model for the short-range interactions does not agree with the numerical results.

9.4.3 Two Field-Theoretical Models

We now consider two models for calculation of the spectral density of Euclidean random matrices. Being more involved as compared to the toy model of previous section, both of these models employ the saddle-point approximation for calculation of functional integrals implied by the assumption of the existence of the thermodynamic limit. In other words, we suppose that the state density as a function of energy is governed exclusively by the density of the Euclidean points seeded in space with the only exception for the uniform displacement of this profile due to repulsion from the cooperative state, which may also depend on the total number of these points in the volume under consideration.

A Nonlinear σ-Model for the Power-Law Interactions

Apart from the simplistic toy model just considered in the previous section, a more refined, although more complicated technically, the so-called nonlinear σ-model can also be proposed for the Euclidean random matrices. It relies on the representation (9.70) for the state density with the allowance for the approximation (9.71). However, instead of (9.72), now one can employ another representation,

$$\exp\left[pF_n\right] \simeq e^{p\frac{iz_n z_n^* + \overline{\chi}_n \chi_n}{\varepsilon}} \int_{-\infty}^{\infty} d\hat{\mathbf{Q}}_n \tag{9.84}$$

$$\exp\left[\frac{\sqrt{p}}{\varepsilon}\left(\overline{\chi}_n \; iz_n^*\right)\hat{\mathbf{Q}}_n \begin{pmatrix} \chi_n \\ z_n \end{pmatrix} - \frac{1}{2}\mathrm{STr}\hat{\mathbf{Q}}_n^2\right],$$

which implies integration over a 2×2 supermatrix $\hat{\mathbf{Q}}_n$ defined on each site n of the lattice. Let us dwell on this formalism.

Actually, without explicitly introducing this terminology, the formalism relying on the integration over a supermatrix has already been employed earlier, in (8.105), for even more complicated case of the population dynamics. A supermatrix

$$\hat{\mathbf{Q}}_n = \begin{pmatrix} {}_nQ_{11} & {}_nQ_{12} \\ {}_nQ_{21} & {}_nQ_{22}i \end{pmatrix}$$

contains regular numbers ${}_nQ_{11}$ and ${}_nQ_{22}i$ on the diagonal, while the off-diagonal matrix elements ${}_nQ_{12}$ and ${}_nQ_{21}$ are Grassmannian. In the convolution

$$\left(\overline{\chi}_n \; iz_n^*\right)\hat{\mathbf{Q}}_n \begin{pmatrix} \chi_n \\ z_n \end{pmatrix},$$

the matrix element ${}_nQ_{11}$ multiplies Grassmannian variables χ_n and $\overline{\chi}_n$, and therefore all such elements form the so-called fermion–fermion block of the entire supermatrix

$$\hat{\mathbf{Q}} = \begin{pmatrix} \cdots & \cdots & \cdots & \cdots \\ \cdots & \hat{\mathbf{Q}}_n & 0 & \cdots \\ \cdots & 0 & \hat{\mathbf{Q}}_{n-1} & \cdots \\ \cdots & \cdots & \cdots & \cdots \end{pmatrix}. \tag{9.85}$$

By the analogy, the matrix elements ${}_nQ_{22}$, ${}_nQ_{12}$, and ${}_nQ_{21}$ belong to the boson–boson, fermion–boson, and boson–fermion blocks, respectively. The terminology simply reflect the fact that Grassmannian variables anticommute, like the fermionic field variables. The notation $\mathrm{STr}\hat{\mathbf{X}}$ means that the diagonal elements corresponding to the fermion–fermion block of the supermatrix $\hat{\mathbf{X}}$ have to be subtracted from ones corresponding to the boson–boson block. Explicitly, the second exponent in (9.84) reads

$$\exp\left\{\frac{-\,{}_nQ_{11}^2 - 2\,{}_nQ_{12}\,{}_nQ_{21} - \,{}_nQ_{22}^2}{2}\right.$$

$$\left.+\frac{\sqrt{p}}{\varepsilon}\left(\overline{\chi}_n\,{}_nQ_{11}\chi_n + \overline{\chi}_n\,{}_nQ_{12}z_n + iz_n^*\,{}_nQ_{21}\chi_n + iz_n^*\,{}_nQ_{22}z_n\right)\right\},$$

and performing now a Gaussian integration over

$$d\hat{\mathbf{Q}}_n = d\,{}_nQ_{11}d\,{}_nQ_{12}d\,{}_nQ_{21}d\,{}_nQ_{22}/\pi$$

according to (8.104) for the Grassmannian part, one obtains

$$\exp\left\{\frac{p}{2\varepsilon^2}\left(\overline{\chi}_n\chi_n + iz_n^*\,z_n\right)^2\right\},$$

which is the second factor in the first line of (9.72).

With the help of the integral over the supermatrix $\hat{\mathbf{Q}}$ of (9.85) one can express the state density as

$$g(\varepsilon) = \frac{1}{\pi}\mathrm{Im}\int d\mathbf{z}^* d\mathbf{z} d\boldsymbol{\chi} d\overline{\boldsymbol{\chi}} d\hat{\mathbf{Q}}\,\left(\overline{\boldsymbol{\chi}}\hat{\mathbf{V}}^{-2}\boldsymbol{\chi}\right)e^{-\mathrm{STr}\hat{\mathbf{Q}}^2}$$

$$\exp\left[-\left(i\mathbf{z}^*\ \overline{\boldsymbol{\chi}}\right)\left(\hat{\mathbf{V}}^{-1} - \frac{p}{\varepsilon} - \frac{\sqrt{p}}{\varepsilon}\hat{\mathbf{Q}}\right)\begin{pmatrix}\mathbf{z}\\ \boldsymbol{\chi}\end{pmatrix}\right]$$

$$= \frac{\partial}{2\pi\partial\lambda}\bigg|_{\lambda\to 0}\mathrm{Im}\int d\mathbf{z}^* d\mathbf{z} d\boldsymbol{\chi} d\overline{\boldsymbol{\chi}} d\hat{\mathbf{Q}}\,e^{-\mathrm{STr}\hat{\mathbf{Q}}^2}$$

$$e^{-\left(i\mathbf{z}^*\ \overline{\boldsymbol{\chi}}\right)\left(\hat{\mathbf{V}}^{-1} - \frac{p}{\varepsilon} - \frac{\sqrt{p}}{\varepsilon}\hat{\mathbf{Q}} + \lambda\hat{\boldsymbol{\sigma}}_{bf}\hat{\mathbf{V}}^{-2}\right)\begin{pmatrix}\mathbf{z}\\ \boldsymbol{\chi}\end{pmatrix}},$$

where $d\mathbf{z}^* d\mathbf{z} d\boldsymbol{\chi} d\overline{\boldsymbol{\chi}} \equiv \prod_n\left\{\frac{dz_n^* dz_n}{\pi}d\chi_n d\overline{\chi}_n\right\}$ and $\hat{\boldsymbol{\sigma}}_{bf}$ is a diagonal matrix with entries 1 for the fermion–fermion block, and -1 for the boson–boson block. Supertrace now implies summation not only over the seeded sites but over all sites. Performing integration over $d\mathbf{z}^* d\mathbf{z} d\boldsymbol{\chi} d\overline{\boldsymbol{\chi}}$ one obtains

$$g(\varepsilon) = \frac{\partial}{2\pi\partial\lambda}\bigg|_{\lambda\to 0}\mathrm{Im}\int d\hat{\mathbf{Q}}\,e^{-\mathrm{STr}\hat{\mathbf{Q}}^2 - \mathrm{STr}\ln\left(\hat{\mathbf{V}}^{-1} + \lambda\hat{\boldsymbol{\sigma}}_{bf}\hat{\mathbf{V}}^{-2} - \frac{p}{\varepsilon} - \frac{\sqrt{p}}{\varepsilon}\hat{\mathbf{Q}}\right)}. \quad (9.86)$$

Formally, this integral is as difficult to calculate as the initial integral of (9.73), since it contains exactly the same large number of variables—two regular and two Grassmannian per each site of the lattice. However, for the regime where a large number of the neighboring Euclidean states get strongly mixed up by the interaction, one can implement the approximate evaluation of this integral by the saddle-point method. In the opposite regime, Euclidean states located close each to the other may not experience strong mixing, since some other states, yet closer located, may move them out of the resonance thus preventing from the coupling. It is a difficult task to say a priori which of the regimes takes place, and therefore we first assume that the saddle-point method is applicable and verify this assumption a posteriori.

We note that the saddle-point condition suggested by the variational principle

$$\frac{\delta}{\delta\hat{\mathbf{Q}}_n}\left[-\mathrm{STr}\hat{\mathbf{Q}}^2 - \mathrm{STr}\ln\left(\hat{\mathbf{V}}^{-1} - \frac{p}{\varepsilon} - \frac{\sqrt{p}}{\varepsilon}\hat{\mathbf{Q}}\right)\right] = 0 \tag{9.87}$$

yields

$$\hat{\mathbf{Q}}_n = \frac{\sqrt{p}}{2\varepsilon}\hat{P}_n\frac{-1}{\frac{p}{\varepsilon} + \frac{\sqrt{p}}{\varepsilon}\hat{\mathbf{Q}} - \hat{\mathbf{V}}^{-1}}, \tag{9.88}$$

and implies that all blocks of the matrix $\hat{\mathbf{Q}}$ of (9.85) are identical, $\hat{\mathbf{Q}}_n = \hat{\mathbf{Q}}_S$, when $\hat{\mathbf{Q}}$ satisfies (9.87). Here $\hat{P}_n$ denotes projection to the site n. Moreover, (9.88) implies that $\hat{\mathbf{Q}}$ has the same matrix symmetry as $\hat{\mathbf{V}}^{-1}$, that is, it has vanishing boson–fermion and fermion–boson matrix elements, while the boson–boson matrix elements differ only in sign from the fermion–fermion one.

Note that (9.88) is the supermatrix version of the renormalization condition, which we have for the first time encountered when deriving (4.22) for the two band quantum systems, although there it was written for the Hamiltonians and energies but not for their inverse values. Moreover, by the replacements $\hat{\mathbf{Q}}_n \to \frac{-\sqrt{p}}{2\varepsilon}\left\langle\hat{\mathbf{G}}_n\right\rangle$, $p/2\varepsilon^2 \to \left\langle V^2\right\rangle$, and $\frac{p}{\varepsilon} \to E$ one obtains the general renormalization equation (9.5) describing transformation of the spectrum perturbed by a random matrix. We therefore can implement the same technique as on p.429 for calculation of the renormalization valid for the case where the perturbation series diagrams have topology of trees. As there, the validity of this approach is implied by a large number of levels that fit to a spectral band of a width given by the coupling size. In the context of the Euclidean random matrices this means a large number of coupled Euclidean states within a typical interaction radius, which may justify the saddle-point method for the evaluation of the integral.

Contributions of the second-order perturbation theory around the saddle point coming from the Grassmannian variables cancel that coming from the regular ones. Therefore from (9.68) one finds the state density

$$g(\varepsilon) = \frac{-1}{\pi}\mathrm{Im}\mathrm{Tr}\frac{\hat{\mathbf{V}}^{-2}}{\frac{p}{\varepsilon} + \frac{\sqrt{p}}{\varepsilon}q - \hat{\mathbf{V}}^{-1}}, \tag{9.89}$$

where the trace can be taken only over the boson–boson block. Here q is the boson–boson matrix element of $\hat{\mathbf{Q}}_S$, to be found for each ε by solving the renormalization equation (9.88) now having the form

$$q = \frac{-\sqrt{p}p}{2\varepsilon M^d}\mathrm{Tr}\frac{1}{\frac{p}{\varepsilon} + \frac{\sqrt{p}}{\varepsilon}q - \hat{\mathbf{V}}^{-1}}. \tag{9.90}$$

Equations (9.89) and (9.90) with the allowance for the requirement $\mathrm{Im}\mathrm{Tr}\hat{V}^{-1} = 0$ result in a system of equations,

$$\varepsilon G(\varepsilon) = (2Q(1+Q)+1)(1+Q) \tag{9.91}$$

$$G(\varepsilon) = \frac{1}{M^d}\mathrm{Tr}\frac{1}{\frac{\varepsilon}{1+Q} - \hat{\mathbf{V}}}, \tag{9.92}$$

where $Q = \frac{q}{\sqrt{p}}$ and the energy is scaled as $\varepsilon \to p\varepsilon$. This system of equations is an analog of (9.3), and it has to be resolved with respect to Q, while the result has to be substituted to the expression for the state density $g(\varepsilon) = -\frac{N}{\pi}\mathrm{Im}G(\varepsilon)$.

In d-dimensional case, for $V(r)\sim r^{-\alpha}$ implied by the interaction $V(k) = k^{\alpha-d}$ written in scaled units of k, (9.92) takes the form

$$G = \frac{d}{k_c^d}\int\limits_{k_0}^{k_c}\frac{k^{d-1}dk}{\frac{\varepsilon}{1+Q} - (k^{\alpha-d} - D)}, \tag{9.93}$$

where the scaled cut-off momentum k_c relates to the number M^d of the lattice sites via (9.80) and the displacement $D = k_c^{\alpha-d}d/\alpha$ ensuring zero trace of the interaction operator $\hat{\mathbf{V}}$ is given by (9.79). After the integration, in the limit $k_o/k_c \to 0$ this yields the relations

$$G(\varepsilon) = \begin{cases} \frac{1}{v-1}\,{}_2F_1\left(1, \frac{v-1}{v}; \frac{2v-1}{v}; \frac{\varepsilon(v+1)+Q+1}{(Q+1)(v+1)}\right); & v < 0 \\ \frac{(Q+1)(v+1)}{Q+1+(v+1)\varepsilon}\,{}_2F_1\left(1, \frac{1}{v}; \frac{v+1}{v}; \frac{(Q+1)(v+1)}{Q+1+(v+1)\varepsilon}\right); & v > 0 \end{cases} \tag{9.94}$$

in complete analogy with the results (9.82, 9.83).

Solutions of (9.91) for Q suggested by the relations (9.94) has to be substituted to the expression $g(\varepsilon) = -\frac{N}{\pi}\mathrm{Im}G(\varepsilon)$ for the density of states. The results are shown in Fig. 9.12 as a function of the parameter v. The main technical difficulty of such calculations is due to the analytical properties of the hypergeometric functions that usually have branching points. Therefore the transcendental equation (9.91) may and usually does have a number of solutions, such that a choice of the one to be substituted to (9.91) has to be done.

Let us illustrate the way to tackle the problem at the example of the short-range interactions. In (9.94) for the case $v > 0$, we take the argument $\frac{(Q+1)(v+1)}{Q+1+(v+1)\varepsilon}$ of the hypergeometric function for a new complex variable e^f, eliminate the variable

$$Q = \frac{\varepsilon(v+1)e^f}{v - e^f + 1} - 1,$$

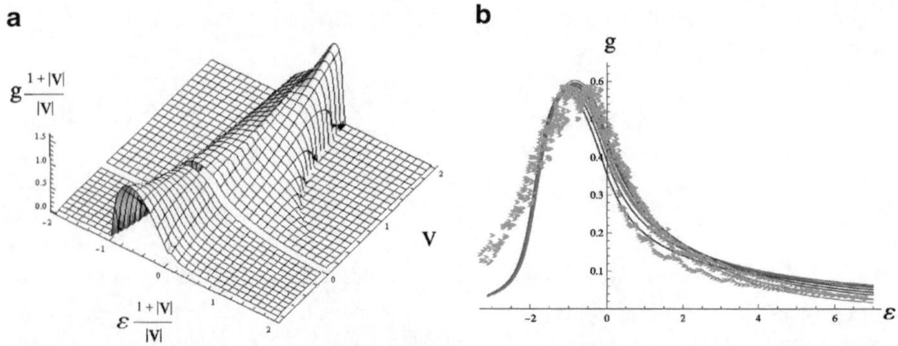

Fig. 9.12 Density of states of Euclidean random matrices suggested by the σ-model, plotted as a function of the parameter $v = \alpha/d - 1$ (**a**). Density of states for the short-range interaction (**b**) for different v scaled such that the curves have the same maximum. Points (>) show the same profiles obtained from the results of Fig. 9.10. One sees the deviations among the numerical results and the profiles suggested by the model at the *red wing* of the *line*. The energy axis ε is shifted by $\gamma = 0.1$ to the complex plane

and arrive at the renormalization equation, quadratic with respect to ε, which yields

$$\varepsilon = \left(\frac{1}{2e^f} - \frac{1}{2\,(v+1)}\right) \left(1 \pm \sqrt{2\left(\frac{e^f}{v+1} - 1\right)\,{}_2F_1\left(1, \frac{1}{v}; \frac{v+1}{v}; e^f\right) - 1}\right), \tag{9.95}$$

The state density

$$g = \frac{N}{\pi}\mathrm{Im}e^f\,{}_2F_1\left(1, \frac{1}{v}; \frac{v+1}{v}; e^f\right),$$

together with the real and the imaginary parts of ε, can be considered as coordinates $\{\mathrm{Re}\varepsilon(f), \mathrm{Im}\varepsilon(f), g(f)\}$ of a hyper surface in three-dimensional space, parametrized by the real and the imaginary parts of f. Intersection of this manifold and the plane $\mathrm{Im}\varepsilon = 0$ gives the required dependence of the state density on the energy. This is illustrated in Fig. 9.13.

Comparison of the state density of Fig. 9.12 obtained with the help of the σ-model in combination with the saddle-point method of evaluation of the functional integral (9.86) and the state density of Fig. 9.10 obtained by numerical simulation shows that though the model reflects some tendencies of the state density profiles and their transformation with the change of the parameter v, still it does not reproduce all the details of the profiles.

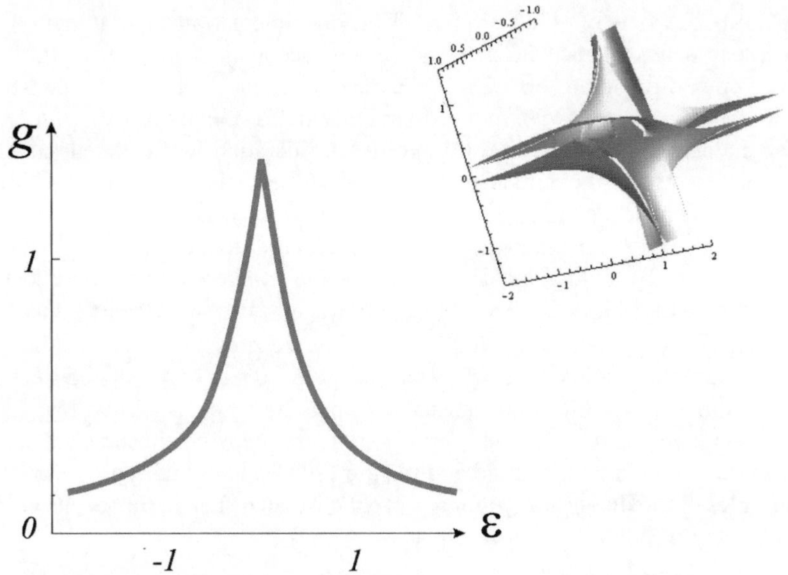

Fig. 9.13 Density of states $g(\varepsilon, v)$ profile for $v = 0.47$. *Inlet* shows the surface $\{\mathrm{Re}\varepsilon(f), \mathrm{Im}\varepsilon(f), g(f)\}$ parametrized by the real and the imaginary parts of f, cut of which by a plane $\mathrm{Im}\varepsilon = 0$ gives the depicted density of state profile

Model of Interacting Clusters

We thus have to admit that the renormalization approach relying on a single parameter Q is insufficient for the detailed description of the Euclidean Random Matrices spectra. In the context of the discussion presented after (9.86) it means that the second regime is more likely than the first one, or in other words, the model based on the idea of renormalization associated with a tree-like topology of the perturbation series diagrams oversimplifies some important properties of such matrices. One of the reasons for this is the contribution of close Euclidean points, which affect each other so strongly that the Taylor expansion (9.71) over the auxiliary fields z_n is no longer valid for the description.

One therefore needs to develop an even more sophisticated model to gain a more adequate description of the spectra of Euclidean random matrices. One of the ways is to single out contributions of the groups of close Euclidean points interacting strongly, such that each group containing a small number α of the points is treated as a new composite unit—a cluster characterized by a Green's operator

$$\hat{G}_\alpha = \frac{1}{\varepsilon - \hat{V}_\alpha}$$

in the Hilbert space of dimension α. Thus the interaction Hamiltonian $\hat{V}_\alpha$ of α Euclidean states in the cluster is taken into account to all-order perturbation theory, while different clusters weakly interact among themselves and with isolated Euclidean points via the weak induced auxiliary fields z_n admitting Taylor expansion of the action. We thus have to take in consideration less isolated Euclidean points, but also have to include in consideration new multilevel point objects—clusters randomly seeded in space. We can call these objects Euclidean clusters, by the analogy. As a consequence, the interaction among Euclidean points and clusters is now described by a block matrix $\hat{V}^{-1}$ where the number N of blocks equals the total number of the objects and each block is a square matrix with the total number of components $\frac{A(A+1)}{2}$ depending on the maximum size A of the cluster that has to be taken into account. For instance, when we take into account only isolated Euclidean states and dimers, each block becomes a 3×3 matrix, which at the first row and the first column contains the variables corresponding to the isolated Euclidean state at the position $\overrightarrow{r}_n$, while the second and the third rows and columns are reserved for the variables of the dimer that might replace the isolated state at this position $\overrightarrow{r}_n$.

The analog of (9.68) for this case reads

$$g(\varepsilon) = \frac{1}{\pi}\mathrm{Im}\mathrm{Tr}\left[\hat{V}^{-2}\left\langle\frac{1}{\hat{V}^{-1} - \sum_\alpha \hat{P}_\alpha \hat{G}_\alpha(\varepsilon)\hat{P}_\alpha}\right\rangle_{\hat{P}_\alpha}\right], \tag{9.96}$$

where the average is taken over projections to all possible arrangements of the Euclidean points and clusters, while (9.70) adopts the form

$$g(\varepsilon) = \frac{1}{\pi}\mathrm{Im}\prod_n\left\{\int_{-\infty}^{\infty}\frac{d\mathbf{z}_n^* d\mathbf{z}_n}{\pi}\int d\boldsymbol{\chi}_n d\overline{\boldsymbol{\chi}}_n\right\} \tag{9.97}$$

$$\sum_{n,m}\overline{\boldsymbol{\chi}}_n\hat{\mathbf{V}}_{nm}^{-2}\chi_m e^{-\sum\limits_{n,m}\left(i\mathbf{z}_n\hat{\mathbf{V}}_{nm}^{-1}\mathbf{z}_m^* + \overline{\boldsymbol{\chi}}_{\alpha n,n}\hat{\mathbf{V}}_{nm}^{-1}\boldsymbol{\chi}_{m,\alpha m}\right)}$$

$$e^{\sum_\alpha p_\alpha \sum_n\left(\exp\left(i\mathbf{z}_n\hat{G}_\alpha(\varepsilon)\mathbf{z}_n^* + \overline{\boldsymbol{\chi}}_n\hat{G}_\alpha(\varepsilon)\boldsymbol{\chi}_n\right) - 1\right)}.$$

The auxiliary fields $\mathbf{z}_n$ and $\boldsymbol{\chi}_n$ now become multicomponent vectors.

We note that in the exponent of the last factor in (9.97) stands the average

$$\overline{\exp\left(i\mathbf{z}_n\hat{G}_\alpha\mathbf{z}_n^* + \overline{\boldsymbol{\chi}}_n\hat{G}_\alpha\boldsymbol{\chi}_n\right) - 1}$$

over eventual structure of the object occupying the point in space. This factor being represented as an integral over auxiliary matrices $\hat{\mathbf{Q}}$ and $\hat{\mathbf{U}}$

$$\int d\hat{\mathbf{Q}}\, d\hat{\mathbf{U}}\, e^{-\mathrm{STr}\overline{\exp\left(\hat{G}_\alpha\hat{\mathbf{U}}\right)} - 1 - i\mathrm{STr}\hat{\mathbf{Q}}\hat{\mathbf{U}} + \left(i\mathbf{z}_n^* - \overline{\boldsymbol{\chi}}_n\right)\hat{\mathbf{Q}}\begin{pmatrix}\mathbf{z}_n\\ \boldsymbol{\chi}_n\end{pmatrix}},$$

after casting the average $\overline{\exp\left(\hat{G}_\alpha \hat{\mathbf{U}}\right) - 1}$ in Taylor series allows one to explicitly write an analog of the functional integral (9.86)

$$g(\varepsilon) = \frac{\partial}{2\pi\partial\lambda}\bigg|_{\lambda\to 0} \operatorname{Im} \int \mathcal{D}\hat{\mathbf{U}}\mathcal{D}\hat{\mathbf{Q}}\, e^S \tag{9.98}$$

$$e^S = e^{-\frac{1}{2}\mathrm{STr}\overline{\hat{\mathbf{U}}\hat{\mathbf{G}}\hat{\mathbf{U}}\hat{\mathbf{G}}} - i\mathrm{STr}\hat{\mathbf{Q}}\hat{\mathbf{U}} - \mathrm{STr}\ln\left(\hat{\mathbf{V}}^{-1} + \lambda\hat{\boldsymbol{\sigma}}_{bf}\hat{\mathbf{V}}^{-2} - \overline{\hat{\mathbf{G}}} - \hat{\mathbf{Q}}\right)}.$$

for the model under consideration.

As earlier, the only practical way to perform functional integration is to invoke the saddle-point method, which yields the average state density per object in the form

$$g(\varepsilon) = \frac{1}{2\pi N}\mathrm{STr}\frac{\hat{\sigma}_{bf}\hat{\mathbf{V}}^{-2}}{\hat{\mathbf{V}}^{-1} - \overline{\hat{\mathbf{G}}} - \hat{\mathbf{Q}}}, \tag{9.99}$$

As long as only the spectral density is addressed, the structures of the boson–boson, fermion–fermion, and boson–fermion blocks of the supermatrix $\hat{\mathbf{Q}}$ are identical, and the calculation can be performed for just the boson–boson block. Therefore in (9.99) one can replace $\frac{1}{2}\mathrm{STr}\ldots$ by $\mathrm{Tr}\ldots$, and with the allowance for the fact that $\mathrm{Tr}\hat{\mathbf{V}}^{-1} = 0$, obtain an equivalent expression,

$$g(\varepsilon) = \frac{-1}{\pi N}\mathrm{Im}\mathrm{Tr}\frac{1}{\left(\overline{\hat{\mathbf{G}}} + \hat{\mathbf{Q}}\right)^{-1} - \hat{\mathbf{V}}},$$

which is also valid for a singular matrix $\hat{\mathbf{V}}$.

The matrix $\hat{\mathbf{Q}}$ entering the equation for the average state density is site independent and it has to be found from an analog of the saddle-point condition (9.90), which for (9.98) reads

$$\frac{\delta S}{\delta\hat{\mathbf{Q}}} = 0; \ \frac{\delta S}{\delta\hat{\mathbf{U}}} = 0$$

and which results in the set of equations

$$\hat{\mathbf{Z}} = \frac{1}{N}\mathrm{Tr}_d\frac{1}{\overline{\hat{\mathbf{G}}} + \hat{\mathbf{Q}} - \hat{\mathbf{V}}^{-1}},$$

$$\hat{\mathbf{Q}} = \overline{e^{\hat{\mathbf{G}}\hat{\mathbf{Z}}}\hat{\mathbf{G}}} - \overline{\hat{\mathbf{G}}} \simeq \overline{\hat{\mathbf{G}}\hat{\mathbf{Z}}\hat{\mathbf{G}}} \tag{9.100}$$

for $\hat{\mathbf{Z}} = i\hat{\mathbf{U}}$ and $\hat{\mathbf{Q}}$. Here the trace operation Tr_d is performed over the Euclidean geometric space exclusively, in contrast to the trace operation for the density, which also involves the block indices. Note that the first, strict, equality in the second equation is exact and does not imply the Taylor expansion, in contrast to the second, approximative, equality resulting from such an expansion.

Hereafter, we concentrate on the case of isolated Euclidean points and dimers, where the calculation can be performed for just boson–boson block, and where $K = 3$. In the basis diagonalizing the average dimer Green's operator $\overline{\hat{\mathbf{G}}}$, one finds

$$\overline{\hat{\mathbf{G}}} = \frac{1}{\varepsilon}\begin{pmatrix} p_1 & 0 & 0 \\ 0 & \overline{g_+} & 0 \\ 0 & 0 & \overline{g_-} \end{pmatrix}$$

$$\overline{\hat{\mathbf{G}}\hat{\mathbf{Z}}\hat{\mathbf{G}}} = \frac{1}{\varepsilon^2}\begin{pmatrix} p_1 Z_{11} & 0 & 0 \\ 0 & \overline{g_+^2} Z_{22} & \overline{g_- g_+} Z_{23} \\ 0 & \overline{g_- g_+} Z_{32} & \overline{g_-^2} Z_{33} \end{pmatrix}$$

where $\overline{G_+} = \overline{g_+}(\varepsilon)/\varepsilon$ and $\overline{G_-} = \overline{g_-}(\varepsilon)/\varepsilon$ stand for the average of contribution of the symmetric and antisymmetric states of dimers, respectively, while $\overline{G_+^2} \equiv \overline{g_+^2}(\varepsilon)/\varepsilon^2, \overline{G_-^2} \equiv \overline{g_-^2}(\varepsilon)/\varepsilon^2, \overline{G_+ G_-} \equiv \overline{g_- g_+}(\varepsilon)/\varepsilon^2$ denote the averages of the corresponding products of the matrix elements of Green's operators expressed in terms of the relevant form-factors. For small dimers, only the symmetric states are coupled with a considerable interaction strength to the smooth fields, while the asymmetries states remain "dark". Therefore the singular interaction matrix in the notation of (9.93) reads

$$\hat{\mathbf{V}} = \left(k^{\alpha-d} - D\right)\begin{pmatrix} 1 & 1 & o_{13}(\varepsilon) \\ 1 & 1 & o_{23}(\varepsilon) \\ o_{13}^*(\varepsilon) & o_{23}^*(\varepsilon) & o_{33}(\varepsilon^2) \end{pmatrix},$$

where $o_{ij}(\varepsilon)$ are infinitesimally small functions, introduced for technical reasons in order to avoid problems with singularity of the interaction matrix, which has to be set to zero at the end of the calculation.

Explicit calculation based on (9.100) modifies the renormalization equation (9.91), which now takes the form

$$\frac{\varepsilon G(\varepsilon)}{1+Q} = 1 + 2(1+Q)\frac{Q - \overline{g_+}(\varepsilon)}{1 + \overline{g_+^2}(\varepsilon)}, \tag{9.101}$$

where $G(\varepsilon)$ is still given by (9.94), while the averages and correlation of the dimer Green's functions are found by the direct calculations of the corresponding form-factors

$$\overline{g_+} = \int \frac{\varepsilon e^{-v(r)/h} d^d r}{\varepsilon - V(r)},$$

$$\overline{g_-} = \int \frac{\varepsilon e^{-v(r)/h} d^d r}{\varepsilon + V(r)},$$

$$\overline{g_+^2} = \int \frac{\varepsilon^2 e^{-v(r)/h} d^d r}{[\varepsilon - V(r)]^2}$$

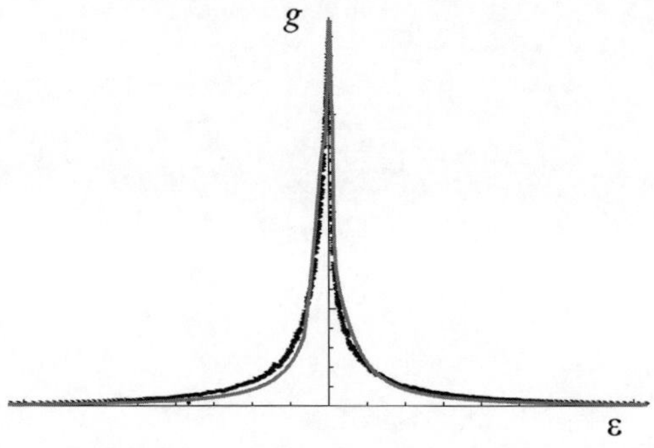

Fig. 9.14 Density of states profile $g(\varepsilon)$ for $v = 0.35$ calculated with the help of the model, which explicitly takes into account contributions of dimers. Parameter h is chosen to be 0.4. *Dots* show the profile obtained by the numerical modeling for $v = 0.35$

of dimers. Here h denotes typical volume per dimer, which is a parameter of the model, while $v(r)$ stands for the volume of sphere of a radius r.

The properly chosen root of (9.101) has to be substituted to the expression for the state density, which now takes the form

$$g(\varepsilon) = \frac{-1}{\pi}\mathrm{Im}\left[\frac{\overline{g_-}}{\varepsilon} + \frac{2AB}{A+B} + \frac{A^2+B^2}{(A+B)^2}G(\varepsilon)\right],$$

where

$$A = \overline{g_+} - \frac{Q-\overline{g_+}}{1+\overline{g_+^2}}\overline{g_+^2},$$

$$B = \frac{2}{\varepsilon}\left(\frac{Q-\overline{g_+}}{1+\overline{g_+^2}} - 1\right).$$

In Fig. 9.14 we show the result of calculation based on this model, which is in an acceptable agreement with the results of direct numerical modeling.

Chapter 10
Entangled States of Composite Quantum Systems

Historically, physical science has always relied on a cognition method known from the antic times as the analytic approach, which is aimed to a deeper understanding of fundamental laws underlying experimentally observed phenomena and to an adequate formulation of these laws in the simplest possible mathematical way. The discovery of quantum mechanics and the development of quantum field theory resulted in a formulation of the natural laws in the form of differential equations, which is one of the best examples of such a method. Addressing complex objects within the framework of statistical physics, scientists have also proceeded in a similar way by identifying and formulating the most fundamental common properties persisting in the absolute majority of qualitatively similar systems, regardless of discrepancy in unimportant details of their numerous and diverse particular realizations. A large number of such realizations of physical systems that have common properties of interest can be associated with the high entropy of the configuration.

There exists a complementary cognition method known as synthetic, which is developing around another fundamental question that can roughly be formulates as follows: how does a combination of parts form an integer whole? In other words, are there a qualitatively new aspects that appear when one puts together two or more components? The rapid development of information sciences during the last half of a century can be considered as the most impressive success of the synthetic method. Being a result of the synthesis of numerous parts, each of which carries the minimum (bit) of information, a computer performs an extremely informative process of calculations by establishing interaction among these parts. Evidently, statistics is not the method to be applied to a description of the computing, and indeed, during the calculation, a particular realization of the computer state contains unique information, thus being a state of zero entropy. One should also note that the main results of informatics, in contrast to that of physics, are not compact mathematical formulas or equations, but volumes of computational codes, with the code complexity rapidly increasing in the course of

V.M. Akulin, *Dynamics of Complex Quantum Systems*, Theoretical and Mathematical Physics, DOI 10.1007/978-94-007-7205-2_10,

progress in this field. From the viewpoint of physics these codes are very elaborated and detailed prescriptions for controlled interventions to complex multipartite systems.

Combination of the analytic approach of quantum mechanics with the synthetic method of informatics gave rise of a new branch of science "Quantum Informatics", which was intensively developing during two last decades. The first basic question that a physicist entering the field of quantum informatics has to address is: what is the new quality which appears as a result of interaction of two quantum systems? The name of this qualitatively new concept is quantum entanglement. It will be the main subject of consideration in this chapter. Entanglement plays a vital role in quantum information science, representing both the defining resource for quantum communication—where it enables, in particular, non-classical protocols such as quantum teleportation and it leads to enhanced security in cryptographic tasks—and a key ingredient for determining the efficiency of quantum algorithms and quantum computation schemes. In addition, studies of entanglement have also proved to be relevant to fields as different as atomic physics, quantum chaos, quantum phase transitions, and quantum networks.

We start the first section with a short historical remark and continue by a few simple examples: entanglement of two spins of $1/2$, and entanglement of two harmonic oscillators subject to squeezing and free evolution. We introduce traditional measures of bipartite entanglement at these simple examples. In the following section we consider the role of entanglement in quantum computation at the example of two main quantum algorithms. At the example of teleportation, we also present some aspects of quantum communications related to entanglement. Next, the third, section describes an approach to exhaustive classification of multipartite entanglement based on a group theory, where we show how to introduce a unique and exhaustive measure of multipartite entanglement with the help of an unusual, but mathematically adequate tool—nilpotent polynomials.

10.1 Product States and Entangled States

In contrast to the majority of the characteristics of physical systems, quantum entanglement is not something that exists as an actor in reality and thus governs the dynamics of natural processes. It is rather a concept convenient for the description and understanding of the relations among an entire system and its parts than a physical quantity. In general, there is no universal and precise commonly accepted definition of a proper measure of entanglement, except for bipartite entanglement. In this section we present the simplest examples of the bipartite entanglement between two spins of $1/2$ and between two linearly coupled harmonic oscillators subject to parametric excitation. The entanglement concept will be further generalized to arbitrary bipartite systems, which are systems comprising just two parts.

10.1.1 Examples of Entangled States

The definition of entangled states is very simple—the state of a system is entangled when the state vector cannot be represented as a direct product of the state vectors of its parts. When the latter is the case, the measurements done with one of the parts are completely independent of the outcomes of the measurement on the other part. For the entangled states, the outcomes of measurements of different parts are correlated. Note that most of the states of a combined system are entangled. We illustrate the structure of the entangled states at a few simple examples.

Two Spins

The simplest example is a pair of spins of $1/2$ given by their spinors

$$\chi_1 = \begin{pmatrix} \varphi \\ \psi \end{pmatrix} \quad \text{and} \quad \chi_2 = \begin{pmatrix} \lambda \\ \mu \end{pmatrix}.$$

Each spin is a two-level system, and a pair of spins is a four-level system. The product state

$$\chi_1 \otimes \chi_2 = \begin{pmatrix} \lambda \\ \mu \end{pmatrix} \otimes \begin{pmatrix} \varphi \\ \psi \end{pmatrix} = \begin{pmatrix} \lambda\varphi \\ \lambda\psi \\ \mu\varphi \\ \mu\psi \end{pmatrix}, \tag{10.1}$$

is not entangled, and a state of the compound four-level system, which cannot be represented in the form (10.1), that is

$$\begin{pmatrix} \Psi_1 \\ \Psi_2 \\ \Psi_3 \\ \Psi_4 \end{pmatrix} \neq \begin{pmatrix} \lambda\varphi \\ \lambda\psi \\ \mu\varphi \\ \mu\psi \end{pmatrix}, \tag{10.2}$$

is entangled. The maximum entangled state

$$\begin{pmatrix} 1 \\ 0 \\ 0 \\ 1 \end{pmatrix} = \frac{1}{\sqrt{2}} \begin{pmatrix} 1 \\ 0 \end{pmatrix} \otimes \begin{pmatrix} 1 \\ 0 \end{pmatrix} + \frac{1}{\sqrt{2}} \begin{pmatrix} 0 \\ 1 \end{pmatrix} \otimes \begin{pmatrix} 0 \\ 1 \end{pmatrix}, \tag{10.3}$$

is known as the Bell state. Later on we are going to give a more precise definition of the expression "maximum entangled", and here we only note that for the Bell state the states of two spins are completely correlated—if one is up, the other is up too.

Dealing with quantum systems it is useful to remember that transformations of quantum states form groups. In particular, the states of an isolated two-level system given by two complex-valued amplitudes satisfying the normalization condition and thus depending on three real parameters is subject to unitary $SU(2)$ transformations in the course of Hamiltonian dynamics. This group has three generators and it is isomorph to the $O(3)$ group of three-dimensional rotations. This very fact was employed for the representation of the quantum states of a spin $1/2$ with the help of the Bloch vector, earlier introduced on Sect. 3.2.3. Unitary transformations of a four-level system belong to a much richer group, $SU(4)$, which has 15 generators, and hence it is described by 15 real parameters. States of such a system are given by four complex-valued amplitudes satisfying the normalization condition, and therefore they depend on seven real parameters. It looks problematic to depict the relevant geometric object as a two-dimensional image. Still the most important features of entanglement can be illustrated graphically when we restrict ourselves to the space of real state vectors, which corresponds to the case where all y-projections of spins are zero.

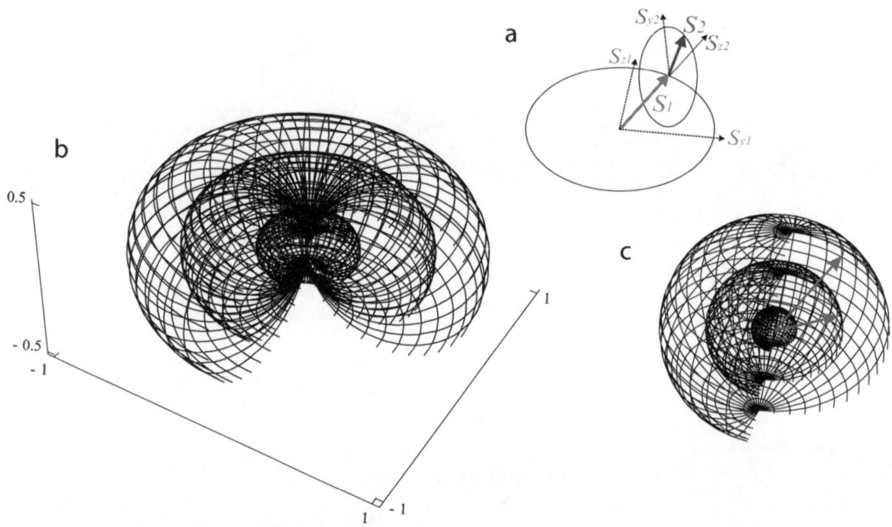

Fig. 10.1 We represent entangled states of two spins depicting the vector sum (**a**) of two spins constructed such that the y direction of the second spin is aligned with the direction of the first spin. The parametric plots (**b**) show the torus surfaces that result from the application of the local operations addressing each spin individually to the total sum of spins, constructed as is shown in (**a**). The representation of the local state with the help of the Bloch sphere (**c**) shows that the length of the Bloch vector decreases with increasing entanglement

Let us parametrize these states as follows:

$$\begin{pmatrix}\Psi_1\\ \Psi_2\\ \Psi_3\\ \Psi_4\end{pmatrix} = \cos\alpha\begin{pmatrix}\cos\beta\\ \sin\beta\end{pmatrix}\otimes\begin{pmatrix}\cos\gamma\\ \sin\gamma\end{pmatrix} + \sin\alpha\begin{pmatrix}-\sin\beta\\ \cos\beta\end{pmatrix}\otimes\begin{pmatrix}-\sin\gamma\\ \cos\gamma\end{pmatrix}$$

$$= \begin{pmatrix}\cos\alpha\cos\beta\cos\gamma + \sin\alpha\sin\beta\sin\gamma\\ \cos\alpha\cos\beta\sin\gamma - \sin\alpha\sin\beta\cos\gamma\\ \cos\alpha\sin\beta\cos\gamma - \sin\alpha\cos\beta\sin\gamma\\ \cos\alpha\sin\beta\sin\gamma + \sin\alpha\cos\beta\cos\gamma\end{pmatrix}, \tag{10.4}$$

and calculate the expectation values of the spin projections

$$S_1^x = \frac{1}{2}\cos(2\alpha)\sin(2\beta)\,;\ S_1^z = \frac{1}{2}\cos(2\alpha)\cos(2\beta)\,;$$
$$S_2^x = \frac{1}{2}\cos(2\alpha)\sin(2\gamma)\,;\ S_2^z = \frac{1}{2}\cos(2\alpha)\cos(2\gamma)\,. \tag{10.5}$$

In Fig. 10.1 we depict the sum of the spin vectors S_1 and S_2 where the z projection of the second spin is aligned with the direction of the first spin.

When one performs the so-called local transformations, which are in our case the unitary transformations $\exp\{i\hat{\sigma}_1^x\beta\}$ or $\exp\{i\hat{\sigma}_2^x\gamma\}$ individually applied to the first or to the second spin, respectively, the end of such a sum vector goes around a circle corresponding to variation either of β or of γ, while the direct product of these circles forms a torus. The size of the torus depends on the parameter α, and for $\alpha = \pi/4$ corresponding to the maximum entangled state the torus degenerates to a point. One can say that the whole family of states corresponds to the points inside the torus of the maximum possible radius $1/2$, while the product states correspond to the surface of this torus. The maximum entangled state corresponds to zero average components of the spins. Note that the parameter α can be defined with the help of the expectation value of the product of the y components of the spin operators, $S_i^y = \frac{1}{2}\sigma_i^y$, via the relation

$$\sin 2\alpha = -4\left\langle S_1^y S_2^y\right\rangle. \tag{10.6}$$

When one considers the general case of the local transformation that involves phases of the amplitudes Ψ_n and moves the Bloch vectors of individual spins along a surface of a three-dimensional spheres, the situation remains qualitatively the same, although an attempt to geometrically draw a torus resulting from the direct product of two such $3D$ spheres on a $2D$ plane looks like a challenge for human imagination. Still one can state that there is a single parameter α defined by (10.6) that marks these tori, while the local transformations are unable to move the state from one torus to another. One can also say that Eq. (10.6) suggests an entanglement measure, which takes the maximum value 1 for the maximum entangled state (10.3).

We have therefore seen that the average projections of the first and the second spin decrease with the increasing of the entanglement, and vanish for the maximum entanglement, while the correlation of the spin projection behaves in the opposite way: being zero for the product state, it increases with the entanglement and takes the maximum value for the maximum entangled state.

Two Oscillators

Among quantum systems, the harmonic oscillator can be considered as the most classic one, since the wave packets of the coherent states of the harmonic oscillators follow classical trajectories and do not experience dispersion. The same is true even for harmonic oscillators driven by an external force. Due to this very fact, two linearly coupled harmonic oscillators, initially in a product state, remain in such a state in spite of their interaction. Entanglement of harmonic oscillators manifests itself only if the parametric interactions are getting involved. Earlier, in Sect. 7.1.3 we have already considered such processes and have shown that parametric excitation of the harmonic oscillator is a problem fully tractable analytically, due to the fact that operators $\hat{a}^\dagger\hat{a},\ \hat{a}^{\dagger^2}$, and $\hat{a}^2$ form a closed subalgebra in the oscillator Hilbert space. The $SU(1,1)$ group generated by these operators in many respects is similar to the $SU(2)$ group of two-level systems generated by the Pauli spin operators, $\hat{\sigma}^z$, $\hat{\sigma}^+$, and $\hat{\sigma}^-$. Still it has an important distinction being non-compact, that is, being characterized by the uniform measure in a space of parameters that may assume infinitely large values.

Wigner Function of the Gaussian States

The non-compactness of the $SU(1,1)$ group implies, in particular, that the quantum averages of the generators $\hat{a}^\dagger\hat{a}$, and $\hat{a}^{\dagger^2} \pm \hat{a}^2$ may take infinitely large values, in contrast to spin operators that are always inferior to $1/2$. Therefore, the states of the parametrically driven oscillators cannot be depicted with the help of the Bloch vector, and require another image, which we present now.

By taking the density matrix $|\Psi\rangle\langle\Psi|$ of a pure state of an arbitrary quantum system, and by projecting this matrix on the coordinate state $\langle x + y/2|$ from the left and the coordinate state $|x - y/2\rangle$ from the right sides, which is followed by the Fourier transformation over the displacement y between these two coordinates, one finds the so-called Wigner function,

$$W_\Psi(x,p) = \int \mathrm{d}y \frac{\langle x + y/2\,|\Psi\rangle\,\langle\Psi|\,x - y/2\rangle}{\pi} e^{-iyp} \tag{10.7}$$

in the phase space p, q. The Wigner function contains all information about the density matrix of the system. In particular, in the phase space where the vacuum oscillation amplitude is chosen for the distance unity, the Wigner function of the harmonic oscillator in the ground state is Gaussian,

$$W_0(x,p) = \frac{1}{\pi} e^{-p^2-x^2}, \tag{10.8}$$

and after being integrated either over the coordinate or over the momentum, it yields the momentum and the coordinate probability distribution, respectively. As long as the Hamiltonian of the system contains only linear and quadratic forms of the phase space vector components p and x, that is,

$$H = (x\ p)\begin{pmatrix} H_{xx} & H_{xp} \\ H_{xp} & H_{pp} \end{pmatrix}\begin{pmatrix} x \\ p \end{pmatrix} + (\mathcal{F}\ \upsilon)\begin{pmatrix} x \\ p \end{pmatrix},$$

which is the case of the oscillator subject to the linear and parametric excitation, the Wigner function satisfies the classical equation of motion

$$\frac{\partial}{\partial t} W_\Psi(x,p) = \{H, W_\Psi(x,p)\}, \tag{10.9}$$

where $\{A,B\} = \frac{\partial A}{\partial p}\frac{\partial B}{\partial q} - \frac{\partial B}{\partial p}\frac{\partial A}{\partial q}$ denotes the Poisson braces. In this section we assume that the force $\mathcal{F}$ and the linear dispersion coefficient υ are zero.

The Wigner function as a mathematical object is a very convenient tool for characterizing the harmonic oscillator subject to the linear and the parametric excitation simultaneously. Indeed, consider a harmonic oscillator initially in the vacuum state, which undergoes a transition to the so-called squeezed vacuum state as a result of the parametric excitation. The corresponding state vector $|\Psi(s,\phi)\rangle$ is given in the basis of the energy eigenstates $|n\rangle$ by the expression

$$|\Psi(s,\phi)\rangle = \sum_m \frac{e^{-2im\phi}(1-s)^m\sqrt{(2m)!}}{(2(1+s))^{m-1/2}\,m!}|2m\rangle, \tag{10.10}$$

where s is the so-called squeezing parameter, and ϕ is the so-called orientation angle in the phase space. Note that $\frac{1}{2}\ln s = Z$ in the notation of Sect. 7.1.3.

The physical meaning of (10.10) remains rather obscure unless one employs the phase space distribution technique, where the meaning of the two parameters (s,ϕ) becomes clear: one immediately finds that the Wigner function of (10.7) for the state (10.10) is a Gaussian

$$W_{s,\phi}(x,p) = \frac{1}{\pi} e^{-s(p\cos\phi + x\sin\phi)^2 - \frac{1}{s}(x\cos\phi - p\sin\phi)^2}, \tag{10.11}$$

like the one of (10.8), but it is compressed by the factor $\sqrt{s}$ in the x direction, and is stretched by the same factor in the p direction, such that the fundamental uncertainty principle of quantum mechanics is respected. After this transformation this compression Gaussian function is rotated by the angle ϕ in the phase space p, q. In other words, in the course of a parametric excitation of the harmonic oscillator, the Gaussian Wigner function of the vacuum state gets squeezed by a factor $\sqrt{s}$ along the direction ϕ, but it still remains a Gaussian one.

The average values of the generators $E = \langle \hat{a}\hat{a}^{\dagger} \rangle$, $L_p = \langle \hat{a}^{\dagger^2} + \hat{a}^2 \rangle$, and $L_m = i\langle \hat{a}^{\dagger^2} - \hat{a}^2 \rangle$ play a role similar to the spin projections $S_x = \frac{1}{2}\langle \hat{\sigma}_x \rangle$, $S_y = \frac{1}{2}\langle \hat{\sigma}_y \rangle$, and $S_z = \frac{1}{2}\langle \hat{\sigma}_z \rangle$ for $SU(2)$—they change as functions of the parameters ϕ and s identifying the state, while the vector

$$\boldsymbol{\mu} = \{E, L_p, L_m\} = \left\{ \cosh\left[\frac{1}{2}\ln s\right], \sinh\left[\frac{1}{2}\ln s\right]\cos\phi, \sinh\left[\frac{1}{2}\ln s\right]\sin\phi \right\}$$

relates the expectation values of the generator with the parameters of the Wigner function. It can serve somewhat as an analog of the Bloch vector for characterization of the transformations of $SU(1,1)$ group, although this group is isomorph not to the group $O(3)$ of $3D$ rotations but to the group of $2D$ rotations of the phase space around the origin and the area preserving uniform compressions of this plane. In accordance with this, instead of the requirement $S_x^2 + S_y^2 + S_s^2 = 1/4$ for spins, one has $E^2 - L_p^2 - L_m^2 = 1$ for the oscillators, which means that the end of the vector $\boldsymbol{\mu}$ remains not at a compact Bloch sphere, but at a non-compact hyperboloid, as shown in Fig. 10.2a. Of course, one can map an infinite hyperboloid surface on the unit sphere according to the relation $\tanh\frac{s}{2} = \frac{1}{\pi}\arctan\left[\sqrt{S_z^2/\left(S_x^2 + S_y^2\right)}\right]$, $\phi = \arctan S_x/S_y$, as shown in Fig. 10.2b, but this mapping will not preserve the uniform Haar measure—the probability distribution of the states. Indeed, a set of states that occupies a domain of the area ΔA at the surface of the Bloch sphere occupies the same area after application of any $SU(2)$ group transformation to all these states, although after the transformation, this area is located in a different position. The states of the hyperboloid projected to the sphere do not have this property, the state density is uniform over the hyperboloid but it changes over the surface of the sphere. Note that the linear excitation of the oscillator results in the displacement of the Wigner function center away from the point of the origin, $x = 0, p = 0$. However, this process is not considered in the present section.

The generators $\hat{E} = \hat{a}\hat{a}^{\dagger}$, $\hat{L}_p = \hat{a}^{\dagger^2} + \hat{a}^2$, and $\hat{L}_m = \hat{a}^{\dagger^2} - \hat{a}^2$ can be related to the uniform transformations of the phase space, where the oscillator quantum states are represented by the Wigner function. The operator $\hat{a}\hat{a}^{\dagger}$ generates rotation of the phase space around the origin, and it can be associated with the operator $i\hat{\sigma}^y$, which generates rotation in the space of two-dimensional vectors via the transformation $\exp[\phi i\hat{\sigma}^y]$, defined by the rotation angle ϕ. The operator $\hat{L}_m$ generates squeezing and corresponds to the transformation $\exp[\sqrt{s}\hat{\sigma}^z]$, while the remaining generator $\hat{L}_p$ can be associated with the operator $\hat{\sigma}^x$, responsible for the transformation $\exp[\eta\hat{\sigma}^x]$ of the phase space plane. Note that in contrast to the $SU(2)$ transformations all the matrices here are real.

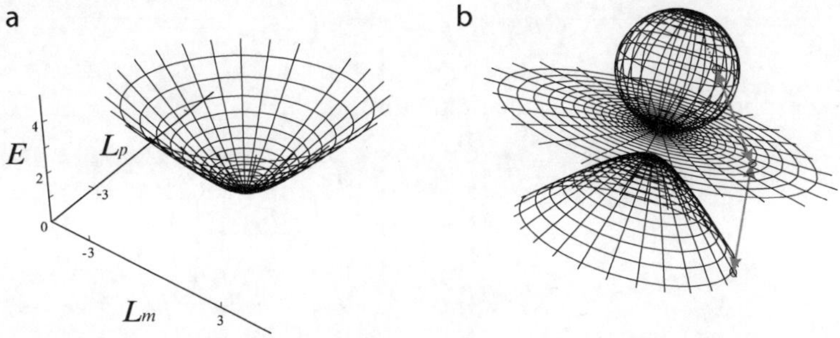

Fig. 10.2 (**a**) Average values of the generator operators of $SU(1,1)$ group that correspond to the squeezing operator $L_p = \left(p^2 - x^2\right)/2$, the sheer operator $L_m = xp$, and the Hamiltonian $E = \left(p^2 + x^2\right)/2$ at a hyperbolic surface is an analog of the average projections of spin 1/2 vector at Bloch sphere. (**b**) Projection of a hyperbolic surface to a sphere. This is a one-to-one mapping, which does not correspond to a uniform Haar measure: the density of states close to the point 0 at the hyperbolic surface is the same as for the spin-down states of the Bloch sphere, while this is not the case for the spin-up states, corresponding to an infinitely high density of the oscillator states

Parametric Excitation of Two Coupled Oscillators as Operation in the 4D Phase Space

We are now in the position to consider an example of the entangled states of two oscillators subject simultaneously to the linear coupling and the individual parametric excitation, namely the two-modes squeezed vacuum states. Let the first of the oscillators correspond to the motion in the direction x and the other oscillates in the orthogonal direction y of the coordinate xy-plane. Let $\hat{b}^\dagger$ and $\hat{b}$ be creation and annihilation operators, respectively, corresponding to the direction y, and q is the relevant momentum. The algebra of all possible transformations of the two-mode phase space, apart from the generators of the $SU(1,1)$ for of each of the oscillators, also contains the mode coupling operator given by the angular momentum $\hat{l} = i\left(\hat{b}\hat{a}^\dagger - \hat{a}\hat{b}^\dagger\right)$—generator of the geometric rotation in xy plane. By commutations, one obtains the other three generators $\hat{l}_{csh} = i\left(\hat{a}^\dagger\hat{b}^\dagger - \hat{a}\hat{b}\right)$, $\hat{l}_{csq} = \hat{a}^\dagger\hat{b}^\dagger + \hat{a}\hat{b}$, and $\hat{l}_{ex} = \hat{a}\hat{b}\dagger + \hat{b}\hat{a}^\dagger$ of the algebra $sp(4)$, of four-dimensional symplectic (the phase volume preserving) uniform transformations. The physical meaning of the cross-shear, the cross-squeezing, and the excitation exchange can be attributed to these generators, respectively. The corresponding Lie group $SP(4)$ depends therefore on ten parameters.

Considering entanglement, it is possible, but extremely inconvenient technically, to work in the energy-eigenstates basis of the oscillators. A much more practical approach relies on the Wigner function in the phase space of two oscillators. The family of states that originate from application of the aforesaid transformations to the two-mode vacuum has a Wigner function of the form

$$W(x,p,y,q) = \frac{1}{\pi^2} \exp\{-F\}, \tag{10.12}$$

where the exponent index F is a quadratic form

$$F = \begin{pmatrix} x\ p\ y\ q \end{pmatrix} \begin{pmatrix} M_{xx} & M_{xp} & M_{xy} & M_{xq} \\ M_{xp} & M_{pp} & M_{py} & M_{pq} \\ M_{xy} & M_{py} & M_{yy} & M_{yq} \\ M_{xq} & M_{pq} & M_{yq} & M_{qq} \end{pmatrix} \begin{pmatrix} x \\ p \\ y \\ q \end{pmatrix} \tag{10.13}$$

of the phase space coordinates (x,p,y,q). The symmetric covariance matrix M given by ten real matrix elements satisfies the normalization condition

$$\det M = 1.$$

From the generalization of (10.9) to $4D$ phase space of the two modes, one finds that any, not necessarily Gaussian, Wigner function satisfies the classical equation of motion for a distribution function $\dot{W} = \{H, W\}$, where the classical Hamilton function H given by the quadratic form $(x,p,y,q)\,\hat{H}\,(x,p,y,q)^T$ over the coordinates and momenta is deafened with the help of the coordinate- and momentum-independent symmetric matrix

$$\hat{H} = \frac{1}{2} \begin{pmatrix} H''_{xx} & H''_{xp} & H''_{xy} & H''_{xq} \\ H''_{xp} & H''_{pp} & H''_{py} & H''_{pq} \\ H''_{xy} & H''_{py} & H''_{yy} & H''_{yq} \\ H''_{xq} & H''_{pq} & H''_{yq} & H''_{qq} \end{pmatrix}. \tag{10.14}$$

Here the superscript T stands for the transposition operation and $\{\ldots,\ldots\}$ for the Poisson braces, as earlier in (10.9).

For the particular case of the Gaussian Wigner function (10.12), one can translate equation of motion for the Wigner function to the equation of motion for its logarithm $\dot{F} = \{H, F\}$, which implies an equation

$$\dot{\hat{M}} = \hat{M}\hat{\Sigma}_y\hat{H} - \hat{H}\hat{\Sigma}_y\hat{M}$$

for the covariance matrix, and brings us to the equation of motion

$$\frac{d}{dt}\hat{M}\hat{\Sigma}_y = \left[\hat{M}\hat{\Sigma}_y, \hat{H}\hat{\Sigma}_y\right], \tag{10.15}$$

resembling that for the density matrix, where

$$\hat{\Sigma}_y = \begin{pmatrix} 0 & 1 & 0 & 0 \\ -1 & 0 & 0 & 0 \\ 0 & 0 & 0 & 1 \\ 0 & 0 & -1 & 0 \end{pmatrix} . \tag{10.16}$$

One notices that the analogy of (10.15) with the equation for the density matrix suggests one to look for the solution in the form of a product

$$\hat{M}(t)\hat{\Sigma}_y = \exp\left[-\hat{H}\hat{\Sigma}_y t\right] \hat{M}(0)\hat{\Sigma}_y \exp\left[\hat{H}\hat{\Sigma}_y t\right],$$

where the mutually inverse evolution operators $\exp\left[-\hat{H}\hat{\Sigma}_y t\right]$ and $\exp\left[\hat{H}\hat{\Sigma}_y t\right]$ multiply the initial condition $\hat{M}(0)\hat{\Sigma}_y$ from the left and the right sides, respectively. By straightforward manipulations one also finds the other forms of this equation

$$\begin{aligned} \hat{M}(t) &= \exp\left[-\hat{H}\hat{\Sigma}_y t\right] \hat{M}(0) \exp\left[\hat{\Sigma}_y \hat{H} t\right] \\ &= \exp\left[-\hat{H}\hat{\Sigma}_y t\right] \hat{M}(0) \exp\left[-\hat{H}\hat{\Sigma}_y t\right]^T . \end{aligned}$$

By the same analogy, the generic transformation $\hat{\Lambda}$ of an arbitrary quantum state of the oscillators can be achieved by application of a sequence of ten standard evolution operations. This also can be viewed as the operations in phase space transforming the covariance matrix $\hat{M}$. In the oscillators Hilbert space, each of these ten operations is characterized by the corresponding time-independent Hamiltonian, say, $\hat{a}^\dagger\hat{a}$ and $\hat{b}^\dagger\hat{b}$ for local rotation of the phase space, local squeezings $i\left(\hat{a}\hat{a} - \hat{a}^\dagger\hat{a}^\dagger\right)$ and $i\left(\hat{b}\hat{b} - \hat{b}^\dagger\hat{b}^\dagger\right)$ along the directions x and y, respectively, and rotation in the x, y-plane, generated by the angular momentum $\hat{L} = i\left(\hat{b}\hat{a}^\dagger - \hat{a}\hat{b}^\dagger\right)$. In the phase space of the oscillators, all these operators correspond to the matrices $\hat{H}$ of (10.14) that read

$$\begin{aligned} &\omega\hat{a}^\dagger\hat{a} + is\left(\hat{a}\hat{a} - \hat{a}^\dagger\hat{a}^\dagger\right) + \\ &+\Omega\hat{b}^\dagger\hat{b} + iS\left(\hat{b}\hat{b} - \hat{b}^\dagger\hat{b}^\dagger\right) \end{aligned} \to \frac{1}{2}\begin{pmatrix} \omega & s & 0 & 0 \\ s & \omega & 0 & 0 \\ 0 & 0 & \Omega & S \\ 0 & 0 & S & \Omega \end{pmatrix} \tag{10.17}$$

$$\begin{aligned} &i\alpha\left(\hat{a}^\dagger\hat{b}^\dagger - \hat{a}\hat{b}\right) + \beta\left(\hat{a}^\dagger\hat{b}^\dagger + \hat{a}\hat{b}\right) + \\ &+\vartheta\left(\hat{a}\hat{b}^\dagger + \hat{b}\hat{a}^\dagger\right) + i\gamma\left(\hat{b}\hat{a}^\dagger - \hat{a}\hat{b}^\dagger\right) \end{aligned} \to \begin{pmatrix} 0 & 0 & \frac{\vartheta+\beta}{2} & \frac{\alpha+\gamma}{2} \\ 0 & 0 & \frac{\alpha-\gamma}{2} & \frac{\vartheta-\beta}{2} \\ \frac{\vartheta+\beta}{2} & \frac{\alpha-\gamma}{2} & 0 & 0 \\ \frac{\alpha+\gamma}{2} & \frac{\vartheta-\beta}{2} & 0 & 0 \end{pmatrix}, \tag{10.18}$$

where the generators $\hat{l}_{csh} = i\left(\hat{a}^\dagger\hat{b}^\dagger - \hat{a}\hat{b}\right)$, $\hat{l}_{csq} = \hat{a}^\dagger\hat{b}^\dagger + \hat{a}\hat{b}$, and $\hat{l}_{ex} = \hat{a}\hat{b}^\dagger + \hat{b}\hat{a}^\dagger$, proportional to the combinations $xq + py$, $xy - pq$, and $xy + pq$ in the Hamilton function, respectively, are also included.

Covariance Matrix of Two-Mode Squeezed Vacuum

We now show that the existence of ten independent parameters of the generic $SP(4)$ transformation does not imply a functional independence of nine matrix elements of the covariance matrix M (10.13) obtained by application of the aforesaid transformations to the vacuum state. In fact, the matrix M depends on only five parameters, in the same way as 15 independent parameters of the unitary $SU(4)$ transformation do not manifest themselves in the state vector (10.2) of two spins. The following demonstrates this fact. We choose a particular parametrization of the generic $SP(4)$ transformation Λ, which is done in three stages: symplectic rotation, local squeezing, and symplectic rotation again.

Let us start with the first stage, the symplectic rotation in the four-dimensional phase space. We start this operation by first applying the local phase space rotations

$$\hat{\Lambda}_1(\phi_1, \phi_2) = e^{i\phi_1\hat{a}^\dagger\hat{a} + i\phi_2\hat{b}^\dagger\hat{b}} \rightarrow \exp\left[\begin{pmatrix} 0 & \frac{\phi_1}{2} & 0 & 0 \\ -\frac{\phi_1}{2} & 0 & 0 & 0 \\ 0 & 0 & 0 & \frac{\phi_2}{2} \\ 0 & 0 & -\frac{\phi_2}{2} & 0 \end{pmatrix}\right] \tag{10.19}$$

given by the application of the operator (10.17) with $s = S = 0$ during a time t, such that the parameters $\phi_1 = \omega t$ and $\phi_2 = \Omega t$ are the required rotation angles in the phase spaces of the first and the second operators, respectively. The arrow shows the correspondence between the operation in the Hilbert space of the harmonic oscillator and the relevant phase space transformation. The transformation $\hat{\Lambda}_1(\phi_1, \phi_2)$ is followed by the second operation,

$$\hat{\Lambda}_2(\varphi) = e^{-i\varphi\left(\hat{a}\hat{b}^\dagger + \hat{b}\hat{a}^\dagger\right)} \rightarrow \exp\left[\begin{pmatrix} 0 & 0 & \frac{\varphi}{2} & 0 \\ 0 & 0 & 0 & \frac{\varphi}{2} \\ -\frac{\varphi}{2} & 0 & 0 & 0 \\ 0 & -\frac{\varphi}{2} & 0 & 0 \end{pmatrix}\right], \tag{10.20}$$

which is a rotation by an angle $\varphi = \gamma t$ in the xy-plane given by the operator (10.18). In order to complete the generic $4D$ symplectic rotation, the latter nonlocal operation can be followed by the second local rotation $\hat{\Lambda}_1(\phi_1', \phi_2')$, characterized by a pair ϕ_1', ϕ_2' of rotation angles, different from the first pair ϕ_1, ϕ_2. We thus have completed the symplectic rotation $\hat{\Lambda}_1(\phi_1', \phi_2')\hat{\Lambda}_2(\varphi)\hat{\Lambda}_1(\phi_1, \phi_2)$ that turns out to be dependent on only five real parameters. Still, not all these five parameters are independent, since the angular momentum operator commutes with the symmetric rotation, that is, rotation of both oscillators in the phase space by the same angle. This yields $\hat{\Lambda}_2(\varphi)\hat{\Lambda}_1(\phi,\phi) = \hat{\Lambda}_1(\phi,\phi)\hat{\Lambda}_2(\varphi)$. We therefore are left with only four independent parameters of the generic symplectic rotation $\hat{\Lambda}_1(\phi_1',\phi_2')\hat{\Lambda}_2(\varphi)\hat{\Lambda}_1(\phi,-\phi)$.

We now come to the next, second, stage, and apply the local squeezing operation

$$\hat{\Lambda}_3(s_1,s_2) = e^{s_1\frac{\hat{a}\hat{a}-\hat{a}^\dagger\hat{a}^\dagger}{2}+s_2\frac{\hat{b}\hat{b}-\hat{b}^\dagger\hat{b}^\dagger}{2}} \to \exp\left[\begin{pmatrix} \frac{s_1}{2} & 0 & 0 & 0 \\ 0 & -\frac{s_1}{2} & 0 & 0 \\ 0 & 0 & \frac{s_2}{2} & 0 \\ 0 & 0 & 0 & -\frac{s_2}{2} \end{pmatrix}\right]. \tag{10.21}$$

Given by application of the operator (10.17) with $\omega = \Omega = 0$, it results in squeezings $s_1 = st$, and $s_2 = St$ of the first and the second modes, respectively. We thus have employed six out of ten transformation parameters. The last, third, stage is again a generic symplectic rotation $\hat{\Lambda}_1(\phi_1'',\phi_2'')\hat{\Lambda}_2(\varphi')\hat{\Lambda}_1(\phi',-\phi')$ depending on four more parameters. The overall transformation can be formally written as

$$\hat{\Lambda} = \hat{\Lambda}_1(\phi_1'',\phi_2'')\hat{\Lambda}_2(\varphi')\hat{\Lambda}_1(\phi',-\phi')\hat{\Lambda}_3(s_1,s_2)\hat{\Lambda}_1(\phi_1',\phi_2')\hat{\Lambda}_2(\varphi)\hat{\Lambda}_1(\phi,-\phi).$$

Consider now the action of the transformation on the vacuum state and calculate the covariance matrix $\hat{M} = \hat{\Lambda}\hat{\Lambda}^T$. The first symplectic rotation $\hat{\Lambda}_1(\phi_1',\phi_2')\hat{\Lambda}_2(\varphi)\hat{\Lambda}_1(\phi,-\phi)$ does not affect the vacuum state with the symmetric Wigner function and the identity covariance matrix $\hat{M} = \hat{I}$, and therefore it can be discarded, leaving us with the transformation

$$\hat{\Lambda}_v = \hat{\Lambda}_1(\phi_1',\phi_2')\hat{\Lambda}_2(\varphi)\hat{\Lambda}_1(\phi,-\phi)\hat{\Lambda}_3(s_1,s_2), \tag{10.22}$$

depending on just six parameters, as it was stated.

It is expedient to give another parametrization of the transformation $\hat{\Lambda}_v$, which starts with a nonlocal operation applied to the vacuum and is followed by local operations. Apparently, the rotation in the xy-plane cannot serve for this purpose, since it leaves the vacuum state Wigner function intact. However, another operation, namely the operation of cross-shear, $\hat{\Lambda}_4(\alpha) = \exp\left[-i\alpha\hat{L}_{csh}\right]$, generated by $\hat{L}_{csh} = i\left(\hat{a}^\dagger\hat{b}^\dagger - \hat{a}\hat{b}\right)$, can do this. We therefore can represent $\hat{\Lambda}_v$ in the form

$$\hat{\Lambda}_v = \hat{\Lambda}_1(\phi_1',\phi_2')\hat{\Lambda}_3(s_1,s_2)\hat{\Lambda}_1(\phi,\phi)\hat{\Lambda}_4(\alpha), \tag{10.23}$$

where we allow for the fact that the rotation $\hat{\Lambda}_1(\theta',-\theta')$ of oscillators in their phase spaces by the same angle but in opposite directions commutes with the $\hat{\Lambda}_4(\alpha)$ operation, can be moved to the right, and discarded as not affecting the vacuum state. Application of $\hat{\Lambda}_4(\alpha)$ alone to the vacuum state yields the entangled state of two modes given by just one parametric covariance matrix,

$$\hat{M}(\alpha) = \begin{pmatrix} \cosh\alpha & 0 & -\sinh\alpha & 0 \\ 0 & \cosh\alpha & 0 & \sinh\alpha \\ -\sinh\alpha & 0 & \cosh\alpha & 0 \\ 0 & \sinh\alpha & 0 & \cosh\alpha \end{pmatrix}, \tag{10.24}$$

which later is subject to exclusively local transformations, squeezings, and rotations.

Effect of Entanglement on Local Measurables

We are now in the position to illustrate the effect of entanglement on the expectation values of the generators of local transformations for squeezed vacuum states of two oscillators, by the analogy to the pair of spins, considered earlier in this section. For this purpose, we first calculate the single-mode Wigner functions for either mode, which is obtained by integrating the two-mode Wigner function over the phase space of the other mode. Since the local operations in a phase space do not change the result of such an integration, one can perform this procedure right after the application of the operation $\hat{\Lambda}_4(\alpha)$, that is, with the covariance matrix (10.24), and find

$$W_1(x,p) = \frac{1}{\pi\cosh\alpha}\exp\left\{-\frac{x^2+p^2}{\cosh\alpha}\right\}, \tag{10.25}$$

and the same for $W_2(y,q)$. One sees that the local Wigner function is a symmetric Gaussian, which at $\alpha \neq 0$ it is broadened as compared to the vacuum Wigner function (10.8). Being subject to local operations it adopts the form

$$W_{1,s_1,\phi_1}(x,p) = \frac{\exp\left\{-\frac{s_1\left(p\cos\phi_1 + x\sin\phi_1\right)^2 - s_1^{-1}\left(x\cos\phi_1 - p\sin\phi_1\right)^2}{\cosh\alpha}\right\}}{\pi\cosh\alpha} \tag{10.26}$$

similar to (10.11).

As in the case of two spins, shown in Fig. 10.1, one can find the average values of the generators

$$\begin{aligned} E_1 &= \langle \hat{a}\hat{a}^\dagger \rangle = \cosh\alpha\cosh\left[\frac{1}{2}\ln s_1\right], \\ L_{p1} &= \left\langle \hat{a}^{\dagger^2} + \hat{a}^2 \right\rangle = \cosh\alpha\sinh\left[\frac{1}{2}\ln s_1\right]\cos\phi_1, \\ L_{m1} &= i\left\langle \hat{a}^{\dagger^2} - \hat{a}^2 \right\rangle = \cosh\alpha\sinh\left[\frac{1}{2}\ln s_1\right]\sin\phi_1 \end{aligned} \tag{10.27}$$

comprising a vector $\mu(\alpha, s_1, \phi_1) = \{E_1, L_{p1}, L_{m1}\}$, which now depends on the entanglement parameter α. One can also find the same vector $\{E_2, L_{p2}, L_{m2}\}$ for the second oscillator, and depict the sum of energies $E = E_1 + E_2$ and the expectation values for of L_{p1} and L_{p2} for the rotation angles $\phi_1 = \phi_2 = 0$, as it is done in Fig. 10.3. Now it can be seen that the total energy as a function of the local squeezing parameters for the entangled and not entangled states behaves in a similar way, although this quantity always increases with increasing entanglement. One also sees that the Wigner function of a single mode gets broader when this mode gets entangled with the other one, such as if the mode is in a thermal state with a positive temperature.

We can also calculate expectation values of the generators responsible for the entanglement $L = \langle l \rangle$, $L_{csh} = \langle l_{csh} \rangle$, $L_{csq} = \langle l_{csq} \rangle$, and $L_{ex} = \langle l_{ex} \rangle$, explicitly given in terms of the transformation parameters as

$$
\begin{aligned}
L &= 4\sinh\alpha\left(\sin\Phi\sinh\frac{s_1}{2}\cosh\frac{s_2}{2} - \sin\Phi'\cosh\frac{s_1}{2}\sinh\frac{s_2}{2}\right),\\
L_{csh} &= 4\sinh\alpha\left(\sin\Theta'\sinh\frac{s_1}{2}\sinh\frac{s_2}{2} - \sin\Theta\cosh\frac{s_1}{2}\cosh\frac{s_2}{2}\right),\\
L_{csq} &= 4\sinh\alpha\left(\cos\Theta\sinh\frac{s_1}{2}\sinh\frac{s_2}{2} + \cos\Theta'\cosh\frac{s_1}{2}\cosh\frac{s_2}{2}\right),\\
L_{ex} &= 4\sinh\alpha\left(\cos\Phi'\cosh\frac{s_1}{2}\sinh\frac{s_2}{2} - \cos\Phi\sinh\frac{s_1}{2}\cosh\frac{s_2}{2}\right),
\end{aligned}
$$

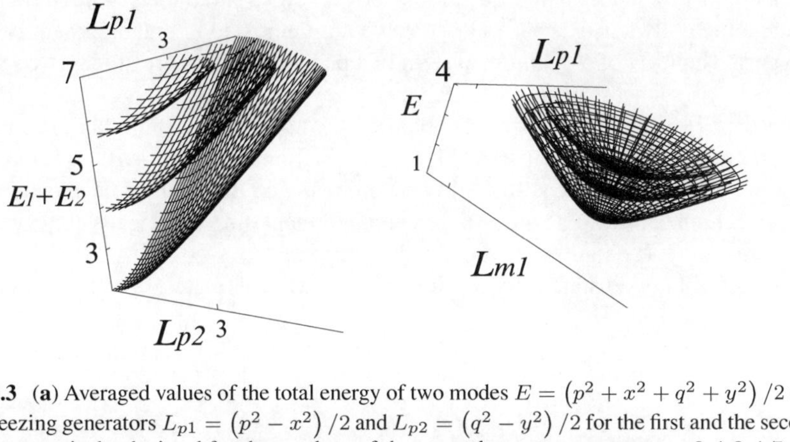

Fig. 10.3 (**a**) Averaged values of the total energy of two modes $E = \left(p^2 + x^2 + q^2 + y^2\right)/2$ and the squeezing generators $L_{p1} = \left(p^2 - x^2\right)/2$ and $L_{p2} = \left(q^2 - y^2\right)/2$ for the first and the second modes, respectively, depicted for three values of the entanglement parameter $\alpha = 0; 1.3; 1.7$. (**b**) Average values of the squeezing operator $L_{p1} = \left(p^2 - x^2\right)/2$, the sheer operator $L_{m1} = xp$, and the Hamiltonian $E = \left(p^2 + x^2\right)/2$ for the first mode. The entanglement parameter α takes the same values as for (**a**)

where $\Phi = \phi - \frac{\phi_1' - \phi_2'}{2}$, $\Phi' = \phi + \frac{\phi_1' - \phi_2'}{2}$, $\Theta = \phi - \frac{\phi_1' + \phi_2'}{2}$, $\Theta' = \phi + \frac{\phi_1' + \phi_2'}{2}$, notice that

$$L_{csh}^2 + L_{csh}^2 - L^2 - L_{ex}^2 = 4\sinh^2\alpha, \tag{10.28}$$

and relate via this equality the entanglement parameter $\sinh\alpha$ with the expectation values of the generators, which is an analog of (10.6) for the case of two entangled spins subject to magnetic fields in the z and y directions. In contrast to the case of spins, this parameter may take an infinite value.

10.1.2 Signatures and Measures of Entanglement for Bipartite Systems

In the examples just considered, of spins and of oscillators, we saw that the inseparability of two parts of a quantum system affects the result of local measurements performed on either of parts, decreasing the expectation values of each of coupled spins components, at in the first example, and increasing the measured energy of each of the two coupled parametrically excited oscillators, as in the second case. In both cases it was possible to associate entanglement with a combination of expectation values of some observables responsible for the interaction of the parts that were comprising a single parameter suitable for the entanglement measure.

One can now address the question: does a single parameter characterizing the entanglement always exists? As long as we are interested in entanglement between two parts A and B of a quantum system in a pure state, there are at least two natural measures of such bipartite entanglement. They are based on the reduced density operator ρ_A of either part, say A, obtained by tracing over the quantum numbers corresponding to the other part B. Two combinations, $S_{vN} = -\mathrm{Tr}[\rho_A \log\rho_A]$ and $S_l = 1 - \mathrm{Tr}[\rho_A^2]$, give the so-called von Neumann and linear entropies, respectively, each of which can serve as the required entanglement measure. Let us consider these measures in more detail in the aforesaid examples.

For the entangled state of two spins of (10.4) we calculate the density matrix

$$\begin{aligned}
&\cos^2\alpha \begin{pmatrix} \cos^2\beta & \frac{\sin 2\beta}{2} \\ \frac{\sin 2\beta}{2} & \sin^2\beta \end{pmatrix} \otimes \begin{pmatrix} \cos^2\gamma & \frac{\sin 2\gamma}{2} \\ \frac{\sin 2\gamma}{2} & \sin^2\gamma \end{pmatrix} \\
&+ \frac{\sin 2\alpha}{2} \begin{pmatrix} -\frac{\sin 2\beta}{2} & \cos^2\beta \\ -\sin^2\beta & \frac{\sin 2\beta}{2} \end{pmatrix} \otimes \begin{pmatrix} -\frac{\sin 2\gamma}{2} & \cos^2\gamma \\ -\sin^2\gamma & \frac{\sin 2\gamma}{2} \end{pmatrix} \\
&+ \frac{\sin 2\alpha}{2} \begin{pmatrix} -\frac{\sin 2\beta}{2} & -\sin^2\beta \\ \cos^2\beta & \frac{\sin 2\beta}{2} \end{pmatrix} \otimes \begin{pmatrix} -\frac{\sin 2\gamma}{2} & -\sin^2\gamma \\ \cos^2\gamma & \frac{\sin 2\gamma}{2} \end{pmatrix} \\
&+ \sin^2\alpha \begin{pmatrix} \sin^2\beta & -\frac{\sin 2\beta}{2} \\ -\frac{\sin 2\beta}{2} & \cos^2\beta \end{pmatrix} \otimes \begin{pmatrix} \sin^2\gamma & -\frac{\sin 2\gamma}{2} \\ -\frac{\sin 2\gamma}{2} & \cos^2\gamma \end{pmatrix},
\end{aligned}$$

trace it over the second spin Hilbert space, and find the density matrix of the first spin,

$$\cos^2\alpha \begin{pmatrix} \cos^2\beta & \frac{\sin 2\beta}{2} \\ \frac{\sin 2\beta}{2} & \sin^2\beta \end{pmatrix} + \sin^2\alpha \begin{pmatrix} \sin^2\beta & -\frac{\sin 2\beta}{2} \\ -\frac{\sin 2\beta}{2} & \cos^2\beta \end{pmatrix}. \qquad (10.29)$$

Now consider the eigenvalues of the density matrix (10.29) $\cos^2\alpha$ and $\sin^2\alpha$. At $\alpha = 0$ the rank of the density matrix is one, since one of the two eigenvalues vanishes. This corresponds to a pure state of the first spin, and hence the quantum state of the system of two spins is a product state. The difference of the density matrix rank from one is the main signature of entanglement. The next question to ask is: how much the system is entangled? The aforesaid entropy measures provide one with two possible answers to this question.

In fact, for the two spin-$1/2$ system, one can immediately calculate the von Neumann and the linear entropies:

$$S_{vN} = -\left(\ln\cos^2\alpha\right)\cos^2\alpha - \left(\ln\sin^2\alpha\right)\sin^2\alpha$$

and

$$S_l = 1 - \cos^4\alpha - \sin^4\alpha = \frac{\sin^2 2\alpha}{2},$$

respectively. Note that up to a numerical factor, the linear entropy is proportional to the square of the expectation values of the product of spins in y-direction, given by (10.6), which was taken as a measure of entanglement. Also note that the linear entropy ranges from zero value for the product state till the maximum value of $1/2$ for the maximum entangled states. It is also easy to calculate the linear entropy for the squeezed vacuum states of a two-mode oscillator. From the Wigner function of one of the modes (10.25) and the universal relation

$$\mathrm{Tr}\rho^2 = 2\pi \int \mathrm{d}p\mathrm{d}x W^2(x,p)$$

one obtains the linear entropy

$$S_l = 1 - \frac{1}{\cosh\alpha} = \frac{2\sinh^2\frac{\alpha}{2}}{\cosh\alpha},$$

which again is related to, but does not directly equal a measure of the entanglement (10.28) composed from the expectation values of measurables responsible for the interaction between two parts of the system. It ranges from zero for the product states till the maximum value of 1 for the maximum entangled states. The latter also correspond to infinitely large oscillator energies and hence squeezings.

The von Neumann entanglement entropy of the two-mode squeezed vacuum state can also be found, although this requires a longer calculation. To this end we calculate the Wigner function of the so-called thermal state given by the density matrix

$$\rho_\beta = \sum_n |n\rangle \frac{e^{-\beta n}}{\left(1-e^{-\beta}\right)^{-1}} \langle n| , \tag{10.30}$$

corresponding to a Hibbs distribution of the energy eigenstate populations at the temperature $T = 1/\beta k$, where k is the Boltzmann constant. In the coordinate representation

$$\begin{aligned}\rho_\beta\left(x,x'\right) &= \sum_n \frac{H_n(x)e^{-x^2/2}}{\pi^{1/4}2^{n/2}\sqrt{n!}} \frac{e^{-\beta n}}{\left(1-e^{-\beta}\right)^{-1}} \frac{H_n(x')e^{-x'^2/2}}{\pi^{1/4}2^{n/2}\sqrt{n!}} \\ &= \frac{e^{-x^2/2-x'^2/2}}{\sqrt{\pi}\left(1-e^{-\beta}\right)^{-1}} \sum_n \frac{H_n(x)H_n(x')e^{-\beta n}}{2^n n!},\end{aligned} \tag{10.31}$$

we employ the integral representations

$$\frac{H_n(x)}{n!} = \frac{1}{2\pi i} \int\limits_{|t|=1} \frac{\mathrm{d}t}{t^{n+1}} e^{-t^2+2tx}$$

$$\frac{H_n(x')}{2^n} = \frac{1}{\sqrt{\pi}} \int\limits_{-\infty}^{\infty} \mathrm{d}r\left(x'+ir\right)^n e^{-r^2}$$

for the Hermite polynomials, where the first integration is performed over the unit circle, change the order of the summation and the integration and find

$$\begin{aligned}\sum_n \frac{H_n(x)H_n(x')e^{-\beta n}}{2^n n!} &= \int\limits_{-\infty}^{\infty} \frac{\mathrm{d}r}{\sqrt{\pi}} \int\limits_0 \frac{\mathrm{d}t}{2\pi i} e^{-r^2-t^2+2tx} \sum_n \frac{e^{-\beta n}\left(x'+ir\right)^n}{t^{n+1}} \\ &= \int\limits_{-\infty}^{\infty} \frac{\mathrm{d}r}{\sqrt{\pi}} \int\limits_{|t|=1} \frac{\mathrm{d}t}{2\pi i} \frac{e^{-r^2-t^2+2tx}}{t-\left(x'+ir\right)e^{-\beta}}.\end{aligned}$$

We next perform integration over $\mathrm{d}t$ by taking the residual at the point $t = \left(x'+ir\right)e^{-\beta}$ and substitute the result into (10.31):

$$\rho_\beta\left(x,x'\right) = \frac{e^{-x^2/2-x'^2/2}}{\pi\left(1-e^{-\beta}\right)^{-1}} \int\limits_{-\infty}^{\infty} \mathrm{d}r e^{-r^2-e^{-2\beta}\left(x'+ir\right)^2+2\left(x'+ir\right)e^{-\beta}x},$$

which yields

$$\rho_\beta\left(x, x'\right) = \frac{1 - e^{-\beta}}{\pi\sqrt{1 - e^{-2\beta}}} \exp\left\{\frac{x^2 + x'^2 - 2e^\beta x x'}{2\left(\coth\beta - 1\right)} - \frac{x^2 + x'^2}{2}\right\},$$

and results in the Wigner function

$$\begin{aligned} W_\beta(x, p) &= \int \mathrm{d}y \frac{\rho_\beta\left(x + y/2, x - y/2\right)}{\pi} e^{-iyp} \\ &= \frac{\tanh\frac{\beta}{2}}{\pi} \exp\left\{-\tanh\frac{\beta}{2}\left(x^2 + p^2\right)\right\}. \end{aligned}$$

We compare this expression with the local Wigner function (10.25) and find that these two expression coincide when

$$\tanh\frac{\beta}{2} = \frac{1}{\cosh\alpha}.$$

Now it is an easy task to find the von Neumann entropy of the entangled modes in the energy representation: by substituting the density matrix (10.30) into the definition of the von Neumann entropy $-\mathrm{Tr}\left[\rho\ln\rho\right]$ one finds

$$\begin{aligned} S_{vN} &= \left(e^{-\beta} - 1\right)\sum_n e^{-\beta n}\ln\left[e^{-\beta n}\left(1 - e^{-\beta}\right)\right] \\ &= \frac{-\beta}{e^\beta - 1} + \ln\left(1 - e^{-\beta}\right) \\ &= 2\ln\cosh\frac{\alpha}{2} - 2\sinh^2\frac{\alpha}{2}\ln\left[\tanh\frac{\alpha}{2}\right], \end{aligned}$$

which is a function monotonously increasing with the increase of the parameter α responsible for the coupling of the modes.

Consider now the general case of a bipartite system, composed of an N-level part A and an M-level part B. In the basis of states composed of the direct product of the basis states for each of the parts $|n\rangle$ for the first part, and $|m\rangle$ for the second one, the state vector

$$|\Psi\rangle = \sum_{n=1,m=1}^{n=N,m=M} \Psi_{n,m}\,|n\rangle\,|m\rangle$$

is defined by a matrix of coefficients $\Psi_{n,m}$. This matrix is not necessarily square, but still, by the analogy to the eigenvalue decomposition, this rectangular matrix can be cast to a sum

$$\Psi_{n,m} = \sum_r \lambda_r a_{rn} b_{rm}$$

with the help of the so-called singular value decomposition. This procedure, also known as Schmidt decomposition, implies

$$|\Psi\rangle = \sum_r \lambda_r |A_r\rangle |B_r\rangle . \tag{10.32}$$

Here $\{\lambda_r\}$ is an analog of the eigenvalue set, and analogs of the eigenvectors

$$|A_r\rangle = a_{rn} |n\rangle$$
$$|B_r\rangle = b_{rm} |m\rangle$$

corresponding to parts A and B, respectively, satisfy the orthogonality conditions

$$\sum_{n=1}^{N} a_{rn} a^*_{r'n} = \delta_{r,r'}$$
$$\sum_{m=1}^{M} b_{rm} b^*_{r'm} = \delta_{r,r'}. \tag{10.33}$$

The rank of the matrix, that is, the number of the elements λ_r, does not exceed the minimum of N and M.

The decomposition (10.32) leads to the explicit expressions for the density matrix

$$|\Psi\rangle \langle\Psi| = \sum_{r,r'} \lambda^*_{r'} \lambda_r |A_r\rangle |B_r\rangle \langle B_{r'}| \langle A_{r'}| \tag{10.34}$$

of the entire system, and hence, with the allowance for (10.33), it yields the local density matrices

$$\rho_A = \mathrm{Tr}_B |\Psi\rangle \langle\Psi| = \sum_r |\lambda_r|^2 |A_r\rangle \langle A_r|$$
$$\rho_B = \mathrm{Tr}_A |\Psi\rangle \langle\Psi| = \sum_{r,r'} |\lambda_r|^2 |B_r\rangle \langle B_r| , \tag{10.35}$$

that both have the same set $\left\{|\lambda_r|^2\right\}$ of positive eigenvalues $|\lambda_r|^2$. In other words, all characteristics of bipartite entanglement that are based on the spectrum of the local density matrices, such as ranks, linear or von Neumann entropies, etc., are identical, and therefore, they can be considered as the characteristics of bipartite entanglement of the entire system. Moreover, the spectrum $\left\{|\lambda_r|^2\right\}$ itself, constrained only by the normalization condition $\sum_r |\lambda_r|^2 = 1$, can serve as a set of entanglement characteristics, whereas the linear and von Neumann entropies are the most important two of a larger number of such characteristics.

10.2 Entanglement and Quantum Informatics

In this section we consider entanglement in the context of problems that belong to the fields of quantum informatics and quantum telecommunication. We start by explaining the physical meaning of some terms widely employed in these fields, and dwell on three main examples: Shor's algorithm relying on bipartite entanglement, Grover's algorithm relying on multipartite entanglement, and quantum teleportation relying on bipartite entanglement among three parts of a quantum system and on classical communication, which helps to transport an unknown quantum state over a large distance.

10.2.1 Computational Basis

Prior to the discussion of the quantum algorithm, we briefly elucidate the physical meanings of the quantum informatic terms, such as number, state of the computer, and the classical code. Quantum codes will be considered as a composite operation, which includes classical codes and essentially quantum unitary transformations as elementary steps.

A natural number x written in the binary code $n_i = 0, 1$ as $x = \sum_{i=1}^{N} n_i 2^{i-1}$ is considered as a specific state of the quantum computer, which corresponds to a product state

$$|x\rangle = \prod_{\otimes i=1}^{N} |n_i\rangle \tag{10.36}$$

of N binary registers, each of which is either in the lower state, $|n_i = 0\rangle$, or in the upper state, $|n_i = 1\rangle$, whereas i numerates the two-level quantum systems comprising the register. Each string $\{n_i\}$ is therefore a state $|x\rangle$ corresponding to the natural number x, whose binary transcription is given by this string.

Unless specified otherwise, we consider a quantum computer as a devise comprising one or several composite registers, which altogether have N two-level elementary registers. Thus a quantum state of the computer is a linear superposition of the computational basis states

$$|\Psi\rangle = \sum_{n_i} \psi_{\{n_i\}} \prod_{\otimes i=1}^{N} |n_i\rangle = \sum_{x=0}^{x=2^N-1} \psi_x \,|x\rangle\,, \tag{10.37}$$

where $\psi_x = \psi_{\{n_i\}}$ stands for the probability amplitude that the register is in the state x given by the binary string $\{n_i\}$. Classical computation implies an operation of state transformation,

$$O_C\,|\Psi\rangle == \sum_{x=0}^{x=2^N-1} \psi_x O_C\,|x\rangle = \sum_{x=0}^{x=2^N-1} \psi_x\,|C(x)\rangle\,, \tag{10.38}$$

where C is a classical code which makes a bijection of natural numbers, that is, sets to one-to-one correspondence the natural input number x and the natural output number $y = C(x)$. In other words, the classical code permutes the computational basis states, but leaves intact the amplitudes in front of these states. Clearly, this permutation operation is a particular case of the unitary operation.

10.2.2 Shor Algorithm

The Shor algorithm is meant to find dividers of a large integer number M. It relies on a mathematical result, which relates periods r of the function

$$f(x) = a^x \bmod M \tag{10.39}$$

with the dividers $a^{r/2} \pm 1$ of the number M. Once a period $r > 0$ satisfying the equations $a^r \bmod M = 1$ for an integer base a is found, one immediately finds the numbers $a^{r/2} \pm 1$ that divide M to integer parts. One, however, needs to satisfy two requirements: r must be even, and $a^{r/2} \bmod M \neq -1$. If either of these requirements is violated, one has to change the integer base a.

For the particular example $M = 15$, by making the choice $a = 2$ not violating the mentioned requirements, and by direct calculations, one finds a string of periodically repeated numbers

$$f(0,1,2,3,4,5,6,7,8\ldots) = 1,2,4,8,1,2,4,8\ldots. \tag{10.40}$$

which implies that the function $f(x) = 2^x \bmod 15$ has a period $r = 4$, while $2^{r/2}\pm 1$ gives two numbers 5 and 3 that both are dividers of 15. The complexity of the problem in the general case comes from the fact that direct verification of the validity of the equation $a^r \bmod M = 1$ requires trying a large ($\sim M$) number of different entries. The main idea of the quantum Shor algorithm is to perform classical calculations not with each individual entry x, but with the equally weighted quantum superposition of all entries at once, and obtain the required period r from the Fourier analysis of the resulting quantum interference of all the equally weighted outcomes.

The Shor algorithm considers the quantum computer as a bipartite system consisting of two parts A and B, each of which is a composite system by itself comprising N and K two-level systems, respectively. These two-level systems numerated by index i are elementary quantum registers, qubits, which can be in the ground state, $|n_i = 0\rangle$, in the excited state, $|n_i = 1\rangle$, or in a linear superposition of these two states. Prior to the Fourier analysis, parts A and B are prepared in a certain state. With a certain probability amplitude ψ_x, part A contains the integer argument $x = \sum_i n_i 2^{n_i}$ with the binary transcription $\{n_i\}$ encoded in the register as the quantum state

$$|x\rangle = \prod_{\otimes i=1}^{N} |n_i\rangle\,, \tag{10.41}$$

while the other part, B, contains the binary transcription of the results of classical calculations $y = f(x)$ for the entry x encoded in part A. The corresponding state vector

$$|\Psi\rangle = \sum_x \psi_x\,|x\rangle\,|y(x)\rangle = \sum_x \psi_x\,|x\rangle\,|a^x \bmod M\rangle \tag{10.42}$$

describes an entangled state of the parts as long as at least two of the amplitudes ψ_x differ from zero.

The first step of the Shor algorithm is meant to make all possible entries x equally probable, such that all the amplitudes ψ_x are equal. To this end, from the initial product ground state

$$|\Psi\rangle = |\mathbf{0}\rangle_A \otimes |\mathbf{0}\rangle_B \tag{10.43}$$

of the parts after applying to each of the qubits comprising the register A the local unitary transformation

$$\hat{H}d \equiv \begin{pmatrix} \frac{1}{\sqrt{2}} & \frac{1}{\sqrt{2}} \\ \frac{1}{\sqrt{2}} & -\frac{1}{\sqrt{2}} \end{pmatrix}, \tag{10.44}$$

known as the Hadamard gate, one obtains a linear combination of equally weighted entries $|x\rangle$ of the register

$$|\Psi\rangle = \left(\frac{1}{2^{N/2}} \sum_{x=0}^{2^N-1} |x\rangle\right) \otimes |\mathbf{0}\rangle_B\,. \tag{10.45}$$

Indeed, the probability amplitude of each string $\{n_i\}$ equals $2^{-N/2}$, since each of the N qubits i can be in the upper and in the lower state with equal amplitudes $2^{-1/2}$. At this step, two parts of the computer still are not entangled. Moreover, even individual qubits belonging to these parts are not entangled either. They all become entangled right after performing the second step of the algorithm, namely the classical calculation, which records in the register B the result of calculation of the function $f(x) = a^x \bmod M$ for each entry x, when the state of the computer reads

$$|\Psi\rangle = \frac{1}{2^{N/2}} \sum_{x=0}^{2^N-1} |x\rangle \otimes |a^x \bmod M\rangle \,. \tag{10.46}$$

Let us trace these transformations for the particular case of $M = 15, a = 2$. Just four qubits are sufficient for the binary transcription of the number 15, and therefore we take $N = K = 4$. The basis of x ranges from 0 through 15. At the beginning

$$|\Psi\rangle = |0,0,0,0\rangle_A \otimes |0,0,0,0\rangle_B = |\mathbf{0}\rangle_A \otimes |\mathbf{0}\rangle_B \,. \tag{10.47}$$

After the first step, the Hadamard transformation,

$$\begin{aligned}
|\Psi\rangle &= \left(\frac{1}{4}|0,0,0,0\rangle_A + \frac{1}{4}|0,0,0,1\rangle_A + \ldots \frac{1}{4}|1,1,1,1\rangle_A\right) \otimes |0,0,0,0\rangle_B \\
&= \left(\frac{1}{4}|0\rangle_A + \frac{1}{4}|1\rangle_A + \frac{1}{4}|2\rangle_A + \ldots \frac{1}{4}|15\rangle_A\right) \otimes |0\rangle_B \\
&= \frac{1}{4}\sum_{x=0}^{15} |x\rangle_A \otimes |0\rangle_B \,,
\end{aligned} \tag{10.48}$$

and after the second step, the classical calculation, according to (10.40), one finds

$$\begin{aligned}
|\Psi\rangle &= \frac{1}{4}\sum_{x=0}^{15} |x\rangle_A \otimes |a^x \bmod M\rangle_B \\
&= \frac{1}{4}|0\rangle_A \otimes |1\rangle_B + \frac{1}{4}|1\rangle_A \otimes |2\rangle_B + \frac{1}{4}|2\rangle_A \otimes |4\rangle_B + \frac{1}{4}|3\rangle_A \otimes |8\rangle_B + \\
&\frac{1}{4}|4\rangle_A \otimes |1\rangle_B + \frac{1}{4}|5\rangle_A \otimes |2\rangle_B + \frac{1}{4}|6\rangle_A \otimes |4\rangle_B + \frac{1}{4}|7\rangle_A \otimes |8\rangle_B + \ldots, \\
&= \frac{|0\rangle_A + |4\rangle_A + |8\rangle_A + |12\rangle_A}{4} \otimes |1\rangle_B + \frac{|1\rangle_A + |5\rangle_A + |9\rangle_A + |13\rangle_A}{4} \otimes |2\rangle_B + \\
&\frac{|2\rangle_A + |6\rangle_A + |10\rangle_A + |14\rangle_A}{4} \otimes |4\rangle_B + \frac{|3\rangle_A + |7\rangle_A + |11\rangle_A + |15\rangle_A}{4} \otimes |8\rangle_B \,,
\end{aligned} \tag{10.49}$$

and hence the two parts A and B are entangled. For this state, the rank of the singular value decomposition (10.32) equals to 4, both density matrices of the parts have all eigenvalues equal to $1/4$, the linear entropy of the parts entanglement amounts to $3/4$, and the von Neumann entropy is $\ln 4$. Note that in the general case, the rank of the density matrix of either part is of the order of the period r, while S_l and S_{vN} can be estimated by the numbers $1 - 1/r$ and $\ln r$, respectively.

We now come to the third step of Shor's algorithm, the determination of the period r. This is the most nontrivial part of the algorithm – an essentially quantum procedure known as discrete quantum Fourier transformation. It relies on essentially nonlocal transformations of the qubit states within part A, which changes not only the order of the states as does the classical computation but also essentially affects the amplitudes ψ_x. However, being a local transformation with respect to the partition of the system to part A and part B, this procedure does not change entanglement between these parts.

We show how to perform the discrete Fourier transform modulo $L = 2^4 = 16$, which is a unitary transformation on $N = 4$ qubits of part A defined by the relations

$$F_L|x\rangle = \frac{1}{\sqrt{L}} \sum_{y=0}^{L-1} \exp(2\pi i x y/L)|y\rangle, \tag{10.50}$$

and then generalize the procedure to the case of a larger number of qubits. Here $|x\rangle$ and $|y\rangle$ are states of the system computational basis, corresponding to the binary strings $\{n_i\}$ and $\{m_i\}$, respectively. The procedure employed to perform the Fourier transformation is based on the representation of the exponent in (10.50) as

$$\exp(2\pi i x y/2^4) = \exp\left\{\frac{2\pi i}{2^4}\left(\sum_{r=1}^{4} n_{4-r+1} 2^{4-r}\right)\left(\sum_{s=1}^{4} m_s 2^{s-1}\right)\right\}$$
$$= \prod_{r=1}^{4}\prod_{s=1}^{r} \exp(i\pi n_{5-r} m_s/2^{r-s}). \tag{10.51}$$

We begin the procedure by reversing the order in which the bits of the input x are stored in our 4-qubit register A, that is, we perform the unitary transformation

$$|n_4\rangle_4 \ldots |n_2\rangle_2|n_1\rangle_1 \;\rightarrow\; |n_1\rangle_4 \ldots |n_3\rangle_2|n_4\rangle_1 \tag{10.52}$$

by applying a sequence of state exchanges, which are operations on pairs of qubits defined as the unitary transformation

$$|0\rangle_i|0\rangle_j \rightarrow |0\rangle_i|0\rangle_j\,;\, |1\rangle_i|1\rangle_j \rightarrow |1\rangle_i|1\rangle_j\,;$$
$$|0\rangle_i|1\rangle_j \rightarrow |1\rangle_i|0\rangle_j\,;\, |1\rangle_i|0\rangle_j \rightarrow |0\rangle_i|1\rangle_j\,; \tag{10.53}$$

applied to qubits i and j. Then we complete the procedure in four steps: (i) We "split" the first qubit by applying the unitary Hadamard transformation (10.44) Note that this would already complete the Fourier transform if we had only one qubit. (ii) Next, we apply to the first and second qubits the conditional phase shift $|a\rangle_2|b\rangle_1 \rightarrow e^{i\pi ab/2}|a\rangle_2|b\rangle_1$ $(a, b = 0, 1)$, given explicitly by

$$\hat{B}_{21} \equiv \begin{pmatrix} 1\,0\,0 & 0 \\ 0\,1\,0 & 0 \\ 0\,0\,1 & 0 \\ 0\,0\,0 & e^{i\pi/2} \end{pmatrix} = \hat{B}(\pi/2), \tag{10.54}$$

where $\hat{B}(\phi)$ is the unitary transformation

$$\hat{B}(\phi) = \exp[-i\phi \begin{pmatrix} 0\,0\,0\,0 \\ 0\,0\,0\,0 \\ 0\,0\,0\,0 \\ 0\,0\,0\,1 \end{pmatrix}]. \tag{10.55}$$

Then we "split" the second qubit by applying to it the transformation Hd. This accounts for the contribution of the second most significant bit of the input x. Similarly, in step (iii) we apply the conditional phase shift $|a\rangle_i|b\rangle_j \to e^{i\pi ab/2^{i-j}}|a\rangle_i|b\rangle_j$ that is $\hat{B}_{ij} \equiv \hat{B}(\pi/2^{i-j})$, to two pairs of qubits $(3,1)$ and $(3,2)$, followed by the Hadamard transformation of the third qubit, and finally (iv) perform three sequentional transformations B_{ij} for the pairs $(4,1)$, $(4,2)$, and $(4,3)$ followed by the gate Hd applied to the fourth qubit. In general, for an N-qubit register A

$$\hat{F}_{2^N} = (\hat{H}d_N \hat{B}_{N;N-1} \cdots \hat{B}_{N,1}) \cdots (\hat{H}d_3 \hat{B}_{32} \hat{B}_{31})(\hat{H}d_2 \hat{B}_{21})(\hat{H}d_1). \tag{10.56}$$

Application of the Fourier transformation (10.56) to the state (10.49) yields the state

$$|\Psi\rangle = \frac{|0\rangle_A + |4\rangle_A + |8\rangle_A + |12\rangle_A}{4}|1\rangle_B + \frac{|0\rangle_A + i\,|4\rangle_A - |8\rangle_A - i\,|12\rangle_A}{4}|2\rangle_B$$
$$+ \frac{|0\rangle_A - |4\rangle_A + |8\rangle_A - |12\rangle_A}{4}|4\rangle_B + \frac{|0\rangle_A - i\,|4\rangle_A - |8\rangle_A + i\,|12\rangle_A}{4}|8\rangle_B,$$

which is a sum of $r = 4$ terms. Each of these terms is given by the direct product of a ket vector $|k\rangle$ of part B, which encodes the remainder $k \in \{2^x \bmod M\} = \{1, 2, 4, 8\}$ resulting from division of the number a^x by the number M, and a ket vector $|y_k\rangle$ of part A, which is equal to the Fourier transform (10.50) of the sum of all states $|x\rangle$ that correspond to this remainder k. Here the symbol of the direct product $\otimes$ is implicit. The sum representing the state $|\Psi\rangle$ can be regrouped differently

$$|\Psi\rangle = \frac{|0\rangle_A}{4}(|1\rangle_B + |2\rangle_B + |4\rangle_B + |8\rangle_B) + \frac{|4\rangle_A}{4}(|1\rangle_B + i\,|2\rangle_B - |4\rangle_B - i\,|8\rangle_B)$$
$$+ \frac{|8\rangle_A}{4}(|1\rangle_B - |2\rangle_B + |4\rangle_B - |8\rangle_B) + \frac{|12\rangle_A}{4}(|1\rangle_B - i\,|2\rangle_B - |4\rangle_B + i\,|8\rangle_B),$$

clearly showing that register A now contains only multiples $|mr\rangle$ of the period $r = 4$, and these states of part A are multiplied by the corresponding states of part B, each of which is given by the sum of the remainder states $|k\rangle$, which enters with a phase factor $e^{im\theta_k}$, where the angles $\theta_0 = 0$, $\theta_2 = \pi/2$, $\theta_4 = \pi$, and $\theta_8 = 3\pi/2$ depend only on the remainder value.

In the general case, the state vector of the computer is a sum of r terms each being given by the direct product of the state $|k\rangle$ of part B, which encodes the remainder

$a^x \bmod M = k$ of the division of the number a^x by the number M, and the state of part A, which is equal to the Fourier transform (10.50) of the properly weighted sum of all states $|x\rangle$ that correspond to the numbers x giving this remainder. Since each remainder periodically appears in the string either with the longest period or with its fraction, the result of the Fourier transformation

$$|\Psi\rangle = \sum_{k \in \text{set of remainders}} \left(\sum_m \psi_m e^{im\theta_k} |mr\rangle_A \right) \otimes |k\rangle_B \tag{10.57}$$

is a sum of all multiples of the periods mr, each of which enters with a phase factor $e^{im\theta_k}$ depending on θ_k—a remainder sensitive angle, whereas the amplitude ψ_m depends exclusively on the number m. Regrouping of this sum yields

$$|\Psi\rangle = \sum_m \psi_m |mr\rangle_A \otimes \sum_{k \in \text{set of remainders}} e^{im\theta_k} |k\rangle_B \tag{10.58}$$

We have therefore arrived at the solution of the most difficult computational part of the factorization problem—finding the period of the function $2^x \bmod M$, while the rest of the computation can be done classically. Of course, finding a common multiple of large numbers coming out as the results of measurements of the state of the register A can also be a difficult numerical job, but this is still a much less challenging task as compared to the initial problem of the number factorization.

10.2.3 Grover's Algorithm

The second fundamental result of quantum informatics is the Grover's search algorithm. In contrast to Shor's algorithm, it relies not on bipartite entanglement discussed thus far, but on essentially multipartite entanglement, which will be considered in detail later on in this chapter. In a very large data base, Grover's algorithm finds an entry which satisfies a predetermined requirement. For a requirement transcripted in an n-bit binary register, the classical search implies verification of its validity for all 2^n different possible entries, while the quantum search algorithm, which makes use of the interference of different entries, can accomplish the task much faster, after just $\sqrt{2^n}$ operations.

The key ingredient of Grover's algorithm is the quantum mechanical formalization of the search criterion. Though given not as an explicit matrix but as a sequence of operational prescriptions, the verification criterion still corresponds to a unitary transformation $\hat{R}_I$ of the quantum computer state, which can be formally associated with a Hermitian Hamiltonian $i\pi^{-1} \ln \hat{R}_I$. The action of this Hamiltonian during a time interval π results in the application of the required transformation. The ideal search corresponds to such a Hamiltonian that has only one unity eigenvalue for the eigenvector $|R\rangle$ satisfying the criterion, while all other eigenvectors $|R'\rangle$ are perpendicular to the latter, $\langle R|R'\rangle = 0$, and have zero eigenvalues. Hence, the verification procedure is a unitary transformation, which simply changes the sign of the projection of the computer state vector at the eigenvector satisfying the criterion. One also calls this procedure the sign inversion with respect to the result. Formally it can be written as $\hat{R}_I = 1 - 2|R\rangle\langle R|$.

Contraction of the algorithm requires another, similar, operation—the sign inversion with respect to the vacuum. The relevant Hamiltonian has only one non-zero eigenvalue equal to 1, which corresponds to the vacuum eigenstate $|0\rangle$. The formal expression for this operation reads $\hat{V}_I = 1 - 2|0\rangle\langle 0|$. Both these sign inversion operations are extremely nonlocal. The remaining necessary operations, the local Hadamard gates, are applied to each qubit of the computer, $\hat{H}_D = \prod_{\otimes i=1}^{N} \hat{H}d$.

The algorithm starts with the Hadamard gate applied to the vacuum state, which results in the equally weighted superposition of all possible entries, in complete analogy to the operation applied to the register A for Shor's algorithm. It is followed by the operation of the sign inversion with respect to the result. The next step is again application of H_D, which is followed by the sign inversion with respect to the vacuum, and by application of $\hat{H}_D$ once again. The circle of four operations $\hat{U} = \hat{V}_I\hat{H}_D\hat{R}_I\hat{H}_D$, or explicitly,

$$\hat{U} = (1 - 2|0\rangle\langle 0|)\,\hat{H}_D\,(1 - 2|R\rangle\langle R|)\,\hat{H}_D \tag{10.59}$$

is repeated again and again, until, after about $\pi 2^{N/2-1}$ circles, the computer arrives at the state $|R\rangle$, satisfying the criterion imposed. Indeed, consider the action of two periods U^2, take into account that $\hat{H}_D\hat{H}_D = 1$ and $\langle R|\,\hat{H}_D\,|0\rangle = 2^{-N/2}$, and find

$$\hat{U}^2 = 1 + \hat{H}_D\,|R\rangle\,\frac{4}{\sqrt{2^N}}\,\langle 0| - |0\rangle\,\frac{4}{\sqrt{2^N}}\,\langle R|\,H_D + O\left(\frac{1}{2^N}\right), \tag{10.60}$$

where $O\,(a)$ is a number of the order of a, which can also be written as an exponent

$$\hat{U}^2 = \exp\left\{-i\frac{4}{\sqrt{2^N}}\left[\hat{H}_D\,|R\rangle\,i\,\langle 0| + \text{c.c.}\right] + O\left(\frac{1}{2^N}\right)\right\}. \tag{10.61}$$

Application of this transformation $L = \mathrm{Ip}\left[2^{N/2-2}\pi\right]$ times, where by $\mathrm{Ip}\,[a]$ we denote the integer part of a real number a, results in the transformation

$$\hat{U}^{2L} = \exp\left\{-i\pi\left[\hat{H}_D\,|R\rangle\,i\,\langle 0| + \text{c.c.}\right]\right\} + O\left(\frac{1}{\sqrt{2^N}}\right). \tag{10.62}$$

The transformation $\hat{U}^{2L}$ is equivalent to a π-pulse, that is, the excitation exchange operation in a two-level system: with the probability $1 - O\left(2^{-N/2}\right)$ very close to unity, it brings the quantum computer, initially in the ground state, to the state

$H_D\,|R\rangle$, which by simple application of the Hadamard gate can be converted to the state $|R\rangle$, which satisfies the search criterion.

It is interesting to note that in half way to the final state, Grover's algorithm yields an equally weighted coherent superposition

$$\frac{\hat{H}_D\,|R\rangle + |0\rangle}{\sqrt{2}} \tag{10.63}$$

of the ground state $|0\rangle$ and the final state $|R\rangle$ transformed by the Hadamard gate. Both of these states are product states of all qubits comprising the quantum computer. Since both $|0\rangle = \prod_{\otimes i} |0_i\rangle$and $|R\rangle = \prod_{\otimes i} |R_i\rangle$; $R_i = 0, 1$ are vectors of the computational basis, the overlap $\langle R|\, H_D\,|0\rangle = 2^{-N/2}$ of two vectors entering the linear combination (10.63) is exponentially small, as was already stated above. Moreover, if we project the operator H_D to a subset A of all qubits comprising the register, the resulting operator

$$\prod_{\otimes i\in A} \langle R_i|\,\hat{H}_D\,|0_i\rangle = 2^{-M/2} \prod_{\otimes i\notin A} \hat{H}d_i \tag{10.64}$$

remains exponentially small as long as the selected subset A contains a large number M of the qubits. If we therefore formally divide the entire register of the computer in two parts A and B, compose the density matrix corresponding to the state (10.63) and trace over either of parts, say B, we arrive at a density matrix of part A, which has just two eigenvalues close to $1/2$ and all the rest eigenvalues are exponentially small in accordance with the smallness of the operator (10.64). As a consequence, the linear and von Neumann entanglement entropies equal $1/2$ and $\ln 2$, respectively, regardless of the way the computer is separated into two big parts. Moreover, each of these eigenstates corresponds to a product state of qubits: one to the product of the vacuum states, which was the initial state of the register, and the other to the product state of qubits containing the final result of calculation, simply transformed locally by the Hadamard gates.

10.2.4 Teleportation

This phenomenon, formulated in the most general way, shown in Fig. 10.4 relies on entanglement of three parts. Two parts A and B of a three partite systems are in a specially prepared entangled state, while the third part C being entangled neither with A nor with B is in an unknown quantum state. The teleportation procedure implies interaction of parts B and C followed by a measurement performed on these two parts in such a way that the outcome of this measurement allows one to uniquely define a local unitary transformation, which being applied to part A brings it to the same unknown state that part C was in initially. The name "teleportation" originates from the fact that the entangled parts A and B can be located at a long distance one from the other, whereas the result of the measurement performed on parts B and C after their interaction can be simply communicated via a classical information channel to the location of part A. This very communication specifies the local transformation to apply to A in order to obtain the desired result.

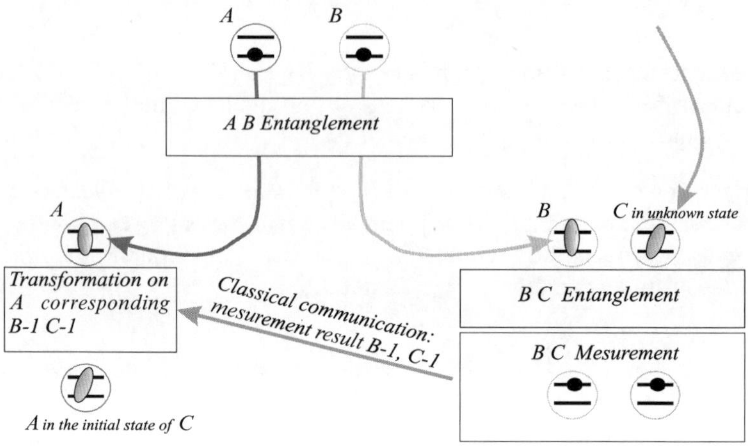

Fig. 10.4 Teleportation scheme. Two qubits A and B initially in the ground state become entangled as a result of an interaction described by a unitary evolution matrix U. After this procedure they are sent do different locations. As a result of the interaction described by a unitary matrix V, the qubit B gets entangled with a qubit C, which initially was in an unknown state $\alpha|0\rangle + \beta|1\rangle$. After the interaction, the state of pair BC is measured and the result of the measurement is communicated to the location of qubit A via a classical communication channel. Depending on the result of the measurement, the qubit A is subject to a local transformation, which puts this qubit into the same unknown state where the qubit C was prior to the procedure

In the simplest case, all three parts are qubits. One takes qubits A and B in the maximum entangled state

$$\frac{|0\rangle_A|0\rangle_B + |1\rangle_A|1\rangle_B}{\sqrt{2}}, \tag{10.65}$$

while part C is in an arbitrary state

$$\alpha|0\rangle_C + \beta|1\rangle_C, \tag{10.66}$$

such that the entire system is in the state

$$\begin{aligned}|\Psi\rangle = &\frac{\alpha}{\sqrt{2}}|0\rangle_A|0\rangle_B|0\rangle_C + \frac{\beta}{\sqrt{2}}|0\rangle_A|0\rangle_B|1\rangle_C \\ &+ \frac{\alpha}{\sqrt{2}}|1\rangle_A|1\rangle_B|0\rangle_C + \frac{\beta}{\sqrt{2}}|1\rangle_A|1\rangle_B|1\rangle_C.\end{aligned} \tag{10.67}$$

We now perform a unitary transformation resulting from the interaction of parts B and C and given by a matrix V^{BC}

$$\begin{pmatrix} |0\rangle_B\,|0\rangle_C \\ |0\rangle_B\,|1\rangle_C \\ |1\rangle_B\,|0\rangle_C \\ |1\rangle_B\,|1\rangle_C \end{pmatrix} \to \begin{pmatrix} V^{BC}_{0,0} & V^{BC}_{0,1} & V^{BC}_{0,2} & V^{BC}_{0,3} \\ V^{BC}_{1,0} & V^{BC}_{1,1} & V^{BC}_{1,2} & V^{BC}_{1,3} \\ V^{BC}_{2,0} & V^{BC}_{2,1} & V^{BC}_{2,2} & V^{BC}_{2,3} \\ V^{BC}_{3,0} & V^{BC}_{3,1} & V^{BC}_{3,2} & V^{BC}_{3,3} \end{pmatrix} \begin{pmatrix} |0\rangle_B\,|0\rangle_C \\ |0\rangle_B\,|1\rangle_C \\ |1\rangle_B\,|0\rangle_C \\ |1\rangle_B\,|1\rangle_C \end{pmatrix}, \tag{10.68}$$

which we choose as

$$V^{BC} = \begin{pmatrix} \frac{1}{\sqrt{2}} & 0 & 0 & \frac{1}{\sqrt{2}} \\ 0 & \frac{1}{\sqrt{2}} & \frac{1}{\sqrt{2}} & 0 \\ 0 & \frac{-1}{\sqrt{2}} & \frac{1}{\sqrt{2}} & 0 \\ \frac{-1}{\sqrt{2}} & 0 & 0 & \frac{1}{\sqrt{2}} \end{pmatrix}, \tag{10.69}$$

such that the state $|\Psi\rangle$ of entire system thus adopts the form

$$|\Psi\rangle = \left(\tfrac{\alpha}{\sqrt{2}}\,|0\rangle_A \;\; \tfrac{\beta}{\sqrt{2}}\,|0\rangle_A \;\; \tfrac{\alpha}{\sqrt{2}}\,|1\rangle_A \;\; \tfrac{\beta}{\sqrt{2}}\,|1\rangle_A \right) \times$$
$$\times \begin{pmatrix} \frac{1}{\sqrt{2}} & 0 & 0 & \frac{1}{\sqrt{2}} \\ 0 & \frac{1}{\sqrt{2}} & \frac{1}{\sqrt{2}} & 0 \\ 0 & \frac{-1}{\sqrt{2}} & \frac{1}{\sqrt{2}} & 0 \\ \frac{-1}{\sqrt{2}} & 0 & 0 & \frac{1}{\sqrt{2}} \end{pmatrix} \begin{pmatrix} |0\rangle_B\,|0\rangle_C \\ |0\rangle_B\,|1\rangle_C \\ |1\rangle_B\,|0\rangle_C \\ |1\rangle_B\,|1\rangle_C \end{pmatrix}. \tag{10.70}$$

Measurements of the states of parts B and C is the next step. If as a result of such a measurement both parts are in the ground state, part A turns to be in the state

$$|\Psi\rangle_A^{00} = \left(\tfrac{\alpha|0\rangle_A}{\sqrt{2}} \;\; \tfrac{\beta|0\rangle_A}{\sqrt{2}} \;\; \tfrac{\alpha|1\rangle_A}{\sqrt{2}} \;\; \tfrac{\beta|1\rangle_A}{\sqrt{2}} \right) \begin{pmatrix} \frac{1}{\sqrt{2}} \\ 0 \\ 0 \\ \frac{-1}{\sqrt{2}} \end{pmatrix} \tag{10.71}$$
$$= \frac{\alpha}{2}\,|0\rangle_A - \frac{\beta}{2}\,|1\rangle_A\,,$$

resulting from the scalar product of the line vector and the first column of the matrix. By the analogy

$$|\Psi\rangle_A^{01} = \frac{\beta}{2}\,|0\rangle_A - \frac{\alpha}{2}\,|1\rangle_A\,,$$
$$|\Psi\rangle_A^{10} = \frac{\beta}{2}\,|0\rangle_A + \frac{\alpha}{2}\,|1\rangle_A\,,$$
$$|\Psi\rangle_A^{11} = \frac{\alpha}{2}\,|0\rangle_A + \frac{\beta}{2}\,|1\rangle_A\,, \tag{10.72}$$

when the qubits B and C are detected in the states $|0\rangle\,|1\rangle$, $|1\rangle\,|0\rangle$, and $|1\rangle\,|1\rangle$, respectively. Evidently, each of these states of the qubit A has to be renormalized in accordance with the general prescription of quantum mechanics, once the corresponding state of part $B+C$ is detected.

After the detection of the state $|i\rangle\,|j\rangle$, the corresponding state $|\Psi\rangle_A^{ij}$ has to be subject to a relevant unitary transformation R^{ij},

$$\begin{pmatrix}|0\rangle_A\\|1\rangle_A\end{pmatrix}=\begin{pmatrix}R_{00}^{ij} & R_{01}^{ij}\\R_{10}^{ij} & R_{11}^{ij}\end{pmatrix}\begin{pmatrix}|0\rangle_A\\|1\rangle_A\end{pmatrix}, \tag{10.73}$$

which has to be chosen such that the result

$$\begin{aligned}
|\Psi\rangle_A^{00}&=\left(\tfrac{\alpha}{2}\ -\tfrac{\beta}{2}\right)\begin{pmatrix}R_{00}^{00} & R_{01}^{00}\\R_{10}^{00} & R_{11}^{00}\end{pmatrix}\begin{pmatrix}|0\rangle_A\\|1\rangle_A\end{pmatrix}\\
|\Psi\rangle_A^{01}&=\left(\tfrac{\beta}{2}\ -\tfrac{\alpha}{2}\right)\begin{pmatrix}R_{00}^{01} & R_{01}^{01}\\R_{10}^{01} & R_{11}^{01}\end{pmatrix}\begin{pmatrix}|0\rangle_A\\|1\rangle_A\end{pmatrix}\\
|\Psi\rangle_A^{10}&=\left(\tfrac{\beta}{2}\ \tfrac{\alpha}{2}\right)\begin{pmatrix}R_{00}^{10} & R_{01}^{10}\\R_{10}^{10} & R_{11}^{10}\end{pmatrix}\begin{pmatrix}|0\rangle_A\\|1\rangle_A\end{pmatrix}\\
|\Psi\rangle_A^{11}&=\left(\tfrac{\alpha}{2}\ \tfrac{\beta}{2}\right)\begin{pmatrix}R_{00}^{11} & R_{01}^{11}\\R_{10}^{11} & R_{11}^{11}\end{pmatrix}\begin{pmatrix}|0\rangle_A\\|1\rangle_A\end{pmatrix}
\end{aligned} \tag{10.74}$$

always yields the same state $|\Psi\rangle_A^{i,j}\sim\alpha\,|0\rangle_A+\beta\,|1\rangle_A$ of part A, which coincides with the initial state of the qubit C. To this end, one has to choose

$$R^{00}=\sigma^z;\,R^{01}=\sigma^y;\,R^{10}=\sigma^x;\,R^{10}=I, \tag{10.75}$$

and, indeed, obtain $|\Psi\rangle_A^{i,j}=\frac{1}{2}\left(\alpha\,|0\rangle_A+\beta\,|1\rangle_A\right)$, which means that any of the four states $|00\rangle$, $|01\rangle$, $|10\rangle$, and $|11\rangle$ with equal probability $1/4=(1/2)^2$ can be detected as the outcome of the measurement performed on part $B+C$, while the state of part A is always the same, $\alpha\,|0\rangle_A+\beta\,|1\rangle_A$.

Now consider the general case of a transformation, which can be employed for the teleportation of qubits. Let the initial entangled state of the qubits A and B results from the application of a unitary transformation $\hat{U}$, given by the matrix $U_{nm}^{kl}=\langle l|_A\,\langle k|_B\,\hat{U}\,|n\rangle_B\,|m\rangle_A$ to the ground state $|0\rangle_B\,|0\rangle_A$. The state vector of the entire system then reads

$$\begin{aligned}
|\Psi\rangle_{ABC}&=\left(\sum_{n,m=0,1}|m\rangle_A\,|n\rangle_B\,U_{nm}^{00}\right)\otimes\left(\alpha\,|0\rangle_C+\beta\,|1\rangle_C\right)\\
&=\sum_{n,m,k}|m\rangle_A\,|n\rangle_B\,|k\rangle_C\,U_{nm}^{00}r_k,
\end{aligned} \tag{10.76}$$

where $r_0 = \alpha$ and $r_1 = \beta$. The second unitary transformation performed on the qubits B and C is given by the matrix V with the elements $V^{kl}_{nm} = \langle l|_B \langle k|_C V |n\rangle_C |m\rangle_B$. It yields the state

$$|\Psi\rangle_{ABC} = \sum_{n,m,k,l,s} |m\rangle_A |l\rangle_B |s\rangle_C V^{kn}_{sl} U^{00}_{nm} r_k, \tag{10.77}$$

which after a measurement performed on the states of the qubits B and C and yielding the result $|l\rangle_B |s\rangle_C$, moves the qubit A to the quantum state

$$|\Psi\rangle_A = \sum_m |m\rangle_A \sum_{n,k} V^{kn}_{sl} U^{00}_{nm} r_k. \tag{10.78}$$

By a unitary transformation ${}^{ls}R$ of the qubit A, which is specific for the outcome ls of the measurement and is given by the matrix elements ${}^{ls}R^m_f = \langle m|_A \, {}^{ls}R \, |m\rangle_A$, the latter state must be moved to the state that was the initial state of the qubit C, that is,

$$\sum_{m,n} {}^{ls}R^m_f V^{kn}_{sl} U^{00}_{nm} \sim \delta^k_f. \tag{10.79}$$

In other words, for each pair l, s, the inverse of the matrix ${}^{ls}M^k_m = \sum_{n=0,1} V^{kn}_{sl} U^{00}_{nm}$ should be proportional to a unitary matrix. This is the very condition, which determines the possibility of teleportation for qubits and allows one to explicitly construct the required transformation matrix

$${}^{ls}R = \frac{1}{a} \begin{pmatrix} U^{00}_{10} V^{01}_{ls} + U^{00}_{00} V^{00}_{ls} & -U^{00}_{01} V^{00}_{ls} - U^{00}_{11} V^{01}_{ls} \\ -U^{00}_{00} V^{10}_{ls} - U^{00}_{10} V^{11}_{ls} & U^{00}_{01} V^{10}_{ls} + U^{00}_{11} V^{11}_{ls} \end{pmatrix},$$

where a is a normalization constant, which makes the determinant of this matrix equal to unity. Note that, for teleportation of quantum systems larger than qubits, the requirements imposed on the transformations U and V are identical to (10.79), with the only difference that the indices l, s, m, f, k, n numerating the states now are not only unities or zeros, but also take larger values, depending on the dimensionality of the corresponding Hilbert space. But again, the requirement is that for each s and l, the matrix $\sum_n V^{kn}_{sl} U^{00}_{nm}$ depending on the indices k and m should be unitary.

10.3 How to Characterize Multipartite Entanglement?

In this section we are going to introduce a way of multipartite entanglement characterization. Just from the beginning, we emphasize that entanglement is not a physical quantity independent of the observer, since it essentially relies on the way the observer divides the system in parts, and on the particular choice of the operators attributed to measurable physical quantities related to different parts. As often happens, some physical concepts and ideas rely on a specific domain of mathematics that offers convenient notions and tools for adequate description of physical phenomena. It turns out that for entanglement, one of the suggestions for such a tool is ring theory, that is, an algebraic construction with well defined addition, subtraction, and multiplication, but which does not incorporate division as a uniquely defined operation—the phenomenon of the so-called dividers of zero, when the product of two non-zero elements equals zero is typical of the rings. One of the objects that form a ring is the rising operator $\hat{\sigma}^+ = \hat{\sigma}^x + i\hat{\sigma}^y$ for a two-level system, which has the property $(\hat{\sigma}^+)^2 = 0$, thus being the divider of zero in the ring of 2×2 matrices. Nontrivial n-th roots of zero are called nilpotent elements of $(n-1)$-th order, and $\hat{\sigma}^+$ is thus a first-order nilpotent element. Nilpotent variables will be widely employed later on in this section for the entanglement characterization, and in order to shorten the notations, the hat signs will be omitted for these objects, $\hat{\sigma}^+ \to \sigma^+$.

10.3.1 Some Group-Theory Aspects of Entanglement

The entanglement description relies on a specific partition of the composite physical system, sketched in Fig. 10.5, which illustrates that the system can often be decomposed into a number of subsystems in many different ways, each of the subsystems possibly being a composite system by itself. In order to avoid an ambiguity related to such a situation, we have to introduce a few definitions. Given a partition of the composite system into n subsystems, we call each of them an "element" and characterize it by a single, possibly collective, quantum number. Thus, the composite system is a collection of the elements. We call this collection an "assembly" in order to avoid confusion with an "ensemble", which is usually understood as a set of all possible realizations of a many-body system with an associated probability distribution over these realizations. The i-th element is assumed to have Hilbert space of dimension d_i. Qubit, qutrit, and qudit are widely used names for two-, three-, and d-level elements (with $d_i = 2$, $d_i = 3$, and $d_i = d$, respectively).

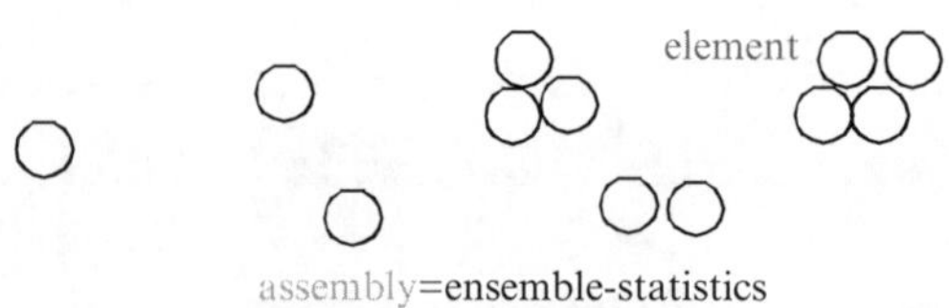

Fig. 10.5 The first step for the entanglement characterization is to choose a partition: an assembly consists of elements each of which can be a composite system by itself

The term "entanglement" has a transparent qualitative meaning: A pure state of an assembly is entangled with respect to a chosen partition when its state vector cannot be represented as a direct product of state vectors of the elements. This notion can also be extended to generic mixed quantum states, whereby entanglement is defined by the inability to express the assembly density operator as a probabilistic combination of direct products of the density operators of the elements, but this issue will not be addressed in this chapter.

We will focus on the question: how can one quantitatively characterize a multipartite entanglement for an assembly of many elements in a pure state? This question can be easily answered for the bipartite setting $n = 2$ discussed in Sect. 10.1.2, that is, for an assembly consisting just of two distinguishable elements A and B, each of arbitrary dimension. In this case, the von Neumann $\mathrm{Tr}[\rho_A \log \rho_A]$ entropy, or linear $\mathrm{Tr}[\rho_A^2]$ entropy, both based on the reduced density operator of either element, e.g., $\rho_A = \mathrm{Tr}_B[|\Psi\rangle\langle\Psi|]$, may be chosen as entanglement measure for a pure state $|\Psi\rangle$ of the assembly. However, already for the tripartite case $n = 3$, characterizing and quantifying entanglement becomes much harder. In fact, there is not one, but there are many different characteristics of entanglement and, apart from the question "How much?", one has also to answer the question: "In which way" are different elements entangled? Therefore the construction of appropriate measures is not unique, and is mostly dictated by convenience.

The main requirement imposed on the construction is that the entanglement characteristics should remain invariant with respect to the local operations that are quantum transformations one applies to each element of the assembly individually. Note that the local operations can be either unitary or simply invertible, depending on the physical situation under consideration. Group theory offers a natural language for the situation: local operations form a subgroup of all possible transformations of the assembly state vector. Under the action of the subgroup, the state vector of the system undergoes changes, still remaining within an orbit, that is, a subset $\mathcal{O}$ of the overall Hilbert space $\mathcal{H}$. The dimension of the coset $\mathcal{H}/\mathcal{O}$ yields the number of independent invariants identifying a generic orbit (see Fig. 10.6). Still, there are singular classes of orbits that require special consideration.

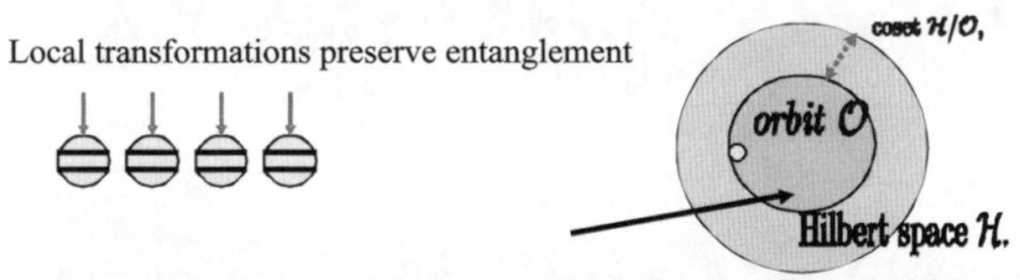

Fig. 10.6 Local transformations move the state vector along an orbit. The coset dimension of the subgroup of the local transformations (dimensionality of the space orthogonal to the orbit) determines how many characteristics of the entanglement are required

One may want to directly construct a set of these invariants under local operations. When dealing with an assembly of k qubits, this task can be accomplished in a regular and straightforward way. One considers the state vector of the assembly

$\psi_{n_1 n_2 \ldots n_k}$ in the representation where each index $n_i = 0, 1$ denotes the state of the qubit i, and its dual vector $\psi^{n'_1 n'_2 \ldots n'_k}$ given by the convolution

$$\psi^{n'_1 n'_2 \ldots n'_k} = \epsilon^{n_1 n'_1} \epsilon^{n_2 n'_2} \ldots \epsilon^{n_k n'_k} \psi_{n_1 n_2 \ldots n_k}, \tag{10.80}$$

with the antisymmetric tensors $\epsilon^{ii'}$ of ranks 2. The required invariants are various convolutions of $\psi_{n_1 n_2 \ldots n_k}$ and $\psi^{n'_1 n'_2 \ldots n'_k}$. The explicit form of these invariants depends critically on the number of elements in the assembly. It has a simple form for 2-qubit assembly

$$I = \psi_{nm} \psi^{nm} = \psi_{00} \psi_{11} - \psi_{01} \psi_{10}. \tag{10.81}$$

For a 3-qubit assembly the situation gets more complicated—five independent local invariants exist, namely three real numbers

$$\begin{aligned} I_1 &= \psi_{kij} \psi^{*pij} \psi_{pmn} \psi^{*kmn}, \\ I_2 &= \psi_{ikj} \psi^{*ipj} \psi_{mpn} \psi^{*mkn}, \\ I_3 &= \psi_{ijk} \psi^{*ijp} \psi_{mnp} \psi^{*mnk}, \end{aligned} \tag{10.82}$$

and the real and the imaginary part of a complex number,

$$I_4 + i I_5 = \psi_{ijk} \psi^{ijp} \psi_{mnp} \psi^{mnk}. \tag{10.83}$$

The quantity $2|I_4 + i I_5|$ is also known by the name residual entanglement or 3-tangle τ.

Similar invariants can still be found for a 4-qubit system. However, with increasing n, the explicit form of the invariants becomes less and less tractable and convenient for practical use, since for $n > 4$ even verification of the functional independence of the chosen invariants may become a computational challenge. Moreover, no explicit physical meaning can be attributed to such invariants, while for the elements larger than qubits no analog of (10.80) for the invariants exists at the moment. Briefly, an explicit construction of the orbit invariants becomes unpractical for assemblies of more than a few qubits.

10.3.2 Canonic States of an Assembly

Fortunately, apart from the invariants, there exists another way to identify an orbit, which the entanglement characterization can rely on. Every orbit can be represented by a "marker", which is one of the orbit state vectors selected according a certain rule. In order to unambiguously attribute a marker to each orbit, we specify a canonic form of an entangled assembly state. To this end, we first identify a reference state

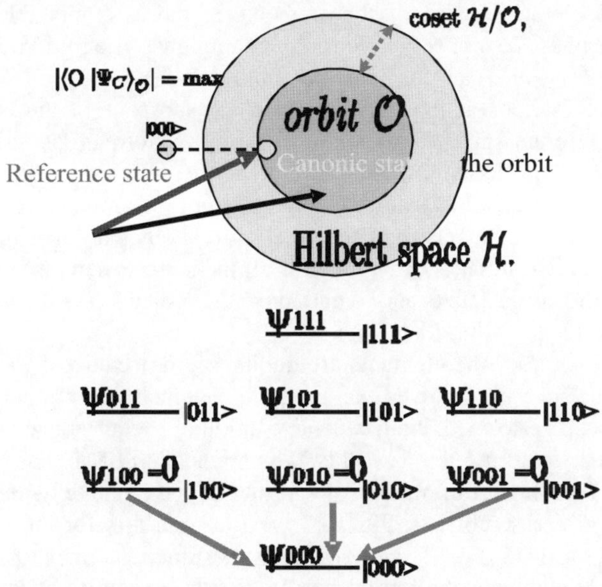

Fig. 10.7 Canonic state determined as the orbit state closest to a chosen product state (say the vacuum state) serves as the orbit marker. Within the orbit, it corresponds to the maximum possible population of the vacuum state, as is shown at the level scheme for a 3-qubit assembly. The canonic form of the entangled state for three qubits. By a local transformation the amplitudes of the lowest excited states *and* the phases of the second highest excited states are set to zero. The amplitude of the lowest state is taken as the common factor determining the normalization and the global phase

$|\mathrm{O}\rangle$ as a direct product of the element states (see Fig. 10.7). The latter can be chosen in an arbitrary way, but for qubits the choice $|0\rangle$, with the lowest energy level occupied, is the one most natural. Thus, the reference state reads $|\mathrm{O}\rangle = |0,\ldots 0\rangle$. Drawing parallels with quantum field theories and spin systems, we will call $|\mathrm{O}\rangle$ the "ground" or "vacuum" state.

By applying local unitary operations to a generic quantum state $|\Psi\rangle$, we can bring it into the "canonic form" $|\Psi_c\rangle$ corresponding to the maximum possible population of the reference state $|\mathrm{O}\rangle$. In other words, we apply a direct product $U_1 \otimes \ldots \otimes U_n$ of local transformations to the state vector $|\Psi\rangle$, and choose the transformation parameters to maximize $|\langle \mathrm{O}| U_1 \otimes \ldots \otimes U_n |\Psi\rangle|^2$. For an assembly of qubits, the local transformation $U(n) = U_1 \otimes \ldots \otimes U_n$ satisfying this requirement,

$$|\langle \mathrm{O}| U_1 \otimes \ldots \otimes U_n |\Psi\rangle|^2 = \max$$

can be seen to be unique up to phase factors multiplying the upper states of each qubit. These factors can further be specified by the requirement that all amplitudes of the states where all but one qubits are in the upper state are real.

However, for a larger size of the elements in the assembly the requirement imposed to the phases is not sufficient for an unambiguous definition of the canonic state. Nevertheless, one can consider the subgroup $U(n,-1) = U_1' \otimes \ldots \otimes U_n'$ of local transformations acting exclusively in the subspace of the element states other than the ground state. Application of these transformations will not change the projection of the assembly state vector at the reference state. In order to uniquely define the canonic state we have to apply a similar procedure to the excited states of the elements, namely, to choose another product reference state (say $|1\rangle = |1,\ldots 1\rangle$) that is orthogonal to all the states where at least one of the elements is in the ground state, and to maximize the projection of the assembly state vector to the state $|1\rangle$ within $U(n,-1)$.

Note that if some of the elements are qubits, the procedure of maximizing the overlap with the state $|1\rangle$ involves exclusively the elements other than these qubits. After having been performed, it allows one to uniquely identify the canonic state for qubits and qutrits entering the assembly. The application of similar procedures to the elements of yet higher dimensions yields at the end a uniquely defined canonic state of the entire assembly. In Fig. 10.7 we illustrate this for the assemblies of qutrits (a) for the local $SU(3)$ and $SL(3)$ transformations, and for an assembly comprising three elements of different size (b). For the assembly shown in Fig. 10.8b

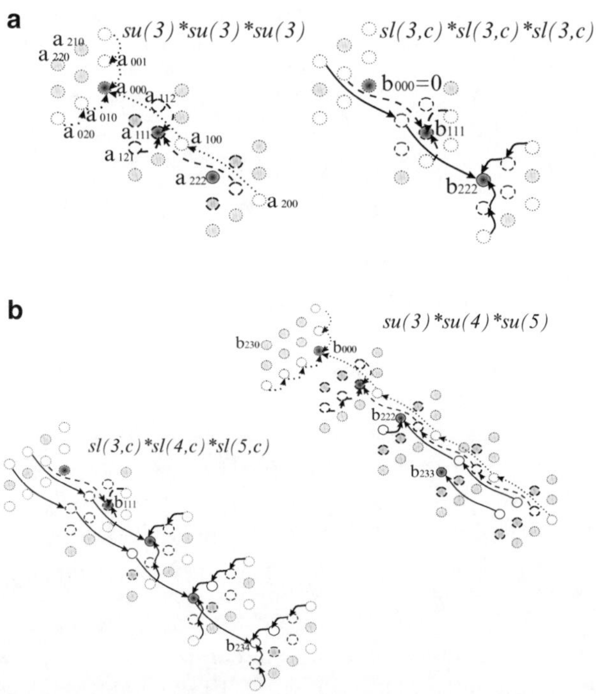

Fig. 10.8 Canonic states for the elements other than quibits require more than one reference product state. One can see this by the example of an assembly of three qutrits (**a**) and an assembly comprising three different qudits of $d = 3, d = 4$, and $d = 5$ levels

of a five-level system, a four-level system, and a qutrit, one should consecutively maximize:

1. the population of the state $|0,0,0\rangle$ by the transformations from $SU(5)\otimes SU(4)\otimes SU(3)$,
2. the population of the state $|1,1,1\rangle$ by the transformations from $SU(4)\otimes SU(3)\otimes SU(2)$,
3. the population of the state $|2,2,2\rangle$ by the transformations from $SU(3)\otimes SU(2)$,
4. the population of the state $|3,3,2\rangle$ by the transformations from the remaining $SU(2)$ mixing the third and the fourth excited states of the five-level system.

The choice of the canonic states uniquely specifies the parameters of local transformations and we thus are left only with the number of parameters of the state vector required exclusively for the entanglement characterization.

10.3.3 Extensive Characteristics Based on Nilpotent Polynomials: Nilpotential and Tanglemeter

Given the canonic state of an orbit, one still needs to characterize entanglement corresponding to this state in an economic and exhaustive way. We believe that the optimum choice for the purpose is the so-called extensive characteristics. Thermodynamic potentials linearly scaling with the number of particles in the system offer examples of extensive characteristics widely employed in statistical physics. The free energy given by the logarithm of the partition function is a specific important example. We will introduce similar characteristics for entangled states in such a way that their values for a product state coincide with the sum of the corresponding values for unentangled groups of elements. However, in contrast to thermodynamical potentials, this technique is based on the notion of nilpotent variables and functions of these variables, whence the logarithm function, transforming products into sums, plays the central role in the construction. As was already stated earlier, an algebraic variable x is called nilpotent if an integer n exists such that $x^n = 0$. In the case of qubits, these variables are naturally provided by the creation operators,

$$\sigma_i^+ = \begin{pmatrix} 0 & 1 \\ 0 & 0 \end{pmatrix}.$$

Evidently, $(\sigma_i^+)^2 = 0$, since the same quantum state cannot be created twice.

Our approach is based on three main ideas: (i) We express the state vector of the assembly in terms of a polynomial of creation operators for elements applied to a reference product state. (ii) Rather than working with the polynomial of nilpotent variables describing the state, we consider its logarithm, which is also a nilpotent polynomial. Due to the important role that this quantity will play throughout the development, we call this quantity the "nilpotential" henceforth, by analogy to thermodynamic potentials. (iii) The nilpotential is not invariant under local transformations, being different in general for different states in the same orbit. We therefore consider nilpotential of the canonic state of an orbit, which contains complete information about the entanglement in the assembly. We therefore call this quantity the "tanglemeter". The latter is, by construction, extremely convenient as an extensive orbit marker: the tanglemeter for an assembly consisting of several unentangled groups of elements equals the sum of tanglemeters of these groups.

Let us explain these ideas, for the simplest example of n qubits subject to the $su(2)_1 \oplus \ldots \oplus su(2)_i \ldots \oplus su(2)_n$ Lie algebra of local transformations. As a reference state, we choose the vacuum, that is, the state $|\mathrm{O}\rangle = |0, 0, \ldots, 0\rangle$ with all the qubits being in the ground state. An arbitrary state $|\Psi\rangle$ of the assembly may be generated via the action of a polynomial $F(\sigma_i^+)$ in the nilpotent operators σ_i^+ on the vacuum:

$$\begin{aligned}
|\Psi\rangle = & \sum_{\{k_i\}=0,1} \psi_{k_n k_{n-1} \ldots k_1} |k_n, k_{n-1}, \ldots, k_1\rangle \\
& = \psi_{00\ldots0} |0, 0, \ldots 0\rangle + \psi_{10\ldots0} |1, 0, \ldots 0\rangle \\
& \quad + \psi_{01\ldots0} |0, 1, \ldots 0\rangle + \ldots + \psi_{11\ldots1} |1, 1, \ldots 1\rangle \\
& = (\psi_{00\ldots0} + \psi_{00\ldots1}\sigma_1^+ + \ldots + \psi_{01\ldots0}\sigma_{n-1}^+ + \psi_{10\ldots0}\sigma_n^+ \\
& \quad + \psi_{00\ldots11}\sigma_2^+\sigma_1^+ \ldots + \psi_{11\ldots0}\sigma_n^+\sigma_{n-1}^+ + \ldots) |\mathrm{O}\rangle \\
& = \sum_{\{k_i\}=0,1} \psi_{k_n k_{n-1} \ldots k_1} \prod_{i=1}^{n} \left(\sigma_i^+\right)^{k_i} |\mathrm{O}\rangle . \qquad (10.84)
\end{aligned}$$

Here, the subscript i enumerates the qubits, and the operator σ_i^+ creates the state $|1\rangle$ out of the state $|0\rangle$ of i-th qubit. The family of all polynomials $F(\sigma_i^+)$ forms a ring, which means that it admits addition and multiplication, but division might be ambiguous. We note that the nilpotent variables introduced here commute with one another, in contrast to the anticommuting nilpotent (Grassmann) variables widely employed in quantum field theory that were discussed on p. 347.

For a technical reason, it is convenient to impose a non-standard normalization condition $|\langle \mathrm{O} \, |\Psi\rangle_{\mathrm{O}}| = 1$, that is,

$$\begin{aligned}
F(\{\sigma_i^+\}) = & \sum_{\{k_i\}=0,1} \alpha_{k_n k_{n-1} \ldots k_1} \prod_{i=1}^{n} \left(\sigma_i^+\right)^{k_i} \\
= & \sum_{\{k_i\}=0,1} \frac{\psi_{k_n k_{n-1} \ldots k_1}}{\psi_{00\ldots0}} \prod_{i=1}^{n} \left(\sigma_i^+\right)^{k_i} . \qquad (10.85)
\end{aligned}$$

Such a normalization is practical, since we mainly work not with F by itself, but with its extensive counterpart—nilpotential—a nilpotent polynomial

$$f(\{\sigma_i^+\}) = \ln\left[F(\{\sigma_i^+\})\right] = \sum_{\{k_i\}=0,1} \beta_{k_n k_{n-1}\dots k_1} \prod_{i=1}^{n} \left(\sigma_i^+\right)^{k_i}, \tag{10.86}$$

where the coefficients $\beta_{k_n k_{n-1}\dots k_1}$ and $\alpha_{k_n k_{n-1}\dots k_1}$ can be explicitly related to each other by expanding $\ln F$ in a Taylor series around 1. Since $(\sigma_i^+)^2 = 0$, this series is a polynomial containing at most $2^n - 1$ terms.

In order to uniquely characterize entanglement, as a convenient orbit marker we take a state $|\Psi_c\rangle_{\mathcal{O}}$ lying in the orbit $\mathcal{O}$, which is the one closest to the reference state $|\mathrm{O}\rangle$ in the inner product sense, that is, $|\langle \mathrm{O}\, |\Psi_c\rangle_{\mathrm{O}}| = \max$. Once the state $|\Psi_c\rangle_{\mathrm{O}}$ is found, we impose the normalization condition $|\langle \mathrm{O}\, |\Psi_c\rangle_{\mathrm{O}}| = 1$. The resulting canonic state $|\Psi_c\rangle_{\mathrm{O}}$ is associated with the canonic form of the polynomial $F_c(\sigma_i^+)$, which begins with a constant term equal to 1. Such a choice yields the tanglemeter—uniquely defined as a nilpotent polynomial $f_c = \ln F_c$. Both the tanglemeter and nilpotential resemble the eikonal, which is the logarithm of the regular *semi-classical* wave function in the position representation, multiplied by $-i$. The difference is that in our case no approximation is made: f represents the logarithm of the exact state vector.

The nilpotential f and the tanglemeter f_c have several remarkable properties: (i) The tanglemeter provides a unique and extensive characterization of entanglement. (ii) A straightforward entanglement criterion can be stated in terms of the cross derivatives $\partial^2 f/\partial\sigma_i^+\partial\sigma_j^+$: **The entanglement criterion:** *The parts A and B of a binary partition of an assembly of n qubits are unentangled iff*

$$\frac{\partial^2 f_c(\{x_i\})}{\partial x_k \partial x_m} = 0, \quad \forall k \in A, : \forall m \in B. \tag{10.87}$$

In other words, the subsystems A and B of the partition are disentangled iff $f_{A\cup B} = f_A(\{x_{\in A}\}) + f_B(\{x_{\in B}\})$, and no cross terms are present in the tanglemeter. Note that this criterion holds not only for the tanglemeter f_c, but for the nilpotential f as well. (iii) The dynamic equation of motion for f can be written explicitly and, suggestively, in the rather general case has the same form as the well-known classical Hamilton–Jacobi equation for the eikonal.

Number of Parameters Characterizing Entanglement

Consider n qubits in a generic pure state $|\Psi\rangle$ of (10.84), specified by 2^n complex amplitudes $\psi_{k_n k_{n-1}\dots k_1}$, i.e. by 2^{n+1} real numbers. When we take normalization into account and disregard the global phase, there are $2^{n+1} - 2$ real parameters characterizing the assembly state. The first relevant question is: What is the maximum number of real numbers needed for the orbit identification, hence, for entanglement characterization?

As has already been stated earlier, it is natural to require that any measure characterizing the intrinsic entanglement in the assembly state remains invariant under unitary transformations changing the state of each qubit. A generic $SU(2)$ transformation is the exponential of an element of the $su(2)$ algebra,

$$U = \exp[i(\sigma^x P^x + \sigma^y P^y + \sigma^z P^z)], \tag{10.88}$$

where σ^x, σ^y, and σ^z are Pauli matrices. It depends on the three real parameters P^x, P^y, and P^z. Therefore, for n qubits the dimension of the coset $\mathcal{H}/\mathcal{O}$, that is, the number of different real parameters invariant under local unitary transformations, reads

$$D_{su} = 2^{n+1} - pn - 2, \tag{10.89}$$

where $p = 3$ the number of the local transformation parameters for single qubit. To be precise, the counting (10.89) is true for $n > 2$ while the case $n = 2$ is special: in spite of the fact that $2^3 - 3 \cdot 2 - 2 = 0$, there is a nontrivial invariant of local transformations for two qubits of the form (10.81).

The coset dimension also depends on the symmetry group of local operations. For example, certain applications involving indirect measurements, and known as SLOCC maps, rely on a class of local transformations constrained only by the requirement of unit determinant. They are described by the complexification $sl(2,\mathbb{C})$ of the $su(2)$ algebra, such that the parameters (P_i^x, P_i^y, P_i^z) specifying the transformation of (10.88) on each qubit are now complex numbers. Therefore in (10.89) $p = 6$ for this case.

10.3.3.1 Examples: Canonic Forms for Two, Three, and Four Qubits

These examples are shown in Fig. 10.9. For two qubits the result is immediate:

$$f_c = \beta_{11}\sigma_2^+\sigma_2^+, \qquad F_c = 1 + \alpha_{11}\sigma_2^+\sigma_2^+ = 1 + \beta_{11}\sigma_2^+\sigma_1^+, \tag{10.90}$$

where the constant $\alpha_{11} = \beta_{11}$ can be chosen real. For three qubits, the canonic forms of F and f also differ only by the unity term,

$$f_c = \beta_3\sigma_2^+\sigma_1^+ + \beta_5\sigma_3^+\sigma_1^+ + \beta_6\sigma_3^+\sigma_2^+ + \beta_7\sigma_3^+\sigma_2^+\sigma_1^+ = F_c - 1. \tag{10.91}$$

Here, we have introduced a shorter notation by considering the indices of β as a binary representation of decimal numbers, $011 \to 3$, etc. One can make use of the fact that the variables σ_i^+ are defined up to phase factors, and set β_3, β_5, and β_6 real.

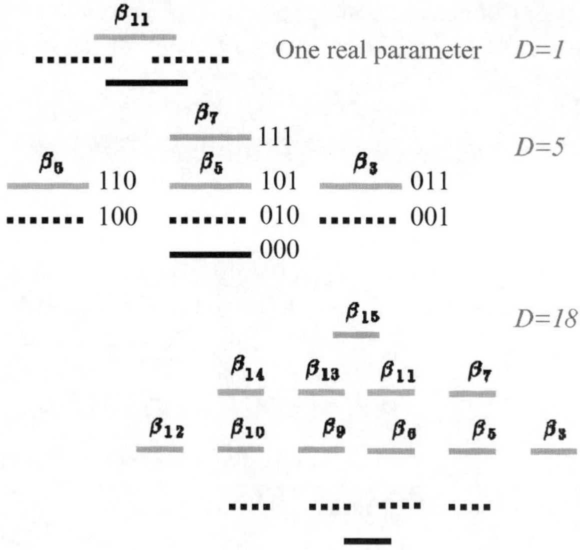

Fig. 10.9 Correspondence between the tanglemeter coefficients and the energy levels for 2-, 3- and 4-qubit cases. The states just below the maximum energy state all have real coefficients β

The tanglemeter for four qubits reads

$$\begin{aligned} f_c = {} & \beta_3\sigma_2^+\sigma_1^+ + \beta_5\sigma_3^+\sigma_1^+ + \beta_9\sigma_4^+\sigma_1^+ + \beta_6\sigma_3^+\sigma_2^+ + \beta_{10}\sigma_4^+\sigma_2^+ + \beta_{12}\sigma_4^+\sigma_3^+ \\ & + \beta_7\sigma_3^+\sigma_2^+\sigma_1^+ + \beta_{13}\sigma_4^+\sigma_3^+\sigma_1^+ + \beta_{11}\sigma_4^+\sigma_2^+\sigma_1^+ \\ & + \beta_{14}\sigma_4^+\sigma_3^+\sigma_2^+ + \beta_{15}\sigma_4^+\sigma_3^+\sigma_2^+\sigma_1^+, \end{aligned} \tag{10.92}$$

while the coefficients α_i of the polynomial F_c differ from β_i only at the last position

$$\alpha_{15} = \beta_{15} + \beta_3\beta_{12} + \beta_5\beta_{10} + \beta_9\beta_6. \tag{10.93}$$

One may note that the sums of the indices of the factors in this expression are equal. The latter is a general feature for the relationship among the coefficients α and β: the coefficients α are given by sums of terms, each of which contains a product of the coefficients β where the sum of the indices equals the index of α.

The next example corresponds to the first nontrivial case of local SL symmetry—the 3-qubit case. The sl-tanglemeter for a generic 3-qubit state reads

$$f_C = \sigma_3^+\sigma_2^+\sigma_1^+, \tag{10.94}$$

where the coefficient is set to unity by the scale freedom in the definition of the nilpotent variables. The corresponding wave function F_C is nothing but the so-called GHZ state $\frac{|000\rangle+|111\rangle}{\sqrt{2}}$. This shows again that all generic states belong to the same sl-orbit, which includes this state. There are, however, also three distinct singular classes of entangled states of measure zero whose tanglemeters do not involve the product $\sigma_1^+\sigma_2^+\sigma_3^+$ and have one of the following forms:

$$f_C = \sigma_2^+\sigma_1^+ + \sigma_3^+\sigma_1^+,$$

$$f_C = \sigma_3^+\sigma_1^+ + \sigma_3^+\sigma_2^+,$$

$$f_C = \sigma_2^+\sigma_1^+ + \sigma_3^+\sigma_2^+.$$

In this classification, we have only taken into account the states whose tanglemeters involve all three σ_i^+ such that no qubit is completely disentangled from the others. For a generic 4-qubit state one finds the sl-tanglemeter

$$f_C = \beta_3\sigma_2^+\sigma_1^+ + \beta_5\sigma_3^+\sigma_1^+ + \beta_9\sigma_4^+\sigma_1^+ + \beta_6\sigma_3^+\sigma_2^+ + \beta_{10}\sigma_4^+\sigma_2^+ + \beta_{12}\sigma_4^+\sigma_3^+ + \beta_{15}\sigma_4^+\sigma_3^+\sigma_2^+\sigma_1^+. \tag{10.95}$$

10.3.4 Quantum Dynamics in Terms of the Nilpotent Polynomials

One can describe the dynamics of an assembly state vector as the evolution of the corresponding polynomials F and f. As an example illustrating the usefulness of the time-dependent tanglemeter description, we present the evolution of the tanglemeter coefficients that emerge from implementation of the Grover's search algorithm. It turns out that the only tanglemeter coefficient different from zero stands in front of the product of the creation operators $\sigma_{i\in R}^+$ corresponding to the qubits that are at the state $|1\rangle$ at the end of the calculation

$$f_c = \beta_G(t)\prod_{i\in R}\sigma_i^+.$$

More specifically, the assembly of qubits that forms the quantum register evolves directly from the initial vacuum state to the final product state $\prod_{i\in R}\sigma_i^+|\text{O}\rangle$, which corresponds to the solution of the search problem: the qubits that belong to a variety R are in the upper state, while the rest of the qubits are in the lower state. The parameter $\beta_G(t)$ linearly increases from zero at the beginning, to its maximum value 1 in the middle of the process and then linearly returns back to zero at the end of the calculation, as the register returns back to a factorizable state, as shown in Fig. 10.10.

Let us now turn to the general situation and derive a dynamic equation for the nilpotential of a quantum assembly. To this end, we focus at infinitesimal transformations, and find the equations of motion describing the dynamics of the nilpotential under continuous local and gate operations. It turns out that for an

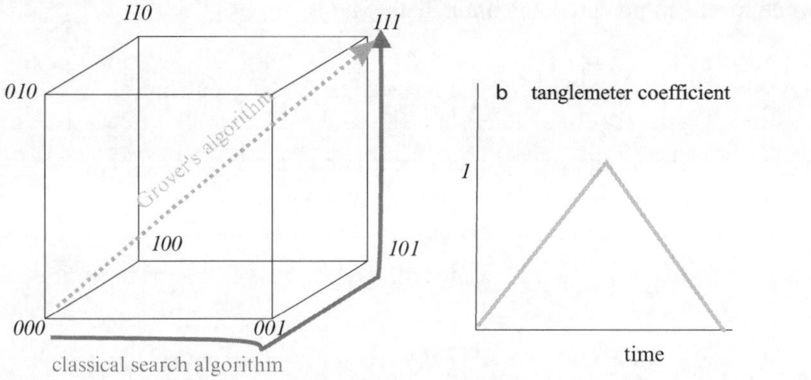

Fig. 10.10 Correspondence between the tanglemeter coefficients and the energy levels for 2-, 3- and 4-qubit cases. The states just below the maximum energy state all have real coefficients β

important class of Hamiltonians supporting universal quantum computation, the dynamic equation for the nilpotential acquires a well-known Hamilton–Jacobi form.

Consider an infinitesimal transformations acting on the state vector $|\Psi\rangle = F(\{\sigma_i^+\})|\mathrm{O}\rangle$ and yielding a transformed state $F'(\{\sigma_i^+\})|\mathrm{O}\rangle$. One can formalize the rules allowing one to obtain F' from F. Bearing in mind that $\sigma^-|0\rangle = 0$ and $\sigma^z|0\rangle = -|0\rangle$, one can represent the action of $\sigma_i^+, \sigma_i^z, \sigma_i^-$ operators on the state vector as appropriate differential operations acting on the polynomials $F(\sigma_i^+)$. The application of the operator σ_i^+ is straightforward—it is a direct multiplication: this operation eliminates the terms that were proportional to σ_i^+ prior to the multiplication. The application of σ_i^- is a kind of inverse: it can be considered as a derivative with respect to the variable σ_i^+, which eliminates the terms independent of σ_i^+ and makes the terms linear in σ_i^+ independent of this variable. Finally, the application of σ_i^z changes the signs of the terms independent of σ_i^+, and leaves intact terms linear in σ_i^+. These actions are summarized by the following formulas:

$$\begin{aligned}
&\sigma_i^+ F \to \sigma_i^+ F, \\
&\sigma_i^- F \to \frac{\partial F}{\partial \sigma_i^+}, \\
&\sigma_i^z F \to -F + 2\sigma_i^+ \frac{\partial F}{\partial \sigma_i^+},
\end{aligned} \tag{10.96}$$

while each unitary operation $U(\sigma_i^x, \sigma_i^y, \sigma_i^z)$ acting on the state vector can be represented by a differential operator $U\left(\sigma_i^+, \frac{\partial}{\partial \sigma_i^+}, 2\sigma_i^+ \frac{\partial}{\partial \sigma_i^+} - 1\right)$ acting on the polynomial

$$F' = U\left(\left\{\sigma_i^+, \frac{\partial}{\partial \sigma_i^+}, 2\sigma_i^+ \frac{\partial}{\partial \sigma_i^+} - 1\right\}\right) F, \tag{10.97}$$

while for the nilpotential one immediately finds

$$f' = \log\left(U\left(\left\{\sigma_i^+, \frac{\partial}{\partial\sigma_i^+}, 2\sigma_i^+\frac{\partial}{\partial\sigma_i^+} - 1\right\}\right)e^f\right). \tag{10.98}$$

Note that a generic transformation (10.98) of an initially canonic polynomial does not necessarily result in another canonic polynomial.

Consider now an infinitesimal unitary transformation $U = 1 - \mathrm{i}\,\mathrm{d}t\,H$ which is not necessarily local. The increment Δf of the nilpotential f suggested by (10.98) reads

$$\Delta f = \log\left(Ue^f\right) - \log\left(e^f\right) = \log\left(1 - \mathrm{i}\;\mathrm{d}t\;\mathrm{e}^{-f}He^f\right). \tag{10.99}$$

This yields the following dynamic equation for f:

$$i\frac{\partial f}{\partial t} = \mathrm{e}^{-f}He^f. \tag{10.100}$$

For the particular case of the local Hamiltonian

$$\begin{aligned} H_L &= \sum_i P_i^x(t)\sigma_i^x + P_i^y(t)\sigma_i^y \\ &= \sum_i P_i^-(t)\sigma_i^+ + P_i^+(t)\sigma_i^-, \end{aligned} \tag{10.101}$$

and the binary interaction

$$H_B = \sum_{i,j} G_{ij}(t)\sigma_i^+\sigma_j^-, \tag{10.102}$$

that together ensure the universal evolution of the qubit assembly, one finds the explicit equation

$$\begin{aligned} i\frac{\partial f}{\partial t} &= \sum_i \left[P_i^-(t)\sigma_i^+ + P_i^+(t)\frac{\partial f}{\partial\sigma_i^+}\left(1 - \sigma_i^+\frac{\partial f}{\partial\sigma_i^+}\right)\right] \\ &+ \sum_{i\neq j} G_{ij}(t)\sigma_j^+\frac{\partial f}{\partial\sigma_i^+}\left(1 - \sigma_i^+\frac{\partial f}{\partial\sigma_i^+}\right). \end{aligned} \tag{10.103}$$

Note that (10.103) for nilpotential formally resembles the Hamilton–Jacobi equation for the mechanical action of classical systems with the Hamiltonian

$$\begin{aligned} H &= \sum_i \left[P_i^-(t)x_i + P_i^+(t)p_i\left(1 - x_ip_i\right)\right] \\ &+ \sum_{i\neq j} G_{ij}(t)\left[x_jp_i\left(1 - x_ip_i\right)\right], \end{aligned} \tag{10.104}$$

where $p_i = \partial f/\partial\sigma_i^+$ play the role of the momenta, while $x_i = \sigma_i^+$ are the coordinates. Comparing with the conventional classical Hamilton–Jacobi equation, the only essential difference is the factor i multiplying the time derivative and the presence of complex parameters that can be interpreted as time-dependent forces and masses.

10.3.5 Entanglement Beyond Qubits

The nilpotent polynomials approach may be extended to describe situations more general than assemblies of qubits. Canonic states of such assemblies have already been discussed, and now the question is: What does one have to employ in place of the Pauli operators σ^+ in order to construct an extensive nilpotent characteristic? The construction of nilpotent polynomials for such systems is based on the so-called Cartan–Weyl decomposition. Here we illustrate this for qutrits corresponding to the local $SU(3)$ symmetry group. There are eight generators represented by the Gell-Mann λ^a matrices that now play the role of the Pauli matrices. In particular the two matrices

$$\lambda^3 = \begin{pmatrix} 1 & 0 & 0 \\ 0 & -1 & 0 \\ 0 & 0 & 0 \end{pmatrix}, \lambda^8 = \frac{1}{\sqrt{3}} \begin{pmatrix} 1 & 0 & 0 \\ 0 & 1 & 0 \\ 0 & 0 & -2 \end{pmatrix} \tag{10.105}$$

play the role of σ^z and compose the so-called Cartan subalgebra L^z. The subalgebra of rising operators L^+ is comprised of three elements,

$$s^+ = \frac{\lambda^1 + i\lambda^2}{2} = \begin{pmatrix} 0 & 1 & 0 \\ 0 & 0 & 0 \\ 0 & 0 & 0 \end{pmatrix},$$

$$u^+ = \frac{\lambda^4 + i\lambda^5}{2} = \begin{pmatrix} 0 & 0 & 1 \\ 0 & 0 & 0 \\ 0 & 0 & 0 \end{pmatrix},$$

$$t^+ = \frac{\lambda^6 + i\lambda^7}{2} = \begin{pmatrix} 0 & 0 & 0 \\ 0 & 0 & 1 \\ 0 & 0 & 0 \end{pmatrix},$$

and only two of them, the commuting pair u^+ and t^+, are required for the construction of the nilpotent polynomials.

A generic pure state of a qutrit may be represented as

$$|\psi\rangle = \psi_0|0\rangle + \psi_1|1\rangle + \psi_2|2\rangle = \left(\psi_0 + \psi_1 t^+ + \psi_2 u^+\right)|0\rangle,$$

with

$$|0\rangle \equiv \begin{pmatrix} 0 \\ 0 \\ 1 \end{pmatrix}, \; |1\rangle \equiv \begin{pmatrix} 0 \\ 1 \\ 0 \end{pmatrix}, \; |2\rangle \equiv \begin{pmatrix} 1 \\ 0 \\ 0 \end{pmatrix},$$

which generalizes a similar representation for the qubit pure states. Note that even for larger d, such a set of $d - 1 = r$ commuting nilpotent operators always exists, allowing one to express a generic qudit state as a first-order polynomial of commuting nilpotent variables.

The state of n qutrits can then be written as

$$|\Psi\rangle = \sum_{\nu_i,\eta_i=0,1} \psi_{2\nu_n+\eta_n\ldots 2\nu_1+\eta_1} \prod_{i=1}^{n} (u_i^+)^{\nu_i} (t_i^+)^{\eta_i} |\mathrm{O}\rangle$$
$$= F(u_i^+, t_i^+)|\mathrm{O}\rangle,$$

where $|\mathrm{O}\rangle = |0, \ldots, 0\rangle$ now denotes the vacuum state of qutrits. For the canonic state the polynomial $F(u_i^+, t_i^+)$ does not have terms linear in u_i^+ and t_i^+ and starts with the bilinear terms. Cubic terms, in turn, do not contain terms linear in u_i^+, which follows from the requirement of the maximum overlap with the state $|1\rangle$, as illustrated in Fig. 10.11. The total number of parameters characterizing the entanglement amounts to

$$D_{su} = 2 \cdot 3^n - 8n - 2,$$

which is valid for $n \geq 3$.

For two qutrits one obtains

$$F_c(u_2^+, t_2^+, u_1^+, t_1^+) = 1 + \alpha_{11} t_2^+ t_1^+ + \alpha_{22} u_2^+ u_1^+$$
$$f_c(u_2^+, t_2^+, u_1^+, t_1^+) = \alpha_{11} t_2^+ t_1^+ + \alpha_{22} u_2^+ u_1^+.$$

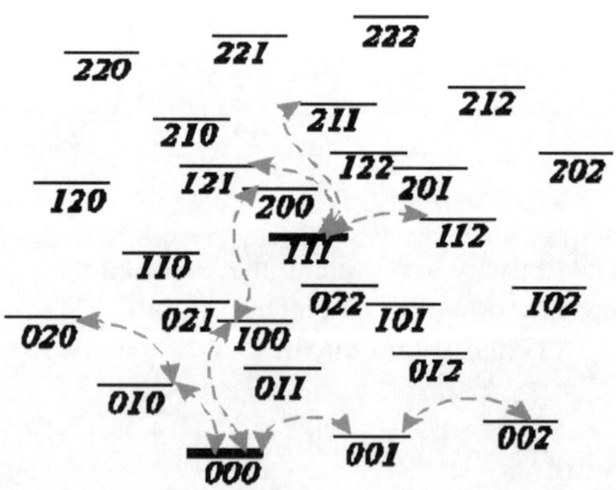

Fig. 10.11 The states $|211\rangle$, $|121\rangle$, and $|112\rangle$ corresponding to qubic terms linear in u_i^+ are depopulated by the requirement of the maximum overlap with the state $|1\rangle$

These polynomials depend only on two complex parameters α_{ii}, which can be set real and positive by a local phase transformation,

$$\exp\left(i\gamma_2\lambda_2^3 + i\delta_2\lambda_2^8 + i\gamma_1\lambda_1^3 + i\delta_1\lambda_1^8\right).$$

Thus, two real parameters are sufficient for characterizing entanglement between two qutrits, which is consistent with the result obtained by a straightforward application of the bipartite Schmidt decomposition.

In order to better understand the pattern, consider two more examples: an assembly of three qutrits, and of two qutrits and a qubit. In the first case, the canonic form is

$$\begin{aligned} F_c = 1 &+ \alpha_{011} t_2^+ t_1^+ + \alpha_{012} t_2^+ u_1^+ + \alpha_{021} u_2^+ t_1^+ + \alpha_{022} u_2^+ u_1^+ + \alpha_{101} t_3^+ t_1^+ \\ &+ \alpha_{102} t_3^+ u_1^+ + \alpha_{201} u_3^+ t_1^+ + \alpha_{202} u_3^+ u_1^+ + \alpha_{110} t_3^+ t_2^+ + \alpha_{120} t_3^+ u_2^+ \\ &+ \alpha_{210} u_3^+ t_2^+ + \alpha_{220} u_3^+ u_2^+ + \alpha_{111} t_3^+ t_2^+ t_1^+ + \alpha_{122} t_3^+ u_2^+ u_1^+ \\ &+ \alpha_{212} u_3^+ t_2^+ u_1^+ + \alpha_{221} u_3^+ u_2^+ t_1^+ + \alpha_{222} u_3^+ u_2^+ u_1^+, \end{aligned}$$

whereas for two qutrits and a qubit (labeled by the index 3), it looks as follows:

$$\begin{aligned} F_c = 1 &+ \alpha_4 t_2^+ t_1^+ + \alpha_5 t_2^+ u_1^+ + \alpha_7 u_2^+ t_1^+ + \alpha_8 u_2^+ u_1^+ + \alpha_{10} t_3^+ t_1^+ \\ &+ \alpha_{11} \sigma_3^+ u_1^+ + \alpha_{12} \sigma_3^+ t_2^+ + \alpha_{15} \sigma_3^+ u_2^+ + \alpha_{13} \sigma_3^+ t_2^+ t_1^+ + \alpha_{17} \sigma_3^+ u_2^+ u_1^+. \end{aligned}$$

10.3.6 Generalized Entanglement

The nilpotent polynomials formalism allows us to consider a special case of generalized entanglement where, due to some constraint, the algebra of local transformations does not include all possible transformations that exist for an element of the dimension d. In this context, of special interest are spin-1 systems, namely three-level systems restricted to evolve under the action of spin operators living in the $so(3) \equiv su(2)$ subalgebra of $su(3)$. In such an assembly, entanglement may exist even within a single element, which still can be characterized via the nilpotent polynomials. For this purpose, one has to invoke the nilpotent variables of higher order whose squares do not vanish and only some higher powers do. With the help of these nilpotent variables one can also consider entanglement among different elements of the assembly.

Though a three-level system corresponds to a full $su(3)$ algebra, we consider it here as a spin-1 system, that is, we concentrate on a situation where the physical observables are restricted to the subalgebra $su(2)$ of spin operators, $S_\pm = S_x \pm iS_y$ and S_z. Note that the latter are equivalent to the $u^\pm + t^\mp$, and λ^3 generators of $su(3)$, respectively. The spin states are characterized by the eigenvalues of

$$\lambda^{(3)} = \begin{pmatrix} 1 & 0 & 0 \\ 0 & 0 & 0 \\ 0 & 0 & -1 \end{pmatrix}, \tag{10.106}$$

and the spin-down state $|-1\rangle$ is chosen as the reference state. Here, in order to make use of the standard spin-1 vector notations, we have interchanged positions of the second and the third lines with respect of (10.105). The operator

$$S_+ = u^+ + t^- = \begin{pmatrix} 0 & 1 & 0 \\ 0 & 0 & 1 \\ 0 & 0 & 0 \end{pmatrix} \tag{10.107}$$

is the only element of the nilpotent subalgebra $su(2) \subset su(3)$, which has therefore to be chosen as the nilpotent variable. It is of the order of two since $S_+^2 \neq 0$, and only $S_+^3 = 0$.

Now we show that even for a single spin, the state $|0\rangle$ is generalized entangled with respect to $SU(2)$. Indeed, by the unitary matrix $e^{-i\pi S_x/\sqrt{2}}$ the state $|0\rangle$ can be transformed to the state with the maximum vacuum population $|\psi_C\rangle = (|-1\rangle + |1\rangle)/\sqrt{2}$, which evidently differs from the reference state $|-1\rangle$. The corresponding canonic state normalized to the unit reference state amplitude reads

$$|\Psi_c\rangle = (1 + S_+^2)\,|-1\rangle\,, \tag{10.108}$$

hence $f_c = S_+^2$ and $F_c = 1 + S_+^2$. The presence of the quadratic term is the signature of generalized entanglement.

One way to understand the meaning of this "self-entanglement" in the state $|0\rangle$ is to see it as a consequence of the fact that the operators $S_{\pm,z}$ of $su(2)$ cannot lift the degeneracy of the two eigenstates of the operator λ_8, which together with $\lambda_3 = S_z$ labels the states in the unrestricted $su(3)$ algebra. In other words, within the group of restricted local transformations, the transition from the state $|0\rangle$ can only access the state $(|-1\rangle + |1\rangle)/\sqrt{2}$ but not $(|-1\rangle - |1\rangle)/\sqrt{2}$; hence the amplitudes of the reference state $|-1\rangle$ and that of the state $|1\rangle$ are fully correlated. One will not be too much surprised by the existence of self-entanglement after paying attention to the fact that here as well as earlier the subgroup of allowed local transformations of the element is also smaller than the entire group of possible transformation of this element (which is, strictly speaking, also local, although not completely allowed).

For a generic state, the canonic nilpotent polynomial and the tanglemeter take the form

$$\begin{aligned} F_c(S_+) &= 1 + \frac{\alpha_s}{2} S_+^2 \\ f_c(S_+) &= \frac{\alpha_s}{2} S_+^2, \end{aligned} \tag{10.109}$$

respectively, where the phase of α'_s can be set to 0, and the factor $\frac{1}{2}$ is introduced for convenience. Therefore, generalized entanglement of a single spin-1 is characterized by a single real parameter.

Consider now entanglement between two spin-1, which we describe as two three-level systems subject to the action of $SU(2) \oplus SU(2)$ local operations. By analogy to the single spin-1 case, we choose $S_{1,2}$ with the implicit subscripts $+$ as nilpotent variables and the state $|-1,-1\rangle$ as the reference. In the canonic form maximizing population of the reference state, one obtains

$$\begin{aligned} F_c(S_1, S_2) &= 1 + \alpha_{s;1}\frac{S_1^2}{2} + \alpha_{s;2}\frac{S_2^2}{2} + \alpha_{s,s}\frac{S_2^2}{2}\frac{S_1^2}{2} \\ &\quad + \alpha_{t,s}\frac{S_2^2}{2}S_1 + \alpha_{st}S_2\frac{S_1^2}{2} + \alpha_{t,t}S_1S_2\,, \\ f_c(S_1, S_2) &= \beta_{s;1}\frac{S_1^2}{2} + \beta_{t,t}S_1S_2 + \beta_{st}\frac{S_1^2}{2}S_2 + \beta_{s;2}\frac{S_2^2}{2} \\ &\quad + \beta_{t,s}S_1\frac{S_2^2}{2} + \beta_{s,s}\frac{S_2^2}{2}\frac{S_1^2}{2}, \end{aligned} \tag{10.110}$$

where $\beta_{s,s} = \alpha_{s,s} - 2\alpha_{t,t}^2$, and all other $\beta = \alpha$. As before, by exploiting the freedom of phase transformations on the nilpotent variables, the parameters $\alpha_{s;1}$ and $\alpha_{s;2}$ (or $\beta_{s;1}$ and $\beta_{s;2}$), characterizing generalized entanglement within each of the three-level systems, can be set real, and we are left with the four complex numbers $\alpha_{s,s}$, $\alpha_{t,s}$, α_{st}, and $\alpha_{t,t}$ characterizing the generalized inter-spin entanglement.

Chapter 11
Dynamics and Entanglement for Open Quantum Systems

The most studied manifestation of entanglement is mixing of quantum states, which results from interaction of an open quantum system with the environment. The quantum state of the open system is a statistical mixture of pure quantum states characterized by the density matrix. In other words, the state vector $|s\rangle$ of an open quantum system is not uniquely defined, but, to the contrary, it is distributed over all Hilbert space of the system with a certain probability $p(s)$, while the results of measurement of a physical observable given by the corresponding operator $\hat{o}$ are distributed according to the weight function $\langle s|\hat{o}|s\rangle p(s)$. The expectation value of the measurement is therefore given by the trace of the product of the operator $\hat{o}$ and the density operator $\int ds\ |s\rangle p(s) \langle s|$.

In Chap. 3, at the simplest example of a two-level system, we have already studied some aspects of a quantum system interacting with the environment, although without invoking the concept of entanglement. Decay and loss of coherence are two processes common to all open quantum systems interacting with an environment. Two different relaxation times T_1 and T_2 characterize these processes in two-level systems as has been mentioned in (5.60). Though both processes represent irreversible relaxation, they are of rather different nature from the point of view of entanglement. The number of occupied quantum states, given by the rank of the system density matrix—an analog of the phase volume in the classical case—may change in an elementary act of relaxation, and usually does it towards diminishing, whereas the act itself essentially involves entanglement with the environment. By contrast, this number does not change in an elementary act of coherence loss but yet increases after being averaged over a sequence of elementary acts, since the probability of a given sequence of elementary acts and the global phase accumulated in the course of this sequence is the result of a random process. The other way to phrase it is to say that the interaction Hamiltonian inducing the coherence loss commutes with the Hamiltonian of the environment, and the phase shift accumulated as a result of such interaction depends on the environment energy. Averaging over the environment energy distribution yields the coherence loss.

V.M. Akulin, *Dynamics of Complex Quantum Systems*, Theoretical and Mathematical Physics, DOI 10.1007/978-94-007-7205-2_11,

In this chapter we consider a similar problem, not just for a two-level system, but for larger systems, also paying attention to the notion of entanglement. The dynamics of the density matrix under some conditions satisfies a linear integro-differential equation, which in certain extreme cases yields a differential master equation that contains an operator of irreversible relaxation. We start by presenting standard descriptions of system interacting with the environment and show several methods that yield the irreversible relaxation rate term in the master equation. Next we demonstrate an exactly soluble model of a quantum system entangled with a bosonic environment, and trace the dynamics of the entanglement entropy. The last part of the chapter is dedicated to the notion of entanglement among the parts of an open system, one of the most challenging concepts that did not receive yet an exhaustive mathematical description. We introduce the commonly accepted measure of such entanglement and discuss possible ways of it generalization.

11.1 Complex Quantum Systems Entangled with the Environment

As long as a system A is not entangled with the environment $\overline{A}$, the density matrix $\hat{\rho}$ of the system is a tensor product of two vectors, a ket vector $|\Psi_A\rangle$ and the dual object—its Hermitian conjugate bra vector $\langle\Psi_A|$. In this case the density matrix has rank 1. This is no longer the case when the system becomes entangled with the environment; in such a situation the density matrix ρ may acquire the maximum possible rank equal to the dimension N of the Hilbert space of the system, and can be represented by the eigenvalue decomposition

$$\hat{\rho} = \sum_i |\Psi_A\rangle_i \, \lambda_i \, {}_i\langle\Psi_A| \, .$$

The eigenvalues λ_i of the density matrix, being positive numbers normalized by the condition $\sum_{i=1}^{N} \lambda_i = 1$, are interpreted as probabilities for the system to be in the corresponding eigenstates $|\Psi_A\rangle_i$.

If a system is entangled with the environment as a result of an interaction that occurred in the past, but does not experience an interaction in the present, the density matrix satisfies the master equation

$$i\dot{\hat{\rho}} = \left[\hat{H}, \hat{\rho}\right], \tag{11.1}$$

where $\hat{H}$ is the Hamiltonian of this part, and $[\ldots,\ldots]$ denotes the commutator. Equation (11.1) is a system of N^2 coupled linear equations, since it relates the matrix elements of a $N \times N$ density matrix. In fact there is no need to solve this equation directly, since it is completely equivalent to the Schrödinger equation for the state vector

$$-i|\dot{\Psi}_A\rangle = \hat{H}\,|\Psi_A\rangle\,,$$

whereas the linearly independent solutions $|\Psi_A\rangle = |E_n\rangle\, e^{-iE_n t}$ of this equation immediately suggest the solution

$$\hat{\rho} = \sum_{i,n,m} |E_n\rangle\, \langle E_n|\Psi_A\rangle_i\, \lambda_i e^{-i(E_n - E_m)t}\, {}_i\langle \Psi_A|E_m\rangle\, \langle E_m|$$

for the density matrix. However, in the case where the parts of the system keep interacting, the equation of motion for the density matrix becomes much more complicated, and the general case of this equation will be discussed later on, in the next section.

In Chap. 3 we have seen that at the time intervals shorter than the Heisenberg time of the environment, the excitation of the system is transferred to the environment and never returns back. In such a situation, one can write down a closed linear master equation for the density matrix not only for the two-level system, but in the general case of a system with a discrete nondegenerate spectrum. The derivation of the master equation for such a situation based on some general assumption will be discussed in the first part of this section. Here we only notice that the most simple form of the master equation is the so-called Lindblad form

$$i\dot{\hat{\rho}} = \left[\hat{H},\hat{\rho}\right] - i\left[\hat{\mathcal{V}},\left[\hat{\mathcal{V}},\hat{\rho}\right]\right], \tag{11.2}$$

which contains a double commutator with a Hermitian operator $\hat{\mathcal{V}}$ responsible for the variation of entanglement of the system with the environment. The second part of the section is dedicated to an exactly soluble model of the environment and relies on the separation of the system and the environment with the help of functional integration, which allows one to make a clearer distinction between the relaxation and the coherence loss. It also suggests a way for description of the entanglement entropy dynamics, which will later be useful for the description of coherence protection.

One more general statement should be made in the context of a system interacting with the environment. Consider the system and the environment as a single compound system, initially in the product state $|\psi_0\rangle\, |\Psi_0\rangle$, where $|\psi\rangle$ is the state vector of the system, and $|\Psi\rangle$ is that of the environment. Let $\hat{\mathbf{H}}$ be the Hamiltonian of the compound system, which allows one to describe the time-dependent state vector in a Fourier representation as

$$|\boldsymbol{\Psi}(\varepsilon)\rangle = \frac{1}{\varepsilon - \hat{\mathbf{H}}}\, |\psi_0\rangle\, |\Psi_0\rangle\,.$$

One immediately finds the density matrix of the compound system

$$|\boldsymbol{\Psi}(\varepsilon)\rangle\, \langle \boldsymbol{\Psi}(\xi)| = \frac{1}{\varepsilon - \hat{\mathbf{H}}}\, |\psi_0\rangle\, |\Psi_0\rangle\, \langle \Psi_0|\, \langle \psi_0|\, \frac{1}{\xi - \hat{\mathbf{H}}},$$

and by tracing it over the environment obtains the density matrix of the system,

$$\hat{\rho}(\varepsilon,\xi) = \mathrm{Tr}_E \left\{ \frac{1}{\varepsilon - \hat{\mathbf{H}}} |\psi_0\rangle |\Psi_0\rangle \langle\Psi_0| \langle\psi_0| \frac{1}{\xi - \hat{\mathbf{H}}} \right\},$$

which should also be averaged over $\underline{\text{possible}}$ initial states of the environment $|\Psi_0\rangle$, that is, over the density matrix $\overline{|\Psi_0\rangle\langle\Psi_0|}$. This procedure is an analog of the ensemble averaging widely employed in this book for description of the dynamics of complex quantum systems, although here the distribution function for the Hamiltonians of the system is not postulated but directly emerges from the averaging over the environment dynamics. Moreover, when the final state of the environment turns out to be different from its initial state, such a procedure does not necessarily yield Hermitian Hamiltonians. This once again indicates the fact that apart from the coherence loss originating from averaging over an ensemble of Hermitian Hamiltonians there exists a qualitatively different process of relaxation associated with the transitions in the system environment.

11.1.1 The Master Equation for Mixed States of an Open System

The concept of mixed quantum states is relevant to the general situation where one deals not with an entire quantum system, but only with a part of it, and one has no means to address the rest of the system. One therefore has to derive equations which contain only the characteristics of this addressable part, which we hereafter call "the system", and we use the name "environment" for the remaining, unaddressed part. The system is thus assumed to be "open", that is, interacting with the unaddressable environment. It is the density matrix $\hat{\rho}$ of the system that accumulates all the required characteristics, since it contains complete information about all the physical measurables relevant to the system. As we have already discussed in the introduction to this chapter, $\hat{\rho}$ is given by a statistical mixture of pure states of the system, that is, by a linear combination of tensor products of different state vectors by their complex conjugates weighted by the positive coefficients, and it satisfies the master equation

$$i\dot{\hat{\rho}} = \left[\hat{H}, \hat{\rho}\right] + \dots . \tag{11.3}$$

Here the commutator on the right hand side allows for the dynamics of the density matrix under the action of the Hamiltonian $\hat{H}$ of the system, while the dots stand for the part resulting from the interaction of the system with the environment, which will be the subject of consideration in this section.

11.1.2 Liouville Formalism

Consider the system and the environment as a composite system that does not interact with anything else, and therefore has a density matrix ρ satisfying the master equation

$$i\dot{\hat{\rho}} = \left[\hat{H}, \hat{\rho}\right] + \left[\hat{\mathcal{H}}, \hat{\rho}\right] + \left[\hat{\mathcal{V}}, \hat{\rho}\right], \tag{11.4}$$

where $\mathcal{H}$ stands for the environment Hamiltonian and $\mathcal{V}$ for the Hamiltonian of interaction between the system and the environment, which can be cast to a sum of tensor products

$$\hat{\mathcal{V}} = \sum \hat{V}_\alpha \otimes \hat{v}_\alpha, \tag{11.5}$$

of the operators $\hat{V}_\alpha$ acting in the Hilbert space of the system, and the operators $\hat{v}_\alpha$ acting on the environment.

Dealing with density matrices, it is worth mentioning that they form a vector space of the dimension given by the square of the dimension of Hilbert space, and therefore all linear operations with the density matrices can be represented by matrices acting in this vector space, though in the regular Hilbert space of the system these operators are higher rank tensors. The trace of the product $\mathrm{Tr}\left(\hat{B}^+\hat{A}\right)$ of a matrix $\hat{A}$ by the Hermitian conjugate of another matrix $\hat{B}$ serves as the scalar product in this vector space. One can employ this formalism in order to find the missing part in (11.3) in a compact way. To this end, consider the density matrix as a vector, and write (11.4) in the form

$$i\dot{\hat{\rho}} = \overset{\cap}{L}\hat{\rho} + \overset{\cap}{\mathcal{L}}\hat{\rho} + \overset{\cap}{\mathcal{L}}_i\hat{\rho} \tag{11.6}$$

where by symbols $\cap$ on the top of the letter we denote linear Liouville operators acting in the linear vector space of the density matrices, while in the regular Hilbert space they act as commutators:

$$\overset{\cap}{L}\hat{\rho} \to \left[\hat{H}, \hat{\rho}\right], \quad \overset{\cap}{\mathcal{L}}\hat{\rho} \to \left[\hat{\mathcal{H}}, \hat{\rho}\right], \quad \overset{\cap}{\mathcal{L}}_i\hat{\rho} \to \left[\hat{\mathcal{V}}, \hat{\rho}\right].$$

We note that in the regular Hilbert space one does not make a distinction between indices related to the system and that related to the environment by employing a cumulative index of states. In the framework of such a description, one considers $\hat{\rho}$-matrix as a second rank tensor, and the Liouville operators are therefore the fourth rank tensors in these notations. When the indices numerating states of the system and the environment are considered independently, $\hat{\rho}$ becomes a fourth-rank tensor, while the Liouville operators are therefore the eighth-rank tensors.

In the general case, no closed set of differential equations can be written for the density matrix of the system $\hat{\rho} = \mathrm{Tr}_{\mathcal{E}}\hat{\rho}$, given by tracing out the variables of the environmental, which in terms of the Liouville formalism we denote by

$$\hat{\rho} = \overset{\cap}{P}_{\mathcal{E}}\hat{\rho}. \tag{11.7}$$

The maximum one can do is to obtain, with the help of the Zwanzig projection method discussed below, an approximate integro-differential equation for a non-orthogonal projection

$$\overset{\cap}{P}\hat{\rho} = \exp[S - \hat{\mathcal{H}}/kT] \otimes \mathrm{Tr}_{\mathcal{E}}\hat{\rho}, \tag{11.8}$$

by assuming the environment density matrix to be close to equilibrium, $\exp[S - \hat{\mathcal{H}}/kT]$, with the environment entropy $S = -\ln \mathrm{Tr}_{\mathcal{E}} \exp[-\hat{\mathcal{H}}/kT]$ guaranteeing the normalization and thereby ensuring the projector property $\overset{\cap}{P}^2 = \overset{\cap}{P}$. When applied in its general form, this method yields an untractable integro-differential equation for the system density matrix, with a memory-type kernel of the integral operator depending exclusively on the difference of the time arguments. A tractable version emerges when one assumes applicability of the second-order time-dependent perturbation theory implied by the weakness of the interaction between the system and the environment, longevity of the environment Heisenberg time implied by the environment's complexity, and shortness of the environment correlation time implied by the broadness of its frequency spectrum.

11.1.3 Zwanzig Projection Method

In order to elucidate the main ideas underlying the derivation of the Liouville equation employing the Zwanzig projection method, we first show at the simplest example how to reduce a system of the first order differential equations to a system of the integro-differential equations of a smaller rank. A system of two differential equations

$$i\dot{X} = aX + bY$$

$$i\dot{Y} = cX + dY \tag{11.9}$$

with the initial condition $X(0)$ and $Y(0)$ is equivalent to a single integro-differential equation

$$i\dot{X} = aX - i\int_0^t be^{-id(t-\tau)}cX(\tau)\mathrm{d}\tau + be^{-idt}Y(0). \tag{11.10}$$

In fact, after the Fourier transformation of (11.9) one finds

$$\zeta X = aX + bY + iX(0)$$
$$\zeta Y = cX + dY + iY(0),$$

from where

$$\zeta X = aX + \frac{bc}{(\varepsilon - d)} X + \frac{ibY(0)}{(\varepsilon - d)} + iX(0),$$

and, finally, the inverse Fourier transformation over ζ indeed yields (11.10).

Generalization of this method can be done for the case of a vector function $\overrightarrow{Z}(t)$ satisfying the linear equation

$$i\dot{\overrightarrow{Z}} = \hat{L}\overrightarrow{Z}$$

where $\hat{L}$ is a linear operator in a vector space of an arbitrary dimension $\mathcal{N}$. Let $\overrightarrow{X}(t) = \hat{P}\overrightarrow{Z}(t)$ and $\overrightarrow{Y}(t) = (\hat{I} - \hat{P})\overrightarrow{Z}(t)$, where $\hat{P}$ projects the $\mathcal{N}$-dimensional vector space to an $\mathcal{M}$-dimensional space, and $\hat{I}$ is the $\mathcal{N}$-dimensional identity operator. The analog of (11.10),

$$i\dot{\overrightarrow{X}}(t) = \hat{P}\hat{L}\overrightarrow{X}(t) - i\int_0^t \hat{P}\hat{L}(\hat{I} - \hat{P})e^{-i(\hat{I}-\hat{P})\hat{L}(\hat{I}-\hat{P})(t-\tau)}(\hat{I} - \hat{P})\hat{L}X(\tau)\mathrm{d}\tau$$
$$+ \hat{P}\hat{L}(\hat{I} - \hat{P})e^{-i(\hat{I}-\hat{P})\hat{L}(\hat{I}-\hat{P})t}\overrightarrow{Y}(0), \quad (11.11)$$

implies $a \to \hat{P}\hat{L}\hat{P}$, $b \to \hat{P}\hat{L}(\hat{I} - \hat{P})$, $c \to (\hat{I} - \hat{P})\hat{L}\hat{P}$, $d \to (\hat{I} - \hat{P})\hat{L}(\hat{I} - \hat{P})$, and also relies on the fact that for all integer powers k one has $(\hat{I} - \hat{P})^k = (\hat{I} - \hat{P})$ and $\hat{P}^k = \hat{P}$.

Let us now apply this generalization to the case of the system density matrix and the projection operator $\overset{\frown}{P}$ of (11.8). The initial density matrix of the system and the environment we take in the form of a tensor product

$$\hat{\rho}(0) = \hat{\rho}(0) \otimes e^{S - \hat{\mathcal{H}}/kT}, \quad (11.12)$$

such that $\overrightarrow{Y}(0) = 0$ and hence (11.11) takes the form

$$i\dot{\hat{\rho}}(t) \otimes e^{-\hat{\mathcal{H}}/kT} = \overset{\frown}{L}\hat{\rho}(t) \otimes e^{-\hat{\mathcal{H}}/kT} - i\int_0^t \overset{\frown}{P}\overset{\frown}{\mathcal{L}}_i\left(\overset{\frown}{I} - \overset{\frown}{P}\right)$$
$$e^{-i(\overset{\frown}{I}-\overset{\frown}{P})\left(\overset{\frown}{L}+\overset{\frown}{\mathcal{L}}+\overset{\frown}{\mathcal{L}}_i\right)(\overset{\frown}{I}-\overset{\frown}{P})(t-\tau)}\left(\overset{\frown}{I} - \overset{\frown}{P}\right)\overset{\frown}{\mathcal{L}}_i\hat{\rho}(\tau) \otimes e^{-\hat{\mathcal{H}}/kT}\mathrm{d}\tau$$

or in the Fourier representation over time

$$\zeta\hat{\rho}(\zeta)\otimes e^{-\mathcal{H}/kT} - i\,\hat{\rho}\big|_{t=0}\otimes e^{-\mathcal{H}/kT} = \overset{\cap}{L}\hat{\rho}(\zeta)\otimes e^{-\mathcal{H}/kT} - \overset{\cap}{P}\overset{\cap}{\mathcal{L}}_i\left(\overset{\cap}{I}-\overset{\cap}{P}\right)$$
$$\frac{1}{\zeta-(\overset{\cap}{I}-\overset{\cap}{P})\left(\overset{\cap}{L}+\overset{\cap}{\mathcal{L}}+\overset{\cap}{\mathcal{L}}_i\right)(\overset{\cap}{I}-\overset{\cap}{P})}\left(\overset{\cap}{I}-\overset{\cap}{P}\right)\overset{\cap}{\mathcal{L}}_i\hat{\rho}(\zeta)\otimes e^{-\mathcal{H}/kT}.$$

We now take the trace over the environment and arrive at

$$\zeta\hat{\rho}(\zeta) - i\,\rho_S|_{t=0} = \overset{\cap}{L}\hat{\rho}(\zeta) - \mathrm{Tr}_E\{\overset{\cap}{P}\overset{\cap}{\mathcal{L}}_i\left(\overset{\cap}{I}-\overset{\cap}{P}\right)$$
$$\frac{1}{\zeta-(\overset{\cap}{I}-\overset{\cap}{P})\left(\overset{\cap}{L}+\overset{\cap}{\mathcal{L}}+\overset{\cap}{\mathcal{L}}_i\right)(\overset{\cap}{I}-\overset{\cap}{P})}\left(\overset{\cap}{I}-\overset{\cap}{P}\right)\overset{\cap}{\mathcal{L}}_i e^{-\hat{\mathcal{H}}/kT}\}\hat{\rho}(\zeta)$$

From this form we can already draw the conclusion that in the time representation the interaction with the environment can be represented as a linear functional of the convolution type over the system density matrix

$$i\dot{\hat{\rho}}(t) = [H,\hat{\rho}(t)] - i\int_0^t \overset{\cap}{\mathcal{G}}(t-\tau)\hat{\rho}(\tau)\mathrm{d}\tau \tag{11.13}$$

with a kernel operator $\overset{\cap}{\mathcal{G}}$ explicitly given by its Fourier transform

$$\overset{\cap}{\mathcal{G}}(\zeta)\ldots = \mathrm{Tr}_E\{\overset{\cap}{P}\overset{\cap}{\mathcal{L}}_i\left(\overset{\cap}{I}-\overset{\cap}{P}\right)$$
$$\frac{1}{\zeta-(\overset{\cap}{I}-\overset{\cap}{P})\left(\overset{\cap}{L}+\overset{\cap}{\mathcal{L}}+\overset{\cap}{\mathcal{L}}_i\right)(\overset{\cap}{I}-\overset{\cap}{P})}\left(\overset{\cap}{I}-\overset{\cap}{P}\right)\overset{\cap}{\mathcal{L}}_i e^{-\hat{\mathcal{H}}/kT}\}. \tag{11.14}$$

If the initial condition is given at $-\infty$ instead of 0, after the replacement $t-\tau\to\tau$, equation (11.13) takes the form of the integro-differential equation

$$i\dot{\hat{\rho}}(t) = \left[\hat{H},\hat{\rho}(t)\right] - i\int_0^\infty \overset{\cap}{\mathcal{G}}(\tau)\hat{\rho}(t-\tau)\mathrm{d}\tau. \tag{11.15}$$

with the same memory kernel $\overset{\cap}{\mathcal{G}}(\tau)$.

11.1.4 The Case of Weak Interaction

Let us now focus on the case when the interaction $\hat{\mathcal{V}}$ between the system and the environment is weak. Physically it implies that the action of the coupling $\hat{\mathcal{V}}$ between the system and the environment accumulated during the correlation time of the environment is small as compared to the quantum of action given by the Planck constant $\hbar = 1$. This assumption allows one to ignore the interaction Liouvillian $\overset{\cap}{\mathcal{L}}_i$ in the denominator at the right hand side of (11.14) for the kernel $\overset{\cap}{\mathcal{G}}(\zeta)$, thus bringing us to the expression of the second-order time-dependent perturbation theory over $\overset{\cap}{\mathcal{L}}_i$, which yields

$$\overset{\cap}{\mathcal{G}}(\zeta)\ldots = \mathrm{Tr}_E\{\overset{\cap}{P}\overset{\cap}{\mathcal{L}}_i\left(\overset{\cap}{I} - \overset{\cap}{P}\right)$$
$$\frac{1}{\zeta - \overset{\cap}{L} - \overset{\cap}{\mathcal{L}}}\left(\overset{\cap}{I} - \overset{\cap}{P}\right)\overset{\cap}{\mathcal{L}}_i e^{-\hat{\mathcal{H}}/kT}\}. \tag{11.16}$$

We have omitted the operators $\left(\overset{\cap}{I} - \overset{\cap}{P}\right)$ in the denominator of (11.16), since the unperturbed Liouvillian commutes with the projection operator. Within this order of approximation, one can also make the replacement

$$\hat{\rho}(t-\tau) \to e^{i\hat{H}\tau}\hat{\rho}(t)e^{-i\hat{H}\tau} \tag{11.17}$$

in (11.15) and arrive at

$$i\dot{\hat{\rho}}(t) = \left[\hat{H}, \hat{\rho}(t)\right] - i\overset{\cap}{\mathcal{R}}\hat{\rho}(t), \tag{11.18}$$

where $\overset{\cap}{\mathcal{R}}\ldots = \int_0^\infty \overset{\cap}{\mathcal{G}}^{(2)}(\tau)\ldots d\tau$ is the relaxation operator given by the explicit expression

$$\overset{\cap}{\mathcal{G}}^{(2)}(\tau)\ldots = \int \frac{d\zeta}{2\pi} e^{i\zeta\tau}\mathrm{Tr}_E\{\overset{\cap}{P}\overset{\cap}{\mathcal{L}}_i\left(\overset{\cap}{I} - \overset{\cap}{P}\right) \tag{11.19}$$
$$\frac{1}{\zeta - \overset{\cap}{L} - \overset{\cap}{\mathcal{L}}}\left(\overset{\cap}{I} - \overset{\cap}{P}\right)\overset{\cap}{\mathcal{L}} e^{-\hat{\mathcal{H}}/kT}\}e^{i\hat{H}\tau}\ldots e^{-i\hat{H}\tau}.$$

For the coupling Lagrangian $\overset{\cap}{\mathcal{L}}_i\ldots = \sum\left[\hat{V}_\alpha \otimes \hat{v}_\alpha, \ldots\right]$ Eq. (11.19) yields

$$\overset{\cap}{\mathcal{R}}\ldots=\sum_{\alpha\beta}\mathrm{Tr}_B\{P[\hat{V}_\alpha\otimes\hat{v}_\alpha,\left(\overset{\cap}{I}-\overset{\cap}{P}\right)\int\limits_0^\infty e^{-i\overset{\cap}{L}\tau-i\overset{\cap}{\mathcal{L}}\tau}$$
$$\left[\hat{V}_\beta\otimes\hat{v}_\beta,e^{i\hat{H}\tau}\ldots e^{-i\hat{H}\tau}\otimes e^{-\mathcal{H}/kT}\right]\mathrm{d}\tau]\}.$$

We note that $e^{-i\overset{\cap}{L}\tau-i\overset{\cap}{\mathcal{L}}\tau}e^{i\hat{H}\tau}\hat{\rho}(t)e^{-i\hat{H}\tau}\otimes e^{-\hat{\mathcal{H}}/kT}=\hat{\rho}(t)\otimes e^{-\hat{\mathcal{H}}/kT}$ and arrive at

$$\overset{\cap}{\mathcal{R}}\ldots=\sum_{\alpha\beta}\mathrm{Tr}_B\{P[\hat{V}_\alpha\otimes\hat{v}_\alpha,\left(\overset{\cap}{I}-\overset{\cap}{P}\right)$$
$$\left[\int\limits_0^\infty\hat{V}_\beta(\tau)\otimes\hat{v}_\beta(\tau)\mathrm{d}\tau,\ldots\otimes e^{-\hat{\mathcal{H}}/kT}\right]]\}$$

where

$$\hat{v}_\beta(\tau)=e^{i\hat{\mathcal{H}}\tau}\hat{v}_\beta e^{-i\hat{\mathcal{H}}\tau}$$
$$\hat{V}_\beta(\tau)=e^{i\hat{H}\tau}\hat{V}_\beta e^{-i\hat{H}\tau}. \tag{11.20}$$

If the operators v_β multiplying the environment density matrix $e^{-\hat{\mathcal{H}}/kT}$ either from the left or from the right sides make the resulting product orthogonal to the projection, that is,

$$\overset{\cap}{P}\hat{v}_\beta(t)e^{-\hat{\mathcal{H}}/kT}=0, \tag{11.21}$$

the projection operators $(\overset{\cap}{I}-\overset{\cap}{P})$ can be omitted in the evolution operator $e^{-i\overset{\cap}{L}_P\tau}=e^{-i\left(\overset{\cap}{L}+\overset{\cap}{\mathcal{L}}\right)\tau}$. If it is not the case, the interaction averaged over the equilibrium state of the environment can be included into $\hat{H}$. We therefore can always assume that $\overset{\cap}{P}\hat{v}_\beta(t)e^{-\hat{\mathcal{H}}/kT}=0$, and we arrive at

$$\overset{\cap}{\mathcal{R}}\ldots=\sum_{\alpha\beta}\mathrm{Tr}_E\{\overset{\cap}{P}[\hat{V}_\alpha\otimes\hat{v}_\alpha,\int\limits_0^\infty\left[\hat{V}_\beta(\tau)\otimes\hat{v}_\beta(\tau),\ldots\otimes e^{-\hat{\mathcal{H}}/kT}\right]\mathrm{d}\tau]\},$$

and evoking explicit expressions for the commutators we obtain

$$\overset{\cap}{\mathcal{R}}\ldots=-\sum_{\alpha\beta}\int\limits_0^\infty\mathrm{d}\tau\{\hat{V}_\alpha\ldots\hat{V}_\beta\left(\tau\right)h_{\alpha\beta}^{II}(\tau)+\hat{V}_\beta\left(\tau\right)\ldots\hat{V}_\alpha h_{\alpha\beta}^{III}(\tau)$$
$$-\hat{V}_\alpha\hat{V}_\beta\left(\tau\right)\ldots h_{\alpha\beta}^{I}(\tau)-\ldots\hat{V}_\beta\left(\tau\right)\hat{V}_\alpha h_{\alpha\beta}^{IV}(\tau)\}. \tag{11.22}$$

The dependencies $h^i_{\alpha\beta}(\tau)$ entering this expression read

$$
\begin{aligned}
h^I_{\alpha\beta}(\tau) &= \mathrm{Tr}\{\overset{\frown}{P}\hat{v}_\alpha\hat{v}_\beta(\tau)e^{-\hat{\mathcal{H}}/kT}\} \\
h^{II}_{\alpha\beta}(\tau) &= \mathrm{Tr}\{\overset{\frown}{P}\hat{v}_\alpha e^{-\hat{\mathcal{H}}/kT}v_\beta(\tau)\} \\
h^{III}_{\alpha\beta}(\tau) &= \mathrm{Tr}\{\overset{\frown}{P}\hat{v}_\beta(\tau)e^{-\hat{\mathcal{H}}/kT}\hat{v}_\alpha\} \\
h^{IV}_{\alpha\beta}(\tau) &= \mathrm{Tr}\{\overset{\frown}{P}e^{-\hat{\mathcal{H}}/kT}\hat{v}_\beta(\tau)\hat{v}_\alpha\},
\end{aligned}
\tag{11.23}
$$

where the projection operator can be omitted. One can also make use of (11.20) and substitute $\hat{v}_\beta(\tau)$ explicitly. For a system interacting with an environment that is close to thermal equilibrium, equation (11.22) with the allowance for (11.23) is the most general final result of the Zwanzig projection method applied at the second-order perturbation theory for calculation of the relaxation operator $\mathcal{R}$ entering (11.18).

11.1.5 Harrovian Environment and the Lindblad Form of Master Equation

In the situation where the correlation time of the environment is much shorter than the inverse of the typical frequency of the system, one can make use of the so-called Markovian approximation by assuming

$$
h^l_{\alpha\beta}(\tau) = w_{\beta\alpha}\delta(\tau), \tag{11.24}
$$

which immediately yields

$$
\begin{aligned}
\overset{\frown}{\mathcal{R}}\hat{\rho}(t) = &-\sum w_{\alpha\beta}\left[\hat{V}_\alpha, \hat{V}_\beta\hat{\rho}(t)\right] \\
&+\sum w_{\beta\alpha}\left[\hat{V}_\alpha, \hat{\rho}(t)\hat{V}_\beta\right].
\end{aligned}
$$

For the specific case $w_{\beta\alpha} = w_{\alpha\beta}$ one obtains the so-called Lindblad form of the master equation,

$$
i\dot{\hat{\rho}} = \left[\hat{H}, \hat{\rho}\right] - i\sum w_{\alpha\beta}\left[\hat{V}_\alpha, \left[\hat{V}_\beta, \hat{\rho}\right]\right]. \tag{11.25}
$$

Due to its simplicity, this form of the master equation is widely used for the description of quantum systems coupled to the environment.

Note that the Markovian approximation is a very strong assumption, and when it is valid, there is no need to impose any other requirements, such as thermal equilibrium of the environment or weakness of the interaction, since this assumption implies that the environment correlation time is short, not only as compared the typical inverse coupling V, but also as compared to the typical frequencies of the system. Also note that the form (11.25) of the equation yields a stationary solution, which commutes with all matrices $\hat{V}_\beta$. It worth mentioning that there exists a generalization of this equation,

$$i\dot{\hat{\rho}} = \left[\hat{H}, \hat{\rho}\right] - i\sum w_{\alpha\beta}\left(\hat{V}_\alpha^+\hat{V}_\beta\hat{\rho} + \hat{V}_\alpha^+\hat{V}_\beta\hat{\rho} - 2\hat{V}_\beta\hat{\rho}\hat{V}_\alpha^+\right), \tag{11.26}$$

where the coupling $\hat{V}_\alpha$ is given by a non-Hermitian matrix. This form of the equation allows one to lift up the assumption of the environment correlation time being short as compared to all periods of the system and to include the asymmetry of the upward and backward transition over the energy scale due to a nonuniform (for example thermal) distribution over the environment energy. Also note that the most simple example is a two-level system, for which the master equation (11.26) reads

$$\begin{aligned} i\frac{\partial}{\partial t}\begin{pmatrix}\rho_{00}(t) & \rho_{01}(t)\\ \rho_{10}(t) & \rho_{11}(t)\end{pmatrix} &= \left[\hat{H}, \hat{\rho}(t)\right] \\ &- i\begin{pmatrix} e^{\frac{-\omega}{kT}}\frac{\rho_{00}(t)}{T_1} - \frac{\rho_{11}(t)}{T_1} & \frac{\rho_{01}(t)}{2T_2} \\ \frac{\rho_{10}(t)}{2T_2} & \frac{\rho_{11}(t)}{T_1} - e^{\frac{-\omega}{kT}}\frac{\rho_{00}(t)}{T_1}\end{pmatrix}, \end{aligned} \tag{11.27}$$

where the relaxation constants W are given in terms of the longitudinal T_1 and the transversal T_2 relaxation times, whereas the Boltzmann factor $e^{\frac{-\omega}{kT}}$ allows for the asymmetry of the upward and downward transition rates.

11.1.6 System and Environment as a Multi-band Composed System

In order to gain deeper insight into the difference between the relaxation and the coherence loss, consider a two-level system interacting with a toy model environment, as shown in Fig. 11.1a. Let this toy environment be a large collection of elements of two kinds. The N_f elements of the first type have just two levels, with the transition frequencies Ω_n close to the frequency Ω_0 of the system, and all these elements initially in the ground state interact with the system via the excitation exchange Hamiltonian $\hat{\sigma}_0^\pm\hat{\sigma}_n^\mp$. The N_s elements of the second kind are harmonic oscillators with frequencies much lower as compared to that of the system, and do not experience the excitation exchange. An interaction between the system and an oscillator from this collection exists due to the fact that the oscillator frequencies ω_m depend on the state of the system. The difference of the frequencies of the oscillator

m for the system in the lower state and in the upper state equals $2\alpha_m$. Moreover, the same interaction couples the elements of the environment of different kinds. The corresponding Hamiltonian $\hat{H} = \hat{H}_0 + \hat{H}_I + \hat{H}_2 + \hat{V}_1 + \hat{V}_2$ consists of five parts that read

$$\hat{H}_0 = \Omega_0 \hat{\sigma}_0^z, \; \hat{H}_I = \sum_{n=1}^{N_f} \Omega_n \hat{\sigma}_n^z, \; \hat{H}_2 = \sum_{m=1}^{N_s} \omega_m \hat{a}_m^\dagger \hat{a}_m$$

$$\hat{V}_1 = \sum_{n=1}^{N_f} v_n \left(\hat{\sigma}_0^+ \hat{\sigma}_n^- + \hat{\sigma}_0^- \hat{\sigma}_n^+ \right), \; \hat{V}_2 = \frac{\hat{\sigma}_0^z}{2} \sum_{m=1}^{N_s} \alpha_m \hat{a}_m^\dagger \hat{a}_m,$$

and correspond to the Hamiltonians of the system, of the environment elements of the first kind, that of the second kind, of the interaction between the system and the elements of the first kind, and that of the second kind, respectively.

The condition $|\Omega_0 - \Omega_m| \ll \Omega_0$ preserves the net number of excitation in the system and in the collection of the environment elements of the first kind, that is, the quantum state of this part can be characterized by the amplitudes $\psi_{k,\{n_m\}}$ with the first index k, showing what the probability amplitude is of the environmental elements $k > 0$ carrying the excitation while the system is in the ground state, or what the amplitude $\psi_{0,\{n_m\}}$ is of the excited state of the system. The amplitude of no excitation either in the system or in the elements of the first kind is denoted as $\phi_{\{n_m\}}$. Apart from the index k all these amplitudes have another cumulative index $\{n_m\}$, which characterizes the distribution of the number of excitations among the oscillators, and the energy of the excited states of the system depends on the total excitation of the oscillators and on the way this excitation is distributed among them. The state of the system and the environment therefore reads

$$\begin{aligned} &\phi_{\{n_m\}} \left|0, 0 \ldots 0, \{n_m\}\right\rangle + \psi_{0,\{n_m\}} \left|1, 0 \ldots 0, \{n_m\}\right\rangle \\ &+ \sum_{k=1}^{N_s} \psi_{k,\{n_m\}} \left|0, 0 \ldots 1_k, \ldots 0, \{n_m\}\right\rangle . \end{aligned} \tag{11.28}$$

In Fig. 11.1b we depict the level scheme of the system and the environment. The corresponding Schrödinger equation

$$\begin{aligned} -i\dot{\psi}_{k,\{n_m\}} &= \left(\Omega_k + \omega_{\{n_m\}}\right) \psi_{k,\{n_m\}} + v_k \psi_{0,\{n_m\}} \\ -i\dot{\psi}_{0,\{n_m\}} &= \left(\Omega_0 + \omega_{\{n_m\}} + \alpha_{\{n_m\}}\right) \psi_{0,\{n_m\}} + \sum_{k=1}^{N_f} v_k \psi_{k,\{n_m\}} \\ -i\dot{\phi}_{\{n_m\}} &= \left(\omega_{\{n_m\}} - \alpha_{\{n_m\}}\right) n_m \phi_{\{n_m\}}, \end{aligned}$$

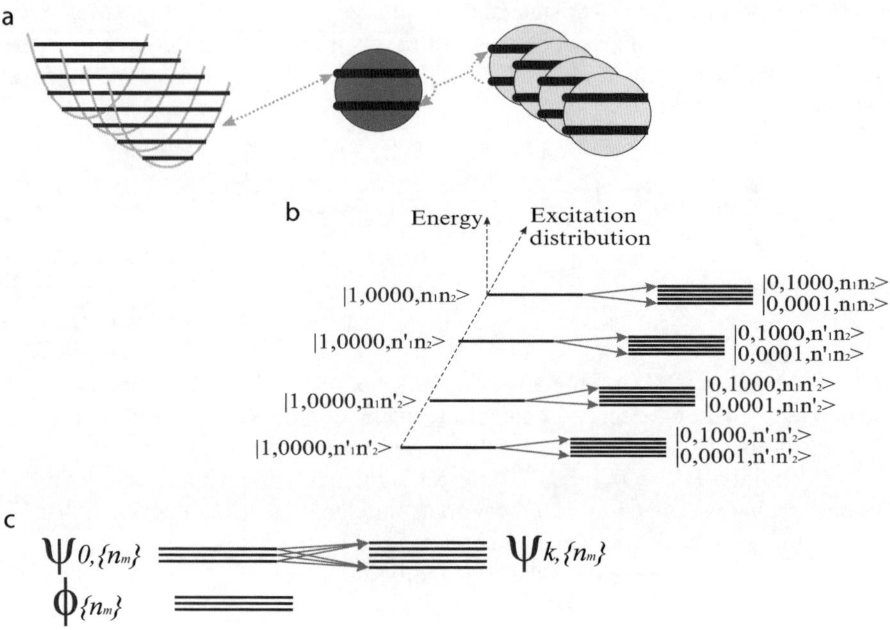

Fig. 11.1 (**a**) A two-level system interacts with the toy environment consisting of two-level elements close to the resonance with the system and of harmonic oscillators with much smaller frequency. There exists an excitation exchange among the two-level system and two-level elements in the environment, while the vibrational excitation of the oscillators remains unchanged but just shifts the system transition energy. (**b**) Level scheme representing the interaction among the levels of the two-level elements and the levels of the system under the conditions when this part contains a single excitation. Detuning of the system depends on the individual distribution of the vibration excitation among the oscillators. (**c**) The combined level scheme

where $\omega_{\{n_m\}} = \sum_{m=1}^{N_s} \omega_m n_m$ and $\alpha_{\{n_m\}} = \sum_{m=1}^{N_s} \frac{\alpha_m}{2} n_m$, has the solution

$$\psi_{k,\{n_m\}} = \frac{v_k e^{-it\omega_{\{n_m\}}}\left(e^{-it\Omega_k} - e^{-tW} e^{-it\Omega_0 - it\alpha_{\{n_m\}}}\right)}{\Omega_k - \Omega_0 - \alpha_{\{n_m\}} - iW} \psi(0)_{0,\{n_m\}}$$

$$\psi_{0,\{n_m\}} = \psi(0)_{0,\{n_m\}} e^{-tW} e^{-it\left(\Omega_0 + \omega_{\{n_m\}} + \alpha_{\{n_m\}}\right)}$$

$$\phi_{\{n_m\}} = \phi(0)_{\{n_m\}} e^{-it\left(\omega_{\{n_m\}} - \alpha_{\{n_m\}}\right)},$$

where $iW = \sum_{k=1}^{N_f} \frac{|v_k|^2}{\varepsilon - \Omega_k - \omega_{\{n_m\}}} = \int \frac{|v_\Omega|^2 g(\Omega)}{\varepsilon - \Omega - \omega_{\{n_m\}}} d\Omega$, while $\psi(0)_{0,\{n_m\}}$ and $\phi(0)_{\{n_m\}}$ stand for the initial conditions for the amplitudes at $t = 0$. In agreement with (11.28) and after performing the calculations in the spirit of that done in Chap. 4, these amplitudes yield the density matrix of the system

$$\hat{\rho} = \sum_{\{n_m\}} \left(\phi_{\{n_m\}} |0\rangle + \psi_{0,\{n_m\}} |1\rangle\right) \left(\phi^*_{\{n_m\}} \langle 0| + \psi^*_{0,\{n_m\}} \langle 1|\right)$$
$$+ \sum_{\{n_m\},k=1}^{N_s} \psi_{k,\{n_m\}} \psi^*_{k,\{n_m\}} |0\rangle \langle 0| ,$$

which can also be presented in the form

$$\hat{\rho} = \sum_{\{n_m\}} [\left(\left|\phi(0)_{\{n_m\}}\right|^2 + \left|\psi(0)_{0,\{n_m\}}\right|^2 \right) |0\rangle \langle 0|$$
$$+ \left|\psi(0)_{0,\{n_m\}}\right|^2 e^{-2tW} \left(|1\rangle \langle 1| - |0\rangle \langle 0|\right)$$
$$+ \phi^*(0)_{\{n_m\}} \psi(0)_{0,\{n_m\}} e^{-2it\alpha_{\{n_m\}} - it\Omega_0} e^{-tW} |1\rangle \langle 0|$$
$$+ \psi(0)_{0,\{n_m\}} \phi(0)_{\{n_m\}} e^{-tW} e^{it\Omega_0 + 2it\alpha_{\{n_m\}}} |0\rangle \langle 1|]. \tag{11.29}$$

The first and the second lines of (11.29) describe the relaxation of population, allowing for transfer of the population $\sum_{\{n_m\}} \left|\psi(0)_{0,\{n_m\}}\right|^2$ of the excited state of the system to its ground state with the rate $2W$, as is typical for the transitions in the level-band systems. The initial population of the ground state $\sum_{\{n_m\}} \left|\phi(0)_{\{n_m\}}\right|^2$ remains there all the time.

The two last lines describe the relaxation of the off-diagonal density matrix elements. Apart from the decay with the rate W related to the population relaxation, the off-diagonal elements experience also so-called coherent damping due to the fact that the transition frequency of the system depends on the energy partition over the oscillators, and for a statistical distribution of the energy among the oscillators this manifests itself in the so-called inhomogeneous broadening of the spectral line. Contributions with different $\alpha_{\{n_m\}}$ interfere destructively and make the off-diagonal elements of the density matrix decay faster than it is suggested by just the population relaxation. One can also interpret it as a loss of coherence, since this kind of damping is related to the fact that we have to make averaging over the distribution of the energy partitions over the oscillators, whereas each of such a partition separately corresponds to a well specified transition frequency and does not result in the coherent damping.

In contrast to the toy model just considered, where two different kinds of elements were responsible for two different types of dampings, in a more general case, the elements of the environment can contribute to both kinds of the damping at the same time. In order to see this phenomenon in more detail, we now consider an exactly soluble model of the environment.

11.2 Exactly Soluble Model of Environment with a Finite Correlation Time

In the previous section we have considered the situation where the environment had an infinitely short correlation time, which resulted in the Markovian master equation for the density matrix of the system. The standard derivation of such an equation relies on a series of qualitative assumptions and is not fully justified by the first principle calculations. In this section we perform a more accurate derivation for a specific model of the environment, which allows one to analytically perform the main part of the calculation with no assumptions. Moreover, this model allows one to address questions related to the quantification of the entanglement between the system and the environment, which was impossible within the framework of the Zwanzig approach based on the projection of the exact density matrix on a tensor product of mixed states that corresponds to zero entanglement.

11.2.1 Hubbard–Stratanovich Separation of a System Parts

Consider a system with Hamiltonian H which is coupled to an environment composed of an infinitely large number of harmonic oscillators with frequencies distributed according a given law g_ω. The total Hamiltonian

$$\hat{H} + \sum_\omega g_\omega \hat{a}^\dagger_\omega \hat{a}_\omega + \hat{V} \otimes \sum_\omega c_\omega \frac{\hat{a}^\dagger_\omega + \hat{a}_\omega}{\sqrt{2}} \tag{11.30}$$

also includes the interaction Hamiltonian given by the tensor product of the system operator $\hat{V}$ and the sum of the environment oscillator displacements weighted by factors c_ω.

Further calculations rely on the expression

$$e^{-iab} = \int \frac{dxdy}{2\pi} e^{-iax+iyb-iyx} \tag{11.31}$$

valid for both a pair of regular numbers $\{a, b\}$ and a pair of commuting operators. This expression can be easily understood, since the integral over $\mathrm{d}y$ is one of the representations of the Dirac delta function, $\delta\,(b - x)$. We employ a functional integral analog of this formula. For an operator $\hat{A} + \hat{B} + \hat{a} \otimes \hat{b}$ of a bipartite system, where operators $\hat{A}$ and $\hat{a}$ act in the Hilbert space of one part while $\hat{B}$ and $\hat{b}$ in that of the other, one has

$$e^{-i\left(\hat{A}+\hat{B}+\hat{a}\otimes\hat{b}\right)t} = \mathrm{T}e^{-i\sum_i\left(\hat{A}+\hat{B}+\hat{a}\otimes\hat{b}\right)\Delta t_i} = \mathrm{T}\prod_i e^{-i\hat{A}\Delta t_i - i\hat{B}\Delta t_i - i\hat{a}\otimes\hat{b}\Delta t_i}.$$

Here T is the time ordering operator, which we will need at a later stage of the calculations. The time intervals Δt_i are taken so short that all operators multiplied by this quantity can be considered as commuting ones. We therefore can apply (11.31) to each of the factors of the product and write

$$e^{-i(\hat{A}+\hat{B}+\hat{a}\otimes\hat{b})t} = \mathrm{T}\prod_i \int \frac{dx_i dy_i}{2\pi/\Delta t_i} e^{-i\hat{A}\Delta t_i - i\hat{B}\Delta t_i - i\hat{a}x_i\Delta t_i - i\hat{b}y_i\Delta t_i - ix_iy_i\Delta t_i}$$

$$= \left(\int\prod_i \frac{dx_i dy_i}{2\pi/\Delta t_i}\right)\left(\mathrm{T}e^{-i\sum_i(\hat{A}-a\hat{x}_i)\Delta t_i}\right)\left(\mathrm{T}e^{-i\sum_i(\hat{B}+\hat{b}y_i)\Delta t_i}\right)e^{-i\sum_i x_iy_i\Delta t_i}.$$

Taking the continuous limit, $\Delta t_i \to 0$, one arrives at the functional integral

$$e^{-i(\hat{A}+\hat{B}+\hat{a}\otimes\hat{b})t} = \int \mathcal{D}x(t)\mathcal{D}y(t)e^{-i\int x(t)y(t)dt}$$
$$\mathrm{T}e^{-i\int(\hat{A}-\hat{a}x(t))dt}\mathrm{T}e^{-i\int(\hat{B}+\hat{b}y(t))dt}.$$

For the Hamiltonian (11.30) the evolution operator $U(t)$ thus reads

$$\int \mathcal{D}x(t)\mathcal{D}y(t)\mathrm{T}e^{-i\int(\hat{H}+\hat{V}x(t)+x(t)y(t))dt}\prod_\omega \mathrm{T}e^{-i\int(\omega\hat{a}^\dagger_\omega\hat{a}_\omega - y(t)c_\omega(\hat{a}^\dagger_\omega+\hat{a}_\omega)/\sqrt{2})dt}.$$

With the help of this technique, the density matrix $\hat{U}(t)\hat{\rho}_{SE}\hat{U}^+(t)$ of the system+environment, initially in the product state $\hat{\rho}_{SE} = \hat{\rho}_0 \otimes \prod_{\otimes\omega}\hat{\rho}_\omega$ of the system given by the initial density matrix ρ_0 and the density matrices ρ_ω of the environment modes, can be represented in the form

$$\int \mathcal{D}x(t)\mathcal{D}y(t)\mathcal{D}X(t)\mathcal{D}Y(t)$$
$$Te^{-i\int(\hat{H}+\hat{V}x(t)+x(t)y(t))dt}\hat{\rho}_0 Te^{i\int(\hat{H}+\hat{V}X(t)+X(t)Y(t))dt}$$
$$\prod_{\otimes\omega}\mathrm{T}e^{-i\int(\omega\hat{a}^\dagger_\omega\hat{a}_\omega - y(t)c_\omega(\hat{a}^\dagger_\omega+\hat{a}_\omega)/\sqrt{2})dt}\rho_\omega \mathrm{T}e^{i\int(\omega\hat{a}^\dagger_\omega\hat{a}_\omega - Y(t)c_\omega(\hat{a}^\dagger_\omega+\hat{a}_\omega)/\sqrt{2})dt}. \tag{11.32}$$

The last line of this expression can be viewed as a weight function $\mathcal{W}$ for the functional integration in the space of the perturbations strengths $x(t)$, $y(t)$, $X(t)$, and $Y(t)$ of the system density matrix.

11.2.2 Environment at a Given Temperature

We are now in the position to take a model for the initial states of the environment modes, which we choose to be in thermal states at a temperature T characterized by the properly normalized density matrices $\rho_\omega = e^{-\frac{\omega}{kT} a^\dagger_\omega a_\omega}$. Tracing out of the environment implies calculation of the traces

$$\mathrm{Tr}\left\{ \mathrm{T} e^{-i\int\left(\omega \hat{a}^\dagger_\omega \hat{a}_\omega - y(t) c_\omega \left(\hat{a}^\dagger_\omega + \hat{a}_\omega\right)/\sqrt{2}\right)dt} \rho_\omega \mathrm{T} e^{i\int\left(\omega \hat{a}^\dagger_\omega \hat{a}_\omega - Y(t) c_\omega \left(\hat{a}^\dagger_\omega + \hat{a}_\omega\right)/\sqrt{2}\right)dt} \right\} \tag{11.33}$$

for all environmental modes individually, which should be performed for arbitrary time-dependent functions $y(t)$ and $Y(t)$.

Fortunately, the dynamics of a quantum harmonic oscillator subject to an arbitrary force $c_\omega y(t)$ can be described analytically. This problem has already been discussed earlier [see (7.16)]. As a result of the evolution, the initial quantum state of the oscillator gets displaced in the phase space, and this displacement can be found from the classical formula and be represented by the phase-space displacement vector

$$\frac{\Delta Q + i\Delta P}{\sqrt{2}} = c_\omega \int^t y(\tau) e^{i(t-\tau)\omega} d\tau, \tag{11.34}$$

where ΔQ and ΔP are properly normalized displacements of the Wigner function in the coordinate and in the momentum, respectively. Note that apart from the displacement operator in the exponent of the evolution operator there is also a scalar part, which yields a phase factor $\exp i\Delta S$. The latter can be adequately associated with the difference of classical actions accumulated by harmonic oscillators subject to the time-dependent forces $y(t)$ and $Y(t)$, as shown in Fig. 11.2.

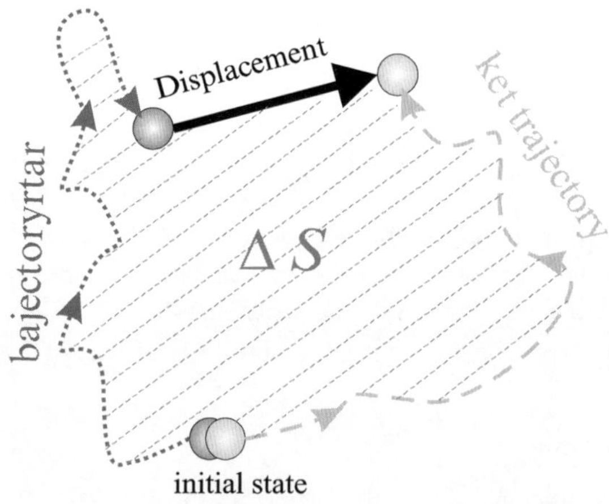

Fig. 11.2 Trajectories of the Wigner function in the phase space of the oscillators that were subject to forces $y(t)$ and $Y(t)$. The *arrow* shows the relative displacement (11.34) and the *shadowed area* corresponds to the difference of the classical actions that give the phase factor ΔS

Consider this description in more detail. To this end we write (11.33) in the interaction representation

$$\mathrm{Tr}\left\{ \mathrm{T}\exp\left[-ic_\omega \int_0^t y(\tau) \frac{\hat{a}^\dagger_\omega e^{-i(t-\tau)\omega} + \hat{a}_\omega e^{i(t-\tau)\omega}}{\sqrt{2}} d\tau \right] \right.$$

$$\left. \hat{\rho}_\omega \mathrm{T}\exp\left[ic_\omega \int_0^t Y(\tau) \frac{\hat{a}^\dagger_\omega e^{-i(t-\tau)\omega} + \hat{a}_\omega e^{i(t-\tau)\omega}}{\sqrt{2}} d\tau \right] \right\}.$$

With the help of the Baker–Campbell–Hausdorff formula applied to the time-ordered exponents entering this equation and the fact that the commutators of the operators at different times are c-numbers, one finds

$$\mathrm{T}\exp\left[i\int dt \frac{c_\omega y(t)}{\sqrt{2}} \hat{a} e^{it\omega} + i\int dt \frac{c_\omega y(t)}{\sqrt{2}} \hat{a}\dagger e^{-it\omega} \right]$$

$$= \exp\left[i\int dt \frac{c_\omega y(t)}{\sqrt{2}} \hat{a}\dagger e^{-it\omega} \right] \exp\left[i\int dt \frac{c_\omega y(t)}{\sqrt{2}} \hat{a} e^{it\omega} \right]$$

$$\exp\left[-\frac{[\hat{a}, \hat{a}\dagger]}{2} \int d\theta dt \Theta_{t-\theta} y(\theta) y(t) c_\omega^2 e^{i(t-\theta)\omega} \right]$$

for the left evolution operator and the similar expression

$$\mathrm{T}\exp\left[-i\int dt \frac{c_\omega Y(t)}{\sqrt{2}} \hat{a}\dagger e^{-it\omega} - i\int dt \frac{c_\omega Y(t)}{\sqrt{2}} \hat{a} e^{it\omega} \right]$$

$$= \exp\left[-i\int dt \frac{c_\omega Y(t)}{\sqrt{2}} \hat{a}\dagger e^{-it\omega} \right] \exp\left[-i\int dt \frac{c_\omega Y(t)}{\sqrt{2}} \hat{a} e^{it\omega} \right]$$

$$\exp\left[-\frac{[\hat{a}, \hat{a}\dagger]}{2} \int d\theta dt \Theta_{\theta-t} Y(\theta) Y(t) c_\omega^2 e^{i(t-\theta)\omega} \right]$$

for the right one, where Θ_x stands for the Heaviside step function. The order of the operators under the trace can be changed by cyclic permutations, and therefore we can interchange order of the first and the last exponential operators under the Tr symbol, combine the operators to the left and to the right sides of the density matrix that now are commuting, and obtain (11.33) in the form

$$\exp\left[-\frac{1}{2} \int\int d\theta dt \left(\Theta_{t-\theta} y(\theta) y(t) + \Theta_{\theta-t} Y(\theta) Y(t) \right) c_\omega^2 e^{i(t-\theta)\omega} \right]$$

$$\exp\left[\frac{1}{2} \int dt d\theta Y(t) y(\theta) c_\omega^2 e^{i(t-\theta)\omega} \right]$$

$$\mathrm{Tr}\left\{ e^{i2^{-1/2}\hat{a}\int (y(t)-Y(t)) c_\omega e^{it\omega} dt} \hat{\rho}_0 e^{i2^{-1/2}\hat{a}\dagger \int (y(t)-Y(t)) c_\omega e^{-it\omega} dt} \right\},$$

where interchange of the factors $e^{-i\int \mathrm{d}t \frac{c_\omega Y(t)}{\sqrt{2}} \hat{a} e^{it\omega}}$ and $e^{i\int \mathrm{d}t \frac{c_\omega y(t)}{\sqrt{2}} \hat{a}\dagger e^{-it\omega}}$ yields the exponential factor in the second line.

For the thermal distribution

$$\hat{\rho}_0 = \left(1 - e^{-\omega/kT}\right) \sum_n^\infty |n\rangle\, e^{-n\omega/kT} \langle n|$$

the trace can be found explicitly, and reads

$$(1 - e^{-\frac{\omega}{kT}}) \exp\left[-\frac{c_\omega^2 e^{-\frac{\omega}{kT}}}{2} \int [y(t) - Y(t)]\, e^{-it\omega} dt \int [y(t) - Y(t)]\, e^{it\omega} dt\right], \tag{11.35}$$

where we have employed the explicit expression for the sum

$$\begin{aligned}
\sum_{m,n}^\infty \frac{\mathrm{Tr}\left\{\hat{a}^m |n\rangle\, e^{-\beta n} \langle n|\, \hat{a}\dagger^m\right\}}{m!m!} X^m &= \sum_{m,n}^\infty \frac{n! e^{-\beta n}}{(n-m)!m!m!} X^m \\
&= \frac{\exp\left(X \frac{e^{-\beta}}{1-e^{-\omega/kT}}\right)}{1 - e^{-\omega/kT}}.
\end{aligned}$$

We now can write explicitly the total weight function entering the last line in (11.32), which is given by the product of contributions of all oscillators with different frequencies

$$\mathcal{W}\left(Y(t), y(t)\right) = \mathcal{N} \exp\left[\int\int \left(Y(\tau)\; y(\tau)\right) \hat{M}_{t-\theta} \begin{pmatrix} Y(\theta) \\ y(\theta) \end{pmatrix} d\tau d\theta\right] \tag{11.36}$$

where the covariance matrix

$$\hat{M}_{t-\theta} = \sum_\omega \frac{c_\omega^2 e^{i(t-\theta)\omega}}{2} \begin{pmatrix} \Theta_{t-\theta} - \frac{e^{-\omega/kT}}{1 - e^{-\omega/kT}} & \frac{e^{-\omega/kT}}{1 - e^{-\omega/kT}} - 1 \\ \frac{e^{-\omega/kT}}{1 - e^{-\omega/kT}} & \Theta_{\theta-t} - \frac{e^{-\omega/kT}}{1-e^{-\omega/kT}} \end{pmatrix} \tag{11.37}$$

determines the probability of the time-dependent perturbations $Y(t)$ and $y(t)$. The integration over time in (11.36) can be extended to $t \to \infty$, since in (11.32) the functional integration over $X(\theta)$ and $x(\tau)$ yields functional delta functions $\delta\left(y(\tau)\right) \delta\left(Y(\theta)\right)$, whereas the functions $X(\theta)$ and $x(\tau)$ do not enter the evolution operators when $\tau, \theta > t$.

Note that the real part of integral in the exponent of (11.36) allows for the overlap of displaced states in the phase space shown in Fig. 11.2, while the imaginary part gives the classical action represented by the corresponding area. Also note that if we assume that each of the environment oscillators is not in a mixed thermal state but in a pure coherent state resulting from the randomly displaced vacuum with the displacements distributed according to a Gaussian law, we obtain the same result. In other words, a mixed state and a complicated pure state are completely equivalent from the viewpoint of the action of a large environment on a much smaller system.

Note that the covariance matrix can be expressed as

$$\hat{M}_{t-\theta} = \begin{pmatrix} \Theta_{t-\theta} h_{t-\theta} - h_{t-\theta,T} & h_{t-\theta,T} - h_{t-\theta} \\ h_{t-\theta,T} & \Theta_{\theta-t} h_{t-\theta} - h_{t-\theta,T} \end{pmatrix} \tag{11.38}$$

in terms of the environment correlation functions

$$h_x = \sum_\omega \frac{1}{2} c_\omega^2 e^{-ix\omega}, \tag{11.39}$$

$$h_{t-\theta,T} = \sum_\omega \frac{1}{2} c_\omega^2 \frac{e^{-ix\omega-\omega/kT}}{1-e^{-\omega/kT}}, \tag{11.40}$$

where (11.39) gives the same physical quantity as the parameters $h^i_{\alpha\beta}(x)$ entering (11.22) for the relaxation matrix, while the typical width of this dependence has the explicit physical meaning of the correlation time. That can be easily understood. In fact, if at the moment $t = 0$ the system makes a transition and sends an excitation to one of the environment oscillators, the probability amplitude that the excitation goes to the oscillator at the frequency ω is proportional to c_ω. In the course of time this probability amplitude evolves in phase, acquiring a phase factor $e^{-it\omega}$ in accordance with the energy of the mode, and after a certain time all the amplitudes together contribute to the amplitude of the backward transition according to their weights c^*_ω. At a time t_c, such that the acquired phases make the interference of the oscillator amplitudes destructive, the backward transition amplitude gets small. Such a time plays the role of the correlation time of the environment, which at times exceeding t_c after the perturbation event coming from the system no longer respond to this perturbation.

11.2.3 Environment as a Source of Langevin Force

Let us focus on the functional weight in the space of the functions $x(t)$ and $X(t)$. Substituting (11.36) into (11.32) traced over the environment in the way shown by (11.35), for the density matrix of the system one obtains

$$\hat{\rho}(t) = \int \mathcal{D}x(t)\mathcal{D}X(t)\mathrm{T}e^{-i\int(\hat{H}+\hat{V}x(t))dt}\hat{\rho}\mathrm{T}e^{i\int(\hat{H}+\hat{V}X(t))dt}\mathcal{W}(\{x(t),X(t)\}) \tag{11.41}$$

where the explicit expression for the functional weight

$$\mathcal{W}\left(\{x(t), X(t)\}\right) = \int \mathcal{D}y(t)\mathcal{D}Y(t)$$

$$e^{-i\int\limits^{\infty}\left(x(\tau)\ X(\tau)\right)\hat{\sigma}^z\begin{pmatrix} y(\tau) \\ Y(\tau)\end{pmatrix}d\tau}$$

$$e^{i\int\limits^{\infty}\int\limits^{\infty}\left(y(\tau)\ Y(\tau)\right)\hat{M}_{\tau-\theta}\begin{pmatrix} y(\theta) \\ Y(\theta)\end{pmatrix}d\theta d\tau} \tag{11.42}$$

depends on the functional correlation covariance matrix (11.37), and σ^z stands for the Pauli matrix.

After performing Gaussian functional integration over $\mathcal{D}y(t)\mathcal{D}Y(t)$, Eq. (11.41) adopts the form

$$\hat{\rho}(t) = \int \mathcal{D}x(t)\mathcal{D}X(t)\mathrm{T}e^{-i\int\limits^{t}\left(\hat{H}+\hat{V}x(t)\right)dt}\hat{\rho}\mathrm{T}e^{i\int\limits^{t}\left(\hat{H}+\hat{V}X(t)\right)dt}$$

$$e^{-\int\int\left(x(\tau)\ X(\tau)\right)\hat{G}_{\tau-\theta}\begin{pmatrix} x(\theta) \\ X(\theta)\end{pmatrix}d\tau d\theta} \tag{11.43}$$

where the functional covariance matrix $\hat{G}_{\tau-\theta}$ for the functions $x(t)$ and $X(t)$ relates to the Fourier transform $\hat{M}_\omega$ of the covariance matrix $\hat{M}_{t-\theta}$ of (11.38) as

$$\hat{G}_t = \frac{1}{2\pi}\int e^{-i\omega t}\hat{\sigma}^z\hat{M}_\omega^{-1}\hat{\sigma}^z d\omega,$$

$$\hat{\sigma}^z\hat{M}_\omega^{-1}\hat{\sigma}^z = \frac{1}{-\left(h_\omega + i\gamma_\omega\right)^2\left(1-2e^{-\omega/kT}\right)+\gamma_\omega^2 e^{-\omega/kT}}$$

$$\begin{pmatrix} \frac{h_\omega+i\gamma_\omega}{e^{\omega/kT}-1}e^{\omega/kT} - h_\omega e^{-\omega/kT} & -h_\omega\left(1-2e^{-\omega/kT}\right) \\ h_\omega e^{-\omega/kT} & \frac{h_\omega-i\gamma_\omega}{e^{\omega/kT}-1}e^{\omega/kT} - h_\omega e^{-\omega/kT}\end{pmatrix}, \tag{11.44}$$

where $\gamma_\omega = \mathrm{vp}\int\frac{c_\mu^2 d\mu}{\pi(\omega+\mu)}$ is given by the Kramers–Kronig relations for the imaginary part of an analytic function $h_\omega + i\gamma_\omega$ defined by the values of its real part $h_\omega = c_\omega^2/2$ at the real axis ω. The presence of γ_ω results from the presence of the step function in (11.38) for $\hat{M}_{t-\theta}$ due to the causality principle.

The quantities $x(t)$ and $X(t)$ entering (11.43) have the explicit physical meaning of Langevin forces acting on the bra and ket parts of the system along the generalized coordinate corresponding to the operator V, whereas (11.44) provides us with the covariance matrix of the Langevin forces. The chosen model allows one to illustrate the fact that the dynamics of a system interacting with its environment and characterized by a density matrix of mixed state can also be viewed as the unitary dynamics of a pure state subject to an external Langevin force, which randomly depends on time, and which is different for the bra and ket state vectors. Each realization of the random force has the probability

$$\mathcal{W}\left(\{x(t), X(t)\}\right) = e^{-\int\int \left(x(\tau)\; X(\tau)\right)\hat{G}_{\tau-\theta}\begin{pmatrix} x(\theta) \\ X(\theta)\end{pmatrix} d\tau d\theta} \tag{11.45}$$

determined via the correlation kernel (11.44), while the net evolution is a statistical mixture of unitary evolutions under the action of different realizations of the random force.

In the interaction representation equation (11.43) reads

$$\hat{\rho}(t) = \int \mathcal{D}X(t)\mathcal{D}x(t)\mathcal{W}\left(\{x(t), X(t)\}\right) \mathrm{T}e^{-i\int\limits_0^t \hat{V}(t)x(t)dt}\hat{\rho}_0 \mathrm{T}e^{i\int\limits_0^t \hat{V}^*(t)X(t)dt}, \tag{11.46}$$

where $\hat{V}(t) = e^{i\hat{H}t}\hat{V}e^{-i\hat{H}t}$. If we now distinguish the operators $\hat{V}(t)$ that act on the density matrix from the left and from the right sides, denote them by indices L and R, respectively, and formally move the right side operators to the left side, we obtain

$$\hat{\rho}(t) = \int \mathcal{D}x(t)\mathcal{D}X(t)\mathcal{W}\left(\{x(t), X(t)\}\right) \mathrm{T}e^{i\int\limits_0^t\left(\hat{V}_R^*(t)X(t) - \hat{V}_L(t)x(t)\right)dt}\hat{\rho}_0.$$

Evaluation of the functional integral is a Fourier transformation, inverse to that of (11.42), and it yields the formal expression

$$\hat{\rho}(t) = \mathrm{T}e^{-\int\limits_0^t\int\limits_0^t \mathrm{d}\tau\mathrm{d}\theta\left(\hat{V}_L(\tau)\; \hat{V}_R^*(\tau)\right)\hat{M}_{\tau-\theta}\begin{pmatrix}\hat{V}_L(\theta) \\ \hat{V}_R^*(\theta)\end{pmatrix}}\hat{\rho}_0 \tag{11.47}$$

determined by the covariance matrix $\hat{M}_{\tau-\theta}$ of (11.37), while the time-ordering operator now implies ordering on both τ and θ.

In the framework of the environment model under consideration, one can derive a differential equation for the density matrix. However, a closed form of the differential equation can be written not for the density matrix $\hat{\rho}(t) = \mathrm{Tr}\left\{|\Psi(t)\rangle\langle\Psi(t)|\right\}$ itself, obtained by tracing over the environment of the tensor product of total state vectors taken at a given time moment, but for a different quantity, namely for $\hat{\rho}(\tau,\theta) = \mathrm{Tr}\left\{|\Psi(\tau)\rangle\langle\Psi(\theta)|\right\}$ where the state vectors are taken at different moments of time. This approach is known as the Keldysh technique. We have already considered many times an analog of this quantity, $\hat{\rho}(\varepsilon,\xi)$, working in the Fourier representation, as, for instance, (4.27) for the population of two bands, although without explicitly presenting the relevant differential equation. The differential equation contains a second derivative and reads

$$\frac{\partial^2}{\partial\tau\partial\theta}\hat{\rho}(\tau,\theta) = -\left(\hat{V}_L(\tau)\ \hat{V}_R^*(\tau)\right)\hat{M}_{\tau-\theta}\begin{pmatrix}\hat{V}_L(\theta)\\ \hat{V}_R^*(\theta)\end{pmatrix}\hat{\rho}(\tau,\theta), \tag{11.48}$$

which is just another way to allow for the integral operator $\overset{\frown}{\mathcal{G}}$ in (11.15) of a particular form corresponding to the model chosen.

11.2.3.1 The Second Order Perturbation Theory

Thus far, we did not make any approximation within the chosen model of the environment. Though the exact result of this consideration given by (11.47) looks rather simple, this simplicity is illusory, since for practical calculations one has to ensure the time ordering for all terms of the time-dependent perturbation theory series for the evolution operator of the density matrix (11.47), and this implies rather involved topological requirements on the corresponding Feynman diagrams representing such terms. Still the form of (11.47) is convenient for description with the help of the second-order time-dependent perturbation theory, since in this order the complicated interplay of time ordering and Gaussian coupling typical of higher orders does not yet manifest itself, or, in other words, one makes the assumption that the leading contribution comes from the Feynman diagrams with the ladder-like topology, such that (11.47) immediately results in

$$\begin{aligned}\Delta\hat{\rho}(t) = -\int_0^t\!\!\int_0^t d\theta d\tau \big(&\hat{V}_L(\tau)\hat{V}_L(\theta)\hat{\rho}_0\left(\Theta_{\tau-\theta}h_{\tau-\theta} - h_{\tau-\theta,T}\right)\\ &+ \hat{\rho}_0\hat{V}_R^*(\theta)\hat{V}_R^*(\tau)\left(\Theta_{\theta-\tau}h_{\tau-\theta} - h_{\tau-\theta,T}\right)\\ &- \hat{V}_L(\tau)\hat{\rho}_0\hat{V}_R^*(\theta)\left(h_{\tau-\theta} - h_{\tau-\theta,T} - h_{\theta-\tau,T}\right)\big)\end{aligned} \tag{11.49}$$

where $\Delta\hat{\rho}(t) = \hat{\rho}(t) - \hat{\rho}_0$.

Let us specify the physical meaning of the assumptions underlying the applicability of the second-order perturbation theory for both the model just considered and the projection method discussed earlier. By doing the second-order perturbation, one assumes that, first, the action of the interaction between the system and the environment accumulated during the time interval equal to the typical correlation time is small, and, second, that the system and the environment were not entangled at the times prior to the moment $t = 0$. The second assumption, also known as the ladder approximation, permits one to write the total density matrix $\hat{\rho}(t = 0)$ as a tensor product of the density matrices of the system and the environment. This assumption implies that for the description of the evolution at $t > 0$, it is sufficient to impose the initial condition at $t = 0$ for the density matrix only. In general, besides the initial condition for the density matrix, the integro-differential equation of the type (11.15) also requires either the values of its first- and higher-order derivatives at $t = 0$ or the values of this matrix at earlier moments of time. The latter is the case where the interaction has started before $t = 0$, whence at the moment $t = 0$ the system and the environment were already entangled. The same is true for (11.47), since formally, the integration interval should be extended till $t = -\infty$, in the case where the system and the environment are entangled at $t = 0$. Another way to say this is to notice that, in general,

$$\mathrm{T}e^{\int_{-t}^{t}\int_{-t}^{t} d\tau d\theta \hat{\Phi}(\tau,\theta)} \neq \mathrm{T}e^{\int_{0}^{t}\int_{0}^{t} d\tau d\theta \hat{\Phi}(\tau,\theta)}\,\mathrm{T}e^{\int_{-t}^{0}\int_{-t}^{0} d\tau d\theta \hat{\Phi}(\tau,\theta)} \tag{11.50}$$

where

$$\hat{\Phi}(\tau,\theta) = -\left(\hat{V}_L(\tau)\;\hat{V}_R^*(\tau)\right)\hat{M}_{\tau-\theta}\begin{pmatrix}\hat{V}_L(\theta)\\ \hat{V}_R^*(\theta)\end{pmatrix},$$

since the left hand side of (11.50) contains a contribution of the domain where $\tau\theta < 0$, which does not contribute to the right hand side of this equation. The assumption that (11.50) holds as an equality implies that no memory effects can last at the time interval longer than the correlation time of the environment, and this hypothesis allows one to describe the evolution of the system by the differential equation for the density matrix containing the relaxation operator $\hat{\mathcal{R}}$.

Fourier Representation for the Relaxation Rates

Consider the second-order expression for the increment of the density matrix (11.49) presented in the Fourier form

$$\begin{aligned}\Delta\hat{\rho} = &\int \hat{V}_\omega\hat{\rho}_0\hat{V}_\omega^*\left(h_{-\omega} + \frac{\omega}{|\omega|}h_{|\omega|,T}\right)d\omega\\ &-\hat{\rho}_0\left(\int\frac{h_\omega - 2h_{\omega,T}}{2}\hat{V}_{-\omega}^*\hat{V}_\omega^* d\omega - i\int\frac{\hat{V}_\omega^*\hat{V}_{-\omega}^* h_{\omega'}}{2\pi(\omega+\omega')}d\omega' d\omega\right)\\ &-\left(\int\frac{h_\omega - 2h_{\omega,T}}{2}\hat{V}_{-\omega}\hat{V}_\omega d\omega + i\int\frac{\hat{V}_\omega\hat{V}_{-\omega}h_{\omega'}}{2\pi(\omega+\omega')}d\omega' d\omega\right)\hat{\rho}_0\end{aligned} \tag{11.51}$$

where the singular integrals over $d\omega'$ are understood as the principal values, the operators and the correlation functions are given by their Fourier decomposition

$$\begin{aligned}\hat{V}(t) &= \int\frac{d\omega}{2\pi}\hat{V}_\omega e^{i\omega t};\ \hat{V}^*(t) = \int\frac{d\omega}{2\pi}\hat{V}_\omega^* e^{-i\omega t};\\ h_\omega &= \frac{\Theta_\omega}{2}c_\omega^2;\ h_{\omega,T} = \frac{\Theta_\omega}{2}c_\omega^2\frac{e^{-\omega/kT}}{1-e^{-\omega/kT}},\end{aligned} \tag{11.52}$$

and where we have omitted the L and R indices, placing the operators to the left and to the right side from $\hat{\rho}_0$, respectively. We have also performed the time integrations, and have made the replacement

$$\frac{e^{i(\omega_1-\omega_2)t}-1}{\omega_1-\omega_2} \to 2i\pi\delta\left(\omega_1-\omega_2\right), \tag{11.53}$$

which is supposed to be still consistent with the second-order perturbation theory, implied by the fact that the action of the interaction accumulated during the correlation time remains small.

Equation (11.51) can also be written as

$$\Delta\hat{\rho} = \int d\omega \hat{V}_\omega \hat{\rho} \hat{V}^*_\omega h^{II}_{-\omega} + i\left[\operatorname{Re}\hat{W}, \hat{\rho}\right] - \left\{\operatorname{Im}\hat{W}, \hat{\rho}\right\}, \tag{11.54}$$

where $h^{II}_{-\omega} = h_\omega + \frac{\omega}{|\omega|} h_{|\omega|,T}$,

$$\hat{W} = \int \frac{\hat{V}^*_\omega \hat{V}^*_{-\omega} h_{\omega'}}{2\pi\left(\omega+\omega'\right)} d\omega' d\omega + i\int \frac{h_\omega - 2h_{\omega,T}}{2} \hat{V}^*_{-\omega}\hat{V}^*_\omega d\omega,$$

and $\hat{\rho}$ stands for $\hat{\rho}_0$. By introducing the quantity

$$h^I_\omega = h_\omega - i\int \frac{h_{\omega'} d\omega'}{2\pi\left(\omega'-\omega\right)} - 2h_{\omega,T}$$

resulting from the standard analytic continuation of the real function h_ω defined for positive arguments ω to the complex plane, one finds

$$\hat{W} = i\int \frac{h^I_\omega}{2} \hat{V}^*_{-\omega}\hat{V}^*_\omega d\omega, \tag{11.55}$$

which allows one to trace an analogy to (11.22) for the correlation functions defining the relaxation operator $\hat{\mathcal{R}}$.

Equation (11.54) clearly shows that the longitudinal relaxation of the density matrix given by the first term on the right hand side does not occur if the emission spectrum of the coupling matrix in the Heisenberg representation does not overlap with the absorption spectrum of the environment modes h^{II}_ω, and vice versa. Even in this case, the transverse relaxation, given by the third term, might still be present, whereas the second term gives a correction to the Hamiltonian dynamics.

Note that the system under consideration contains a finite number of energy levels, and therefore the frequency decomposition V_ω of the time-dependent operators $V(t)$ in the interaction representation has a discrete spectrum, unless this system is subject to a time-dependent external perturbation. Therefore the integrals entering (11.51) contain contributions linearly increasing in time, which arise from the terms $\sim V(\omega)V(\omega)$ with coinciding arguments. In fact, for $V^n_m(t) = V^n_m e^{i\Omega_{mn}t}$ one has the explicit form $(V_\omega)_m^n = V_m^n \delta(\omega - \Omega_{mn})$, which being substituted into (11.49) yields the terms on the right hand side of the form

$$V_m^n \rho_n^k V_k^{*r} \int d\omega \int^t \int^t h_\omega e^{i\Omega_{mn}\tau - i\Omega_{kr}\theta + i\omega(\tau-\theta)} d\theta d\tau,$$

$$V_m^n V_n^k \rho_k^r \int d\omega \int^t \int^\tau h_\omega e^{i\Omega_{mn}\tau - i\Omega_{kn}\theta + i\omega(\tau-\theta)} d\theta d\tau,$$

$$\rho_m^n V_n^{*k} V_k^{*r} \int d\omega \int^t \int^\tau h_\omega e^{i\Omega_{rk}\tau - i\Omega_{nk}\theta + i\omega(\tau-\theta)} d\theta d\tau,$$

and for coinciding frequencies Ω_{ij} in the exponents of the integrands results in

$$V_m^n \rho_n^k V_k^{*r} \int d\omega h_\omega \frac{2\left(\cos\left(\omega + \Omega_{mn}\right)t - 1\right)}{\left(\omega + \Omega_{mn}\right)^2},$$

$$V_m^n V_n^k \rho_k^r \int d\omega h_\omega \left[\frac{it}{\omega + \Omega_{mn}} - \frac{e^{i(\omega + \Omega_{mn})t} - 1}{\left(\omega + \Omega_{mn}\right)^2}\right],$$

$$\rho_m^n V_n^{*k} V_k^{*r} \int d\omega h_\omega \left[\frac{it}{\omega + \Omega_{rk}} - \frac{e^{i(\omega + \Omega_{rk})t} - 1}{\left(\omega + \Omega_{rk}\right)^2}\right].$$

Here we have set $T \to 0$ for the simplicity of the presentation. For long times, where (11.53) still holds, these contributions after the replacement $\omega \to \alpha/t - \Omega$ give the expressions

$$2tV_m^n \rho_n^k V_k^{*r} \int h_{\alpha/t - \Omega_{mn}} \frac{\cos\alpha - 1}{\alpha^2} d\alpha,$$

$$tV_m^n V_n^k \rho_k^r \int h_{\alpha/t - \Omega_{mn}} \left[\frac{i}{\alpha} - \frac{e^{i\alpha} - 1}{\alpha^2}\right] d\alpha,$$

$$t\rho_m^n V_n^{*k} V_k^{*r} \int h_{\alpha/t - \Omega_{rk}} \left[\frac{i}{\alpha} - \frac{e^{i\alpha} - 1}{\alpha^2}\right] d\alpha, \tag{11.56}$$

linearly increasing in time, and for times much longer than the correlation time of h_ω result in the master equation

$$\dot{\rho_m^r} = \frac{\Delta\rho_m^n(t)}{-t} = 2\pi\delta_m^r \sum_k h_{-\Omega_{mk}} \left|V_m^k\right|^2 \rho_k^k,$$
$$-\pi \sum_n \left(h_{-\Omega_{mn}} \left|V_m^n\right|^2 + \left|V_r^n\right|^2 h_{-\Omega_{rn}}\right) \rho_m^r. \tag{11.57}$$

Here we have assumed that the frequencies Ω_{mn} and Ω_{kr} coincide only when both the first and the second indices coincide, which is typical of the finite systems with discrete energy spectrum.

The right hand side of (11.57) has the physical meaning of the relaxation operator $\mathcal{R}$, which does not depend on time, and moreover, it does not change after returning back to the Schrödinger representation. We therefore have arrived at the master equation of the form (11.18), where the relaxation operator allows for the transversal and longitudinal relaxation. The longitudinal relaxation from level k to level m goes with the rate $2\pi\delta_m^r h_{-\Omega_{mk}} \left|V_m^k\right|^2$, provided level k has an energy higher than level m, and $\Omega_{mk} < 0$. This process involves the diagonal elements of the density matrix, and it allows for the population cascading down along the energy scale. Off-diagonal matrix elements corresponding to the energy levels m and k experience exponential transverse relaxation, which has a rate given by the half sum of the rates of the longitudinal relaxation from these levels towards all the energy levels below.

At finite temperatures, one has to perform the same calculations, retaining the term $h_{\tau-\theta,T}$ in (11.49). However, as a result of these calculations, the transverse relaxation rates of the off-diagonal elements of density matrix do not change, while the longitudinal relaxation rates for the diagonal matrix elements do. The transitions upward the energy scale become allowed, just being attenuated by the Bose distribution prefactors, such that the longitudinal transition probability reads

$$2\pi\delta_m^r \sum_k h_{-\Omega_{mk}} \left|V_m^k\right|^2 + 2\pi\delta_m^r \sum_k \frac{e^{-\Omega_{mk}/kT}}{1-e^{-\Omega_{mk}/kT}} h_{\Omega_{mk}} \left|V_m^k\right|^2, \tag{11.58}$$

where the coefficients h are taken at zero temperature.

We should also mention another possibility: for a time-dependent Hamiltonian $\hat{H}(t)$ of the system, the frequency spectra $V_m^n(\omega)$ of the matrix elements in the interaction representation are not necessarily discrete and therefore not necessarily yield a contribution linearly increasing in time. It means that one can rely on (11.55) and think of a considerable inhibition of the relaxation rates by a proper choice of these spectra with the help of a control procedure.

11.2.4 The Increment of the Linear Entropy of Entanglement

Let us now focus on the entanglement between the system and the environment, which we characterize by the linear entanglement entropy $S_l = 1 - \mathrm{Tr}\left\{\hat{\rho}^2(t)\right\}$. Augmentation of this quantity in the course of time is given by the derivative

$$\dot{S}_l = -\mathrm{Tr}\left\{2\hat{\rho}(t)\dot{\hat{\rho}}(t)\right\}, \tag{11.59}$$

while the exact expression (11.47) for the density matrix, and the auxiliary quantity $\rho(t,\theta)$ satisfying the differential equation (11.48) allows one to find an approximative expression for the density matrix derivative $\dot{\hat{\rho}}(t)$ required for the calculation. To this end one has to multiply both sides of (11.48) by the step function $\Theta_{\tau-\theta}$ and integrate over the time θ. In the integrand on the right hand side, the derivative over θ can be moved from the matrix $\rho(t,\theta)$ to the step function thus resulting in the Dirac delta function, which after the integration yields the required derivative $\int d\theta \delta_{\tau-\theta} \frac{\partial}{\partial \tau} \hat{\rho}(\tau,\theta) = \dot{\hat{\rho}}(\tau)$, such that we arrive at the exact expression

$$\dot{\hat{\rho}}(\tau) = \int d\theta \Theta_{t-\theta} \left(\hat{V}_L(t)\ \hat{V}_R^*(t) \right) \hat{M}_{t-\theta} \begin{pmatrix} \hat{V}_L(\theta) \\ \hat{V}_R^*(\theta) \end{pmatrix} \hat{\rho}(t,\theta).$$

The much simpler approximate expression

$$\frac{\partial}{\partial t}\rho(t) \simeq \left\{ \int \left(\hat{V}_L(t)\ \hat{V}_R^*(t) \right) \hat{M}_x \begin{pmatrix} \hat{V}_L(t-x) \\ \hat{V}_R^*(t-x) \end{pmatrix} dx \right\} \hat{\rho}(t) \tag{11.60}$$

emerges from the assumption that the density matrix $\rho(t,\theta)$ experiences just a small change during the time interval given by the correlation time of the environment. In order to be consistent within the approximation, the integral in parentheses at the right hand side of (11.60) should be averaged over an interval T of the time t, which is either longer than or of the order of the correlation time of the environment, still remaining shorter than any of the relaxation times. We denote such an averaging by a bar on top of the expression and arrive at

$$\dot{S}_l = -2\mathrm{Tr} \left\{ \hat{\rho}(t) \overline{\int_0^\infty \left(\hat{V}_L(t)\ \hat{V}_R^*(t) \right) \hat{M}_x \begin{pmatrix} \hat{V}_L(t-x) \\ \hat{V}_R^*(t-x) \end{pmatrix} dx} \hat{\rho}(t) \right\} \tag{11.61}$$

or in the Fourier representation

$$\dot{S}_l = -2\mathrm{Tr} \left\{ \hat{\rho}(t) \int d\omega \left(\hat{V}_L(\omega)\ \hat{V}_R^*(\omega) \right) \hat{M}_\omega \begin{pmatrix} \hat{V}_L(-\omega) \\ \hat{V}_R^*(-\omega) \end{pmatrix} \hat{\rho}(t) \right\}, \tag{11.62}$$

where we have assumed that the time averaging interval T is long enough to yield the Dirac delta functions for the overlaps of harmonics. It is evident that the assumptions made are equivalent to the second-order approximation discussed earlier.

Allowing for (11.38) and expressing the superoperators $\hat{V}_L$ and $\hat{V}_R$ via the regular operator $\hat{V}$, from (11.61) in the Fourier representation we obtain

$$\dot{S}_l = 2\int d\omega \mathrm{Tr}\{\left(h_\omega - 2h_{\omega,\omega/kT}\right)\hat{V}(\omega)\hat{\rho}(t)\hat{V}^*(-\omega)\hat{\rho}(t) \tag{11.63}$$

$$-\left(h_\omega - h_{\omega,\omega/kT}\right)\left(\hat{V}(-\omega)\hat{V}(\omega) + \hat{V}^*(\omega)\hat{V}^*(-\omega)\right)\hat{\rho}^2(t)\}, \tag{11.64}$$

where $h_\omega = \frac{1}{2}c_\omega^2$ and $h_{\omega,T} = \frac{1}{2}c_\omega^2\frac{e^{-\omega/kT}}{1-e^{-\omega/kT}}$. In the case of a discrete spectrum of $V(\omega)$ and zero temperature, the leading contribution reads

$$\dot{S}_l = 2\pi\left(\sum_{m,k} V_m^k\rho_k^k V_k^{*m}\rho_m^m h_{-\Omega_{mk}} + \sum_{n,k,r} V_k^n V_n^k \rho_k^r \rho_r^k h_{-\Omega_{kn}}\right).$$

One sees that both the transverse and the longitudinal relaxations contribute to the variation of the entanglement entropy.

The entropy variation stops only when a stationary distribution attains. However, in contrast to the usual entropy, the entanglement entropy is not necessarily a monotonous function of time. In order to see this, one can consider the simplest case of a two-level system satisfying the master equation (11.27) with the initial condition given by the density matrix

$$\hat{\rho}(0) = \frac{1}{2}\begin{pmatrix} 1 & 1 \\ 1 & 1 \end{pmatrix}$$

of the pure state $\left(|0\rangle + |1\rangle\right)/\sqrt{2}$. The solution

$$\hat{\rho}(t) = \begin{pmatrix} \frac{1/2}{e^{-\frac{\omega}{T}}+1} + \frac{\tanh\left(\frac{\omega}{2T}\right)}{2}e^{-t\left(1+e^{-\frac{\omega}{T}}\right)/T_1} & \frac{1}{2}e^{-t/2T_2} \\ \frac{1}{2}e^{-t/2T_2} & \frac{1/2}{e^{\frac{\omega}{T}}+1} - \frac{\tanh\left(\frac{\omega}{2T}\right)}{2}e^{-t\left(1+e^{-\frac{\omega}{T}}\right)/T_1} \end{pmatrix}$$

of (11.27), satisfying this initial condition, gives the linear entropy of entanglement

$$\begin{aligned} S_l &= 1 - \mathrm{Tr}\left\{\hat{\rho}^2(t)\right\} \\ &= \frac{1}{2} - \frac{1}{2}e^{-\frac{t}{T_2}} - \frac{\tanh^2\left(\frac{\omega}{2T}\right)}{2}\left(1 - e^{-t\left(1+e^{-\frac{\omega}{T}}\right)/T_1}\right)^2 \end{aligned}$$

shown in Fig. 11.3. In the long-time limit, the linear entropy tends to its stationary value $\frac{1}{2}\left(1-\tanh^2\left(\frac{\omega}{2T}\right)\right)$ determined by the thermal distribution, and for zero temperature reaches the pure ground state, although at the beginning of the process it increases from zero value till a maximum value, which amounts to $\frac{1}{4}\left(1+\coth\left(\frac{\omega}{T}\right)\right)$ at the time $t = 2T_1\ln\left(1+\coth^2\left(\frac{\omega}{2T}\right)\right)$, for the case $T_2 = T_1$.

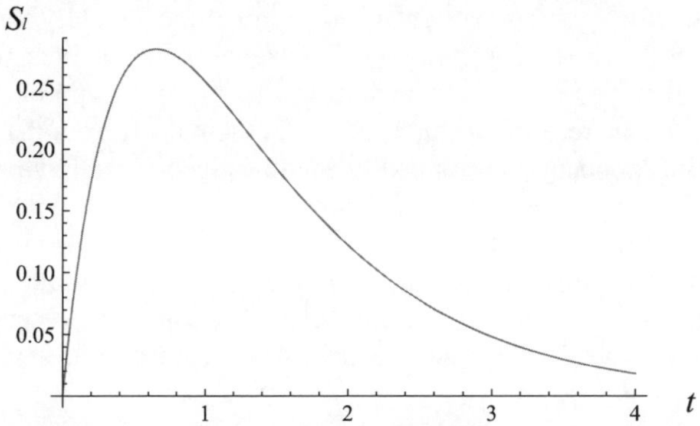

Fig. 11.3 Linear entropy of entanglement of a two-level system initially in the state $(|0\rangle + |1\rangle)/\sqrt{2}$ interacting with environment according to Eq. (11.27). An increase of the entropy at the initial stage of the relaxation process is followed by a decrease, which for low temperature moves the two-level system towards the pure ground state $|0\rangle$

11.3 Entanglement of Mixed Quantum States

Multipartite entanglement among elements of an assembly in a mixed quantum state cannot be straightforwardly characterized with the technique described in Sect. 10.3. Moreover, not only such a refined procedure as characterization but even the definition of entanglement by itself needs to be modified considerably for the mixed states. In this section we introduce this definition, present a widely adopted entanglement measure for two qubits in a mixed state, show how this definition and this measure lead to a paradox known as the sudden death of entanglement, and discuss an alternative approach for multipartite entanglement characterization for mixed states based on the nilpotent polynomial technique.

11.3.1 Definition of Entanglement for Mixed States

Thus far we have addressed only pure states of an assembly, where all manipulations with the corresponding state vector were formulated in terms of group theory as unitary rotations in the relevant Hilbert space. The relevant algebraic technique relies on two basic operations in the exponent—summation and the algebraic binary product that is commutation. The situation is different when one addresses mixed states. The definition specifies the property of the unentangled mixed states density operator. The elements of the assembly are considered not entangled when the density matrix can be cast to a statistically weighted sum of the tensor products of different unentangled bra and ket state vectors. Moreover, the product states vectors contributing to this sum are not necessarily orthogonal. Transformations of the density operators suggested by group theory do not offer full access to manipulation of these statistical weights. In particular, the group of local unitary transformations does not change them at all. Trying to characterize entanglement for the mixed states, we therefore go beyond the domain where the algebraic approach seems to be an adequate tool. The reason for this is in the fact that apart from the two standard operations in the exponent, the summation and the algebraic binary product typical of Lie group algebras, a third operation, namely the summation of the exponents, is indispensable for the manipulation of mixed states.

Let us formally introduce the definition of mixed separable (unentangled) states, – a state is considered as not entangled if it can be represented as a sum of statistically weighted product states, or, more specifically, if a system density matrix can be cast in the sum

$$\rho = \sum_{\{s_i\}} g\left(\{s_i\}\right) \prod_{\otimes i} \left|s_i\right\rangle_i \left\langle s_i\right|_i \,. \tag{11.65}$$

Here index i numerates the elements, the bra and ket vectors describing states of each elements $\langle s_i|_i \;\; |s_i\rangle_i$ entering the tensor product can take different values that are not necessarily orthogonal, $\langle s_i'|s_i\rangle_i \neq 0$, while a positive function of the element states $g\left(\{s_i\}\right)$ satisfies the normalization condition

$$\sum_{\{s_i\}} g\left(\{s_i\}\right) = 1, \tag{11.66}$$

where the summation is performed over all possible combinations $\{s_i\}$ of the element product states.

Hence description of entanglement encounters a difficulty—the same density operator can result from various statistical mixtures of different states. In particular, from a naive point of view, one may expect that an equally weighted statistical mixture of the maximum entangled pure states that all belong to the same orbit is entangled. Nevertheless, this mixture may and does yield a density operator identical to that obtained by a statistical sum of the product states, and hence, according to the definition, it must be considered separable. For example, the statistical mixture with equal weights $1/4$ of four maximum entangled states given by (10.4) with $\alpha = \beta = \gamma = \pi/4$, with $\alpha = -\beta = \gamma = \pi/4$, with $\alpha = \beta = \gamma = -\pi/4$, and with $\alpha = \beta = -\gamma = -\pi/4$ indeed gives the density matrix $\rho = 1/4I$ proportional to the identity matrix I, which can be also obtained as a statistical mixture of four product states $|0\rangle\,|0\rangle\,\langle 0|\,\langle 0|$, $|0\rangle\,|1\rangle\,\langle 1|\,\langle 0|$, $|1\rangle\,|0\rangle\,\langle 0|\,\langle 1|$, and $|1\rangle\,|1\rangle\,\langle 1|\,\langle 1|$ with the same statistical weights $1/4$. This example suggests a metaphoric interpretation: statistical mixtures of the product states produce a sort of shadow, in which many different statistical mixtures of the entangled states cannot be seen.

Note that the set of all product states form a complete basis in the space of Hermitian matrices, and therefore any density matrix, including the density matrices of entangled states, can be represented as their linear combination (11.65). The requirement, which is difficult to verify, is that all the expansion coefficients $g\left(\{s_i\}\right)$ can be made positive by a proper choice of a basis of the product states that are not necessarily orthogonal. A linear combination satisfying this requirement is called a convex sum. Therefore the definition of entanglement for mixed states is also known as the impossibility to cast the density matrix of the entire system to a convex sum of the product states of its parts.

11.3.2 An Entanglement Measure for Mixed States

There exists one commonly accepted measure of entanglement for a bipartite system composed of two 2-level systems in a mixed state—the so-called Wootters concurrence

$$C = \max\left(0, 2\max\left[\lambda_i\right] - \sum_{k=1}^{4}\lambda_k\right). \quad (11.67)$$

In order to evaluate this quantity, one has to find the difference between the largest eigenvalue λ_1 and the sum $\sum_{k=2}^{4}\lambda_k$ of all the rest of the eigenvalues of the concurrence operator

$$|\lambda_k\rangle\, \lambda_k\, \langle\lambda_k| = \sqrt{\hat{\rho}\hat{\epsilon}_1 \otimes \hat{\epsilon}_2 \hat{\rho}^* \hat{\epsilon}_1 \otimes \hat{\epsilon}_2}, \quad (11.68)$$

where $\otimes$ stands for the tensor product, and $\hat{\epsilon}_j = i\hat{\sigma}_j^y$ is the antisymmetric tensor for qubit j. For $C = 0$, the two-level systems are proven to be not entangled and can be cast in the form of a convex sum (11.65). Let us consider how the concurrence appears in a natural way in the context of entanglement.

In the matrix standing under the square root one recognizes a product of the system density matrix

$$\hat{\rho} = \sum_i |i\rangle\, p_i\, \langle i| = \sum_{i,n,m,k,l} |n,m\rangle\, {}^{(i)}\psi_{n,m} p_i\, {}^{(i)}\psi^*_{k,l} \quad \langle k,l|\,, \quad (11.69)$$

with the eigenstates $|i\rangle = \sum_{n,m}\, {}^{(i)}\psi_{n,m}\, |n,m\rangle$ written in the computational basis, by the complex conjugate of its counterpart

$$\hat{\epsilon}_1 \otimes \hat{\epsilon}_2 \cdot \hat{\rho} \cdot \hat{\epsilon}_1 \otimes \hat{\epsilon}_2 = \sum_{i,n,m,k,l} |n,m\rangle\, {}^{(i)}\psi^{n,m} p_i \left({}^{(i)}\psi^{k,l}\right)^* \langle k,l|\,, \quad (11.70)$$

where the indices are raised according to the rule (10.80). For a pure quantum state (${}^{(i=1)}\psi_{n,m} = \psi_{n,m}$, ${}^{(i>1)}\psi_{n,m} = 0$), the trace of the operator under the square root on the right hand side of (11.68)

$$\sum_k |\lambda_k\rangle\, \lambda_k^2\, \langle\lambda_k| = \sum_{n,m,k,l,v,w} |n,m\rangle\, \psi_{n,m} \left(\psi^{k,l}\psi_{k,l}\right)^* \psi^{v,w}\, \langle v,w| \quad (11.71)$$

coincides with the square of the absolute value of the invariant suggested by (10.81)

$$\sum_k \lambda_k^2 = \left(\psi^{k,l}\psi_{k,l}\right)^* \psi^{n,m}\psi_{n,m} = \left|\psi^{n,m}\psi_{n,m}\right|^2, \tag{11.72}$$

which is one of the possible measures of entanglement for pure states. It vanishes for the product states and equals unity for the maximum entangled state. Note that the choice of ϵ_i for the index rising operation is dictated by the fact that these tensors remain invariant under the local transformations $\hat{\epsilon}_i \to \hat{U}_i^T \; \hat{\epsilon}_i \hat{U}_i$, where $\hat{U}_i$ is a unitary operator and $\hat{U}_i^T$ its transpose. By introducing the index rising operation and the relevant index convolution, we go beyond the operations normally employed in quantum mechanics, since the rule $\hat{\epsilon}_i \to \hat{U}_i^T \; \hat{\epsilon}_i \hat{U}_i$ suggests to consider a transformation typical of quadratic forms instead of that typical of quantum mechanical operators representing physical observables in Hilbert space. This is one more indication of the fact that entanglement is not a physical observable, but a tool for analysis of the structure of quantum states of composed systems.

Let us now turn to the mixed states of two-qubit systems given by the density matrix $\hat{\rho} = \sum_i |i\rangle \, p_i \, \langle i|$ of full rank, that is, $p_i \neq 0$ for all i, and rewrite it as

$$\hat{\rho} = \sum_{i=1}^{4} |\iota_i\rangle \langle \iota_i| \, , = \hat{\Upsilon}\hat{\Upsilon}^+, \tag{11.73}$$

in terms of the eigenvectors $|\iota_i\rangle = \sqrt{p_i}\,|i\rangle$, which are not normalized to unity and serve as rows of a non-unitary matrix Υ. The representation (11.73) of a positive Hermitian matrix, known as Cholesky decomposition, is unique as long as we require orthogonality of the matrix columns. If we lift up this requirement, the representation of density matrix in the form of a product of two non-unitary Hermitian conjugate matrices becomes ambiguous, and can be given as

$$\hat{\rho} = \hat{\Upsilon}\hat{\mathcal{U}}\hat{\mathcal{U}}^+\hat{\Upsilon}^+, \tag{11.74}$$

where $\hat{\mathcal{U}}$ is a unitary matrix of general type, that is, not necessarily a square matrix. For the case under consideration, this is a $4 \times M$ matrix with $M \geq 4$ satisfying the condition $\hat{\mathcal{U}} \cdot \hat{\mathcal{U}}^+ = 1$. In other words the rows of the matrix $\hat{\mathcal{U}}$ are orthonormal vectors of length M. By contrast, the vectors $|\varrho_j\rangle$ of length 4 that form columns of the non-unitary product of matrices $\hat{\Upsilon}\hat{\mathcal{U}}$ are not necessary orthogonal, neither are they normalized, and therefore the corresponding normalized vectors $|j\rangle = |\varrho_j\rangle / \sqrt{\langle \varrho_j|\varrho_j\rangle}$ give a representation of the density matrix as a sum

$$\hat{\rho} = \sum_{j=1}^{M} |\varrho_j\rangle \langle \varrho_j| = \sum_{j=1}^{M} |j\rangle \langle \varrho_j|\varrho_j\rangle \langle j| \tag{11.75}$$

of tensor products $|j\rangle \otimes \langle j|$ of M non-orthogonal unit vectors weighted by the positive coefficients $\langle \varrho_j|\varrho_j\rangle$. The entanglement criterion for mixed states can now be formulated in terms of the matrices $\hat{\Upsilon}$ and $\hat{\mathcal{U}}$—the state is not entangled if for a matrix $\hat{\Upsilon}$ given by the decomposition of the density matrix $\hat{\rho}$ one can find a rectangular unitary matrix $\hat{\mathcal{U}}$ such that the column-vectors $|\varrho_j\rangle$ of the matrix $\hat{\Upsilon}\hat{\mathcal{U}}$ are the product states, that is,

$$\langle\varrho_j|^* \,\hat{\epsilon}_1 \otimes \hat{\epsilon}_2\, |\varrho_j\rangle = 0, \tag{11.76}$$

for all $j = 1, \ldots, M$.

In order to understand when (11.76) holds, consider a matrix $\hat{C}$ with the elements

$$C^i_j = \langle\varrho_i|^* \,\hat{\epsilon}_1 \otimes \hat{\epsilon}_2\, |\varrho_j\rangle = \frac{\langle i|^* \,\hat{\epsilon}_1 \otimes \hat{\epsilon}_2\, |j\rangle}{\sqrt{p_i p_j}} = \sum_{n,m} \frac{{}^{(i)}\psi^{n,m} \quad {}^{(j)}\psi_{n,m}}{\sqrt{p_i p_j}} \tag{11.77}$$

evaluated for the eigenstates i and j of the density matrix, where $\langle i|^*$ is the transpose of $|i\rangle$. If there exists an orthogonal transformation $\hat{\mathcal{U}}^T\hat{C}\hat{\mathcal{U}}$ performed by application of a $4 \times M$ unitary matrix $\hat{\mathcal{U}}$, which from the matrix $\hat{C}$ can generate an $M \times M$ anti-diagonal matrix, that is, a matrix with zero diagonal

$$\left(\hat{\mathcal{U}}^T \cdot \hat{C} \cdot \hat{\mathcal{U}}\right)_{j,j} = 0, \tag{11.78}$$

the states $|\varrho_j\rangle = \sum_{i=1}^{4} \mathcal{U}_{j,i}\, |\iota_i\rangle$ do indeed satisfy (11.76). Thus they are the very product states that realize the required representation (11.75).

We now have to find the conditions under which (11.78) holds. To this end we note that by construction $\hat{C}$ is not a Hermitian, but a symmetric matrix $C^i_j = C^j_i$, and therefore, according to the general rules of linear algebra, it can be set to the diagonal form by an orthogonal transformation

$$\hat{u}^T \cdot \hat{C} \cdot \hat{u} = \begin{pmatrix} \lambda_1 e^{i\phi_1} & 0 & 0 & 0 \\ 0 & \lambda_2 e^{i\phi_2} & 0 & 0 \\ 0 & 0 & \lambda_3 e^{i\phi_3} & 0 \\ 0 & 0 & 0 & \lambda_4 e^{i\phi_4} \end{pmatrix} \tag{11.79}$$

performed with the help of a 4×4 unitary matrix $\hat{u}$, although the spectrum of the singular values of $\hat{C}$ is not necessarily real and may contain phase factors $e^{i\phi_k}$. Since the product $\hat{u} \cdot \hat{\mathcal{U}}$ is a unitary $4 \times M$ matrix, the transformation $\hat{u}$ can be absorbed by $\hat{\mathcal{U}}$ such that the requirement (11.78) puts a constraint exclusively on the spectrum $\lambda_k e^{i\phi_k}$ of the matrix (11.79): this spectrum should allow a transformation $\hat{\mathcal{U}}^T \cdot \hat{C} \cdot \hat{\mathcal{U}}$, which makes a diagonal matrix anti-diagonal. The latter implies the existence of the solution of the system of second-order equations

$$\sum_{j=1}^{M} \lambda_j e^{i\phi_j} \left(\mathcal{U}^i_j\right)^2 = 0, \tag{11.80}$$

$$\sum_{j=1}^{M} \left(\mathcal{U}_j^i\right)^* \mathcal{U}_j^r = \delta_{ir}. \tag{11.81}$$

with respect to the matrix elements of $\hat{\mathcal{U}}$.

Let us first consider a unit vector of the form

$$\{\mathcal{U}_1^1, \mathcal{U}_2^1, \mathcal{U}_3^1, \mathcal{U}_4^1\} = \frac{1}{2}\left\{e^{i\varphi_1}, e^{i\varphi_2}, e^{i\varphi_3}, e^{i\varphi_4}\right\}$$

composed by the matrix elements of $\hat{\mathcal{U}}$, which satisfies (11.80) provided the moduli λ_j of the eigenvalues of $\hat{C}$ are such that they can be the lengths of the sides of a quadrangle, implied by the condition

$$C = \max\left(0, \lambda_1 - \sum_{k=2}^{4} \lambda_k\right) = 0$$

that is, the maximum length λ_1 of a side is shorter as compared to the sum of lengths of three other sides. The required phases φ_i are such that the differences $\phi_{i+1} + 2\varphi_{i+1} - \phi_i - 2\varphi_i$ amount to the angles between two successive sides $i+1$ and i of this quadrangle. Three other vectors of unit length,

$$\{\mathcal{U}_1^2, \mathcal{U}_2^2, \mathcal{U}_3^2, \mathcal{U}_4^2\} = \frac{1}{2}\left\{e^{i\varphi_1}, e^{i\varphi_2}, -e^{i\varphi_3}, -e^{i\varphi_4}\right\}$$

$$\{\mathcal{U}_1^3, \mathcal{U}_2^3, \mathcal{U}_3^3, \mathcal{U}_4^3\} = \frac{1}{2}\left\{e^{i\varphi_1}, -e^{i\varphi_2}, -e^{i\varphi_3}, e^{i\varphi_4}\right\}$$

$$\{\mathcal{U}_1^4, \mathcal{U}_2^4, \mathcal{U}_3^4, \mathcal{U}_4^4\} = \frac{1}{2}\left\{-e^{i\varphi_1}, e^{i\varphi_2}, -e^{i\varphi_3}, e^{i\varphi_4}\right\},$$

are orthogonal both to $\{\mathcal{U}_1^1, \mathcal{U}_2^1, \mathcal{U}_3^1, \mathcal{U}_4^1\}$ and among themselves. They immediately suggest entries of the other rows of the required unitary matrix, for the particular case $M = 4$, since each of them also satisfies (11.80). Thus, by the direct construction of the 4×4 matrix $\hat{\mathcal{U}}$, we have shown that the condition $C = 0$ for the concurrence defined by (11.67) is sufficient for the possibility to cast the density matrix in the form of a sum of four statistically weighted product states. The necessity of this condition emerges from the fact that the anti-diagonal $M \times M$ matrix $\hat{\mathcal{U}}^T \hat{C} \hat{\mathcal{U}}$ must satisfy the condition $\mathrm{Tr}\hat{\mathcal{U}}^T \hat{C} \hat{\mathcal{U}} \hat{A} = 0$ for an arbitrary diagonal $M \times M$ matrix $\hat{A}$, which implies that the matrix $\hat{C}$ in the diagonal representation of (11.79) and a generic symmetric 4×4 matrix $\hat{\mathcal{U}} \hat{A} \hat{\mathcal{U}}^T$ should satisfy the condition $\mathrm{Tr}\hat{C} \hat{\mathcal{U}} \hat{A} \hat{\mathcal{U}}^T = 0$. But for an arbitrary diagonal matrix $\hat{A}$, the trace of $\hat{\mathcal{U}} \hat{A} \hat{\mathcal{U}}^T$ can have an arbitrary value, regardless of the choice of $\hat{\mathcal{U}}$ and hence the requirement $C = 0$ implying the possibility to set the trace $\sum_j \lambda_j e^{i\phi_j}$ of $\hat{C}$ equal to zero is necessary.

To complete the consideration we have to show that the λ_i entering (11.79) and (11.68) are the same quantities. For this purpose it is sufficient to note that the product $\hat{C} \cdot \hat{C}^{+}$ is a Hermitian matrix with the spectrum $(\lambda_i)^2$, while

$$\hat{C} = \sqrt{\hat{\rho}} \cdot \hat{\epsilon}_1 \otimes \hat{\epsilon}_2 \cdot \sqrt{\hat{\rho}^*},$$

in the accordance with (11.77), and hence

$$\hat{\rho} \cdot \hat{\epsilon}_1 \otimes \hat{\epsilon}_2 \cdot \hat{\rho}^* \cdot \hat{\epsilon}_1 \otimes \hat{\epsilon}_2 = \sqrt{\hat{\rho}} \cdot \hat{C} \cdot \hat{C}^{+} \cdot \left(\sqrt{\hat{\rho}}\right)^{-1}.$$

The cyclic permutations of non-degenerate rectangular matrix factors do not change the characteristic polynomial of the product matrix, and hence the spectrum of the left hand side of the last equality is also given by $(\lambda_i)^2$.

We conclude this section by noting that (10.82) and (10.83), giving invariants for three two-level systems, contain probability amplitudes in the fourth power, which does not allow one to easily generalize the concurrence technique on this, the simplest multipartite case.

11.3.3 Sudden Death and Resurrection of Entanglement for Mixed States

The definition of entanglement for mixed states as the impossibility to represent a system density matrix as a statistical mixture of the product states of its parts may and does result in physically counterintuitive regimes of entanglement dynamics. One of them results in a phenomenon named "sudden death of entanglement", where entangled pure states entering with certain probabilities the mixed state density matrix of the system experience individual natural evolution, such that at a certain moment of time the density matrix turns to be representable in the form of (11.65), although such a representation was impossible before. The entangled system becomes not-entangled although no physical action or event occur at this point; simply the density matrix still composed by the statistical mixture of the entangled states changes its mathematical structure by allowing a different representation as a statistical sum of the product states. A similar phenomenon, sudden birth of entanglement, has exactly the same nature—entangled eigenstates of the density matrix in the course of their dynamics move out of the "shadow" of the separable states and become entangled.

Let us consider this phenomenon at the simplest example of a pair of two-level systems with the Hamiltonian $\hat{H}_0 = \Omega\left(\hat{\sigma}_1^z + \hat{\sigma}_2^z\right)$, which, in complete analogy to (11.30), interacts with the environing radiation field via the dipole moment operator $\hat{V} = d\left(\hat{\sigma}_1^x + \hat{\sigma}_2^x\right)$ multiplied by the operator of electromagnetic field strength $\hat{\mathcal{E}} = \sum_\omega c_\omega \left(\hat{a}_\omega^\dagger + \hat{a}_\omega\right)/\sqrt{2}$. The 4×4 density matrix of the pair satisfies the Lindblad master equation (11.18) $\dot{\hat{\rho}} = -i\left[\hat{H}_0, \hat{\rho}\right] - \overset{\cap}{\mathcal{R}}\hat{\rho}$ with the relaxation operator given by (11.57,11.58)

$$\left(\overset{\cap}{\mathcal{R}}\hat{\rho}\right)^r_m = 2\pi\delta^r_m \sum_k h_{\Omega_{km}} \left|V^k_m\right|^2 \rho^k_k, -\pi \sum_n \left(h_{\Omega_{nm}} \left|V^n_m\right|^2 + \left|V^n_r\right|^2 h_{\Omega_{nr}}\right)\rho^r_m,$$

where the environment correlation spectra given by Eq. (11.52) for the particular case read $h_\omega = \frac{\Theta_\omega}{2}c^2_\omega - \frac{\Theta_\omega}{2}c^2_\omega \frac{e^{-\omega/kT}}{1-e^{-\omega/kT}} + \frac{\Theta_{-\omega}}{2}c^2_{-\omega}\frac{e^{\omega/kT}}{1-e^{\omega/kT}}$. It is convenient to write the explicit expression for the master equation in the basis of Dicke states, which in the computational basis can be expressed as

$$|-1\rangle = |0,0\rangle\,,\ |0\rangle = \frac{|0,1\rangle + |1,0\rangle}{\sqrt{2}},$$
$$|1\rangle = |1,1\rangle\,,\ |D\rangle = \frac{|0,1\rangle - |1,0\rangle}{\sqrt{2}}.$$

Here the dark state $|D\rangle$ does not interact with the environing radiation field for symmetry reasons. Pure states of the three-level bright part $\{|-1\rangle\,,|0\rangle\,,|1\rangle\}$ of the pair of two-level systems have been considered in Sect. 10.3.6 in the context of generalized entanglement, where we have introduced the operators $\hat{\lambda}^{(3)}$ of (10.106) and $\hat{S}^+$ of (10.107), which allow us to write $\hat{H}_0 = \Omega\hat{\lambda}^{(3)}$ and $\hat{V} = d\left(\hat{S}^+ + \hat{S}\right)$.

The dynamics of the matrix elements evolving the dark state given by the master equation splits into four independent equations:

$$\dot{\rho}^D_D = 0,$$
$$\dot{\rho}^{-1}_D = \left(i\Omega - \pi\left|V^0_{-1}\right|^2 h_{-\Omega}\right)\rho^{-1}_D,$$
$$\dot{\rho}^1_D = -\left(i\Omega + \pi\left|V^0_1\right|^2 h_\Omega\right)\rho^1_D,$$
$$\dot{\rho}^0_D = -\pi\left(\left|V^1_0\right|^2 h_{-\Omega} + \left|V^{-1}_0\right|^2 h_\Omega\right)\rho^0_D,$$

and so do the equations for the non-diagonal terms for the bright part

$$\dot{\rho}^{-1}_0 = \left(i\Omega - \pi\left|V^0_{-1}\right|^2 h_{-\Omega} - \pi\left|V^{-1}_0\right|^2 h_\Omega - \pi\left|V^1_0\right|^2 h_{-\Omega}\right)\rho^{-1}_0,$$
$$\dot{\rho}^1_0 = -\left(i\Omega + \pi\left|V^0_1\right|^2 h_\Omega + \pi\left|V^{-1}_0\right|^2 h_\Omega + \pi\left|V^1_0\right|^2 h_{-\Omega}\right)\rho^1_0,$$
$$\dot{\rho}^{-1}_1 = \left(2i\Omega - \pi\left|V^0_{-1}\right|^2 h_{-\Omega} - \pi\left|V^0_1\right|^2 h_\Omega\right)\rho^{-1}_1.$$

The equation for the transposed density matrix elements are complex conjugate, as is dictated by the Hermiticity of this matrix. The diagonal matrix elements for the bright part satisfy a system of coupled equations

$$
\begin{aligned}
\dot{\rho}_{-1}^{-1} &= -2\pi h_{-\Omega}\left|V_{-1}^{0}\right|^{2}\rho_{-1}^{-1} + 2\pi h_{\Omega}\left|V_{0}^{-1}\right|^{2}\rho_{0}^{0},\\
\dot{\rho}_{0}^{0} &= 2\pi h_{-\Omega}\left|V_{-1}^{0}\right|^{2}\rho_{-1}^{-1} - 2\pi h_{\Omega}\left|V_{0}^{-1}\right|^{2}\rho_{0}^{0}\\
&\quad + 2\pi h_{\Omega}\left|V_{1}^{0}\right|^{2}\rho_{1}^{1} - 2\pi h_{-\Omega}\left|V_{0}^{1}\right|^{2}\rho_{0}^{0},\\
\dot{\rho}_{1}^{1} &= -2\pi h_{\Omega}\left|V_{1}^{0}\right|^{2}\rho_{1}^{1} + 2\pi h_{-\Omega}\left|V_{0}^{1}\right|^{2}\rho_{0}^{0}.
\end{aligned}
\tag{11.82}
$$

One can introduce the upward $w_{\uparrow} = \pi\left|V_{0}^{-1}\right|^{2} h_{-\Omega} = \pi c_{\Omega}^{2} d^{2}\frac{e^{-\omega/kT}}{2(1-e^{-\omega/kT})}$ and downward $w_{\downarrow} = \pi\left|V_{0}^{-1}\right|^{2} h_{\Omega} = \pi d^{2} c_{\Omega}^{2}\frac{1}{2(1-e^{-\omega/kT})}$ relaxation rates for the probability amplitudes and by straightforward calculations explicitly obtain the solution $\hat{\rho}(t)$ of the master equation in the form

$$
\begin{pmatrix}
\sum_k M_k^1 \rho_k^k & \rho_0^1 e^{-(i\Omega+2w_\downarrow+w_\uparrow)t} & \rho_{-1}^1 e^{-(2i\Omega+w_\downarrow+w_\uparrow)t} & \rho_D^1 e^{-(i\Omega+w_\downarrow)t}\\
\rho_1^0 e^{(i\Omega-2w_\downarrow-w_\uparrow)t} & \sum_k M_k^0 \rho_k^k & \rho_{-1}^0 e^{-(i\Omega=w_\downarrow+2w_\uparrow)t} & \rho_D^0 e^{-(w_\downarrow+w_\uparrow)t}\\
\rho_1^{-1} e^{(2i\Omega-w_\downarrow-w_\uparrow)t} & \rho_0^{-1} e^{(i\Omega-w_\downarrow-2w_\uparrow)t} & \sum_k M_k^{-1}\rho_k^k & \rho_D^{-1} e^{(i\Omega-w_\uparrow)t}\\
\rho_1^D e^{(i\Omega-w_\downarrow)t} & \rho_0^D e^{-(w_\downarrow+w_\uparrow)t} & \rho_{-1}^D e^{-(i\Omega+w_\uparrow)t} & \rho_D^D
\end{pmatrix}
\tag{11.83}
$$

where the density matrix elements entering (11.83) correspond to the initial moment of time $t=0$, while M_k^l with $k,l=-1,0,1$ is a 3×3 matrix

$$
\hat{M} = \exp\left\{2t\begin{pmatrix}-w_\downarrow & w_\uparrow & 0\\ w_\downarrow & -w_\downarrow - w_\uparrow & w_\uparrow\\ 0 & w_\downarrow & -w_\uparrow\end{pmatrix}\right\}
\tag{11.84}
$$

representing the fundamental solution of (11.82) for the diagonal matrix elements.

Let us now consider the particular case of the initial product state

$$
\begin{pmatrix}\Psi_{11}\\ \Psi_{10}\\ \Psi_{01}\\ \Psi_{00}\end{pmatrix} = \begin{pmatrix}\sin\beta\sin\gamma\\ \cos\beta\sin\gamma\\ \sin\beta\cos\gamma\\ \cos\beta\cos\gamma\end{pmatrix}
\tag{11.85}
$$

with the initial density matrix $\hat{\rho}(0)$ in the basis $\{|D\rangle, |-1\rangle, |0\rangle, |1\rangle\}$ of the form

$$
\begin{pmatrix}
\sin^2\beta\sin^2\gamma & \frac{\sin\beta\sin\gamma\sin(\beta+\gamma)}{\sqrt{2}} & \frac{\sin 2\beta\sin 2\gamma}{4} & \frac{\sin\beta\sin\gamma\sin(\beta-\gamma)}{\sqrt{2}}\\
\frac{\sin\beta\sin\gamma\sin(\beta+\gamma)}{\sqrt{2}} & \frac{\sin^2(\beta+\gamma)}{2} & \frac{\cos\beta\cos\gamma\sin(\beta+\gamma)}{\sqrt{2}} & \frac{\sin(\beta-\gamma)\sin(\beta+\gamma)}{2}\\
\frac{\sin 2\beta\sin 2\gamma}{4} & \frac{\cos\beta\cos\gamma\sin(\beta+\gamma)}{\sqrt{2}} & \cos^2\beta\cos^2\gamma & \frac{\cos\beta\cos\gamma\sin(\beta-\gamma)}{\sqrt{2}}\\
\frac{\sin\beta\sin\gamma\sin(\beta-\gamma)}{\sqrt{2}} & \frac{\sin(\beta-\gamma)\sin(\beta+\gamma)}{2} & \frac{\cos\beta\cos\gamma\sin(\beta-\gamma)}{\sqrt{2}} & \frac{\sin^2(\beta-\gamma)}{2}
\end{pmatrix}.
\tag{11.86}
$$

At zero temperature $w_\uparrow = 1$ and the matrix of (11.84) reads

$$\hat{M} = \begin{pmatrix} e^{-4t} & 0 & 0 \\ 4e^{-4t}t & e^{-4t} & 0 \\ -4e^{-4t}t - e^{-4t} + 1 & 1 - e^{-4t} & 1 \end{pmatrix},$$

where we have set $w_\downarrow = 1/4$. The time-dependent system density matrix has the explicit form

$$\hat{\rho}(t) = \begin{pmatrix} e^{-t}\rho_1^1 & \rho_0^1 e^{-i\Omega t - \frac{t}{2}} & \rho_{-1}^1 e^{-2i\Omega t - \frac{t}{4}} & \rho_D^1 e^{-i\Omega t - \frac{t}{4}} \\ \rho_1^0 e^{i\Omega t - \frac{t}{2}} & te^{-t}\rho_1^1 + e^{-t}\rho_0^0 & \rho_{-1}^0 e^{-i\Omega t - \frac{t}{4}} & \rho_D^0 e^{-\frac{t}{4}} \\ \rho_1^{-1} e^{2i\Omega t - \frac{t}{4}} & \rho_0^{-1} e^{i\Omega t - \frac{t}{4}} & \Lambda & \rho_D^{-1} e^{i\Omega t} \\ \rho_1^D e^{i\Omega t - \frac{t}{4}} & \rho_0^D e^{-\frac{t}{4}} & \rho_{-1}^D e^{-i\Omega t} & \rho_D^D \end{pmatrix}$$

$$\Lambda = \left(1 - e^{-t} - te^{-t}\right)\rho_1^1 + \left(1 - e^{-t}\right)\rho_0^0 + \rho_{-1}^{-1} \tag{11.87}$$

where the ρ_j^i are given by (11.86). With the help of this density matrix, one can calculate the concurrence (11.67) paying attention to the fact that in the chosen Dicke representation the tensor $\hat{\epsilon}_1 \otimes \hat{\epsilon}_2$ entering (11.68) reads

$$\hat{\epsilon}_1 \otimes \hat{\epsilon}_2 = \begin{pmatrix} 0 & 0 & 1 & 0 \\ 0 & -1 & 0 & 0 \\ 1 & 0 & 0 & 0 \\ 0 & 0 & 0 & 1 \end{pmatrix}. \tag{11.88}$$

The result is shown in Fig. 11.4, where one can identify the time moments when the concurrence becomes positive, which means that the density matrix Eq. (11.87) does not have an alternative representation as a statistical mixture of non-orthogonal product states.

11.3.4 Nilpotent Variables for Entanglement of Mixed States

Classification of entanglement for mixed states can rely on a conceptually transparent idea where bra $\langle|$ and ket $|\rangle$ vectors of the same element are considered as two independent ket vectors belonging to two distinct but tween elements. That is, an $|a\rangle_i \langle b|_i$ element of the density matrix can be unambiguously associated with the quantum state $|a\rangle_i \langle b|_i^+ \equiv |a\rangle_i \left|\overline{b}\right\rangle_i$ of a system containing twice more elements: initial elements in states $|a\rangle_i$ and their fictitious twins in states $\left|\overline{b}\right\rangle_i$, where a bar marks the twin element. In this way, one arrives at a pure state of an assembly with twice more elements, which can be described as a pure state with the help of the technique of nilpotent polynomials of Sect. 10.3, with the only constraint that the Hermitian symmetry of the density operator must be respected by taking into

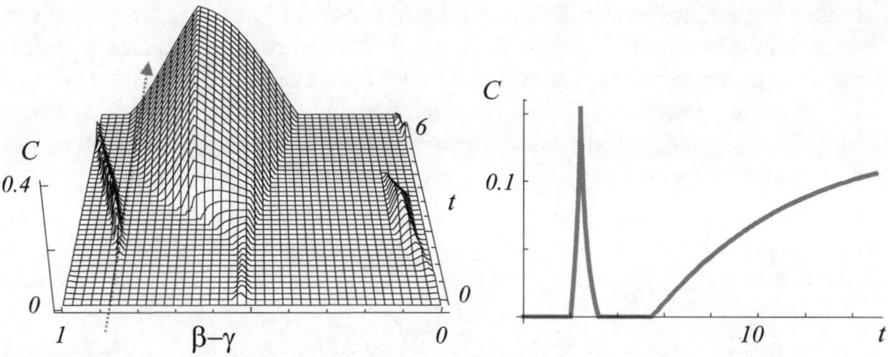

Fig. 11.4 Sudden birth and sudden death of entanglement for a two-qubit system at a mixed state. A pair of two-level atoms initially in a product state $|\Psi\rangle = |\psi_1\rangle \otimes |\psi_2\rangle$, where $|\psi_1\rangle = \cos\beta\,|0\rangle + \sin\beta\,|1\rangle$ and $|\psi_2\rangle = \cos\gamma\,|0\rangle + \sin\gamma\,|1\rangle$, interacts with modes of the electromagnetic field in the vacuum states, such that only the symmetric configuration has non-zero dipole moment. Density matrix of the pair given by (11.87) with the initial condition (11.86), with the help of (11.88) for the product $-\hat{\sigma}_1^y \otimes \hat{\sigma}_2^y$, allows one to calculate the concurrence C of (11.67). At the *left panel*, the concurrence is depicted as a function of the time t and the parameter $\beta - \gamma$ for the second atom initially in the state $(|0\rangle + |1\rangle)/\sqrt{2}$, that is, $\gamma = \pi/4$. One sees that in spite of the fact that the eigenvectors of the density matrix evolve smoothly in time, the entanglement suddenly appears and disappears, which simply means that between these time points the density matrix cannot have an alternative representation as a statistical sum of product states. By the *arrow* we indicate the regime where entanglement appears twice and disappears in between of these events, which is illustrated at the *right panel*

account the symmetry with respect to interchange of twins elements. One of the possibilities to allow for this symmetry is to consider the Hermitian conjugate of bra vector of the assembly as a ket state vector of the twin assembly as was done while considering the concurrence. In such a framework, entanglement among the states of the original assembly and the same states of the twin assembly accounts for the state mixing. Evidently, performing a local transformation on the elements of the assembly, one has to perform exactly the same operations with the elements of the twin assembly. This technique can easily identify factorizable density matrices of open systems, but the question whether or not it is equally efficient for the identification of separable states is more difficult.

Let us consider an assembly of qubits and denote by σ_i^+ and $\overline{\sigma}_i^+$ the raising operator for i-th qubit and for its twin, respectively. When applying the technique of nilpotent polynomials, one can immediately identify three limiting cases: (i) the tanglemeter has no cross terms containing both variables σ_i^+ of the assembly and the variables $\overline{\sigma}_j^+$ of the twin assembly, which means that the state is pure; (ii) the tanglemeter has no cross terms among the variables (no matter with or without bar) with indices $i \in A$ and $j \in \overline{A}$, which means that the density operator is a tensor product of the density operators of the group of qubits A and that of the rest of the qubits $\overline{A}$; and (iii) the tanglemeter contains only the second-order cross terms $\sigma_i^+\overline{\sigma}_j^+$, which means that this is a statistical mixture of product states, as one can infer by applying the Hubbard–Stratanovich separation of the operator products in the exponent. The condition (iii) is sufficient for the mixed state to be unentangled. However, one cannot yet be sure whether or not it is necessary.

Let us go one step further in the consideration and focus on the mixed states of two qubits. By analogy to the tanglemeter (10.92) of a 4-qubit system in the pure canonic state, for the tanglemeter $\mathbf{f}_c$ of a mixed state one can write

$$\begin{aligned}\mathbf{f}_c = {} & \beta_{0,3}\left(\sigma_2^+\sigma_1^+ + \overline{\sigma}_2^+\overline{\sigma}_1^+\right) + \beta_{3,3}\overline{\sigma}_2^+\overline{\sigma}_1^+\sigma_2^+\sigma_1^+ \\ & + \beta_{1,1}\overline{\sigma}_1^+\sigma_1^+ + \beta_{1,2}\left(\overline{\sigma}_2^+\sigma_1^+ + \overline{\sigma}_1^+\sigma_2^+\right) + \beta_{2,2}\overline{\sigma}_2^+\sigma_2^+ \\ & + \beta_{1,3}\left(\overline{\sigma}_1^+\sigma_2^+\sigma_1^+ + \overline{\sigma}_2^+\overline{\sigma}_1^+\sigma_1^+\right) \\ & + \beta_{2,3}\left(\overline{\sigma}_2^+\sigma_2^+\sigma_1^+ + \overline{\sigma}_2^+\overline{\sigma}_1^+\sigma_2^+\right),\end{aligned} \tag{11.89}$$

where the Hermiticity of the density operator is taken into account by the fact that the tanglemeter coefficients $\beta_{i,j}$ compose a symmetric matrix. The fact that this tanglemeter originates not from a generic pure state of a four-qubit system but from the density matrix of a two-qubit system can be clearly seen when one represents it as a sum

$$\mathbf{f}_c = \left(\overline{\sigma}_2^+, \overline{\sigma}_1^+\right)\hat{M}_{\text{nil}}\begin{pmatrix}\sigma_2^+\\ \sigma_1^+\end{pmatrix} + f_c\left(\sigma_1^+, \sigma_2^+\right) + f_c\left(\overline{\sigma}_1^+, \overline{\sigma}_2^+\right),$$

of two identical tanglemeters of the dual assemblies $f_c\left(\sigma_1^+, \sigma_2^+\right) = \beta_{0,3}\sigma_2^+\sigma_1^+$ and $f_c\left(\overline{\sigma}_1^+, \overline{\sigma}_2^+\right) = \beta_{0,3}\overline{\sigma}_2^+\overline{\sigma}_1^+$ and a convolution term which allows for entanglement of the dual assemblies. The matrix $\hat{M}_{\text{nil}}$ with nilpotent polynomial entries

$$\begin{pmatrix}\beta_{2,2} + \frac{\beta_{2,3}\left(\sigma_1^+ + \overline{\sigma}_1^+\right)}{2} + \frac{\beta_{3,3}}{4}\overline{\sigma}_1^+\sigma_1^+ & \beta_{1,2} + \frac{\beta_{1,3}\overline{\sigma}_1^+ + \beta_{2,3}\sigma_2^+}{2} + \frac{\beta_{3,3}}{4}\overline{\sigma}_1^+\sigma_2^+ \\ \beta_{1,2} + \frac{\beta_{1,3}\sigma_1^+ + \beta_{2,3}\overline{\sigma}_2^+}{2} + \frac{\beta_{3,3}}{4}\overline{\sigma}_2^+\sigma_1^+ & \beta_{1,1} + \frac{\beta_{1,3}\left(\sigma_2^+ + \overline{\sigma}_2^+\right)}{2} + \frac{\beta_{3,3}}{4}\overline{\sigma}_2^+\sigma_2^+\end{pmatrix}$$

determines the convolution term. It resembles a Hermitian matrix, although it has an important difference—instead of the complex conjugation one performs permutation of the dual nilpotent variables $\overline{\sigma}_i^+ \rightleftharpoons \sigma_i^+$.

With the help of (11.89) after straightforward calculations one finds the density operator of two qubits in the canonic state:

$$\hat{\rho}(\hat{\beta}) = \begin{pmatrix}\beta_{0,3}^2 + \beta_{22}\beta_{1,1} + \beta_{1,2}^2 + |\beta_{3,3}| & \beta_{2,3} & \beta_{1,3} & \beta_{0,3} \\ \beta_{2,3} & \beta_{2,2} & \beta_{1,2} & 0 \\ \beta_{1,3} & \beta_{1,2}^* & \beta_{1,1} & 0 \\ \beta_{0,3}^* & 0 & 0 & 1\end{pmatrix}$$

normalized by the condition $\rho_{0,0} = 1$ natural for the approach based on the nilpotent variables. Here all $\beta_{i,j}$ except two, say $\beta_{1,2}$ and $\beta_{0,3}$, can be made real and positive by the properly chosen phases of the qubits eigenstates. One of possible decompositions $\hat{\rho} = \hat{\Upsilon}\hat{\mathcal{U}}_{\text{pos}}\hat{\mathcal{U}}^{+}_{\text{pos}}\hat{\Upsilon}^{+}$ of (11.74) for this matrix is given by

$$\hat{\Upsilon}\hat{\mathcal{U}}_{\text{pos}} = \begin{pmatrix} \beta_{0,3} & \frac{\beta_{1,3}}{\sqrt{\beta_{1,1}}} & \frac{\beta_{1,1}\beta_{2,3}-\beta^{*}_{1,2}\beta_{1,3}}{\sqrt{\beta_{1,1}}\sqrt{\beta_{1,1}\beta_{2,2}-|\beta_{1,2}|^2}} & \Gamma \\ 0 & \frac{\beta^{*}_{1,2}}{\sqrt{\beta_{1,1}}} & \frac{\sqrt{\beta_{1,1}\beta_{2,2}-|\beta_{1,2}|^2}}{\sqrt{\beta_{1,1}}} & 0 \\ 0 & \sqrt{\beta_{1,1}} & 0 & 0 \\ 1 & 0 & 0 & 0 \end{pmatrix}$$

where

$$\Gamma = \left[\left(|\beta_{1,2}|^2 + \beta_{1,1}\beta_{2,2} + \beta_{3,3}\right)\left(|\beta_{1,2}|^2 - \beta_{2,2}\beta_{1,1}\right)\right.$$
$$\left. -2\beta_{1,3}\beta_{2,3}\,\mathrm{Re}\,\beta_{1,2} + \beta^2_{1,3}\beta_{2,2} + \beta_{1,1}\beta^2_{2,3}\right]^{\frac{1}{2}} \left[|\beta_{1,2}|^2 - \beta_{1,1}\beta_{2,2}\right]^{-\frac{1}{2}},$$

and this suggests a representation

$$\begin{pmatrix} 2\beta_{0,3} & \frac{\beta_{1,3}}{\sqrt{\beta_{1,1}}} & \frac{\beta_{1,1}\beta_{2,3}-\beta^{*}_{1,2}\beta_{1,3}}{\sqrt{\beta_{1,1}(\beta_{1,1}\beta_{2,2}-|\beta_{1,2}|^2)}} & \Gamma \\ \frac{\beta_{1,3}}{\sqrt{\beta_{1,1}}} & -2\beta_{1,2} & -\sqrt{\beta_{1,1}\beta_{2,2} - |\beta_{1,2}|^2} & 0 \\ \frac{\beta_{1,1}\beta_{2,3}-\beta^{*}_{1,2}\beta_{1,3}}{\sqrt{\beta_{1,1}(\beta_{1,1}\beta_{2,2}-|\beta_{1,2}|^2)}} & -\sqrt{\beta_{1,1}\beta_{2,2} - |\beta_{1,2}|^2} & 0 & 0 \\ \Gamma & 0 & 0 & 0 \end{pmatrix} \tag{11.90}$$

for the matrix $\hat{\mathcal{U}}^{T}_{\text{pos}}\hat{C}\hat{\mathcal{U}}_{\text{pos}} = \hat{\mathcal{U}}^{T}_{\text{pos}}\hat{\Upsilon}^{T}\hat{\epsilon}_1 \otimes \hat{\epsilon}_2\hat{\Upsilon}\hat{\mathcal{U}}_{\text{pos}}$, which determines the spectrum $\lambda_i e^{i\phi_i}$ of (11.79).

As we have already discussed in Sect. 11.3.2 the mixed state is separable if by a transformation $\hat{\mathcal{U}}^{T}\hat{C}\hat{\mathcal{U}}$ performed by a unitary matrix $\hat{\mathcal{U}}$ one can reduce the matrix (11.90) to the anti-diagonal form $\left(\hat{\mathcal{U}}^{T}\hat{C}\hat{\mathcal{U}}\right)_{jj} = 0$. As one can see, this matrix is already close to such a form, and indeed it can have all diagonal matrix elements zero after transforming the left upper block,

$$e^{-id\hat{\sigma}^y} e^{ic\hat{\sigma}^x} e^{ib\hat{\sigma}^z} \begin{pmatrix} 2\beta_{0,3} & \frac{\beta_{1,3}}{\sqrt{\gamma_{1,1}}} \\ \frac{\beta_{1,3}}{\sqrt{\beta_{1,1}}} & -2\beta_{1,2} \end{pmatrix} e^{ib\hat{\sigma}^z} e^{ic\hat{\sigma}^x} e^{id\hat{\sigma}^y}$$

to the anti-diagonal form. By direct calculations one finds the solution for the real b, c, and d, which exists only if $\beta_{1,2} = -\beta^{*}_{0,3}$ is real. One immediately finds the corresponding tanglemeter,

$$\mathbf{f}_c = i\mathrm{Im}\beta_{0,3}\left(\sigma_1^+ - \overline{\sigma}_1^+\right)\left(\sigma_2^+ - \overline{\sigma}_2^+\right) + \mathrm{Re}\beta_{0,3}\left(\sigma_1^+ + \overline{\sigma}_1^+\right)\left(\sigma_2^+ + \overline{\sigma}_2^+\right)$$
$$+ \left(\beta_{1,1} + \beta_{1,3}\left(\sigma_2^+ + \overline{\sigma}_2^+\right) + \frac{\beta_{3,3}}{2}\overline{\sigma}_2^+\sigma_2^+\right)\overline{\sigma}_1^+\sigma_1^+$$
$$+ \left(\beta_{2,2} + \beta_{2,3}\left(\sigma_1^+ + \overline{\sigma}_1^+\right) + \frac{\beta_{3,3}}{2}\overline{\sigma}_1^+\sigma_1^+\right)\overline{\sigma}_2^+\sigma_2^+, \tag{11.91}$$

which illustrates that separability manifests itself in vanishing of the imaginary part of the tanglemeter coefficient in front of the symmetric cross product $\left(\sigma_1^+ + \overline{\sigma}_1^+\right)\left(\sigma_2^+ + \overline{\sigma}_2^+\right)$ and the real part in front of $\left(\sigma_1^+ - \overline{\sigma}_1^+\right)\left(\sigma_2^+ - \overline{\sigma}_2^+\right)$. This result shows that a composed system may have a rather general type of entanglement among its parts, which, however, can be completely shadowed by the existence of the possibility to cast the system density matrix to a statistical mixture of the product states of the parts.

Chapter 12
Coherence Protection and Control Over Complex Quantum Systems

This chapter is dedicated to consideration of complex quantum systems with time-dependent Hamiltonians. However, in contrast to what was discussed in Chap. 6, here we focus on a completely different, and in a sense the opposite, aspect of the problem—our aim will be to construct an external action that would force a complex quantum system to behave in a desired way, that is, to undergo predetermined desired evolution and not react to undesired external interventions. We start the consideration with the simplest case of the two-level system and by this example we introduce the main ideas and concepts of unitary quantum control, quantum state engineering, and Zeno state protection. We next turn to the mathematical aspects of the control and discuss the procedure of finding the control sequence on semigroups. We finally focus on the methods that permit one to protect a system from the undesired action of the environment, and at the same time to exert complete control over its unitary evolution.

12.1 Control of Complex Quantum Systems

Thus far we have been considering the universal properties common to almost all realizations of complex quantum systems, regardless of the details of the unperturbed Hamiltonian $\hat{H}_0$ and the interaction $\hat{V}$. We now consider another aspect of such systems, namely their controllability, the possibility to make a complex system follow a predescribed evolution by choosing the strengths and types of applied external perturbations. The control goals can be of two types. One can either place the system at a certain time in a well-defined particular quantum state, beginning from another well-defined initial state, or perform a well-defined transformation of an unknown initial quantum state. In the first case we speak of quantum state manipulation, whereas the second case implies a more involved control over the unitary evolution of the entire system, given by a specific unitary

V.M. Akulin, *Dynamics of Complex Quantum Systems*, Theoretical and Mathematical Physics, DOI 10.1007/978-94-007-7205-2_12,

transformation matrix. The latter also implies a predetermined evolution of each basis state in the Hilbert space of the system.

Though it is not evident at first glance, the justification of the statistical description of complex quantum systems and the requirement necessary for the controllability of a multilevel system have common algebraic and group-theory roots. In spite of this, the ways of describing these two processes are very different. Considering the dynamics of a complex system, one ignores individual details of the spectrum and the interaction performing the ensemble average, thus assuming that the particular realization of a complex system has a small influence on the universal character of its behavior, given by a few average characteristics. For the problem of control the situation is quite the opposite: one needs to know all of the spectral and interaction details in order to achieve the desired system evolution. The techniques for solving these two problems, as well as the final results obtained, are accordingly distinct. In order to describe the universal dynamics of a complex system, one tries to find a finite algebraic expression for the ensemble averaged physical quantity of interest in terms of standard functions, and to check, whenever it is possible, whether the dispersion of the results is small. The solution of the control problem is usually an algorithm which enables one to compute the control functions $f_i(t)$, by employing detailed information about the Hamiltonian $\hat{H}_0$, the perturbations $\hat{V}_j$, and the desired transformation matrix $\hat{U}_d$.

The most general control problem is formulated for a Hamiltonian

$$\hat{H}(t) = \hat{H}_0 + \sum\nolimits_j f_j(t)\hat{V}_j \tag{12.1}$$

comprising a given time-independent unperturbed part $\hat{H}_0$ and a given set of different perturbation operators $\hat{V}_j$. Each of the perturbations may have an individual time-dependent strength $f_j(t)$ and the solution implies that one can specify the particular forms of $f_j(t)$ such that the evolution matrix

$$\begin{aligned}\hat{U}(t) = \hat{T}e^{-i\int_0^t \hat{H}(t)dt} &= 1 - i\int_0^t \hat{H}(t)dt + i^2\int_0^t \hat{H}(t)\int_0^t \hat{H}(t')dt'dt + \ldots \\ &+(-i)^n\int_0^t \hat{H}(t)\int_0^t \hat{H}(t')\int_0^{t'} \hat{H}(t'')\ldots\int_0^{t^{(n-1)}} \hat{H}(t^{(n)})dt^{(n)}\ldots dt''dt'dt + \ldots\end{aligned} \tag{12.2}$$

assumes a predetermined form $\hat{U}(T) = \hat{U}_d$ at a given time moment $t = T$. When this is possible the system is called controllable. Here $\hat{T}$ is the time-ordering operator that has been encountered earlier in (5.2). A particular case, where all functions $f_j(t)$ take only two values $f_j(t) = 0$ and $f_j(t) = 1$, is called bang–bang control.

The necessary condition of the system's controllability is that the Hamiltonian $\hat{H}_0$ and perturbation operators $\hat{V}_j$ together with their commutators of different orders form an operator basis set which spans the entire linear space of the operators in the Hilbert space of the system. In other words, it should be possible to represent any operator in the Hilbert space of the considered quantum system as a linear combination of $\hat{H}_0$, operators $\hat{V}_j$, and their commutators of various orders.

We consider several aspects of the control problem, starting with the simplest examples of the two-level system. Though the consideration of these examples can be performed directly in exact analytical form, we discuss several algorithms for the numerical solution and introduce the main concepts and images convenient for the description of more sophisticated situations.

12.1.1 Control of Two-Level Systems

A two-level quantum system within two-dimensional Hilbert space corresponds to the Lie group $SU(2)$ and can be described with the help of the three-element algebra of Pauli matrices $\hat{\sigma}_x, \hat{\sigma}_y, \hat{\sigma}_z$, that compose the set of generators of $SU(2)$ supplemented by the identity matrix $\hat{I}$. This group is isomorphic to the group of three-dimensional rotations, and therefore the main ideas of the control technique can be illustrated by three-dimensional images, convenient for intuitive perception.

Indeed, each Hermitian operator $\hat{H}$ can be represented as a linear combination of Pauli matrices

$$\hat{H} = X\hat{\sigma}_x + Y\hat{\sigma}_y + Z\hat{\sigma}_z + E_0\hat{I}, \tag{12.3}$$

where the constant E_0 determining the mean energy responsible for the global phase can be omitted. Therefore the vector $\overline{H} = (X, Y, Z)$ formed by three real numbers determines the Hermitian operator with zero trace. The quantum state of the two-level system can also be represented as a real three-dimensional vector of unit length. Two eigenvectors of an operator $\hat{H}$ are given by the three-dimensional vectors (x, y, z) of length 1 that are parallel and antiparallel to the vector $\overline{H}$. Each state vector (x, y, z) also corresponds to a unit rank density matrix

$$\rho = |\psi\rangle\langle\psi| = \frac{x}{2}\hat{\sigma}_x + \frac{y}{2}\hat{\sigma}_y + \frac{z}{2}\hat{\sigma}_z + \frac{1}{2}\hat{I} \tag{12.4}$$

of the pure state. The coordinates (x, y, z) of the state $|\psi\rangle$ and the state $\left|\tilde{\psi}\right\rangle$ orthogonal to it, thus correspond to opposite points on the unit sphere, as shown in Fig. 12.1a. These coordinates are also known as the components of the Bloch vector.

Each unitary transformation

$$\hat{U} = e^{-it\hat{H}} = \exp\left[-it\left(X\hat{\sigma}_x + Y\hat{\sigma}_y + Z\hat{\sigma}_z\right)\right] \tag{12.5}$$

of a state vector (x, y, z) under the action of a time-independent Hamiltonian is a counterclockwise rotation around the axis $\overline{H}$ by the angle $\phi = t\left(X^2 + Y^2 + Z^2\right)^{1/2}$

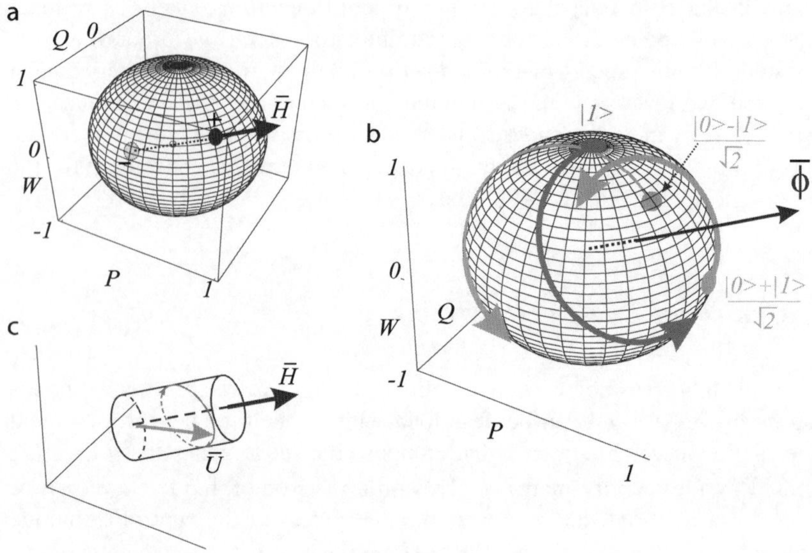

Fig. 12.1 Representation of the states and operators of a two-level system by three-dimensional vectors. (**a**) Each Hamiltonian $\hat{H}$ corresponds to a vector $\overline{H}$ with the coordinates (X, Y, Z) given by the coefficients in the representation (12.3) of this Hamiltonian as the sum of Pauli matrices σ_x, σ_y, and σ_z. Two eigenstates of the Hamiltonian, with positive (+) and negative (−) eigenenergies, correspond to the points given by the intersection of the vector axis with the unit sphere. Coordinates $\pm(x, y, z)$ of these points are given by the components P, Q, and W of the Bloch vector (12.4), respectively. (**b**) A unitary transformation $\exp(-it\hat{H})$ can be represented by an axial vector $\overline{\phi}$, directed parallel to $\overline{H}$, which has a length equal to the rotation angle. The points representing the states move on the unit sphere. (**c**) Vector $\overline{U}$ representing the unitary transformation as a sum (12.6) does not conserve its length, moving under the action of a different transformation around the vector $\overline{H}$, along an elliptic trajectory

represented by the axial vector $\overline{\phi} = (Xt, Yt, Zt)$, as shown in Fig. 12.1b. One can also derive a differential equation describing the evolution of the unitary operator $\hat{U}$ under the action of a time-dependent Hamiltonian. Indeed, by representing the evolution operator as a linear combination

$$\hat{U} = U_0\hat{I} - iU_X\hat{\sigma}_x - iU_Y\hat{\sigma}_y - iU_Z\hat{\sigma}_z \tag{12.6}$$

and by calculating the products of the Pauli matrices, one can rewrite the Schrödinger equation $i\frac{\partial}{\partial t}\hat{U} = \hat{H}\hat{U}$ in the form

$$\frac{\partial}{\partial t}U_0 = -(\overline{H}, \overline{U})$$
$$\frac{\partial}{\partial t}\overline{U} = \overline{H}U_0 + \left[\overline{H} \times \overline{U}\right] \tag{12.7}$$

employing the vector notation $\overline{U} = (U_X, U_Y, U_Z)$ and $\overline{H} = (X, Y, Z)$. Here $(\ldots, \ldots)$ stands for the scalar product and $[\ldots \times \ldots]$ for the vector product of two vectors.

The unitarity of $\hat{U}$ implies that $U_0^2 + U_X^2 + U_Y^2 + U_Z^2 = 1$, which means that the length of the three-dimensional part $\overline{U}$ of the four-dimensional vector (U_0, U_X, U_Y, U_Z) is not conserved. Inspection of (12.7) shows that $\overline{U}$ moves along an elliptic trajectory over the surface of a cylinder directed along the vector $\overline{H}$ as shown in Fig. 12.1c, since the component of $\overline{U}$ rotates around the axis given by the direction of $\overline{H}$ with the angular velocity $|\overline{H}|$, whereas the component $U_{||}$ parallel to $\overline{H}$ oscillates with the same frequency as that of the component U_0. The latter is a manifestation of the rotation of the vector $(U_{||}, U_0)$ in four-dimensional space. However, due to the unitarity condition, the three-dimensional vector $\overline{U}$ describes the evolution operator completely.

The fact that the quantum states of the system, the Hamiltonians, and the unitary transformation matrices can all be represented in the same three-dimensional real vector space is a unique property of the two-level systems corresponding to the $SU(2)$ symmetry group. For other systems, the dimensionality of the real vector spaces corresponding to operators are larger as compared to the vector spaces representing the states. When we ignore the global phase of the state and allow for the normalization condition, each state of an N-level system given by a normalized complex-valued vector in N-dimensional Hilbert space corresponds to a $2N-2$ real vector. The Hamiltonians, being Hermitian matrices, can be described by $N^2 - 1$ real parameters when we set their mean energy responsible for the global phase to zero, thus assuming that all Hamiltonians have zero trace. This is also valid for the unitary matrices generated by such Hamiltonians via the relation $\hat{U} = \exp(-it\hat{H})$.

Unitary Quantum State Engineering by Iterative Transformations of States

The control of two-level systems is not a challenging task, and usually solutions can be found with the help of relatively simple algebraic calculations. However, it is useful to consider several control algorithms as operations in the vector space, since this illustrates the main principle of finding the control amplitudes $f_j(t)$ for large systems. We mainly restrict ourselves to the so-called "bang–bang" control when the amplitudes assume only the values 0 or 1 during certain time intervals.

Let us assume that $\hat{H}_0 = 0$ and that we have only two different perturbations $\hat{V}_1 = \hat{\sigma}_x$ and $\hat{V}_2 = \hat{\sigma}_z$ at our disposal. The commutator $\left[\hat{V}_1, \hat{V}_2\right] = 2i\hat{\sigma}_y$ provides us with the missing third operator and the operator basis becomes complete and the system is therefore controllable. For this simplest of examples, we will illustrate the geometrical meaning of the algorithms that allow one to control the system.

The simplest iterative algorithm allows one to approach a given quantum state $|f\rangle$ starting from an initial state $|i\rangle$. We apply to the system initially in the state $|i\rangle$, the perturbations $\hat{V}_1$ and $\hat{V}_2$ in sequence, trying to approach the desired state $|f\rangle$ with each iteration step. As a result, the initial state experiences the series of transformations $|i_N\rangle = \exp\left(-it_N\hat{V}_{1,2}\right)|i_{N-1}\rangle$ where $\hat{V}_{1,2}$ can be either $\hat{V}_1$ or $\hat{V}_2$, depending on the parity of N.

At each step of the iteration we try to maximize the scalar product

$$\langle f|i_N\rangle = \langle f|\exp\left(-it_N\hat{V}_{1,2}\right)|i_{N-1}\rangle \tag{12.8}$$

by the proper choice of the parameter t_N denoting the exposure time to the perturbation. After N iterations we therefore achieve the state

$$|i_N\rangle = \hat{U}\,|i\rangle \tag{12.9}$$

which approaches the desired state $|f\rangle$ as a result of the application of the iteratively constructed operator

$$\hat{U} = \exp\left(-it_N\hat{V}_{1,2}\right)\ldots\exp\left(-it_3\hat{V}_1\right)\exp\left(-it_2\hat{V}_2\right)\exp\left(-it_1\hat{V}_1\right)|i\rangle\,. \tag{12.10}$$

as illustrated in Fig. 12.2a.

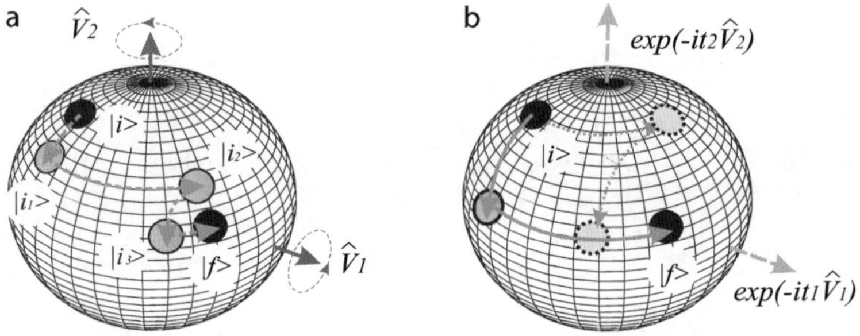

Fig. 12.2 The iterative procedure compared to the direct transition. (**a**) By successive application of two different transformations, $\exp(-i\hat{V}_1 t)$ and $\exp(-i\hat{V}_2 t)$, suggested by the two Hamiltonians $\hat{V}_1 = \hat{\sigma}_x$ and $\hat{V}_2 = \hat{\sigma}_z$ one can reach the final state $|f\rangle$ from the initial state $|i\rangle$ by choosing the variable t_N at each step in such a way that the scalar product of the transformed state and the target state is maximum. (**b**) The same transformation can be performed in two steps only by finding the maximum scalar product as a function of two variables, t_1 and t_2. However, this approach may not result in the coincidence of the transformed state with the target state and can result only in the closest approach if the order of the applied transformations is wrong, as shown by the spots with the *dashed line* contour. The scalar product then has a local maximum, different from the global maximum value 1

The iterative procedure is not a very efficient way to perform the desired transformation. Indeed, it depends on a large number of parameters t_N, whereas only two parameters are required for determining the position of a point on the unit sphere representing a state of the two-level system. Therefore the objective can be achieved by application of only two transformations

$$|i_2\rangle = \exp\left(-iT_2\hat{V}_2\right)\exp\left(-iT_1\hat{V}_1\right)|i\rangle\,. \tag{12.11}$$

However, for this purpose, one needs to perform a simultaneous search for the parameters T_1 and T_2 that yield the maximum scalar product $\langle f|i_2\rangle$, as illustrated in Fig. 12.2b. The main obstacle in this approach is that the problem may not always have a solution for a given transformation sequence. The natural way to overcome this obstacle is to add an extra transformation

$$|i_3\rangle = \exp\left(-iT_3\hat{V}_1\right)\exp\left(-iT_2\hat{V}_2\right)\exp\left(-iT_1\hat{V}_1\right)|i\rangle \tag{12.12}$$

and look for the maximum in the space of three variables instead of two. The solution of the latter problem is not unique, and the set of solutions form a one-dimensional curve in the three-dimensional space of the parameters (T_1, T_2, T_3).

Iterative Construction of a Unitary Transformation

Prior to the consideration of an algorithm allowing one to construct a desired unitary transformation, we take advantage of the simplicity of the two-level system and solve the problem directly by calculation of the most direct transformation. For the case $\hat{V}_1 = \hat{\sigma}_x$ and $\hat{V}_2 = \hat{\sigma}_z$ under consideration, the direct multiplication of matrices yields

$$\begin{aligned}\hat{U} &= \exp\left(-iT_3\hat{V}_1\right)\exp\left(-iT_2\hat{V}_2\right)\exp\left(-iT_1\hat{V}_1\right)\\ &= \cos\left(T_3+T_1\right)\cos(T_2)\hat{I} - i\sin\left(T_3+T_1\right)\cos(T_2)\hat{\sigma}_x\\ &\quad + i\sin\left(T_3-T_1\right)\sin(T_2)\hat{\sigma}_y - i\cos\left(T_3-T_1\right)\sin(T_2)\hat{\sigma}_z.\end{aligned} \tag{12.13}$$

Comparison of (12.13) and (12.6) allows us to find the vector (U_0, U_X, U_Y, U_Z). The solution of the equation $\hat{U}(T_1, T_2, T_3) = \hat{U}_d$ can be found for any desired unitary matrix $\hat{U}_d = U_0\hat{I} - iU_X\hat{\sigma}_x - iU_Y\hat{\sigma}_y - iU_Z\hat{\sigma}_z$, and reads

$$\begin{aligned}T_1 &= \frac{1}{2}\arcsin\frac{U_Y}{\sqrt{U_X^2+U_Y^2}} + \frac{1}{2}\arcsin\frac{U_Z}{\sqrt{U_0^2+U_Z^2}},\\ T_2 &= \arccos\left(-\sqrt{U_X^2+U_Y^2}\right),\\ T_3 &= \frac{1}{2}\arccos\frac{-U_Y}{\sqrt{U_X^2+U_Y^2}} + \frac{1}{2}\arccos\frac{U_Z}{\sqrt{U_0^2+U_Z^2}}.\end{aligned} \tag{12.14}$$

One can verify directly that the coefficients U given by the trigonometric expressions do indeed satisfy the requirement $U_0^2 + U_X^2 + U_Y^2 + U_Z^2 = 1$.

The solution (12.14) can also be found numerically by minimizing the function

$$\mathrm{Tr}[(\hat{U}(T_1,T_2,T_3)-\hat{U}_d)(\hat{U}(T_1,T_2,T_3)-\hat{U}_d)^*] \tag{12.15}$$

which gives the square of the distance between two matrices vanishing for the desired solution. However, if by analogy to the state control we try to find a solution minimizing successively the function

$$\mathrm{Tr}\left\{\left[\exp\left(-it_N\hat{V}_{1,2}\right)\ldots\exp\left(-it_2\hat{V}_2\right)\exp\left(-it_1\hat{V}_1\right)-\hat{U}_d\right]\times c.c\right\} \tag{12.16}$$

for each new added operator $\exp\left(-it_N\hat{V}_{1,2}\right)$, the procedure may converge to a local minimum different from zero and not give the desired result. We illustrate this

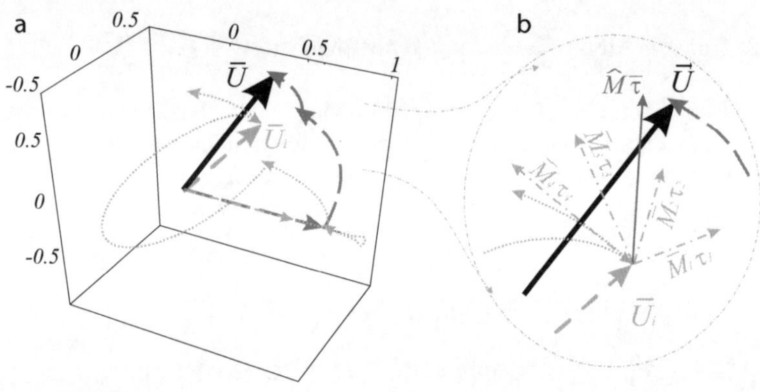

Fig. 12.3 The iterative procedure for a predescribed unitary transformation. The trajectories of the vector (12.6), changing with the course of the transformations, are shown by the *long dashed* and *dotted arrows* for the direct and iterative transformations, respectively. (**a**) The desired transformation represented by the vector $\overline{U}$ (*bold solid arrow*) can be achieved directly by three successive transformations that are found by minimizing the functional (12.14). The iterative approach (12.15) may result in a transformation $\overline{U}_i$ which cannot be improved, since the next iteration corresponds to displacement in the direction perpendicular to that required. (**b**) The situation can be improved by linear iterations based on the linear expansion $\overline{T}=\overline{T}^{(0)}+\overline{\tau}$ of the operator (12.9) near the point $(T_1^{(0)},T_2^{(0)},T_3^{(0)},T_4^{(0)})$ corresponding to $\overline{U}_i$. Linear corrections (*dash-dot arrows*) allow one to iteratively achieve the solution $\overline{U}$

in Fig. 12.3a, showing the situation where the last added operator moves the vector $\overline{U}$ in a direction perpendicular to that going towards the solution. Note that a similar complication resulting from the presence of local minima for the function (12.15) can also be encountered for systems with a large Hilbert space dimension.

If the iteration procedure given by the function (12.16) or any other iterative algorithm based on an incomplete set of the variables $\{t\}$ has stopped close enough to the global minimum of the function (12.15), the exact solution can still be found by successive application of the linear approximation which involves all variables. Though for low-dimensional systems this technique does not give a substantial advantage as compared to the direct minimization of (12.15), for large systems it is much more convenient and requires much less numerical work.

Let us consider this approach in more detail. For a set of variables $\left\{T_n^{(0)}\right\}$ close to the solution $\{T_n\}$, the difference

$$\exp\left(-iT_N^{(0)}\hat{V}_{1,2}\right)\ldots\exp\left(-iT_2^{(0)}\hat{V}_2\right)\exp\left(-iT_1^{(0)}\hat{V}_1\right)-\hat{U}_d \qquad (12.17)$$

can be expanded into a Taylor series over the variations $\{\delta T_n\}$ of the variables $\{T_n\}$ up to the first order near the point $\left\{T_n^{(0)}\right\}$, that is,

$$\hat{U}_{\delta T}=\exp\left(-iT_N\hat{V}_{1,2}\right)\ldots\exp\left(-iT_2\hat{V}_2\right)\exp\left(-iT_1\hat{V}_1\right)-\hat{U}_d$$
$$+\sum_n^N\delta T_n\frac{\partial}{\partial T_n^{(0)}}\exp\left(-iT_N^{(0)}\hat{V}_{1,2}\right)\ldots\exp\left(-iT_2^{(0)}\hat{V}_2\right)\exp\left(-iT_1^{(0)}\hat{V}_1\right). \qquad (12.18)$$

Calculation of the derivatives is the most time-consuming part of the numerical work but must be done only once. With the help of these derivatives, we construct an algorithm which allows one to find the exact solution $T_n = T_n^{(0)} + \delta T_n$ as a result of the iterations over δT_n. Once the derivatives are found, one can find the first approximation of the variables δT_n solving the system of linear equations $\hat{U}_{\delta T} = 0$ with $\hat{U}_{\delta T}$ given by (12.18).

For the two-level system under consideration, the matrix equations $\hat{U}_{\delta T} = 0$ formally correspond to a system of four linear equations. However, due to the unitarity of $\hat{U}$, the actual number of linearly independent equations is three, as can be shown when we employ the Pauli matrices as the basis set in the matrix space according to (12.6) and take into account the unitarity requirement $U_0^2 + U_X^2 + U_Y^2 + U_Z^2 = 1$. Hence for $N > 3$ the number of equations is less than the number of variables. In this representation for $N = 4$ the equation $\hat{U}_{\delta T} = 0$ adopts the matrix form

$$\begin{pmatrix} M_{X,1} & M_{X,2} & M_{X,3} & M_{X,4} \\ M_{Y,1} & M_{Y,2} & M_{Y,3} & M_{Y,1} \\ M_{Z,1} & M_{Z,2} & M_{Z,3} & M_{Z,3} \end{pmatrix}\begin{pmatrix}\delta T_1\\ \delta T_2\\ \delta T_3\\ \delta T_4\end{pmatrix}=\begin{pmatrix}\delta U_X\\ \delta U_Y\\ \delta U_Z\end{pmatrix} \qquad (12.19)$$

with

$$M_{jn} = \frac{\partial}{\partial T_n^{(0)}} \mathrm{Tr}\left[\hat{\sigma}_j e^{-iT_4^{(0)}\hat{V}_2} e^{-iT_3^{(0)}\hat{V}_1} e^{-iT_2^{(0)}\hat{V}_2} e^{-iT_1^{(0)}\hat{V}_1}\right]$$
$$\delta U_j = \mathrm{Tr}\left[\hat{\sigma}_j e^{-iT_4\hat{V}_2} e^{-iT_3\hat{V}_1} e^{-iT_2\hat{V}_2} e^{-iT_1\hat{V}_1} - \hat{\sigma}_j\hat{U}_d\right], \tag{12.20}$$

where the Pauli matrices have been employed for a basis of the operator space.

As the first step we solve (12.19) with respect to $\delta T_1, \delta T_2$, and δT_3 since the fourth column of the matrix is perpendicular to the vector δU, which was precisely the reason for the inefficiency of the iterative variations (12.16). Then we substitute the new values of $T_n = T_n^{(0)} + \delta T_n$ into the second equation (12.20), find new $\delta\overline{U}$ and repeat the iteration. We note that the deviations $\delta\overline{U}$ have changed as a result of the iteration, whereas the matrix $\hat{M}$ is retained in its original form. This is the main difference between the employed method and the steepest descent technique, which requires recalculation of all of the derivatives at each step of the iteration. Nevertheless, if we are close enough to the exact solution, at each step of the iteration the displacement of variables δT_n occurs at a small angle to the direction of the steepest descent and hence this procedure will converge, since for $\delta\overline{U} \to 0$ the parameters $\delta T_n \to 0$ as well. We illustrate this in Fig. 12.3b.

Control by a Periodic Field

One can control the two-level system completely by applying a harmonic external field of a given polarization, frequency, and amplitude. These three parameters determine the three parameters of the effective Hamiltonian. For an external field $V_x = \mathcal{E}_x d/2$ polarized in the x-direction we have obtained (3.9) describing the evolution of the state vector (ψ_0, ψ_1) in the resonant approximation. For an arbitrary polarization, this equation adopts the form

$$i\frac{\partial}{\partial t}\begin{pmatrix}\psi_1\\ \psi_0\end{pmatrix} = \begin{pmatrix}\frac{\Delta}{2} & V_x - iV_y\\ V_x + iV_y & -\frac{\Delta}{2}\end{pmatrix}\begin{pmatrix}\psi_1\\ \psi_0\end{pmatrix}, \tag{12.21}$$

where $\Delta = E_1 - E_0 - \omega$ is the detuning of the field quantum energy from the energy of the transition. Here we do not include the initial conditions in the form of δ-functions and shift the energy scale such that the mean value of the interaction vanishes.

The Hamiltonian of the system has an evident representation (12.3)

$$\hat{H} = \frac{\Delta}{2}\hat{\sigma}_z + V_x\hat{\sigma}_x + V_y\hat{\sigma}_y$$

with $X = V_x$, $Y = V_y$, and $Z = \Delta/2$. Therefore any unitary evolution $\hat{U} = \exp(-it\hat{H})$ can be constructed directly by choosing the strength and the orientation of the external field polarization along with the detuning. The only restriction of this method is the applicability of the resonant approximation, which requires that the Rabi frequency $\Omega = (\Delta^2 + \mathcal{E}^2 d^2)^{1/2}$ should be much smaller than the field frequency ω. This implies that the detuning and the interactions should be small, whereas the desired transformation can be achieved on a time-scale much longer than the field period. This type of control can also be employed for larger systems, provided they satisfy some requirements that are in close relation with the general algebraic requirements necessary for the controllability of generic quantum systems.

12.1.2 Holonom and Non-Holonom Systems

The universal character of complex quantum system dynamics relies on the fact that it is possible for the generic system with an unperturbed Hamiltonian $\hat{H}_0$ and a perturbation $\hat{V}$ to occupy any of the levels close in energy to the initially populated state, while the state vector of the system, evolving with the course of time, can approach any linear combination of the energy eigenfunctions corresponding to these levels. Mathematically this implies the validity of the bracket generation condition: for almost any pair $\hat{V}$ and $\hat{H}_0$ the set of all possible commutators $\left[\hat{V}, \left[\hat{V}, \left[\ldots, \left[\hat{H}_0\right]\right] \ldots\right]\right], \left[\hat{H}_0, \left[\hat{H}_0, \left[\ldots, \left[\hat{V}\right]\right] \ldots\right]\right], \ldots$ spans the entire N^2-dimensional space of the operators, which is restricted only by the condition of Hermiticity and anti-Hermiticity for the even and odd orders of the commutation, respectively. Roughly speaking, the bracket generation condition is necessary for the complexity of a system and it becomes sufficient condition when all the linearly independent commutators are of the same order of magnitude. If the latter is not the case, the universality manifests itself only on the energy scale corresponding to the smallest typical size of the commutators and in the time domain given by the inverse of this value. At larger energy scales, or for shorter time intervals, the system may manifest individual behavior.

In special cases, when all of the commutators are restricted to a lower-dimensional subspace and form there a closed subalgebra, with a number of elements smaller than N^2, the system may exhibit individual behavior at any energy and at all time-scales. In the extreme case of a low-dimensional subalgebra, the multilevel system may have an exact analytical solution, thus being equivalent to a simpler quantum system. The two-level system with the three-element algebra $su(2)$ of the Pauli matrices $\sigma_x, \sigma_y, \sigma_z$, along with the harmonic oscillator with the closed three-dimensional $su(1,1)$ algebra of $a\dagger$, a, $aa\dagger$ are two examples of such simple systems.

The bracket generation condition is also necessary for complete control over the dynamics of quantum systems. Indeed, considering the time-dependent systems in Sect. 5.1, we have seen that even in the simplest case of the evolution $\hat{U} = e^{-it_B \hat{B}} e^{-it_A \hat{A}}$ under the action of two Hamiltonians $\hat{A}$ and $\hat{B}$ applied in sequence during the time intervals t_A and t_B, respectively, all series of different-order commutators appeared in the Baker–Campbell–Hausdorff formula (5.4). The coefficients with which these commutators enter the linear combination in the exponent, were powers of the exposition times t_A and t_B. For a time-dependent perturbation $f(t)\hat{V}$ acting together with the unperturbed Hamiltonian $\hat{H}_0$ the operator structure of the evolution matrix is also of the type suggested by the Baker–Campbell–Hausdorff formula: it is the exponent of a linear combination of different commutators

$$\hat{U}(t) = \exp\Bigg\{ - i\hat{H}_0 t - i\sum_j \hat{V}_j \int_0^t f_j(t)dt + \sum_j c_{j0}[f_i(t)]\Big[\hat{V}_j, \hat{H}_0\Big]$$
$$+ \sum_{jk} c_{kj0}[f_i(t)]\Big[\hat{V}_k, \Big[\hat{V}_j, \hat{H}_0\Big]\Big] + \dots$$
$$+ \sum_{\dots j} c_{00\dots j}[f_i(t)]\Big[\hat{H}_0, \Big[\hat{H}_0, \Big[\dots, \hat{V}_j\Big]\dots\Big]\Big] + \dots \Bigg\}. \qquad (12.22)$$

However, in this case, the corresponding coefficients c are complex functionals of all functions $f_j(t)$. If the commutators of $\hat{V}_j$ and $\hat{H}_0$ span all N^2-dimensional operator space corresponding to the N-dimensional Hilbert space of the system, one can try to find the functions $f_j(t)$ that give the desired evolution of the system. Usually this requires much more effort since the functional relations between set of the coefficients $c_{\dots}$ and the set of the functions $f_j(t)$ are complicated and unknown explicitly in most cases.

The bracket generation condition is equivalent to the property known in analytical mechanics as holonomicity. The essence of this concept is the following. The infinitesimal evolution of a mechanical system given by a set of local differential Hamilton equations relates the coordinates and momenta of the system in two sequential moments of time. One can often find integrals of these equations that are quantities conserved throughout the evolution. For a conservative system, the most well-known integral is the total energy, whereas the energy conservation law is the global constraint following from the local equations of motion. A local constraint, resulting in a global one, is designated as integrable or holonom. If the number of holonom constraints equals the number of degrees of freedom, the system is completely integrable or completely holonom. The motion of the completely holonom system can be separated to independent motions of different degrees of freedom. In this case, a set of coordinates can be introduced in such a way that the motion of the system is equivalent to an independent evolution of different coordinates, and the motion of a completely holonom mechanical system with n degrees of freedom occurs along the surface of an n-dimensional torus in $2n$-dimensional phase space. A system is called partially holonom when the number of holonom constraints is less than the number of degrees of freedom, and for no constraints the system is non-holonom. The evolution of a non-holonom system occurs in entire (unconstrained) phase space.

There is a simple, sufficient criterion ensuring that the bracket generation condition holds for a pair of operators $\hat{A}$ and $\hat{B}$. Each of the operators should have non-degenerate eigenvalues $A_j \neq A_k$ with all frequencies different $A_j - A_l \neq A_k - A_s$, and in the representation where one operator, say $\hat{A}$, is diagonal, the other one should not have zero matrix elements $B_{j,k} \neq 0$. Note that this very condition allows one to construct any desired evolution with the Hamiltonian $\hat{H}$ by applying harmonic external fields in complete analogy with the case of a two-level system (12.21). In fact, considering each pair of eigenvectors of $\hat{A}$ as an independent two-level system, and applying periodic fields at the frequencies $\omega_{j,k}$ detuned from the corresponding transition frequency $A_j - A_k$ by the values $H_{j,j} - H_{k,k}$ and with the amplitudes $v_{j,k} = H_{j,k}/B_{j,k}$ one can construct any desired $\hat{H}$ in the rotating wave approximation.

Partially Controlled Systems

We now demonstrate an analogy between the holonom dynamics of a mechanical system and the uncontrollable dynamics of the simplest compound quantum system comprising a pair of two-level systems. At this example we draw a parallel between non-controllability and holonomicity by introducing quantum variables that behave like classical coordinates of holonom mechanical systems moving along tori surfaces. We start with the simplest case of two non-interacting two-level parts, each of which can be exposed to the action of just one Hamiltonian, $\hat{\sigma}_x$ for the first part and $\hat{\Sigma}_z$ for the second part. We will employ capital letters to designate the Pauli operators of the second two-level system. The system has a four-dimensional Hilbert space spanned by the orthogonal states $|0,0\rangle, |0,1\rangle, |1,0\rangle$, and $|1,1\rangle$, whereas the space of Hermitian operators consists of 16 elements, including the identity operator. The operators $\hat{I}, \hat{\sigma}_x, \hat{\sigma}_y, \hat{\sigma}_z, \hat{\Sigma}_x, \hat{\Sigma}_y, \hat{\Sigma}_z$ along with nine binary direct products $\hat{\sigma}_j \otimes \hat{\Sigma}_k$ with $j,k = x,y,z$ can be selected as the basis spanning the operator space.

This system is apparently non-controllable since the algebra consists of only two operators $\hat{\sigma}_x$ and $\hat{\Sigma}_z$ that moreover commute corresponding to different Hilbert subspaces. Therefore, one cannot create new linearly independent operators via bracket generation, and the creation of controllability cannot be satisfied, since two operators $\hat{\sigma}_x$ and $\hat{\Sigma}_z$ alone cannot span the entire 16-dimensional operator space. Furthermore, neither of the subsystems is controllable individually, since the motion of the Bloch vectors representing these subsystems is just a simple rotation around the x-axis for the first part and around the z-axis for the second part of the system; they do not span the corresponding Bloch spheres. Let us construct an adequate image corresponding to this motion by representing the quantum state geometrically.

A complex vector in four-dimensional Hilbert space depends on six real parameters, provided we ignore the global phase of the state vector and allow for the normalization condition. Therefore it is impossible to depict this object as a single vector in three-dimensional space, and at least a pair of such vectors is required. When two parts of the systems are completely independent, the state vector is a direct product of the state vectors of the parts $|\Psi\rangle = |\psi_1\rangle\,|\psi_2\rangle$. Each part then has its own Bloch vector of unit length. By simply appending one of the vectors to the other, scaled by a certain factor for convenience of presentation, we can construct a composition representing a quantum state of the system. The sum of the Bloch vectors contains only part of the information about the state, in contrast to the composition. Taken just by itself, the sum is insufficient for a complete description, and supplementary parameters of the composition are required for this purpose. We specify these parameters by considering different trajectories of the sum.

The initial condition $|\psi_1\rangle = |1\rangle\,, |\psi_2\rangle = (|1\rangle + |0\rangle)/\sqrt{2}$ corresponds to the eigenstates of the operators $\hat{\sigma}_z$ and $\hat{\Sigma}_x$, respectively, that are given by the vectors $(0,0,1)$ and $(1,0,0)$ on the Bloch spheres related to the first and second parts. Under the action of the Hamiltonians $\hat{\sigma}_x$ and $\hat{\Sigma}_z$ these vectors start to rotate in the (z,y) and in the (x,y) planes, respectively, and the entire evolution can be represented as a motion on the surface of a torus, as shown in Fig. 12.4 Note that each torus corresponds to a certain initial condition, which is not necessarily of the type $|\psi_1\rangle\,|\psi_2\rangle$, but can also include the entangled states $\alpha\,|\psi_1\rangle\,|\psi_2\rangle + \beta\left|\tilde{\psi}_1\right\rangle\left|\tilde{\psi}_2\right\rangle$ with $\langle\psi_i\left|\tilde{\psi}_i\right\rangle = 0$. Whatever the strengths $f(t)$ and $g(t)$ of the Hamiltonian $g(t)\hat{\sigma}_x + f(t)\hat{\Sigma}_z$ are, the evolution cannot bring the system off the surface of the torus, which is the holonom constraint of the system motion. Therefore in order to specify the state of the system, one has to indicate not only the position of the composition of the Bloch vectors, but also to specify the parameters of the torus to which it belongs. The parameters of tori, namely the vertical position of the equatorial plane, the larger radius, the eccentricity, and the smallest radius are independent integrals of motion that specify the holonom constraint as shown in Fig. 12.4b–d.

Let us consider a more involved example of an uncontrollable system. We take two Hamiltonians $\hat{\sigma}_x$ and $\hat{\Sigma}_z \otimes \hat{\sigma}_z$ and find that together with their commutator $\hat{\Sigma}_z \otimes \hat{\sigma}_y$ they form a closed, three-element algebra and together with the identity matrix span a four-dimensional subspace of the entire 16-dimensional space of all 4×4 Hermitian matrices. The latter also includes the operators $\hat{\sigma}_y$, $\hat{\sigma}_z$, $\hat{\Sigma}_x$, $\hat{\Sigma}_y$, $\hat{\Sigma}_z$, $\hat{\Sigma}_z \otimes \hat{\sigma}_x$, $\hat{\Sigma}_y \otimes \hat{\sigma}_j$, $\hat{\Sigma}_z \otimes \hat{\sigma}_j$, $(j = x, y, z)$ that can never be obtained by the commutation of $\hat{\sigma}_z$ and $\hat{\Sigma}_z \otimes \hat{\sigma}_x$.

Under the action of the Hamiltonian $\hat{\Sigma}_z \otimes \hat{\sigma}_x$, the trajectory of the composition of the Bloch vectors moves along a torus surface in such a way that a 2π rotation of the large radius corresponds to the same rotation of the small one. After the application of the Hamiltonian $\hat{\sigma}_z$ the composition moves to another torus with the same larger radius and the same position of the equatorial plane, but with a different small radius, as illustrated in Fig. 12.5a. Having returned to the original torus by another application of the Hamiltonian $\hat{\sigma}_z$, the trajectory can never return precisely to its initial position, but acquires a shift. Therefore all of the inner part of the largest torus for a given larger radius and equatorial plane can be filled by

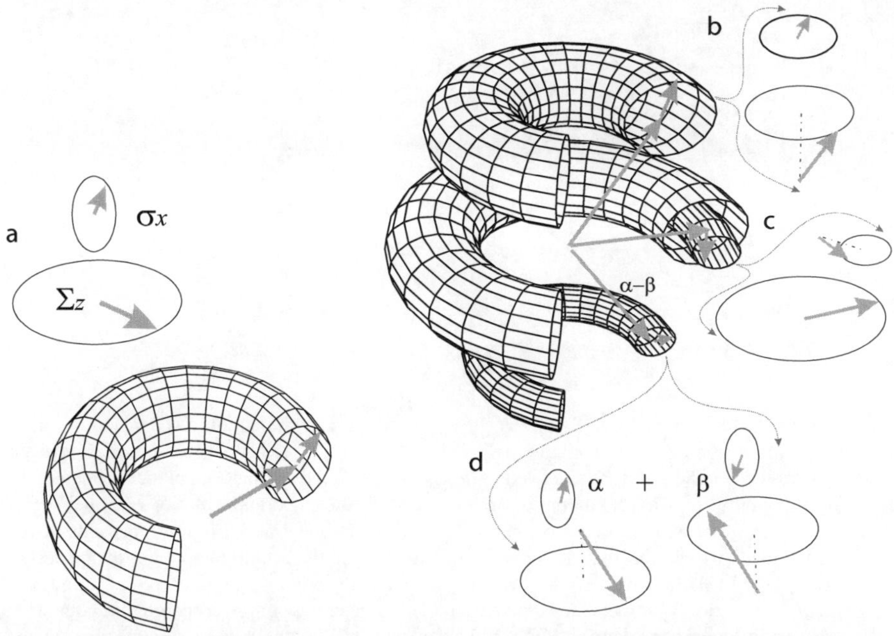

Fig. 12.4 A pair of two-level systems moving independently under the action of a single operator for each system is not completely controllable. The evolution of the pair, considered as a single uncontrollable compound quantum system, can be represented as a motion along the surface of a torus with the parameters depending on the initial condition. The torus is formed by the rotation of the Bloch vector $\overline{b}$ of the first part around the end of the Bloch vector $\overline{B}$ of the second part, whereas the plane $(\overline{Z}, \overline{B})$ is considered as the (zy) plane of the first part. This image is valid even for the case when the Bloch vectors of the second (**b**) or the first (**c**) are not in the planes (z, y) and (x, y), respectively. The image can also be extended to the case of entangled states that cannot be factorized as a direct product. In this case, $\overline{B}$ is not in the unit sphere, and has a length equal to the population difference of the factorizable components. The phase difference can also be visualized by the ellipticity of the cross-sections of the tori or by inclination of the equatorial plane with respect to the Z axis

the trajectories when we apply two Hamiltonians in sequence. In other words, the set of states accessible from an initial state by control over the operators $\hat{\sigma}_x$ and $\hat{\Sigma}_z \otimes \hat{\sigma}_z$ comprising a three-element subalgebra is a three-dimensional variety. By employing more complicated Hamiltonians, one can obtain subalgebras with more elements which imposes fewer holonom constraints on the system dynamics and results in a variety of accessible states of a larger spatial dimensionality.

Now let us take the following Hamiltonians:

$$\hat{A} = \hat{\sigma}_z + \hat{\Sigma}_x + \hat{\Sigma}_x \otimes \hat{\sigma}_x + \hat{\Sigma}_z \otimes \hat{\sigma}_y$$
$$\hat{B} = \hat{\sigma}_x + \hat{\sigma}_z + \hat{\Sigma}_y + \hat{\Sigma}_x \otimes \hat{\sigma}_x + \hat{\Sigma}_y \otimes \hat{\sigma}_x \tag{12.23}$$

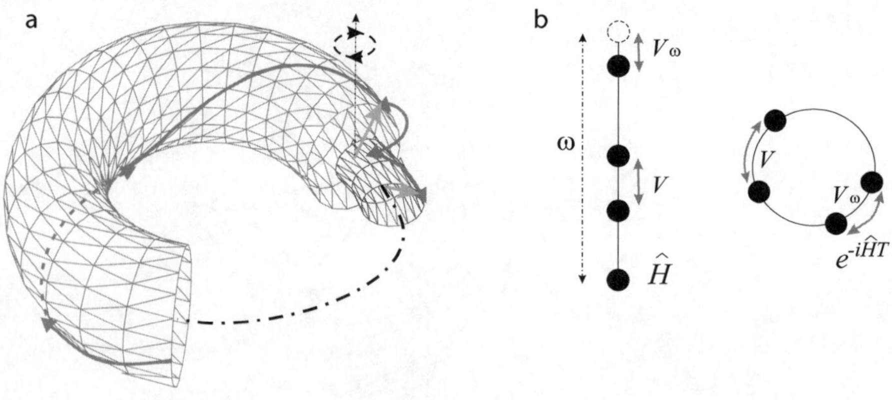

Fig. 12.5 (**a**) Evolution of a couple of two-level systems under the action of two Hamiltonians that belong to a three-element subalgebra. The composition $\overline{b}$ of the Bloch vectors may now go from one torus to another with the same axis (the big radius, and the equatorial plane are identical). Holonom constraints now have only three parameters, while the constrained motion occupies three-dimensional sets, that is, inner parts of the tori of smaller radius 1 and different positions of the equatorial planes. (**b**) The spectrum of the Hamiltonian $\hat{H}_0$ and the evolution operator over a period $\hat{U} = \exp(-i\hat{H}T)$. Static perturbation V repels the levels close in energy. A harmonic of a time-dependent perturbation V_ω repels the levels separated by the energy of the corresponding quantum and leads to a more uniform distribution of the eigenstates of $\hat{U}$ along the unit circle

chosen more or less randomly among the other combinations of a rather general structure. Their commutator gives a new, linearly independent combination which does not commute with the Hamiltonians yielding two more linearly independent combinations, and so on, until after eight cycles of commutation the number of all possible linearly independent combinations reaches the maximum possible number 15. This is an example of non-holonom system, which does not have constraints and can be controlled completely. The trajectory of the sum of the Bloch vectors can occupy the entire space restricted by the normalization requirement limiting the maximum length, and moreover each of two Bloch vectors comprising the composition occupies a three-dimensional set, and thus the state vector moves in a six-dimensional space.

Control of Non-Holonom Systems

Let us try to construct a bang–bang control for the two interacting two-level systems with the Hamiltonians (12.23) assuming that the interaction $\hat{H}_0 = \hat{\sigma}_z + \hat{\Sigma}_x \otimes \hat{\sigma}_x$ is always present, whereas two different perturbations $\hat{V}_A = \hat{\Sigma}_x + \hat{\Sigma}_z \otimes \hat{\sigma}_y$ and $\hat{V}_B = \hat{\sigma}_x + \hat{\Sigma}_y + \hat{\Sigma}_y \otimes \hat{\sigma}_x$ are switched on and off in sequence with the time-independent amplitudes $f_{A,B}(t) = 1$ during the exposition time intervals $T_1, T_2, \ldots, T_{16}$. Here, the even subscripts correspond to the perturbation $\hat{V}_B$ with $f_A(t) = 0,\ f_B(t) = 1$ and the odd to $\hat{V}_A$ with $f_B(t) = 0,\ f_A(t) = 1$. Evidently $\int f_A(t) f_B(t) dt = 0$

and the total number for the switching required for complete control in the case of four-dimensional Hilbert space amounts either to 16, or to 15 if we ignore the global phase. By analogy to a single two-level system with the evolution operator (12.13) we write

$$\hat{U}(\{T_i\}) = e^{-iT_{16}\hat{B}}e^{-iT_{15}\hat{A}}e^{-iT_{14}\hat{B}}e^{-iT_{13}\hat{A}}e^{-iT_{12}\hat{B}}e^{-iT_{11}\hat{A}}e^{-iT_{10}\hat{B}}$$

$$e^{-iT_9\hat{A}}e^{-iT_8\hat{B}}e^{-iT_7\hat{A}}e^{-iT_6\hat{B}}e^{-iT_5\hat{A}}e^{-iT_4\hat{B}}e^{-iT_3\hat{A}}e^{-iT_2\hat{B}}e^{-iT_1\hat{A}} \quad (12.24)$$

for the generic evolution operator of this type. Our aim is to find the timings T_i that would result in a desired evolution of the system $\hat{U}_d$.

The identity operator $\hat{U}_d = \hat{I}$ that corresponds to a non-trivial timing $\sum_i T_i = T > 0$ is the simplest example of a desired evolution. Such an evolution allows one to return the system to its initial state without even knowing this state explicitly. The straightforward way to find this evolution by minimizing the function

$$\mathrm{Tr}[(\hat{U}(T_1,\ldots,T_{16}) - \hat{I})(\hat{U}(T_1,\ldots,T_{16}) - \hat{I})^*] \quad (12.25)$$

in the 16-dimensional space of the parameters T_i in complete analogy with (12.15) requires an already appreciable amount of numerical work. Moreover, this procedure, started from an arbitrary point in the parameter space $\{T_i\}$, usually converges to a local minimum and yields a matrix $\hat{U}$ which is far from the desired one. In order to be successful, one should start already in a very close vicinity of the solution—for each dimension the value of T_i should be within at least a half of the shortest period corresponding to the evolution operators $e^{-iT\hat{B}}$ or $e^{-iT\hat{A}}$. For the N-dimensional Hilbert space this vicinity has an exponentially small $\sim 2^{-N^2}$ volume relative to the total volume $\sim (2\pi)^{N^2}$ of the parameter space. Therefore the functional (12.25) is not a very useful tool, unless we have some other means to gain an insight into the dynamics of the system by narrowing the domain of the solution search.

An alternative algorithm relies on the natural property of complex quantum systems to have a "rigid" spectrum resulting from the interaction-induced repulsion of the energy eigenstates. Let us discuss this in more detail. Consider a time-dependent Hamiltonian $\hat{H}(t) = \hat{H}_0 + \hat{V}(t)$ that changes periodically in time with a period T. As we have already seen in the context of (5.12), the time-dependent operator can be replaced by a time-independent one $\hat{H}(\tau) - i\frac{\partial}{\partial\tau}$ containing one additional direction τ in geometric space. Let us perform the Fourier transformation over the variable τ and make use of the quasienergy representation (6.147) by complete analogy to the quasimomentum representation for a quantum system, periodic in space. The Hamiltonian adopts the form $\sum_k \hat{H}(k\omega)\delta_{n+k,n'} + n\omega\delta_{n,n'}$ where n and n' are the quasienergy indices and $\omega = 2\pi/T$. For the operator of the evolution over a period we therefore obtain $\hat{U}(T) = \exp[-iT\sum_k \hat{H}(k\omega)\delta_{n+k,n'}]$ whereas the term $-iTn\omega\delta_{n,n'}$ in the Hamiltonian just gives a multiple of 2π phase for each n and can be omitted. This implies that the energy eigenstates of the effective Hamiltonian $iT^{-1}\ln\hat{U}(T)$ separated by the energy ω of a quantum, or by a multiple of this value, have to be considered as neighboring states from the point of view of the contributions they give to the operator of evolution over a period. In other words, the eigenstates of $\hat{H}_0$ located at an energy distance $n\omega$ are coupled by the n-th harmonics of the time-dependent part $\hat{V}(t)$ and repel each other in the same way as the stationary coupled states that lie close together in energy. Mutual repulsion of the levels causes the eigenvalues of $\hat{U}(T)$ to be distributed over the unit circle in the complex plane with the maximum possible distance among the neighboring eigenvalues. We illustrate this in Fig. 12.5b.

In the context of two interacting, two-level systems we now consider the truncated evolution operator

$$\hat{U}_0(\{T_i\}) = e^{-iT_4\hat{B}}e^{-iT_3\hat{A}}e^{-iT_2\hat{B}}e^{-iT_1\hat{A}} \tag{12.26}$$

which depends on $N = 4$ parameters. All four eigenvalues of this 4×4 matrix are on the unit circle in the complex plane. Let us try to choose the four exposure times $\{T_i\}$ such that the eigenvalues of $\hat{U}_0$ become equally spaced on the unit circle, thus aiding this natural tendency, typical of strong and complex perturbations. For this purpose we calculate the coefficients $c_n(\{T_i\})$ of the characteristic polynomial

$$\begin{aligned}\det\left|\hat{U}_0(\{T_i\}) - \lambda\hat{I}\right| &= \lambda^4 + \lambda^3 c_3(\{T_i\}) + \lambda^2 c_2(\{T_i\}) \\ &\quad + \lambda c_1(\{T_i\}) + c_0(\{T_i\})\end{aligned} \tag{12.27}$$

and minimize the sum

$$F(\{T_i\}) = \sum_{n=1}^{n=N-1} |c_n(\{T_i\})|^2 \tag{12.28}$$

of its coefficients. When the function (12.28) assumes a zero value, the roots of the resulting characteristic polynomial $\lambda^N + c_0 = 0$ are equidistantly distributed on the unit circle. Due to the fact that a homogeneous distribution of the eigenvalues of $\hat{U}_0$ is a natural property of complex systems, the function (12.28) reaches the global minimum much more often, as compared to the function (12.25).

Being repeated $N = 4$ times, the truncated evolution yields the identity transformation $\hat{U}_0^4 = \hat{I}$. Starting from this non-trivial identity $\hat{U}(\{T_i\})$ with $\{T_i\} = T_{1,\dots,}T_4, T_{1,\dots,}T_4, T_{1,\dots,}T_4, T_{1,\dots,}T_4$ one finds the desired transformation $\hat{U}_d$ by solving iteratively (with the help of (12.19), (12.20)) the problem $\hat{U}(\{T_i\}) = \exp(\mu\ln\hat{U}_d)$ for μ gradually increasing from 0 to 1.

12.1.3 Control on Lie Semigroups

We now discuss some mathematical aspects of quantum control. As we have already seen at the previous example, one can control a two-level system by sequential applications of two non-commuting operators. How general is such a situation? The answer of this question is the following. Given two operators satisfying the bracket generation condition presented on the p. 12.1.2, and given a possibility to completely control the coefficients in front of these operators entering the Hamiltonian, the system can be completely controlled, which means that it can be forced to follow a predetermined trajectory in the Hilbert space with any prescribed accuracy. In the case where the control over the coefficients is not complete, in particular, when one cannot change the signs of the coefficients, the situation is more complicated. Still one can perform any desired transformation, provided the algebra corresponding to the unitary transformation of the system is compact. However, this transformation cannot follow a predetermined trajectory, and just can coincide with such a trajectory stroboscopically, that is, at a sequence of equally spaced moments of time.

In fact, consider the result of application of four operators,

$$e^{i\Delta tb\hat{B}}e^{i\Delta ta\hat{A}}e^{-i\Delta tb\hat{B}}e^{-i\Delta ta\hat{A}} = e^{(\Delta t)^2ab\left[\hat{A},\hat{B}\right]+O\left(\Delta t^3\right)}.$$

According to the Baker–Campbell–Hausdorff formula, for $\Delta t \to 0$ and $(\Delta t)^2\,ab \to c$ it yields the third operator $e^{c\left[\hat{A},\hat{B}\right]}$ with any prescribed accuracy and any predetermined parameter c. In this way one has in hands not just two, but three operators $\hat{A}$, $\hat{B}$, and $\hat{C} = \left[\hat{A},\hat{B}\right]$. The same operation can be continued, increasing the number of the operators in hands, such that one will have access to all operators comprising the algebra, and thus one is able to apply any desired infinitesimal transformation at each moment of time. The key requirement here is that one can implement an operation, say $e^{-i\Delta ta\hat{A}}$, and its inverse $e^{i\Delta ta\hat{A}}$ at any moment of time, when needed. In other words, one can perform locally in time all infinitesimal transformations of the Lie group.

The latter in not always the case. For instance, in the presence of a nonzero unperturbed Hamiltonian $\hat{H}_0$ and small controlled perturbations of the form $f(t)\hat{V}_A + g(t)\hat{V}_B$, one cannot implement an infinitesimal operation inverse to the natural evolution $e^{-i\Delta t\hat{H}_0}$, since the forces f and g might not be strong enough to detract the system far from its natural evolution. The only way to get the inverse operation is to control the system trajectory in such a way that at the end of some rather long time interval it returns back to the initial point, that is, the evolution operator turns out to be close to the unity matrix, still having a small mismatch corresponding to the desired inverse infinitesimal operation. For the two-level system this can be easily done by simply waiting for the moment when the natural evolution $\hat{H}_0$ completes the Rabi circle, bringing the system back to the starting point. The inverse of the natural evolution $e^{-i\Delta t\hat{H}_0}$ results from the direct evolution during the time given by the Rabi period minus Δt. Detraction from the natural evolution resulting from the application of the small forces ensure other infinitesimal transformations, at times close to the Rabi period.

Now it becomes evident why the requirement of compactness of the group is crucial—for a noncompact group a generic trajectory never returns back close enough to the initial point, as is the case for compact groups, in virtue of the Poincaré theorem. The latter states that with any predetermined accuracy a system will pass close to the initial state in the course of free evolution. The problem is that this process may take an infinitely long time, and the control is meant to ensure the return as fast as is possible. We therefore encounter a situation where for any transformation there exists an inverse one, but this does not hold locally in time. In other words, our control forms a group, but not a Lie group, remaining just a Lie semigroup.

In contrast to the case of a two-level system, for a general N-level system, finding a control bringing it close to the initial state, whatever this state was, is a nontrivial task. Still it can be accomplished with the help of the procedure of (12.27) and (12.28) generalized to the case of N-dimensional Hilbert space. Given two operators $\hat{A}$ and $\hat{B}$ satisfying the bracket generation condition, one performs a sequence of N unitary transformations

$$\hat{U}_0(\{T_i\}) = e^{-iT_N\hat{B}} \ldots e^{-iT_2\hat{B}} e^{-iT_1\hat{A}} \tag{12.29}$$

with arbitrary chosen parameters $\{T_i\}$ and calculates the characteristic polynomial of this transformation,

$$\det\left|\hat{U}_0(\{T_i\}) - \lambda\hat{I}\right| = \lambda^N + \lambda^{N-1}c_{N-1}(\{T_i\}) + \ldots + c_0(\{T_i\}).$$

Then one constructs the sum of the coefficients (12.28) and searches the coordinates $\{t_i\}$ of the global minimum of this sum, which equals 0. The corresponding characteristic polynomial $\lambda^N + c_0(\{t_i\})$, with $|c_0(\{t_i\})| = 1$ in virtue of the unitarity of transformations, has roots uniformly distributed on the unit circle, and hence the operator $\hat{U}_0(\{t_i\})$ is the N-th root of the identity operator, up to an unimportant phase factor.

The main surprise is that the global minima of $F(\{T_i\})$ can be found with reasonably high probability by the regular steepest descent method, which is usually employed for finding the local minima. In other words, the function $F(\{T_i\})$ is such that a big fraction of its minima are the global ones. This situation becomes clearer when one remembers that complex systems without hidden integrals of motion have rigid spectra of the unitary evolution matrices satisfying Dyson statistics (1.3) with $\beta = 2$. In other words, by simply taking random entries for the timings $\{T_i\}$ we generate a random matrix $\hat{U}_0(\{T_i\})$ with a spectrum, which is already rather close to the equidistant one, and hence the steepest descent method starts at the point rather close to the global minimum. Application of a sequence of N transformations $\hat{U}_0(\{T_i\})$ results in the identity transformation, and small variation

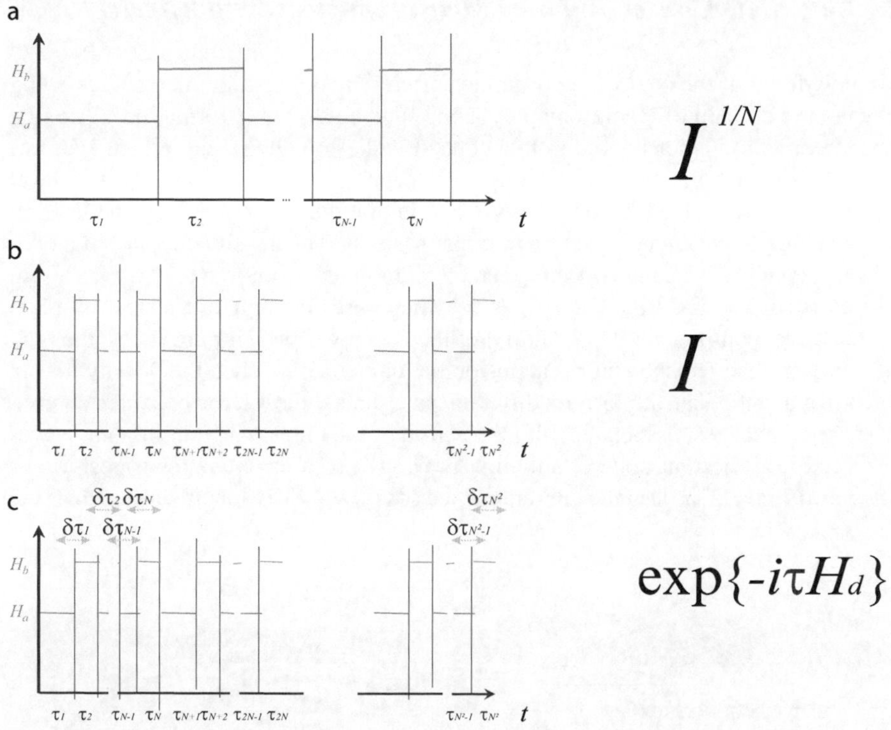

Fig. 12.6 (**a**) Control sequence with $K = N$ switching, which yields a non-degenerate N-th root of the identity operator. (**b**) Repetition of this sequence N times yields a control sequence of length $K = N^2$, corresponding to the non-infinitesimal identity operator, which allows one to construct a group out of a compact Lie semigroup. (**c**) Small variations $\delta\tau_i$ of the control timings allows one to construct any desired operation in a finite vicinity of the identity operator

of the parameters T_i in this sequence of N^2 transformations with $i = 1, \ldots, N^2$ generates a set of N^2 infinitesimal operations that can be employed for construction of the relevant Lie group. Moreover, for any H_d, the operation $\exp[i\alpha H_d]$ with a small but finite parameter α can be found by rapidly converging series of sequential linear approximations. The situation is illustrated in Fig. 12.6. Note that, in practice, parameters $\{T_i\}$ for any desired unitary transformation $\exp[iH_d]$ can be found with the help of the steepest descent method by minimizing the mismatch between the unitary transformation

$$\hat{U}(\{T_i\}) = e^{-iT_{N^2}\hat{B}} \ldots e^{-iT_2\hat{B}} e^{-iT_1\hat{A}}$$

and the scaled transformation $\exp[\alpha i H_d]$, starting from the identity transformation at $\alpha = 0$ and gradually increasing the scaling parameter α till it takes the desired value 1. After each step of the augmentation of α, one has to repeat the steepest descent calculations.

12.1.4 State Conservation by Measurements: Zeno Effect

We now focus at the control opportunities offered by the measurements, thus going beyond the domain of Hamiltonian dynamics and unitary transformations. The first phenomenon to be considered in this context is the so-called Zeno effect. We start with a geometric illustration of the regular Zeno effect for the simplest two-level system, show how this method allows one to inhibit the coherence loss, and later on we concentrate on a generalization to larger systems. The density operator of a two-level system is a 2×2 matrix which can be represented according to (12.4) as a sum of four parts $\hat{\rho} = \hat{I} + P\hat{\sigma}_x + Q\hat{\sigma}_y + W\hat{\sigma}_z$ where three Pauli matrices are multiplied by three real components P, Q, and W of the Bloch vector. For pure states, the rank of $\hat{\rho}$ is 1 and the Bloch vector remains on the unit sphere. When the density matrix becomes a statistical mixture of different pure states, the vector becomes shorter. The coherence loss, associated with the action of the unknown Hamiltonian, leaves the vector on the unit sphere, although moving it to a new, unknown position as shown in Fig. 12.7a. Usually the coherence loss time T_2 is much shorter than the

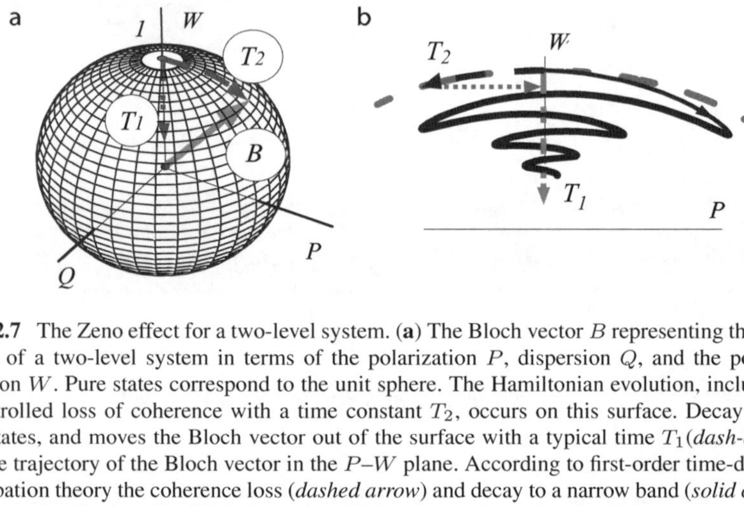

Fig. 12.7 The Zeno effect for a two-level system. (**a**) The Bloch vector B representing the density matrix of a two-level system in terms of the polarization P, dispersion Q, and the population inversion W. Pure states correspond to the unit sphere. The Hamiltonian evolution, including the uncontrolled loss of coherence with a time constant T_2, occurs on this surface. Decay destroys pure states, and moves the Bloch vector out of the surface with a typical time T_1 (*dash-dot line*). (**b**) The trajectory of the Bloch vector in the P–W plane. According to first-order time-dependent perturbation theory the coherence loss (*dashed arrow*) and decay to a narrow band (*solid curve*) do not move the Bloch vector away from the unit sphere, and can therefore be corrected to second order by frequent measurement-induced projections (*dotted arrow*) onto the initial upper state $W = 1$. The decay (*dash-dot line*) moves B out of the sphere with a typical time T_1

longitudinal relaxation time T_1 which is associated with spontaneous decay and entanglement, with an escaping photon thus moving the Bloch vector out of the unit sphere. In the case of spontaneous decay to an infinite continuum, this process results in a decrease, linear in time, of the length of the Bloch vector. For a band-like continuum, a short period of quadratic decrease of the length may exist prior to this linear regime. This persists for a time equal to the inverse width of the band. These three possibilities are shown in Fig. 12.7b.

Measurement of a physical quantity results in the projection of the Bloch vector to a vector representing the measured eigenstate of the corresponding operator. In Fig. 12.7 we illustrate this for the measurement of the population of the upper level, corresponding to the positive eigenvalue of the operator $\hat{\sigma}_z$. Being subjected to the action of a Hamiltonian of type $\hat{\sigma}_y$, for unknown time T, the Bloch vector turns from its initial upper position in the direction P to an unknown angle $\sim T/T_2$, which is small, provided that $T \ll T_2$. The linear increment of the vector, resulting from the evolution, is therefore orthogonal to the vector itself. Being projected after the measurement to the W-axis, it returns to the initial position almost entirely. In other words, the regular Zeno effect allows one to restore the initial state vector up to the second-order terms in T/T_2. Though for a multilevel system, the Bloch representation is less convenient, the essence of the Zeno effect relies on the same geometric idea. We consider the real and the imaginary parts of the probability amplitudes as independent components of a real state vector v, which rotates in the course of an uncontrolled unitary evolution at a small angle, and returns after the measurement to the initial position, up to the second-order corrections.

12.1.5 *Quantum State Engineering by Indirect Measurements*

The Zeno effect considered above is based on direct projective measurements, performed right on the system under consideration. There exists another, indirect, type of successive measurements that are performed not on the system itself but on auxiliary systems after they have interacted with the system of interest. Prior to the interaction, the auxiliary systems are supposed to be in known initial states, either all in the same state, or each one in an individually prepared state, depending on the aim and the chosen strategy of such a manipulation. As a result of the projective measurement on the auxiliary systems, the quantum state of the considered system changes, but the way it changes depends essentially on the outcomes of the measurements. One can design a series of indirect measurements such that in the case where they yield a predetermined sequence of the outcomes, the system of interest undergoes a desired transformation. However, if even for a single measurement in this series the outcome turns out to be different from the one prescribed, the system does not experience the desired transformation. One can say that implementation of the series of indirect measurements exerts a control over the system, however, this control is probabilistic, since it gives the desired system transformation only if the measurement results are exactly as required, while the probability of such an outcome might be rather small.

Let us look at such a process in more detail. A quantum system, composed of a part A under consideration and of an auxiliary part B, being initially in a product state $|\Psi\rangle_A |\Phi\rangle_B$, experience a unitary transformation $\hat{U}$ and thereby passes to the state $\hat{U}\,|\Psi\rangle_A |\Phi\rangle_B$ as a result of the interaction between the parts. If the measurement of system B yields the required result characterized by a state vector

$|\psi\rangle_B$, the part A undergoes transformation $\hat{R}\,|\Psi\rangle_A$ where a nonunitary matrix $\hat{R} = {}_B\langle\psi|\,\hat{U}\,|\Phi\rangle_B$ is given by the projection of the unitary matrix $\hat{U}$ to a subspace determined by the initial and the final states of the part B. Since the one-to-one correspondence between the initial and the final states typical of the unitary transformation $\hat{U}$ is preserved even after the projection, the matrix $\hat{R}$ must satisfy only one requirement—to be invertible. Matrices of this type can be normalized such that their determinant equals unity, and then they form a group $SL(N)$, where N stands for the dimension of the Hilbert space of the system A. The Lie algebra corresponding to this group has the same set of generators as the unitary group $SU(N)$, but the parameters multiplying these generators are no longer purely imaginary, being generic complex numbers.

Consider now an example of an infinitesimal unitary transformation $\hat{U} = 1 - iH\Delta t$ induced by application of a Hamiltonian $\hat{H} = \hat{V}_A \otimes \hat{V}_B$ to the composed system, which is followed by a projective measurement on the part B for the case of identical initial and final states, $|\Phi\rangle_B = |\psi\rangle_B$. In other words, we perform a Zeno-like type of the measurement on the part B, which always yields the same result with a probability up to the second-order correction $O(\Delta t^2)$ equal to unity. The part A, in turn, experiences the infinitesimal transformation $\hat{R} = 1 - i\hat{V}_A\Delta t$, which is unitary for this particular situation. If another, second, type of Hamiltonian $\hat{H} = \hat{v}_A \otimes \hat{v}_B$ is also available, then by implementing the strategy of infinitesimal transformations and Zeno measurements we reproduce the situation typical of the control on Lie semigroups discussed in Sect.12.1.3. This example shows that the control by indirect measurements can to some extent be equivalent to the unitary control, although it relies on application of the transformations comprising a noncompact group $SL(N)$ group, which still contains the compact $SU(N)$ as a subgroup. An important discrepancy between $SL(N)$ and $SU(N)$ appears while attempting to perform a predetermined transformation of a state in a finite number of steps: typically, the probability of the desired result in the case of $SL(N)$ is much smaller than unity.

Let us illustrate the general situation for the simplest example of a pair of two-level systems, and choose as the unitary transformation applied to both parts

$$\hat{U} = e^{-i\frac{\alpha}{2}\left(\hat{\sigma}_1^y\hat{\sigma}_2^y + \hat{\sigma}_1^x\hat{\sigma}_2^x\right)} = \begin{pmatrix} 1 & 0 & 0 & 0 \\ 0 & \cos\alpha & i\sin\alpha & 0 \\ 0 & i\sin\alpha & \cos\alpha & 0 \\ 0 & 0 & 0 & 1 \end{pmatrix}. \tag{12.30}$$

We assume that the two-level system B initially in the state $|i\rangle = \Phi_0\,|0\rangle + \Phi_1\,|1\rangle$ is detected after the interaction in the final state $|f\rangle = \Psi_0\,|0\rangle + \Psi_1\,|1\rangle$, such that the part A experiences the transformation

$$\hat{R} = \Phi_0\Psi_0^* \begin{pmatrix} \cos\alpha & 0 \\ 0 & 1 \end{pmatrix} + \Phi_0\Psi_1^* \begin{pmatrix} 0 & 0 \\ i\sin\alpha & 0 \end{pmatrix} \tag{12.31}$$

$$+ \Phi_1 \Psi_0^* \begin{pmatrix} 0 & i \sin\alpha \\ 0 & 0 \end{pmatrix} + \Phi_1 \Psi_1^* \begin{pmatrix} 1 & 0 \\ 0 & \cos\alpha \end{pmatrix}.$$

We now address the question: under what conditions is the transformation

$$\hat{R} = \begin{pmatrix} \Phi_0 \Psi_0^* \cos\alpha + \Phi_1 \Psi_1^* & i\Phi_1 \Psi_0^* \sin\alpha \\ i\Phi_0 \Psi_1^* \sin\alpha & \Phi_0 \Psi_0^* + \Phi_1 \Psi_1^* \cos\alpha \end{pmatrix} \quad (12.32)$$

proportional to a unitary matrix

$$\hat{u} = \begin{pmatrix} \cos\beta & ie^{i\gamma} \sin\beta \\ ie^{-i\gamma} \sin\beta & \cos\beta \end{pmatrix}? \quad (12.33)$$

Applying consecutively two such unitary transformations $\hat{u}$ with different parameters β and γ, one can construct any predetermined unitary transformation of the part A. From the diagonal part of the matrix relation $\hat{R} = \hat{u}$, after lifting up for a moment the normalization requirement posed on the amplitudes Ψ_i and Φ_i, one obtains

$$\begin{aligned} \Phi_0 \Psi_0^* \cos\alpha + \Phi_1 \Psi_1^* &= \cos\beta \\ \Phi_0 \Psi_0^* + \Phi_1 \Psi_1^* \cos\alpha &= \cos\beta, \end{aligned} \quad (12.34)$$

which yields $\Phi_0 = \frac{\cos\beta}{\Psi_0^*(\cos\alpha+1)}$ and $\Phi_1 = \frac{\cos\beta}{\Psi_1^*(\cos\alpha+1)}$, and after substitution into the off-diagonal part of the matrix relation $\hat{R} = \hat{u}$ results in the equations

$$\begin{aligned} \Psi_0^* \frac{\sin\alpha}{\cos\alpha + 1} - \Psi_1^* e^{i\gamma} \tan\beta &= 0 \\ \Psi_1^* \frac{\sin\alpha}{\cos\alpha + 1} - \Psi_0^* e^{-i\gamma} \tan\beta &= 0. \end{aligned} \quad (12.35)$$

Provided the corresponding determinant $\tan^2 \frac{\alpha}{2} - \tan^2 \beta$ vanishes, these equations have a normalizable non-trivial solution $\Psi_0^* = \mathcal{A} e^{i\gamma}$, $\Psi_1^* = \mathcal{A}$ depending on a complex normalization constant $\mathcal{A}$, which immediately results in $\Phi_0 = e^{-i\gamma} \mathcal{A}'$ and $\Phi_1 = \mathcal{A}'$ with another normalization constant $\mathcal{A}'$. The proportionality coefficient between the matrices $\hat{R}$ and $\hat{u}$ thus amounts to $2\mathcal{A}\mathcal{A}' \cos\frac{\alpha}{2}$. The absolute values of the normalizations are $|\mathcal{A}| = |\mathcal{A}'| = 1/\sqrt{2}$, and therefore the probability of the required outcome of the measurement is $\left|2\mathcal{A}\mathcal{A}' \cos\frac{\alpha}{2}\right|^2 = \cos^2 \beta$. Thus the problem is resolved for the simplest case.

In order to mimic a unitary transformation in larger systems with N-dimensional Hilbert space, a set of many sequential transformations $\left\{\hat{U}_i\right\}$ is required, as is clear from a comparison of the number $2N$ of the complex parameters required for a single transformation with the number of requirements N^2 imposed by the matrix equation $\hat{R} = \hat{u}$. As a general rule, one can state that for every desired final transformation $\hat{u}$, specific parameters for each of the transformations $\hat{U}_i$ have to be identified, such that there exists a non-trivial solution of the problem. Search for these parameters and for the corresponding choices of states of the part B prior to and after the measurement is a matter of numerical work.

The situation gets much simpler when one tries to control not the evolution, but only the known state of the part A. For example, as the part A one can take a harmonic oscillator and as the part B a two-level system. Let the oscillator be initially in the vacuum state $|0\rangle$, and our aim is to set it to a target state, which is given as a sum of a finite number of energy eigenstates,

$$|\psi\rangle = \sum_{n=0}^{n_{\max}} \psi_n |n\rangle \,, \tag{12.36}$$

by making the parts consecutively interacting via the Jaynes–Cummings Hamiltonian

$$V = g\left(\hat{\sigma}^+ \hat{a} + \hat{\sigma}^- \hat{a}\dagger\right), \tag{12.37}$$

while the system B is set to $|\Phi_k\rangle = \Phi_k^0 |0\rangle + \Phi_k^1 |1\rangle$ prior to the k-th interaction and is registered in the state $|\Psi_k\rangle = \Psi_k^0 |0\rangle + \Psi_k^1 |1\rangle$ right after this interaction. As a result of the k-th, which has lasted during a time interval τ_k, the entire system experiences the evolution $e^{-i\tau_k g\left(\hat{\sigma}^+ \hat{a} + \hat{\sigma}^- \hat{a}\dagger\right)}$, which after projection yields the matrix $\hat{R} = \langle \Psi_k | \, e^{-i\tau_k g\left(\hat{\sigma}^+ \hat{a} + \hat{\sigma}^- \hat{a}\dagger\right)} \, |\Phi_k\rangle$ acting in the Hilbert space of the oscillator. The matrix R_{nm} is tridiagonal, and it can be explicitly written as

$$\begin{aligned} R_m^n = {} & \left[\Psi_k^{*0}\Phi_k^0 \cos\left(\tau_k g\sqrt{n}\right) + \Psi_k^{*1}\Phi_k^1 \cos\left(\tau_k g\sqrt{n+1}\right)\right] \delta_m^n \\ & - i\Psi_k^{*1}\Phi_k^0 \sin\left(\tau_k g\sqrt{n}\right) \delta_{m+1}^n - i\Psi_k^{*0}\Phi_k^1 \sin\left(\tau_k g\sqrt{n+1}\right) \delta_{m-1}^n . \end{aligned} \tag{12.38}$$

By setting $\Psi_k^1 = 0$, which implies that the two-level system is always registered in the ground state, one makes the matrix R_m^n only two diagonal, containing the principal diagonal $a_n \delta_m^n$ with $a_n = \Phi_k^0 \cos\left(\tau_k g\sqrt{n}\right)$ and the next upper diagonal

$$R_m^n = a_n \delta_m^n + b_n \delta_{m-1}^n, \tag{12.39}$$

with $b_n = -i\Phi_k^1 \sin\left(\tau_k g\sqrt{n+1}\right)$, while a product of $n_{\max}$ of such matrices with different parameters Φ contains $n_{\max}$ upper diagonals next to the principal one, since as a result of each multiplication, the number of upper diagonals increases by unity. Application of such a product to the vacuum state yields a linear combination of $n_{\max} + 1$ energy eigenstates, ranging from $|0\rangle$ till the maximum excited state $|n_{\max}\rangle$.

Determination of the parameters that lead to the desired linear combination (12.36) can be done recursively—one finds a state $|\phi\rangle = \sum_{n=0}^{n_{\max}-1} \phi_n |n\rangle$ that after multiplication by the matrix (12.39) gives the desired state $|\psi\rangle$ of (12.36) simply with the help of the inverse matrix,

$$|\phi\rangle = \hat{R}^{-1}|\psi\rangle \tag{12.40}$$

and the condition that $|\phi\rangle$ does not contain contribution of the state $|n_{\max}\rangle$. The matrix inverse to $\hat{R}$ of (12.39) reads

$$\left(\hat{R}^{-1}\right)_m^n = \sum_{l=0}^{n} (-1)^l \frac{\prod_{k=1}^{l} b_{n+1-k}}{\prod_{r=0}^{l} a_{n-k}} \delta_{m+l}^n$$

where $\prod_{k=1}^{0} \dots \equiv 1$, and therefore the ratio $\mathcal{Q} = i\Phi_{n_{\max}}^1/\Phi_{n_{\max}}^0$ can be numerically found from the explicit condition $\langle n_{\max}|\phi\rangle = 0$, which reads

$$\sum_{l=0}^{n_{\max}} \mathcal{Q}^l \frac{\prod_{r=0}^{l-1} \sin\left(\tau_k g\sqrt{n-l+1}\right)}{\prod_{r=0}^{l} \cos\left(\tau_k g\sqrt{n-r}\right)} \psi_{n-l} = 0.$$

The normalization condition determines $\Phi_{n_{\max}}^1$ and $\Phi_{n_{\max}}^0$ up to an irrelevant common phase factor, and hence allows one to obtain explicitly all the matrix elements of $\hat{R}^{-1}$ and by means of (12.40) the new target state $|\psi\rangle = |\phi\rangle$. In the next step one determines $\Phi_{n_{\max}-1}^1$ and $\Phi_{n_{\max}-1}^0$ and continues the procedure till all the amplitudes Φ_j^i are found.

We summarize the results of this section, by noting that the control of the quantum evolution of a system can be exerted in many different manners and often it yields the desired result, although the way the system approaches this result is usually far from intuitively evident—the system trajectory in the Hilbert space is not smooth and straightforward, and on its way it typically visits domains located far both from the final and from the initial positions. Engineering of the interventions required for pushing the system towards the goal usually relies on numerical algorithms, which pick out one of many possible solutions of the problem.

12.2 Coherence Control of Open Quantum Systems

Decay and loss of coherence are two processes common to all open quantum systems interacting with the environment.

Considering irreversible processes such as decay or measurement, one can always employ certain physical models, describing these phenomena as a coherent process in a large compound system, which, apart from the system of interest, also includes a much larger part with a continuous spectrum; they were called "the environment" and "the meter", respectively, in the previous sections. The most relevant model for the measurement and for the decay is the level–continuum problem considered on p. 78. Here we will not dwell on this aspect of the process, simply assuming that after a given time considered as short relative to other typical times, the system of interest simply moves to a certain state (either known of unknown) with an exponentially vanishing probability of the opposite result.

On the contrary, loss of coherence does not require a quantum model of the environment. It can be viewed as a systematic and uncontrolled classical external intervention in the system, resulting in the application of a known Hamiltonian during unknown time intervals. Without performing a measurement, one can never know whether such an intervention has occurred. Coherence loss protection is therefore a special kind of measurement which produces the action of canceling all of the results of coherence loss without affecting the quantum state of the system.

12.2.1 Dynamical Decoupling from the Environment

In Sect. 11.1 we have considered the relaxation processes that occur in a system as a result of interaction with the environment. Now we address the question: how could one minimize the rates of these processes and eventually completely exclude the influence of the environment on the system? There are several tools one can employ for the purpose, and in this section we consider one of them, based on the same strategy as quantum control, namely on the special choice of an external intervention to the system via a time-dependent controlled interaction Hamiltonian.

Stabilizing Interventions

Consider the interaction Hamiltonian (11.5) and assume that the system experiences a periodic series of unitary transformations $\hat{U}_j$, each of which is a result of a short application of strong controlled Hamiltonians at the time moment t_j. Let the period contain J interventions each of which is so short that at the scale of typical frequencies of the system this transformation can be considered as instantaneous. The other way to phrase this is to say that the system is subject to actions of strong external Hamiltonians $\hat{H}_j$ that, however, last during very short time intervals Δt_j, such that the action $\hat{H}_j \Delta t_j$ takes some finite, not necessarily small, values. One can employ the interaction representation with respect of these Hamiltonians and write the remaining time-dependent Hamiltonian (11.5) in the form

$$\hat{\mathcal{V}} = \sum_{\alpha,j} \Theta(t_{j+1} - t)\Theta(t - t_j)\hat{\mathbf{U}}_j \hat{V}_\alpha \hat{\mathbf{U}}_j^{-1} \otimes \hat{v}_\alpha, \tag{12.41}$$

where

$$\hat{\mathbf{U}}_j^{-1} = \prod_{k \le j} \hat{U}_j = \prod_{k \le j} e^{-i\hat{H}_j \Delta t_j}.$$

If during the intervention time t_J the action accumulated due to the coupling $\hat{\mathcal{V}}$ is small, and if all the intervals $t_{j+1} - t_j$ between the sequential interventions are equal, one can average the interaction over the period and write

$$\hat{\mathcal{V}} = \sum_{\alpha,j} \overline{\hat{V}_\alpha} \otimes \hat{v}_\alpha, \tag{12.42}$$

where

$$\overline{\hat{V}_\alpha} = \sum_{j=1}^{J} \hat{\mathbf{U}}_j \hat{V}_\alpha \hat{\mathbf{U}}_j^{-1}. \tag{12.43}$$

Note that the same averaging has to be applied to the unperturbed Hamiltonian of the system

$$\overline{\hat{H}} = \sum_{j=1}^{J} \hat{\mathbf{U}}_j \hat{H} \hat{\mathbf{U}}_j^{-1}. \tag{12.44}$$

The unitary transformations $\hat{U}_j$ (and hence $\hat{\mathbf{U}}_j$) may form a subgroup of the group of all possible unitary operations in the Hilbert space of the system. In this case, the operators $\overline{\hat{V}_\alpha}$ are the results of averaging of the operators $\hat{V}_\alpha$ over this subgroup. Otherwise, the averaging is performed over the entire unitary group and usually yields zeros for both the coupling (12.43) and the Hamiltonian (12.44). The aim is to find a subgroup of transformations, which would decouple the system from the environment by averaging out the coupling Hamiltonian but would still leave intact a subspace of the Hilbert space where non-trivial evolution may occur. The latter is often called the decoherence-free subspace. The most economical way is to find a discrete subgroup of standard operations containing the minimum number of elements, which could suggest the shortest possible sequence of the transformations required for the decoupling. The elements of the subgroup may have common eigenvectors x_k corresponding to the same eigenvalues,

$$\hat{\mathbf{U}}_j x_k = e^{i\phi_j} x_k.$$

In this case, the space spanned by such vectors is the decoherence-free subspace of the Hilbert space, while the subgroup composed of the operations $\hat{\mathbf{U}}_j$ is called the "stabilizer" of this subspace, according to group-theory terminology.

Let us illustrate this method of coherence protection by the simple example shown in Fig. 12.8a. A four-level system $A + B$ composed of two interacting two-level systems A and B is characterized by the Hamiltonian

$$H = \omega_A \hat{\sigma}_A^z + \omega_B \hat{\sigma}_B^z + \alpha \hat{\sigma}_A^x \otimes \hat{\sigma}_B^x.$$

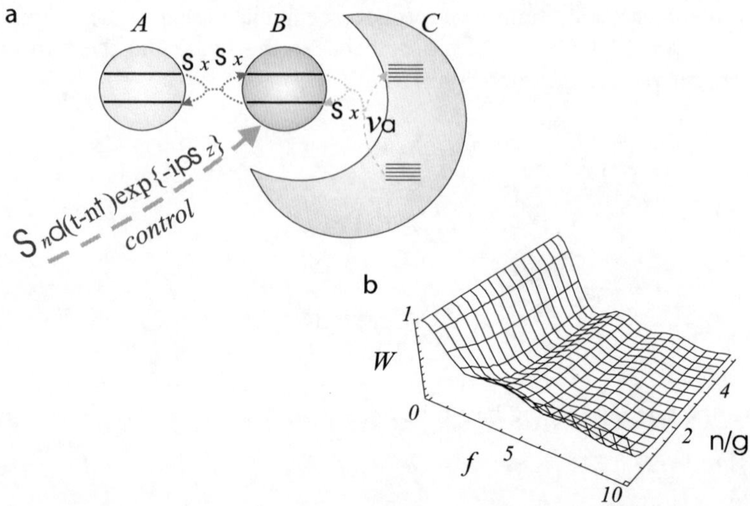

Fig. 12.8 (**a**) A four-level system $A + B$ is composed of two two-level systems one of which is coupled to the environment C. Rapid intervention into the system B decouples less the system A from the environment C, such that the Hilbert space of the system A becomes the decoherence-free subspace of the Hilbert space of $A + B$. (**b**) Decrease of the relaxation rate W as a result of phase modulation $f \cos \nu t$ for the environment absorption spectrum of the width γ centered at the resonance frequency of the part B

One of the two-level systems, say B, interacts with the environment via the interaction Hamiltonian

$$\hat{\mathcal{V}} = \hat{\sigma}_B^x \otimes \hat{v}_C.$$

A periodic application of the transformations $\hat{U}_j = \exp\left[-i\pi\hat{\sigma}_B^z\right]$ separated by a time interval $T \ll 1/\omega_B, 1/\alpha, 1/\left|\hat{\mathcal{V}}\right|$ results in the two-element subgroup of transformations $\hat{\mathbf{U}}_j = \exp\left[-ij\pi\hat{\sigma}_B^z\right]$ with $j = 1, 2$, and yields the Hamiltonians

$$\begin{aligned}\overline{\hat{H}}_{2T} &= \omega_A\hat{\sigma}_A^z + \omega_B\hat{\sigma}_B^z + \alpha\hat{\sigma}_A^x \otimes \sum_{j=1}^{2} \exp\left[-ij\pi\hat{\sigma}_B^z\right]\hat{\sigma}_B^x \exp\left[ij\pi\hat{\sigma}_B^z\right] \\ &= \omega_A\hat{\sigma}_A^z + \omega_B\hat{\sigma}_B^z, \\ \overline{\hat{\mathcal{V}}}_{2T} &= \sum_{j=1}^{2} \exp\left[-ij\pi\hat{\sigma}_B^z\right]\hat{\sigma}_B^x \exp\left[ij\pi\hat{\sigma}_B^z\right] \otimes \hat{v}_C = 0,\end{aligned}$$

averaged over the period of $2T$. Eigenvectors of $\hat{\mathbf{U}}_j$ corresponding to the same eigenvalues have the form of either $|x\rangle_A \otimes |0\rangle_B$ or $|x\rangle_A \otimes |1\rangle_B$ and compose the decoherence-free subspace coinciding with the Hilbert space of the part A

multiplied either by state $|0\rangle$ or by state $|1\rangle$ of the part B. The two-element subgroup $\left\{\hat{\mathrm{U}}_1, \hat{\mathrm{U}}_2\right\}$ is therefore a stabilizer of the Hilbert space of the part A, while the transformation in this decoherence-free subspace can be done by application of the corresponding Pauli matrices $\hat{\sigma}_A^{x;y;z}$.

Note that the existence of non-trivial decoherence-free subspaces and the corresponding stabilizers is not a universal property of complex systems, and hence, one has to straightforwardly verify for each particular case of the system, the coupling to the environment, and the suggested sequence of transformations $\hat{U}_j$.

Decoupling from Rapid Noise

In the previous section we focused on the group-theory aspects of the interaction between a system and the environment related to the possibility of existence of the decoherence-free subspaces. The possibility to perform the control as fast as required was taken for granted. Now we turn to the situation where the correlation time of the environment and the typical time of the coherence protecting interventions are of the same order of magnitude, and our goal is to considerably decrease the rate of relaxation and coherence loss by a proper time-dependent perturbation $f(t)\hat{H}_c$ of the system.

Equation (11.51) serves as a good starting point for the consideration, where the spectral decomposition

$$\hat{V}_\omega = \int e^{-i\omega t} \mathrm{T}\exp\left\{-i\int^{t} \hat{H}(\tau)d\tau\right\} \hat{V}\mathrm{T}\exp\left\{i\int^{t} \hat{H}(\tau)d\tau\right\} dt \qquad (12.45)$$

of the coupling operator in the interaction representation is now affected by the time-dependent control $f(t)\hat{H}_c$ entering the system Hamiltonian $\hat{H}(t) = \hat{H} + f(t)\hat{H}_c$. As follows from (11.54) and the discussion after that equation, for an arbitrary density matrix $\hat{\rho}$ the goal is achieved, provided $\mathrm{Im}\left[i\int h^I_\omega \hat{V}^*_{-\omega}\hat{V}^*_\omega d\omega\right]$ and $\int d\omega \hat{V}_\omega \hat{\rho} \hat{V}^*_\omega h^{II}_{-\omega}$, take the minimum possible values. The second requirement cannot be achieved globally, since it is governed by the density matrix, which is different at each given moment of time, and therefore one has to choose a certain combination of the individual rates written for every pair of matrix elements

$$\int d\omega \left(\hat{V}_\omega\right)^k_m \left(\hat{V}^*_\omega\right)^s_r h^{II}_{-\omega} \qquad (12.46)$$

to be minimized. The sum $W_{tot} = \sum_{r,s} W_{r,s}$ of all possible relaxation rates $W_{r,s}$ given by (12.46) for $k = r$, $m = s$, is one of the possible choices. Note that being chosen in this way, the rate W_{tot} does not include the dephasing rate $\mathrm{Re}\int h^I_\omega \hat{V}^*_{-\omega}\hat{V}^*_\omega d\omega$ responsible for the decay of the off-diagonal elements of the density matrix.

When $f(t)$ is periodic function with period T, the frequency spectrum $\left(\hat{V}_\omega\right)_m^k$ of matrix elements of the time-dependent operators $\hat{V}(t)$ is discrete, since the evolution operator of the system for this case reads

$$\mathrm{T}\exp\left\{-i\int^{t}\hat{H}(\tau)d\tau\right\} = \hat{F}(t)e^{-it\hat{H}_T}. \tag{12.47}$$

Here $\hat{H}_T$ is defined as

$$e^{-iT\hat{H}_T} = \mathrm{T}\exp\left\{-i\int_0^T\hat{H}(\tau)d\tau\right\}$$

and has the physical meaning of an effective Hamiltonian that gives an exact evolution over the period, while the matrix $\hat{F}(t)$ satisfies the periodicity requirement $\hat{F}(t) = \hat{F}(t+T)$. The periodicity requirement implies

$$\hat{F}(t) = \sum_n e^{in\frac{2\pi}{T}t}\hat{F}_n,$$

and hence

$$\hat{V}_\omega = \sum_{n,m}\int dt e^{i(n-m)\frac{2\pi}{T}t+i\omega t}\hat{F}_n e^{-it\hat{H}_T}\hat{V}e^{it\hat{H}_T}\hat{F}_m^* \tag{12.48}$$

has indeed a discrete frequency spectrum. Equation (12.47), known as the Floquet–Lyapunov theorem for a set of differential equations with periodic coefficients, is a generalization of (5.25) written at $\tau = t$ for the state vector of a periodically driven quantum system.

For a given matrix element, say r, s, Eq. (12.48) yields

$$\begin{aligned}\left(\hat{V}_\omega\right)_r^s &= \sum_{n,m,f,j}\int dt e^{i(n-m)\nu t+i\omega t-it\Omega_{fj}}\left(\hat{F}_n\right)_r^f V_f^j\left(\hat{F}_m^*\right)_j^s \\ &= 2\pi\sum_{k,f,j}\delta\left(\frac{2\pi}{T}k+\omega-\Omega_{fj}\right)\sum_m\left(\hat{F}_{m+k}\right)_r^f V_f^j\left(\hat{F}_m^*\right)_j^s,\end{aligned} \tag{12.49}$$

where $\Omega_{kl} = E_{T;k} - E_{T;l}$ stands for the frequency spectrum suggested by the eigenvalues $E_{T;n}$ of the effective Hamiltonian $\hat{H}_T$ in the interaction representation. Equation (12.49) substitutes the expression $V_r^s\delta\left(\omega-\Omega_{rs}\right)$ in the derivation of (11.57), such that in the expressions for the relaxation matrix elements one has to make the replacement

$$2\pi h^{II}_{-\Omega_{rs}} \left|V^s_r\right|^2 \to \sum_{k,f,j} 2\pi h^{II}_{\frac{2\pi}{T}k-\Omega_{fj}} \left|\sum_m \left(\hat{F}_{m+k}\right)^f_r V^j_f \left(\hat{F}^*_m\right)^s_j\right|^2, \tag{12.50}$$

and after summation of the right hand side over r and s one obtains the quantity W_{tot} to be minimized by a proper choice of the control $f(t)$. Note that this quantity is a sum of essentially positive contributions, and therefore cannot be set to zero exactly. Though within the second-order perturbation theory the coupling of system and environment resulting in relaxation cannot be completely suppressed by a periodic control, still it might be considerably reduced in the case where the overlap $\sum_{rs} 2\pi h^{II}_{-\Omega_{rs}} \left|V^s_r\right|^2$of the environment absorption spectrum $h^{II}_{-\omega}$ and the coupling power spectrum $2\pi \left|V^s_r\right|^2 \delta\left(\omega - \Omega_{rs}\right)$ gets decreased as a result of such an intervention.

Let us illustrate this by the example of the same system as shown in Fig. 12.8a and for the simplest version of phase control $f(t)\hat{\sigma}^z_B$. The function $f(t)$ is taken as periodic with zero average over the period T. The choice $f(t) = \nu f \sin \nu t$ admits an exact analytic solution. The system Hamiltonian thus adopts the form

$$H = \omega_A \hat{\sigma}^z_A + \left(\omega_B - \nu f \sin \nu t\right) \hat{\sigma}^z_B + \alpha \hat{\sigma}^x_A \otimes \hat{\sigma}^x_B,$$

while the coupling to the environment in the interaction representation

$$\hat{\mathcal{V}} = \begin{pmatrix} 0 & Ve^{-i\omega_B t - if\cos\nu t} \\ Ve^{i\omega_B t + if\cos\nu t} & 0 \end{pmatrix}_B \otimes \hat{v}_C \tag{12.51}$$

experiences the phase modulation $f \cos \nu t$ and suggests the explicit expression

$$\hat{V}_\omega = \sum_k \begin{pmatrix} 0 & V\delta_{\omega-\omega_B-k\nu} i^k J_k(f) \\ V\delta_{\omega+\omega_B-k\nu}(-i)^n J_k(f) & 0 \end{pmatrix}$$

for the spectral decomposition of the operator $\hat{V}(t)$ in terms of the Bessel functions $J_k(f)$. Hence, according to (12.49,12.50)

$$W_{tot} = \sum_k 2\pi h^{II}_{k\nu-\omega_B} \left|VJ_k(f)\right|^2 + \sum_k 2\pi h^{II}_{k\nu+\omega_B} \left|VJ_k(f)\right|^2$$

is the quantity to be minimized.

Let us consider zero temperatures and choose a model where

$$h^{II}_\omega = \frac{c^2_{\omega_B}\pi\gamma}{\left(\omega - \omega_B\right)^2 + \gamma^2}, \tag{12.52}$$

which implies that the environment absorption spectrum is centered at the frequency of the part B of the system and has a width γ. The latter we choose to be small, $\gamma \ll \omega_B$, and thus we obtain the expression

$$W_{tot} = \pi^2 c_{\omega_B}^2 \gamma^{-1} |V|^2 \sum_k \frac{|J_k(f)|^2}{(k\nu/\gamma)^2 + 1} \tag{12.53}$$

to be minimized by a proper choice of the frequency ν and the amplitude f of the phase modulation. In Fig. 12.8b we show the dependence (12.53) where the constant factor in front of the sum is set to unity. One sees that the relaxation rate indeed decreases with increasing of both the amplitude and the frequency modulation when the spectrum of the coupling matrix element $V_{01}(t)$ gets larger than the width γ of the environment absorption spectrum. This result shows an example of the general situation—in order to protect a quantum system from rapid noise one has to act on the system faster than this noise.

Minimization of the Entropy Growth Rate

Though the choice of a global characteristic of the coupling between the system and the environment as a sum of all contributions (12.46) is physically motivated, from the formal point of view it is a rather arbitrary choice. We now introduce a formal measure of the coupling reduction due to an intervention, which relies on the minimization of the linear entanglement entropy growth rate.

The definition of the linear entropy together with (11.46) suggests one to consider the functional average

$$\mathrm{Tr}\hat{\rho}^2(t) = \int \mathcal{D}X(t)\mathcal{D}x(t)\mathcal{D}X'(t)\mathcal{D}x'(t) W\left(x(t), X(t)\right) W\left(x'(t), X'(t)\right)$$

$$\mathrm{Tr}\left\{ \mathrm{T}e^{-i\int_0^t \hat{V}(t)x(t)dt} \hat{\rho}_0 \mathrm{T}e^{i\int_0^t \hat{V}^*(t)X(t)dt} \mathrm{T}e^{-i\int_0^t \hat{V}(t)x'(t)dt} \hat{\rho}_0 \mathrm{T}e^{i\int_0^t \hat{V}^*(t)X'(t)dt} \right\},$$

which depends on the initial density matrix $\hat{\rho}_0$ of the system. One can take the average over all possible initial density matrices of rank 1 and arrive at

$$\overline{\mathrm{Tr}\hat{\rho}^2(t)} = \int \frac{\mathcal{D}X(t)\mathcal{D}x(t)\mathcal{D}X'(t)\mathcal{D}x'(t)}{N(N+1)} W\left(x(t), X(t)\right) W\left(x'(t), X'(t)\right)$$

$$\left(\mathrm{Tr}\,\mathrm{T}e^{i\int_0^t \hat{V}^*(t)X(t)dt} \mathrm{T}e^{-i\int_0^t \hat{V}(t)x'(t)dt} \mathrm{T}e^{i\int_0^t \hat{V}^*(t)X'(t)dt} \mathrm{T}e^{-i\int_0^t \hat{V}(t)x(t)dt} + \right.$$

$$\left. \mathrm{Tr}\,\mathrm{T}e^{i\int_0^t \hat{V}^*(t)X(t)dt} \mathrm{T}e^{-i\int_0^t \hat{V}(t)x'(t)dt} \mathrm{Tr}\,\mathrm{T}e^{i\int_0^t \hat{V}^*(t)X'(t)dt} \mathrm{T}e^{-i\int_0^t \hat{V}(t)x(t)dt} \right),$$

where N is the dimension of the system Hilbert space.

We now assume that as the result of the controlled intervention, the action of this coupling

$$\int_{t}^{t+\tau_C} \hat{V}(t)x(t)dt$$

accumulated during a time interval of the order of the environment correlation time τ_C, is small and therefore performing the Magnus expansion,

$$\mathrm{T}e^{-i\int_0^t \hat{V}(t)x'(t)dt} = e^{-i\int_0^t \hat{V}(\tau)x(\tau)d\tau - \frac{1}{2}\int_0^t \Theta_{\tau-\theta}\left[\hat{V}(\tau),\hat{V}(\theta)\right]x(\tau)x(\theta)d\theta d\tau + \ldots},$$

in the time-ordered exponents one can keep only the second-order commutators that will be considered as c-numbers in the following calculations. We substitute this relation to the expression for $\overline{\mathrm{Tr}\hat{\rho}^2(t)}$ and obtain

$$\overline{\mathrm{Tr}\hat{\rho}^2(t)} = \int \frac{\mathcal{D}X(t)\mathcal{D}x(t)\mathcal{D}X'(t)\mathcal{D}x'(t)}{N(N+1)} W\left(x(t),X(t)\right) W\left(x'(t),X'(t)\right)$$

$$\left(\mathrm{T}e^{i\int_0^t\left(\hat{V}^*(\tau)X(\tau)-\hat{V}(\tau)x'(\tau)\right)d\tau+\Xi_{X,x'}}\, e^{i\int_0^t\left(\hat{V}^*(\tau)X'(\tau)-\hat{V}(\tau)x(\tau)\right)d\tau+\Xi_{X',x}} + \right.$$

$$\left.\mathrm{T}e^{i\int_0^t\left(\hat{V}^*(\tau)X(\tau)-\hat{V}(\tau)x'(\tau)\right)d\tau+\Xi_{X,x'}}\,\mathrm{T}e^{i\int_0^t\left(\hat{V}^*(t)X'(t)-\hat{V}(t)x(t)\right)d\tau+\Xi_{X',x}}\right),$$

where

$$\Xi_{X,x'}(t) = -\int_0^t \left(X(\tau)\ x'(\tau)\right)\hat{m}(\tau,\theta)\begin{pmatrix} X(\theta) \\ x'(\theta)\end{pmatrix} d\theta d\tau$$

$$\Xi_{X',x}(t) = -\int_0^t \left(X'(\tau)\ x(\tau)\right)\hat{m}(\tau,\theta)\begin{pmatrix} X'(\theta) \\ x(\theta)\end{pmatrix} d\theta d\tau$$

$$\hat{m}(\tau,\theta) = \begin{pmatrix} \frac{\left[\hat{V}^*(\tau),\hat{V}^*(\theta)\right]\Theta_{\tau-\theta}}{2} & -\frac{\left[\hat{V}^*(\tau),\hat{V}(\theta)\right]}{2} \\ 0 & \frac{\left[\hat{V}(\tau),\hat{V}(\theta)\right]\Theta_{\tau-\theta}}{2}\end{pmatrix}.$$

We now assume that the perturbation is small, cast the evolution operators to the second-order perturbation series, ignore the terms linear in random functions, since they vanish after the functional average, and obtain

$$1-\overline{\mathrm{Tr}\hat{\rho}^2(t)}=\int\frac{\mathcal{D}X(t)\mathcal{D}x(t)\mathcal{D}X'(t)\mathcal{D}x'(t)}{N\left(N+1\right)}W\left(x(t),X(t)\right)W\left(x'(t),X'(t)\right)$$
$$\left(X(\tau)\;x'(\tau)\;X'(\tau)\;x(\tau)\right)\hat{C}(\tau,\theta)\begin{pmatrix}X(\theta)\\x'(\theta)\\X'(\theta)\\x(\theta)\end{pmatrix}, \tag{12.54}$$

where the 4×4 matrix $\hat{C}(\tau,\theta)$ has the block structure

$$\hat{C}(\tau,\theta)=\begin{pmatrix}\hat{A}(\tau,\theta) & \hat{A}(\tau,\theta)+\hat{B}(\tau,\theta)\\ \hat{0} & \hat{A}(\tau,\theta)\end{pmatrix}$$

with 2×2 blocks

$$\hat{A}(\tau,\theta)=\begin{pmatrix}\mathrm{Tr}\left\{\hat{V}_\tau^*\hat{V}_\theta^*\right\} & -\mathrm{Tr}\left\{\hat{V}_\tau^*\hat{V}_\theta\right\}\\ -\mathrm{Tr}\left\{\hat{V}_\tau\hat{V}_\theta^*\right\} & \mathrm{Tr}\left\{\hat{V}_\tau\hat{V}_\theta\right\}\end{pmatrix}$$

$$\hat{B}(\tau,\theta)=\begin{pmatrix}\mathrm{Tr}\left\{\hat{V}_\tau^*\right\}\mathrm{Tr}\left\{\hat{V}_\theta^*\right\} & -\mathrm{Tr}\left\{\hat{V}_\tau^*\right\}\mathrm{Tr}\left\{\hat{V}_\theta\right\}\\ -\mathrm{Tr}\left\{\hat{V}_\tau\right\}\mathrm{Tr}\left\{\hat{V}_\theta^*\right\} & \mathrm{Tr}\left\{\hat{V}_\tau\right\}\mathrm{Tr}\left\{\hat{V}_\theta\right\}\end{pmatrix}.$$

We now are in the position to derive an explicit equation for the linear entropy of entanglement (12.54) in terms of the correlation functional (11.38) of the thermostat. To this end we note that the relation

$$\frac{\int X_iC_{ij}X_j\exp[-X_iG_{ij}X_j]\prod_i dX_i}{\int\exp[-X_iG_{ij}X_j]\prod_i dX_i}=\frac{1}{2}\mathrm{Tr}\left\{\hat{C}\hat{G}^{-1}\right\}$$

holds for a Gaussian average of a quadratic form given by a normalized n-fold integral over the real variables X_i. The functional analog of this relation written for the average linear entropy yields

$$\int\mathrm{Tr}\left\{\begin{pmatrix}\hat{A}(\theta,\tau) & \hat{A}(\theta,\tau)+\hat{B}(\theta,\tau)\\ \hat{0} & \hat{A}(\theta,\tau)\end{pmatrix}\begin{pmatrix}\hat{M}(\tau,\theta) & 0\\ \hat{0} & \hat{M}(\tau,\theta)\end{pmatrix}\right\}d\theta d\tau,$$

or explicitly

$$\overline{S_l(t)} = 1 - \overline{\mathrm{Tr}\hat{\rho}^2(t)} = \int d\theta d\tau \sum_{\omega} \frac{c_\omega^2 e^{i(\tau-\theta)\omega}}{2}$$
$$(\mathrm{Tr}\left\{\hat{V}_\tau \hat{V}_\theta^*\right\} + \Theta_{\tau-\theta}\mathrm{Tr}\left\{\hat{V}_\tau^* \hat{V}_\theta^* + \hat{V}_\tau \hat{V}_\theta\right\}$$
$$- \mathrm{Tr}\left\{\left(\hat{V}_\tau^* + \hat{V}_\tau\right)\left(\hat{V}_\theta^* + \hat{V}_\theta\right)\right\} \frac{e^{-\omega/kT}}{1 - e^{-\omega/kT}}). \tag{12.55}$$

As earlier, the coinciding frequencies in the Fourier decompositions of the operators $\hat{V}_\theta, \hat{V}_\tau, \hat{V}_\theta^*$, and $\hat{V}_\tau^*$ result in the terms on the right hand side that linearly increase in time. This linearly increasing increment emerging from (12.55) is the very quantity to be minimized when one looks for the maximum possible decoupling of the system from the environment by a proper choice of the control.

Note, however, that setting $\frac{\partial}{\partial t}\overline{S_l(t)}$ to zero via a proper choice of the control does not necessary mean that the system is completely decoupled from the environment. In fact, the example shown in Fig. 11.3 illustrates that in contrast to the regular entropy, the linear entropy of entanglement cannot only increase, but also decrease as the result of relaxation. Therefore the linear entropy averaged over all possible density matrices may remain unchanged under conditions where this quantity increases for some density matrices and decreases for the other. One therefore needs to consider the dispersion $D_{S_l} = \overline{(1-\mathrm{Tr}\hat{\rho}^2)^2} - \left(\overline{1-\mathrm{Tr}\hat{\rho}^2}\right)^2$ of the linear entropy over the distribution of the density matrices as well.

From the exact expression for $\left(1-\mathrm{Tr}\hat{\rho}^2\right)^2$

$$\int \mathcal{D}^4\overrightarrow{X}\mathcal{D}^4\overrightarrow{\tilde{X}} d\left|\psi\right\rangle W\left(x,X\right) W\left(x',X'\right) W\left(\tilde{x},\tilde{X}\right) W\left(\tilde{x}',\tilde{X}'\right)$$
$$(1 - \langle\psi|\,\mathrm{Te}^{i\int_0^t \hat{V}_t^* X'(t)dt}\,\mathrm{Te}^{-i\int_0^t \hat{V}_t x(t)dt}\,|\psi\rangle\langle\psi|\,\mathrm{Te}^{i\int_0^t \hat{V}_t^* X(t)dt}\,\mathrm{Te}^{-i\int_0^t \hat{V}_t x'(t)dt}\,|\psi\rangle)$$
$$(1 - \langle\psi|\,\hat{\rho}_0\mathrm{Te}^{i\int_0^t \hat{V}_t^* \tilde{X}(t)dt}\,\mathrm{Te}^{-i\int_0^t \hat{V}_t \tilde{x}'(t)dt}\,|\psi\rangle\langle\psi|\,\mathrm{Te}^{i\int_0^t \hat{V}_t^* \tilde{X}'(t)dt}\,\mathrm{Te}^{-i\int_0^t \hat{V}_t \tilde{x}(t)dt}\,|\psi\rangle),$$

where $\overrightarrow{X}$ stands for the vector $\left(X(\tau)\; x(\tau)\; X'(\tau)\; x'(\tau)\right)$ and $\overrightarrow{\tilde{X}}$ for the similar vector of tilded functions, with the help of the second-order time-dependent perturbation theory, one obtains the integrand

$$\left(\langle\psi|\left(\left(\hat{V}_\tau\hat{V}_\theta + \hat{V}_\theta^*\hat{V}_\tau^*\right)M_d^{\tau-\theta} + \left(\hat{V}_{\tau'}\hat{V}_{\theta'} + \hat{V}_{\theta'}^*\hat{V}_{\tau'}^*\right)M_d^{\tau'-\theta'}\right)|\psi\rangle\right.$$
$$\left.+ \langle\psi|\hat{V}_\tau|\psi\rangle\langle\psi|\hat{V}_\theta^*|\psi\rangle M_o^{\tau-\theta} + \langle\psi|\hat{V}_{\tau'}|\psi\rangle\langle\psi|\hat{V}_{\theta'}^*|\psi\rangle M_o^{\tau'-\theta'}\right)$$
$$\left(\langle\psi|\hat{V}_{\tilde{\tau}}|\psi\rangle\langle\psi|\hat{V}_{\tilde{\theta}}^*|\psi\rangle M_o^{\tilde{\tau}-\tilde{\theta}} + \langle\psi|\hat{V}_{\tilde{\tau}'}|\psi\rangle\langle\psi|\hat{V}_{\tilde{\theta}'}^*|\psi\rangle M_o^{\tilde{\tau}'-\tilde{\theta}'}\right.$$
$$\left.+ \langle\psi|\left(\left(\hat{V}_{\tilde{\tau}}\hat{V}_{\tilde{\theta}} + \hat{V}_{\tilde{\theta}}^*\hat{V}_{\tilde{\tau}}^*\right)M_d^{\tilde{\tau}-\tilde{\theta}} + \left(\hat{V}_{\tilde{\tau}'}\hat{V}_{\tilde{\theta}'} + \hat{V}_{\tilde{\theta}'}^*\hat{V}_{\tilde{\tau}'}^*\right)M_d^{\tilde{\tau}'-\tilde{\theta}'}\right)|\psi\rangle\right),$$

of the eightfold integral over the times $\tau, \theta, \tau', \theta', \tilde{\tau}, \tilde{\theta}, \tilde{\tau}'$, and $\tilde{\theta}$, yielding the square of the linear entropy. Here $M_d^x = M_{11}^x = M_{22}^x$ and $M_o^x = M_{12}^x + M_{21}^x$ stand for the combinations of elements of the covariance matrix $\hat{M}_x$ of (11.38). The required dispersion D_{S_l} results from the averaging of this expression over all possible state vectors and subtraction of the square of the average entropy.

Allowing for the symmetry with respect to interchange of times with and without primes, as well as that with respect to the tilde sign, one obtains a shorter expression for the integrand:

$$
\begin{aligned}
&4(\langle\psi|\,\hat{V}_\tau\hat{V}_\theta + \hat{V}_\theta^*\hat{V}_\tau^*\,|\psi\rangle\,\langle\psi|\,\hat{V}_{\tilde{\tau}}\hat{V}_{\tilde{\theta}} + \hat{V}_{\tilde{\theta}}^*\hat{V}_{\tilde{\tau}}^*\,|\psi\rangle\, M_d^{\tau-\theta}M_d^{\tilde{\tau}-\tilde{\theta}} \\
&+ 2\,\langle\psi|\,\hat{V}_\tau\hat{V}_\theta + \hat{V}_\theta^*\hat{V}_\tau^*\,|\psi\rangle\,\langle\psi|\,\hat{V}_{\tilde{\tau}}\,|\psi\rangle\,\langle\psi|\,\hat{V}_{\tilde{\theta}}^*\,|\psi\rangle\, M_d^{\tau-\theta}M_o^{\tilde{\tau}-\tilde{\theta}} \\
&+ \langle\psi|\,\hat{V}_\tau\,|\psi\rangle\,\langle\psi|\,\hat{V}_\theta^*\,|\psi\rangle\,\langle\psi|\,\hat{V}_{\tilde{\tau}}\,|\psi\rangle\,\langle\psi|\,\hat{V}_{\tilde{\theta}}^*\,|\psi\rangle\, M_o^{\tau-\theta}M_o^{\tilde{\tau}-\tilde{\theta}})
\end{aligned}
$$

of the expression for the entropy squared, which allows one to write the entropy dispersion as

$$
\begin{aligned}
D_{S_l} = 4\int & d\tau d\theta d\tilde{\tau} d\tilde{\theta}((\overline{\psi_i^*\psi_j\psi_k^*\psi_l} - \overline{\psi_i^*\psi_j}\,\overline{\psi_k^*\psi_l}) \\
&\langle i|\,\hat{V}_\tau\hat{V}_\theta + \hat{V}_\theta^*\hat{V}_\tau^*\,|j\rangle\,\langle k|\,\hat{V}_{\tilde{\tau}}\hat{V}_{\tilde{\theta}} + \hat{V}_{\tilde{\theta}}^*\hat{V}_{\tilde{\tau}}^*\,|l\rangle\, M_d^{\tau-\theta}M_d^{\tilde{\tau}-\tilde{\theta}} \\
&+ 2\,(\overline{\psi_i^*\psi_j\psi_k^*\psi_l\psi_r^*\psi_s} - \overline{\psi_i^*\psi_j}\,\overline{\psi_k^*\psi_l\psi_r^*\psi_s}) \\
&\langle i|\,\hat{V}_\tau\hat{V}_\theta + \hat{V}_\theta^*\hat{V}_\tau^*\,|j\rangle\,\langle k|\,\hat{V}_{\tilde{\tau}}\,|l\rangle\,\langle r|\,\hat{V}_{\tilde{\theta}}^*\,|s\rangle\, M_d^{\tau-\theta}M_o^{\tilde{\tau}-\tilde{\theta}} \\
&+ (\overline{\psi_i^*\psi_j\psi_m^*\psi_n\psi_k^*\psi_l\psi_r^*\psi_s} - \overline{\psi_i^*\psi_j\psi_m^*\psi_n}\,\overline{\psi_k^*\psi_l\psi_r^*\psi_s}) \\
&\langle i|\,\hat{V}_\tau\,|j\rangle\,\langle m|\,\hat{V}_\theta^*\,|n\rangle\,\langle k|\,\hat{V}_{\tilde{\tau}}\,|l\rangle\,\langle r|\,\hat{V}_{\tilde{\theta}}^*\,|s\rangle\, M_o^{\tau-\theta}M_o^{\tilde{\tau}-\tilde{\theta}}),
\end{aligned}
$$

employing the decomposition $|\psi\rangle = \sum_j \psi_j\,|j\rangle$ of the state vector over a basis $\{|j\rangle\}$.

With the help of the representation

$$
\underbrace{\overline{\psi_i^*\dots\psi_s}}_{2K} = \int_{-\infty}^{\infty} \underbrace{\frac{\partial^{2K}}{\partial\lambda_i\dots\partial\lambda_s^*}}_{2K}\Bigg|_{\lambda\to 0} e^{i\sum\limits_j \psi_j\psi_j^* u - iu + i\sum\limits_j(\lambda_j^*\psi_j + \psi_j^*\lambda_j)}\, du\prod_j d\psi_j d\psi_j^*
$$

for the average of amplitudes products over $SU(N)$ group, where

$$
\frac{\partial^{2K}}{\partial\lambda_i\dots\partial\lambda_s^*}\Bigg|_{\lambda\to 0} e^{i\sum\limits_j(\lambda_j^*\psi_j + \psi_j^*\lambda_j)}
$$

stands for $\psi_i^* \dots \psi_s$,

$$\delta\left(\sum_j \psi_j \psi_j^* - 1\right) \prod_j d\psi_j d\psi_j^*$$

stands for the Haar measure of $SU(N)$ group acting on a state vector $|\psi\rangle$, and the integral representation

$$\delta\left(\sum_j \psi_j \psi_j^* - 1\right) = \int_{-\infty}^{\infty} e^{i\sum_j \psi_j \psi_j^* u - iu} du$$

of the Dirac δ-function is employed, one finds

$$\begin{aligned}
\overline{\psi_i^* \psi_j} &= \frac{(N-1)!}{N!} \delta_{ij} \\
\overline{\psi_i^* \psi_j \psi_k^* \psi_l} &= \frac{(N-1)!}{(N+1)!} \left(\delta_{ij}\delta_{kl} + \delta_{il}\delta_{kj}\right) \\
\overline{\psi_i^* \psi_j \psi_k^* \psi_l \psi_r^* \psi_s} &= \frac{(N-1)!}{(N+2)!} \left(\delta_{ij}\delta_{kl}\delta_{rs} + \{j, l, s\}\,\text{permutations}\right) \\
&\dots \\
\underbrace{\overline{\psi_i^* \dots \psi_s}}_{2K} &= \frac{(N-1)!}{(N+K-1)!} (\underbrace{\delta_{ij} \dots \delta_{rs}}_{2K} + \{j, \dots, s\}\,\text{permutations}),
\end{aligned}$$

where all permutations of the second indices of δ-functions enters with equal weights. For the case where $\mathrm{Tr}\left\{\hat{V}_t\right\} = 0$, one explicitly finds

$$\begin{aligned}
D_{S_l} = \int d\tau d\theta d\tilde{\tau} d\tilde{\theta} \Bigg[4M_d^{\tau-\theta} M_d^{\tilde{\tau}-\tilde{\theta}} \Bigg(&\frac{\mathrm{Tr}\left\{\hat{V}_\tau \hat{V}_\theta + \hat{V}_\theta^* \hat{V}_\tau^*\right\} \mathrm{Tr}\left\{\hat{V}_{\tilde{\tau}} \hat{V}_{\tilde{\theta}} + \hat{V}_{\tilde{\theta}}^* \hat{V}_{\tilde{\tau}}^*\right\}}{N(N+1)} \\
&- \frac{\mathrm{Tr}\left\{\left(\hat{V}_\tau \hat{V}_\theta + \hat{V}_\theta^* \hat{V}_\tau^*\right)\left(\hat{V}_{\tilde{\tau}} \hat{V}_{\tilde{\theta}} + \hat{V}_{\tilde{\theta}}^* \hat{V}_{\tilde{\tau}}^*\right)\right\}}{N^2(N+1)} \Bigg) \\
- 8M_d^{\tau-\theta} M_o^{\tilde{\tau}-\tilde{\theta}} \Bigg(&\frac{2\mathrm{Tr}\left\{\hat{V}_\tau \hat{V}_\theta + \hat{V}_\theta^* \hat{V}_\tau^*\right\} \mathrm{Tr}\left\{\hat{V}_{\tilde{\tau}} \hat{V}_{\tilde{\theta}}^*\right\}}{N^2(N+1)(N+2)}
\end{aligned}$$

$$
\left.-\frac{\mathrm{Tr}\left\{\left(\hat{V}_{\tau}\hat{V}_{\theta}+\hat{V}_{\theta}^{*}\hat{V}_{\tau}^{*}\right)\left(\hat{V}_{\tilde{\tau}}\hat{V}_{\tilde{\theta}}^{*}+\hat{V}_{\tilde{\theta}}^{*}\hat{V}_{\tilde{\tau}}\right)\right\}}{N\left(N+1\right)\left(N+2\right)}\right)
$$

$$
+4M_{o}^{\tau-\theta}M_{o}^{\tilde{\tau}-\tilde{\theta}}\left(-\frac{2\left(2N+3\right)\mathrm{Tr}\left\{\hat{V}_{\tau}\hat{V}_{\theta}^{*}\right\}\mathrm{Tr}\left\{\hat{V}_{\tilde{\tau}}\hat{V}_{\tilde{\theta}}^{*}\right\}}{N^{2}\left(N+1\right)^{2}\left(N+2\right)\left(N+3\right)}\right.
$$

$$
+\frac{\mathrm{Tr}\left\{\hat{V}_{\tau}\hat{V}_{\theta}^{*}\hat{V}_{\tilde{\tau}}\hat{V}_{\tilde{\theta}}^{*}\right\}+\mathrm{Tr}\left\{\hat{V}_{\tau}\hat{V}_{\tilde{\tau}}\hat{V}_{\tilde{\theta}}^{*}\hat{V}_{\theta}^{*}\right\}+\mathrm{Tr}\left\{\hat{V}_{\tau}\hat{V}_{\tilde{\theta}}^{*}\hat{V}_{\theta}^{*}\hat{V}_{\tilde{\tau}}\right\}}{N\left(N+1\right)\left(N+2\right)\left(N+3\right)}
$$

$$
+\frac{\mathrm{Tr}\left\{\hat{V}_{\tau}\hat{V}_{\tilde{\tau}}\hat{V}_{\theta}^{*}\hat{V}_{\tilde{\theta}}^{*}\right\}+\mathrm{Tr}\left\{\hat{V}_{\tau}\hat{V}_{\theta}^{*}\hat{V}_{\tilde{\theta}}^{*}\hat{V}_{\tilde{\tau}}\right\}+\mathrm{Tr}\left\{\hat{V}_{\tau}\hat{V}_{\tilde{\theta}}^{*}\hat{V}_{\tilde{\tau}}\hat{V}_{\theta}^{*}\right\}}{N\left(N+1\right)\left(N+2\right)\left(N+3\right)}
$$

$$
\left.\left.+\frac{\mathrm{Tr}\left\{\hat{V}_{\tilde{\tau}}\hat{V}_{\theta}^{*}\right\}\mathrm{Tr}\left\{\hat{V}_{\tau}\hat{V}_{\tilde{\theta}}^{*}\right\}+\mathrm{Tr}\left\{\hat{V}_{\tilde{\theta}}^{*}\hat{V}_{\theta}^{*}\right\}\mathrm{Tr}\left\{\hat{V}_{\tilde{\tau}}\hat{V}_{\tau}\right\}}{N\left(N+1\right)\left(N+2\right)\left(N+3\right)}\right)\right]. \tag{12.56}
$$

Let us now calculate the average entropy (12.55) and its dispersion (12.56) for the particular example of the system shown in Fig. 12.8 with the phase modulated interaction

$$
\hat{V}_t=\begin{pmatrix}0 & Ve^{-i\omega_B t-if\cos\nu t}\\ Ve^{i\omega_B t+if\cos\nu t} & 0\end{pmatrix}
$$

corresponding to (12.51), at zero temperature. For the environment model (12.52) resulting in the correlation functions $M_o^{\tau-\theta}=-e^{-i(\tau-\theta)\omega-\gamma|\tau-\theta|}$ and $M_d^{\tau-\theta}=\Theta_{\tau-\theta}e^{-i(\tau-\theta)\omega-\gamma|\tau-\theta|}$, this yields the average entanglement entropy

$$
\overline{S_l(t)}=2V^2\int d\theta d\tau e^{-i(\tau-\theta)\omega-\gamma|\tau-\theta|}\left(\cos\left(\phi_\theta-\phi_\tau\right)+\cos\left(\phi_\theta+\phi_\tau\right)\right),
$$

and the dispersion

$$
D_{S_l}(t)=V^4\int^t d\tau d\theta d\tilde{\tau}d\tilde{\theta}e^{-i(\tau-\theta)\omega-\gamma|\tau-\theta|}e^{-i(\tilde{\tau}-\tilde{\theta})\omega-\gamma|\tilde{\tau}-\tilde{\theta}|}
$$

$$
\left(\frac{4}{15}\cos\left(\phi_\theta-\phi_\tau\right)\cos\left(\phi_{\tilde{\theta}}-\phi_{\tilde{\tau}}\right)+\frac{4}{15}\sin\left(\phi_\theta-\phi_\tau\right)\sin\left(\phi_{\tilde{\theta}}-\phi_{\tilde{\tau}}\right)-\right.
$$

$$
\left.\frac{22}{15}\cos\left(\phi_\theta+\phi_\tau\right)\cos\left(\phi_{\tilde{\theta}}+\phi_{\tilde{\tau}}\right)+8\Theta_{\tau-\theta}\cos\left(\phi_\theta-\phi_\tau\right)\Theta_{\tilde{\theta}-\tilde{\tau}}\cos\left(\phi_{\tilde{\theta}}-\phi_{\tilde{\tau}}\right)\right),
$$

where $\phi_t=\omega_B t+f\cos\nu t$. Note that this time we do not assume that the system frequency ω_B coincides with the environment absorption frequency ω. We concentrate on the leading terms of these expressions emerging from the combinations independent on $\tau + \theta$ and $\tilde{\tau} + \tilde{\theta}$, which results in the average entropy linearly increasing in time and given by the expression

$$\overline{S_l(t)} = 2tV^2 \sum_{n=-\infty}^{\infty} \frac{2\gamma\left((\omega_B + n\nu)^2 + \gamma^2 + \omega^2\right) J_n^2(f)}{\left((\omega - \omega_B - n\nu)^2 + \gamma^2\right)\left((\omega_B + \omega + n\nu)^2 + \gamma^2\right)}$$
$$+ 2tV^2 \sum_{n=-\infty}^{\infty} (-1)^n \frac{2\gamma\left(n^2\nu^2 + \gamma^2 + \omega^2\right) J_{n-\frac{\omega_B}{\nu}}(f)\, J_{n+\frac{\omega_B}{\nu}}(f)}{\left((\omega - n\nu)^2 + \gamma^2\right)\left((n\nu + \omega)^2 + \gamma^2\right)},$$

in the case where $\frac{2\omega_B}{\nu}$ is an even integer p. Otherwise the second sum vanishes. The corresponding dispersion

$$D_{S_l}(t) = \frac{4V^4t^2}{15}\left(\sum_{n=-\infty}^{\infty} \frac{2\gamma\left((\omega_B + n\nu)^2 + \gamma^2 + \omega^2\right) J_n^2(f)}{\left((\omega - \omega_B - n\nu)^2 + \gamma^2\right)\left((\omega_B + \omega + n\nu)^2 + \gamma^2\right)}\right)^2$$
$$- \frac{4V^4t^2}{15}\left(\sum_{n=-\infty}^{\infty} \frac{4\gamma\omega\left(\omega_B + n\nu\right) J_n^2(f)}{\left((\omega_B - \omega + n\nu)^2 + \gamma^2\right)\left((\omega_B + \omega + n\nu)^2 + \gamma^2\right)}\right)^2$$
$$- \frac{22V^4t^2}{15}\left(\sum_{n=-\infty}^{\infty} (-1)^n \frac{2\gamma\left(n^2\nu^2 + \gamma^2 + \omega^2\right) J_{n-\frac{\omega_B}{\nu}}(f)\, J_{n+\frac{\omega_B}{\nu}}(f)}{\left((\omega - n\nu)^2 + \gamma^2\right)\left((n\nu + \omega)^2 + \gamma^2\right)}\right)^2$$
$$+ 8V^4t^2\left|\sum_{n=-\infty}^{\infty} \frac{J_n^2(f)\left(\gamma + i\omega\right)}{(\omega_B + n\nu)^2 + (\gamma + i\omega)^2}\right|^2$$

quadratically increases in time. In Fig. 12.9 we show dependencies $\overline{S_l(t)}/2tV^2$ and $D_{S_l}(t)/V^4t^2$ as functions of the environment absorption line width and the modulation amplitude. One sees that for the chosen example of the system, at zero

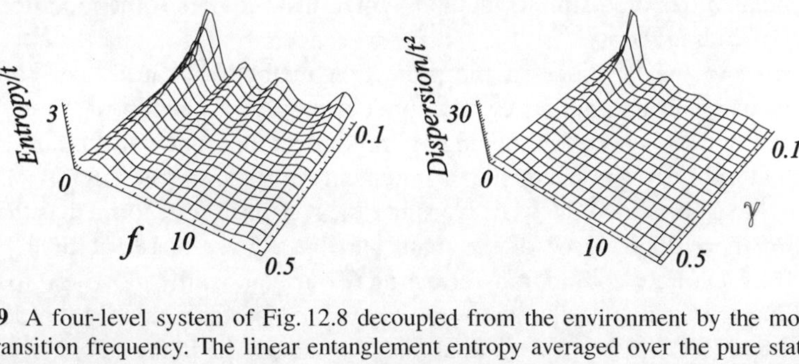

Fig. 12.9 A four-level system of Fig. 12.8 decoupled from the environment by the modulation of the transition frequency. The linear entanglement entropy averaged over the pure state of the system and its dispersion are shown as a function of the modulation amplitude f and the width γ of the environment absorption line. Transition frequency $\omega_B = 6$, the environment resonance frequency $\omega = 5$, and the modulation frequency $\nu = 1$

temperature, both these quantities remaining positive and decrease on average with the increasing frequency modulation amplitude.

12.2.2 Decoupling by Zeno Effect

We now show how one can completely suppress coherence loss in another universal way. The suggested method relies on the basic ideas of classical error-correcting codes and on the idea of the Zeno effect, discussed earlier in Sect. 12.1.4. The latter implies that periodic measurements allow one to keep a quantum system in the initial state for as long as needed, provided that the period T is short relative to the relaxation time T_2. In order to protect a quantum system from coherence loss, we do not intervene on the entire system, but only affect the auxiliary part, the ancilla, which we have to add for the purpose of protection. We perform encoding in the entire Hilbert space of the system and the ancilla by employing non-holonom control.

In the N-dimensional Hilbert space $\mathcal{H}_{\mathcal{N}}$ of a compound quantum system $\mathcal{N}$ comprising the main part $\mathcal{K}$ and an axillary part, the ancilla $\mathcal{A} = \mathcal{N}/\mathcal{K}$, a measurement of the ancilla state projects the compound state vector onto a multidimensional subspace $\mathcal{H}_{\mathcal{K}}$. The dimension K of this subspace coincides with the dimension of the Hilbert space of the main part. If the evolution induced by the error Hamiltonians always results in a linear increment of the state vector, perpendicular to $\mathcal{H}_{\mathcal{K}}$, then such a projecting measurement returns the system back to its initial state. This is sketched in Fig. 12.10a in complete analogy to the regular Zeno effect. In other words, the errors do not affect the main part of the system. Though this assumption corresponds to an unphysical situation, when all errors affect only the ancilla and no error Hamiltonians act on the main system, it nevertheless allows us to draw the important conclusion that the total number M of different error-inducing Hamiltonians may not exceed the dimensionality A of the ancilla Hilbert space. Universal protection against this number of generic Hamiltonians relies on a similar method, which just requires some modification in the spirit of coding theory.

We come to the key idea of the protection method. For a system composed of a main part and an ancilla, eventually subjected to an action of a number M of different but known error-inducing Hamiltonians E_m, one can find a subspace $\mathcal{C}$ such that any error-induced linear increment δv of any state vector $v \in \mathcal{C}$ is orthogonal to $\mathcal{C}$, provided $M \leq A$. Consider the subspace $\mathcal{H}_{\mathcal{K}}$, formed as the direct product of the Hilbert space of the main part by a given state vector $|\tilde{\alpha}\rangle$ of the ancilla. If by a unitary coding transformation $\hat{C}$ one can rapidly move $\mathcal{H}_{\mathcal{K}}$ to $\mathcal{C}$ prior to exposing the system to the action of errors, and perform rapidly the decoding transformation $\hat{C}^{-1}$ just before measurement of the ancilla in the state $|\tilde{\alpha}\rangle$, then the resulting dynamics is equivalent to the errors acting solely on the ancilla. Hereafter we call $\mathcal{C}$ a code space. A system $\mathcal{N}$ may have many $\mathcal{C}$.

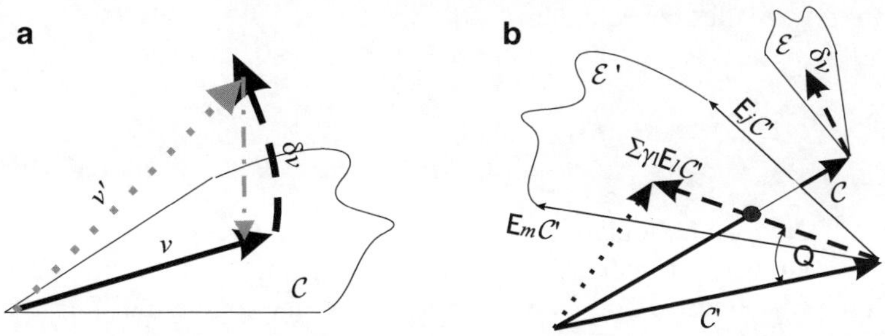

Fig. 12.10 (**a**) The multidimensional Zeno effect. If all the linear increments $\delta v \sim \hat{E}_j v$ of the state vectors v from the code subspace $\mathcal{C}$ are orthogonal to this subspace for any error-induced transformation $v' \to v$, the projection to $\mathcal{C}$ induced by the measurement compensates for the errors up to $O((T/T_2)^2)$. (**b**) The iterative algorithm for finding the code space. The code space $\mathcal{C}$ is represented here by a single vector. The application of all possible errors $\hat{E}_m\mathcal{C}'$ to a code space guess $\mathcal{C}'$ forms the space of errors $\mathcal{E}'$. A linear combination $\sum \gamma_m \hat{E}_m \mathcal{C}'$ shows a direction in $\mathcal{E}'$ in which the length of the sum $\mathcal{C}' + \sum \gamma_m \hat{E}_m \mathcal{C}'$ assumes the minimum value. Displacement in this direction increases the angle Θ between $\mathcal{C}'$ and $\mathcal{E}'$, which approaches the maximum value $\pi/2$ in the vicinity of the "median" point. Repeating this exponentially converging iteration one obtains the code space $\mathcal{C}$ orthogonal to the error space $\mathcal{E}$

Let us perform formal calculations illustrating this idea. Consider a state vector $|S\rangle = |s\rangle \otimes |\tilde{\alpha}\rangle$ of $\mathcal{N}$ formed by $\mathcal{K}$ in a state $|s\rangle$ and $\mathcal{A}$ in the state $|\tilde{\alpha}\rangle$. We treat the errors $\hat{E}_m$ as matrix parts of interaction Hamiltonians with external random fields $f_m(t)$ that produce an uncontrolled evolution of the system. For $\hat{E}_m \sim 1$ one can estimate $f_m \sim 1/T_2$. Under the action of the error-inducing Hamiltonians $\hat{E}_m f_m(t)$, the vector $|S\rangle$ undergoes an uncontrolled evolution given by the unitary operator

$$\hat{U} = \prod_{m=1}^{M} e^{-i\hat{E}_m \int f_m(t)dt} \simeq \hat{I} - i\sum_{m=1}^{M} \hat{E}_m \int f_m(t)dt, \tag{12.57}$$

where the time ordering of the product is implicit. Actions over the Zeno period T of the M different fields are small, $\int_t^{t+T} f_m(x)dx \left|\hat{E}_m\right| \ll 1$, so that only the identity operator $\hat{I}$ and the first-order terms $\sim \int f_m(t)dt$ are important in the Taylor series for the evolution operator.

Let $\hat{\rho}_{sc} = |s\rangle \otimes \langle s|$ be the density matrix of the main part before the action of errors. Then $\hat{\rho} = |S\rangle \otimes \langle S| = |\tilde{\alpha}\rangle \hat{\rho}_{sc} \langle\tilde{\alpha}|$ is the density matrix of the entire system. After coding by the unitary transformation $\hat{C}$, this matrix takes the form $\hat{C}\hat{\rho}\hat{C}^{-1}$, and reads $\hat{U}\hat{C}\hat{\rho}\hat{C}^{-1}\hat{U}^*$ at the end of the action of errors. Decoding yields $\hat{\rho}' = \hat{C}^{-1}\hat{U}\hat{C}\hat{\rho}\hat{C}^{-1}\hat{U}^*\hat{C}$ whereas the direct substitution shows that the variation $\delta\hat{\rho}_{sc} = \langle\tilde{\alpha}|\,\hat{\rho}' - \hat{\rho}\,|\tilde{\alpha}\rangle$ of the density matrix of the main system after the perturbation (12.57) and the measurement of the ancilla is given by the commutator

$$\delta\hat{\rho}_{sc} = -i\left[\sum_{m=1}^{M} \int f_m(t)dt\, \langle\tilde{\alpha}|\, \hat{C}^{-1}\hat{E}_m\hat{C}\, |\tilde{\alpha}\rangle\, , \hat{\rho}_{sc}\right]. \tag{12.58}$$

Therefore $\hat{\rho}_{sc}$ satisfies the equation

$$i\frac{d\hat{\rho}_{sc}}{dt} = \left[\hat{h}_e, \hat{\rho}_{sc}\right]; \quad \hat{h}_e = \sum_{m=1}^{M} f_m \langle\tilde{\alpha}|\, \hat{C}^{-1}\hat{E}_m\hat{C}\, |\tilde{\alpha}\rangle \tag{12.59}$$

with the effective Hamiltonian $\hat{h}_e$ made up of error-inducing Hamiltonians transformed by coding and decoding and projected to the state of the ancilla. Our aim is to set these projections to zero by a proper choice of the coding transformation $\hat{C}$.

The relation $|v\rangle = \hat{C}\,|s\rangle \otimes |\tilde{\alpha}\rangle$ holds for the code vectors and the states of the system. If $\hat{h}_e = 0$, the errors do not affect the main part of the system and $\hat{\rho}_{sc} =$ const, which implies

$$\langle v'|\, \hat{E}_m\, |v\rangle = 0; \ \ \forall\, |v\rangle\,, |v'\rangle \in \mathcal{C}; \ \ \forall \hat{E}_m \in \left\{\hat{E}_m\right\}. \tag{12.60}$$

In other words, a linear increment $\delta v = \hat{E}_m v$ resulting from any error $\hat{E}_m$ applied to a vector v of the coding space is orthogonal to any other vector v' of the coding space. Note that all Hamiltonians $\hat{E}_m$ in (12.60) projected onto the code space have zero traces, or should be set in this form by subtracting a scalar operator $\lambda\hat{I}$ that does not change the Hamiltonian dynamics. We note the physical meaning of (12.60). A corrupted state vector of the code space does not overlap with another vector of the code space if they were orthogonal before the corruption. This corresponds to the so-called Hamming error detection condition in classical information theory. It simply means that the variety of all of the results obtained by the corruption of one message should not include another possible message. The only difference is that in quantum mechanics we interpret the overlap as a scalar product of the state vectors in the Hilbert space, whereas in classical information theory this concept is formulated in terms of "distance", a metrical property which denotes the number of distinct information bits for two messages.

Now we show an algorithm which allows one to design such codes. For $N = KA$ dimensional Hilbert space $\mathcal{H}_{\mathcal{N}}$ formed by the main part $\mathcal{K}$ and ancilla $\mathcal{A}$, condition (12.60) is a set of MK^2 equations taken over AK^2 matrix elements $\langle\alpha|\otimes\langle s'|\, \hat{C}\, |s\rangle\otimes |\tilde{\alpha}\rangle$ of the $N \times N$ encoding operator $\hat{C}$. This defines a rectangular $K \times N$ part $\hat{C}\,|\tilde{\alpha}\rangle$ of $\hat{C}$. To specify (12.60) further, let the states $|s\rangle \otimes |\tilde{\alpha}\rangle$ correspond to the first K positions of the state vector, which implies that each of the M matrices $\hat{C}^{-1}\hat{E}_m\hat{C}$ has an all-zero upper left $K \times K$ corner. The standard methods of linear algebra do not give a recipe for finding a linear transformation $\hat{C}$ that simultaneously sets to zero the corners of M different matrices. The following iterative algorithm accomplishes this task—we find a code space $\mathcal{C}$ composed of all linear combinations of K first columns of $\hat{C}$ which satisfies the conditions of (12.60) given M arbitrary error Hamiltonians $\hat{E}_m$.

Though it can directly be done in the Hilbert space, in order to elucidate the geometrical meaning of the algorithm we write the code basis as a single real vector

C of length $2KN$ by appending vectors of length N corresponding to real and imaginary parts of the first K columns of $\hat{C}$. We rewrite (12.60) in the form

$$\left(\mathsf{C}\hat{\mathsf{E}}_m\mathsf{C}\right) = 0, \tag{12.61}$$

where $2KN \times 2KN$ error matrices $\hat{\mathsf{E}}_m$ have a $K \times K$ block structure. Each diagonal and upper-diagonal block repeats real $2N \times 2N$ matrices $\hat{\mathsf{E}}_m$ corresponding to Hamiltonians $\hat{E}_m$, whereas the lower blocks are formed by the unity operator $\hat{I}$ in the Hilbert space $\mathcal{H}_N$ rewritten in the chosen real representation. The first type of blocks corresponds to the condition (12.60) whereas the second part ensures the orthogonality of different columns of $\hat{C}\,|\tilde{\alpha}\rangle$. In other words, the vector C should satisfy M quadratic equations (12.61) for each m.

The input of our algorithm is a randomly selected and properly normalized vector C'. If by chance C' is orthogonal to all vectors $\hat{\mathsf{E}}_m\mathsf{C}'$, the problem is solved. Otherwise, there exists a linear combination $\sum_{m=1}^{M}\gamma_m\hat{\mathsf{E}}_m\mathsf{C}'$ that minimizes the length of the vector $\sum_{m=1}^{M}\gamma_m\hat{\mathsf{E}}_m\mathsf{C}' + \mathsf{C}'$. This is schematically shown in Fig. 12.10b. To find it, we employ the standard variational principle and solve the corresponding set of linear equations for γ_m. When we displace the end of our vector C' in this direction, the angle Θ between this vector and the plane $\left\{\hat{\mathsf{E}}_m\mathsf{C}'\right\}$ increases. Since the orientation of the plane $\left\{\hat{\mathsf{E}}_m\mathsf{C}'\right\}$ may also change with the displacement, the latter should not be too big. To achieve the best convergency of our further iterations we direct the new vector to the "median" point $\mathsf{C}' \to \sum_{m=1}^{M}\frac{1}{2}\gamma_m\hat{\mathsf{E}}_m\mathsf{C}' + \mathsf{C}'$, then normalize the new vector, and repeat this procedure until the required accuracy is achieved.

Summarizing, we state that the discussed protecting codes give a universal means to protect a quantum system from coherence loss, although they require a physical procedure of rapid coding and frequent measurement, which determines the efficiency of the error protection.

12.2.3 Coherence Protection by Random Coding and Measurements

The protection method described in the previous section relies on the coding transformation that makes error-induced increments of the state vector orthogonal to the coding space. Finding such transformations implies an exponentially complex calculation, and therefore it cannot be practically applied for rather large systems. Still a similar approach can be suggested, such that it does allow if not to completely exclude then to drastically diminish the action of errors on a large system. It is based on the very fact that the system is large, with exponentially high dimensionality of the corresponding Hilbert space, whereas two randomly chosen vectors in a space of high dimension, though not orthogonal, have an exponentially small overlap given by the inverse square root of the Hilbert space dimension. This phenomenon is often referred to as the orthogonality catastrophe. The main idea of the method is to perform a random quantum coding, which is an analog of classical random codes.

Classical random coding is a method for producing efficient linear codes $[n, k, d]$ where k-bit words are encoded as randomly chosen n-bit sequences. The minimum distance d between any two code words, that is, the minimum number of bits to be flipped in order to obtain a meaningful code word from another meaningful code word, known as Hamming distance, meets the so-called Varshamov–Gilbert bound $k/n \leq 1 - H(d/n)$ in the limit $n \to \infty$, where $H(x) = -x\log_2 x - (1-x)\log_2(1-x)$ is the binary entropy function. The idea of random coding is also applicable to quantum codes and allows for efficient production, encoding, decoding, and protection of the information contained in a state vector of a large quantum system. This can be achieved by a manipulation similar to that described in the previous section, with the one important difference—the coding is done not by a certain unitary transformation found as a result of exponentially complex calculation but by a rather short sequence of simple unitary transformations with randomly chosen parameters.

Strong mixing occurring in the phase space of complex classical systems and its quantum counterpart, entanglement, occurring in the Hilbert space of composed quantum systems, are the very phenomena the random coding relies on. In practice, however, the dynamics of classical complex systems is not reversible, and therefore subsequent decoding will not be possible. And yet, the dynamics of quantum multi-dimensional systems can be reversible, provided the underlying physical system is simple. In this section we show how to harness such high dimensionality to provide coherence protection of generic applicability.

To combine strong mixing with reversibility, we consider a quantum system with a large number of separate energy levels and two simple interactions that can be fully controlled in the sense of a Lie group—together with unitary transformations produced by either of the interaction the inverse operations are supposed to be equally available. In such a system we can use non-holonomic control to encode quantum data into strongly mixed states by a unitary transformation $\hat{C}$ and subsequently to apply the inverse transformation $\hat{C}^{-1}$ to decode the data. Encoding the data into many levels allows us to strongly reduce the rate at which the data is corrupted by arbitrary physical errors, as will be shown. In turn, by applying the Zeno effect we can restore the slightly corrupted data back to its original value with high probability.

To encode k qubits of data in a high-dimensional Hilbert space of n qubits, we introduce $n-k$ ancilla qubits in addition to the data qubits. This results in an increased number of possible errors, which depends polynomially on n, but the error rate decreases at will because the infinitesimal errors are semi-orthogonal to the encoded data to a degree exponential in $n-k$. The degree of semi-orthogonality reflects the error-correction efficacy of the coding. For coding in minimal dimensions, efficacy requires a careful choice of the code, but in high dimensions it is trivially achieved by random coding. In physical terms, random coding translates into strong mixing, or full population of all energy levels, which is attainable by non-holonomic control with a number of interaction switchings depending only polynomially on n. Thus the random coding approach complements the non-holonomic Zeno decoherence suppression scheme of the previous section, which demands an exponential effort to find and effect an exact code, and which is therefore restricted to low-dimension systems.

The state of the k data qubits is a continuous complex-valued vector in the entire 2^k-dimensional Hilbert space $\mathcal{H}_k$. When the original system is expanded by $n-k$ ancilla qubits in some state $|\tilde{\alpha}\rangle$, the extended state vectors form a subspace $\mathcal{H}_{k,\tilde{\alpha}} = |\tilde{\alpha}\rangle \langle\tilde{\alpha}| \mathcal{H}_n$ in the larger space $\mathcal{H}_n$. The quantum encoding $\hat{C}$ is a unitary transformation that maps the subspace $\mathcal{H}_{k,\tilde{\alpha}}$ onto a 2^k-dimensional code subspace $\mathcal{C} = \hat{C}\mathcal{H}_{k,\tilde{\alpha}}$ in $\mathcal{H}_n$. Note that $\mathcal{C}$ is spanned by the columns of the rectangular matrix $\hat{C}\,|\tilde{\alpha}\rangle$, and has zero projections on the $2^n - 2^k$ column vectors of the matrices $\hat{C}\,|\alpha\rangle$ for states $|\alpha\rangle \perp |\tilde{\alpha}\rangle$ completing a basis for the ancilla Hilbert space.

The regular Zeno effect of Sect.12.1.4 allows to preserve a quantum system in a known initial state by periodic projective measurements, provided that the period T is short relative to the relaxation time T_2. Here, however, in complete analogy to the situation shown in Fig. 12.10, we only perform measurements on the ancilla part, in order to protect the unknown quantum state of the data part, restoring it to the initial state up to the second-order terms in T/T_2. To see the results of the Zeno effect, consider first the state vector $|S\rangle = |\tilde{\alpha}\rangle\,|s\rangle$ of the compound system formed by the ancilla in the state $|\tilde{\alpha}\rangle$ and the data part in some state $|s\rangle$. The encoding $\hat{C}$ maps the state $|S\rangle$ to the code vector $|v\rangle = \hat{C}\,|\tilde{\alpha}\rangle\,|s\rangle \in \mathcal{C}$, which then undergoes an error-induced, uncontrolled unitary evolution of the form of (12.57), where, as earlier, the error sources are assumed to be described by Hamiltonians $\hat{E}_m$ of interactions with external random fields $f_m(t)$ that produce an uncontrolled evolution of the system. We also assume, as earlier, that over the Zeno period T the different fields have small actions, $|\hat{E}_m \int_t^{t+T} f_m(x)dx| \ll 1$, so that only the identity operator $\hat{I}$ and the first-order terms are important in the Taylor series.

The evolution of the system under the action of errors is, again, given by (12.59) where effective Hamiltonian $\hat{h}_e = \sum_{m=1}^{M} f_m \langle\tilde{\alpha}|\,\hat{C}^{-1}\hat{E}_m\hat{C}\,|\tilde{\alpha}\rangle$ now should not vanish completely for arbitrary field strengths f_m, as suggested by (12.60) in the low-dimension scheme, but just diminished considerably by random encoding $\hat{C}$ in the Hilbert space of extremely high dimension thus suppressing the error accumulation in the density matrix. For systems with binary interaction an exponential reduction of the error-accumulation rate can be achieved if the projection rate $1/T$ grows faster than n^2. Moreover, the code rate $R = k/n$ tends to unity as $1-o(n^2T)$.

In particular, one can reduce all effective Hamiltonians

$$\hat{h}_m = \langle\tilde{\alpha}|\,\hat{C}^{-1}\hat{E}_m\hat{C}\,|\tilde{\alpha}\rangle$$

by applying a random encoding $\hat{C}$, which allows to reduce the rate of decoherence by a factor of the order of ancilla dimension $A = 2^{n-k}$. Indeed, for individual qubits, a typical single-particle or binary interaction $\hat{E}_m$ is given by a sparse matrix that strongly couples each state to only few other states, with most matrix elements being zero. A generic unitary encoding $\hat{C}$ smears out this coupling over all 2^n states of the entire system, thus producing a typical matrix element of $\hat{C}^{-1}\hat{E}_m\hat{C}$ that is 2^n times smaller than the typical nonzero matrix element of $\hat{E}_m$. Since only 2^k of these states contribute to the projected interactions $\hat{h}_m$, the overall reduction is equal to 2^{n-k}. From a more general viewpoint, the effectiveness of the random encoding rests on the fact that the decoherence rate is only polynomial in the number of particles n, whereas the dimension of the entire Hilbert space is exponential in n. The random encoding allows to store quantum information in strongly entangled states of many-body systems that are only weakly affected by physically realistic perturbations.

This random encoding also allows to specify a relation between the dimension $K = 2^k$ of the information system and the maximum number t of possible errors caused by binary interactions. Given D single-particle quantum states, there exist $M = \sum_{l=1}^{t} \binom{n}{l} \left(D^2 - 1\right)^l$ different possible error matrices (here $D = 2$ for qubits). Since the energy of binary interaction scales as n^2, each error matrix yields a decoherence rate of $W = n^2/T_2$. This rate is reduced $A = 2^{n-k}$ times after the matrix elements are smeared out over the entire system. Thus, given $n \to \infty$ identical particles, one can correct t errors with high fidelity if the total decoherence rate $W2^{k-n}M$ vanishes. This condition, $W2^{k-n}M \to 0$, gives the Hamming bound

$$1 - k/n - \log_2 M^{1/n} \geq 0. \tag{12.62}$$

Hence, to asymptotically meet this classical bound, we can randomly choose an (n, k)-code thus leaving the error patterns of weight t or less uncorrected with a vanishing probability.

The question arises now as to how to perform a random encoding in an efficient way. One can achieve this feasibility with the help of non-holonomic control over the Lie group, by applying two different natural Hamiltonians $\hat{H}_1$ and $\hat{H}_2$ that satisfy certain conditions. A generic choice of a sequence of non-holonomic control timings $t_1, t_2, t_3, \cdots, t_{M'}$ of a reasonably short length M' will result in a random encoding transformation

$$\hat{C} = e^{-it_{M'}\hat{H}_2} \cdots e^{-it_3\hat{H}_1} e^{-it_2\hat{H}_2} e^{-it_1\hat{H}_1}. \tag{12.63}$$

Indeed, the sequential application of two different Hamiltonians suggested by (12.63) results in a completely generic, random transformation provided that: (a) the number M' of the control timings t_l is of the order of $n = \log_2(A \times K)$; (b) the operators $\hat{H}_1$ and $\hat{H}_2$ satisfy the conditions of non-holonomic control; and (c) the time intervals t_l are long enough to ensure big acquired actions, $|\hat{H}_{1,2}\, t_l| \gg \hbar$. Therefore, given a physical system, the main problem is to find two controllable interactions that satisfy the conditions needed for non-holonomic control, and allow to perform the decoding transformation $\hat{C}^{-1}$ at the same level of complexity by simply changing the sign of the interaction.

12.2.4 Topological Decoupling from the Environment

Consideration of protection of a quantum system against uncontrolled interventions presented in Sects. 12.2.2 and 12.2.3 relies on the general idea of classical Coding Theory, which suggests to give the system "an additional space" by combining it with an ancilla, such that two meaningful code words in this new combined system get separated by a certain Hamming distance d after the coding procedure. The uncontrolled interventions that in the classical case are simple flips of individual bits can be identified and corrected as long as the number of such flips is inferior of d. The other way to formulate this, is to say that the space of the code words embedded in the space of all possible states of the combined system is sparse enough, such that the errors are unable to mix up two different code words. In the analogous quantum situation, the state vector $|v\rangle$ of the combined system belonging to a code subspace $\mathcal{C}$ of the entire Hilbert space carries the quantum information as long as the error-inducing perturbations $\hat{E}_i$ are unable to provoke transitions among different basis vectors spanning the code space, as is formalized in (12.60).

The quantum coding considered thus far was a unitary transformation to perform as a physical control action after the system and the ancilla were combined together. However, a quantum system in the code space $\mathcal{C}$ is not necessarily a result of such an action—it may exist naturally as a multiply degenerate ground state of a certain many-body Hamiltonian. Designed in a special way, such a manifold is protected against errors, if it remains insensitive to the single-body error Hamiltonians $\hat{E}_i$ up to a certain order of the perturbation theory. In other words, the coding space exists as a degenerate lowest energy subspace of a much larger Hilbert space of the many-body system, while $\hat{E}_i$ together with the products $\underbrace{\hat{E}_i \dots \hat{E}_j}_{d}$ up to a certain order d have zero projections to this subspace.

Consider the 4-qubit system shown in Fig. 12.11a as an example. Four two-level systems, say spins $1/2$, are placed at the vertices of a square and labeled by pairs of indices $\{i, j\}$ each of which may take values 0 or 1. The tight-binding Hamiltonian

$$\hat{H} = \frac{\beta}{2}\left(\hat{\sigma}^x_{0,0}\hat{\sigma}^x_{0,1} + \hat{\sigma}^x_{1,0}\hat{\sigma}^x_{1,1}\right) + \frac{\alpha}{2}\left(\hat{\sigma}^z_{0,0}\hat{\sigma}^z_{1,0} + \hat{\sigma}^z_{0,1}\hat{\sigma}^z_{1,1}\right) \tag{12.64}$$

responsible for the interaction of these spins has different nearest neighbor couplings in different direction: in one directions the spins are coupled via x-components with a binding parameter $\frac{\alpha}{2}$, while in the other direction via z-components with a coupling $\frac{\beta}{2}$. Provided $\alpha > \beta$, one can explicitly find the energy spectrum

$$\mathcal{E}_i = \left\{-\sqrt{\alpha^2 + \beta^2}, -\alpha, -\beta, 0, \beta, \alpha, \sqrt{\alpha^2 + \beta^2}\right\},$$

where all but zero energy levels are doubly degenerate. The level at zero energy is fourfold degenerate.

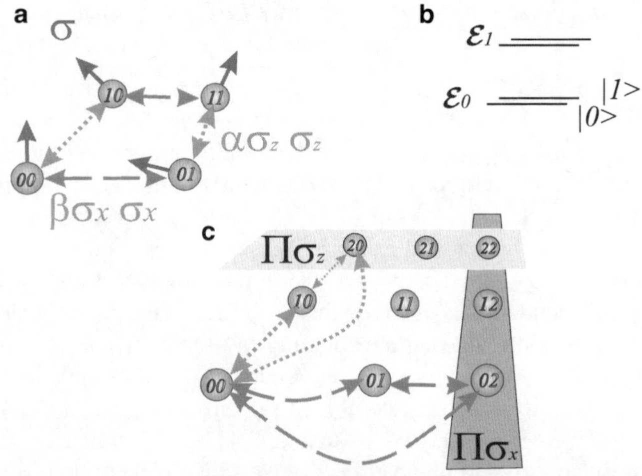

Fig. 12.11 (**a**) Four qubits arranged in a $2D$ lattice. Along one direction the tight-binding interaction between the qubits is given by the Hamiltonian $\frac{\alpha}{2}\hat{\sigma}^x_{i,j}\hat{\sigma}^x_{i,j+1}$, and along the other direction it reads $\frac{\beta}{2}\hat{\sigma}^z_{i,j}\hat{\sigma}^z_{i+1,j}$. (**b**) Two lowest energy manifolds of the eigenstates of this system at the energies $\mathcal{E}_0 = -\sqrt{\alpha^2+\beta^2}$ and $\mathcal{E}_1 = -\alpha$ are doubly degenerate. (**c**) Nine qubits arranged in 3×3 lattice structure with interactions of the torus symmetry also coupling the sites at different edges. The energy spectrum is sixfold degenerate

One can also find the energy eigenstates and check explicitly that the two lowest states

$$|0\rangle = \left|\begin{matrix}0\,0\\0\,0\end{matrix}\right\rangle - a\left|\begin{matrix}0\,0\\1\,1\end{matrix}\right\rangle - a\left|\begin{matrix}1\,1\\0\,0\end{matrix}\right\rangle + \left|\begin{matrix}1\,1\\1\,1\end{matrix}\right\rangle$$

$$|1\rangle = \left|\begin{matrix}0\,1\\0\,1\end{matrix}\right\rangle - a\left|\begin{matrix}0\,1\\1\,0\end{matrix}\right\rangle - a\left|\begin{matrix}1\,0\\0\,1\end{matrix}\right\rangle + \left|\begin{matrix}1\,0\\1\,0\end{matrix}\right\rangle, \tag{12.65}$$

where $a = \frac{\alpha-\sqrt{\alpha^2+\beta^2}}{\beta}$, indeed have the same energy, and moreover in the first order of the perturbation theory these states are not affected by the single spin operators $\hat{E}_{i,j,k} \sim \hat{\sigma}^{k=x,y,z}_{i,j}$ since

$$\langle 1|\,\hat{E}_{i,j,k}\,|1\rangle = \langle 0|\,\hat{E}_{i,j,k}\,|1\rangle = \langle 0|\,\hat{E}_{i,j,k}\,|0\rangle = 0.$$

Here, by

$$\left|\begin{matrix}a_{1,1}\;a_{1,2}\\a_{2,1}\;a_{2,2}\end{matrix}\right\rangle$$

with $a_{i,j} = 0, 1$ we denote the basis states, where 0 at the position $\{i, j\}$ of the ket parentheses denotes the spin z- projection down at the site with $x = i$, and $y = j$, while 1 stands for the corresponding spin up state. In the ground state duplet of

(12.65), the state $|1\rangle$ differs from the state $|0\rangle$ by flips of the spins in the second column. Coupling of the states of the duplet takes place only in the second-order perturbation theory over the local operators $\hat{E}_{i,j,k}$.

The double degeneracy of the energy eigenstates is not occasional—it follows from the symmetry of the Hamiltonian (12.64). The part of the Hamiltonian proportional to α and the part proportional to β do not satisfy the bracket generation condition, and therefore it turns out to be possible to identify operators that commute with both parts. Indeed, one finds at least two operators

$$\hat{Q} = \hat{\sigma}^z_{0,0}\hat{\sigma}^z_{0,1}$$
$$\hat{P} = \hat{\sigma}^x_{0,1}\hat{\sigma}^x_{1,1}$$

not commuting each with the other, although both commuting with the Hamiltonian (12.64) for any α and β. This implies that $\hat{Q}$ and $\hat{H}$ can be simultaneously diagonalized, $\hat{P}$ and $\hat{H}$ can be simultaneously diagonalized, but not $\hat{P}$ and $\hat{Q}$, which is only possible when the spectrum of $\hat{H}$ is doubly degenerate. Note that the state $|0\rangle$ is an eigenstate of the operator $\hat{Q}$ with the eigenvalue 1, while the state $|1\rangle$ is the eigenstate with the eigenvalue -1. In other words, $\hat{Q}$ works as $\hat{\sigma}^z$ operator in the code space of the doubly degenerate ground state. The operator $\hat{P}$ flips the "protected qubit" since $|0\rangle = \hat{P}\,|1\rangle$ and $|1\rangle = \hat{P}\,|0\rangle$ thus working as $\hat{\sigma}^x$ operator in the code space.

Consider now a 3×3 lattice of spins, shown in Fig. 12.11c, that have the Hamiltonian

$$\hat{H} = \sum_{i,j=0}^{2} \frac{\alpha}{2}\hat{\sigma}^z_{i,j}\hat{\sigma}^z_{i+1,j} + \frac{\beta}{2}\hat{\sigma}^x_{i,j}\hat{\sigma}^x_{i,j+1}, \tag{12.66}$$

where the index $i = 3$ is identified with $i = 0$. In other words, the lattice is implied to be at a torus surface. One finds six operators

$$\hat{Q}_i = \hat{\sigma}^z_{i,0}\hat{\sigma}^z_{i,1}\hat{\sigma}^z_{i,2},$$
$$\hat{P}_i = \hat{\sigma}^x_{0,i}\hat{\sigma}^x_{1,i}\hat{\sigma}^x_{2,i} \tag{12.67}$$

commuting with this Hamiltonian and not commuting among themselves. This system with 512 eigenstates is on the border of complexity where the complete and straightforward analysis is still feasible. The ground state can have a degeneracy of either 2 or 6, depending on the relation between α and $\beta \neq \alpha$. The double degeneracy is universal and does not depend on the boundary conditions. In particular, it persists even when instead of the torus topology one sets the plane topology of the Hamiltonian (12.66), by assuming that when at least one of the indices i or j assumes the value 3, the corresponding spin operator vanishes.

The doubly degenerate ground state of the Hamiltonian (12.66) is spanned by two energy eigenstates $|0\rangle$ and $|1\rangle$, each of which is a linear combination of 2^6 states of the form

$$\left|\begin{matrix} a_{11} & a_{12} & a_{12} \\ a_{21} & a_{22} & a_{22} \\ a_{31} & a_{32} & a_{32} \end{matrix}\right\rangle = \prod_{\otimes i,j} \left(\hat{\sigma}_{ij}^{+}\right)^{a_{ij}} |\mathbf{0}\rangle\,. \tag{12.68}$$

Here, as earlier, the numbers $a_{i,j} = 1$ specify which are the nods $\{i, j\}$ of the lattice that carry the spin up single-particle states while $a_{i,j} = 0$ do that for the spin down states. The right hand side of (12.68) presents these states via the spin rising operators $\hat{\sigma}_{ij}^{+}$, in the spirit of the notations that have been employed in Chap. 9 for the entanglement description with the help of nilpotent variables σ_{ij}^{+}. Direct calculation for $\beta < \alpha < 0$ gives explicitly two states $|0\rangle$ and $|1\rangle$ of the doubly degenerate ground state in terms of linear combinations of the basis state (12.68) with just three real coefficients a, b, c ($|a| > |b| > |c|$) that depend on the ratio β/α exclusively. One of the states reads

$$\begin{aligned}
|0\rangle =&[a+\\
&a(\sigma_{11}^{+}\sigma_{12}^{+}\sigma_{13}^{+}\sigma_{21}^{+}\sigma_{22}^{+}\sigma_{23}^{+} + \sigma_{21}^{+}\sigma_{22}^{+}\sigma_{31}^{+}\sigma_{32}^{+}\sigma_{33}^{+}\sigma_{23}^{+} + \sigma_{11}^{+}\sigma_{12}^{+}\sigma_{13}^{+}\sigma_{31}^{+}\sigma_{32}^{+}\sigma_{33}^{+})+\\
&b(\sigma_{11}^{+}\sigma_{12}^{+}\sigma_{13}^{+}\sigma_{22}^{+}\sigma_{23}^{+}\sigma_{31}^{+} + \sigma_{12}^{+}\sigma_{13}^{+}\sigma_{21}^{+}\sigma_{22}^{+}\sigma_{23}^{+}\sigma_{31}^{+} + \sigma_{11}^{+}\sigma_{12}^{+}\sigma_{13}^{+}\sigma_{23}^{+}\sigma_{32}^{+}\sigma_{31}^{+}+\\
&\sigma_{13}^{+}\sigma_{21}^{+}\sigma_{22}^{+}\sigma_{23}^{+}\sigma_{32}^{+}\sigma_{31}^{+} + \sigma_{11}^{+}\sigma_{12}^{+}\sigma_{13}^{+}\sigma_{22}^{+}\sigma_{33}^{+}\sigma_{31}^{+} + \sigma_{12}^{+}\sigma_{21}^{+}\sigma_{22}^{+}\sigma_{23}^{+}\sigma_{33}^{+}\sigma_{31}^{+}+\\
&\sigma_{12}^{+}\sigma_{13}^{+}\sigma_{21}^{+}\sigma_{32}^{+}\sigma_{33}^{+}\sigma_{31}^{+} + \sigma_{11}^{+}\sigma_{13}^{+}\sigma_{22}^{+}\sigma_{32}^{+}\sigma_{33}^{+}\sigma_{31}^{+} + \sigma_{13}^{+}\sigma_{21}^{+}\sigma_{22}^{+}\sigma_{32}^{+}\sigma_{33}^{+}\sigma_{31}^{+}+\\
&\sigma_{11}^{+}\sigma_{12}^{+}\sigma_{23}^{+}\sigma_{32}^{+}\sigma_{33}^{+}\sigma_{31}^{+} + \sigma_{12}^{+}\sigma_{21}^{+}\sigma_{23}^{+}\sigma_{32}^{+}\sigma_{33}^{+}\sigma_{31}^{+} + \sigma_{11}^{+}\sigma_{22}^{+}\sigma_{23}^{+}\sigma_{32}^{+}\sigma_{33}^{+}\sigma_{31}^{+}+\\
&\sigma_{11}^{+}\sigma_{12}^{+}\sigma_{13}^{+}\sigma_{21}^{+}\sigma_{23}^{+}\sigma_{32}^{+} + \sigma_{11}^{+}\sigma_{13}^{+}\sigma_{21}^{+}\sigma_{22}^{+}\sigma_{23}^{+}\sigma_{32}^{+} + \sigma_{11}^{+}\sigma_{12}^{+}\sigma_{13}^{+}\sigma_{21}^{+}\sigma_{22}^{+}\sigma_{33}^{+}+\\
&\sigma_{11}^{+}\sigma_{12}^{+}\sigma_{21}^{+}\sigma_{22}^{+}\sigma_{23}^{+}\sigma_{33}^{+} + \sigma_{11}^{+}\sigma_{12}^{+}\sigma_{13}^{+}\sigma_{21}^{+}\sigma_{32}^{+}\sigma_{33}^{+} + \sigma_{11}^{+}\sigma_{21}^{+}\sigma_{22}^{+}\sigma_{23}^{+}\sigma_{32}^{+}\sigma_{33}^{+})+\\
&b(\sigma_{11}^{+}\sigma_{21}^{+} + \sigma_{31}^{+}\sigma_{21}^{+} + \sigma_{12}^{+}\sigma_{22}^{+} + \sigma_{13}^{+}\sigma_{23}^{+} + \sigma_{11}^{+}\sigma_{31}^{+} + \sigma_{12}^{+}\sigma_{32}^{+} + \sigma_{22}^{+}\sigma_{32}^{+}+\\
&\sigma_{13}^{+}\sigma_{33}^{+} + \sigma_{23}^{+}\sigma_{33}^{+}) + b(\sigma_{11}^{+}\sigma_{12}^{+}\sigma_{21}^{+}\sigma_{22}^{+} + \sigma_{12}^{+}\sigma_{13}^{+}\sigma_{23}^{+}\sigma_{22}^{+} + \sigma_{21}^{+}\sigma_{31}^{+}\sigma_{32}^{+}\sigma_{22}^{+}+\\
&\sigma_{23}^{+}\sigma_{32}^{+}\sigma_{33}^{+}\sigma_{22}^{+} + \sigma_{11}^{+}\sigma_{13}^{+}\sigma_{21}^{+}\sigma_{23}^{+} + \sigma_{11}^{+}\sigma_{12}^{+}\sigma_{31}^{+}\sigma_{32}^{+} + \sigma_{11}^{+}\sigma_{13}^{+}\sigma_{31}^{+}\sigma_{33}^{+}+\\
&\sigma_{21}^{+}\sigma_{23}^{+}\sigma_{31}^{+}\sigma_{33}^{+} + \sigma_{12}^{+}\sigma_{13}^{+}\sigma_{32}^{+}\sigma_{33}^{+}) + c(\sigma_{11}^{+}\sigma_{12}^{+}\sigma_{22}^{+}\sigma_{31}^{+} + \sigma_{12}^{+}\sigma_{21}\sigma_{22}^{+}\sigma_{31}^{+}\\
&+ \sigma_{11}^{+}\sigma_{13}^{+}\sigma_{23}^{+}\sigma_{31} + \sigma_{13}^{+}\sigma_{21}^{+}\sigma_{23}^{+}\sigma_{31}^{+} + \sigma_{12}^{+}\sigma_{21}^{+}\sigma_{32}^{+}\sigma_{31}^{+} + \sigma_{11}^{+}\sigma_{22}^{+}\sigma_{32}^{+}\sigma_{31}^{+}+\\
&\sigma_{13}^{+}\sigma_{21}^{+}\sigma_{33}^{+}\sigma_{31}^{+} + \sigma_{11}^{+}\sigma_{23}^{+}\sigma_{33}^{+}\sigma_{31}^{+} + \sigma_{11}^{+}\sigma_{12}^{+}\sigma_{21}^{+}\sigma_{32}^{+} + \sigma_{11}^{+}\sigma_{21}^{+}\sigma_{22}^{+}\sigma_{32}^{+}\\
&+ \sigma_{12}^{+}\sigma_{13}^{+}\sigma_{23}^{+}\sigma_{32}^{+} + \sigma_{13}^{+}\sigma_{22}^{+}\sigma_{23}^{+}\sigma_{32}^{+} + \sigma_{11}^{+}\sigma_{13}^{+}\sigma_{21}^{+}\sigma_{33}^{+} + \sigma_{12}^{+}\sigma_{13}^{+}\sigma_{22}^{+}\sigma_{33}^{+}+\\
&\sigma_{11}^{+}\sigma_{21}^{+}\sigma_{23}^{+}\sigma_{33}^{+} + \sigma_{12}^{+}\sigma_{22}^{+}\sigma_{23}^{+}\sigma_{33}^{+} + \sigma_{13}^{+}\sigma_{22}^{+}\sigma_{32}^{+}\sigma_{33}^{+} + \sigma_{12}^{+}\sigma_{23}^{+}\sigma_{32}^{+}\sigma_{33}^{+})+\\
&c(\sigma_{12}^{+}\sigma_{13}^{+}\sigma_{21}^{+}\sigma_{23}^{+}\sigma_{31}^{+}\sigma_{32}^{+} + \sigma_{11}^{+}\sigma_{13}^{+}\sigma_{22}^{+}\sigma_{23}^{+}\sigma_{31}^{+}\sigma_{32}^{+} + \sigma_{11}^{+}\sigma_{13}^{+}\sigma_{21}^{+}\sigma_{22}^{+}\sigma_{33}^{+}\sigma_{32}^{+}+\\
&\sigma_{11}^{+}\sigma_{12}^{+}\sigma_{21}^{+}\sigma_{23}^{+}\sigma_{33}^{+}\sigma_{32}^{+} + \sigma_{12}^{+}\sigma_{13}^{+}\sigma_{21}^{+}\sigma_{22}^{+}\sigma_{31}^{+}\sigma_{33}^{+} + \sigma_{11}^{+}\sigma_{12}^{+}\sigma_{22}^{+}\sigma_{23}^{+}\sigma_{31}^{+}\sigma_{33}^{+})]\,|\mathbf{0}\rangle\,,
\end{aligned} \tag{12.69}$$

while the other state, $|1\rangle$, results from the applications of the total flip operator $\prod_{\otimes i,j} \hat{\sigma}^x_{ij}$ to the state $|0\rangle$. Note that the states $|0\rangle$ and $|1\rangle$ are again the eigenstates of $\hat{Q}_k$ operators with $\hat{Q}_k |0\rangle = (-1)^k |0\rangle$ and $\hat{Q}_k |1\rangle = -|1\rangle$, which means that either of the operators $\hat{Q}_0$ and $\hat{Q}_2$ can be considered as $\hat{\sigma}^z$ operator for the protected qubit. Direct analysis suggests that the role of $\hat{\sigma}^x$ operator can be played by the product $\hat{P}_0\hat{Q}_1$, as one of the possibilities.

Inspection of (12.69) for the state $|0\rangle$ and its counterpart for expressions for $|1\rangle$ shows that the minimum Hamming distance d among them equals 2, that is, all the basis vectors (12.68) contributing to either $|0\rangle$ or $|1\rangle$ differ at least in two positions $\{i, j\}$, or, in other words, each of the terms entering (12.69) differs from each of the terms entering the analogous expression for $|1\rangle$ at least by two factors σ^+. This implies that the Hilbert subspace corresponding to the doubly degenerate ground state is insensitive to the single-particle perturbations $\hat{E}_{i,j,k}$. Moreover, it turns out that the amplitudes in front of the basis states are such that the ground state is also insensitive to $\hat{E}_{i,j,k}$ in the second-order perturbations theory, whereas the first nonvanishing contribution $\sim \left[\left\|\hat{E}_{i,j,k}\right\| / \min(\alpha, \beta)\right]^3$ comes from the third-order perturbation and correspond to flips of all spins in a same column of the lattice. Note that here we count only the non-trivial contributions that are not proportional to the identity.

One can conjecture that for a larger $M \times M$ lattice the first nonvanishing contribution $\left[\left\|\hat{E}_{i,j,k}\right\| / \min(\alpha, \beta)\right]^M$ appears in the M-th order of the perturbation theory, which means that the efficiency of the protection against single spin perturbations scales exponentially with the size of the lattice. Note, however, that the fact that the ground state is well-protected against single spin perturbation also means that the quantum transformations in this subspace must rely on an essentially nonlocal interaction. One can state that this kind of protection also complicates the way how the desired quantum operations have to be performed in the error-protected subspace. There is a certain physical reason for this, which is hidden in the fact that a Hamiltonian of even more general type

$$\hat{H} = \sum_{i,j=0}^{M} \frac{\alpha_{ij}}{2} \hat{\sigma}^z_{i,j} \hat{\sigma}^z_{i,j+1} + \frac{\beta_{ij}}{2} \hat{\sigma}^x_{i,j} \hat{\sigma}^x_{i+1,j}, \tag{12.70}$$

exhibits a number of holonomic constrains, whereas the local symmetry imposes a certain global invariance. This follows from the commutativity of all the local contributions to the Hamiltonian $\hat{\sigma}^x_{i,j}\hat{\sigma}^x_{i,j+1}$ and $\hat{\sigma}^z_{i,j}\hat{\sigma}^z_{i+1,j}$ with all the global products of the types $\prod_i^M \hat{\sigma}^x_{i,j}$ and $\prod_j^M \hat{\sigma}^z_{i,j}$. The degenerate eigenstates of the Hamiltonian are therefore numbered by the eigenvalues of the functionally independent global invariants, and in order to change these invariants one has to apply either a global transformation or a combination of local transformations in a very high order of the perturbation theory.

A natural question arises: can the dimension of the subspace protected against local error systematically be made larger than two? The answer to this question is positive: the structure of the Hamiltonian can be designed in such a way that the ground state has multiple degeneracy, which depends exclusively on the topological properties of the spin coupling network. One can see it at the simplest example comparing the spectrum of 8 spin system with the Hamiltonian

$$\begin{aligned}\hat{H} = {} & \frac{\alpha}{2}\hat{\sigma}^z_{1,1}\hat{\sigma}^z_{1,2} + \frac{\alpha}{2}\hat{\sigma}^z_{1,2}\hat{\sigma}^z_{\boxed{1,1}} + \frac{\alpha}{2}\hat{\sigma}^z_{1,3}\hat{\sigma}^z_{1,4} + \frac{\alpha}{2}\hat{\sigma}^z_{1,4}\hat{\sigma}^z_{\boxed{1,3}} + \\ & \frac{\alpha}{2}\hat{\sigma}^z_{2,1}\hat{\sigma}^z_{2,2} + \frac{\alpha}{2}\hat{\sigma}^z_{2,2}\hat{\sigma}^z_{2,3} + \frac{\alpha}{2}\hat{\sigma}^z_{2,3}\hat{\sigma}^z_{2,4} + \frac{\alpha}{2}\hat{\sigma}^z_{2,4}\hat{\sigma}^z_{2,1} + \\ & \frac{\beta}{2}\hat{\sigma}^x_{1,1}\hat{\sigma}^x_{2,1} + \frac{\beta}{2}\hat{\sigma}^x_{1,2}\hat{\sigma}^x_{2,2} + \frac{\beta}{2}\hat{\sigma}^x_{1,3}\hat{\sigma}^x_{2,3} + \frac{\beta}{2}\hat{\sigma}^x_{1,4}\hat{\sigma}^x_{2,4}, \end{aligned} \tag{12.71}$$

the Hamiltonian

$$\begin{aligned}\hat{H} = {} & \frac{\alpha}{2}\hat{\sigma}^z_{1,1}\hat{\sigma}^z_{1,2} + \frac{\alpha}{2}\hat{\sigma}^z_{1,2}\hat{\sigma}^z_{\boxed{1,3}} + \frac{\alpha}{2}\hat{\sigma}^z_{1,3}\hat{\sigma}^z_{1,4} + \frac{\alpha}{2}\hat{\sigma}^z_{1,4}\hat{\sigma}^z_{\boxed{1,1}} + \\ & \frac{\alpha}{2}\hat{\sigma}^z_{2,1}\hat{\sigma}^z_{2,2} + \frac{\alpha}{2}\hat{\sigma}^z_{2,2}\hat{\sigma}^z_{2,3} + \frac{\alpha}{2}\hat{\sigma}^z_{2,3}\hat{\sigma}^z_{2,4} + \frac{\alpha}{2}\hat{\sigma}^z_{2,4}\hat{\sigma}^z_{2,1} + \\ & \frac{\beta}{2}\hat{\sigma}^x_{1,1}\hat{\sigma}^x_{2,1} + \frac{\beta}{2}\hat{\sigma}^x_{1,2}\hat{\sigma}^x_{2,2} + \frac{\beta}{2}\hat{\sigma}^x_{1,3}\hat{\sigma}^x_{2,3} + \frac{\beta}{2}\hat{\sigma}^x_{1,4}\hat{\sigma}^x_{2,4}, \end{aligned} \tag{12.72}$$

and the Hamiltonian

$$\begin{aligned}\hat{H} = {} & \frac{\alpha}{2}\hat{\sigma}^z_{1,1}\hat{\sigma}^z_{1,2} + \frac{\alpha}{2}\hat{\sigma}^z_{1,2}\hat{\sigma}^z_{\boxed{1,4}} + \frac{\alpha}{2}\hat{\sigma}^z_{1,3}\hat{\sigma}^z_{1,4} + \frac{\alpha}{2}\hat{\sigma}^z_{\boxed{1,3}}\hat{\sigma}^z_{1,1} + \\ & \frac{\alpha}{2}\hat{\sigma}^z_{2,1}\hat{\sigma}^z_{2,2} + \frac{\alpha}{2}\hat{\sigma}^z_{2,2}\hat{\sigma}^z_{\boxed{2,4}} + \frac{\alpha}{2}\hat{\sigma}^z_{2,3}\hat{\sigma}^z_{2,4} + \frac{\alpha}{2}\hat{\sigma}^z_{\boxed{2,3}}\hat{\sigma}^z_{2,1} + \\ & \frac{\beta}{4}\hat{\sigma}^x_{1,1}\hat{\sigma}^x_{2,1} + \frac{\beta}{4}\hat{\sigma}^x_{1,2}\hat{\sigma}^x_{2,2} + \frac{\beta}{4}\hat{\sigma}^x_{1,3}\hat{\sigma}^x_{2,3} + \frac{\beta}{4}\hat{\sigma}^x_{1,4}\hat{\sigma}^x_{2,4} + \\ & \frac{\beta}{4}\hat{\sigma}^x_{\boxed{1,1}}\hat{\sigma}^x_{\boxed{1,3}} + \frac{\beta}{4}\hat{\sigma}^x_{\boxed{1,2}}\hat{\sigma}^x_{\boxed{1,4}} + \frac{\beta}{4}\hat{\sigma}^x_{\boxed{2,1}}\hat{\sigma}^x_{\boxed{2,3}} + \frac{\beta}{4}\hat{\sigma}^x_{\boxed{2,2}}\hat{\sigma}^x_{\boxed{2,4}}, \end{aligned} \tag{12.73}$$

where indices in the frames mark the difference in the topology of the interactions illustrated in Fig. 12.12. The Hamiltonian of (12.71) has connectivity corresponding to topological genus 2 in contrast to that of (12.72), which has toric topology of genus 1, and also in contrast with that of (12.73), which is topologically equivalent to a sphere thus having genus 0. As a consequence the minimum degeneracy of eigenstates for the first Hamiltonian amounts to 4, while that of the second one is 2. The third Hamiltonian has a number of non-degenerate eigenstates, including the ground state.

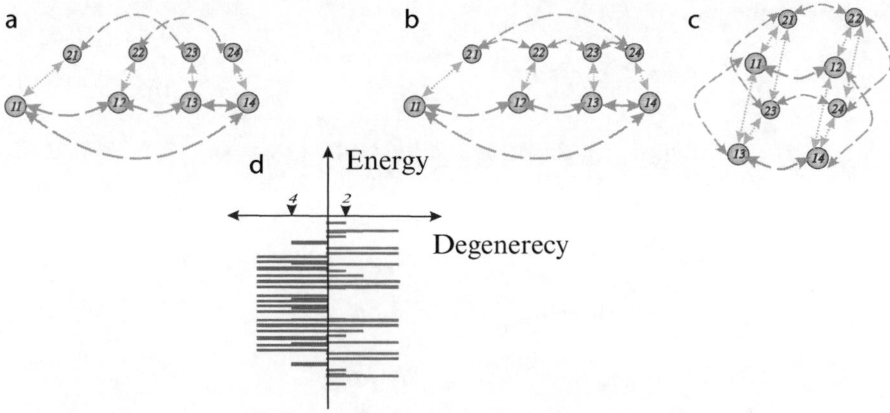

Fig. 12.12 (**a**) Eight qubit arranged in a $2D$ lattice with $\alpha = 2, \beta = 2.4$. The interaction topology corresponds to the Hamiltonian Eq. (12.71). (**b**) Same for Eq. (12.72) topology is equivalent to a torus. (**c**) Same for Eq. (12.73) topology is equivalent to a sphere. (**d**) Energy spectrum of the Hamiltonian Eq. (12.71) (*left*) has at least fourfold degeneracy of all levels. For the Hamiltonian Eq. (12.72) (*right*) the minimum degeneracy is 2. Energy levels of the Hamiltonian Eq. (12.73), including the ground state, have minimum degeneracy 1. The very high degeneracy of the states at zero energy is not shown

In the case $\alpha \gg \beta$ one can get short expressions for the ground state vectors:

$$|0\rangle = |\mathbf{0}\rangle + \sigma_{21}^{+}\sigma_{22}^{+}\sigma_{23}^{+}\sigma_{24}^{+}|\mathbf{0}\rangle + O(1)\left(\sigma_{11}^{+}\sigma_{12}^{+}\sigma_{13}^{+}\sigma_{14}^{+} + \sigma_{11}^{+}\sigma_{12}^{+}\sigma_{13}^{+}\sigma_{21}^{+}\sigma_{22}^{+}\sigma_{23}^{+}\right.$$
$$\left.\sigma_{24}^{+}\sigma_{14}^{+}\right)|\mathbf{0}\rangle$$

for the Hamiltonian (12.73),

$$|0\rangle = |\mathbf{0}\rangle + O\left(\frac{\beta^2}{\alpha^2}\right)\left(\sigma_{11}^{+}\sigma_{21}^{+} + \sigma_{12}^{+}\sigma_{22}^{+} + \sigma_{13}^{+}\sigma_{23}^{+} + \sigma_{24}^{+}\sigma_{14}^{+}\right)|\mathbf{0}\rangle$$

$$|1\rangle = \sigma_{21}^{+}\sigma_{22}^{+}\sigma_{23}^{+}\sigma_{24}^{+}|\mathbf{0}\rangle + O\left(\frac{\beta^2}{\alpha^2}\right)\left(\sigma_{13}^{+}\sigma_{21}^{+}\sigma_{22}^{+}\sigma_{24}^{+} +\right.$$
$$\left.\sigma_{12}^{+}\sigma_{21}^{+}\sigma_{23}^{+}\sigma_{24}^{+} + \sigma_{11}^{+}\sigma_{22}^{+}\sigma_{23}^{+}\sigma_{24}^{+} + \sigma_{21}^{+}\sigma_{22}^{+}\sigma_{23}^{+}\sigma_{14}^{+}\right)|\mathbf{0}\rangle$$

for the Hamiltonian (12.72), and

$$|0\rangle = |\mathbf{0}\rangle + O\left(\frac{\beta^2}{\alpha^2}\right)(\sigma_{11}^+\sigma_{21} + \sigma_{12}^+\sigma_{22}^+ + \sigma_{13}^+\sigma_{23}^+ + \sigma_{24}^+\sigma_{14}^+)\,|\mathbf{0}\rangle$$

$$|1\rangle = \sigma_{12}^+\sigma_{13}^+\,|\mathbf{0}\rangle + O\left(\frac{\beta^2}{\alpha^2}\right)(\sigma_{11}^+\sigma_{12}^+\sigma_{13}^+\sigma_{21}^+ + \sigma_{13}^+\sigma_{22}^+ +$$
$$\sigma_{12}^+\sigma_{23}^+ + \sigma_{12}^+\sigma_{13}^+\sigma_{24}^+\sigma_{24}^+)\,|\mathbf{0}\rangle$$

$$|2\rangle = \sigma_{21}^+\sigma_{22}^+\sigma_{23}^+\sigma_{24}^+\,|\mathbf{0}\rangle + O\left(\frac{\beta^2}{\alpha^2}\right)(\sigma_{13}^+\sigma_{21}^+\sigma_{22}^+\sigma_{24}^+ +$$
$$\sigma_{12}^+\sigma_{21}^+\sigma_{23}^+\sigma_{24}^+ + \sigma_{11}^+\sigma_{22}^+\sigma_{23}^+\sigma_{24}^+ + \sigma_{21}^+\sigma_{22}^+\sigma_{23}^+\sigma_{14}^+)\,|\mathbf{0}\rangle$$

$$|3\rangle = \sigma_{12}^+\sigma_{13}^+\sigma_{21}^+\sigma_{22}^+\sigma_{23}^+\sigma_{24}^+\,|\mathbf{0}\rangle + O\left(\frac{\beta^2}{\alpha^2}\right)(\sigma_{12}^+\sigma_{21}^+\sigma_{22}^+\sigma_{24}^+ +$$
$$\sigma_{13}^+\sigma_{21}^+\sigma_{23}^+\sigma_{24}^+ + \sigma_{11}^+\sigma_{12}^+\sigma_{13}^+\sigma_{22}^+\sigma_{23}^+\sigma_{24}^+ + \sigma_{12}^+\sigma_{13}^+\sigma_{14}^+\sigma_{21}^+\sigma_{22}^+\sigma_{23}^+)\,|\mathbf{0}\rangle$$

for the Hamiltonian (12.71). The non-degenerate state corresponding to the spherical topology is sensitive to the first-order local perturbations, as one can conclude from the presence of the contribution $\sim O(1)$ of the order of unity, while the toric and higher genus topologies are more robust and require at least third-order perturbations, $\sim O(\frac{\beta^2}{\alpha^2})$.

In a larger spin lattice, one can form a more sophisticated topology of the interaction network thus arriving at the multiply degenerate ground state exponentially protected against local spin errors. Note that the topologically guaranteed state degeneracy persists for any particular choice of the parameters α_{ij} and β_{ij} entering the Hamiltonian (12.70), yet an additional (occasional) degeneracy may occur for some particular choices of these parameters. The latter is the degeneracy to be avoided, since it may destroy the exponential protection against the local noises. A description of large spin lattices in terms of explicit entangled local spin states of the type (12.69) is extremely unpractical, since the vast majority of the quantum states of such systems does not belong to the protected code subspace of interest, which we associate with the degenerate ground state. One has to invoke another language where the topological aspects of the problem become more evident.

To this end we focus on the properties of the operators P_i and Q_j, which have already been introduced by (12.67) in the context of 3×3 spin lattice. For larger lattices one has to generalize these operators,

$$\hat{P}_i = \prod_j \hat{\sigma}_{ij}^x$$
$$\hat{Q}_i = \prod_j \hat{\sigma}_{ji}^z, \qquad (12.74)$$

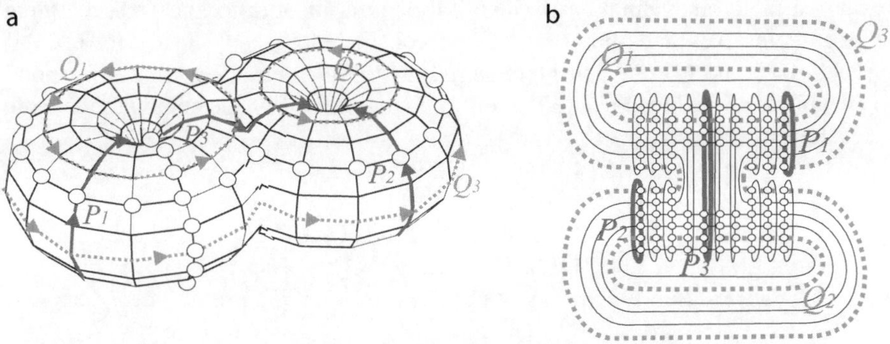

Fig. 12.13 Genus 2 surface (**a**) and its representation as a correspondence of the plane borders (**b**). A rectangular spin network on a surface of such topology has as a minimum fourfold degenerate states, as follows from the commutation properties of the operators P_i and Q_i of Eq. (12.74) given by the products of the operators along the topologically inequivalent trajectories. Six inequivalent trajectories are marked as P_1, P_2, P_3, Q_1, Q_2, and Q_3

such that the index j runs along a closed trajectory—a loop P_i or Q_i, which is not necessarily connected, and which we denote by the same symbol as the corresponding operator of (12.74), just without the hat sign. Loops of the same type do not intersect, hence the relevant products do not contain the Pauli matrices acting on the same spins, and therefore the operators $\hat{P}$ and $\hat{Q}$ corresponding to these products of the Pauli matrices commute. Moreover, even if a pair of loops intersects an even number of times, the corresponding operators also commute. For the model of a rectangular lattice under consideration the topologically equivalent loops do not intersect, as one sees in Fig. 12.13b, although in more sophisticated models they may intersect an even number of times. The crucial role is played by the pairs of topologically inequivalent loops that intersect an odd number of times since these very loops result in non-commuting pairs of the operators. Examples of the topologically inequivalent loops are shown in Fig. 12.13 for the particular choice of a genus 2 surface.

The topological degeneracy depends on the minimum number of the distinct operators $\hat{P}_i$ that can tolerate the non-commutativity relations suggested by the collection of topologically distinct loops that intersect just once. Let us consider which pairs of the operators commute, and which do not. All P_j operators commute among themselves, and so do Q_j operators, while some operators P_j do not commute with some operators Q_j, according the table

$$\begin{array}{llll} NO & YES & NO & Q_1 \\ YES & NO & NO & Q_2 \\ NO & NO & YES & Q_3 \\ P_1 & P_2 & P_3 & \end{array}$$

In such a table one cannot easily identify a subset of operators that all commute among themselves and therefore can be considered as equivalent. However, if instead of $\hat{P}_2$ and $\hat{Q}_2$ one considers the products $\hat{P}_1\hat{P}_2$ and $\hat{Q}_1\hat{Q}_2$ that correspond to the unions of the loops $P_1 \cup P_2$ and $Q_1 \cup Q_2$, respectively, the commutation table takes the form

$$\begin{array}{cccc} NO & NO & NO & Q_1 \\ NO & YES & YES & Q_1 \cup Q_2 \\ NO & YES & YES & Q_3 \\ P_1 & P_1 \cup P_2 & P_3 & \end{array}$$

since the union of the loops intersects twice with the loops with index 3. The last table clearly shows that the operators $\hat{P}_3$, $\hat{Q}_3$, $\hat{P}_1\hat{P}_2$, and $\hat{Q}_1\hat{Q}_2$ all commute among themselves, do not commute with the operators $\hat{P}_1$ and $\hat{Q}_1$, and hence $\hat{P}_3$ can be considered as an operator topologically equivalent to $\hat{P}_1\hat{P}_2$, while $\hat{Q}_3$ can be considered as equivalent to $\hat{Q}_1\hat{Q}_2$. If now one notices that, since $\left(\sigma_{ij}^{\alpha}\right)^2 = 1$, all operators $\hat{Q}_j$ and $\hat{P}_j$ are square roots of the unity operator and have eigenvalues ± 1, the topologically degenerate states can be classified by the signs of the eigenvalues of topologically inequivalent operators $\hat{Q}_1$ and $\hat{Q}_3$ (or, alternatively, $\hat{P}_1$ and $\hat{P}_3$). There are four opportunities to have these signs: $++$, $+-$, $-+$, and $--$, and therefore all the quantum states, including the ground state, are at least fourfold degenerate. Note that the operators corresponding to different topologically equivalent loops may but do not need to have the same eigenvalues for the same eigenstates of the degenerate energy level, and therefore the degeneracy of the level may occasionally be higher than the topologically guaranteed one.

One can further increase the dimensionality of the topologically protected quantum subspace by "gluing" borders of rectangular spin lattice in such a way that it forms a surface of higher genus g. The dimensionality of the protected Hilbert subspace then scales as 2^g. Note that the rectangular spin lattice is not the unique system that has invariants of the type (12.74) and thus gives a possibility to construct the topologically protected subspaces. A few other known examples all manifest a common property—the possibility to construct operators that commute with the Hamiltonian and have the structure of products of local operators taken along loop trajectories consecutively visiting neighboring lattice sites.

12.2.5 Universal Suppression of Slow Noise

The conception employed thus far for the dynamical protection of a quantum system against unknown and unwanted noise-induced interventions was based on the idea that has been originally developed for classical systems. It prescribes one to give the system an additional space and perform coding in such a way that every two different states carrying information get separated in space by a certain Hamming distance, which cannot be overcome by the error induced dynamics. Now we demonstrate that the action of slow unwanted interventions can be suppressed just by a properly chosen control protocol, without increasing the dimension of the system Hilbert space. We show a universal, that is, applicable to any quantum system, technique that allows one to simultaneously perform universal control of the evolution operator and compensate for the first-order contribution of an arbitrary Hermitian slow noise. The higher-order perturbations can also be suppressed, if needed.

Earlier, in Sect. 12.1.3, we have already discussed the possibility to control a quantum system by two-valued Hamiltonian. Here we show that, at least, a three-valued Hamiltonian is needed in order to perform any arbitrarily chosen evolution of a quantum system while compensating for all Hermitian static noises. To be more specific, our goal here is to show how to design a time-dependent control Hamiltonian, which can, at the same time, impose a chosen evolution to the system and eliminate the first-order action of any Hermitian static noise. The interesting result is that for the purpose two out of three values of the Hamiltonian have to satisfy the bracket generation condition presented on p. 583 while the third value can be arbitrary, even a null operator, such that during the time intervals where the Hamiltonian takes this value, the system evolves under the action of noise only.

In this section, we first formulate the problem to solve, and we give the explicit conditions the evolution matrix must fulfill. Then we show that these conditions are impossible to meet by just a two-valued control Hamiltonian. However, by slight modification of the two-valued procedure allowing for extra control steps during which the control Hamiltonian is set to zero, an algorithm of finding the appropriate protected control sequence can be constructed. This algorithm is illustrated at the example of a controlled error-protected three-level system (qutrit).

Conditions for First-Order Noise Elimination

Consider an N-level quantum system the Hamiltonian of which comprises a controllable part $\hat{H}_c(t)$ and an unknown static contribution, the noise $\hat{\mathcal{N}} = \sum_i \varepsilon_i \hat{G}_i$ given in terms of the Hermitian traceless generators $\hat{G}_i$ of the algebra $su(N)$ numerated by the index $i = 1, \ldots, N^2 - 1$. The evolution matrix satisfies the dynamical equation

$$\begin{aligned} \imath \frac{\partial \hat{U}(t)}{\partial t} &= \left(\hat{H}_c(t) + \hat{\mathcal{N}}\right) \hat{U}(t) \\ \hat{U}(0) &= I. \end{aligned} \tag{12.75}$$

Upon transforming to the interaction picture relative to $\hat{H}_c(t)$ with the help of the unperturbed evolution operator

$$\hat{U}_c(t) \equiv \mathrm{T}\exp\left\{\frac{1}{\imath}\int_0^t \hat{H}_c(s)\,ds\right\}, \tag{12.76}$$

where T denotes the chronological product, one isolates the evolution $\tilde{\hat{U}}(t)$ due to the noise only, which satisfies the dynamical system

$$\imath \frac{\partial \tilde{\hat{U}}}{\partial t} = \left[\sum_i \hbar \varepsilon_i \hat{U}_c^\dagger (t) \hat{G}_i \hat{U}_c (t) \right] \tilde{\hat{U}} (t)$$
$$\tilde{\hat{U}} (0) = I. \tag{12.77}$$

The first-order contribution of the noise to the evolution is thus given by

$$\tilde{\hat{U}}^{(1)} (t) = \sum_i \varepsilon_i \int_0^t \hat{U}_c^\dagger (s) \hat{G}_i \hat{U}_c (s) \, ds. \tag{12.78}$$

Our aim is to design a control Hamiltonian $\hat{H}_c(t)$, such that, at the end of the control sequence, say at time T_c, the evolution operator takes an arbitrarily prescribed value $\hat{U}_d \in SU(N)$ while the first-order contribution of any constant noise vanishes, that is,

$$\hat{U}_c (T_c) = \hat{U}_d \tag{12.79}$$
$$\tilde{\hat{U}}^{(1)} (T_c) = \sum_i \varepsilon_i \int_0^{T_c} \hat{U}_c^\dagger (t) \hat{G}_i \hat{U}_c (t) \, dt = 0 \tag{12.80}$$

for any set of constants $\{\varepsilon_i\}$, which implies

$$\forall i, \quad \int_0^{T_c} \hat{U}_c^\dagger(t) \hat{G}_i \hat{U}_c(t) dt = 0. \tag{12.81}$$

Two-Valued Control Is Insufficient

The conditions (12.79) alone can be met if the control Hamiltonian $\hat{H}_c(t)$ alternates between two values $\hat{A}$ and $\hat{B}$ (satisfying the bracket generation condition) during time intervals T_i which play the role of adjustable control parameters, and can be found with the help of the strategy presented in Sect. 12.1.3. The overall evolution operator induced by a K-step control sequence ($K \propto N^2$) has the form of (12.29),

$$\hat{U}_c (T_c) = e^{-iT_K \hat{H}_K} \times \ldots \times e^{-iT_2 \hat{H}_2} \times e^{-iT_1 \hat{H}_1}, \tag{12.82}$$

where $\hat{H}_k \equiv \hat{A}$ when k is even, $\hat{B}$ when k is odd, $\sum_{k=1}^K T_k = T_c$, and with $\hat{U}_c(T_c)$ satisfying (12.79). However, the number of equations (12.81) and conditions

(12.79) amounts to N^4, which implies that the control sequence should have at least $K \propto N^4$ steps in order to simply provide the required number of the parameters. Still the correspondence between the number of equations and the number of the variable timings is not enough, since a control based on just a two-valued Hamiltonian, though yielding any desired unitary transformation in the absence of noise, cannot simultaneously ensure universal protection against some errors. In particular, N operators of the form

$$\hat{G}_i \sim C_n = \frac{1}{2}\left[\hat{A} + \hat{B}, \left(\hat{B} - \hat{A}\right)^n\right] \tag{12.83}$$

never will satisfy (12.80), no matter how many parameters contains the time-dependent evolution operator

$$\hat{U}_c(t) = e^{-i\tau_K(t)\hat{H}_K} \times \ldots \times e^{-i\tau_2(t)\hat{H}_2} \times e^{-i\tau_1(t)\hat{H}_1}, \tag{12.84}$$

with

$$\tau_j(t) = \begin{cases} 0; & t < \sum_{n=1}^{j-1} T_n \\ t - \sum_{n=1}^{j-1} T_n; & \sum_{n=1}^{j-1} T_n < t < \sum_{n=1}^{j} T_n \\ T_j; & \sum_{n=1}^{j} T_n < t \end{cases}$$

corresponding to the chosen two-valued control sequence (12.82).

In order to demonstrate this we introduce the superoperators $\hat{\mathcal{H}}_i$, whose action on any operator $\hat{X}$ is defined by $\hat{\mathcal{H}}_i\hat{X} \equiv \left[\hat{H}_i, \hat{X}\right]$, and hence $e^{+\varepsilon\hat{H}_i}\hat{X}e^{-\varepsilon\hat{H}_i} = \sum_{k=0}^{\infty}\frac{\varepsilon^k}{k!}\hat{\mathcal{H}}_i^k\hat{X}$. Since $\left[\hat{A}, (\hat{B} - \hat{A})^n\right] = \left[\hat{B}, (\hat{B} - \hat{A})^n\right] = \left[\hat{H}_i, (\hat{B} - \hat{A})^n\right]$ for every i, one can set $\hat{C}_n = \hat{\mathcal{H}}_i(\hat{B} - \hat{A})^n$. Now by evaluating the first-order effect of $\hat{C}_n$ according to (12.81,12.83) one finds

$$\begin{aligned}
&\int_0^{T_c} \hat{U}_c^\dagger(t)\hat{C}_n\hat{U}_c(t)ds \\
=&\sum_i \hat{U}^\dagger\left(\sum_{n=1}^{i-1} T_n\right)\int_0^{T_i} dt\sum_{k=0}^{\infty}\frac{(it)^k}{k!}\hat{\mathcal{H}}_i^{k+1}(\hat{B} - \hat{A})^n\hat{U}\left(\sum_{n=1}^{i-1} T_n\right) \\
=&-i\sum_i \hat{U}^\dagger\left(\sum_{n=1}^{i-1} T_n\right)\left[e^{iT_i\hat{H}_i}(\hat{B} - \hat{A})^n e^{-iT_i\hat{H}_i} - (\hat{B} - \hat{A})^n\right]\hat{U}\left(\sum_{n=1}^{i-1} T_n\right) \\
=&-i\sum_i \hat{U}_{i+1}^\dagger(\hat{B} - \hat{A})^n\hat{U}_{i+1} + i\sum_i \hat{U}_i^\dagger(\hat{B} - \hat{A})^n\hat{U}_i \\
=&-i\hat{U}^\dagger(T_c)(\hat{B} - \hat{A})^n\hat{U}(T_c) + i(\hat{B} - \hat{A})^n.
\end{aligned} \tag{12.85}$$

The contribution obtained does not depend on the specific timing parameters T_i but only on the final operation $\hat{U}(T_c) = \hat{U}_d$. Moreover, if $\hat{U}_d$ and $(\hat{B} - \hat{A})$ do not commute, this contribution is nonzero. This completes the proof: universal control and universal protection cannot always be simultaneously performed in the two-valued control Hamiltonian approach.

Three-Valued Control

Two-valued Hamiltonian thus cannot solve the control problem along with the simultaneous suppression of slow noise. We therefore choose a control Hamiltonian $\hat{H}_c(t)$, which alternates between three values $\hat{A}$, $\hat{B}$, and $\hat{C}$ during time intervals T_i such that the overall evolution operator induced by such a three-valued K-step control sequence has the form

$$\hat{U}_c(T_c) = e^{-\imath T_K \hat{H}_k} \times \ldots \times e^{-\imath T_3 \hat{C}} \times e^{-\imath T_2 \hat{B}} \times e^{-\imath T_1 \hat{A}}, \tag{12.86}$$

where $\sum_{k=1}^{K} T_k = T_c$ and $\hat{H}_k \equiv \hat{A}$ for $k_{\mathrm{mod}\,3} \equiv 1$, $\hat{H}_k \equiv \hat{B}$ for $k_{\mathrm{mod}\,3} \equiv 2$, and $\hat{H}_k \equiv \hat{C}$ for $k_{\mathrm{mod}\,3} \equiv 0$.

As we have already stated, the minimum length required to make the control sequence robust against first-order perturbations is $K = N^4$, which is consistent with the number of equations (12.81) and requirements (12.79). Let us consider the particular case of $\hat{C} = 0$ and construct this long control sequence by inserting $K - K_0 \propto N^4$ time intervals with zero Hamiltonian into a sequence of $K_0 = N^2$ timings, which is found as solving the control problem by the method of Sect. 12.1.3. The way these time intervals are inserted into the original time intervals is illustrated in Fig. 12.14.

By the construction, the control aim formulated in (12.79) is fulfilled, while (12.80) takes the form

$$\begin{aligned} 0 = & \sum_{j \in \{j : \hat{H}_j = \hat{A}, \hat{B}\}} \int_0^{T_j} \hat{U}_c^\dagger \left(\sum_{n=1}^{j-1} T_j \right) e^{\imath t \hat{H}_k} \hat{G}_i e^{-\imath t \hat{H}_k} \hat{U}_c \left(\sum_{n=1}^{j-1} T_j \right) dt \\ & + \sum_{j \in \{j : \hat{H}_j = 0\}} T_j \hat{U}_c^\dagger \left(\sum_{n=1}^{j-1} T_n \right) e^{\imath t \hat{H}_k} \hat{G}_i e^{-\imath t \hat{H}_k} \hat{U}_c \left(\sum_{n=1}^{j-1} T_n \right), \end{aligned} \tag{12.87}$$

where we separate contributions of the time intervals with nonzero (first sum) and zero (second sum) control Hamiltonian. One finds the durations T_j of the $N^4 - N^2$ inserted time intervals by solving this system of $N^4 - N^2$ linear equations. In the case some of T_j turn out to be negative, the number of the inserted intervals can be increased and made larger than $N^4 - N^2$, such that the system of $K > N^4$ equations has nonnegative solutions for the duration of these time intervals. The same strategy can be chosen for the arbitrary third control, although in this case

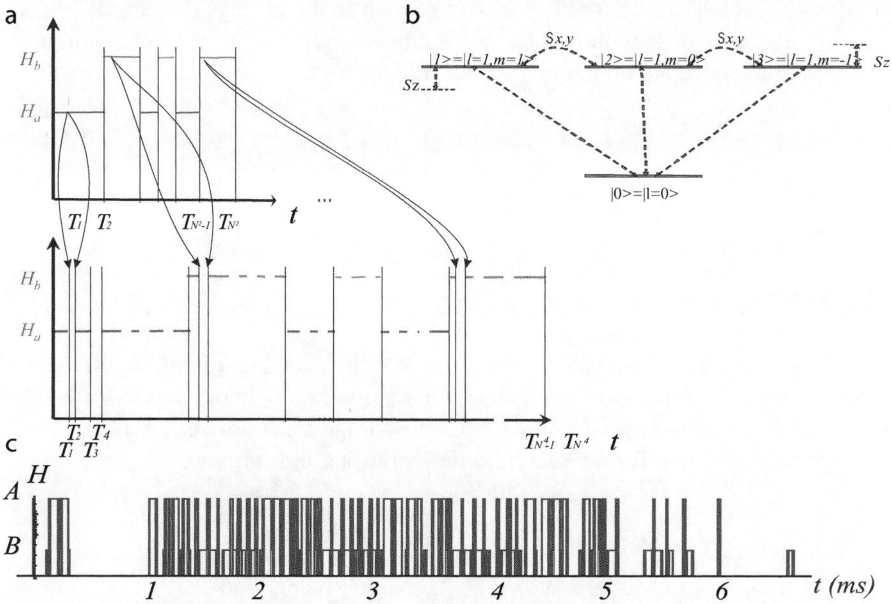

Fig. 12.14 (**a**) In order to achieve an arbitrary predetermined unitary evolution and simultaneously to eliminate the action of an unknown slow error-inducing Hamiltonian one can take an existing N^2 interval control sequence, and insert $N^4 - N^2$ additional time intervals where the control Hamiltonian takes the value $\hat{C} = 0$. The duration of these additional time intervals should be chosen such that $N^2 - 1$ requirements Eq. (12.80) for $N \times N$ matrices are satisfied. These durations are found from the solution of the linear equation (12.87). In the case $\hat{C} \neq 0$, the corresponding equations are nonlinear. (**b**) The completely protected atomic qutrit gate performing the operation Eq. (12.88) consists of the three hyperfine levels ($|0,1,2\rangle \equiv |n=1; L=0; S=1/2; I=1/2; F=1, m_F=1,0,-1\rangle$) of the hydrogen atom. The couplings $|0\rangle \leftrightarrow |1\rangle$ and $|1\rangle \leftrightarrow |2\rangle$ are due to a static magnetic field, while the coupling $|0\rangle \leftrightarrow |2\rangle$ consists in an effective second-order Hamiltonian, due to a time-dependent magnetic field, slightly detuned from the transitions $|0,2\rangle \leftrightarrow |g\rangle$ where $|g\rangle \equiv |n=1; L=0; S=1/2; I=1/2; F=0\rangle$ serves as a relay state. (**c**) Control sequence for realistic parameters. The plot is done for $\hat{H} = \frac{(g_e \mu_B \mathcal{B})^2}{8\hbar\Delta}\hat{\Gamma} - \frac{g_e \mu_B}{\hbar}\left(\hat{\boldsymbol{S}} \cdot \boldsymbol{B}\right)$ where $g_e \simeq 2$ and $\mu_B \simeq -9.274084 \times 10^{-24} J.T^{-1}$ are the electronic gyromagnetic factor and Bohr magneton, respectively, $\hat{\boldsymbol{S}}$ is the vector of spin 1 operators of the form $\hat{S}_x = \frac{1}{\sqrt{2}}\left(s^+ + t^+ + s^- + t^-\right)$, $\hat{S}_y = \frac{i}{\sqrt{2}}\left(t^+ + s^- - t^- - s^+\right)$, $\hat{S}_z = \lambda_3$, and the operator $\hat{\Gamma} = \lambda_8 - s^- - s^+$ in the notations of Eq. (10.105), while the static magnetic field $B_{x,y,z}$ and the magnetic field of the amplitude $\mathcal{B}$ oscillating along the axis x at a frequency detuned by Δ from the hyperfine transition frequency of the hydrogen ground state take the values $B_x = -0.203\,\text{mT}$, $B_y = 0.101\,\text{mT}$, $B_z = -0.051\,\text{mT}$, $\mathcal{B} = 1.757\,\text{mT}$, $\Delta = 0.892\,\text{GHz}$, for $\hat{A}$, and $B_x = -0.203\,\text{mT}$, $B_y = -0.203\,\text{mT}$, $B_z = 0.101\,\text{mT}$, $\mathcal{B} = 1.015\,\text{mT}$, $\Delta = -0.892\,\text{GHz}$ for $\hat{B}$. The third value of the Hamiltonian $\hat{C} = 0$. The control sequence of a total control period $T_c \simeq 6.58\,\text{ms}$ comprises $208 > 81$ pulses of durations ranging from $t_{min} \simeq 2.5\mu s$ to $t_{max} \simeq 689.7\mu s$

the resulting equations are nonlinear with respect to the timings. In Fig. 12.14c we show an example of completely protected three-level system of a hydrogen atom performing a qutrit gate operation

$$\hat{U}_d = \begin{bmatrix} 1 & 0 & 0 \\ 0 & 0 & -i \\ 0 & -i & 0 \end{bmatrix}. \tag{12.88}$$

Optimum Order of the Correction

When the control is introduced in such a way that, during a control time T_c, it compensates a static noise $\mathcal{N}$ in the first-order perturbation theory, the second-order term $S^{(2)} = (\mathcal{N}T_c)^2/2$ gives a good estimate for the remaining error, even though the explicit expression for the second-order contribution is different

$$S^{(2)} = \int_0^{T_c} dt \int_0^t ds \hat{U}_c^\dagger(t)\mathcal{N}\hat{U}_c(t)\hat{U}_c^\dagger(s)\mathcal{N}\hat{U}_c(s). \tag{12.89}$$

One can explore the feasibility of a control Hamiltonian canceling the noise in the second-order perturbation theory as well, which is given by the requirement $S^{(2)} = 0$ implied by vanishing of the $(N^2-1)^2$ matrix expressions

$$\int_0^{T_c} dt \int_0^t ds \hat{U}_c^\dagger(t)\hat{G}_m\hat{U}_c(t)\hat{U}_c^\dagger(s)\hat{G}_n\hat{U}_c(s) = 0, \tag{12.90}$$

for the generators $\hat{G}_n$ of the group $SU(N)$, and equivalent to $N^2 \cdot (N^2-1)^2$ scalar equations. Together with the first-order requirements and the condition (12.79), this imposes N^6 requirements on the control sequence. The error suppression up to the second order thus requires N^6 control intervals, while the protection scheme extended to n-th order of the perturbation theory implies $N^{2(n+1)}$ control intervals. The similar requirements $S^{(n)} = 0$ can be formulated for yet higher orders.

How far is it reasonable to go in the higher-order corrections? Estimations show that for systems whose Hilbert space dimensionality exceeds 4 one needs at most to correct the second-order contribution. Indeed, let us turn attention to higher orders of the perturbation expansion in the static noise and first estimate the size of the high-order perturbation. If the control were absent, the contribution of the noise to the evolution would be given by

$$e^{-i\mathcal{N}T_c} \simeq 1 - i\mathcal{N}T_c - \frac{(\mathcal{N}T_c)^2}{2} + \ldots + \frac{(-i\mathcal{N}T_c)^n}{n!} + S^{(n+1)}. \tag{12.91}$$

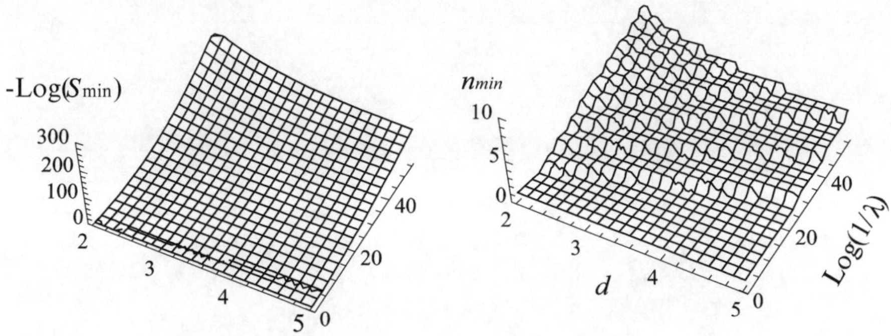

Fig. 12.15 Minimum error $S_{\min}(\lambda, d) = \min S^{(n)}$ (*left*) and the corresponding order of the perturbation (*right*) as functions of the typical accumulated error during the control step λ and the dimension of the Hilbert space of the system $N = 2^d$

Provided that errors are suppressed by a control sequence of length $K \simeq N^{2(n+1)}$ up to the order n, the remaining error is

$$S^{(n+1)} \sim (|\mathcal{N}|\, T_c)^{n+1} / (n+1)!, \tag{12.92}$$

where the control time T_c can be estimated by the minimum time τ required for one step of the sequence multiplied by the number of steps K, that is,

$$S^{(n+1)} \sim \frac{\left(|\mathcal{N}|\, \tau N^{2(n+1)}\right)^{n+1}}{(n+1)!}. \tag{12.93}$$

The combination $\lambda = |\mathcal{N}|\, \tau$ serves as an external parameter of the problem, corresponding to the typical action of the errors during one control step. For an ensemble of d qubits, the Hilbert space dimension reads $N = 2^d$.

Given λ and d, the remaining error $S^{(n+1)}$of (12.93), as a function of n, reaches a minimum $S_{\min}^{(n+1)}(\lambda, d)$, for a certain order $n_{\min}(\lambda, d)$. In Fig. 12.15 we show the minimum errors, and the corresponding orders of error correction as functions of the size of the system N and the parameter λ, corresponding to the error probability amplitude during one control step. One sees that the second-order noise correction of a system consisting of more than three qubits becomes reasonable only for $\lambda \sim e^{-20} \sim 2\ 10^{-9}$, which is beyond the physical limit of feasibility for the moment.

Chapter 13
Bibliography and Problems

Even considering particular aspects of problems that have been encountered through many decades in different branches of physics and chemistry, it is impossible to give a comprehensive review of all relevant publications. This is particularly the case for the dynamics of multilevel quantum systems, which usually is just a technical aspect and not the main concern of research papers and textbooks dedicated mostly to physical phenomena, rather than to the technical details of their description. In considering a bibliographic search in such a situation, one has to make a difficult choice amongst the numerous possibilities. The main objective of the bibliography presented here was to find a balance between the pioneering papers, topical reviews, monographs, textbooks concerning particular physical domains, and the original publications most closely related to the dynamical processes in complex quantum systems. This review therefore consists of two parts: the "Bookshelf" comprising a selection of classical original papers and monographs, and the "Problems" where a selection of more recent publications from different domains is suggested to the reader for consideration from the point of view of the dynamics of multilevel systems developed in the previous chapters. Though some old classical papers cited below might be more difficult to access than their subsequent interpretations, we believe that the pleasure of following the development of the original ideas of the great physicists should definitely compensate for this trouble.

Among numerous modern publications, only a limited number of papers from a number of different fields have been selected. The majority of the problems considered there have been formulated in terms traditional to the corresponding physical domains and we suggest that they be approached from the point of view of the dynamics of complex quantum systems, since we believe that the effort aimed at such a diversification of approaches will help the reader to attain a deeper insight into the nature of these processes. Consideration of some of the problems requires a certain advanced experience in adjacent fields, and that can be achieved by consulting the cited monographs and reviews. In particular, this relates to the field-theoretical models in combination with the saddle-point approximation that have not been considered here in sufficient detail.

V.M. Akulin, *Dynamics of Complex Quantum Systems*, Theoretical and Mathematical Physics, DOI 10.1007/978-94-007-7205-2_13,

13.1 Bookshelf

Cascades and Exponential Decay

"Consider an ensemble of two-level systems interacting with a quantized oscillator of a field mode in a highly excited Fock state in the presence of an infinite number of other field modes in the vacuum state"—that is how we would formulate today the problem considered by A. Einstein [85] in his famous paper *Zur Quantentheorie der Strahlung* (1917). Solving this problem, we would draw the level scheme, Fig. 8.1d, consider the corresponding Schrödinger equation, and find that the system is described by the master equation with the probability amplitudes Bn coupling the closest states of the excited field oscillator and the atomic ensemble, and the probabilities A coupling states of the atomic ensemble with the bands of states describing the reservoir of the other field oscillators. Of course Einstein did not formulate this problem in terms of the Schrödinger equation or Dirac formalism that were introduced later, but this problem can still be viewed as the first example of the dynamics of complex quantum systems. This approach was first adopted a decade later by L. D. Landau [145] in his paper *Das Dämpfungsproblem in der Wellenmechanik* (1927).

The description of the interaction between radiation and matter in terms of multilevel quantum systems is already found in the paper *Berechnung der natürlichen Linienbreite auf Grund der Diracschen Lichttheorie* (1930) by V. Weisskopf and E. Wigner [214], dedicated to the analogy between the width of the resonance of a classical harmonic oscillator with the friction and the spectral line widths. The dynamical approach had been employed by E. Fermi [93] in his paper *Quantum theory of radiation* (1932), where the level–band problem had been formulated and solved in the context of atomic spontaneous emission. The radiative widths of resonances were the main physical quantities of interest. Later on he generalized the result and obtained the "Fermi golden rule" for the transition rate in an arbitrary level–band system, valid for times shorter than the Heisenberg time.

The dynamical consideration of multilevel quantum systems for a long time remained an auxiliary tool in the investigation of the spectral properties of quantum objects that manifest themselves in scattering processes. G. Breit and E. Wigner [33] in their paper *Capture of slow neutrons* (1936) had demonstrated that nuclear interactions can result in scattering profiles much broader than those suggested by the radiation lifetimes. The development of the microwave technique had offered an opportunity to control the strength of interaction between radiation and matter, such that the rates of radiative transitions became of the order of the interactions in quantum systems. R. Karplus and J. Schwinger [129] in their paper *A note on saturation in microwave spectroscopy* (1948) have shown that the width of the resonance can depend on the radiation field intensity.

Alternative Line Shapes

For open quantum systems, there always exists a time parameter which gives the typical time of an exponential decay. However, some important spectral features may result from another, non-exponential type of dynamics, which yields different line shapes. One of the most striking theoretical results is the semicircular profile of the energy eigenstate density obtained by E. Wigner [217] in his paper *On the distribution of the roots of certain symmetric matrices* (1957). This distribution emerges from the model of a Gaussian random matrix based on the central limit theorem of statistics. Though the physical validity of this model for the description of complex nuclei has been questioned by F. Dyson [81] in his series of papers *Statistical theory of the energy levels of complex systems* (1962), it nevertheless remains an elegant example of a complex system being adequate and useful in other domains of physics, in situations where the spectral complexity truly results from the randomness of the system parameters.

Corrections to the semicircular Wigner distributions near the spectrum edges had been explored by Ya. G. Sinai and A. B. Soshnikov [205] in their paper *A refinement of Wigner's semi-circle law in a neighborhood of the spectrum edge for random symmetric matrices* (1998), while the general properties of the Hamiltonians perturbed by a random matrix had been discussed by E. Brezin, S. Hikami, and A. Zee [38] in the paper *Universal correlations for deterministic plus random Hamiltonians* (1995).

The other famous example of a spectral profile different from a Lorentzian one is given by U. Fano [89] in the paper *Effects of configuration interaction on intensities and phase shifts* (1961), which was a development of the ideas formulated in his earlier publication [88] *Sullo spettro di assorbimento dei gas nobili presso il limite dello spettro d'arco* (1935). Formulated initially for purely intra-atomic interactions, this example became of extreme importance with the development of lasers in the situations when laser-induced interactions strongly perturb atomic spectra. The dynamics corresponding to these spectra also differ from a simple exponential decay, as illustrated by P. L. Knight, M. A. Lauder and B. J. Dalton [131] in the review paper *Laser-induced continuum structure* (1990).

Non-Lorentzian spectral profiles are typical of solid state physics. J. Hubbard [114] in his series of works entitled *Electron correlations in narrow energy bands* (1963–1964) had demonstrated that under certain conditions, spectra of random alloys have a semicircular structure resulting from electron correlation. The Hubbard model Hamiltonian for the electron liquid does not have any random parameters, however, the resulting spectra so resemble those obtained via the random perturbation of a system of two degenerate levels, shown in Fig. 9.1, that this invites one to think about the fundamental reasons for such a similarity.

Non-trivial line shapes are also typical of glassy media and excitons in cold gases, which conform the model proposed by M. Mesard, G. Parisi and A. Zee [165] in the paper entitled *Spectra of Euclidean random matrices* (1999), and presented in detail in the work [177] *Euclidean random matrices: solved and open problems* written by G. Parisi (2006).

Dynamics

The general relation between the dynamical and spectral properties of quantum systems had been established by R. Kubo [139] in his paper *A general expression for the conductivity tensor* (1956), which gives the linear response of the system to a weak perturbation. This expression implies that the system is subject to relaxation, such that the nonlinear effects do not have time to manifest themselves.

Hamiltonian systems behave differently. The first example of such behavior had been given by E. L. Hahn [112] in his paper *Spin Echoes* (1950), where it had been shown that an ensemble of spins in a slightly inhomogeneous field experiences a coherent damping which can be reversed by sending a π-pulse which inverts the phases of the probability amplitudes. A very convenient tool for the consideration of this kind of problem had been introduced in the paper by I. I. Rabi, N. F. Ramsey, and J. Schwinger [186] entitled *Use of rotating coordinates in magnetic resonance problems* (1954).

The interaction of the ensemble of two-level systems with radiation in the regime of developed coherent dynamics also differs considerably from the linear response regime. R. H. Dicke [70] in his paper *Coherence in spontaneous radiation processes* (1954) had demonstrated that the spontaneous decay rate resulting from the interaction of the ensemble with a photon vacuum increases strongly for the highly correlated states of individual atoms achieved in the course of their coherent dynamics. These states also change the propagation of the electromagnetic radiation in a medium composed of the ensemble of two-level atoms as shown by C. K. Rhodes, A. Szöke, and A. Javan [188] in the paper *The influence of level degeneracy on the self-induced transparency effect* (1968), and by S. L. McCall and E. L. Hahn [161] in the paper *Self-induced transparency* (1969).

Localization and Chains

The dynamics of complex quantum systems are strongly related to the structure of the energy eigenfunctions. P. W. Anderson [8] in his famous paper *Absence of diffusion in certain random lattices* (1957) had demonstrated that in a perturbed infinite complex system with random parameters, the diffusion process can be partially or completely suppressed, such that the energy eigenfunctions turn out to be localized on a limited number of unperturbed states. In the one-dimensional case, the corresponding energy eigenfunctions are localized exponentially, as had been numerically demonstrated in the paper by R. E. Borland [32] entitled *The nature of the electronic states in disordered one-dimensional systems* (1963). An analytically soluble model had been introduced by P. Lloyd [155], in his paper *Exactly solvable model of electronic states in a three-dimensional disordered Hamiltonian: non-existence of localized states* (1969). D. J. Thouless [210] in his paper *Maximum metallic resistance in thin wires* (1977) had discussed the localization conditions for the one-dimensional case.

Exhaustive numerical analysis of the one-dimensional case had been presented in the paper by K. Ishii [119] entitled *Localization of eigenstates and transport phenomena in the one-dimensional disordered systems* (1972) whereas the analytical consideration had been summarized in the monograph by I.M. Lifshits, S.A. Gredeskul and L.A. Pastur [152] *Introduction to the theory of disordered systems* (1988). It turns out that the localization phenomenon is not something specific to disordered systems, but has deeper roots in the algebraic structure of operators, as had been shown in the paper by Ya. G. Sinai [203] entitled *Anderson localization for one-dimensional difference Schroedinger operator with quasi-periodic potential* (1987).

Quantum dynamics in chains had been considered for the first time in the paper by V. L. Berezinskii [19] entitled *Kinetics of a quantum particle in a one-dimensional random potential* (1974). This work had demonstrated that the all-order correlations of the Green's function are important in the one-dimensional case, contrary to the Bogolyubov method for the description of complex dynamics.

Ensembles: Levels Statistics

The idea of a statistical description of complex quantum systems had first been employed by E. Wigner [216] in his paper *On a class of analytic functions from the quantum theory of collisions* (1951) for the neutron scattering amplitudes that have been assumed to possess certain statistical properties in the complex plane of energies. Much later, F. Dyson [81] in his paper *Statistical theory of the energy levels of complex systems* (1962) had proposed the ensemble average as a universal approach to complex quantum systems, which was relevant even in the case where no real physical ensemble had been associated with the complex system, whereas the statistical approach emerged from the "renounce of the exact knowledge... of the nature of the system...". This very concept, called, following F. Dyson, "a new kind of statistical mechanics" remains the cornerstone of the modern approach to complex quantum systems.

The dynamical aspects of the behavior of multilevel systems had been restricted to the statistics of resonance widths in open multilevel quantum systems, which had been related in the paper by C.E. Porter and R.G. Thomas [184] entitled *Fluctuations of nuclear reaction widths* (1956) to the statistics of eigenvectors of complex systems. One finds more details of the statistical approach to nuclear and atomic spectra in the book by C. E. Porter [183] *Statistical theories of spectra: fluctuations* (1965). The mathematical background of the statistical approach to complex spectra along with the fundamental results of the level correlation statistics is given in the book by M. L. Metha [163] *Random matrices and the statistical theory of energy levels* (1967).

In contemporary research, statistical models are widely employed in condensed matter physics. One can consult the review by C. W. J. Beenakker [16] *Random matrix theory of quantum transport* (1997), the book by J. Imry [117] *Introduction to mesoscopic physics* (2002), and the special issue of *The Journal of Physics* [125] (2003) for a modern view on the statistical description of complex quantum systems in the context of mesoscopic condensed matter physics.

Classical and Quantum Chaos, Level Statistics

Quantum complexity and chaos do not only have a long history of parallel development and mutual enrichment via the exchange of basic ideas but also have common roots in ergodic theory. The first demonstration of the relevance of the statistical approach to the description of dynamical systems has been performed by Ya. G. Sinai [201] in his pioneering paper *On the concept of entropy for a dynamic system* (1959), which got further development in the later monographs [202] *Introduction to ergodic theory* (1976), and *Topics in ergodic theory* (1994) [204] by the same author. About the same time, B. V. Chirikov [55] in his paper *Resonance processes in magnetic traps* (1960) had introduced the resonance overlap criterion necessary for the destruction of the separability of the motion, which results in the chaotic classical dynamics of mechanical systems. These ideas were further developed in detail in the monographs by A.J. Lichtenberg, M.A. Liberman [150] entitled *Regular and stochastic motion* (1982), and by G. M. Zaslavski, R. Z. Sagdeev, D. A. Usikov, and A. A. Chernikov [222] entitled *Weak chaos and quasi-regular patterns* (1991).

The quantum mechanical counterpart of classically regular and chaotic systems has been studied numerically. For regular dynamics, the Poissonian statistics of distances to the closest neighboring level and level clustering had been found in the paper by M. V. Bery and M. Tabor [20], entitled *Level clustering in the regular spectrum* (1977), whereas the level repulsion results were typical of chaotic dynamics, as had been shown in the well-known publication of O. Bohigas, M. J. Giannoni, and C. Schmit [28] entitled *Characterization of chaotic quantum spectra and universality of level fluctuation laws* (1984). For more details concerning the manifestation of the chaotic dynamics at the quantum level of consideration, one can consult the reviews by B. Eckhardt [84] entitled *Quantum mechanics of classically non-integrable systems* (1988), and F. M. Izrailev [121], *Simple models of quantum chaos: spectrum and eigenfunctions* (1990) where some dynamical aspects related to Anderson localization in kicked rotors had been considered, as well as the monographs by M. Gutzwiller [110] *Chaos in classical and quantum mechanics* (1990), by L. E. Reichl [187] *The transition to chaos* (1992), and by F. Haake [111] *Quantum signatures of chaos* (2001) for a general point of view of this field.

An interesting example of level statistics had been discussed in the papers by E. B. Bogomolny, U. Gerland, and C. Schmit [29] entitled *Models of intermediate spectral statistics* (1999), and [30] *Singular statistics* (2001), where it had been demonstrated that for essentially quantum systems that have no classical analogs, one could observe intermediate statistics, corresponding neither to a Poissonian nor to a Wigner–Dyson distribution on the closest neighboring levels spacing.

Supersymmetry, Unification Disorder and Chaos

Considerable progress in the understanding and a better description of disordered quantum systems has been achieved after the application of field theory methods to disordered quantum systems. This supersymmetry technique, along with the representative references to earlier publications, had been presented in detail in the book by K. Efetov [87] entitled *Supersymmetry in disorder and chaos* (1997). One should append to this exhaustive publication the review by M. Zirnbauer [225] *Anderson localization and non linear sigma-model with graded symmetry* (1986) and the recent paper [37] *New correlation functions for random matrices and integrals over supergroups* (2003) by E. Brezin and S. Hikami.

In the paper *Quantum billiards with surface scattering: Ballistic sigma-model approach* (1998) by Ya. M. Blanter, A. D. Mirlin, and B. A. Muzykantskii [23], the supersymmetry methods have been generalized to the case of ballistically propagating particles which allowed one to apply field theory methods to the ballistic billiards typically considered in the models of quantum chaos.

Dynamics of Atoms and Molecules

Contrary to the case of nuclear collisions, for the collision of atoms one has the small Born–Oppenheimer parameter $\Lambda = (m/M)^{1/4}$ which allows for the fact that the electrons of mass m move much faster relative to the nuclei of mass M. The adiabaticity parameter Λ is widely employed for the description of atomic collisions, as one can see from the classical monograph by H.S.W. Massey and N. F. Mott [159] entitled *The theory of atomic collisions* (1997) and the monograph by I. I. Sobelman, L. A. Vainshtain, and E. A. Yukov [207] entitled *Excitation of atoms and broadening of spectral lines* (1981). When the trajectories of the center-of-mass motion of colliding atoms or molecules can be considered as given, the electron Hamiltonian acquires an explicit time-dependence. Numerous models of such collisions have been considered in the monograph [173] *Theory elementary atomic and molecular processes* (1985) by E. E. Nikitin and in the monograph [174] *Theory of slow atomic collisions* (1984) by E. E. Nikitin and S. Ya. Umanskii.

Among the pioneering papers related to molecular dynamics, one should cite the paper by A. M. Dykhne [79], entitled *Adiabatic perturbation of the discrete spectrum states* (1962), where the general semiclassical expression for the transition probability had been obtained, the paper by Yu. N. Demkov and V. I. Osherov [67] entitled *Stationary and nonstationary problems in quantum mechanics that can be solved by means of contour integration* (1968) where the famous example of an exactly soluble problem had been suggested, and the papers by M. Bixon and J. Jortner [21] entitled *Intramolecular radiationless transitions* (1968) and [22] *Long radiative lifetimes of small molecules* (1969) where the multilevel quantum model had been introduced for the description of intramolecular conversion and spontaneous emission. High efficiency of the excitation transfer in biological systems known as "photon harvesting" was discovered due to the experimental development of the two-time technique presented in the paper by T. Brixner, J. Stenger, H.M. Vaswani, M. Cho, R.E. Blankenship, and G.R. Fleming [40], entitled *Two-dimensional spectroscopy of electronic couplings in photosynthesis* (2005).

Complex multilevel systems corresponding to the Rydberg states of atoms and small molecules had been exhaustively described in the monographs by E. W. Schlag [192] entitled *ZEKE spectroscopy* (1998), the book [126] *Molecular applications of quantum defect theory* (1996), edited by Ch. Jungen, and the monograph by T. F. Gallagher [101] entitled *Rydberg atoms* (1994).

Laser Theory, Quantum Optics

Though Albert Einstein in his work [85] (1917) did not consider the situation where the population of the excited atomic states exceeds the population of the ground state, the master equation derived by him had been employed for the prediction and explanation of induced coherent emission of masers and lasers. Later on it became clear that the master equation did not include information relating to the phase of the electromagnetic field and other, more general, characteristics, such as the photon statistics of the emitted radiation. A consistent description of the interaction of quantized electromagnetic fields with matter had been presented in the review by S. Stenholm [209] entitled *Quantum theory of electromagnetic fields interacting with atoms and molecules* (1973), where the dynamical description of the quantum evolution of multilevel systems composed of atomic ensembles and multimode fields, arose naturally and described all of the dynamical details of the process, including irreversible relaxation. For more details, one can consult the book by M. Sargent, W. E. Lamb, and M. O. Scully [190] *Laser physics* (1977) whereas for the modern view on the problems of quantum optics the monograph W. Schleich [193] *Quantum optics in phase space* (2001) can be recommended.

Atoms and Molecules in Strong Fields

Atoms and molecules interacting with strong laser fields are the classical example of multilevel quantum systems that experience coherent quantum evolution. F. V. Bunkin and A. M. Prokhorov [42] in their paper *The excitation and ionization of atoms in a strong radiation field* (1964) had employed the rotating wave approach, earlier introduced by I. I. Rabi, N. F. Ramsey, and J. Schwinger in the already cited paper [186]. It was the first application of the seminal concept of quasienergy, which was introduced some years later by Ya. B. Zeldovich [223] in his papers *The quasienergy of a quantum-mechanical system subjected to a periodic action* (1967) and *Scattering and emission of a quantum system in a strong electromagnetic wave* (1973), and by V. I. Ritus [189] in the paper *Shift and splitting of atomic energy levels by the field of an electromagnetic wave* (1967), which had yielded the method of the effective Hamiltonian describing the time evolution of a multilevel quantum system under a strong time-dependent perturbation. The mathematical formalism of this approach relied on the classical paper by G. Floquet [94] *Sur les équations différentialles linéaires à coefficients périodiques* (1889). For the modern applications of the Floquet theory to dynamics of atoms in strong fields, one can consult the review by S. Guérin and H. R. Jauslin [109] entitled *Control of quantum dynamics by laser pulses: adiabatic Floquet theory* (2003). The dynamics of molecules in laser fields had first been considered by G. A. Askaryan [11] in the paper *Excitation and dissociation of molecules in an intense light field* (1964) and by F. V. Bunkin, A. M. Prokhorov, and R. V. Karapetyan [43] in the paper *Dissociation of molecules in strong radiation field* (1965).

Fundamental aspects of the interaction between intense radiation and matter are widely presented in monographs and textbooks. One can consult the book [27] *Nonlinear optics* by N. Blombergen (1965) for the details of the coupling of photons in atomic media, the book by L. Allen, J. H. Eberly [5] entitled *Optical resonance and two-level atoms* (1975) for fundamentals of the interaction between strong radiation and matter, the monograph by N.B. Delone, V.P. Krainov [64] entitled *Atoms in strong light fields* (1985) for a review of multiphoton processes in atoms, and the textbook by C. Cohen-Tannoudji, J. Dupont-Roc, and G. Grynberg [59] *Atom–photon interactions: basic processes and applications* (1998). The consistent description of atoms and molecules in strong laser fields in terms of multilevel quantum systems is also given in the textbook by N. V. Karlov and V.M. Akulin [4] *Intense resonant interactions in quantum electronics* (1991), which evidently is a publication that is conceptually closest to the present text.

Interatomic Forces in Resonators and Laser Fields, Ionic Traps, Atomic Optics

The development of lasers in the 1990s had opened up new possibilities for the observation of quantum phenomena. On the one hand, the photon measurement technique had reached the sensibility level of a single photon that had allowed one to address questions about atomic interaction mediated by a single photon in resonators, and the manifestations of such interactions in the dynamics of individual photons. One of the examples of such systems had been presented in the paper by G. Kurizki, A. G. Kofman, and V. Yudson [140] entitled *Resonant photon exchange by atom pairs in high-Q cavities* (1996). In this context the important role played by photonic band structures had been described in detail in the monograph [124] by J. D. Joannopoulos, R. D. Meade, and J. N. Winn *Photonic crystals* (1995). The novel aspect of the problem was the coupling of internal atomic degrees of freedom with translational atomic motion.

On the other hand, the powerful and stable sources of electromagnetic radiation had enabled experimentalists to control and confine the translational motion of atoms and atomic ions. In the situation where the quantum levels corresponding to the coupled internal and translational motion were addressed individually, the multilevel quantum system had been proposed for quantum computations by J. I. Cirac and P. Zoller [57] in their pioneering paper *Quantum computation with cold trapped ions* (1995). Note that the quantization of the translational motion of atomic ions results in this case from the effective Kapitsa–Dirac potential imposed by rapidly oscillating radio-frequency fields, as had been described in detail in the monograph [105] *Ion traps* by P. K. Ghosh (1997). This quantization of ionic vibrational modes in ionic traps suggests the comparison of the ensemble of trapped ions with a polyatomic molecule, as has been done in Sect. 2.5.1.

The quantization of the translational motion of neutral atoms can also be achieved by strong laser fields. The fundamentals of such processes are presented in the monograph [164] by P. Meystre entitled *Atom optics* (2001). The consideration of the aforesaid systems by the methods developed in the previous sections for complex multilevel systems is a matter for future research. For such systems, the complex translational quantum dynamics of atoms results from the spatial structure of the laser field modes.

Long-Term Dynamics

The main difference between discrete and continuous complex spectra manifests itself in the long-term behavior of the populations. Coherent returns of the population to the initial state were discussed for the first time by J. Eberly, N. B. Narozhny, and J. J. Sanchez-Mondregon [83] in their pioneering paper *Periodic spontaneous collapse and revivals in a simple quantum model* (1980). Later on, I. Sh. Averbukh and N. F. Perelman [12] in their cornerstone paper *Fractional revivals: universality in the long-term evolution of quantum wave packets beyond the correspondence principle dynamics* (1989) had elucidated the role of anharmonicity on the violation of the Ehrenfest principle, and had revealed a rich domain of possible intermediate asymptotics of quantum mechanics resulting from strong and long-lasting interference phenomena in multilevel systems.

The manifestation of such interference in the dynamical properties of multilevel relay-like systems had been numerically shown in the example of a hydrogen atom by G. Casati, B. V. Chirikov, and D. L. Shepelyansky [50,56] in the paper *Quantum limitation for chaotic excitation of the hydrogen atom in a monochromatic field* (1984) and further discussed by D. L. Shepelyansky [197] in his paper *Localization of diffusive excitation in multi-level systems* (1987) where the relations between Anderson localization and the long-time asymptotic behavior of complex systems had been demonstrated.

The long-time limit of the coherent dynamics of a strongly perturbed continuum was the main concern of the paper by M. V. Fedorov, M. Yu. Ivanov, and A. M. Movsesian [90] entitled *Strong-field photoionization of an initially excited hydrogen atom: formation of Rydberg wavepacket, its structure and trapping of population at Rydberg levels* (1990), where it had been demonstrated that the lifetime of the atomic states increases with the increase of the ionizing laser field intensity. This counterintuitive result is a natural consequence of correlations between the interaction matrix elements, typical of atomic systems.

In the context of polyatomic molecules, the non-exponential, long-time behavior of complex quantum systems had been discussed by S. A. Schofield and P. G.

Wolynes [194] in the paper *A scaling perspective on quantum energy flow in molecules* (1992), while for metals, a similar approach had been developed by J. T. Chalker, I. V. Lerner and R. A. Smith [54] in the paper *Random walks through the ensemble: linking spectral statistics with wave-function correlations in disordered metals* (1996).

Open Quantum Systems

Though the idea that interaction of an open quantum system with the environment results in relaxation was widely accepted from the very beginning of Quantum Mechanics, as was the case for the Einstein coefficient A describing spontaneous emission of photons and for the relaxation times T_1 and T_2 in NMR, the exhaustive technical tool for the relaxation description has been proposed only in 1961 in the paper by R. Zwanzig [226] entitled *Memory effects in irreversible thermodynamics*, where the projection method has been introduced. In the context of Quantum Optics this formalism has been further specified and developed in the book by W. Gardiner and P. Zoller [103] *Quantum noise.*

Quantum Entanglement

Inseparability of quantum states of composite systems, discovered in the early days of quantum mechanics by A. Einstein, B. Podolsky, and N. Rosen [86] in their article *Can Quantum-Mechanical Description of Physical reality Be Considered Complete?* (1935) and named "entanglement" ("Verschränkung") by E. Schrödinger [195] in the paper *Die gegenwärtige Situation in der Quantenmechanik* (1935), became one of the central concepts of physics during the last decade. Example of the simplest maximum entangled state of two two-level systems given in the pioneering paper by J.S. Bell [17] published in 1964 and entitled *On the Einstein–Podolsky–Rosen Paradox,* had been generalized by D.M. Greenberger, M.A. Horne, and A. Zeilinger [107] in the paper entitled *Going Beyond Bell's Theorem* (1989) to the states of three two-level systems. The group theory aspects of multipartite entanglement had been formulated by N. Linden and S. Popescu in the article *On Multi-Particle Entanglement* (1998) where the concept of orbits of the local transformation and polynomial invariants of these transformations had been discussed. A technique for exhaustive characterization of multipartite entanglement of systems in pure state based on an algebraic approach had been proposed in the paper [157] (2006).

Exhaustive characterization of quantum entanglement in open systems still remains an open problem in the general case. The definition of the entangled mixed states given by R.F. Werner [215] in the article *Quantum states with Einstein–Podolsky–Rosen correlations admitting a hidden-variable model* (1989) and the necessary condition of the entanglement proposed by A. Peres [182] in his paper entitled *Separability Criterion for Density Matrices* (1996), had been summarized in the concept of distillated entanglement by M. Horodecki, P. Horodecki, and R. Horodecki in their paper [113] entitled *Mixed-State Entanglement and Distillation: Is there a "Bound" Entanglement in Nature?* (1998). For the particular case of two two-level systems the nessecery and sufficient entanglement criterion had been suggested by W. K. Wootters [218] in his paper *Entanglement of Formation of an Arbitrary State of Two Qubits* (1998). More details one can find in the book [18] *Geometry of Quantum States* by I. Bengtsson and K. Zyczkowski (2006). Phenomenon of entanglement sudden death, typical of open quantum systems, had been discovered in 2004 in the paper [219] by T. Yu and J. H. Eberly, entitled *Finite-time disentanglement via spontaneous emission.*

Control

The possibility to control a multilevel quantum system by applying a time-dependent perturbation is based on the existence theorem proved by V. Jurdevic and H. J. Sussmann [127] in their paper *Control systems on Lie groups* (1972), where it had been shown that the compactness of the group generated by the unperturbed Hamiltonian and the perturbations plays a crucial role and permits one to exert complete control if the group coincides with the group of all possible unitary transformations of the system. Application of this general result to quantum systems has been discussed by A. G. Butkovskiy and Yu. I. Samoilenko [45] in the monograph *Control of quantum-mechanical processes and systems* (1990). Control by measurement has been suggested in the paper [213] by K. Vogel et al., *Quantum State Engineering of the Radiation Field* (1993). Practical methods of control are presented in the monograph by P. W. Brumer, M. Shapiro [41] entitled *Principles of the quantum control of molecular processes* (2003). Mathematical Aspects of the control theory, are presented in the book [60] *Introduction to Quantum Control and Dynamics* by D. D'Alessandro (2008).

Error Corrections and Coherence Protection

For basic principles of classical error corrections one can consult the monograph by R. G. Gallager [100] *Information theory and reliable communications* (1968). First suggestions for the quantum error corrections proposed by P.W. Shor in the paper [199] entitled *Scheme for reducing decoherence in quantum computer memory* (1995) and by A.R. Calderbank and P.W. Shor [46] in the paper *Good quantum error-correcting codes exist* (1996) were based on the ideas conceptually very close to the classical error correction. The extreme of this approach is given in the article [144] *Perfect Quantum Error Correcting Code* by R. Laflamme C. Miquel, J.P. Paz, and W.H. Zurek (1996).

However, very soon it became clear that there exist other, essentially quantum, approaches to the protection of quantum systems against errors. One of them has been suggested in the paper [151] by D.A. Lidar, I.L. Chuang, and K.B. Whaley entitled *Decoherence-Free Subspaces for Quantum Computation* (1998). Another approach rely on the construction of multiparticle Hamiltonians that have multiply degenerate ground state not affected by a few-particle perturbations. This is the case of topological protection suggested in the article [130] by A.Yu. Kitaev *Fault-tolerant quantum computation by anyons Annals Phys. 303 (2003) 2–30* (1997). For coherence protection by non-holonom quantum control one can consult paper [39] (2011).

13.2 Problems

Two Levels

1. Describe the dynamics of spin classically (F. Bloch [24], *Nuclear induction* (1946)) and in terms of two-level systems and Bloch vectors (F. Bloch [25], *Generalized theory of relaxation* (1956)).

Level–Band

1. Consider the level–band problem in terms of scattering theory (G. Breit and E. Wigner [33] *Capture of slow neutrons* (1936)).
2. Consider the level–band model of intramolecular transitions and find the conversion rate of the electronic excitation to vibrational energy (M. Bixon and J. Jortner [21] *Intramolecular radiationless transitions* (1968)).
3. Consider the transition rates in atoms as a function of the frequency of the probe radiation in the presence of a strong laser field.(P. L. Knight, M. A. Lauder and B. J. Dalton [131] *Laser-induced continuum structure* (1990)).
4. Consider the interaction of two atoms via a narrow resonator mode (G. Kurizki, A. G. Kofman, and V. Yudson [140] *Resonant photon exchange by atom pairs in high-Q cavities* (1996)) and generalize the result to the case of a photonic band.
5. Consider the population dynamics corresponding to the Fano spectral profile (Ph. Durand, I. Paladrova and F. X. Cadea [78] *Theory of Fano profiles* (2001)).
6. Consider the time evolution of an atomic doublet coupled to an ionization continuum (Yu. V. Dubrovskii, M. V. Fedorov, and M. Yu. Ivanov [76] *Rydberg atom ionization by an intense laser pulse with smooth time envelope: A model of two closely spaced levels coupled to continuum* (1992)). Compare this to the method of projections (L. Mower [169] *Decay theory of closely coupled unstable states* (1966)).
7. Consider the dynamics of level decay to a band edge, with the density of states increasing according to the power law (A. G. Kofman [133] *Theory of single-photon bound-free transitions: Extension of the pole approximation* (1997)).

Revivals Wavepackets

1. Consider the regime of recurrences in the molecular quasicontinuum (E. Kyrölä and J. Eberly [142], *Quasicontinuum effects in molecular excitation* (1985)).
2. Consider the long-time asymptotics of the quasicontinuum and continuous band models (J. Javanainen and E. Kyrölä [123] *Long-time limit of a quasicontinuum model* (1985)).
3. Consider the time-dependence of the fractional revivals spikes (C. Lechtle, I. Sh. Averbukh, and W. P. Schleich [147] *Generic structure of multilevel quantum beats* (1996)).
4. Consider a band of atomic Rydberg states in terms of wavepackets (M. V. Fedorov, M. Yu. Ivanov and A. M. Movsesian [90] *Strong-field photoionization of an initially excited hydrogen atom: Formation of Rydberg wavepacket, its structure and trapping of population at Rydberg levels* (1990)).
5. Consider the excitation of an atom by a laser field and its second harmonics in terms of a multilevel system. Interpret the phase dependence of the numerically observed structure (K. J. Schafer and K. C. Kulander [191] *Phase dependent effects in multiphoton ionization induced by a laser field and its second harmonic* (1992)).
6. Consider recurrences in a spectrum with the Poisson statistics of levels (P. Pechukas [179] *Quantum chaos in the irregular spectrum* (1982)).
7. Trace relation between fractional revivals and factorization of large numbers (M. Mehring, K. Muller, I. Sh. Averbukh, W. Merkel, and W.P. Schleich, [162] *NMR Experiment Factors Numbers with Gauss Sums* (2007)).

Singular Statistics and Power Decay

1. Consider the singular billiard (P. Seba [196] *Wave chaos in singular quantum billiard* (1990)) and describe it in terms of a level interacting with a band. Compare this to the system perturbed by a rank-1 matrix (E. B. Bogomolny, U. Gerland, and C. Schmit [30] *Singular statistics* (2001)), and relate the singular statistics to the non-exponential decay of the level to a continuum.
2. Consider a perturbation imposed by a magnetic flux line on a band corresponding to a billiard (R. Narevich, R. E. Prange, O. Zaitsev [172] *Square billiard with an Aharonov–Bohm flux line* (2001)). Formulate this problem in terms of a rank-1 perturbation.
3. Relate the spectral correlations at the mobility edge of a disordered metal (V. E. Kravtsov, I. V. Lerner, B. L. Altshuler and A. G. Aronov [138] *Universal spectral correlations at the mobility edge* (1994)) with the power-law decay law (J. T. Chalker, I. V. Lerner and R. A. Smith [54] *Random walks through the ensemble: Linking spectral statistics with wave-function correlations in disordered metals* (1996)).

4. Consider the vibrational dynamics of polyatomic molecules in the regime of power-law decay (M. Gruebele [108] *Intramolecular vibrational dephasing obeys a power law at intermediate times* (1998)).
5. Consider ensembles suggested by hidden integrability of the motion (E. Bogomolny, O. Giraud, and C. Schmit [31] *Random Matrix Ensembles Associated with Lax Matrices* (2009)).
6. Consider Eucledean random matrices with the help of the supersymmetry technique (C.R. Offer, B.D. Simons *Field Theory of Euclidean Random Matrix Ensembles* (2000)).

Two Bands

1. Consider an example of an individual addressing the transitions in a two-band system (I. J. Cirac and P. Zoller [57] *Quantum computing with cold trapped ions* (1995)).
2. Compare the coherent damping and revivals in the Jannes–Cumming model (J. H. Eberly, N. B. Narozhny, and J. J. Sanchez-Mondregon [83], *Periodic spontaneous collapse and revival in a simple quantum model* (1980)) with that for an ensemble of spins subjected to a phase conjugating action (E. L. Hahn [112], *Spin Echoes* (1950)).
3. Consider the two-band model of metallic wires (D. J. Thouless [210], *Maximum metallic resistance in thin wires* (1977)).
4. Consider the two-band model of unimolecular reactions in laser fields (M. Quack [185] *Theory of unimolecular reactions induced by monochromatic infrared radiation* (1978)). Identify the master equation regime, the role of correlations of the dipole matrix elements, and consider the case of several privileged transitions.
5. Determine the probability of multiphoton transition as a function of laser intensity in the presence of removed degeneracy over the momentum projections (V. M. Akulin [2] *Contribution of weak and multiphoton transitions to the excitation of polyatomic molecules in an intense laser field* (1982)).
6. Consider a Rydberg atom in a strong laser field as a discrete band coupled to a continuum (M. Yu. Ivanov [120] *Suppression of resonant multiphoton ionization via Rydberg states* (1994)). Demonstrate the role of the rank-1 interaction matrix in the narrowing of the decay resonances and as an inhibitor of the decay rates. Determine the positions of the quasienergy eigenstates in the complex plane for this problem.

Quasienergy

1. Derive the time-independent Schrödinger equation for an atom in a strong, circularly polarized laser field (F. V. Bunkin, A. M. Prokhorov [42] *The excitation and ionization of atoms in a strong radiation field* (1964)).

2. Consider the role of photon quantum states in the concept of the quasienergy (Ya. B. Zeldovich [223, 224] *The quasienergy of a quantum-mechanical system subjected to a periodic action* (1967) and *Scattering and emission of a quantum system in a strong electromagnetic wave* (1973)).
3. Derive the Schrödinger equation for an atom in an arbitrary periodic external field (M. Gavrila and J. Z. Kaminski [104] *Free–free transitions in intense high-frequency laser fields* (1984)).
4. Consider the dynamics of Rydberg wave packets in the representation of quasienergy eigenstates (D. Delande and J. Zakrzewski [61] *Spontaneous emission of nondispersive Rydberg wave packets* (1998)).

Time-Dependent Systems

1. Consider a quantum particle scattered by the potential $-U_0/\cosh^2(x)$ (Landau and Lifshits *Quantum mechanics*, problem for § 23) and trace the analogy with the two-level system excited by the pulse interaction n.5.2.4.
2. Reproduce the result of the paper (Yu. N. Demkov [66] *Charge transfer at small resonance defects* (1964)) for the exponentially rising coupling between two detuned levels.
3. Consider two coupled levels with a detuning changing parabollically in time (E. E. Nikitin and S. Ya. Umanskii [174] *Theory of slow atomic collisions* § 8.3 (1984)). Compare the result with the approximate one given by the Dykhne formula (A. M. Dykhne [79] *Adiabatic perturbation of the discrete spectrum states* (1962)).
4. Estimate the probability of a non-adiabatic transition corresponding to the level–band system of (5.169) with the help of the Dykhne formula.
5. Allow for the effect of variation of the level velocity in the Landau–Zener Problem (D.A. Garanin and R. Schilling [102] Effect of nonlinear sweep in the Landau–Zener–Stueckelberg effect (2002)).
6. Derive the Magnus series approximation for multilevel systems with time-dependent Hamiltonians [118] (A. Iserles A. Marthinsen, S. P. Norsett (1999)).

Time-Dependent Complex Systems

1. Consider the role of the degeneracy of band states in the Demkov–Osherov problem (V. A. Yurovsky, A. Ben-Reuven, P. S. Julienne, and Y. B. Brand [221] *Counterintuitive transitions in multistate curve crossing involving linear potentials* (1999)).
2. Consider two-state curve crossing of a band (V. A. Yurovsky, A. Ben-Reuven [220] *Saturated transition in exactly soluble models of two-state curve crossing with time-dependent potentials* (1999)).

3. Estimate the rate of population diffusion over the energy scale for a complex system, adiabatically perturbed by a strong time-dependent perturbation.
4. In the quasienergy representation, consider the decay of a single level to a continuous band of finite width, for the case of harmonic modulation of the coupling (G. S. Agarwal [1] *Control of decoherence and relaxation by frequency modulation of a heat bath* (2000)).
5. Perform the same for the case of phase modulation (A. G. Kofman and G. Kurizki [135] *Universal dynamical control of quantum mechanical decay: Modulation of the coupling to the continuum* (2001), A. G. Kofman [134] *Relaxation of a two-level system strongly coupled to a reservoir: Anomalously slow decoherence* (2001)).
6. Compare the universal population distribution in a quantum system perturbed by a random matrix for colliding molecules (V.M. Akulin [3] *Singularities of collisional vibrational relaxation processes in strongly excited spectrally complex polyatomic molecules* (1983)) and for quantum dots in an external field (V. E. Kravtsov [137], *Time-reversal symmetry breaking by ac field: Effect of commensurability in frequency domain* (2002)).
7. Consider a quantum particle evolving in the presence of a chaotic background (D. Kuznetsov, A. Bulgak and G. D. Dang [141] *Quantum Lévy process and fractional kinetics* (1999)).

Group Symmetry Underlying Exact Solubility

1. Show that the system of n two-level atoms which has 2^n quantum levels can be separated into subsystems, each of which is a high-dimensional representation of a SU(2) group (F. Bloch and I. I. Rabi [26] *Atoms in variable magnetic fields* (1945)).
2. Consider a harmonic oscillator with time-dependent parameters (K. Husimi [116] *Miscellanea in elementary quantum mechanics* (1953)). Relate the possibility to find an exact solution with the algebraic properties of the operators entering the Hamiltonian.
3. Consider the underlying group symmetry for the Morse oscillator (R. D. Levine [148] *Representation of one-dimensional motion in a Morse potential by a quadratic Hamiltonian* (1983)).
4. Generalize the time-dependent oscillator problem to a multidimensional case (V. V. Dodonov and V. I. Man'ko [72] *Phase space eigenfunctions of multidimensional quadratic Hamiltonians* (1986)).
5. Consider the underlying group symmetry for the oscillator with a $1/x$ perturbation (V. V. Dodonov, V. I. Manko, and D. E. Nikonov [73] *Exact propagators for time-dependent Coulomb, delta and other potentials* (1992)), and for the Landau levels (V. A. Kostelecky, V. I. Manko, M. M. Nieto, and D. R. Traux [136] *Supersymmetry and a time-dependent Landau system* (1993)).

Relay of Isolated Levels

1. Derive the relay-like Schrödinger equation for an ensemble of two-level atoms interacting with the quantized field mode for one-photon and multi-photon cases (S. Stenholm [209] *Quantum theory of electromagnetic fields interacting with atoms and molecules* (1973)).
2. Perform the same derivation for the photon down-conversion, that is, for a field mode nonlinearly coupled to another field mode at half the frequency (G. Drobny, I. Jex, and V. Buzek [75] *Mode entanglement in nondegenerate down-conversion with quantized pump* (1993)).
3. Consider an exactly soluble relay and determine the role of relay truncation on the recurrence dynamics (M. Lindberg and E. Kyrölä [153] *Infinite multilevel system coupled to a quantum-field mode* (1992)).
4. Find the analytic description of the previous problem (B. W. Shore and J. H. Eberly [200] *Analytic approximations in multi-level excitation theory* (1978)).
5. Determine the population distribution evolution in a perfect lattice with a single defect (P. Lloyd [155] *Exactly solvable model of electronic states in a three-dimensional disordered Hamiltonian: non-existence of localized states* (1969)).
6. Consider the soluble, semi-infinite relay of levels with a coupling decreasing with the level number and with an arbitrary coupling for the first transition (A. Makarov [156] *Coherent excitation of equidistant multilevel systems in a resonant monochromatic field* (1977)).
7. Consider the relay with quadratically changing detunings (S. Mukamel and J. Jortner [171] *A model for isotope separation via molecular multiphoton photodissociation* (1976)) and compare the numerical results of the paper with the predictions of the WKB method.

Disordered Chains

1. Consider resonant levels, consecutively coupled by matrix elements randomly distributed according to Poisson statistics, by analogy to the one-dimensional elastic chain of randomly distributed masses (F. J. Dyson [80] *The dynamics of a disordered linear chain* (1953)) shown in Fig. 2.21 and find the mean density of states.
2. Consider the mean Green's functions for the disordered chains (P. Lloyd [155] *Exactly solvable model of electronic states in a three-dimensional disordered Hamiltonian: non-existence of localized states* (1969)).
3. Consider quantum localization with the help of the scaling idea applied to transparencies and reflectivities of groups of consecutive levels in the disordered chain (P. W. Anderson, D. J. Thouless, E. Abrahams, and D. S. Fisher [9] *New method for a scaling theory of localization* (1980)).

4. For the same problem consider the role of phase relations in the scaling (L. I. Deych, A. A. Lisyansky, and B. L. Altshuler [69] *Single parameter scaling in one-dimensional localization revisited* (2000)).
5. Consider the density of states for a non-Hermitian chain (C. Mudry, P. W. Brouwer, B. I. Halperin, V. Gurarie, A. Zee [170] *Density of states in the non-Hermitian Lloyd model* (1998)).
6. Consider the effect of a time-dependent perturbation on the localized electron states in a random one-dimensional lattice (N.F. Mott and W. D. Twose [167] *The theory of impurity conduction* § 7 (1961)).

Composite Systems: Relay of Bands

1. Consider the possibility of representing generic, multilevel systems as a level–band system and as a relay of isolated levels (E. Kyrölä, and M. Lindberg [143] *Serial and parallel multilevel systems* (1987)).
2. Represent the Schrödinger equation describing the Hall effect (B. Huckestein [115] *Scaling theory of the integer quantum Hall effect* (1995)) as a relay of degenerate bands of Landau levels.
3. Derive the master equation for randomly coupled, multi-band systems (Sect. 8.1.2) with the help of a standard Feynman diagram approach (B. Carmeli, A. Nitzan [47] *Kinetic equation for collisionless multiphoton excitation of large molecules* (1978) and [48] *Random coupling model for intramolecular dynamics* (1980)).
4. Consider the limit of applicability of the master equation (D. L. Shepelyansky [197] *Localization of diffusive excitation in multi-level systems* (1987)).
5. Consider the one-dimensional chain of levels with time-dependent level positions (J. Karczmarek, M. Scott, and M. Ivanov [128] *Two-color control of localization: From lattices to spin systems* (1999)) in the quasienergy representation, as a relay of bands.
6. Describe a quantum dot in a microwave field (D. M. Basko, M. A. Skvortsov, and V. E. Kravtsov [15] *Dynamic localization in quantum dots: analytical theory* (2003)) as a relay of quasienergy bands.
7. Describe an electron in a nanotube (L.S. Levitov and A. M. Tsvelik [149] *Narrow-gap Luttinger liquid in carbon nanotubes* (2003)) in terms of a multi-band system with four levels in each band.
8. Describe Arnold diffusion (V. Ya. Demikhovskii, F. M. Izrailev, and A. I. Malyshev [65] *Manifestation of Arnold diffusion in quantum systems* (2002)) as a process in a multiband system.
9. Consider multiphoton excitation of atoms above the ionization limit (N. B. Delone and M. V. Fedorov [62] *Above-Threshold Ionization* (1989), and [63] *Multiphoton ionization of atoms: New effects* (1989)).

Effect of Correlations and Relay with Decay

1. Consider a relay of isolated levels coupled to a continuum (R. S. Burkey and C. D. Cantrell [44] *Multichannel excitation of the quasi-continuum* (1985)) as a multiband system with correlated matrix elements.
2. Consider the above-threshold ionization of atoms (Z. Deng and J. H. Eberly [68] *Multiphoton absorption above ionization threshold by atoms in strong laser field* (1985)) as a process in a multiband relay with completely correlated matrix elements.
3. Consider a Rydberg atom in a strong laser field (M. V. Fedorov and A. M. Movsesian [91] *Interference suppression of photoionization of Rydberg atoms in a strong electromagnetic field* (1988) and [92] *Field-induced effects of narrowing of photoelectron spectra and stabilization of Rydberg atoms* (1988)) as continuum–continuum transitions with a rank-1 interaction.
4. Consider a hydrogen atom in an intense RF field (G. Casati, B. V. Chirikov, and D. L. Shepelyansky [50] *Quantum limitation for chaotic excitation of the hydrogen atom in a monochromatic field* (1984)) as a relay of bands with correlated matrix elements. Estimate the typical time for ballistic propagation for this model.

Localization in Relay of Bands, Field-Theory Methods

1. Determine the quantum localization border for Rydberg atoms in an RF field (G. Casati, B. V. Chirikov, I. Guarneri, and D. L. Shepelyansky [51] *Dynamical stability of quantum "chaotic" motion in a hydrogen atom* (1986)) and find the parameter governing the localization (G. Casati, I. Guarneri, F. Izrailev, and R. Scharf [52] *Scaling behavior of localization in quantum chaos* (1990)). Express this in terms of band random matrices (G. Casati, L. Molinari, and F. Izrailev [53] *Scaling properties of band random matrices* (1990)) and compare it to molecular models (G. Grahm and D. Shepelyansky [106] *A solid-state model for photonic localization in molecular quasi-continua* (1991)).
2. Consider complex systems with the help of the σ-model technique and the saddle-point approximation (K. Efetov [87] *Supersymmetry in disorder and chaos* (1996), Chap. 4).
3. Consider random band matrices in terms of a σ-model (Y. V. Fyodorov and A. D. Mirlin [96] *Scaling properties of localization in random band matrices: A σ -model approach* (1991). Consider the transition to localization in such systems (A. D. Mirlin, Y. V. Fyodorov, F.-M. Dittes J. Quezada, and T. H. Seligman [166] *Transition from localized to extended eigenstates in the ensembles of power-law random banded matrices* (1996)).
4. Consider the role of particle collisions on the localization (B. L. Altshuler, A. G. Aronov and D. E. Khmelnitsky [6] *Effects of electron-electron collisions with small energy transfers on quantum localisation* (1982)). Formulate the problem of binary collisions in terms of random matrices (D. L. Shepelyansky

[198] *Coherent propagation of two interacting particles in a random potential* (1994)). Consider this problem with the help of a σ-model (Y. V. Fyodorov, and A. Mirlin [97] *Statistical properties of random banded matrices with strongly fluctuating diagonal elements* (1995)).

5. Consider the transition to localization for sparse band matrices (Y. V. Fyodorov, A. D. Mirlin, and H-J. Sommers [99] *A novel field theoretical approach to the Anderson localization: sparse random hopping model* (1992)).
6. Consider dynamic localization in driven fermionic systems (M.A. Skvortsov, D.M. Basko, and V.E. Kravtsov [206], *Energy absorption in time-dependent unitary random matrix ensembles: dynamic vs. Anderson localization* (2004))
7. Consider the dynamics of elementary excitations in cold Rydberg atoms as a level coupled to a band of random walks (I. Mourachko et al. [168] *Many-body effects in a frozen Rydberg gas* (1998)).

Transformation of Complex Spectra

1. Consider the transformation of a non-degenerate spectrum perturbed by a small random matrix (P. Pechukas [180] *Distribution of energy eigenvalues in the irregular spectrum* (1983)). Derive the diffusion equation describing this process.
2. Derive a functional equation describing the transformation of a general multilevel system perturbed by a random matrix (L.A. Pastur [178] *On the spectrum of random matrices* (1972)).
3. Derive the Pastur equation via a direct average over a random matrix perturbation of a deterministic quantum system (E. Brezin, S. Hikami, and A. Zee [38] *Universal correlations for deterministic plus random Hamiltonians* (1995)), and find the correlations of the level positions. Consider the corresponding time evolution (E. Brezin and S. Hikami [34] *Spectral form factor in a random matrix theory* (1997)). Consider the general case of level correlations (E. Brezin and S. Hakami [35] *Extension of level-spacing universality* (1997).
4. Trace the transition from the Breit–Wigner density of states to a Wigner semicircular profile for sparse band random matrices (Y. V. Fyodorov, O. A. Chubykalo, F. M. Izrailev and G. Casati [98] *Wigner random banded matrices with sparse structure: local spectral density of states* (1996)).
5. Derive a differential equation of motion for the mean density of states considering the mean square coupling of a random perturbation as a variable (F. Dyson [82] *A class of matrix ensembles* (1972)).
6. Consider the behavior near the gap closing (E. Brezin, S. Hikami [36] *Universal singularity at the closure of a gap in a random matrix theory* (1998)).
7. Find the transformation under perturbation by a random matrix of the following spectral profiles: (a) Density of states of a one-dimensional free particle $g(E) = 2\pi(1-E^2)^{-1/2}$; (b) Hubbard random alloys; (c) two-dimensional free particle $g(E) = \pi\Theta(E)$; (d) a particle in the conduction band of a semiconductor with the kinetic energy $T(p) = \cos p$; (e) the Rydberg spectrum $E = 1/n^2$ for $n = 2, 3, 4$ as given by the solution of the fourth-order algebraic equation.

Entanglement

1. Find inequivalent entanglement classes for three two-level elements (W. Dur, G. Vidal, and I.J. Cirac [77] *Three qubits can be entangled in two inequivalent ways* (2000)).
2. Find inequivalent entanglement classes for four two-level elements (F. Verstraete, J. Dehaene, B. De Moor, and H. Verschelde [212], *Four qubits can be entangled in nine different ways* (2002)).
3. Consider generalization of entanglement in terms of observables (H. Barnum, E. Knill, G. Ortiz, R. Somma, and L. Viola, [13] *A Subsystem-Independent Generalization of Entanglement* (2004)).
4. Derive algebraic rules for nilpotent polynomials for the entanglement description (A.M. Frydryszak [95], *Pure state entanglement in terms of nilpotent variables: η-toolbox* (2011)).
5. Consider general conditions for the entanglement measures of mixed states (V. Vedral, M.P. Plenio, M.A. Rippin, and P.L. Knight [211], *Quantifying Entanglement* (1997)).
6. Consider entanglement of mixed states in Ising model (P. Stelmachovi and V. Buzek [208], *Quantum-information approach to Ising model: Entanglement in chains of qubits* (2004)).
7. Consider dynamics of entanglement in chains of two-level systems (L. Amico, A. Osterloh, F. Plastina, R. Fazio, G.M. Palma [7], *Dynamics of entanglement in one-dimensional spin systems* (2004)).
8. Trace behavior of entanglement in phase transitions (T.J. Osborne and M. A. Nielsen [176], *Entanglement in a simple quantum phase transition* (2002)).
9. Trace the relation between entanglement and high efficiency of the photon harvesting (F. Caruso, A.W. Chin, A. Datta, S.F. Huelga, and M.B. Plenio [49], *Entanglement and entangling power of the dynamics in light-harvesting complexes* (2010)).

Control and Coherence Protection

1. Determine a way to perform a given transformation via controlled collision (D. Jaksch, H. Briegel, J. I. Cirac, C. W. Gardiner, and P. Zoller [122] *Entanglement of atoms via cold controlled collisions* (1999)).
2. Trace the equivalence of the sufficient criterion of the bracket generation condition and the requirement given in the book by P. W. Brumer, M. Shapiro [41] entitled *Principles of the quantum control of molecular processes* (2003).
3. Check the possibility to control a couple of dipole–dipole interacting spins by application of an external time-dependent magnetic field.
4. Consider control of tanglemeter by multiparticle manipulations (A. Mandilara, V. M. Akulin, M. Kolar and G. Kurizki [158], *Control of multiatom entanglement in a cavity* (2007)).

5. Consider universal protection of a pair of two-level system by control (K. Khodjasteh, and L. Viola [132], *Dynamically Error-Corrected Gates for Universal Quantum Computation* (2009)).
6. Study a physical implementation of the topological protection scheme (B. Doucot, M.V. Feigel'man, L.B. Ioffe, A.S. Ioselevich [74], *Protected qubits and Chern–Simons theories in Josephson junction arrays* (2005)).

References

1. Agarwal, G.S.: Control of decoherence and relaxation by frequency modulation of a heat bath. Phys. Rev. A **61**, 013809 (2000)
2. Akulin, V.M.: Contribution of weak and multiphoton transitions to the excitation of polyatomic molecules in an intense laser field. Sov. Phys. JETP **56**, 533–538 (1982)
3. Akulin, V.M.: Singularities of collisional vibrational relaxation processes in strongly excited spectrally complex polyatomic molecules. Sov. Phys. JETP **57**, 774–780 (1983)
4. Akulin, V.M., Karlov, N.V.: Intense Resonant Interactions in Quantum Electronics. Springer, Berlin (1991)
5. Allen, L., Eberly, J.H.: Optical Resonance and Two-Level Atoms. Wiley, Chichester (1975)
6. Altshuler, B.L., Aronov, A.G., Khmelnitsky, D.E.: Effects of electron-electron collisions with small energy transfers on quantum localisation. J. Phys. C Solid State Phys. **15**, 7367–7386 (1982)
7. Amico, L., Osterloh, A., Plastina, F., Fazio, R., Palma, G.M.: Dynamics of entanglement in one-dimensional spin systems. Phys. Rev. A **69**, 022304 (2004)
8. Anderson, P.W.: Absence of diffusion in certain random lattices. Phys. Rev. **109**, 1492–1505 (1957)
9. Anderson, P.W., Thouless, D.J., Abrahams, E., Fisher, D.S.: New method for a scaling theory of localization. Phys. Rev. B **22**, 3519–3526 (1980)
10. Arnold, V.I., Weinstein, A., Vogtmann, K.: Mathematical Methods of Classical Mechanics. Springer, New York (1989)
11. Askaryan, G.A.: Excitation and dissociation of molecules in intense light field. Sov. Phys. JETP **19**, 273–274 (1964)
12. Averbukh, S., Perelman, N.F.: Fractional revivals: universality in the long-term evolution of quantum wave packets beyond the correspondence principle dynamics. Physics Letters A **139**, 449–454 (1989)
13. Barnum, H., Knill, E., Ortiz, G., Somma, R., Viola, L.: A subsystem-independent generalization of entanglement. Phys. Rev. Lett. **92**, 107902 (2004)
14. Barut, A., Raczka, R.: Theory of Group Representations and Applications. Polish Scientific Publisher, Warszava (1977)
15. Basko, D.M., Skvortsov, M.A., Kravtsov, V.E.: Dynamic localization in quantum dots: analytical theory. Phys. Rev. Lett. **90**, 096801–096804 (2003)
16. Beenakker, C.W.J.: Random matrix theory of quantum transport. Rev. Mod. Phys. **69**, 731–808 (1997)
17. Bell, J.S.: On the Einstein–Podolsky–Rosen Paradox. Physics **1**, 195–200 (1964)

V.M. Akulin, *Dynamics of Complex Quantum Systems*, Theoretical and Mathematical Physics, DOI 10.1007/978-94-007-7205-2,

18. Bengtsson, I., Zyczkowski, K.: Geometry of Quantum States. Cambridge University Press, Cambridge (2006)
19. Berezinskii, V.L.: Kinetics of a quantum particle in a one-dimensional random potential. Sov. Phys. JETP **38**, 620–627 (1974)
20. Bery, M.V., Tabor, M.: Poissonian statistics and clustering. Proc. R. Soc. Lond. **356**, 375–394 (1977)
21. Bixon, M., Jortner, J.: Intramolecular radiationless transitions. J. Chem. Phys. **48**, 715–726 (1968)
22. Bixon, M., Jortner, J.: Long radiative lifetimes of small molecules. J. Chem. Phys. **50**, 3284–3290 (1969)
23. Blanter, Y.M., Mirlin, A.D., Muzykantskii, B.A.: Quantum billiards with surface scattering: ballistic sigma-model approach. Phys. Rev. Lett. **80**, 4161 (1998)
24. Bloch, F.: Nuclear induction. Phys. Rev. **70**, 460 (1946)
25. Bloch, F.: Generalized theory of relaxation. Phys. Rev. **105**, 1206–1222 (1957)
26. Bloch, F., Rabi, I.I.: Atoms in variable magnetic fields. Rev. Mod. Phys. **17**, 237–244 (1945)
27. Blombergen, N.: Nonlinear Optics. Benjamin, New York (1965)
28. Bohigas, O., Giannoni, M.J., Schmit, C.: Characterization of chaotic quantum spectra and universality of level fluctuation laws. Phys. Rev. Lett. **52**, 1–4 (1984)
29. Bogomolny, E.B., Gerland, U., Schmit, C.: Models of intermediate spectral statistics. Phys. Rev. E **59**, R1315–R1318 (1999)
30. Bogomolny, E.B., Gerland, U., Schmit, C.: Singular statistics. Phys. Rev. E **63**, 036206 (2001)
31. Bogomolny, E., Giraud, O., Schmit, C.: Random matrix ensembles associated with lax matrices. Phys. Rev. Lett. **103**, 054103 (2009)
32. Borland, R.E.: The nature of the electronic states in disordered one-dimensional systems. Proc. Roy. Soc. Lond. A **274**, 529–545 (1963)
33. Breit, G., Wigner, E.: Capture of slow neutrons. Phys. Rev. **40**, 519–531 (1936)
34. Brezin, E., Hikami, S.: Spectral form factor in a random matrix theory. Phys. Rev. E **55**, 4067–4083 (1997)
35. Brezin, E., Hakami, S.: Extension of level-spacing universality. Phys. Rev. E **56**, 264–269 (1997)
36. Brezin, E., Hikami, S.: Universal singularity at the closure of a gap in a random matrix theory. Phys. Rev. E **57**, 4140–4149 (1998)
37. Brezin, E., Hikami, S.: New correlation functions for random matrices and integrals over supergroups. J. Phys. A **36**, 711–751 (2003)
38. Brezin, E., Hikami, S., Zee, A.: Universal correlations for the deterministic plus random Hamiltonians. Phys. Rev. E **51**, 5442–5452 (1995)
39. Brion, E., Carlier, F., Harel, G., Akulin, V.M.: Non-holonomic quantum control. J. Phys. B At. Mol. Opt. Phys. **44**, 154001 (2011)
40. Brixner, T., Stenger, J., Vaswani, H.M., Cho, M., Blankenship, R.E., Fleming, G.R.: Two-dimensional spectroscopy of electronic couplings in photosynthesis. Nature **434**, 625–628 (2005)
41. Brumer, P.W., Shapiro, M.: Principles of the Quantum Control of Molecular Processes. Wiley, New York (2003)
42. Bunkin, F.V., Prokhorov, A.M.: The excitation and ionization of atoms in a strong radiation field. Sov. Phys. JETP **19**, 739–743 (1964)
43. Bunkin, F.V., Prokhorov, A.M., Karapetyan, R.V.: Dissociation of molecules in a strong radiation field. Sov. Phys. JETP **20**, 145–148 (1965)
44. Burkey, R.S., Cantrell, C.D.: Multichannel excitation of the quasi-continuum. J. Opt. Soc. Am. B **2**, 451–457 (1985)
45. Butkovskiy, G., Samoilenko, Y.I.: Control of Quantum-Mechanical Processes and Systems. Kluwer Academic, Dordrecht (1990)
46. Calderbank, A.R., Shor, P.W.: Good quantum error-correcting codes exist. Phys. Rev. A **54**, 1098–1105 (1996)

47. Carmeli, B., Nitzan, A.: Kinetic equation for collisionless multiphoton excitation of large molecules. Chem. Phys. Lett. **62**, 457–460 (1978)
48. Carmeli, B., Nitzan, A.: Random coupling model for intramolecular dynamics. J. Chem. Phys. **72**, 2054–2080 (1980)
49. Caruso, F., Chin, A.W., Datta, A., Huelga, S.F., Plenio, M.B.: Entanglement and entangling power of the dynamics in light-harvesting complexes. Phys. Rev. A **81**, 062346 (2010)
50. Casati, G., Chiricov, B.V., Shepelyansky, D.L.: Quantum limitation for chaotic excitation of the hydrogen atom in a monochromatic field. Phys. Rev. Lett. **53**, 2525–2528 (1984)
51. Casati, G., Chirikov, B.V., Guarneri, I., Shepelyansky, D.L.: Dynamical stability of quantum chaotic motion in a hydrogen atom. Phys. Rev. Lett. **23**, 2437–2440 (1986)
52. Casati, G., Guarneri, I., Izrailev, F., Scharf, R.: Scaling behavior of localization in quantum chaos. Phys. Rev. Lett. **64**, 5–8 (1990)
53. Casati, G., Molinari, L., Izrailev, F.: Scaling properties of band random matrices. Phys. Rev. Lett. **64**, 1851–1854 (1990)
54. Chalker, J.T., Lerner, I.V., Smith, R.A.: Random walks through the ensemble: linking spectral statistics with wave-function correlations in disordered metals. Phys. Rev. Lett. **77**, 554–557 (1996)
55. Chirikov, B.V.: Resonance processes in magnetic traps. J. Nucl. Energy C Plasma Phys. **1**, 253–260 (1960)
56. Chirikov, B.V., Shepelyansky, D.L.: Correlation properties of dynamical chaos in Hamiltonian systems. Physica D **13**, 395–400 (1984)
57. Cirac, J.I., Zoller, P.: Quantum computation with cold trapped ions. Phys. Rev. Lett. **74**, 4091–4094 (1995)
58. Cohen-Tanoudji, C., Diu, B., Laloe, F.: Quantum Mechanics. Wiley, Paris (1977)
59. Cohen-Tannoudji, C., Dupont-Roc, J., Grynberg, G.: Atom-Photon Interactions: Basic Processes and Applications. Wiley-Interscience, New York (1998)
60. D'Alessandro, D.: Introduction to Quantum Control and Dynamics. Chapman & Hall/CRC Applied Mathematics & Nonlinear Science, London (2008)
61. Delande, D., Zakrzewski, J.: Spontaneous emission of nondispersive Rydberg wave packets. Phys. Rev. A **58**, 446–477 (1998)
62. Delone, N.B., Fedorov, M.V.: Above-threshold ionization. Progr. Quant. Electron. **13**, 267–298 (1989)
63. Delone, N.B., Fedorov, M.V.: New effects in the multiphoton ionization of atoms. Sov. Phys. Uspekhi **32**, 500–520 (1989)
64. Delone, N.B., Krainov, V.P.: Atoms in Strong Light Fields. Springer, Berlin (1985)
65. Demikhovskii, V.Y., Izrailev, F.M., Malyshev, A.I.: Manifestation of Arnold diffusion in quantum systems. Phys. Rev. Lett. **88**, 154101 (2002)
66. Demkov, Y.N.: Charge transfer at small resonance defects. JETP **18**, 138–142 (1964)
67. Demkov, Y.N., Osherov, V.I.: Stationary and nonstationary problems in quantum mechanics that can be solved by means of contour integration. Sov. Phys. JETP **26**, 916–921 (1968)
68. Deng, Z., Eberly, J.H.: Multiphoton absorption above ionization threshold by atoms in strong laser field. J. Opt. Soc. Am. B **2**, 486–493 (1985)
69. Deych, L.I., Lisyansky, A.A., Altshuler, B.L.: Single parameter scaling in one-dimensional localization revisited. Phys. Rev. Lett. **84**, 2678–2681 (2000)
70. Dicke, R.H.: Coherence in spontaneous radiation processes. Phys. Rev. **93**, 99 (1954)
71. Dodonov, V.V.: Universal integrals of motion and universal invariants of quantum systems. J. Phys. A **33**, 7721–7738 (2000)
72. Dodonov, V.V., Man'ko, V.I.: Phase space eigenfunctions of multidimensional quadratic Hamiltonians. Physica A **137**, 306–316 (1986)
73. Dodonov, V.V., Manko, V.I., Nikonov, D.E.: Exact propagators for time-dependent Coulomb, delta and other potentials. Phys. Lett. A **162**, 359–364 (1992)
74. Doucot, B., Feigel'man, M.V., Ioffe, L.B., Ioselevich, A.S.: Protected qubits and Chern–Simons theories in Josephson junction arrays. Phys. Rev. B **71**, 024505 (2005)

75. Drobny, G., Jex, I., Buzek, V.: Mode entanglement in nondegenerate down-conversion with quantized pump. Phys. Rev. A **48**, 569–579 (1993)
76. Dubrovskii, Y.V., Fedorov, M.V., Ivanov, M.Y.: Rydberg atom ionization by an intense laser pulse with smooth time envelope: a model of two closely spaced levels coupled to continuum. Laser Physics **2**, 288–298 (1992)
77. Dur, W., Vidal, G., Cirac, I.J.: Three qubits can be entangled in two inequivalent ways. Phys. Rev. A **62**, 062314 (2000)
78. Durand, P., Paladrova, I., Cadea, F.X.: Theory of Fano profiles. J. Phys. B **34**, 1953–1986 (2001)
79. Dykhne, M.: Adiabatic perturbation of the discrete spectrum states. Sov. Phys. JETP **14**, 941–943 (1962)
80. Dyson, F.J.: The dynamics of a disordered linear chain. Phys. Rev. **92**, 1331–1338 (1953)
81. Dyson, F.: Statistical theory of the energy levels of complex systems. J. Math. Phys. **3**, 140–175 (1962)
82. Dyson, F.: A class of matrix ensembles. J. Math. Phys. **13**, 90–97 (1972)
83. Eberly, J., Narozhny, N.B., Sanchez-Mondregon, J.J.: Periodic spontaneous collapse and revivals in a simple quantum model. Phys. Rev. Lett. **44**, 1323–1325 (1980)
84. Eckhardt, B.: Quantum mechanics of classically non-integrable systems. Phys. Rep. **163**, 205–297 (1988)
85. Einstein, A.: Zur Quantentheorie der Strahlung. Physikalische Zeitschrift **18**, 121–128 (1917). The Collected Papers of Albert Einstein, Volume 6: The Berlin Years: Writings, 1914–1917 (English translation supplement). University Press, Princeton
86. Einstein, A., Podolsky, B., Rosen, N.: Can quantum-mechanical description of physical reality be considered complete? Phys. Rev. **47**, 777 (1935)
87. Efetov, K.: Supersymmetry in Disorder and Chaos. Cambridge University Press, Cambridge (1997)
88. Fano, U.: Sullo spettro di assorbimento dei gas nobili presso il limite dello spettro d'arco. Nuovo Cimento N. S. **12**, 154–161 (1935)
89. Fano, U.: Effects of configuration interaction on intensities and phase shifts. Phys. Rev. **124**, 1866–1878 (1961)
90. Fedorov, M.V., Ivanov, M.Y., Movsesian, A.M.: Strong-field photoionization of an initially excited hydrogen atom: formation of Rydberg wavepacket, its structure and trapping of population at Rydberg levels. J. Phys. B **23**, 2245S–2257S (1990)
91. Fedorov, M.V., Movsesian, A.M.: Interference suppression of photoionization of Rydberg atoms in strong electromagnetic field. J. Opt. Soc. Am. B **6**, 928–936 (1988)
92. Fedorov, M.V., Movsesian, A.M.: Field-induced effects of narrowing of photoelectron spectra and stabilization of Rydberg atoms. J. Phys. B **21**, L155 (1988)
93. Fermi, E.: Quantum theory of radiation. Rev. Mod. Phys. **4**, 87–133 (1932)
94. Floquet, G.: Sur leséquations différentialles linéairesà coefficients périodiques. Ann. de l'Ecole Normale 2-e serie **12**, 47–88 (1889)
95. Frydryszak, A.M.: Pure state entanglement in terms of nilpotent variables: eta-toolbox. In: Olkiewicz, R., Cegla, W., Frydryszak, A., Garbaczewski, P., Jakobczyk, L. (eds.) Quantum Dynamics and Information. World Scientific, Singapore (2011)
96. Fyodorov, Y.V., Mirlin, A.D.: Scaling properties of localization in random band matrices: a $sigma$-model approach. Phys. Rev. Lett. **67**, 2405–2409 (1991)
97. Fyodorov, Y.V., Mirlin, A.: Statistical properties of random banded matrices with strongly fluctuating diagonal elements. Phys. Rev. B **52**, R11580–R11583 (1995)
98. Fyodorov, Y.V., Chubykalo, O.A., Izrailev, F.M., Casati, G.: Wigner random banded matrices with sparse structure: local spectral density of states. Phys. Rev. Lett. **76**, 1603–1606 (1996)
99. Fyodorov, Y.V., Mirlin, A.D., Sommers, H.-J.: A novel field theoretical approach to the Anderson localization: sparse random hopping model. J. Phys. I France **2**, 1571–1605 (1992)
100. Gallager, R.G.: Information Theory and Reliable Communications. Wiley, New York (1968)
101. Gallagher, T.F.: Rydberg Atoms. Cambridge University Press, Cambridge (1994)

102. Garanin, D.A., Schilling, R.: Effect of nonlinear sweep in the Landau-Zener-Stueckelberg effect. Phys. Rev. B **66**, 174438 (2002)
103. Gardiner, W., Zoller, P.: Quantum Noise. Springer, Berlin (2004)
104. Gavrila, M., Kaminski, J.Z.: Free-free transitions in intense high-frequency laser fields. Phys. Rev. Lett. **52**, 613–616 (1984)
105. Ghosh, P.K.: Ion Traps. Clarendon, Oxford (1995)
106. Grahm, G., Shepelyansky, D.: A solid-state model for photonic localization in molecular quasi-continua. Europhys. Lett. **14**, 211–215 (1991)
107. Greenberger, D.M., Horn, M.A., Zeilinger, A.: Going beyond Bell's theorem. In: Kafatos, M. (ed.) Bell's Theorem, Quantum Theory, and Conceptions of the Universe, pp. 73–76. Kluwer Academic, Dordrecht (1989)
108. Gruebele, M.: Intramolecular vibrational dephasing obeys a power law at intermediate times. Proc. Natl. Acad. Sci. USA **95**, 5965–5970 (1998)
109. Guérin, S., Jauslin, H.R.: Control of quantum systems by pulses: adiabatic Floquet theory. Adv. Chem. Phys. **125**, 1–75 (2003)
110. Gutzwiller, M.: Chaos in Classical and Quantum Mechanics. Springer, New York (1990)
111. Haake, F.: Quantum Signatures of Chaos. Springer, Berlin (2001)
112. Hahn, E.L.: Spin echoes. Phys. Rev. **80**, 580–594 (1950)
113. Horodecki, M., Horodecki, P., Horodecki, R.: Mixed-state entanglement and distillation: is there a "bound" entanglement in nature? Phys. Rev. Lett. **80**, 5239–5242 (1998)
114. Hubbard, J.: Electron correlations in narrow energy bands. Proc. R. Soc. Lond. Ser. A **276**, 238–257(1963); **277**, 237–259 (1964); **281**, 401–419 (1964)
115. Huckestein, B.: Scaling theory of the integer quantum Hall effect. Rev. Mod. Phys. **67**, 375–396 (1995)
116. Husimi, K.: Miscellanea in elementary quantum mechanics. Prog. Theor. Phys. **9**, 238–244, 381–402 (1953)
117. Imry, J.: Introduction to Mesoscopic Physics. Oxford University Press, Oxford (2002)
118. Iserles, A., Marthinsenet, A., Nörsett, S.P.: On the implementation of the method of Magnus series for linear differential equations. BIT Numer. Math. **39**, 281–304 (1999)
119. Ishii, K.: Localization of eigenstates and transport phenomena in the one-dimensional disordered systems. Supl. Prog. Theor. Phys. **53**, 77–138 (1972)
120. Ivanov, M.Y.: Suppression of resonant multiphoton ionization via Rydberg states. Phys. Rev. A **49**, 1165–1170 (1994)
121. Izrailev, F.M.: Simple models of quantum chaos: spectrum and eigenfunctions. Physics Reports **5–6**, 299–392 (1990)
122. Jaksch, D., Briegel, H., Cirac, J.I., Gardiner, C.W., Zoller, P.: Entanglement of atoms via cold controlled collisions. Phys. Rev. Lett. **82**, 1975–1978 (1999)
123. Javanainen, J., Kyrölä, E.: Long-time limit of a quasicontinuum model. Opt. Commun. **56**, 17–21 (1985)
124. Joannopoulos, J.D., Meade, R.D., Winn, J.N.: Photonic Crystals. Princeton University Press, Princeton (1995)
125. J. Phys. A **36**(12) (28 March 2003)
126. Jungen, C.: Molecular Applications of Quantum Defect Theory. Institute of Physics Publisher, Philadelphia (1996)
127. Jurdevic, V., Sussmann, H.J.: Control systems on Lie groups. J. Differ. Eq. **12**, 313–329 (1972)
128. Karczmarek, J., Scott, M., Ivanov, M.: Two-color control of localization: from lattices to spin systems. Phys. Rev. A **60**, R4225–R4228 (1999)
129. Karplus, R., Schwinger, J.: A note on saturation in microwave spectroscopy. Phys. Rev. **73**, 1020–1026 (1948)
130. Kitaev, Y.: Fault-tolerant quantum computation by anyons. Ann. Phys. **303**, 2–30 (2003)
131. Knight, P.L., Lauder, M.A., Dalton, B.J.: Laser-induced continuum structure. Phys. Rep. **190**, 1–61 (1990)
132. Khodjasteh, K., Viola, L.: Dynamically error-corrected gates for universal quantum computation. Phys. Rev. Lett. **102**, 080501 (2009)

133. Kofman, G.: Theory of single-photon bound-free transitions: extension of the pole approximation. J. Phys. B **30**, 5141–5156 (1997)
134. Kofman, G.: Relaxation of a two-level system strongly coupled to a reservoir: anomalously slow decoherence. Phys. Rev. A **64**, 033809 (2001)
135. Kofman, G., Kurizki, G.: Universal dynamical control of quantum mechanical decay: modulation of the coupling to the continuum. Phys. Rev. Lett. **87**, 270405 (2001)
136. Kostelecky, V.A., Manko, V.I., Nieto, M.M., Traux, D.R.: Supersymmetry and time-dependent Landau system. Phys. Rev. A **48**, 951–963 (1993)
137. Kravtsov, V.E.: Time reversal symmetry breaking by ac field: effect of commensurability in frequency domain. Pramana J. Phys. **58**, 183–193 (2002)
138. Kravtsov, V.E., Lerner, I.V., Altshuler, B.L., Aronov, A.G.: Universal spectral correlations at the mobility edge. Phys. Rev. Lett. **72**, 888–891 (1994)
139. Kubo, R.: A general expression for the conductivity tensor. Can. J. Phys. **34**, 1274–1277 (1956)
140. Kurizki, G., Kofman, A.G., Yudson, V.: Resonant photon exchange by atom pairs in high-Q cavities. Phys. Rev. A **53**, R35–R38 (1996)
141. Kuznetsov, D., Bulgak, A., Dang, G.D.: Quantum Lévy process and fractional kinetics. Phys. Rev. Lett. **82**, 1136–1139 (1999)
142. Kyrölä, E., Eberly, J.: Quasicontinuum effects in molecular excitation. J. Chem. Phys. **82**, 1841–1853 (1985)
143. Kyrölä, E., Lindberg, M.: Serial and parallel multilevel systems. Phys. Rev. A **35**, 4207–4225 (1987)
144. Laflamme, R., Miquel, C., Paz, J.P., Zurek, W.H.: Perfect quantum error correcting code. Phys. Rev. Lett. **77**, 198–201 (1996)
145. Landau, L.D.: Das Dämpfungsproblem in der Wellenmechanik. Zeitschrift Phys. Bd **XLV**, 430–441 (1927)
146. Landau, L.D., Lifshits, E.M.: Quantum mechanics. Butterworth-Heinemann, Oxford (1975–1999)
147. Lechtle, C., Averbukh, I.S., Schleich, W.P.: Generic structure of multilevel quantum beats. Phys. Rev. Lett. **77**, 3999–4002 (1996)
148. Levine, R.D.: Representation of one-dimensional motion in a Morse potential by a quadratic Hamiltonian. Chem. Phys. Lett. **95**, 87–90 (1983)
149. Levitov, L.S., Tsvelik, A.M.: Narrow-gap Luttinger liquid in carbon nanotubes. Phys. Rev. Lett. **90**, 016401 (2003)
150. Lichtenberg, A.J., Liberman, M.A.: Regular and Stochastic Motion. Springer, Berlin (1982)
151. Lidar, D.A., Chuang, I.L., Whaley, K.B.: Decoherence-free subspaces for quantum computation. Phys. Rev. Lett. **81**, 2594–2597 (1998)
152. Lifshits, I.M., Gredeskul, S.A., Pastur, L.A.: Introduction to the Theory of Disordered Systems. Wiley, New York (1988)
153. Lindberg, M., Kyrölä, E.: Infinite multilevel system coupled to a quantum-field mode. J. Opt. Soc. Am. B **9**, 595–604 (1992)
154. Linden, N., Popescu, S.: On multi-particle entanglement. Fortschr. Phys. **4–5**, 567–578 (1998)
155. Lloyd, P.: Exactly solvable model of electronic states in a three-dimensional disordered Hamiltonian: non-existence of localized states. J. Phys. C Solid State Phys. **2**, 1717–1725 (1969)
156. Makarov, A.: Coherent excitation of equidistant multilevel systems in a resonant monochromatic field. Sov. Phys. JETP **45**, 918–924 (1977)
157. Mandilara, A., Akulin, V.M., Smilga, A., Viola, L.: Description of quantum entanglement with nilpotent polynomials. Phys. Rev. A **74**, 022331 (2006)
158. Mandilara, A., Akulin, V.M., Kolar, M., Kurizki, G.: Control of multiatom entanglement in a cavity. Phys. Rev. A **75**, 022327 (2007)
159. Massey, H.S.W., Mott, N.F.: The Theory of Atomic Collisions. Clarendon, Oxford (1997)
160. Mathews, J., Walker, R.L.: Mathematical Methods of Physics. Addison-Wesley, New York (1971)

161. McCall, S.L., Hahn, E.L.: Self-induced transparency. Phys. Rev. **183**, 457–485 (1969)
162. Mehring, M., Muller, K., Averbukh, I.Sh., Merkel, W., Schleich, W.P.: NMR experiment factors numbers with gauss sums. Phys. Rev. Lett. **98**, 120502 (2007)
163. Metha, M.L.: Random Matrices and Statistical Theory of Energy Levels. Academic, New York (1967)
164. Meystre, P.: Atom Optics. Springer, New York (2001)
165. Mezard, M., Parisi, G., Zee, A.: Spectra of Euclidean random matrices. Nucl. Phys. B **59**, 689–701 (1999)
166. Mirlin, D., Fyodorov, Y.V., Dittes, F.-M., Quezada, J., Seligman, T.H.: Transition from localized to extended eigenstates in the ensembles of power-law random banded matrices. Phys. Rev. E **54**, 3221–3230 (1996)
167. Mott, N.F., Twose, W.D.: The theory of impurity conduction. Adv. Phys. **10**, 107–163 (1961)
168. Mourachko, I., Comparat, D., de Tomasi, F., Fioretti, A., Nosbaum, P., Akulin, V.M., Pillet, P.: Many-body effects in a frozen Rydberg gas. Phys. Rev. Lett. **80**, 253–256 (1998)
169. Mower, L.: Decay theory of closely coupled unstable states. Phys. Rev. **142**, 799–816 (1966)
170. Mudry, P.W., Brouwer, B.I., Halperin, V., Gurarie, A.Z.: Density of states in the non-Hermitian Lloyd model. Phys. Rev. B **58**, 13539 (1998)
171. Mukamel, S., Jortner, J.: A model for isotope separation via molecular multiphoton dissociation. Chem. Phys. Lett. **40**, 150–156 (1976)
172. Narevich, R., Prange, R.E., Zaitsev, O.: Square billiard with an Aharonov-Bohm flux line. Physica E **9**, 578–582 (2001)
173. Nikitin, E.E.: Theory Elementary Atomic and Molecular Processes. Oxford University Press, Oxford (1985)
174. Nikitin, E.E., Umanskii, S.Y.: Theory of Slow Atomic Collisions. Springer, New York (1984)
175. Offer, C.R., Simons, B.D.: Field theory of Euclidean random matrix ensembles. J. Phys. A Math. Gen. **33**, 7567 (2000)
176. Osborne, T.J., Nielsen, M.A.: Entanglement in a simple quantum phase transition. Phys. Rev. A **66**, 032110 (2002)
177. Parisi, G.: Euclidean random matrices: solved and open problems. In: Brezin, E., Kazakov, V., Serban, D., Wiegmann, P. (Eds.), Application of Random Matrices in Physics. NATO Science Series 206, vol. 21, pp. 219–260. Springer, Berlin (2006)
178. Pastur, L.A.: On the spectrum of random matrices. Theor. Math. Phys. (USSR) **10**, 67–74 (1972)
179. Pechukas, P.: Quantum chaos in the irregular spectrum. Chem. Phys. Lett. **86**, 553–557 (1982)
180. Pechukas, P.: Distribution of energy eigenvalues in the irregular spectrum. Phys. Rev. Lett. **51**, 943–946 (1983)
181. Perelomov, M.: Integrable Systems of Classical Mechanics and Lie Algebras. Springer, Berlin (1990); Generalized Coherent States and Their Applications, Springer, Berlin (1986)
182. Peres, A.: Separability criterion for density matrices. Phys. Rev. Lett. **77**, 1413–1418 (1996)
183. Porter, E.: Statistical Theories of Spectra: Fluctuations. Academic, New York (1965)
184. Porter, C.E., Thomas, R.G.: Fluctuation of nuclear reaction width. Phys. Rev. **104**, 483–491 (1956)
185. Quack, M.: Theory of unimolecular reactions induced by monochromatic infrared radiation. J. Chem. Phys. **69**, 1282–1307 (1978)
186. Rabi, I.I., Ramsey, N.F., Schwinger, J.: Use of rotating coordinates in magnetic resonance problem. Rev. Mod. Phys. **26**, 167–171 (1954)
187. Reichl, L.E.: The Transition to Chaos. Springer, New York (1992)
188. Rhodes, K., Szöke, A., Javan, A.: The influence of level degeneracy on the self-induced transparency effect. Phys. Rev. Lett. **21**, 1151–1155 (1968)
189. Ritus, V.I.: Shift and splitting of atomic energy levels by the field of an electromagnetic wave. Sov. Phys. JETP **24**, 1041–1044 (1967)
190. Sargent, M., Lamb, W.E., Scully, M.O.: Laser Physics. Addison-Wesley, Reading (1987)
191. Schafer, K.J., Kulander, K.C.: Phase dependent effects in multiphoton ionization induced by a laser field and its second harmonic. Phys. Rev. A **45**, 8026–8033 (1992)

192. Schlag, W.: ZEKE Spectroscopy. Cambridge University Press, Cambridge (1998)
193. Schleich, W.: Quantum Optics in Phase Space. Wiley, Weinheim (2001)
194. Schofield, S.A., Wolynes, P.G.: A scaling perspective of quantum energy flow in molecules. J. Chem. Phys. **98**, 1123–1131 (1992)
195. Schrödinger, E.: Die gegenwärtige Situation in der Quantenmechanik. Naturwissenschaften **23**, 807–812, 823–828, 844–849 (1935). English translation: Schrödinger, E.: Discussion of probability relations between separated systems. Proc. Camb. Philosoph. Soc. **31**, 555–563 (1935)
196. Seba, P.: Wave chaos in singular quantum billiard. Phys. Rev. Lett. **64**, 1855–1858 (1990)
197. Shepelyansky, L.: Localization of diffusive excitation in multi-level systems. Phys. D Nonlinear Phenom. **28**, 103–114 (1987)
198. Shepelyansky, L.: Coherent propagation of two interacting particles in a random potential. Phys. Rev. Lett. **73**, 2607–2610 (1994)
199. Shor, P.W.: Scheme for reducing decoherence in quantum computer memory. Phys. Rev. A **52**, R2493–R2496 (1995)
200. Shore, W., Eberly, J.H.: Analytic approximations in multi-level excitation theory. Opt. Comm. **24**, 83–88 (1978)
201. Sinai, Y.G.: On the concept of entropy for a dynamic system. Sov. Phys. Doklady **124**, 768–771 (1959)
202. Sinai, Y.G.: Introduction to Ergodic Theory. Princeton University Press, Princeton (1976)
203. Sinai, Y.G.: Anderson localization for one-dimensional difference Schrödinger operator with quasi-periodic potential. J. Stat. Phys. **46**, 861–909 (1987)
204. Sinai, Y.G.: Topics in Ergodic Theory. Princeton University Press, Princeton (1994)
205. Sinai, Y.G., Soshnikov, A.B.: A refinement of Wigner's semicircle in a neighbourhood of the spectrum edge for random symmetric matrices. Funct. Anal. Appl. **32**, 114–131 (1998)
206. Skvortsov, M.A., Basko, D.M., Kravtsov, V.E.: Energy absorption in time-dependent unitary random matrix ensembles: dynamic vs Anderson localization. JETP Lett. **80**, 60–66 (2004)
207. Sobelman, I.I., Vainshtain, L.A., Yukov, E.A.: Excitation of Atoms and Broadening of Spectral Lines. Springer, New York (1981)
208. Stelmachovi, P., Buzek, V.: Quantum-information approach to Ising model: entanglement in chains of qubits. Phys. Rev. A **70**, 032313 (2004)
209. Stenholm, S.: Quantum theory of electromagnetic fields interacting with atoms and molecules. Physics Reports **6**, 1–122 (1973)
210. Thouless, J.: Maximum metallic resistance in thin wires. Phys. Rev. Lett. **39**, 1167–1169 (1977)
211. Vedral, V., Plenio, M.P., Rippin, M.A., Knight, P.L.: Quantifying entanglement. Phys. Rev. Lett. **78**, 2275–2279 (1997)
212. Verstraete, F., Dehaene, J., De Moor, B., Verschelde, H.: Four qubits can be entangled in nine different ways. Phys. Rev. A **65**, 052112 (2002)
213. Vogel, K., Akulin, V.M., Schleich, W.P.: Quantum state engineering of the radiation field. Phys. Rev. Lett. **71**, 1816–1819 (1993)
214. Weisskopf, V., Wigner, E.: Berechnung der natürlichen Linienbreite auf Grund der Diracschen Lichttheorie. Z. Phys. **63**, 54–73 (1930)
215. Werner, R.F.: Quantum states with Einstein–Podolsky–Rosen correlations admitting a hidden-variable model. Phys. Rev. A **40**, 4277–4281 (1989)
216. Wigner, E.: On a class of analytic functions from the quantum theory of collisions. Ann. Math. **53**(36) (1951)
217. Wigner, E.: On the distribution of the roots of certain symmetric matrices. Ann. Math **65**, 203–207 (1957)
218. Wootters, W.K.: Entanglement of formation of an arbitrary state of two qubits. Phys. Rev. Lett. **80**, 2245–2248 (1998)
219. Yu, T., Eberly, J.H.: Finite-time disentanglement via spontaneous emission. Phys. Rev. Lett. **93**, 140404 (2004)

220. Yurovsky, V.A., Ben-Reuven, A.: Saturated transition in exactly soluble models of two-state curve crossing with time-dependent potentials. Phys. Rev. A **60**, 4561–4566 (1999)
221. Yurovsky, V.A., Ben-Reuven, A., Julienne, P.S., Brand, Y.B.: Counterintuitive transitions in multistate curve crossing involving linear potentials. J. Phys. B **32**, 1845–1857 (1999)
222. Zaslavski, M., Sagdeev, R.Z., Usikov, D.A., Chernikov, A.A.: Weak Chaos and Quasi-Regular Patterns. Cambridge University Press, Cambridge (1991)
223. Zeldovich, Y.B.: The quasienergy of a quantum mechanical system subjected to a periodic action. Sov. Phys. JETP **24**, 1006–1008 (1967)
224. Zeldovich, Y.B.: Scattering and emission of a quantum system in a strong electromagnetic wave. Sov. Phys. Uspekhi **16**, 427–433 (1973)
225. Zirnbauer, M.: Anderson localization and non linear sigma-model with graded symmetry. Nucl. Phys. B **265**, 375–408 (1986)
226. Zwanzig, R.: Memory effects in irreversible thermodynamics. Phys. Rev. **124**, 983–992 (1961). See also the book by Zwanzig, R. (Ed.), Nonequilibrium Statistical Mechanics. Oxford University Press, Oxford (2001)

Index

V.M. Akulin, *Dynamics of Complex Quantum Systems*, Theoretical and Mathematical Physics, DOI 10.1007/978-94-007-7205-2,

Printed by Printforce, the Netherlands